OILFIELD SCALE MITIGATION

From Prediction to Prevention: Tackling Challenges in Energy Transition and Sustainability

Qiwei Wang and Tao Chen

AMPP™

AMPP
Association for Materials Protection and Performance

Printed in the United States of America

Print ISBN: 978-1-57590-010-0
eBook ISBN: 978-1-57590-011-7

AMPP
Association for Materials Protection and Performance
15835 Park Ten Place
Houston, TX 77084
ampp.org

Table of Contents

SECTION I. FUNDAMENTALS

Chapter 1: Overview 1

1.1 Introduction 1
1.2 Produced Waters from Oil and Gas Production 2
1.3 Typical Scales Encountered in Oil and Gas Production 4
1.3.1 Acid-Soluble Scales 4
1.3.1.1 pH of Produced Water Inside the Production Tubing 5
1.3.1.2 Carbonate Scales 6
1.3.1.3 Sulfide Scales 8
1.3.1.4 Silica and Silicates 11
1.3.2 Acid-Insoluble Scales 11
1.3.2.1 Barite 11
1.3.2.2 Celestite 16
1.3.2.3 Calcium Sulfates 17
1.3.2.4 Halite 18
1.4 Scale Prediction 22
1.5 Chemical Treatments for Scales 29
1.5.1 Common Scale Inhibitors 29
1.5.2 Calcite and Sulfate Scales Inhibition 30
1.5.3 Gypsum Scale Inhibition 36
1.5.4 The Effects of High Temperature and Anhydrite Scale Inhibition 38
1.5.5 Halite Scale Inhibition 41
1.5.6 Sulfide Scale Control 41
1.6 Summary 43
Nomenclature 44
References 45

Chapter 2: Scale Prediction: Thermodynamics and Kinetics 53

2.1 Introduction 53
2.2 Thermodynamic Models 55
2.2.1 Common Thermodynamic Models 57
2.2.1.1 D-H Limiting Law 58
2.2.1.2 The Extended D-H Equation and Its Modifications 59
2.2.1.3 Robinson-Stokes equation 61
2.2.1.4 Meissner Model 62
2.2.1.5 Specific Ion Interaction Theory 63
2.2.1.6 Pitzer Theory 63

2.2.2 Common Thermodynamic Software ... 67
2.2.2.1 Visual MINTEQ ... 67
2.2.2.2 PHREEQC ... 68
2.2.2.3 SOLMINEQ.88 ... 68
2.2.2.4 MultiScale ... 68
2.2.2.5 ScaleSoftPitzer ... 68
2.2.2.6 Other Software ... 69
2.2.3 Semiempirical Saturation Indices ... 69
2.2.3.1 Langelier Saturation Index ... 69
2.2.3.2 Larson-Buswell Saturation Index ... 71
2.2.3.3. SOLMINEQ.88 ... 71
2.2.3.4 Stiff-Davis Stability Index ... 71
2.2.3.5 Ryznar Stability Index ... 71
2.2.3.6 Feitler Critical pH ... 72
2.2.3.7 Oddo-Tomson index ... 72
2.2.4 Challenges in Scale Predictions ... 74
2.3 Kinetics of Scale Formation ... 75
2.3.1 Nucleation ... 77
2.3.2 Surface Growth ... 79
2.3.3 Aggregation ... 79
2.3.4 Ostwald Ripening ... 81
2.3.5 Induction Period ... 81
2.3.6 Surface Deposition ... 86
2.3.7 Challenges in Scale Rate Predictions ... 88
2.4 Scale Inhibition Mechanisms and Models ... 89
2.4.1 Inhibition via Adsorption ... 89
2.4.2 Inhibition via Structure Interference and Matrix Incorporation ... 91
2.4.3 I nhibition Model for Deposition Rate ... 92
2.5 Monitoring Methods for Scaling Processes ... 93
2.5.1 I on Concentration Methods ... 93
2.5.2 Scattering Methods ... 94
2.5.3 Partial Pressure Methods ... 96
2.5.4 Microscopic Methods ... 96
2.5.5 Mass Methods ... 96
2.6 Conclusions ... 97
Nomenclature ... 98
References ... 99

Chapter 3: Role of PVT in Scale Modeling ... 109

3.1 Introduction ... 109
3.2 Multiphase Equilibria ... 111
3.2.1 Multiphase Systems ... 111
3.2.2 Phase Equilibrium ... 112
3.3 Equation of State ... 113
3.3.1 Ideal Gas ... 114
3.3.2 Real Gases ... 114
3.3.3 Cubic Equation of State ... 114
3.3.4 Solving the Equation of State ... 115
3.3.5 Flash Calculations ... 116
3.4 Multiphase Scale Prediction Models ... 117
3.5 Input Data to PVT Models ... 119

3.5.1 PVT Sampling ... 120
3.5.2 PVT Analysis and Reports ... 121
3.5.3 PVT Input and Fluid Characterization ... 123
3.6 Scale Modeling in Oil and Gas Wells ... 125
3.6.1 Effect of GOR ... 128
3.6.2 Effect of Water Cut ... 129
3.6.3 The Effect of Temperature and Water Evaporation/Condensation ... 129
3.6.4 Effect of Lift Gas Injection ... 130
3.7 Scale Modeling in Topside Processes ... 131
3.8 Summary ... 132
Nomenclature ... 133
References ... 134

Chapter 4: Scale Modeling for Gas Wells ... 135
4.1 Introduction ... 135
4.2 Gas Reservoirs ... 136
4.3 Thermodynamics ... 137
4.3.1 Water Evaporation and Condensation ... 137
4.3.2 Temperature Change During Inflow to the Well ... 140
4.4 Scale Modeling for Gas Wells ... 142
4.4.1 Input Data and Fluid Tuning ... 143
4.4.2 Scale Prediction During Well Inflow ... 145
4.4.3 Temperature Profile from Bottom Hole to Wellhead ... 147
4.5 Halite Formation in Gas Wells ... 150
4.6 Gas Well Water Sampling and Estimation of Formation Water Composition ... 154
4.7 Summary ... 156
Nomenclature ... 158
References ... 159

SECTION II. SCALE INHIBITOR TREATMENT

Chapter 5: Scale Inhibitor and Evaluation ... 161
5.1 Introduction ... 161
5.2 Scale Inhibitor Chemistries ... 162
5.2.1 Polyphosphates ... 163
5.2.2 Phosphate Esters ... 163
5.2.3 Phosphonates ... 164
5.2.4 Phosphorus-Based Polymers ... 165
5.2.5 Sulfonated Polymers ... 167
5.2.6 Biodegradable Polymeric Scale Inhibitors ... 169
5.3 Laboratory Evaluation of Scale Inhibitors ... 171
5.3.1 Inhibition Performance Evaluation of Scale Inhibitors ... 171
5.3.2 Brine, Solvents, and Chemical Compatibility Evaluation ... 176
5.3.3 Material Compatibility Evaluation ... 177
5.3.4 Thermal Stability Testing ... 178
5.3.5 Emulsion Tendency and Foaming Tendency Testing ... 179
5.3.6 Subsea/Deepwater Umbilical Application Qualification Testing ... 180
5.3.7 Core-flood Test ... 181
5.4 Final Remarks ... 184

Nomenclature......185
References......186

Chapter 6: Squeeze Treatment: Theory, Design, and Optimization......191
6.1 Introduction......191
6.2 Squeeze Treatment Basics......192
6.2.1 Preflush......193
6.2.2 Main Treatment......194
6.2.3 Overflush......194
6.2.4 Inhibitor Retention Mechanism......194
6.2.5 Placement Considerations......195
6.2.6 Treatment Monitoring......196
6.2.7 Treatment Design Optimization......197
6.3 Treatment Modeling......197
6.3.1 Scale Inhibitor Transport and Adsorption......198
6.3.2 Inhibitor Adsorption Characteristics......201
6.3.3 Field Treatment Design Lifecycle......202
6.3.4 Core Flood Isotherm Derivation......203
6.4 Operational Constraints......205
6.5 Squeeze Treatment Lifetime......206
6.5.1 Fixed Target Lifetime......206
6.5.2 Global Treatment Optimization......206
6.6 Nonaqueous Squeeze Treatments......208
6.6.1 Oil-Soluble......209
6.6.2 Invert/Reverse Emulsions......209
6.6.3 Amphiphilic Solvent Systems......209
6.6.4 Microemulsions......209
6.6.5 Water-Free Materials......209
6.6.6 Part-Aqueous Systems......210
6.7 Combined Treatments......210
6.7.1 Squimulation......210
6.7.2 Squeeze Treatment and Tracer Program......210
6.8 Summary......211
Nomenclature......211
References......211

Chapter 7: Advanced Squeeze Treatment......217
7.1 Introduction......217
7.2 Tagged Polymer Scale Inhibitor......218
7.3 Special Additives Developed to Extend Scale Inhibitor Treatment Life......222
7.3.1 Precipitation Squeeze......222
7.3.2 Mutual Solvents as Preflush......225
7.3.3 Kaolinite......226
7.4 Enhanced Squeeze Treatments through the Surface Charge Alternation and Inhibitor Surface Precipitation......227
7.4.1 Enhanced Inhibitor Retention via Bridging......227
7.4.2 Enhance Inhibitor Retention via Surface Precipitation......229
7.5 Foamed Scale Inhibitor Squeeze Treatment......232
7.5.1 Laboratory Test to Qualify the Foamed Scale Inhibitor for Downhole Squeeze Treatments......234
7.5.1.1 Scale Inhibitor Compatibility with the Foam Agent......234

7.5.1.2 Foam Half-Life and Foam Friction Loop Tests 235
7.5.1.3 Scale Inhibitor Core Flood 236
7.5.2 Field Trial Results 236
7.5.3 Summary 237
7.6 Gelled Scale Inhibitor Squeeze Treatment 238
7.6.1 Qualification of Gelled Scale Inhibitor 240
7.6.1.1 Polymer Gel/Scale Inhibitor Compatibility and Temperature Stability 240
7.6.1.2 Gelled Scale Inhibitor-Breaking Characteristics and Shear Thinning Properties 240
7.6.1.3 Gelled Scale Inhibitor by Core Flooding 240
7.6.2 Field Applications 241
7.6.3 Summary 242
7.7 Nonaqueous Scale Inhibitors 242
7.7.1 Mutual Solvent-Based Package for Low-Pressure Reservoir 242
7.7.1.1 Requirements for the Nonaqueous Scale Inhibitor Package 242
7.7.1.2 Laboratory Qualification Tests 243
7.7.1.3 Field Applications 245
7.7.2 Mutual Solvent-Based Package for Water-Sensitive/Fine-Rich Reservoir 245
7.7.2.1 Laboratory Evaluations 246
7.7.2.2 Field Applications 247
7.7.3 Summary 247
7.8 Conclusions 247
Nomenclature 249
References 249

Chapter 8: Solid Scale Inhibitor Treatment 255

8.1 Introduction 255
8.2 Solid Scale Inhibitors for Hydraulic Fracturing 256
8.2.1 Encapsulated Bivalent Metal Salt Scale Inhibitor 256
8.2.2 Scale Inhibitor Impregnated Porous Proppants 258
8.2.3 Comparison Studies 259
8.2.4 Latest Development on Impregnated Porous Proppants 260
8.3 Proppant-Sized Solid Scale Inhibitor 261
8.3.1 Case Studies 261
8.3.2 Combined Application with Paraffin and Asphaltene Solid Inhibitors 266
8.3.3 Solid Inhibitor Loading Effect 267
8.3.4 Environment-Friendly Proppant-Sized Solid Inhibitor 270
8.3.5 Application for Metal Cross-Linked Fluid 271
8.3.6 Application in Prepacked Screen and Rechargeable Annular Space 274
8.3.7 Ultra-High-Strength Proppant-Sized Solid Scale Inhibitor 275
8.4 Other Solid Inhibitors for Hydraulic Fracturing 278
8.5 Encapsulated Scale Inhibitor for Batch Treatment 278
8.6 Final Remarks 280
Nomenclature 282
References 282

Chapter 9: Scale Control in EOR Processes 287

9.1 Introduction 287
9.2 Scale Formation and Control in Chemical-Enhanced Oil Recovery Process 289
9.2.1 Chemical Injection-Enhanced Oil Recovery 289

9.2.2 Mechanism of Scale Deposition ... 291
9.2.3 Scale Control Method ... 293
9.2.4 Case Studies in China ... 297
9.3 Scale Formation and Control in Gas-Enhanced Oil Recovery ... 300
9.3.1 Gas Injection-Enhanced Oil Recovery ... 300
9.3.2 Mechanism of Scale Deposition ... 301
9.3.3 Scale Control Method ... 301
9.3.4 Case Studies ... 304
9.4 Scale Formation and Control in Thermal-Enhanced Oil Recovery ... 304
9.4.1 Thermal Injection-Enhanced Oil Recovery ... 304
9.4.2 Mechanism of Scale Deposition ... 305
9.4.3 Scale Control Method ... 306
9.3.4 Case Studies ... 307
9.5 Summary ... 308
Nomenclature ... 308
References ... 309

Chapter 10: Scale Prevention in Shale Production ... 315

10.1 Introduction ... 315
10.2 Challenges for Mineral Scaling in Shale Production ... 316
10.2.1 Varieties of Shale Formation Characteristics ... 316
10.2.2 Intensified Rock-Fluid Interactions during Shale Development ... 318
10.2.3 Sensitivity to Scaling for Low-Permeability Shale Formation ... 320
10.2.4. Complexity of Water Management in Unconventional Operation ... 321
10.2.4.1 Elevated Water Footprint ... 321
10.2.4.2 Water-Commingling Scenario ... 323
10.2.4.3 Uncertainty in Flowback Water Chemistry ... 325
10.2.4.4 Diverse Water Chemistries ... 326
10.3 Challenges for Scale Inhibitor Application in Fracturing Fluids ... 329
10.3.1 Scale Inhibitor Tolerance with Iron-Rich Water ... 329
10.3.2 Scale Inhibitor Compatibility with Cross-Linkers ... 331
10.3.3 Scale Inhibitor Interaction with Friction Reducers ... 334
10.4 Scale Control Innovations for Unconventional Operation ... 335
10.4.1 Solid Scale Inhibitors ... 335
10.4.2 Environment-Friendly Inhibitor Package ... 335
10.4.2.1 Chemical Disclosure ... 335
10.4.2.2 Green Chemistries for Scale Inhibitors ... 335
10.4.2.3 Ranking of Entire Inhibitor Package ... 336
10.4.3 Scale Squeeze while Fracturing ... 336
10.4.4 Digitalization ... 337
10.5 Summary ... 337
Nomenclature ... 338
References ... 339

Chapter 11: Multifunctional Production Chemicals ... 343

11.1 Introduction ... 343
11.2 Types of Multifunctional Chemicals ... 345
11.2.1 Binary Combinations for Flow Assurance ... 345
11.2.1.1 Scale Inhibitors and Asphaltene Dissolvers ... 346
11.2.1.2 Scale Inhibitors and Paraffin Inhibitors ... 346
11.2.1.3 Scale Inhibitors and Hydrate Inhibitors ... 346

11.2.1.4 Scale Inhibitors and Dispersant ... 347
11.2.1.5 Scale Inhibitor with Dissolvers ... 347
11.2.1.6 Asphaltene and Paraffin Inhibitors ... 347
11.2.1.7 Asphaltene Dispersant and Solvent ... 347
11.2.1.8 Paraffin Inhibitor and Solvent ... 347
11.2.1.9 Paraffin dispersant and solvent ... 347
11.2.2 Binary Combinations for Integrity Management ... 348
11.2.2.1 Corrosion Inhibitors and H_2S Scavengers ... 348
11.2.2.2 Corrosion Inhibitors and Oxygen Scavengers ... 348
11.2.2.3 Corrosion Inhibitors and Biocides ... 348
11.2.3 Binary Combinations for Flow Assurance and Integrity Management ... 349
11.2.3.1 Scale and Corrosion Inhibitors ... 349
11.2.3.2 Scale Inhibitors and H_2S Scavenger ... 349
11.2.3.3 Hydrate and Corrosion Inhibitors ... 349
11.2.3.4 Dispersants and Corrosion Inhibitors ... 350
11.2.4 Binary Combinations for Flow Assurance and Production Optimization ... 350
11.2.4.1 Scale Inhibitors and Foamers ... 350
11.2.5 Binary Combinations for Integrity Management and Production Optimization ... 350
11.2.5.1 Corrosion Inhibitors and Drag Reducers ... 351
11.2.5.2 Corrosion Inhibitors and Foamers ... 351
11.2.5.3 Corrosion Inhibitors and Demulsifiers ... 351
11.2.6 Ternary Combinations ... 351
11.2.6.1 Scale Inhibitor, Scale Dissolver, and Corrosion Inhibitor ... 351
11.2.6.2 Scale Inhibitor, Corrosion Inhibitor, and Foamer ... 352
11.3 Challenges in Chemical Formulation ... 352
11.3.1 Reactivity of Different Chemicals ... 352
11.3.2 Performance Decreases due to Component Interference ... 352
11.3.3 Difficulty in Creating a Homogeneous Solution ... 353
11.3.4 Thermal Stability ... 353
11.3.5 Metal Compatibility ... 354
11.3.6 Elastomeric Compatibility ... 354
11.3.7 Gas Hydrate Formation in Chemical Delivery Line ... 354
11.3.8 Overdosing of Production Chemicals ... 354
11.4 Laboratory Assessment ... 354
11.4.1 TGA ... 355
11.4.2 Capillary Screening ... 355
11.4.3 Screening of Aged Intermediates ... 356
11.4.4 High-Pressure Viscosity ... 356
11.4.5 Chemical Compatibility with Metals ... 356
11.4.6 Chemical Compatibility with Elastomers ... 356
11.5 Field Applications ... 357
11.5.1 Scale and Asphaltene Control ... 357
11.5.2 Scale and Corrosion Inhibition ... 357
11.5.3 Scale and H_2S Control ... 357
11.5.4 Scale Control and Corrosion Inhibition ... 357
11.5.6 Scale and Corrosion Inhibition in Low-Pressure Gas Wells ... 358
11.5.7 Corrosion and H_2S Control ... 360
11.5.8 Corrosion Inhibitor and Biocide Treatment ... 360
11.5.9 Hydrate and Corrosion Prevention ... 361
11.5.10 Dispersion and Corrosion Inhibition ... 361

11.5.11 Corrosion Inhibition and Demulsification ... 361
11.5.12 Corrosion Inhibitor Treatment of Low-Pressure Wells ... 361
11.6. Summary ... 361
Nomenclature ... 362
References ... 362

Chapter 12: Scale Inhibitor Analysis ... 369

12.1 Introduction ... 369
12.2 Overview ... 371
12.3 Interferences ... 373
12.4 Sample Preservation ... 374
12.4.1 Case Studies ... 376
12.4.1.1 Wells with High Fe in Produced Water ... 376
12.4.1.2 Well with Moderate-Barium Sulfate Scale ... 377
12.4.1.3 Wells with Harsh Barium Sulfate Scale ... 378
12.4.2 Guidelines ... 378
12.4.2.1 Preservation of Samples for Phosphonate ... 378
12.4.2.2 Preservation of Samples for Polymers ... 379
12.5 Detectable Tags ... 379
12.6 Inhibitor Speciation ... 382
12.6.1 Importance of P Speciation ... 382
12.6.1.1 Phosphorus Speciation in Deta-Phosphonate, Diethylenetriaminepenta Phosphonic Acid, by Ion Chromatography ... 383
12.6.2 Polymer Speciation ... 384
12.6.2.1 Isolate Polyacrylate from Polyacrylate/Sulphonated Polymers Mix (Simple Approach) ... 385
12.6.2.2 Isolate and Quantify P-Tagged and Non-P-Containing Polyacrylate Polymers ... 385
12.6.2.3 Isolate and Quantify P-Tagged and Non-P-Containing Sulphonated Copolymers ... 385
12.6.2.4 High-Performance Liquid Chromatography and Mass Spectroscopy ... 386
12.7 Fluorescence ... 388
12.8 Time-Resolved Fluorescence ... 392
12.9 Squeezing Comingled Wells in the Same Flowline ... 396
12.10 Summary of Options for Scale Inhibitor Analysis: Benefits, Limitations, and Common Interferences ... 397
12.11 Conclusions ... 403
Nomenclature ... 404
References ... 405

Chapter 13: Scale Monitoring Techniques ... 409

13.1 Introduction ... 409
13.2 Scale Monitoring ... 410
13.2.1 Proactive Scale Monitoring ... 410
13.2.1.1 Scale Modeling ... 411
13.2.1.2 Water Ion Monitoring ... 411
13.2.1.3 Inhibitor Residual Measurements ... 411
13.2.1.4 Suspended Solid Analysis ... 412

13.2.2 Reactive Scale Monitoring ... 412
13.2.2.1 Thickness Shear-Mode Resonator ... 412
13.2.2.2 Gamma Ray Transmission ... 413
13.2.2.3 Heat-Based Sensor ... 414
13.2.2.4 Scale Coupon ... 415
13.3. Suspended Solid Analysis ... 416
13.3.1 Bulk Scale Analysis Principle ... 416
13.3.2 Differentiation Between Scale in Reservoir and Wellbore ... 417
13.3.3 Zonal Resolution ... 419
13.3.4 Squeeze Timing Optimization ... 420
13.3.5 Solid Morphology ... 420
13.3.5.1 Overlapping ... 422
13.3.5.2 Coverage vs. Morphology ... 424
13.3.5.3 Inhibition after Scale Formation ... 424
13.3.5.4 Static vs. Dynamic Scale Conditions ... 424
13.4. Summary ... 425
Nomenclature ... 426
References ... 426

SECTION III. EXOTIC SCALES

Chapter 14: Sulfide Scales ... 429
14.1 Introduction ... 429
14.2 Mineral Chemistry ... 430
14.2.1 Lead Sulfide ... 430
14.2.2 Zinc Sulfide ... 431
14.2.3 Iron Sulfide ... 431
14.2.3.1 (Amorphous) Ferric Sulfide and Mackinawite ... 432
14.2.3.2 Pyrite and Marcasite ... 432
14.2.3.3 Greigite ... 432
14.2.3.4 Pyrrhotite and Troilite ... 433
14.2.3.5 Kansite ... 433
14.2.3.6 Smythite ... 433
14.3 Origin and Formation ... 433
14.4 Occurrence in the Field ... 435
14.4.1 Lead and Zinc Sulfides ... 435
14.4.2 Iron Sulfide ... 436
14.4.3 Black Powder ... 439
14.5 Operational Challenges Caused by Sulfide Scale ... 442
14.6 Prevention and Remediation ... 443
14.6.1 Prevention by Chemical Inhibition ... 444
14.6.2 Prevention by Chelation Chemistry ... 445
14.6.3 Remediation with Mechanical Methods ... 446
14.6.4 Remediation Using Scale Dissolvers ... 447
14.7 Summary ... 448
Nomenclature ... 448
References ... 449

Chapter 15: Halite Scale 455
15.1 Introduction 455
15.2 Halite Scale Form and Occurrence 456
15.2.1 Halite Mineralogy 456
15.2.2 Field Occurrence 456
15.2.3 Formation Mechanisms 457
15.3 Halite Scale Prediction 460
15.3.1 Modeling of Halite Scale 460
15.3.2 Field Monitoring: Leading and Lagging Indicators 461
15.4 Halite Scale Prevention 462
15.4.1 Nonchemical Methods 462
15.4.2 Continuous Injection of Wash Water 462
15.5 Use of Prevention Chemicals 463
15.5.1 Halite Inhibitors 463
15.5.1.1 Ferrocyanide (Hexacyanoferrate) Salts 464
15.5.1.2 Nitrilotrialkanamides and Quaternary Salts 465
15.5.1.3 Other Chemistries for Halite Control 466
15.5.2 Mechanisms of Halite Inhibition 467
15.5.3 Chemical Application Methods 468
15.6 Halite Scale Testing 468
15.6.1 Static Testing 469
15.6.2 Dynamic Testing 469
15.6.3 Kinetic Turbidity Testing 470
15.6.4 Test Method Summary 471
15.7 Halite Removal 471
15.8 Case Histories 472
15.8.1 Green Multifunctional Scale Inhibitor in the U.S. Rockies 472
15.8.2 Squeezing Halite Inhibitor into High-Temperature Reservoirs 473
15.8.3 Deepwater Halite Scale Deposition in the Gulf of Mexico 474
15.9 Final Remarks 477
Acknowledgments 477
Nomenclature 478
References 478

Chapter 16: Silica and Silicate Deposits 483
16.1 Introduction 483
16.2 Silica Scale 485
16.3 Metal Silicates 487
16.3.1 Magnesium Silicate 487
16.3.2 Iron Silicate 488
16.3.3 Aluminum Silicate 489
16.3.4 Other Metal Silicates 489
16.4 Occurrences 490
16.4.1 Water Purification and Clarification Systems 490
16.4.2 Oil and Gas Production Systems 491
16.4.3 Geothermal Systems 492
16.5 Mitigation 493
16.5.1 pH Control 493
16.5.2 Silica Removal by Precipitation 493
16.5.3 Ion Exchange Technologies 494

16.5.4 Chemical Inhibition ... 495
16.6 Detection and Evaluation Methods ... 497
16.7 Final Remarks ... 498
Nomenclature ... 500
References ... 500

SECTION IV. NEW TECHNOLOGIES FOR SUSTAINABLE PRODUCTION

Chapter 17: Green Scale Inhibitors ... 505

17.1 Introduction ... 505
17.2 Environmental Regulations ... 507
17.2.1 Oslo and Paris Convention Regulations ... 507
17.2.1.1 Biodegradation ... 512
17.2.1.2 Toxicity ... 512
17.2.1.3 Bioaccumulation Factor Tests ... 512
17.2.1.4 Bioaccumulation (Log Pow) ... 513
17.2.2 USA Regulations ... 515
17.2.3 Canada Regulations ... 515
17.2.4 Australia Regulations ... 516
17.2.5 Summary of the Regulations ... 517
17.3 Chemistries and Applications ... 517
17.3.1 Polyaspartate ... 518
17.3.2 Carboxymethyl Inulin ... 521
17.3.3 Hybrid Polymer ... 523
17.3.4 Biodegradable Phosphonate-Based Scale Inhibitors ... 526
17.3.5 Other Green Scale Inhibitors ... 530
17.4 Summary ... 532
Nomenclature ... 534
References ... 535

Chapter 18: Nanomaterials for Scale Inhibition ... 541

18.1 Introduction ... 541
18.2 Potential for Scale Inhibition ... 542
18.3 Common Scale Inhibitor Nanomaterials ... 543
18.3.1 Amorphous Ca-Diethylenetriamine Pentakis (Methylene Phosphonic Acid) Scale Inhibitor Nanomaterials ... 543
18.3.2 Crystalline Ca-Diethylenetriamine Pentakis (Methylene Phosphonic Acid) Scale Inhibitor Nanomaterials ... 547
18.3.3 Carbon-Based Scale Inhibitor Nanomaterials ... 550
18.3.4 Metal/Metal Oxide Scale Inhibitor Nanomaterials ... 552
18.3.5 Silicon-Based Scale Inhibitor Nanomaterials ... 554
18.3.6 Polymer Scale Inhibitor Nanomaterials ... 556
18.4 Summary ... 558
Nomenclature ... 559
References ... 560

Chapter 19: Engineering Coatings and Surfaces ... 565

19.1 Introduction ... 565
19.2 Solid Deposition Mechanisms ... 569

19.3 Coatings and Surface Treatment Engineering 574
19.3.1 Engineering Requirements 574
19.3.2 Polymer Coatings 575
19.3.3 Metallic Coatings 576
19.3.4 Ceramic Coatings 578
19.4 Coating Qualification Test and Qualification Procedures 580
19.5 Coating Performances against Specific Solid Deposits 584
19.6 Conclusions 592
Nomenclature 594
References 595

Chapter 20: Hardware Technologies for Subsea Scale Control 601

20.1 Introduction 601
20.2 Challenges Associated with Subsea Fields 602
20.3 Hardware Technologies for Scale Control 605
20.3.1 Subsea Access and Fluid Injection Systems 606
20.3.1.1 Subsea Well and Manifold Access Technology 606
20.3.1.2 Subsea Modular Injection System with Hydraulic Access 608
20.3.2 Subsea Sampling 609
20.3.3 Control Subsea Chemical Injection Rates—Chemical Injection Metering Valves 612
20.4 Case Studies and Associated Benefits 615
20.4.1 Case Study 1: Offshore Sabah, Malaysia 616
20.4.1.1 Problem 616
20.4.1.2 Fluid Delivery Options 616
20.4.1.3 Subsea Modular Injection System with Hydraulic Access 617
20.4.1.4 Results 617
20.4.2 Case Study 2: Schiehallion Field 618
20.4.2.1 Problem 618
20.4.2.2 Fluid Delivery Options 618
20.4.2.3 Subsea Modular Injection System with Hydraulic Access 619
20.4.2.4 Results 619
20.4.3 Economical Considerations—Cost Analysis 620
20.4.4 Multiaccess-Enabled Subsea Sampling Case Study—Offshore Angola 622
20.4.5 Inflow Control Equipment: Inflow Control Valve Scaling Case Study—Presalt Carbonate Reservoir 626
20.5 Conclusions 629
Nomenclature 630
References 631

1 Overview

Amy T. Kan[1] and Mason B. Tomson[1]
[1]Brine Chemistry Consortium, Rice University, Houston, Texas, USA

1.1 Introduction

Oil and natural gas have been the backbone of global energy production for many years and are expected to remain. America has been the leading oil producer since 2008, thanks to hydraulic fracturing and offshore technology advancements that have allowed the extraction of massive oil and gas resources from onshore shale formations and offshore deep water.[1] While petroleum products have made life easier, finding, producing, and transporting crude oil can adversely affect the environment. For instance, hydraulic fracturing requires large quantities of water, which can result in the depletion of freshwater sources, the potential contamination of chemicals used in the process, and increasing demand for scale and corrosion controls. However, the industry has made significant progress in reducing the environmental impact of hydraulic fracturing by reusing produced water instead of fresh water. Indeed, the demand for more sustainable and environmentally friendly oil and gas production has increased.

It has been discovered that the water used in drilling and completion operations pales compared to the amount of water produced from shale plays. The water produced from conventional fields is usually reinjected to enhance oil production, but this is not possible with the water produced from shale plays.[2] Instead, it must be disposed of through other means, such as hauling, storage, and saltwater disposal wells. Only a small portion (13%) of the produced water is recycled for reuse in hydraulic fracturing operations.

Deepwater production often encounters produced water at ultra-high temperatures (150–200°C) and pressure (1,000–1,500 bars) and high levels of total dissolved solids (TDS;

>300,000 mg/L).[3, 4] The produced water from different zones and wells is often mixed in the production tubing, subsea tieback, or common facilities, resulting in varying oil, gas, and water compositions that can lead to significant scaling issues.

Various measures are employed to mobilize residual or heavy oil, including steam, supercritical carbon dioxide, seawater, or alkaline surfactant flooding.[5] However, these measures can result in significant scaling problems and require careful consideration to mitigate their impact. Industry must continue developing strategies and technologies to address these scaling issues and minimize their negative impact on the production process and the environment.

The root cause of the scale problem encountered during oil and gas production is the co-production of produced water from the reservoir and the oil and gas.[6] In 2022, the United States alone produced over 4 billion barrels of oil and 21 billion barrels of produced water.[2] Scale buildup in production systems can reduce production efficiency. Effective scale control can help increase production efficiency to reduce the environmental impact of production. Additionally, using environmentally friendly scale inhibitors can help minimize the potential harm caused by chemical releases into the environment.[7]

1.2 Produced Waters from Oil and Gas Production

Formation water is assumed to be in equilibrium with the reservoir rocks.[8] The equilibrium condition changes as water flows near the wellbore and up the production tubing due to the changes in temperature and pressure.[9, 10] The changes in temperature and pressure also lead to the evaporation of CO_2, H_2S, and H_2O from the aqueous phase.[11, 12] Some production systems may allow the mixing of produced waters from different production zones or wells. The produced water can also mix with other injected chemicals, such as scale and corrosion inhibitors, surfactants, or methanol and ethylene glycol, which are added to aid production.[13–15] These factors can cause a change in pH, alkalinity, redox condition, and precipitation of scale-forming minerals, leading to blockages and reduced production rates. Moreover, the altered produced water composition collected after production may have implications for disposal or reuse.

The percentile distribution of the produced water ion composition in the United States is plotted in Figure 1.1 based on the U.S. Geological Survey (USGS) produced water data of 1905–2019.[16] The number of data points for each ion is included in Figure 1.2. Dai *et al.*[17] studied water produced from the Permian Basin of West Texas. They concluded that most produced water originated from seawater and evolved by evaporation and precipitation. Thus, most seawater ion components are also present in the produced water. The produced water predominantly comprises sodium (Na^+), chloride (Cl^-), calcium (Ca^{2+}), magnesium (Mg^{2+}), bicarbonate (HCO_3^-), and sulfate (SO_4^{2-}) ions. At the 50 percentiles, the produced water contains approximately 1 M Na^+ and Cl^- and 0.05 M Ca^{2+} ions. These three ions (Na^+, Cl^-, and Ca^{2+}) are the major constituents of the TDS (mg/L). Ca^{2+} and Mg^{2+} cations are concentrated for the last five percentile of the highly saline-produced water. Due to the stronger ion-ion interaction with the divalent cation, they may cause a nonlinear effect on the ion activity in the highly saline-produced water. In addition to the first five most predominant cations and anions, bicarbonate (HCO_3^-), sulfate (SO_4^{2-}), and potassium (K^+) are also present in the produced water at significant concentrations (above 1 mM concentrations at the 50 percentile). Bicarbonate and sulfate are the constituents of several major carbonate and sulfate scales, while potassium rarely forms scales. The last group of ions, including barium

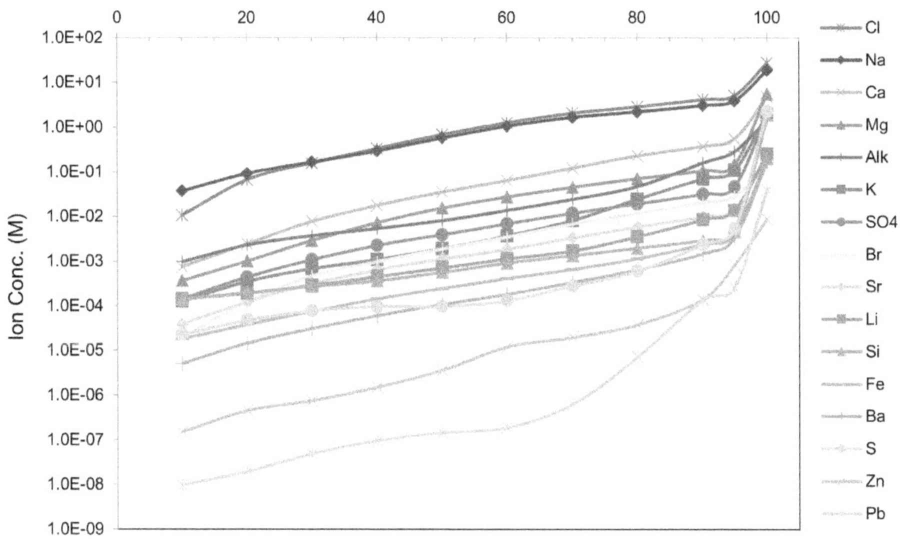

Figure 1.1. The percentile distribution of produced water composition. Produced with data from the USGS National Produced Waters Geochemical Database.[16]

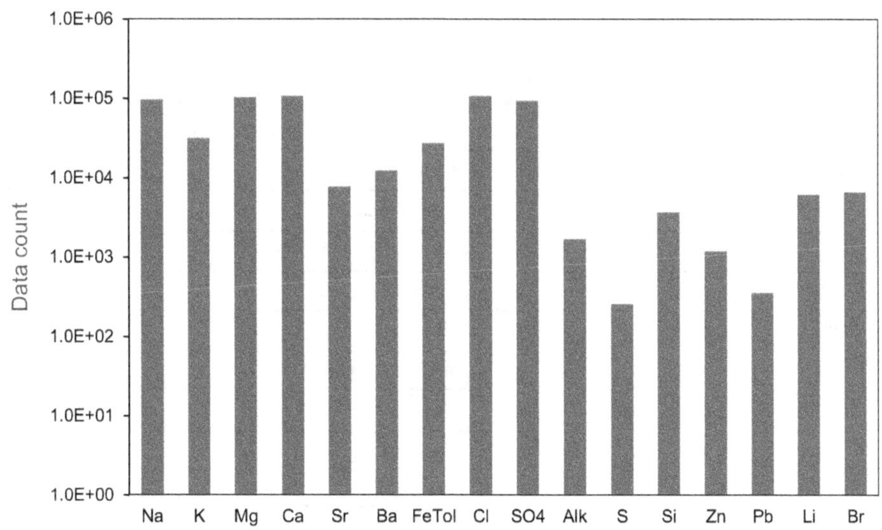

Figure 1.2. The number of data used in the percentile analysis. Produced with data from the USGS National Produced Waters Geochemical Database.[16]

(Ba^{2+}), strontium (Sr^{2+}), iron (Fe^{2+}), zinc (Zn^{2+}), lead (Pb^{2+}), bisulfide (HS^-), and silicon (Si), may or may not exist in the produced water. If they are present, they typically are the minor components at a concentration below 1 mM at the 50 percentile. Many of these ions, such as Ba^{2+}, Sr^{2+}, Fe^{2+}, and HS^-, play major roles in scale forming and will be discussed below. Lithium (Li+) and bromide (Br^-) concentrations are also plotted in this graph. Li^+ and Br^- are not known to contribute to scaling issues, but both have high-economic values. Li^+ in the produced water is of interest in the context of evaluating its potential as a valuable resource for industrial processes, particularly in the manufacturing of batteries.

1.3 Typical Scales Encountered in Oil and Gas Production

The pictures in Figure 1.3 are typical scales observed in the production equipment. The most common types of oilfield scales include carbonates, sulfates, sulfides, silicates, and halite scales. Table 1.1 lists the typical scale classification as acid-soluble or insoluble, since the solubility of scale in acid is commonly used in scale identification and removal in the field.

1.3.1 Acid-Soluble Scales

This group of minerals strongly depends on the pH changes inside the production tubing. All minerals in a reservoir are in equilibrium with the surrounding, undisturbed "connate" or natural water. Due to the reduction in partial pressure of CO_2 up the production tubing, the pH increases significantly at lower pressure with the associated increase in solution pH. The change in solution pH is often the root cause of the problem with carbonate and sulfide scales.

Table 1.1: A list of common mineral scales observed in the oil production equipment.

Acid Insoluble	Acid Soluble
Halite, NaCl	Calcite/aragonite/vaterite, $CaCO_3$
Gypsum, $CaSO_4{\times}2H_2O$	Dolomite, $CaMg(CO_3)_2$
Anhydrite, $CaSO_4$	Siderite, $FeCO_3$
Celestite, $SrSO_4$	Ankerite, $Ca(Fe,Mg,Mn)(CO_3)_2$
Barite, $BaSO_4$	Strontianite, $SrCO_3$
Pyrite, FeS_2	Witherite, $BaCO_3$
	Mackinawite, FeS
	Troilite/pyrrhotite, Fe_xS
	Sphalerite, ZnS
	Galena, PbS
	Magnetite, Fe_3O_4
	Metal silicates

Figure 1.3. Common scales observed in oil production systems: (a) a calcite scale buildup from a well in South Texas, (b) SrSO4 scale buildup on a well in Quata, (c) anhydrite scale from a well in Kuwait, and (d) gypsum scale from a well in Permian basin.

1.3.1.1 pH of Produced Water Inside the Production Tubing

The pH of the aqueous phase inside the production tubing can fluctuate due to the partitioning of acid gas components plus water (CO_2, H_2S, and H_2O) when temperature and pressure change. The carbonates, sulfides, and silicates are sensitive to pH changes. Changing the pH of the aqueous phase can significantly impact the formation of carbonate and sulfide scales. The pressure decreases as the production fluid flows out of the formation and up the well. Particularly, if the confining pressure is below the bubble point, lower pressures cause dissolved CO_2 to escape the solution, and the pH of the solution generally increases. The rise in pH results in aqueous bicarbonate (HCO_3^-) being converted to carbonate (CO_3^{2-}), leading to calcite precipitation in the formation pore throats, on the production tubing walls, or in surface handling equipment. Additionally, the decreased pressure also reduces the solubility of calcite in salt water, in that nearly all minerals are more soluble at higher pressures, assuming other variables remain unchanged. Similarly, the pressure drop could cause dissolved hydrogen sulfide to come out of the solution, converting bisulfide (HS^-) to sulfide (S^{2-}) and forming metal sulfide precipitates.

In addition to the changes in temperature and pressure, the pH of the produced water can also be influenced by the organic acids in the water. Organic acids, predominantly C2 to C5 organic acids (acetate, propionate, butyrate, and valerate) in produced water, are buffers at typical produced water pH ranges, naturally occurring and biogenically formed in those reservoirs at 180–320°F.[18] Figure 1.4 plots the percentile distribution of C2 to C5 organic acid concentrations based on the USGS produced water database values.[16] Due to the constraint of charge balance and buffering capacity of organic acids, the concentrations of bicarbonate, carbonate, bisulfide, and sulfide and the corresponding scaling tendency would be strongly influenced when the organic acids are present in significant concentrations in the produced water. Generally, the pH will decrease as the pressure increases down a well with a gas phase. Still, since the total alkalinity must remain constant, this pH drop will cause a decrease in the fraction of alkalinity due to carbonates and a concomitant increase in the fraction of alkalinity due to carboxylic acids—that is, more of a corrosion concern than a scale problem.

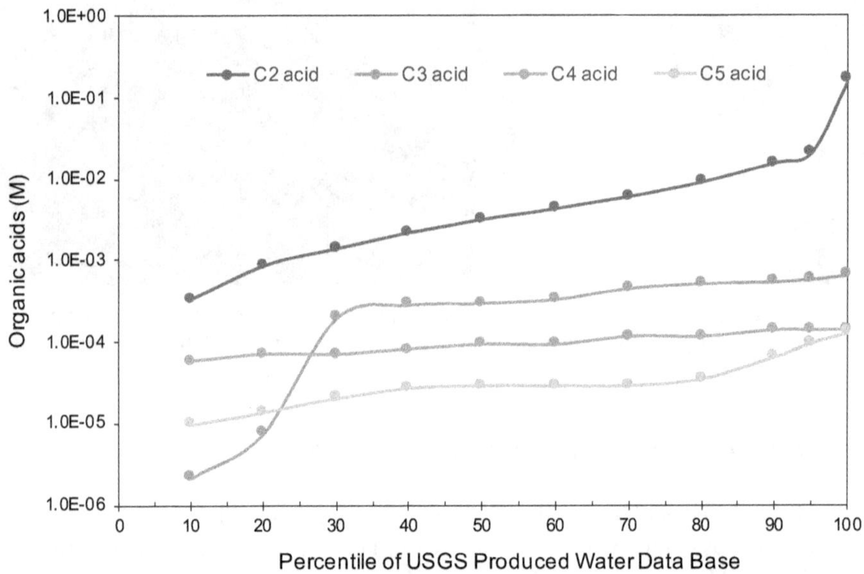

Figure 1.4. Plot of the percentile distribution of organic acids concentrations. Produced with data from the USGS National Produced Waters Geochemical Database.[16]

1.3.1.2 Carbonate Scales

While calcite ($CaCO_3$) and dolomite ($CaMg(CO_3)_2$) are the two most common carbonate minerals found in oil-bearing formations, the dolomite scale is rarely formed in the production tubing, possibly due to the slow reaction kinetics of dolomite formation. Calcite is the most commonly observed scale during oil and gas production, and the chemistry of calcite formation has been well understood.[8] The scale formed in production tubing often contains up to 20% iron or magnesium carbonate ($Ca_{0.8-1.0}(Fe, Mg)_{0.2-0.0}CO_3$). Ferrous carbonate (siderite) has gained significant attention to control corrosion.[19,20] The oil industry is particularly concerned with CO_2 corrosion of steel due to the presence of CO_2 in produced gases. Corrosion is often less severe at the bottom of deep gas wells, where scale seems to cover the tubing uniformly and is frequently composed primarily of ferrous carbonate. Laboratory tests have shown that a thin, adherent layer of ferrous carbonate on the surface of a corroding material can significantly reduce corrosion rates. Li *et al.*[21] observed that ankerite, $Ca(Fe_xMg_{1-x})(CO_3)_2$, is likely the mineral phase that forms at the interface of carbon steel tubing and produced water, as shown in the conceptual drawing of Figure 1.5(a), with observations shown in Figures 1.5(b) through (g). The apparatus comprises a flow-through glass reactor with a continuous deoxygenated CO_2/bicarbonate buffered calcite saturated solution. A stainless steel coupon and a carbon steel coupon were placed in series in the glass reactor.[21] While no scale was formed on the stainless steel coupon, a visible amount

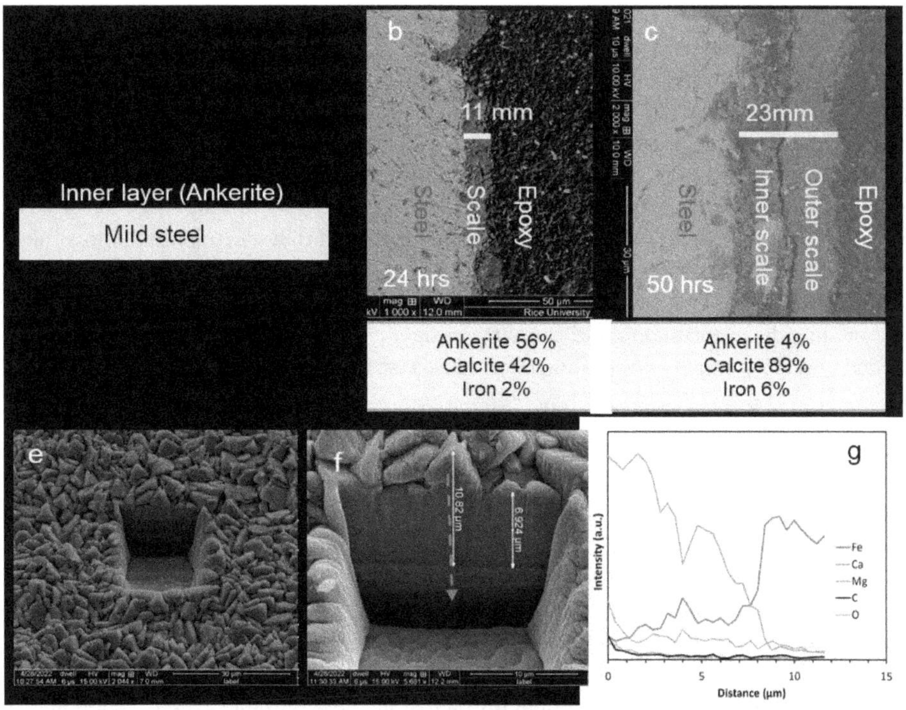

Figure 1.5. The plot of (a) a schematic diagram of how the carbonate scale formed on top of a 1018 carbon steel coupon; (b and c) the cross-sectional scanning electron microscope (SEM) pictures of the scale layers grew on top of the mild steel after 24 h and 50 h. Epoxy was used to fix the scale in place for specimen cutting and SEM analysis; (d) the scale composition estimated by X-ray diffraction analysis of scale formed after 24 h and 50 h; (e) the top view of SEM picture of the scale covered coupon with a square hole cut by the focused ion beam (FIB). (f) FIB prepared the scale on the side surface. The red line indicates the direction of the energy dispersive spectroscopy (EDS) line scan, and the white line measures the thickness of the scale; (g) the EDS line scan results. The experiments were done by flowing a deoxygenated synthetic produced water containing 1 M NaCl, 200 mg/L Ca, and 67 mg/L Mg, 3,000 mg/L bicarbonate for 24 h and 50 h. The solutions were purged with 100% CO_2 gas to yield a pH of 6.09 and supersaturated with respect to calcite (SI = 0.34).

of scale was observed with the carbon steel coupon. Figure 1.5(b) is the cross-sectional scanning electronic microscopy (SEM) picture of the scale layer on the carbon steel coupon after 24 h and 50 h of reaction, where two mineral layers were observed. XRD analysis of the scale layer shows the presence of ankerite and calcite minerals (Figures 1.5[b] and [c]). Figure 1.5(e) is the SEM image of a rectangular shape cutting of the scale layer by the focus ion beam (FIB). Figure 1.5(f) shows the direction and depth of the energy dispersive spectroscopy (EDS) line scan alone on the FIB surface of the scale layer. Figure 1.5(g) shows the EDS line scan results. The study confirmed that calcite crystals would build on top of the ankerite layer to form the protective layer and prevent corrosion.

There seems to be a link between the adherence of protective films and the temperature at which they form. Siderite ($FeCO_3$) is formed without Ca in the produced water. The exact relationship between siderite, ankerite, and its solid solution is currently being studied, and it may have implications for both the thermodynamic and kinetic aspects of the initial scale layer formed in the carbonate system. Due to a high activation energy to growth, $FeCO_3$ is slow to form at low temperatures but so is corrosion. As the temperature increases to around 100°C, the rate of $FeCO_3$ formation accelerates. Between 90°C and 110°C, $FeCO_3$ decomposes and forms black magnetite (Fe_3O_4), a better corrosion–prevention film.[22] At higher temperatures, the corrosion rates tend to decrease to a lower limiting value. As the need for high-efficiency corrosion control increases with saltwater production, more relationships between scale and corrosion are expected to be developed. Li *et al.*[21] observed that corrosion inhibitors that inhibit carbon steel corrosion while allowing a surface coating of siderite or ankerite formation would also inhibit the calcite scale buildup. However, an excess scale inhibitor would inhibit ankerite or siderite formation and cause more significant corrosion. Further investigation is needed on the potentially devastating effect of excessive scale inhibitor use.

1.3.1.3 Sulfide Scales

Hydrogen sulfide (H_2S), naturally occurring and biologically generated from sulfate-reducing bacteria, is a common and toxic gas found in some oil and gas production facilities.[23, 24] Approximately 1% of USGS produced water data reported a sulfide concentration.[16] Sulfide can cause corrosion in production facilities and other equipment, such as transportation pipes, pumps, and valves.[25] The corrosion can lead to reduced transportation efficiency, asset integrity failure, and environmental contamination. Iron sulfide (FeS) formation due to H_2S corrosion can cause operational problems, such as equipment malfunction and stabilized oil–water emulsions. In addition, exposure to H_2S can be dangerous to humans, causing symptoms such as coughing, eye irritation, unconsciousness, and death at high concentrations. Sulfide deposition can also occur at depth, even in low sulfide brines, forming zinc sulfide (ZnS) or lead sulfide (PbS).

Despite the importance of sulfide scales in oil and gas production, there is limited knowledge of the prediction and control of metal sulfide deposits. Iron sulfide scale deposits from oil and gas wells often include a mixture of pyrrhotite, troilite, mackinawite, greigite, pyrite, and marcasite.[26] Figure 1.6 lists several ferrous sulfide mineral phases and their formation pathways. Aqueous iron-sulfide complexes (FeS^0) play a potentially important role in the chemistry of iron sulfides.[27] The aqueous iron–sulfide complex can form nanoparticles of

Fe_2S_2 to $Fe_{150}S_{150}$. This nanoparticle may be the precursor for the formation of mackinawite and additional mineral phases. Figure 1.7 is the Eh–pH diagram of the Fe-S-H_2O system at 25°C, 10^{-5} M Fe^{2+}, and S^{2-}.[28] While mackinawite, troilite, and pyrrhotite have a stoichiometry of very close to 1:1, greigite (Fe_3S_4) has a stoichiometry of 1:1.33, pyrite and marcasite have a stoichiometry of 1:2, which implies that slight oxidation must accompany.[29] Unfortunately, there is little understanding of how these solid phases form and transform.

The thermodynamics of sulfides and their mineral salts are poorly understood, especially at high temperatures, pressures, and salinity. Recent studies by Liu *et al.*[30] and Wang *et al.*[31] extended the thermodynamic solubility prediction of mackinawite (FeS) over a wider range of conditions (0–250,000 mg/L TDS, 77–250°F, pH 3–9) by measuring the solubility

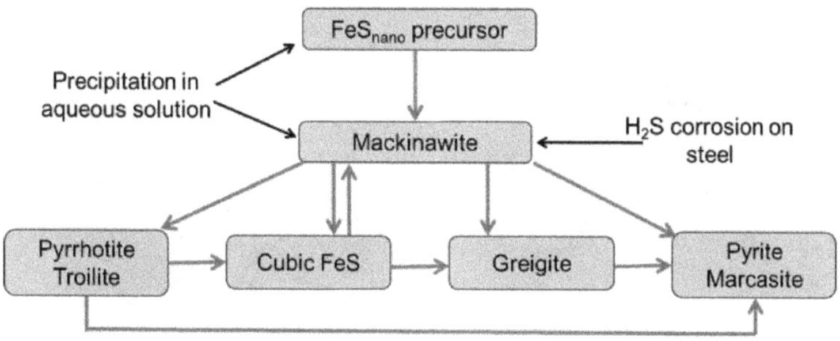

Figure 1.6. A proposed diagram of formation pathways of different ferrous sulfide solid phases.

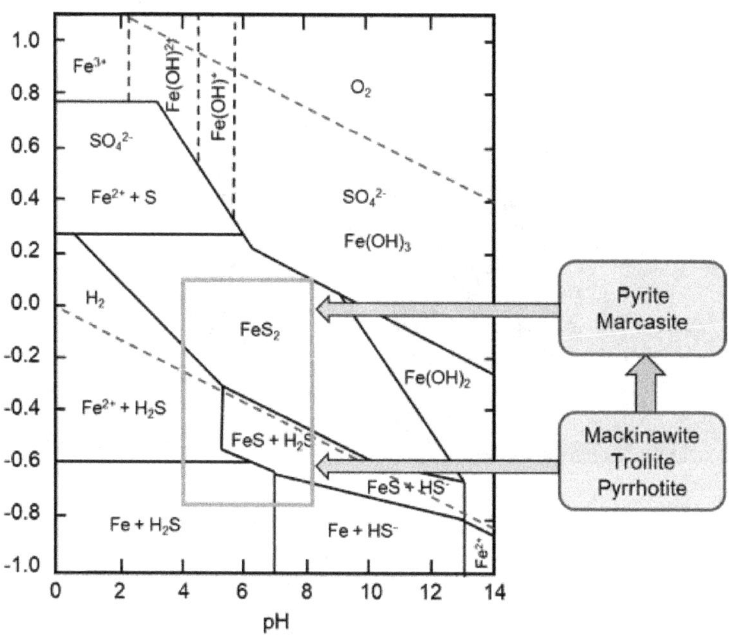

Figure 1.7. The Eh-pH diagram of Fe-S-H_2O system at 25°C, 10^{-5} M Fe^{2+} and S^{2-}. Reproduced with permission from Ref. (28).

of mackinawite and FeS_{aq}^{0} stability constants at different temperature and pressure. The thermodynamic model for FeS scale prediction is included in the latest version of ScaleSoft-Pitzer software, SSP2023.[32]

Mackinawite is the first detectable FeS crystalline phase at low temperatures (<212°F). It was proposed that mackinawite formation kinetics is faster than other stable phases, and mackinawite is the most soluble FeS phase. There is limited understanding of converting acid-soluble FeS to other iron sulfide solid phases, such as troilite, pyrrhotite, or pyrite. Liu *et al.*[30] observed that the formation of troilite depends on reaction time, temperature, and solution ionic strength. The conversion of initial precipitate mackinawite to troilite is more pronounced at low ionic strength and higher temperature (Figure 1.8).

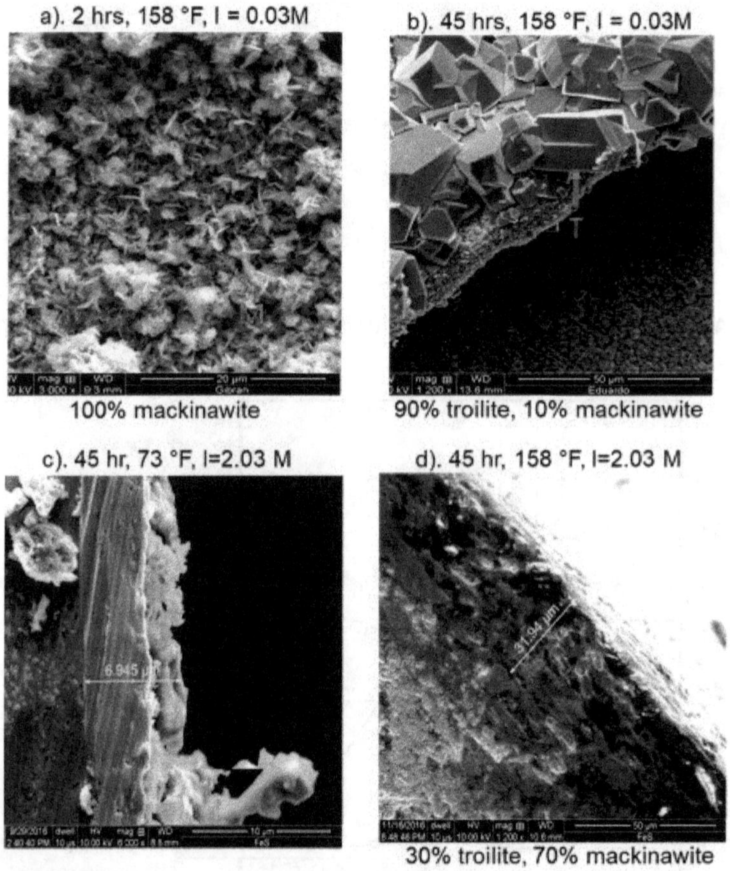

Figure 1.8. Plots of the SEM pictures of the FeS scales formed on 1018 carbon steel coupons at the following conditions overflowed with solutions of Fe(II) and sulfide supersaturated solution at ~0.3 mackinawites SI, at different reaction times, temperature, and ionic strength controlled by the addition of NaCl: (a) SEM of a coupon that was reacted for 2 h, where M is identified as mackinawite, (b) SEM of a coupon that was reacted for 45 h, where T stands for troilite; (c) the cross-sectional SEM of the scale layer at 73°F; (d) the cross-sectional SEM of the scale later at 158°F. The % shown below the SEM graphs are the results of XRD analyses.

Some research has been done on pyritization in marine sediments.[33] However, how pyrite forms in the oil and gas production well is poorly understood.[23, 25, 34] Pyrite must be physically removed once formed on the production tubing. More research is needed to improve the understanding of the formation and behavior of metal sulfides in oil and gas production.

1.3.1.4 Silica and Silicates

Silica and silicates are the two water scale types. Silica scale problems may occur in high-temperature boilers for steam floods or in water associated with geothermal energy production.[6, 35, 36] Silicate scales may become problematic during alkaline surfactant polymer (ASP) flood processes as silica becomes more soluble at higher pH levels.[13] As the ASP water flows to the production well, it encounters neutral pH connate water near the wellbore or in the well.

Due to its limited solubility, silica typically does not exist in natural water at substantial concentrations. The solubility of silica increases with temperature. The temperature range where silica problems arise is circa 400–500°F (204–260°C). Nearly all produced water becomes supersaturated with silicates as temperature decreases, but silicate precipitation is rare. Dissolved silica begins to polymerize and forms colloidal silica.[37] The process of silica monomer species in solution transforming into larger polymers takes time and occurs at a slower pace at lower temperatures. Cations like Mg^{2+}, Ca^{2+}, or Fe^{2+} and Fe^{3+} can bridge the colloidal silicate particles and form magnesium, calcium, or iron silicate scales. Silicate can also coprecipitate with calcium carbonate. No threshold chemical treatment option exists if silica and silicate scales are the problem. Also, silica and silicates do not appear to affect the precipitation kinetics of other common scales.[38]

1.3.2 Acid-Insoluble Scales

The formation of sulfates (such as calcium sulfates, barite, and celestite) and halite scale is not influenced by fluctuations in fluid pH. However, the inhibitors that prevent sulfate scale formation can often be pH-sensitive.

1.3.2.1 Barite

Barite or barium sulfate scale is one of the most common and problematic scales. Under most conditions, barite is among the least soluble scales. Unlike other scales, the chemical removal of the barite scale is extremely difficult. Many oilfield waters contain up to several hundred ppm of barium. Such water usually contains very little sulfate due to the low solubility of barite. When production occurs from a single zone and the well is not being flooded by water from an injection well, the produced water should be at saturation to barite in the reservoir. Barite scale is not expected to form from such "single-zone" wells since the barite is typically more soluble up the production tubing at the typical operation temperature zone of 150–400°F. Barite scale usually forms when water from multiple zones or wells is mixed or when incompatible water is injected for pressure maintenance.[8] For example, when multiple zones are produced, one of the brines high in sulfate and the other high in barium, a barite scale can form when they mix at the bottom of the well. In this scenario, the saturation index (SI) value for barite can be exceptionally high, resulting in the rapid formation of large amounts of scale.[39] The most significant issues of concern generally relate to the few feet around the injection well and up the production well, where conditions can change rapidly.[40, 41] In subsea completions, methanol is often injected to prevent gas hydrates, and these hydrate inhibitors can exacerbate the barite scale problem.[14, 42, 43]

The thermodynamics of barite solubility and the mechanistic understanding of barite nucleation and growth process have been studied extensively. Dai *et al.*[44] proposed the sequence of barite nucleation and growth involved a precluster formation stage,[45] followed by nucleus formation and crystal growth. Figure 1.9 shows the schematic procedure to measure the detectable particles using cryogenic transmission electron microscopy (TEM) and the TEM images. A laser scattering signal was used to detect the formation of detectable particles, where the time it takes to form the detectable particles is often termed the induction time (t_{ind}). The particle sizes were uniformly observed with a radius of 10–35 nm from six supersaturated barite solution experiments (SI = 1.57–2.6) at 25°C. Assuming aggregation is the primary reaction between nucleation and the appearance of crystal laser scattering, Dai *et al.*[44] combined the Smoluchowski theory for aggregation with the Mie light theory to model the laser signal versus time to deduce the nucleation time (t_n). Such a fundamental mechanistic understanding of barite nucleation and growth would establish a common model to predict the kinetics of all scale formation and benefit the study of how chemicals may inhibit scale formation.[46]

Rapidly formed barite (relative to geological times) from produced water often contains up to one strontium ion (Sr^{2+}) for every seven barium ions (i.e., approximately $Ba_{0.88-1}Sr_{0.12-}$

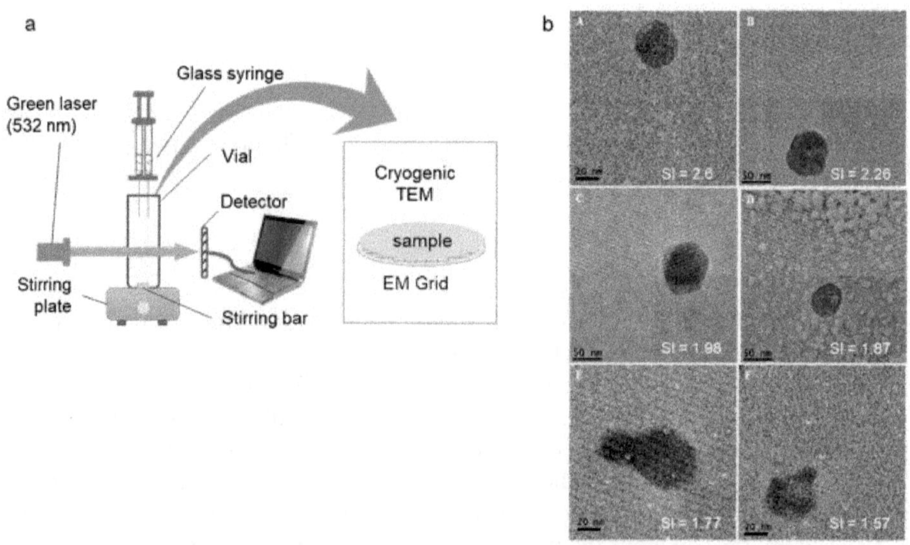

Figure 1.9. (a) The schematic of the experimental procedure for the particle measurement and (b) the measured barite particle sizes with six supersaturated barium sulfate solutions at a saturation index (SI) of 1.57-2.6 and room temperature. Reproduced with permission from Ref. (44).

$_0SO_4$).[47] In some cases, radioactive radium ions ($^{226}Ra^{2+}$) can also substitute into the barite crystal lattice, making them a major source of "naturally occurring radioactive materials" or NORMs.[48] The $Ba_xSr_{1-x}SO_4$ solid solution formed under nonequilibrium conditions has a significantly higher Sr incorporation than those formed under equilibrium conditions, even at the same aqueous phase compositions. Such mixed ion effects on the kinetics of scale formation are common among the different minerals with divalent cations. Figure 1.10 plots the log induction time (t_{ind}) of barite scale formation as a function of the second metal ion (Mg^{2+}, Ca^{2+}, or Sr^{2+}) concentration.

The barium and sulfate concentrations are ~0.97 mm (dotted line) at the experimental condition of Figure 1.10. Note that the based 10 logarithms of the sulfate scale solubility products ($\log_{10} Ksp$) are –4.60 for gypsum, –6.61 for celestite, and –9.97 for barite at 25°C, that is, gypsum is 100 times more soluble than celestite, and celestite is greater than 2,200 times more soluble than barite. In this plot, the solution is fairly undersaturated with respect to gypsum, but the Sr^{2+} data on the right side of the dashed line would exceed the celestite solubility limit. Thus, the data show that both Ca^{2+} and Sr^{2+} in the produced water can slow the rate of barite precipitation, which is consistent with the solid solution formation of Ba_x-$Sr_{(1-x)}SO_4$.[47] Zhao *et al.*[49] observed a change in the inhibitor concentration needed to inhibit the formation of $Ba_xSr_{(1-x)}SO_4$ solid solution. Ca^{2+} concentrations never exceed the solubility of gypsum in the barite. However, Ca^{2+} also slows down the kinetics of barite precipitation, possibly due to the formation of solution phase stable complex of $CaSO_4^0$. Lastly, Bhandari et al.[50] observed increased sulfate nucleation and precipitation rate due to pressure increase. Barite, anhydrite, and celestite nucleation kinetics highly depend on hydrostatic pressure

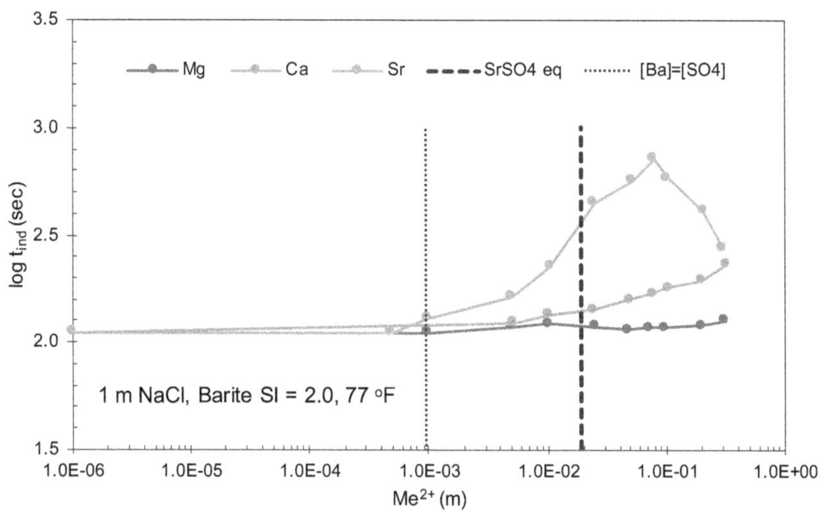

Figure 1.10. The plot of the log induction time (seconds) versus the secondary ion (Me = Mg, Ca, or Sr) concentrations of barite scale formation, where the solution matrix contains 1 M NaCl, equal molar of Ba^{2+} and SO_4^{2-} ions at barite SI = 2.0 and one of the secondary ions at 77°F. The dotted line indicates where [Ba^{2+}] and [SO_4^{2-}] concentrations are at SI = 2.0, and the dashed line indicates where $SrSO_4$ is at equilibrium.

applied even under constant thermodynamic driving force at the same supersaturation level. The pressure effect was attributed to an intrinsic volume collapse in the rate-determining step of the nucleation and crystal growth processes.

Previous discussions on the kinetics of scale formation focused on the nucleation and growth of scale particles. The nucleation kinetics is one of the most important issues related to chemical scale inhibition mechanism and prediction. Recent studies also focused on the mechanism of scale growth and deposition that govern the deposition of scale on the production tubing. Some field observations show that the mineral scale deposition is found in the production well even at conditions predicted to be kinetically stable.[51] The contradiction suggests a missing step of mineral deposition in the flow. Recent research on nucleation in the boundary layer can be used to explain the contradiction,[52] as illustrated in Figure 1.11. Since the solution flow velocity inside the tubing at a laminar flow condition approaches zero near the steel surface, the barium sulfate can nucleate inside the boundary layer at the surface. At the same time, the average reaction time in the reactor is much shorter than the induction time (t_{ind}) predicted by the nucleation theory, where the average reaction time is the time for the bulk of the fluid to travel through the reactor, that is, $t_{reactor}$ (s) = L/v, where L is the reactor length in cm and v is the linear velocity in cm/s.

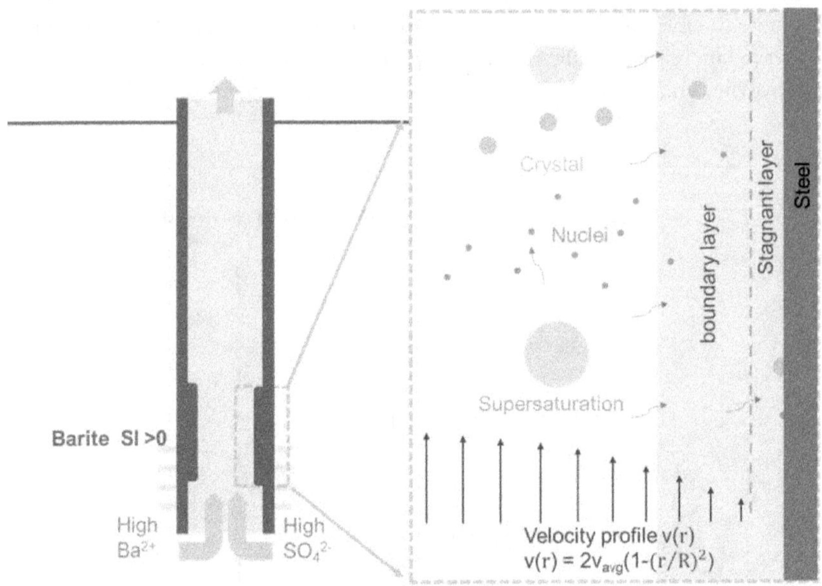

Figure 1.11. Left drawing illustration of barite scale deposition in a downhole location of a production tubing where barite supersaturation may be induced by mixing a high Ba^{2+} produced water with a high sulfate water from a different zone or due to seawater breakthrough. The drawing on the right illustrates how the solution inside the pipe forms $BaSO_4$ clusters, nuclei, and crystals. Based on the fluid dynamic calculation, the liquid flow's linear velocity should reduce from the tubing's center to the edge of a stagnant layer adjacent to the steel surface, where the Ba^{2+} and SO_4^{2-} would be undisturbed and allow the scale to build up.

This conclusion is supported by observing barite scale deposition in a flow-through apparatus and crystal growth in the microfluidic (MF) channels. Due to the negligibly small linear velocity near the tubing wall of MF channels, this stagnant boundary layer offers enough time for the crystals to form without being flushed away, as shown in Figures 1.12(a) and (b). In Figure 1.12(a), six flow-through experiments were tested with $t_{reactor}$ = 5.18 min, and the calculated induction time (t_{ind}) ranged from 8.7 min to 107 h. Barite scale deposition was evaluated after 48 h of flow. Since the resident time in the reactor should be 5.18 min, no scale is expected to form since (5.18 min) if the stagnant layer does not exist. If scale forms within the stagnant layer, one may expect scale formation for all experiments where t_{ind} < 48 h, and no scale should be observed in the experiments where t_{ind} > 48 h. The results shown in Figure 1.12(a) supported the above hypothesis, except for the experiment with t_{ind} = 72 h, wherein a trace amount of scale was still observed. The hypothesis was further validated by four experiments (barite SI = 1.57–1.9, 25°C) using the MF device ($t_{reactor}$ = 0.278 s). Figure 1.12(c) shows four micrographs of the barite SI = 1.57 experiment at four different times, where the particle sizes grew with time. The time for a visible particle appearing in MF is typically similar to, slightly shorter than the predicted t_{ind} (Figure 1.12[b]). The nucleation time (t_n) extrapolated from MF particle sizes versus experimental time matches the theoretical nucleation time well (Figure 1.12[d]), supporting the hypothesis of Figure 1.11.

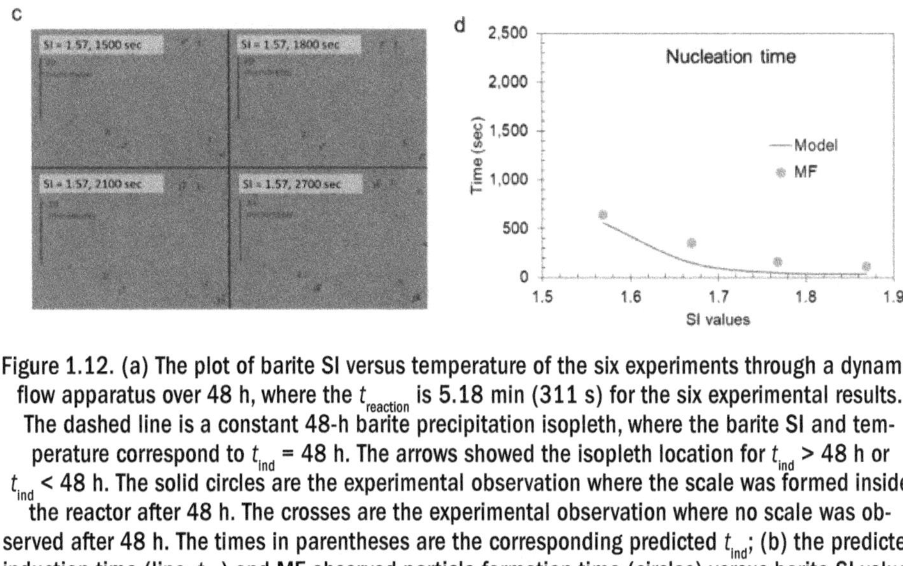

Figure 1.12. (a) The plot of barite SI versus temperature of the six experiments through a dynamic flow apparatus over 48 h, where the $t_{reaction}$ is 5.18 min (311 s) for the six experimental results. The dashed line is a constant 48-h barite precipitation isopleth, where the barite SI and temperature correspond to t_{ind} = 48 h. The arrows showed the isopleth location for t_{ind} > 48 h or t_{ind} < 48 h. The solid circles are the experimental observation where the scale was formed inside the reactor after 48 h. The crosses are the experimental observation where no scale was observed after 48 h. The times in parentheses are the corresponding predicted t_{ind}; (b) the predicted induction time (line, t_{ind}) and MF observed particle formation time (circles) versus barite SI values for four MF experiments at 25°C; (c) the micrograph of the MF experiment at four different times where $t_{reactor}$ is 0.278 s. The barite particles are observed for barite SI = 1.57, and the particle sizes increase with time; and (d) the calculated (line) and observed nucleation time (t_n, circle), extrapolated from MF data. Reproduced with permission from Ref. (52).

1.3.2.2 Celestite

The formation of the celestite scale has become a problem during oil and gas production, especially in the Middle East region.[53–58] To treat these celestite scales efficiently during production, it is important to quantitatively study the kinetics of celestite scale crystallization with and without scale inhibitors. Several studies have investigated celestite crystal growth or spontaneous precipitation rates at various experimental conditions with or without scale inhibitors.[46, 59–61] Nancollas[60] showed that 10^{-9} M of sodium pyrophosphate significantly reduced the celestite crystal growth rate with seeded growth experiments. He *et al.*[62] and Zhao *et al.*[63] studied the nucleation kinetics of celestite in the presence and absence of a scale inhibitor. Most common threshold inhibitors are effective in inhibiting celestites.

Celestite in water-formed scale deposits is often coprecipitated with other alkaline earth sulfate deposits, such as barite. Literature reports on celestite solubility vary greatly, with a review by Monnin and Galinier's[64] finding that its solubility in NaCl follows the Pitzer model. Yet, the observed and predicted barite solubility discrepancy is often due to $Ba_xSr_{(1-x)}SO_4$ solid solution formation. Figure 1.13 plots the log induction time of celestite precipitation at different secondary ion concentrations. The solution contains strontium and sulfate at equal molar (0.024 m) and different second cation concentrations. Note that the solution is supersaturated with celestite (SI = 1.5) at room temperature. Adding Ca^{2+} slows celestite precipitation, while adding a trace level of Ba^{2+} accelerates celestite precipitation. Note that 0.5 μm Ba^{2+} would be saturated in the high-sulfate environment. Therefore, the accelerated precipitation would likely be due to the barite-induced celestite precipitation. Unfortunately, no analyses of the solid samples were available.

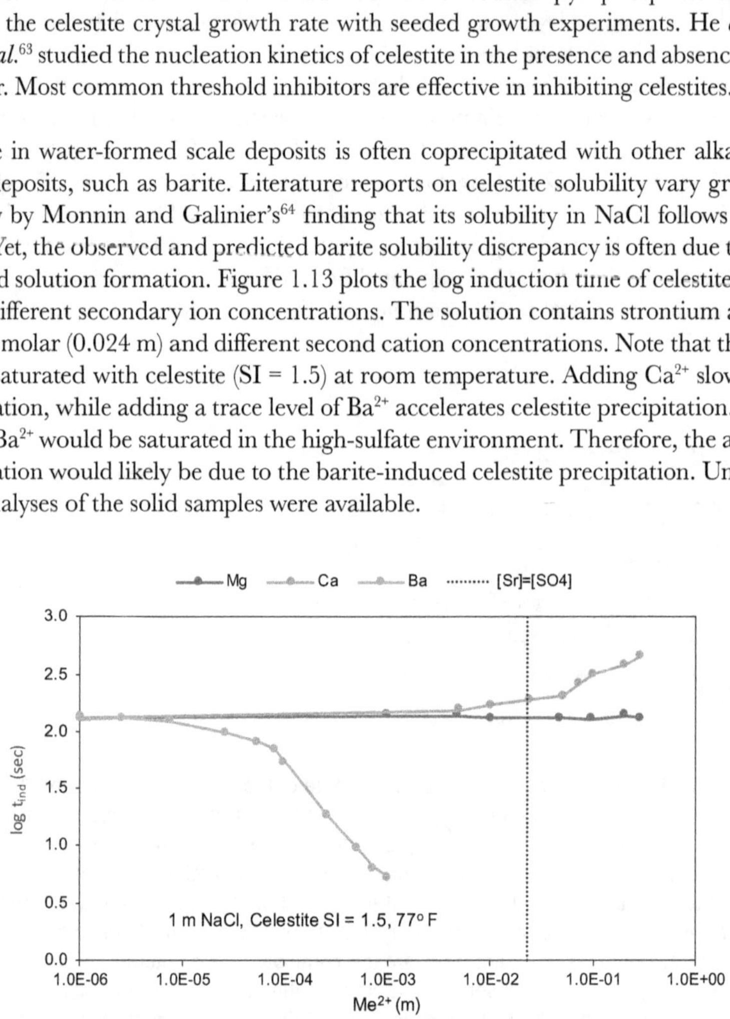

Figure 1.13. The plot of the log induction time (seconds) of celestite formation versus the secondary ion (Mg^{2+}, Ca^{2+}, or Ba^{2+}) concentrations, where the solution matrix contains 1 M NaCl, equal molar of Sr and SO_4 ions at celestite SI = 1.5 and one of the secondary ions at 77°F. The dotted line indicates where [Sr^{2+}] and [SO_4^{2-}] are at SI = 1.5.

1.3.2.3 Calcium Sulfates

Gypsum, hemihydrate (bassanite), and anhydrite are the three solid phases of calcium sulfates that can form in aqueous solutions.[65] The three phases of calcium sulfate differ in the number of water molecules in the crystal formula. Gypsum ($CaSO_4 \cdot 2H_2O$) is stable at low temperatures, while anhydrite ($CaSO_4$) is the commonly observed phase at higher temperatures. Hemihydrate ($CaSO_4 \cdot \frac{1}{2}H_2O$) is a phase that may form in high TDS water at intermediate temperatures (100–121°C). Anhydrite is thermodynamically more stable above about 42–45°C in pure water. However, the transformation from metastable gypsum to anhydrite is very slow. The transition temperatures depend on the salt content of the brine and other chemical species, with higher salt concentrations tending to favor the formation of a solid calcium sulfate phase with fewer water molecules in the crystal formula.

The precipitation of calcium sulfate scale phases is often caused by mixing incompatible waters. It can also occur due to increased temperature during water processing on the surface, such as during membrane filtration, steam flood, or when large quantities of thermodynamic hydrate inhibitors are used for hydrate control. In the presence of hydrate inhibitors such as methanol or MEG, anhydrite was observed to form at much lower temperatures due to the lowering of water activity in their presence.[66]

Anhydrite has been reported in steam flood operations, where the reservoir was heated with steam to facilitate heavy oil production, and the reservoir contains dolomite or anhydrite nodules.[67] It was postulated that steam raising the reservoir temperature would cause the lesser soluble anhydrite precipitation in connate water at equilibrium with the more soluble calcium sulfate phase of a reservoir from the same zone or different zone during mixing.[68]

Gypsum scale is uncommon in conventional oil and gas production since it forms at a low temperature and has high solubility. However, gypsum scale is often problematic in processes such as the CO_2-treated enhanced oil recovery (EOR), where gypsum scale formation leads to production shutting down because of formation damage or equipment failure. In water desalination using a membrane system, gypsum scale fouling is a severe challenge of the membrane system. It can decrease filtration efficiency, compromise permeation quality, and cause membrane damage.[69] Since gypsum is not acid soluble, it is one of the more troublesome scaling minerals in membrane desalination processes.

In CO_2-WAG flooding, carbonate/carbonic acid equilibrium is altered by dissolving carbon dioxide into water to form carbonic acid (H_2CO_3).[70] When the carbonic acid dissociates in favorable pH, bicarbonate and carbonate ions are released into the water, calcite- or dolomite-rich formation rock in the reservoir dissolves. The formation water with Ca^{2+} ions potentially mixes with other produced water from different zones rich in sulfate ions, such as the connate water at equilibrium with anhydrite nodules. Such gypsum scale problems have been noticed in CO_2-WAG flooded wells around West Texas.

The impact of the other alkaline earth divalent cations on gypsum nucleation kinetics was also studied. Figure 1.14 plots the log induction time of gypsum precipitation at different secondary ion concentrations. The solution contains calcium and sulfate at equal molar

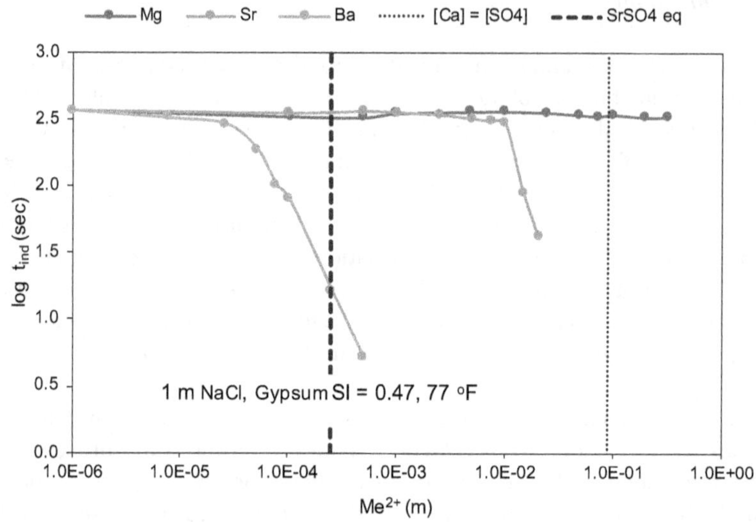

Figure 1.14. The plot of the log induction time (seconds) of gypsum formation versus the secondary ion (Mg^{2+}, Ca^{2+}, or Ba^{2+}) concentrations, where the solution matrix contains 1 M NaCl, equal molar of Ca^{2+} and SO_4^{2-} ions at gypsum SI = 0.47 and one of the secondary ions at 77°F. The dotted line indicates where [Ca^{2+}] and [SO_4^{2-}] are at SI = 0.47, and the dashed line indicates where the solution is at equilibrium concerning celestite.

(0.09 m) and different secondary ion concentrations. Note that the solution is supersaturated with celestite when the Sr^{2+} concentration exceeds 0.025 mm, while a Ba^{2+} concentration of 0.015 μm would be at equilibrium with respect to barite. Both Ba^{2+} and Sr^{2+} accelerated gypsum scale formation, possibly due to the presence of barite or celestite nuclei in the solution to accelerate gypsum precipitation.

1.3.2.4 Halite

Halite precipitation can become a concern in oil and gas production when the ions are near saturation levels. A slight decrease in temperature or evaporation of water causes halite to precipitate. In the most recent USGS-produced water database, over 5,000 data points contain TDS concentrations of over 300,000 mg/L (Figure 1.15). These systems have the potential to have a halite scaling problem. A temperature drop (ΔT) of about 3°C (~5°F) from the saturation state (T_0) would result in $\Delta SI \approx +0.005$ SI changes or approximately 500 lb of precipitate per 1,000 barrels of water produced. Another condition of concern is wherein a production well that is at, or near, equilibrium at T and P is shut in for any reason, and the water in the tubing begins to cool. Halite will begin nucleation almost instantaneously to form a halite scale on the steel tubing, and the scale may not be readily reversed upon rebeginning flow. It is also important to note that the solubility of halite is reduced in the presence of methanol, and halite scale can become a problem at much lower TDS values, around 180,000 mg/L and above, when methanol is present.[42]

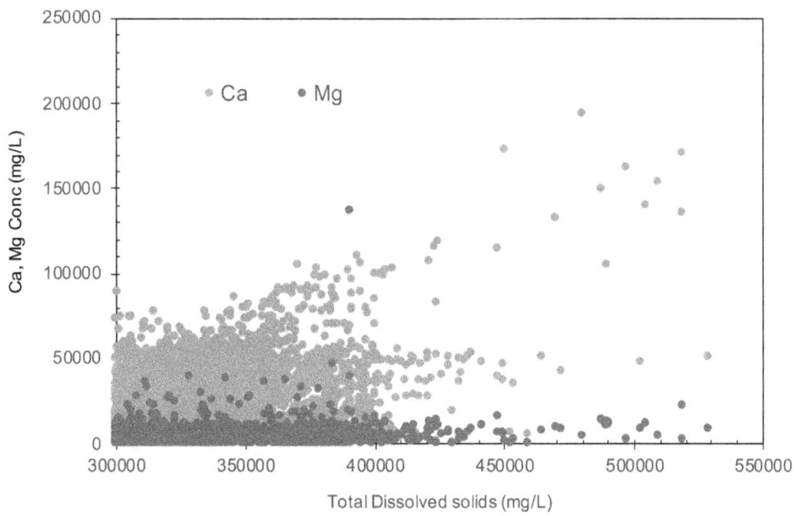

Figure 1.15. A plot of the Ca^{2+} and Mg^{2+} concentration in over 5,000 data points in the USGS produced water database with a total dissolved solid concentration of over 300,000 mg/L.

Additionally, halite scale can be induced to form due to water evaporation with pressure decrease, especially for wells with small water cuts and large gas volume. Note that the Ca and Mg concentrations are often very high in the high-TDS water, which would significantly complicate the halite SI prediction. The halite problem is easily resolved if fresh water is available for injection. A small volume of freshwater can dissolve a large mass of halite precipitate. Still, the halite SI prediction must be accurate. If too little water is added, the scale will form; if too much water is added, excess produced water must be disposed of. The problem is more severe when fresh water is unavailable for injection since few chemicals are effective halite inhibitors.

The halite precipitate formed in a dynamic flow loop is shown in Figure 1.16, where the three pictures on the left to right are halite crystals in the tubing from the fluid entrance, the middle of the tubing, and the tubing toward the fluid exit point. The flow loop was set up to saturate the NaCl water with 20,000 mg/L $CaCl_2$ in the matrix at 100°C and 3,000 psi. The supersaturation is created by cooling the saturated solution in the tubing through the 25°C (77°F) temperature zone, that is, the fluid experienced a $\Delta T = 75$°C (135°F) temperature drop corresponding to 0.15 SI changes. Large halite crystals were formed at the inlet of the tubing, and lesser halite crystals were formed at the end of the tubing, but both SEM and EDX (Energy Dispersive X-ray analyzer) showed that the crystals were predominantly halite with halite crystal habit.[71] It is unclear if Ca^{2+} is incorporated into the halite crystals.

Halite supersaturation can also be induced in the laboratory by adding a small amount of $MgCl_2$ and $CaCl_2$ solution. Figure 1.17 is a plot of the changes in laser intensity versus time when a halite-saturated solution at 25°C (77°F) is brought to a supersaturation state (halite SI = +0.006) by adding a small amount of $MgCl_2$ and $CaCl_2$ solution to a final concentration

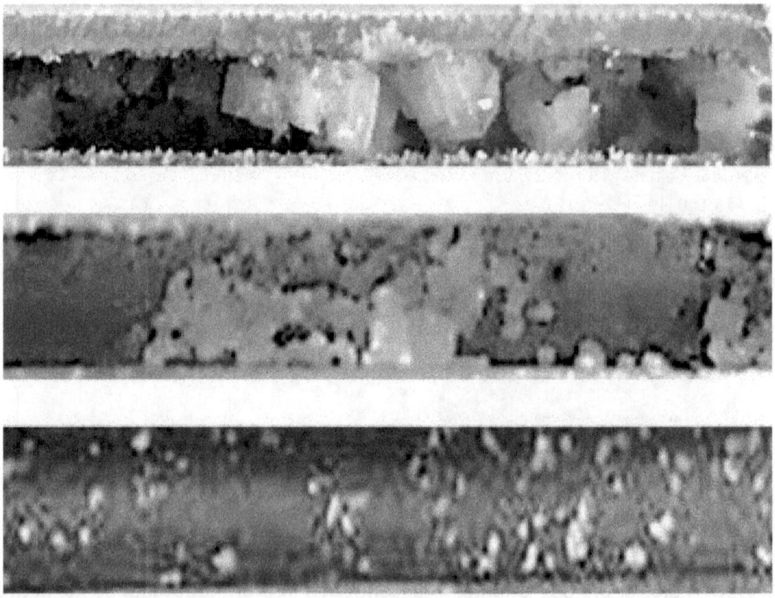

Figure 1.16. Halite scale formation in a dynamic flow loop after the solution was saturated concerning halite at 100°C was brought to a second temperature zone of 25°C to generate a supersaturation solution (halite SI = 0.15) due to a temperature drop of 75°C. The feed solution is a Na-Ca-Cl solution that contains 20,000 mg/L Ca and is saturated at 100°C by flowing over a halite column.

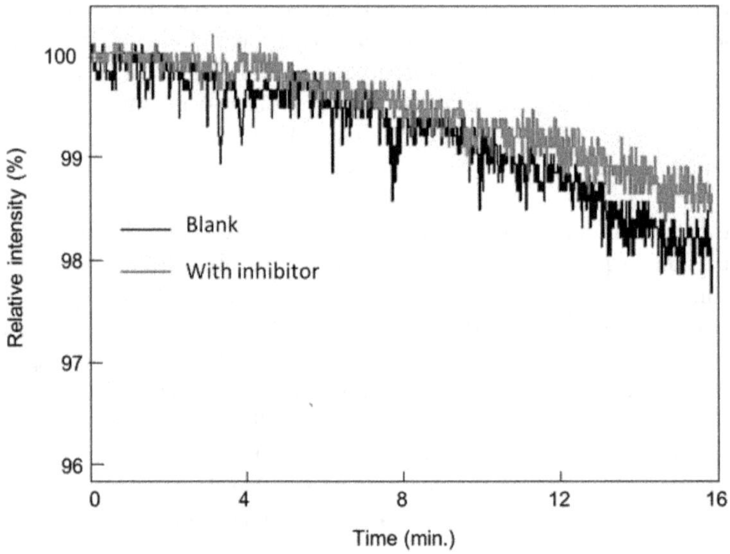

Figure 1.17. A plot of the laser intensity changes of a solution supersaturated with halite at 77°F in the presence and absence of a scale inhibitor. The solution is supersaturated with respect to halite at SI = 0.006 by dosing with the $MgCl_2$ and $CaCl_2$ solution to a halite-saturated solution at 77°F to final concentrations of 480 ppm Ca^{2+} and 174 ppm Mg^{2+}.

of 480 ppm [Ca^{2+}], 174 ppm [Mg^{2+}] in the presence and absence of a scale inhibitor which is a polymer containing carbonate, amide and sulfonate functional groups, where the polyamide is known to have halite crystal modifying properties and sulfonate is known to provide better calcium tolerance at high temperature. The laser signal was used to detect the turbidity of the solution. The laser signal would decrease due to light scattering upon the presence of scale particles in the solution. The data in Figure 1.17 indicates that the laser signal started to decrease at about 1 min after the injection of the $MgCl_2$ and $CaCl_2$ solution, that is, the precipitates are visible within 1 min after the increase in halite SI value to 0.006.[72] A polymeric scale inhibitor cannot control halite scale formation under a low SI value.

Navarathna *et al.*[73] developed a unique experimental design to monitor the halite precipitation by laser scattering as a function of linear changes in the solution temperature (Figure 1.18). The experimental design mimics the dynamic temperature changes in a typical production tubing from downhole to surface. Figure 1.19 illustrates the dynamic relationship of temperature changes to halite SI and the observed nucleation induction time. The reactor temperature was set to be about 5°C higher than the theoretical equilibrium temperature to dissolve the salts in the reactor. The temperature was reduced linearly with time from a temperature above the equilibrium temperature (SI < 0) to a temperature when the solution is supersaturated with respect to halite (SI > 0) using an oven with a programable temperature ramp (Figure 1.18). In Figure 1.19, T_0 is the thermodynamic temperature where halite is at saturation (SI = 0). T_n is the temperature where the laser signal drops, that is, the observed induction time. ΔT (=$T_n - T_0$) is the deviation between the induction and the thermodynamic temperatures. If one assumes that the kinetics of halite precipitation is fast, then a negative ΔT may imply the true halite solubility is lower than the predicted thermodynamic solubility and vice versa for a positive ΔT. However, a positive ΔT may also indicate that the kinetics of halite formation is measurably slower.

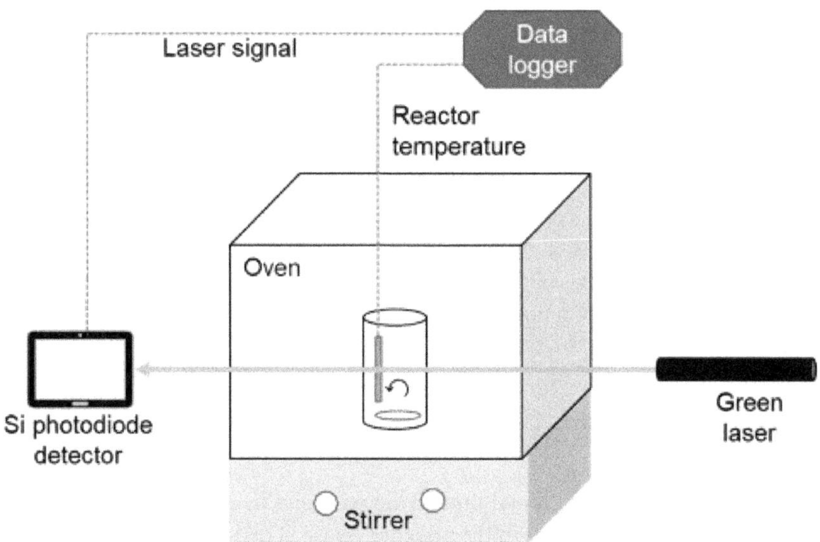

Figure 1.18. Experimental design to study the precipitation of halite as a function of solution temperature.

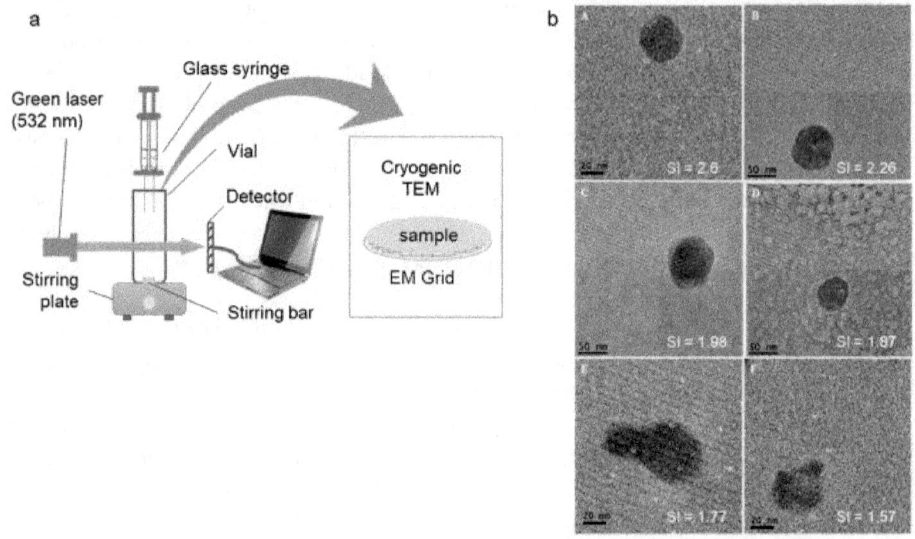

Figure 1.19. Schematic diagram of the relationship of temperature, SI and laser signal for the experimental setup, where T_0 is the calculated equilibrium temperature and the corresponding to SI = 0, T_n is the temperature when halite particles were observed at a corresponding SI > 0, ΔT is the differences in temperature of the theoretical versus observed temperature.

The laser signal changes of three experiments with varying concentrations of $CaCl_2$ and NaCl in solution, where the solutions contain 0–1.47 m $CaCl_2$ (i.e., 0 to ~60,000 mg/L Ca), are plotted in Figure 1.20. All three solutions shared the same equilibrium temperature of 115°C (T_0). The theoretical equilibrium temperature (T_0) is calculated by the Pitzer model (SSP2023). Note that ΔTs for all three experiments are positive and $\Delta T < 1.2$°C. The nucleation induction temperatures where particles were observed (T_n) are less than 1.2°C deviation from the theoretical equilibrium temperature of 115°C (T_0), which is equivalent to a deviation of less than 0.0037 halite SI value (Figure 1.20). The results prove that our Pitzer model can predict the theoretical temperature of the halite solution containing 0–60,000 mg/L $CaCl_2$ with a deviation of less than 0.9% from the true saturation. Such an accurate prediction has never previously been reported. The results establish this method for halite inhibitor evaluation. Navarathna *et al.*[73] showed a significant increase in ΔT with several chemical inhibitors as the chemicals slow the halite precipitation kinetics. More research is underway to document the accuracy of such predictions at higher pressures relevant to oil and gas production.

1.4 Scale Prediction

Scale prediction is primarily based on thermodynamics by calculating the saturation state with respect to various minerals in the aqueous phase. This section covers the fundamental principles. The next three chapters (Chapters 2, 3, and 4) provide more detailed discussions on this subject. The saturation state is typically represented by the saturation ratio (SR) or the SI, where the SR is the ratio of the ion activity product divided by the solubility prod-

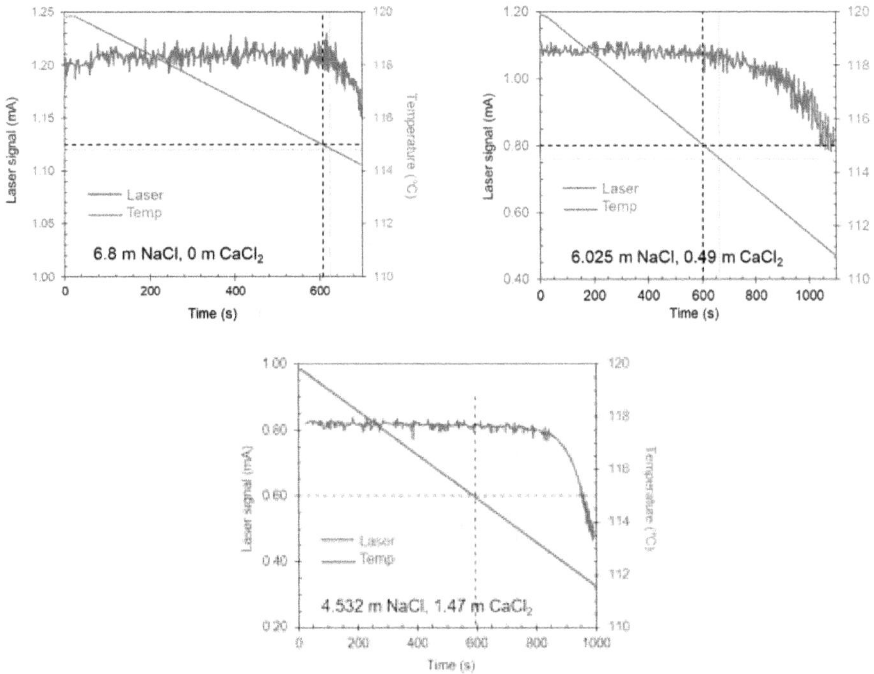

Figure 1.20. The experimental results of three halite precipitation experiments of decreasing NaCl and increasing $CaCl_2$ concentrations so that the predicted halite solution is at the equilibrium temperature (T_0 = 115°C, the top horizontal lines). The lower horizontal line indicates the observed temperatures (T_n) when the laser signal began to decrease.

uct, and SI is the 10-based logarithm of the SR value, as listed in Equation (1.2) with barite as an example.

$$BaSO_4 \rightarrow Ba^{2+} + SO_4^{\ 2-} \tag{1.1}$$

$$\begin{aligned} \text{SI (barite)} &= \log_{10}(\text{SR}) = \log_{10}\frac{\{Ba^{2+}\}\cdot\{SO_4^{\ 2-}\}}{K_{\text{sp,barite}}} \\ &= \log_{10}\frac{[Ba^{2+}]\cdot\gamma_{Ba^{2+}}\cdot[SO_4^{\ 2-}]\cdot\gamma_{SO_4^{\ 2-}}}{K_{\text{sp,barite}}} \end{aligned} \tag{1.2}$$

where [ ] represents ion concentration, { } denotes ion activity, γ_i is the activity coefficient and $K_{sp\text{'}barite}$ is the thermodynamic solubility product of the barite scale.

Equation (1.1) can be adapted for other acid-insoluble mineral phases by substituting the activity of Ba^{2+} and $SO_4^{\ 2+}$ with the corresponding lattice ion activity and their corresponding solubility products. The SR or SI equation for the carbonate and sulfide system needs to consider the speciation of the acid gases in water, and it is more complicated. Below are the equations for calcite:

$$CaCO_{3,s} \rightarrow Ca^{2+} + CO_3^{2-} \tag{1.3}$$

$$\text{SI (calcite)} = \log_{10} \frac{\{Ca^{2+}\}\cdot\{CO_3^{2-}\}}{K_{\text{sp, calcite}}} \tag{1.4}$$

Carbonate ions (CO_3^{2-}) are related to bicarbonate (HCO_3^-) and dissolved CO_2 ($CO_{2,aq}$) species by following equilibrium equations:

$$CO_{2,aq} + H_2O \rightarrow H^+ + HCO_3^-, \qquad K_1 = \frac{\{HCO_3^-\}\cdot\{H^+\}}{\{CO_{2,aq}\}\cdot\{H_2O\}} \tag{1.5}$$

$$HCO_3^- \rightarrow H^+ + CO_3^{2-}, \qquad K_2 = \frac{\{CO_3^{2-}\}\cdot\{H^+\}}{\{HCO_3^-\}} \tag{1.6}$$

By combining the above equations, the overall reaction for calcite scale formation can be described as below:

$$Ca^{2+} + 2HCO_3^- \rightarrow CaCO_{3,s} + CO_{2,aq} + H_2O \tag{1.7}$$

Thus, the SI for calcite can be expressed as:

$$\text{SI (calcite)} = \log_{10}\left[\frac{\{Ca^{2+}\}\cdot\{HCO_3^-\}^2}{\{CO_{2,aq}\}\cdot\{H_2O\}}\cdot\frac{K_2}{K_{\text{sp, calcite}}\cdot K_1}\right] \tag{1.8}$$

The Pitzer model can calculate both the activity coefficients and water activity. Suppose the brine is also at equilibrium with CO_2 in the gas phase:

$$CO_2^{gas} \rightarrow CO_{2,aq} \quad K_H = \frac{\{CO_{2,aq}\}}{P_{CO_2}\varphi_{CO_2}} \tag{1.9}$$

Equation (1.8) can be replaced by Equation (1.10):

$$\text{SI (calcite)} = \log_{10}\left[\frac{[Ca^{2+}]\cdot[HCO_3^-]^2}{P_{CO_2}}\cdot\frac{\gamma_{Ca^{2+}}\cdot\gamma_{HCO_3^-}^2}{\varphi_{CO_2}}\cdot\frac{K_2}{K_{\text{sp, calcite}}\cdot K_1\cdot K_H}\right] \tag{1.10}$$

where P_{CO_2} and φ_{CO_2} are CO_2 partial pressure and fugacity coefficient, respectively.

Sulfide equations can be derived by substituting the metal ions for Ca^{2+} and the sulfide equilibrium constants for carbonate equilibrium constants, using FeS as an example:

$$FeS \rightarrow Fe^{2+} + S^{2-} \tag{1.11}$$

$$\text{SI (FeS)} = \log_{10} \frac{\{Fe^{2+}\}\cdot\{S^{2-}\}}{K_{\text{sp,FeS}}} \tag{1.12}$$

$$H_2S_{aq} \rightarrow H^+ + HS^- \quad K_{1,H_2S} = \frac{\{H^+\}\cdot\{HS^-\}}{\{H_2S_{aq}\}} \tag{1.13}$$

$$HS^- \rightarrow H^+ + S^{2-} \quad K_{2,HS^-} = \frac{\{H^+\}\cdot\{S^{2-}\}}{\{HS^-\}} \tag{1.14}$$

However, the second ionization constant of sulfide varied considerably in the literature (*e.g.*, pK_2 = 13.8, 15.3, 17.4 at 25°C).[74–76] The conventional solubility product for sulfide is defined as in Equation (1.15), using mackinawite as an example. The solubility products for ZnS and PbS are defined similarly.[18]

$$FeS + H^+ \rightarrow Fe^{2+} + HS^- \tag{1.15}$$

where $K_{sp,mackinawite} = K_{sp}/ K_{2,HS^-}$.

Thus, the SI calculations for the sulfide scale are defined below:

$$\text{SI (mackinawite)} = \log_{10}\left[\frac{\{Fe^{2+}\}\cdot\{HS^-\}}{\{H^+\}}\cdot\frac{1}{K_{sp,mackinawite}}\right] \tag{1.16}$$

For the overall reaction:

$$FeS_s + H_2S_{aq} \rightarrow Fe^{2+} + 2HS^- \tag{1.17}$$

$$\text{SI (mackinawite)} = \log_{10}\left[\frac{\{Fe^{2+}\}\cdot\{HS^-\}^2}{\{H_2S_{aq}\}}\cdot\frac{1}{K_{sp,mackinawite}\cdot K_{1,H_2S}}\right] \tag{1.18}$$

In the SI equation, the SI is calculated from the measured ion concentrations multiplied by a calculated activity coefficient (γ_i) to account for the impact of ion interaction on the activity of the ions and divided by the equilibrium constants, such as the solubility product for the acid insoluble scales or the product of equilibrium constants for the acid-soluble scales. The fugacity of gas is calculated using the Peng–Robinson cubic equation of state with Peneloux volume shift and activity coefficients calculated by the Pitzer model.

A comparison of the solubility products of different mineral scales showed a span of 20 orders of magnitude among the different scales mentioned above (Figure 1.21).

Figure 1.21 plots the logarithm of the solubility products of acid-insoluble and acid-soluble minerals over various temperatures. The temperature dependence of such equilibrium constants is that of Kharaka *et al.*[18] For highly soluble scales, such as halite and gypsum, a large amount of scale can form when the solution is barely supersaturated. In contrast, the amount of scale formed is less detectable if the solubility product of the mineral, such as barite, is of concern. The solubility product is usually derived in two methods. The first method is to calculate the solubility product from the Gibbs free energy approach, where the net free energy change can be derived based on other thermodynamic properties of

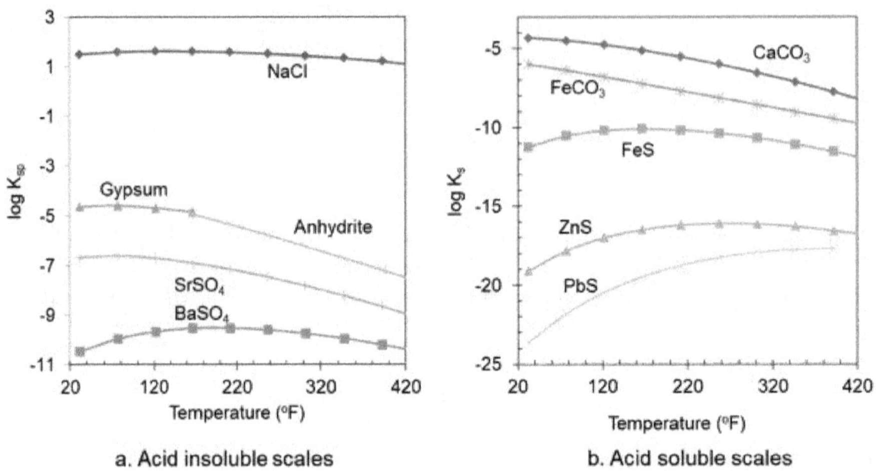

Figure 1.21. Comparison of the solubility products of common oilfield scales where (a) is the temperature dependence of log K_{sp} for the acid-insoluble scales and (b) is the temperature dependence of the log K_s values of the acid-soluble scales, where $K_s = K_{sp}K_1/K_2$ for both the carbonates and sulfides.

this reaction.[18, 77] The second method is to extrapolate the mineral solubility to zero ionic strength based on the measured mineral solubility. At zero ionic strength, the ion activity coefficients are assumed to be 1, and thus, the solubility product can be calculated as the product of ion molalities.[10, 78] However, strong ion pairs (*e.g.*, $CaSO_4^0$) might impact the second method. If the solubility product and ion activity coefficients can be accurately predicted, the mineral solubilities can be calculated at any temperature, pressure, and solution composition. Wang[79] has summarized a significant uncertainty of K_{sp} values of the sulfate scale from the uncertainty of the chemical potentials of the cation (Me^{2+}) and anion (SO_4^{2-}) at infinite dilution, at high temperature. Such uncertainty in K_{sp} values may require further investigation for the geothermal energy applications.

Many thermodynamic models have been proposed to predict the thermodynamic properties of electrolyte solutions. Milner[80, 81] tried to use statistical thermodynamics to approximate the excess free energy and concluded that both the osmotic coefficient of solvent (water) and the activity coefficient of ion are proportional to the square root of electrolyte molality when the solution is very dilute. Debye and Hückel derived the famous Debye–Hückel (D–H) limiting law based on Boltzmann's law and Poisson equation.[82] It approximates the exponential ion density radial distribution function with the first two terms of its polynomial expansion. It exactly reflects the nonideality due to long-range electrostatic interionic attraction in the limit of infinite dilution.[83] When electrolyte concentration is larger, the statistical thermodynamics suggests the appropriate functional forms to account for the short-range intermolecular potential and fluid properties, especially at high ionic strength.[84–88] Among them, Pitzer's theory is one of the most widely adopted models for mixed electrolyte solutions, adopting the appropriate form and the fitted virial coefficients to represent the nonideality due to short-range intra-ion interactions. The Pitzer theory correlates the nonideal intermolecular or ionic interactions to the macroscopic properties of the system with a more solid theoretical basis. It provides the flexibility and capability of tackling electrolyte solutions over much wider temperature, pressure, and ionic strength ranges with the virial coefficients. As a result, it has become one of the most accurate and widely adopted thermodynamic models in geochemical and industrial applications. It has been widely adopted by different geochemical software, such as Geochemist's Workbench, PHRQPITZ, PHREEQC, and ScaleSoftPitzer.[4, 89–92]

Oil and gas production at ultra-high temperatures, pressure, and TDS water poses significant challenges to scaling prediction and control. Several researchers[33, 91, 93–96] have conducted experimental research to establish comprehensive thermodynamic data relevant to the oil and gas production regime. For example, the anhydrite solubility in pure water[97–99] and NaCl solution[93, 100, 101] were studied at high temperatures and pressures. The solubility of calcium sulfates in calcium chloride solution had only been studied recently at a temperature up to 90°C.[102] Azimi and Papangelakis[103] and Deng *et al.*[104] presented some anhydrite solubility data in a mixed electrolyte system at 25–250°C. These solubility data have been included in scale prediction evaluation for high-TDS water.

From adopting a carefully selected set of virial coefficients, critical constants, and experimental data from the literature, additional virial coefficients and constants were fitted in a simultaneous regression model. In this way, this model overcomes the inconsistency problem that usually occurs in developing Pitzer theory and extends the applicable ranges of Pitzer theory. Table 1.2 summarizes the ranges of conditions for the experimental data included in the thermodynamic modeling. To compile this table, about 10,000 measurements from circa 1850 to present were examined, and about 6,000 results were kept after quality control screening. A few of the measurements were from the authors' laboratory.

Table 1.2: The range of literature available data for scale prediction modeling.

Parameter	*T* (°C)	*P* (bar)	Na^+ (M)	K^+ (M)	Mg^{2+} (M)	Ca^{2+} (M)	Cl^- (M)	SO_4^{2-} (M)	SI Mean	Std Error
Calcite[A]	0-250	1-1,500	<6.2	<1.14	-	<0.66	<6.2	<0.05	0.006	0.0035
Barite	0-250	1-1,500	<6.0	<0.25	<0.16	<0.55	<6.0	<0.03	-0.007	0.0039
Gypsum	0-100	1-1,400	<6.4	-	<0.05	<0.11	<6.4	<0.12	-0.016	0.0028
Anhydrite	95-250	1-1,400	<6.3	-	-	<0.04	<6.3	<0.04	0.023	0.0098
Celestite	25-250	1-575	<5.0	<1.4	<0.12	<0.05	<5.0	<0.05	-0.020	0.0074
Halite	0-250	1-1,500	5.6-9.0	-	-	<0.41	6.1-9.0		0.000	0.0017
FeS[B]	23-120	1-5	<5	-	-	-	<5		-0.005	0.016

[A]Alkalinity range 0-0.07 m; H_2S < 3% in gas or dissolved sulfide <0.04 m.
[B]FeS refers to mackinawite solubility.

Due to the nature of the virial coefficient parameter, the parameter range shown in Table 1.2 should be considered as the constraint for such a thermodynamic model. Saturation indices for applications outside the range in Table 1.2 may post more uncertainty. Most produced water shows a high Ca/SO_4 ratio, either because the brine with high Ca concentration was mixed with low SO_4 solution after water injection or because SO_4 was reduced by sulfate-reducing bacteria. About 95% of the produced water in the USGS database has a Ca concentration of less than 0.44 m.[4] However, extreme Ca concentrations are also reported in some North Dakota fields and Marcellus shale. The produced water from Marcellus shale contains 19,000–39,200 mg/L Ca, with a TDS range of 200,000–300,000 mg/L. Such high Ca concentrations are attributed to the dolomitization process, that when dolomite is formed by recrystallization of calcium carbonate in an Mg-rich brine, Ca is replaced by Mg in its lattice position and is released into the aqueous phase.[105] High Ca and TDS brine could also originate from deep gas wells, where gradual evaporation of water from the solution into the gas phase concentrates the connate water.

Figure 1.22 plots the calculated halite SI values of a set of experimental data where the solution matrix contains high $CaCl_2$ or $MgCl_2$. $CaCl_2$ and $MgCl_2$ affected the solubility of NaCl due to the common ion effect. However, the theoretical saturation indices of these solutions should be 0.00, assuming the solutions are at equilibrium with halite. The dotted horizontal lines are the quality lines at ±0.005 halite SI values, showing the maximum halite SI error tolerance. The thermodynamic model fails when the Ca^{2+} or Mg^{2+} concentrations exceed 0.5 m. Further research in the thermodynamic modeling for high-TDS water with >20,000 mg/L Ca^{2+} is needed, and it is one of the authors' future research thrusts.

Lastly, accurately interpreting the partitioning of H_2O in different phases is important to account for the scaling problem in a well due to evaporation and condensation for low water

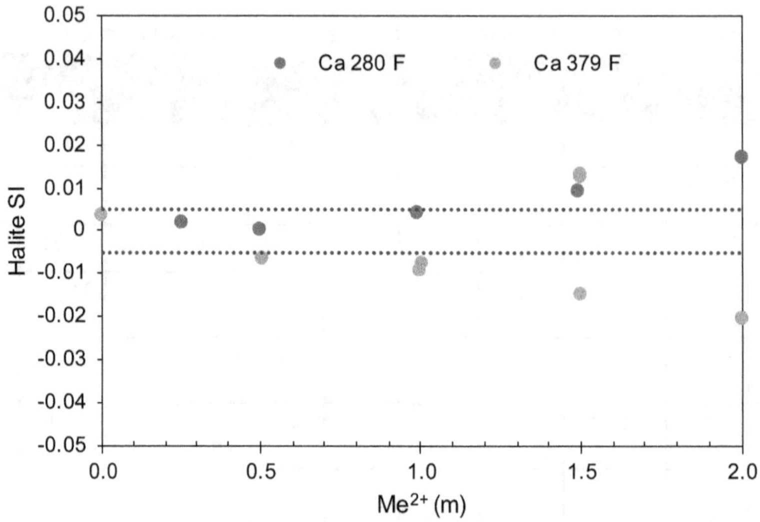

Figure 1.22. A plot of calculated halite SI values versus the Ca^{2+} or Mg^{2+} concentrations. The dotted lines are the quality lines at ±0.005 SI.

cut production. Also, the partitions and speciation of CO_2 and H_2S in different phases and the phase behavior are important to describe the scaling of acidic scales accurately. Kan *et al.*[12] summarize a significant improvement in models to accurately model the partitioning of H_2O, CO_2, and H_2S in hydrocarbon/aqueous phases and the aqueous ion activities at ultra-high temperature, pressure, and mixed electrolyte conditions. The equation of state used to model the phase behavior was Peng–Robinson with Peneloux volume shift (PR-P). Still, some important modifications were introduced to calculate the fugacity of the acid gases and H_2O.

1.5 Chemical Treatments for Scales

Many chemical scale inhibitors are effective in controlling both calcite and barite scales. Treatment of calcite scale can easily be achieved by low dosage continuous injection of a scale inhibitor or via a squeeze treatment. Also, the calcite scale can be remediated by acidizing the well to dissolve the deposit. However, the barite scale is harder to remediate if it forms. Scale inhibitor may be applied by continuous injection into the surface equipment or downhole via a small diameter treat string or the annulus. Inhibitors can also be forced into the formation via a squeeze treatment to release the inhibitor when production commences slowly. Injection of inhibitors into the surface equipment would not protect the production tubing, and installing a downhole treat string can be prohibitively expensive and is often a source of failure. An inhibitor squeeze, which protects the near wellbore formation, the production tubing, and the surface equipment, is often the preferred scale control method.[106–109] All treatment methods require a good knowledge of the chemical choices and minimum treatment concentrations.

1.5.1 Common Scale Inhibitors

The common scale inhibitors used in oil fields include phosphonates, phosphate esters, polymers of acrylic acids, sulfonic acid or maleic acid copolymers, and environmentally safe inulin or aspartate-based polymers. Table 1.3 lists the chemical names and abbreviations of 18 common scale inhibitors.

Table 1.3: Common scale inhibitors for calcite and the sulfates scales.

Abbreviation	Chemical Name
Phosphonate	
ATMP	Amino tris(methylene phosphonic acid)
BHPMP	Bis-hexamethylene triamine-penta(methylene phosphonic) acid
DETPMP	Diethylene triamine-penta(methylene phosphonic) acid
HDTMP	Hexamethylene diamine-tetra(methylene phosphonic) acid
EDTMP	Ethylenediamine tetra (methylene phosphonic) acid
HEDP	1-Hydroxy ethylidene-1,1-diphosphonic Acid
HEBMP	Hydroxyethyl amino bis(methylene phosphonic) acid
PhnAA	Phosphonated amino acid
PAPEMP	Polyamino polyether methylene phosphonic acid
PBTC	2-Phosphonobutane-1,2,4-tricarboxylic acid

Table 1.3: Common scale inhibitors for calcite and the sulfates scales (continued).

Abbreviation	Chemical Name
Phosphate ester	
TEAPE	Triethanolamine phosphate ester
Polycarboxylate	
PPCA	Phosphinopolycarboxylic acid
MAC	Phosphorous incorporated maleic acid polymer
Polysulfonate/polycarboxylate	
SPCA	Sulfonated polycarboxylate
PVS	Polyvinyl sulfonate
AA/AMPS	Sulfonated polyacrylic acid copolymer
Environmentally Friendly Inhibitor	
PASP	Polyaspartic acid
CMI	Carboxymethyl inulin (0.25 substitution)

1.5.2 Calcite and Sulfate Scales Inhibition

Calcite, barite, celestite, and gypsum scales can normally be controlled with chemical scale inhibitors at normal operating conditions. The scale inhibitors are typically evaluated using a dynamic scale loop (DSL) or a laser batch reactor,[110, 111] as shown in Figures 1.23 and 1.24. Both figures are for calcite scale inhibition, where a combination of (a) CO_2/alkalinity or (b) alkalinity/HCl is used for pH and SI control.[112–114] For the acid-insoluble scales, such as barite, pH can be controlled with a pH buffer agent, and therefore, the apparatus can be significantly simplified. Fan *et al.*[115] developed a coaxial tubing for the DSL apparatus to minimize the excess adsorption of scale inhibitors on the steel tubing at high temperatures. Adsorption of scale inhibitors on steel tubing could cause erroneous results unless the equipment is thoroughly cleaned. The DSL is typically used to simulate a specific produced brine under production conditions. The result is difficult to apply to other well conditions.

The batch reactor shown in Figure 1.24 can simulate temperature and SI conditions to develop a quantitative profile of the minimum effective inhibitor concentration (MIC) needed for scale control at a given T, SI conditions. It has been demonstrated that the MIC generated with this approach provides quantitative MIC values that agree well with that produced by the DSL test method.[116] Note that the inhibitor concentrations discussed in this chapter refer to active component concentrations.

He *et al.*[117] derived a semiempirical equation to predict the minimum inhibitor concentration to inhibit scale (MIC) from a large number of nucleation kinetic data of barite and calcite (Equations [1.19] through [1.21]).

$$\log_{10}(t_0) = \alpha_0 + \frac{\alpha_1}{\mathrm{SI}} + \frac{\alpha_2}{T} + \frac{\alpha_3}{\mathrm{SI}\cdot T} \tag{1.19}$$

$$b_{\text{inh}} = 10^{\beta_0 + \beta_1 \cdot \text{SI} + \frac{\beta_2}{T} + \beta_3 \cdot \text{pH} + \beta_4 \cdot \log R_{\text{inh}}} \tag{1.20}$$

$$C_{\text{inh}}\ (\text{mg/L}) = \frac{1}{b_{\text{inh}}\ (\text{L/mg})} \log\left(\frac{t_{\text{inh}}\ (\text{s})}{t_0\ (\text{s})}\right) \tag{1.21}$$

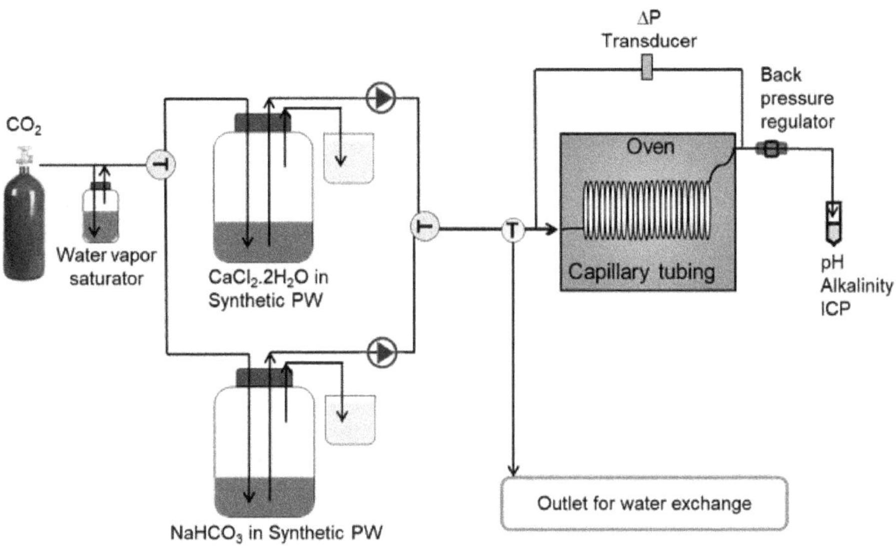

Figure 1.23. Plots of a dynamic flow-through apparatus for calcite scale inhibitor evaluation.

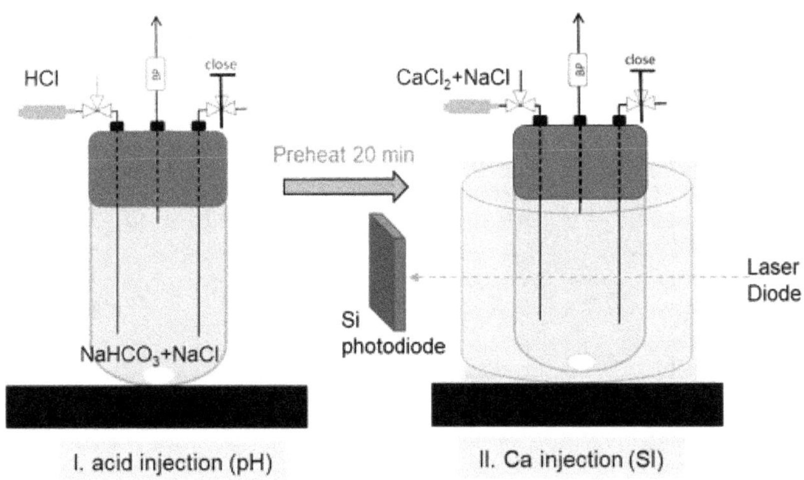

Figure 1.24. Plots of a batch reactor for calcite scale inhibitor evaluation.

where t_0 (s) is the time required for scale to form a precipitate, t_{inh} (s) is the time required for scale inhibition at a specific inhibitor concentration (C_{inh}), and b_{inh} (L/mg) is the inhibitor-specific inhibition coefficient.

Figure 1.25 shows a typical observation using the batch reactor shown in Figure 1.24, where the data were for calcite nucleation inhibition study at a calcite SI of 1.31 at pH 6.5 and 90°C with DETPMP as the scale inhibitor. Note that the nucleation may not occur immediately and is related to the mineral type, degree of supersaturation, and temperature. The induction time without an inhibitor (t_0) typically varies with the degree of supersaturation and temperature (Equation [1.19]). With the presence of inhibitors, the nucleation induction time (t_{inh}, s) can be exponentially increased with higher inhibitor concentration. A linear relationship between the log(t_{inh}) and inhibitor concentration (Equation [1.21]) was observed over a wide range of conditions.[63, 115, 117–119]

The slope of the linear correlation shown in Figure 1.26 corresponds to a specific b_{inh}, and the intercept is $\log_{10}(t_0)$. Both t_0 and b_{inh} have been correlated to temperature (T, K) and SI values (Equations [1.19] and [1.20]) with α_x and β_x are semiempirical parameters for the correlation from a large set of experimental data. Furthermore, b_{inh} is a function of solution pH and lattice ion concentrations (R_{inh}). Therefore, t_0 is specific to a production condition (temperature and produced water composition). MIC is the minimum inhibitor concentration required to protect the well. Thus, MIC would equal C_{inh} at a user-defined inhibition time (t_{inh}), typically the time for the fluid to travel from downhole to surface equipment. We often assign a protection time of 2 h, which seems suitable for most applications.

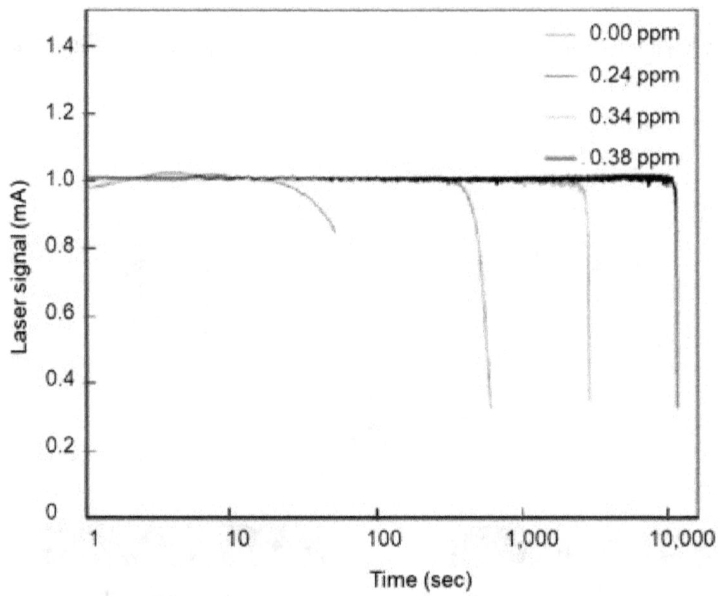

Figure 1.25. Plots of the typical results of laser signal versus time of four batch scale experiments using the apparatus of Figure 1.24 with 0 mg/L, 0.24 mg/L, 0.34 mg/L, and 0.38 mg/L DETPMP.

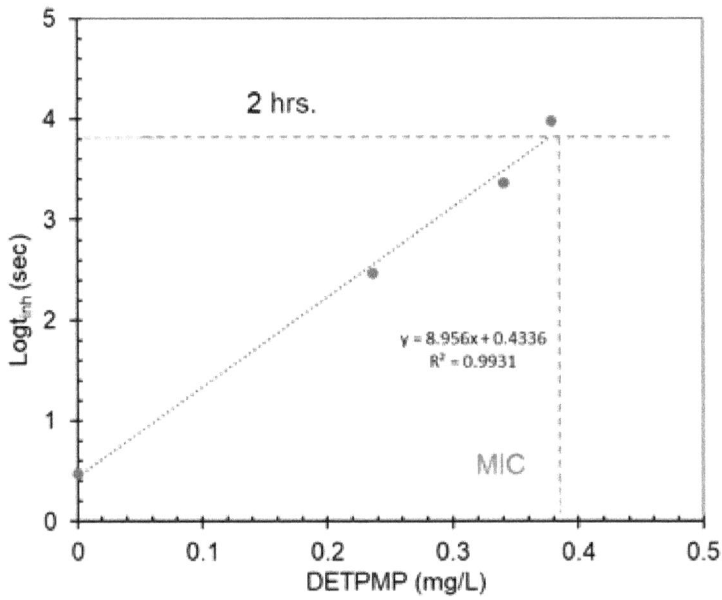

Figure 1.26. Plot of the based 10 logarithms of time versus DETPMP concentration. The experimental solution is supersaturated with calcite (calcite SI = 1.31, 90°C, pH 6.5, background electrolyte 1 m NaCl).

Dai *et al.*[120] advanced He's model to predict the kinetics of nucleation based on the classical nucleation theory and inhibition effect as an impact on changes in Gibbs free energy or interfacial tension. Figure 1.27 plots the agreement of the measured and predicted barite $\log_{10}(t_{inh})$ for five common scale inhibitors using Dai's model, where the black solid lines show that measured and predicted induction time are equal; the black dashed lines show that the predicted $\log_{10}(t_{ind})$ = predicted $\log_{10}(t_{ind}) \pm 0.5$.

Dissolved iron in water can have detrimental effects on scale inhibition. Both corrosion and Fe(II) containing minerals in the formation could be the source of Fe(II) in produced water. Fe(III) may form in the water produced due to proton reduction at extreme anoxic conditions or exposure to air by chemical injection or in surface equipment. The presence of ferric ions can have a detrimental impact on the efficiency of scale inhibitors due to the strong interaction of Fe(III) with the scale inhibitor (Figure 1.28), where the inhibition efficiency is expressed as the ratio in the percentage of the measured inhibition time in the presence of a fixed concentration of Fe to that in the absence of Fe. Ferric ion has a detrimental impact on both phosphonates and polymeric scale inhibitors.[121] It was hypothesized that the rapid nucleation of ferric oxide nanoparticles in solution was a nucleation aid to induce heterogeneous barite nucleation. Such a detrimental effect on barite inhibition may be overcome by adding a chelating agent, such as ethylenediaminetetraacetic acid or citrate, to chelate Fe(III). The main iron species in water is expected to be Fe(II) ion. Fortunately, the Fe(II) ion has a less detrimental impact on several polymeric inhibitors, such as SPCA and maleic acid polymer.[121] At the same time, Fe(II) also had a detrimental effect on barite inhibition by a phosphonate, possibly due to the extremely low solubility of a Fe(II) phosphonate

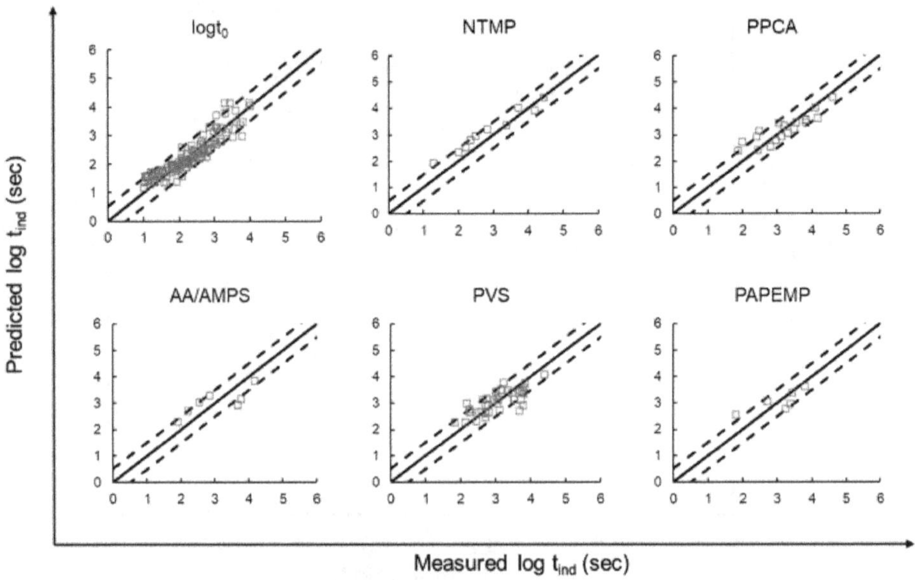

Figure 1.27. Plots of the measured and predicted induction times for barite by Dai's mechanistic inhibition model for the induction time in the absence of a scale inhibitor and that with five other common scale inhibitors. The measured induction times represent many experimental nucleation inhibition studies at different barite SI, temperature, pH, and background electrolyte composition.

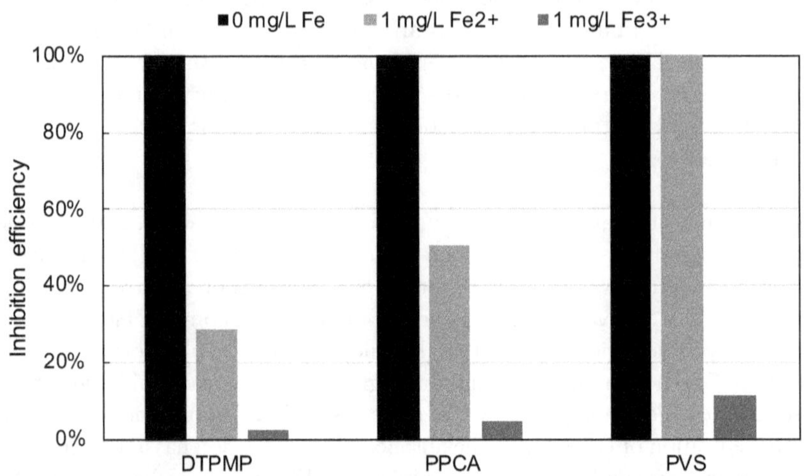

Figure 1.28. Comparison of three different inhibitors at 1 mg/L concentration in the presence and absence of 1 mg/L ferrous and ferric ions. Test conditions: 70°C, 1 M NaCl, barite SI = 2.0, pH 6.74, 400 mg/L Ca^{2+}, inhibitor conc. = 1.22 mg/L.

scale.[122] Figures 1.29 and 1.30 show the effects of Fe(II) on DETPMP at two temperature and pH values. The trend of inhibition reduction due to temperature and pH is consistent with the hypothesis that the impact of Fe(II) with phosphonate is due to the Fe-phosphonate precipitation.[122]

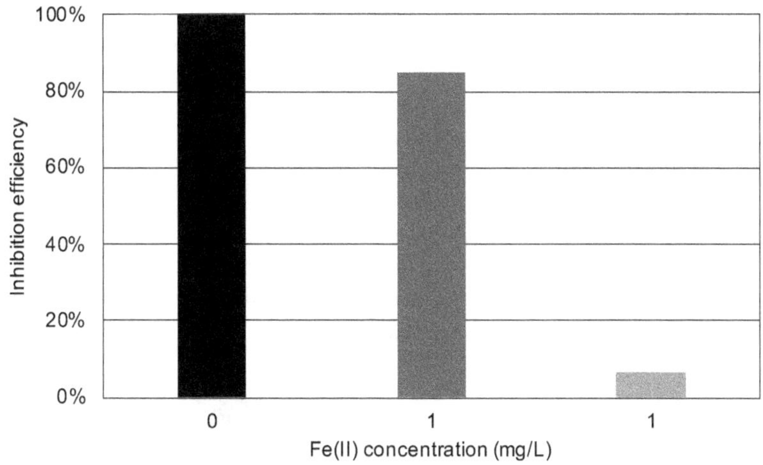

Figure 1.29. The impact of ferrous ion Fe(II) on barite inhibition by DETPMP at 25°C versus 70°C. Test conditions: 1 M NaCl, pH = 5.5, barite SI = 2.0.

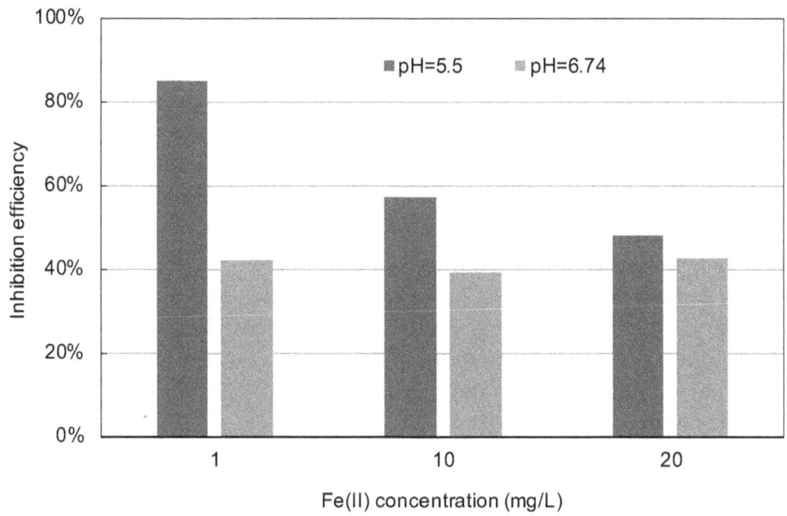

Figure 1.30. The impact of Fe(II) on barite inhibition by DETPMP at pH 5.5 and 6.74. Test conditions: 25°C, 1 M NaCl, barite SI = 2.0.

1.5.3 Gypsum Scale Inhibition

The above-mentioned semiempirical equation to predict the induction time in the presence and absence of inhibitor (Equations [1.4] through [1.6]) has been developed for calcite, barite, and celestite with or without scale inhibitors.[63, 117, 123] However, the induction times of gypsum do not show a similar trend and cannot be described by the semiempirical models. Dai *et al.*[124] adopted the classical nucleation theory to model the induction time of gypsum. It was proposed that gypsum nucleation may gradually transit from homogeneous to heterogeneous nucleation when the gypsum SI decreases. Such transition is represented by a gradual decrease of surface tension at smaller SI values.

The traditional DSL or batch laser tests are significantly limited by their testing efficiency under many conditions. A new modified CSTR was developed to circumvent such a problem, and the method was tested for the inhibition of the gypsum scale.[124, 125] Figure 1.31 is the setup of the apparatus. The apparatus was modified from the batch reactor developed previously (Figure 1.24) using a bath for temperature control and a magnetic stir bar for mixing.

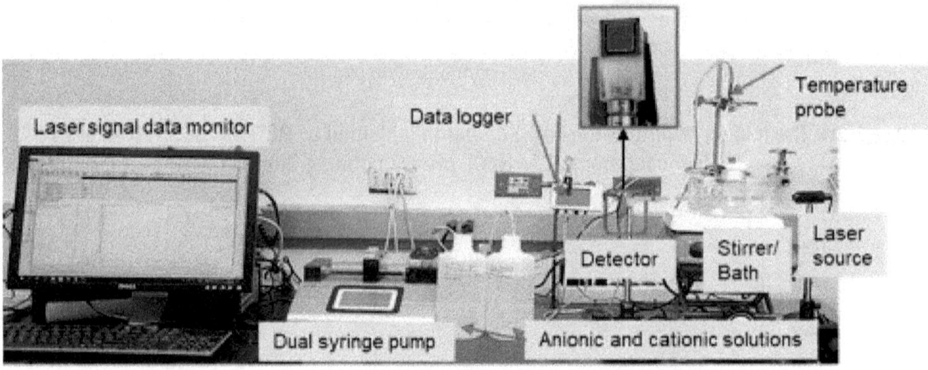

Figure 1.31. A modified CSTR test setup.

At time zero, equal volumes of cationic and anionic solutions were injected into the reactor along with scale inhibitors of a certain initial concentration (C_s) via preheating loops. After mixing the initially injected solutions, the cationic and anionic solutions (without scale inhibitors) were simultaneously pumped into the reactor with the same flow rate and no headspace. CSTR procedure reduces the original inhibitor concentration (C_s) exponentially with time for the inhibitor concentration in the reactor at a given time (C_{inh}^{CSTR}) according to the equation below:

$$C_{inh}^{CSTR} = C_s e^{-k_w t} \tag{1.22}$$

where k_w is the mass exchange rate of the reaction vial and is equal to Q/V, the reciprocal of the residence time (τ). The critical time is picked when the laser intensity drops (about 1%, or 0.004 mA in this study). Thus, the time to complete inhibitor testing is significantly reduced due to the reactor's exponential reduction in inhibitor concentration at constant SI. Figure 1.32 compares seven inhibitor screenings of gypsum supersaturated solutions at

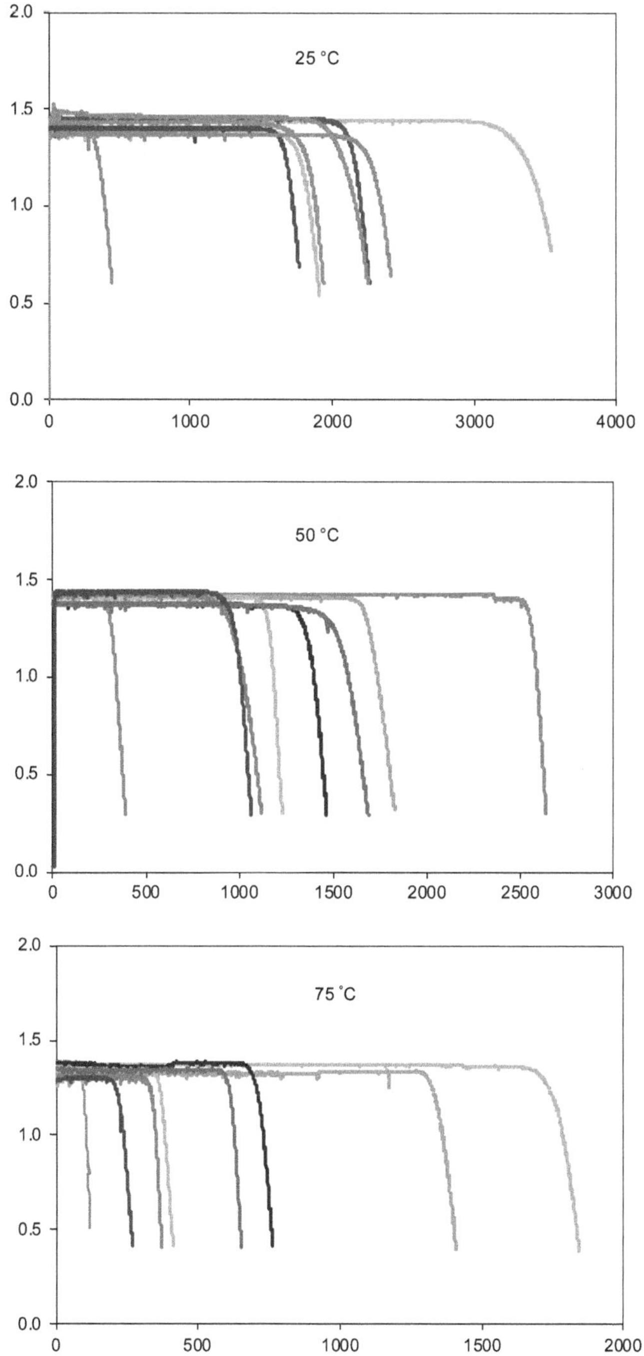

Figure 1.32. Gypsum nucleation inhibition by 1 mg/L (active concentration) of seven scale inhibitors at 25°C, 50°C, and 75°C. The synthetic produced water contains 0.1 m NaCl and 0.07 m, 0.06 m, and 0.052 m of Ca^{2+} and SO_4^{2-}ions and the gypsum SIs correspond to 0.65, 0.56, 0.56, respectively, at 25°C, 50°C, and 75°C.

25–75°C and a similar gypsum SI value (SI = 0.65, 0.56, 0.56 for 25°C, 50°C, and 75°C experiments, respectively). HDTMP was the most effective gypsum inhibitor at all three temperatures. The second best-ranked scale inhibitors are BHPMP, PPCA, and DETPMP at 25°C and 50°C. However, BHPMP alone performs better than DETPMP and PPCA at 75°C. It is proposed that the limited solubility of Ca-DETPMP and Ca-PPCA at 75°C may impair the performance of DETPMP at higher temperatures.[126, 127] However, the true cause of such discrepancy has not been carefully evaluated. Interestingly, three inhibitors (PA-PEMP, PE, SPCA) consistently perform worse than the other four inhibitors for gypsum. More chemical/structure/reaction studies may be very useful to guide future understanding of inhibition mechanisms and inhibitor selection.

1.5.4 The Effects of High Temperature and Anhydrite Scale Inhibition

Screening effective inhibitors for high-temperature (HT) environments is important for scale control of HPHT wells. Potential scale inhibitors for HT must be thermally stable and inhibition effective at $T > 150$°C. Deng[128] developed a hydrothermal bomb for high-temperature inhibitor evaluation to $T < 250$°C, $P \leq 2{,}000$ psi (Figure 1.33). The HT environment would facilitate corrosion to parts with the contact of steam or salt solution. Proper injection and mixing of cation/anion solutions at the vapor pressure of HT require special equipment and materials. The hydrothermal cell is customized with (1) a thin layer of Teflon coating on the surface of the interior, (2) a borosilicate glass and Teflon cover as the reactor, (3) one-time used PFA (perfluoroalkoxy plastic) tubing for injecting reagents to avoid corrosion problems with metal tubing, (4) Hastelloy C-276 or titanium tubing at the heated zone, and (5) two fused borosilicate glass sight windows. The high-performance liquid chromatography (HPLC) switching valve with a fixed volume sampling loop is used for cation/anion injection and sample withdrawer. The HPLC pump was used to inject de-

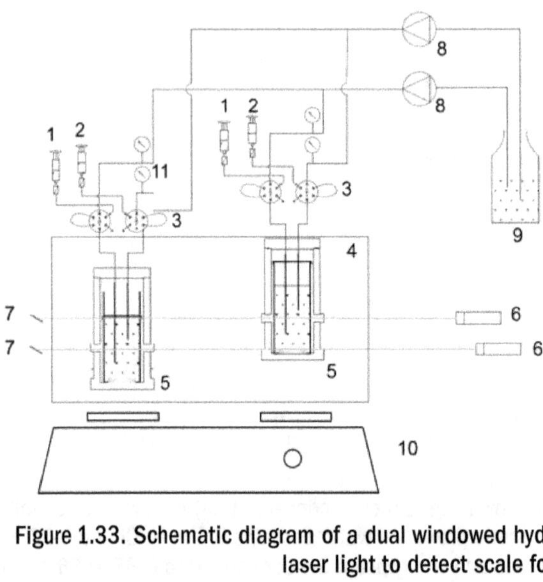

No	Description
1.	Anion feed
2.	Cation feed
3.	HPLC switch valve w/ sample loop
4.	Oven
5.	Modified static test bomb (see text for details)
6.	Green laser diode
7.	Si photodiode
8.	HPLC pump with 0.5 mm in-line filter
9.	DI water
10.	NeFeB magnetic stirrer
11.	Pressure gauge

Figure 1.33. Schematic diagram of a dual windowed hydrothermal apparatus equipped with a laser light to detect scale formation.

ionized [water] to deliver the reagents and clean the sampling loop between injection and sampling. A magnetic stir plate with Ne-Fe-B magnets generates strong magnetic coupling between the stirrer and stir bar to facilitate effective mixing. The nucleation induction time was measured by a laser-turbidity setup similar to those of low-temperature experiments.

Wang studied the thermal stability of several phosphonate and polymeric scale inhibitors.[129] The 1 M NaCl solution scale inhibitors were thermally treated at 95–200°C under an argon atmosphere for 0–72 h. Their chemical structure was evaluated by both 1H nuclear magnetic resonance (NMR; after 24 h of thermal treatment at various temperatures; Figure 1.34) and by scale inhibition testing (Figure 1.35). The *f*(active) is expressed as the fractional inhibition efficiency (b_{inh}) of the thermally treated scale inhibitor relative to that of the untreated (fresh) inhibitor efficiency:

$$f(\text{active}) = \frac{b_{\text{inh,heated}}}{b_{\text{inh,fresh}}} \tag{1.23}$$

The data showed that phosphonate scale inhibitors degraded more than polymeric scale inhibitors at high temperatures (>100°C or 212°F), and faster degradation was observed.

Due to the difficulty of inhibition in the presence of high Ca and temperature, inhibition of the anhydrite scale has proven difficult. Zhang *et al.* discussed the successful selection of an anhydrite scale inhibitor for a steam flood operation. Most known common scale inhibitors failed the high temperature and calcium tolerance challenge for anhydrite inhibition. Even though HDTMP is a good scale inhibitor for gypsum at lower temperatures, it cannot inhibit anhydrite scale formation at high temperatures, presumably due to its inferior thermal stability and solubility in Ca. None of the four gypsum scale inhibitors perform well in anhydrite inhibition. Surprisingly, SPCA is an excellent inhibitor for anhydrite inhibition at 200°C (392°F) temperature (Figure 1.36). The data from Figure 1.36 are the results of the dynamic flow loop experiments. The inhibitor was injected into the produced water at 10 mg/L, and the concentration was reduced to 5 mg/L, 3 mg/L, and 1 mg/L every 25 min

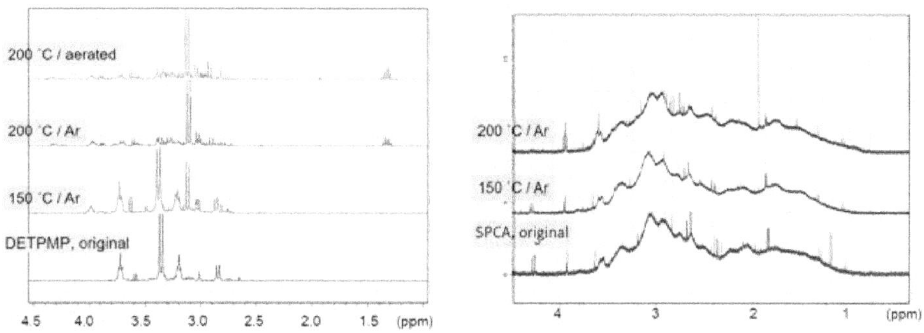

Figure 1.34. Plots of the 1H nuclear magnetic resonance spectrum of DETPMP and SPCA scale inhibitors after different degrees of thermal treatment in both aerated and under argon atmosphere. All inhibitor solutions (1,000 mg/L, active) were prepared in 1 M NaCl solution at pH 7.

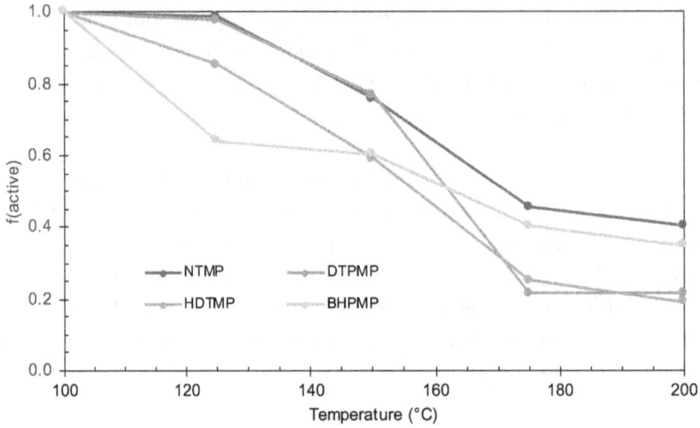

Figure 1.35. Inhibition activity of the thermally treated inhibitor, *f*(active), of four phosphonates versus different thermal treatment temperature after 24 h of heating time.

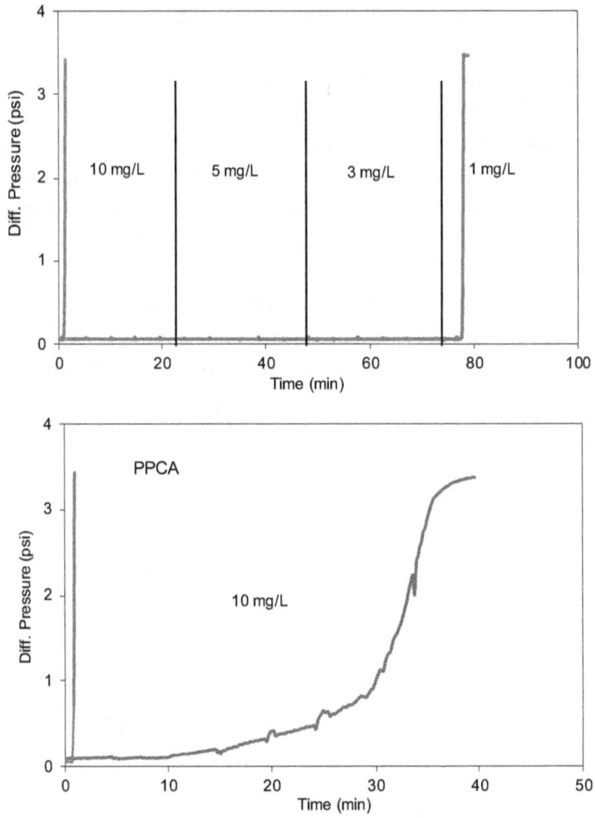

Figure 1.36. Testing anhydrite scale inhibition with a dynamic scale loop with two inhibitors, that is, sulfonated polycarboxylate (SPCA) and phosphinopolycarboxylate (PPCA). The testing was conducted with a synthetic produced water containing 1 m NaCl, ~20 mm Ca^{2+}, and sulfate ions at anhydrite SI = 1.0 at 200°C (392°F).

unless the scale was formed. The result showed that SPCA could inhibit the anhydrite scale for 25 min in the scaling loop at a concentration above 1 mg/L. Still, anhydrite scale is formed when the inhibitor concentration is reduced to 1 mg/L. However, PPCA failed to protect the DSL at 10 mg/L concentration. Zhang *et al.*[130] studied the speciation chemistry of SPCA in acid-base and calcium complex from 0.01 M to 2 M ionic strength and from 25°C to 90°C. The results showed that SPCA forms a stable aqueous complex with Ca^{2+}. Combining the thermostability and calcium complex stability may be why SPCA is one of the few good inhibitors for anhydrite inhibition. It also may be why a sulfonated scale inhibitor is not a chemical for inhibitor squeeze operation.[131] Yao *et al.*[132] proposed an interesting idea to recycle nonadsorbing scale inhibitors, such as SPCA, to protect the reservoir from the injection well to production well—further research on this new inhibitor treatment technique is in progress.

1.5.5 Halite Scale Inhibition

No effective chemistry can prevent halite precipitation at a large temperature drop. Figure 1.37 shows two batch nucleation inhibition studies at 25°C and 70°C (77°F and 158°F), where the supersaturation was induced by adding a salt solution containing $CaCl_2$ and $MgCl_2$ as background electrolytes. A common carboxylate/sulfonate copolymeric scale inhibitor can control the halite scale at 100 mg/L concentration and with a halite SI = 0.006 at 25°C. However, 1,700 mg/L of the same scale inhibitor is needed to control halite precipitation for about 50 min (~3,000 s) at a halite SI = 0.0139 and 70°C. The chemical efficiency dropped significantly at high SI and temperatures. Most high-salinity brine contains very high Ca concentrations. Few chemicals effectively inhibit halite at high Ca, salinity, and high-temperature environments. Navarathna *et al.*[73] observed that most salt anticaking chemicals, such as nitrilotriacetamide, polyacrylamide, or ferrocyanide ($[Fe(CN)_6]^{4-}$, FCN) are amide or cyanide-based. However, FCN cannot tolerate high temperature and calcium conditions. More work is still underway to identify an effective inhibitor/combination to inhibit halite at low concentrations. The novel experimental design mimics the temperature changes in the production tubing. It certainly would be more relevant to mimic the halite precipitation/inhibition condition than the conventional batch inhibitor testing methods at fixed SI.

1.5.6 Sulfide Scale Control

Due to the instantaneous formation of Fe_xS_x nanoparticles in the aqueous phase, chemical inhibition of iron sulfide formation is nearly impossible, while dispersants might be considered a sulfide control option. Mackinawite scale is hydrophobic and would partition into the oil phase, creating a nuisance for oil transport and refinery. Ko *et al.*[133] tested a series of polymeric dispersants. FeS particles partitioned into the oil phase when no dispersant or ineffective polymers, for example, polyaspartate (PASP) or dextran (DEX), were applied; while they remained in the aqueous phase when either polyvinyl pyrrolidone (PVP) or carboxymethyl cellulose (CMC) was added to the solution before FeS formation. Ko *et al.* found that CMC at a molecular weight of 90,000–200,000 Dalton performed well as a dispersant for FeS mackinawite. Figure 1.38 illustrates a proposed mechanism where

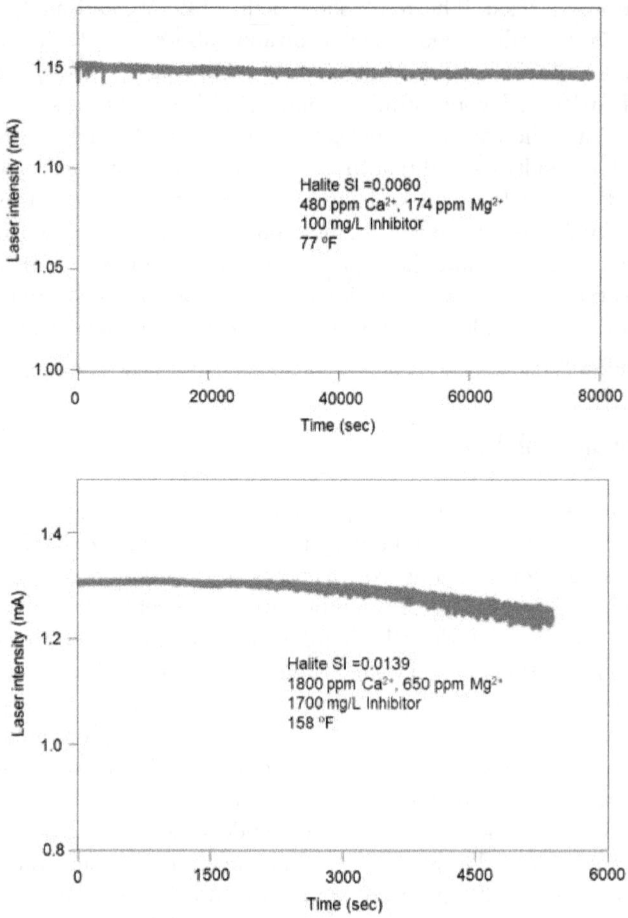

Figure 1.37. Inhibition of the halite scale at halite SI of 0.006 and 0.014 with a polymeric scale inhibitor. The synthetic produced water is a halite-saturated solution with various concentrations of Ca^{2+} and Mg^{2+} chloride salt to raise the halite SI at 25°C (77°F) and 70°C (158°F). Reproduced with permission from Ref. (72).

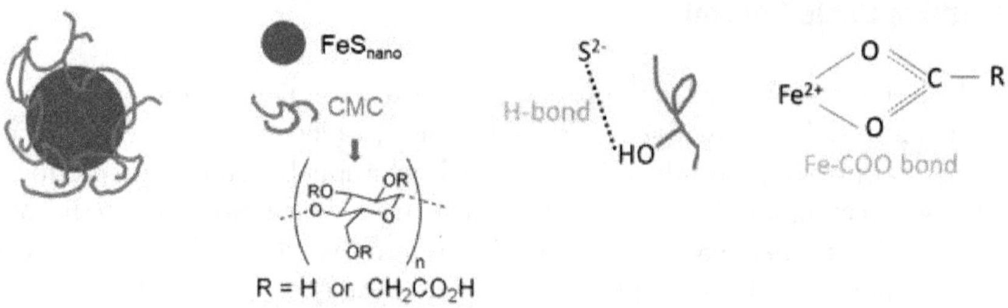

Figure 1.38. A schematic representation of dispersion mechanism based on Fourier transform infrared spectroscopy spectrum and transmission electron microscopy analyses.

CMC interacts with FeS at the particulate level. CMC-dispersed FeS particles convert the hydrophobic FeS surface to hydrophilic; the dispersed FeS particles remain in the aqueous phase. The dispersed FeS is in nanosize (5–10 nm), as observed in the transmission electron microscope.[133] The TEM image suggested that CMC stabilizes the particles, and this coating prevents the FeS particles from agglomerating and growing. CMC is also an effective dispersant for ZnS and other mineral scales such as barite, gypsum, calcite, and siderite.[134] Wang *et al.*[135] developed a mechanistic model to calculate the effective dispersant concentration based on the FeS SI, CMC molecular weight, solution pH, and ionic strength.

1.6 Summary

Scaling is a complex issue that can significantly impact oil and gas production. It can negatively impact the carbon footprint of well operations. This chapter aims to provide an overview of scale formation and control, new methodologies for scale inhibitor studies, and key research challenges that might improve scale control technologies.

The challenges of scale control can be oriented to both thermodynamic and kinetic origins. Common scale inhibitors are effective in controlling calcite and barite scales. Earlier literature on scale inhibition is mainly focused on the most common scales, that is, calcite and barite. A similar group of inhibitor chemistry can inhibit both calcite and barite. However, it does not apply to other scale types, such as the highly soluble halite or the least soluble sulfide scales. The inhibition of gypsum, anhydrite, halite, sulfides, silica, and silicates presents more challenges and requires new thinking and approaches.

The kinetics of various scale types, mixed scale, or solid solutions are discussed. Mg^{2+} and Fe^{2+} significantly affect carbonate scale deposition and scale inhibition. In contrast, the Group IIA cations, Mg^{2+}, Ca^{2+}, Sr^{2+}, and Ba^{2+}, all have some impact on the kinetics of barite, celestite, and gypsum precipitation. The inhibitor's thermal and Ca complex stability and metal ion solid solubility may strongly influence scale inhibition and squeeze treatment efficiency. At the sub-ppm level, scale inhibitors may inhibit barite and calcite scales at relatively high SI values. The same level of scale inhibitors would only effectively control the scaling of a more soluble scale at much lower SI values. Halite scale inhibition would require a different chemistry. The nitrilotriacetamide or ferrous cyanide chemistry may inhibit halite at lower temperatures, but its efficiency decreases at higher temperatures.

Scale prediction is based on comparing lattice ion activity versus the thermodynamic solubility limits. The mineral scale solubility at low ionic strength is well known. Predicting the activity of ions at high TDS, temperature, and pressure still requires additional focus, especially in the context of produced water from deep water, shale gas, and tight oil-producing fields, where mixed ion concentrations, high temperature, and pressure pose significant challenges. The kinetics of scale formation, deposition, and inhibition efficiency have received less attention, while they are ultimately important. A system may have low-scale risk if the scale doesn't form over the time required for a liquid to travel in the production tubing. Such conditions may exist at low SI and temperature or in the presence of a kinetic scale inhibitor. A system could also have very high-scale risk where no scale inhibitor can control at very high concentration. Such conditions may occur at very high SI values and high temperature and pressure, where the inhibitor becomes incompatible with the produced water. Thus, understanding the kinetics of scale formation and inhibition is also very important, but it is an area of lesser investigation.

This book aims to comprehensively examine all aspects of the scale problem, including a fundamental understanding of common and exotic scale risk and chemical treatment, and explore various new scale mitigation approaches. The partitioning of acid gases and the phase behavior of H_2O can significantly impact the scaling tendency of the aqueous phase inside the production tubing. A thorough discussion of the kinetics, thermodynamics, and phase behavior of gas/liquid phases on acid-soluble scale prediction is presented in Chapters 2–4. Methods for controlling scales include injecting inhibitors, squeezing inhibitors into the formation, and using a downhole treat string, which will be further explored in the following chapters.

Nomenclature

a_i	Activity of an ion
b_{inh}	Inhibition efficiency coefficient (L/mg)
C_{inh}	Inhibitor concentration (mg/L), as active composition in this chapter
CMC	Carboxy methyl cellulose
DEX	Dextran
EDS	Energy dispersive X-ray spectroscopy
EOR	Enhanced oil recovery
FCN	Ferrocyanide
FIB	Focused ion beam
FT-IR	Fourier transform infrared spectroscopy
HPHT	High pressure high temperature
I	Ionic strength
m	Molality concentration (= moles of dissolved ions in one kilogram of water)
M	Morality concentration (= moles of dissolved ions in one liter of water)
mm	Millimolal (a molality concentration)
mM	Millimolar (a morality concentration)
MF	Microfluidics
MIC	Minimum inhibitor concentration, as active composition in this chapter
nm	Nanometer
NORM	Naturally occurring radioactive materials
SEM	Scanning electron microscopy
SI	Saturation index
SR	Saturation ratio
SSP	ScaleSoftPitzer software
T_0	Equilibrium temperature (°C)

T_n	Halite precipitation temperature (°C)
ΔT	Differential temperature ($=T_n - T_0$, °C)
t_0	Induction time of scale in the absence of a scale inhibitor (s)
t_{inh}	Induction time of scale in the presence of a scale inhibitor (s)
$t_{reactor}$	Retention time of a reactor (s)
TDS	Total dissolved solids (mg/L)
TEM	Transmission electron microscopy
USGS	U.S. Geological Survey
WAG	Water alternating gas
XRD	X-ray diffraction analysis
γ_i	Activity coefficient of an ion
φ_i	Fugacity coefficient of a gas
μm	Micromolal (a molality concentration)

References

1. U.S. Energy Information Administration, *Oil and Petroleum Products Explained: Oil and the Environment* (Washington DC, 2022).
2. IHS Markit, *Produced Water from Onshore US Oil and Gas Activities to Decline to Nearly 20 Billion Barrels Annually; Reach $28 Billion in Value by 2022* (New York: S&P Global, 2020).
3. J.J. Wylde, S.F. McKay, M. Falla, "Development of a Suitable $CaCO_3$ Scale Inhibitor System for use in ultrahigh temperature and pressure wells," SPE-80398-MS, International Symposium on Oilfield Scale, held January 29–30, 2003 (Aberdeen, United Kingdom).
4. A.T. Kan, J.Z. Dai, G. Deng, et al., "Recent Advances in Scale Prediction: Approach and Limitations," *SPE Journal* 24 (2019): pp. 2209–2220.
5. M. Davies, P.J.B. Scott, *Oilfield Water Technology* (Houston: NACE International, 2006), pp. 614.
6. J.C. Cowan, D.J. Weintritt, *Water-formed Scale Deposits* (Houston: Gulf Publishing Company, 1976), pp. 596.
7. M. Jordan, E. Sorhaug, D. Marlow, et al., "'Red' vs. 'Green' Scale Inhibitors for Extending Squeeze Life—A Case Study From North Sea, Norwegian Sector," NACE-10137, NACE Annual Conference, held March 14–18, 2010 (San Antonio, TX, USA).
8. J.E. Oddo, M.B. Tomson, "Why Scale Forms in the Oil Field and Methods To Predict It," *Society of Petroleum Engineers* 9 (1994): pp. 47–54.
9. A.T. Kan, M.B. Tomson, "Scale Prediction for Oil and Gas Production," *SPE Journal* 17 (2012): pp. 362–378.
10. B. Kaasa, "Prediction of pH, Mineral Precipitation and Multiphase Equilibria During Oil Recovery," (Ph.D. diss., Norges Teknisk-Naturvitenskapelige Universitet Trondheim, 1998).
11. K.S. Pedersen, P.L. Christensen, J.A. Shaikh, et al., *Phase Behavior of Petroleum Reservoir Fluids* (Boca Raton: CRC Press, 2006), pp. 422.
12. A.T. Kan, M. Garcia-Bermudes, Z.J. Dai, et al., "Modeling H_2S Partitioning in Deep Water Production Systems," SPE-184517-MS, SPE International Conference on Oilfield Chemistry, held April 3–5, 2017 (Montgomery, TX, USA).
13. J. Arensdorf, S. Kerr, K. Miner, et al., "Mitigating Silicate Scale in Production Wells in an Oilfield in Alberta," SPE-141422-MS, SPE International Symposium on Oilfield Chemistry, held April 11–13, 2011 (The Woodlands, TX, USA).

14. A.T. Kan, G. Fu, M.B. Tomson, "Effect of Methanol on Carbonate Equilibrium and Calcite Solubility in a Gas/Methanol/Water/Salt Mixed System," *Langmuir* 18 (2002): pp. 9713–9725.
15. M.A. Kelland, *Production Chemicals for the Oil and Gas Industry*, 2nd ed. (Boca Raton: CRC Press, 2014), pp. 454.
16. M.S. Blondes, K.D. Gans, M.A. Engle, et al., *National Produced Waters Geochemical Database v2.3* (Reston: US Geological Survey, 2019), pp. 41.
17. Z. Dai, B. Ouyang, X. Wang, et al., "From Water Analysis To Scale and Corrosion Control Improvements: A Permian Basin Example," *Chemical Geology* 610 (2022): p. 121077.
18. Y.K. Kharaka, W.D. Gunter, P.K. Aggarwal, et al., *SOLMINEQ.88: A Computer Program for Geochemical Modeling of Water-Rock Interactions* (Reston: US Geological Survey, 1988), pp. 420.
19. R. Barker, D. Burkle, T. Charpentier, et al., "A Review of Iron Carbonate ($FeCO_3$) Formation in the Oil and Gas Industry," *Corrosion Science* 142 (2018): pp. 312–341.
20. J. Greenberg, M. Tomson, "Precipitation and Dissolution Kinetics and Equilibria of Aqueous Ferrous Carbonate vs Temperature," *Applied Geochemistry* 7 (1992): pp. 185–190.
21. W. Li, Z. Dai, X. Wang, et al., "Observations of CO_2 Corrosion-Induced Carbonate Scale Formation and Inhibition on Mild Steel," *SPE Journal* 27 (2022): pp. 1–12.
22. T. Tanupabrungsun, B. Brown, S. Nesic, "Effect of pH on CO_2 Corrosion of Mild Steel at Elevated Temperatures," NACE-2013-2348, NACE Annual Conference, held March 17–21, 2013 (Orlando, FL, USA).
23. J.L. Pryzybylinski, "Iron Sulfide Deposit Formation and Prevention Under Anaerobic Conditions Typically Found in the Oilfield," SPE-65030-MS, SPE International Symposium on Oilfield Chemistry, held February 13–16, 2001 (Houston, TX, USA).
24. S. Nešić, "Key Issues Related To Modelling of Internal Corrosion of Oil and Gas Pipelines—A Review," *Corrosion Science* 49 (2007): pp. 4308–4338.
25. H.A. Nasr-El-Din, A.S. Metcalf, H.B. Luft, "Sour Environments: Protecting Production and Equipment-Case Studies," PETSOC-2003-025, Canadian International Petroleum Conference, held June 10–12, 2003 (Calgary, Canada).
26. Q. Wang, H. Ajwad, T. Shafai, et al., "Iron Sulfide Scale Dissolvers: How Effective are They?," SPE-168063-MS, SPE Saudi Arabia Section Technical Symposium and Exhibition, held May 19–22, 2013 (Al-Khobar, Saudi Arabia).
27. D. Rickard, G.W. Luther, "Chemistry of Iron Sulfides," *Chemical Reviews* 107 (2007): pp. 514–562.
28. D. Kocabag, H.L. Shergold, G.H. Kelsall, "Natural Oleophilicity/Hydrophobicity of Sulphide Minerals, II. Pyrite," *International Journal of Mineral Processing* 29 (1990): pp. 211–219.
29. Q. Wang, J.W. Morse, "Pyrite Formation Under Conditions Approximating Those in Anoxic Sediments I. Pathway and Morphology," *Marine Chemistry* 52 (1996): pp. 99–121.
30. Y. Liu, Z. Zhang, N. Bhandari, et al., "New Approach To Study Iron Sulfide Precipitation Kinetics, Solubility, and Phase Transformation," *Industrial and Engineering Chemistry Research* 56 (2017): pp. 9016–9027.
31. X. Wang, S. Ko, Y. Liu, et al., "Kinetics and Thermodynamics of Iron Sulfide, Precipitation, Deposition and Control," SPE-193630-MS, SPE International Conference on Oilfield Chemistry, held April 8–9, 2019, Galveston, TX, USA.
32. X. Wang, Z. Dai, S. Ko, et al., "Iron Sulfide Solubility Measurement and Modeling Over Wide Ranges of Temperatures, Ionic Strength, and pH," *SPE Journal* 27 (2022): pp. 1263–1274.
33. M. Huerta-Diaz, J. Morse, "Pyritization of Trace Metals in Anoxic Sediments," *Geochimica et Cosmochimica Acta* 56 (1992): pp. 2681–2702.
34. H. Nasr-El-Din, A. Al-Humaidan, "Iron Sulfide Scale: Formation, Removal and Prevention," SPE-68315-MS, International Symposium on Oilfield Scale, held January 30–31, 2001 (Aberdeen, UK).

35. Z. Amjad, R.W. Zuhl, "An Evaluation of Silica Scale Control Additives for Industrial Water Systems," NACE-08368, NACE Annual Conference, held March 16–20, 2008 (New Orleans, LA, USA).
36. J.S. Gill, "Silica Scale Control," NACE-98226, NACE Annual Conference, held March 22–27, 1998 (San Diego, CA, USA).
37. W. Wang, W. Wei, "Silica and Silicate Scale Formation and Control: Scale Modeling, Lab Testing, Scale Characterization, and Field Observation," SPE-179897-MS, SPE International Oilfield Scale Conference and Exhibition, held May 11–12, 2016 (Aberdeen, UK).
38. Y. Zhao, H.D.M Sriyarathne, K. Harouaka, et al., "Evaluation of Silica Matrix Ion Effects on Barite Scale Inhibitors," SPE-193548-MS, SPE International Conference on Oilfield Chemistry, held April 8–9, 2019 (Galveston, TX, USA).
39. J.E. McElhiney, M.B. Tomson, A.T. Kan, "Design of Low Sulphate Seawater Injection Based Upon Kinetic Limits," SPE-100480-MS, SPE International Oilfield Scale Symposium, held May 31–June 1, 2006 (Aberdeen, UK).
40. K.S. Sorbie, E.J. Mackay, "Mixing of Injected, Connate and Aquifer Brines in Waterflooding and Its Relevance To Oilfield Scaling," *Journal of Petroleum Science and Engineering* 27 (2000): pp. 85–106.
41. I. Collins, "Predicting the Location of Barium Sulfate Scale Formation in Production Systems," SPE-94366-MS, SPE International Symposium on Oilfield Scale, held May 11–12, 2005 (Aberdeen, UK).
42. A.T. Kan, G. Fu, M.B. Tomson, "Effect of Methanol and Ethylene Glycol on Sulfates and Halite Scale Formation," *Industrial and Engineering Chemistry Research* 42 (2003): pp. 2399–2408.
43. E. Odeigah, T. Pojtanabuntoeng, "Regeneration and Reclamation of Mono-Ethylene Glycol (MEG) Used as a Hydrate Inhibitor: A Review," *American Journal of Chemical Engineering* 10 (2022): pp. 32–45.
44. Z. Dai, F. Zhang, A.T. Kan, et al., "Two-Stage Model Reveals Barite Crystallization Kinetics From Solution Turbidity," *Industrial and Engineering Chemistry Research* 58 (2019): pp. 10864–10874.
45. D. Gebauer, A. Völkel, H. Cölfen, "Stable Prenucleation Calcium Carbonate Clusters," *Science* 322 (2008): pp. 1819–1822.
46. L.T. MacHale, R.G. Finke, "Solid Barium Sulfate Formation From Aqueous Solution: Re-Examination of Key Literature and Kinetics of This Classic System in Search of a Minimum Mechanism of Formation," *Industrial and Engineering Chemistry Research* 62 (2023): pp. 9639–9661.
47. Y. Zhao, Z. Dai, X. Wang, et al., "Non-Equilibrium $Ba_xSr_{1-x}SO_4$ Solid Solution Compositions at Elevated Sr2+ Concentration, Ionic Strength, and Temperature," *Geochimica et Cosmochimica Acta* 336 (2022): pp. 113–130.
48. J. Oddo, X. Zhou, D. Linz, et al., "The Mitigation of NORM Scale in the Gulf Coast Regions of Texas and Louisiana: A Laboratory and Field Study," *SPE Advanced Technology Series* 4 (1996): pp. 84–93.
49. Y. Zhao, Z. Dai, C. Dai, et al., "A Semiempirical Model for Barium-Strontium-Sulfate Solid Solution Scale Crystallization and Inhibition Kinetics at Oilfield Conditions," *SPE Journal* 26 (2021): pp. 4037–4050.
50. N. Bhandari, A.T. Kan, F. Zhang, et al., "Mineral Precipitation Kinetics: Assessing the Effect of Hydrostatic Pressure and Its Implication on the Nucleation Mechanism," *Crystal Growth & Design* 16 (2016): pp. 4846–4854.
51. K. Ramstad, T. Tydal, K.M. Askvik, et al., "Predicting Carbonate Scale in Oil Producers From High-Temperature Reservoirs," *SPE Journal* 10 (2005): pp. 363–373.

52. Y. Lu, A.T. Kan, M.B. Tomson, "Nucleation and Crystallization Kinetics of Barium Sulfate in the Hydrodynamic Boundary Layer: An Explanation of Mineral Deposition," *Crystal Growth & Design* 21 (2021): pp. 1443–1450.
53. M. Nassivera, A. Essel, "Fateh Field Sea Water Injection—Water Treatment, Corrosion, and Scale Control," SPE-7765-MS, Middle East Technical Conference and Exhibition, held February 25–28, 1979 (Manama, Bahrain).
54. D.F. Jacques, B.I. Bourland, "A Study of Solubility of Strontium Sulfate," *SPE Journal* 23 (1983): pp. 292–300.
55. J. Shen, C.C. Crosby, "Insight into Strontium and Calcium Sulfate Scaling Mechanisms in a Wet Producer," *Journal of Petroleum Technology* 35 (1983): pp. 1249–1255.
56. J. Al-Ashhab, H. Al-Matar, S. Mokhtar, "Techniques Used to Monitor and Remove Strontium Sulfate Scale in UZ Producing Wells," SPE-101401-MS, Abu Dhabi International Petroleum Exhibition and Conference, held November 5–8, 2006 (Abu Dhabi, UAE).
57. C. Johnston, L. Sutherland, "The Influence of Turbulence (or Hydrodynamic Effects) on Strontium Sulphate Scale Formation and Inhibitor Performance," SPE-169760-MS, SPE International Oilfield Scale Conference and Exhibition, held May 14–15, 2014 (Aberdeen, UK).
58. A. Essel, B. Carlberg, "Strontium Sulfate Scale Control by Inhibitor Squeeze Treatment in the Fateh Field," *Journal of Petroleum Technology* 34 (1982): pp. 1302–1306.
59. I. Malollari, P.G. Klepetsanis, P.G. Koutsoukos, "Precipitation of Strontium Sulfate in Aqueous Solutions at 25°C," *Journal of Crystal Growth* 155 (1995): pp. 240–246.
60. G.H. Nancollas, "The Kinetics of Growth of Seed Crystals in Solution," *Croatica Chemica Acta* 42 (1970): pp. 311–318.
61. S.K. Hamdona, S.M. Hamza, A.H. Mangood, "The Influence of Polyphosphonates on the Precipitation of Strontium Sulfate (Celestite) From Aqueous Solutions," *Desalination and Water Treatment* 24 (2010): pp. 55–60.
62. S. He, J.E. Oddo, M.B. Tomson, "The Nucleation Kinetics of Strontium Sulfate in NACL Solutions Up To 6 m and 90°C With or Without Inhibitors," *Journal of Colloid and Interface Science* 174 (1995): pp. 327–335.
63. Y. Zhao, Z. Dai, C. Dai, et al., "Semiempirical Model for Predicting Celestite Scale Formation and Inhibition in Oilfield Conditions," *Energy and Fuels* 35 (2021): pp. 11285–11294.
64. C. Monnin, C. Galinier, "The Solubility of Celestite and Barite in Electrolyte Solutions and Natural Waters at 25°C: A Thermodynamic Study," *Chemical Geology* 71 (1988): pp. 283–296.
65. D. Freyer, W. Voigt, "Crystallization and Phase Stability of $CaSO_4$ and $CaSO_4$ Based Salts," *Monatshefte* für *Chemie - Chemical Monthly* 134 (2003): pp. 693–719.
66. H. Lu, A.T. Kan, P. Zhang, et al., "Phase Stability and Inhibition of Calcium Sulfate in the System NaCl/Monoethylene Glycol/H_2O," *SPE Journal* 17 (2012): pp. 187–197.
67. D. Barge, F. Al-Yami, D. Uphold, et al., "Steamflood Piloting the Wafra Field Eocene Reservoir in the Partitioned Neutral Zone, Between Saudi Arabia and Kuwait," SPE-120205-MS, SPE Middle East Oil and Gas Show and Conference, held March 15–18, 2009 (Manama, Bahrain).
68. F. Zhang, C.J. Hinrichsen, A.T. Kan, et al., "Calcium Sulfate Scaling Risk and Inhibition for a Steamflood Project," *SPE Journal* 22 (2017): pp. 881–891.
69. D.M. Warsinger, J. Swaminathan, E. Guillen-Burrieza, et al., "Scaling and Fouling in Membrane Distillation for Desalination Applications: A Review," *Desalination* 356 (2015): pp. 294–313.
70. S. Mateen, "Modeling of Mineral Scaling in a West Texas CO_2-WAG EOR Field Using Produced Water Analysis and a 1-D Reactive Mixing-Material Balance Coupled Approach," (Master's thesis, Rice University, 2020).
71. S. Paudyal, G. Ruan, J. Lee, et al., "Development of Dynamic Tube Blocking Test Method to Study Halite Scale Deposition and Inhibition," SPE-204389-MS, SPE International Conference on Oilfield Chemistry, held December 6–7, 2021 (The Woodlands, TX, USA).

72. G. Ruan, Y. Liu, A.T. Kan, et al., "Sodium Chloride (Halite) Mineral Scale Threat Assessment and Scale Inhibitor Evaluation By Two CoμμonJar Test Based Methods," *Journal of Water Process Engineering* 43 (2021): pp. 102241.
73. C. Navarathna, C. Leschied, X. Wang, et al., "A Novel Experiment Setup To Model the Effects of Temperature on Halite Scaling and Inhibition," SPE-213849-MS, SPE International Conference on Oilfield Chemistry, held June 28–29, 2023 (The Woodlands, TX, USA).
74. D.J. Phillips, S.L. Phillips, "High Temperature Dissociation Constants of HS– and the Standard Thermodynamic Values for S2–," *Journal of Chemical and Engineering Data* 45 (2000): pp. 981–987.
75. F.W. Dickson, "Solubilities of Metallic Sulfides and Quartz in Hydrothermal Sulfide Solutions," *Bulletin of Volcanology* 29 (1966): pp. 605–628.
76. J.W. Cobble, H.P. Stephens, "Thermodynamic Properties of the Aqueous Sulfide and Bisulfide Ions and the Second Ionization Constant of Hydrogen Sulfide Over Extended Temperatures," *Inorganic Chemistry* 10 (1971): pp. 619–625.
77. F.J. Millero, *Chemical Oceanography*, 4th ed. (Boca Raton: CRC Press, 2013), pp. 591.
78. G. Atkinson, M. Mecik, "The Chemistry of Scale Prediction," *Journal of Petroleum Science and Engineering* 17 (1997): pp. 113–121.
79. X. Wang, G. Deng, S. Ko, et al., "Improved Scale Prediction for High Calcium Containing Produced Brine and Sulfide Scales," SPE-200699-MS, SPE International Oilfield Scale Conference and Exhibition, held June 24–25, 2020 (Virtual).
80. S.R. Milner, "The Current-Potential Curves of the Oscillating Spark and the Mechanism of Spark Conduction," *The London, Edinburgh, and Dublin Philosophical Magazine and Journal of Science* 24 (1912): pp. 709–721.
81. S.R. Milner, "The Virial of a Mixture of Ions," *The London, Edinburgh, and Dublin Philosophical Magazine and Journal of Science* 23 (1912): pp. 551–578.
82. P. Debye, E. Hückel, "Zur Theorie der Elektrolyte. I. Gefrierpunktserniedrigung und Verwandte Erscheinungen," *Physikalishce Zeitschrift* 24 (1923): p. 305.
83. K.S. Pitzer, *Thermodynamics*, 3rd ed. (New York: McGraw-Hill, 1995), pp. 626.
84. M. Margules, "Über die Zusammensetzung der gesättigten Dämpfe von Mischungen," *Sitzungsberichte der Kaiserlichen Akademie der Wissenschaften, Mathematisch-Naturwissenschaftliche Classe* 104 (1895): pp. 1243–1278.
85. G. Scatchard, "Concentrated Solutions of Strong Electrolytes," *Chemical Reviews* 19 (1936): pp. 309–327.
86. E.A. Guggenheim, J.C. Turgeon, "Specific Interaction of Ions," *Transactions of the Faraday Society* 51 (1955): pp. 747–761.
87. K.S. Pitzer, "Thermodynamics of Electrolytes. I. Theoretical Basis and General Equations," *Journal of Physical Chemistry* 77 (1973): pp. 268–277.
88. K.S. Pitzer, G. Mayorga, "Thermodynamics of Electrolytes. II. Activity and Osmotic Coefficients for Strong Electrolytes With One or Both Ions Univalent," *Journal of Physical Chemistry* 77 (1973): pp. 2300–2308.
89. N. Plummer, D.L. Parkhurst, G.W. Fleming, et al., *A computer program incorporating Pitzer's equations for calculation of geochemical reactions in brines*. Water-Resources Investigations Report 88-4153, US Geological Survey, 1988.
90. C.M. Bethke, *Geochemical Reaction Modeling: Concepts and Applications* (New York: Oxford University Press, 1996), pp. 397.
91. Z. Dai, A.T. Kan, W. Shi, et al., "Solubility Measurements and Predictions of Gypsum, Anhydrite, and Calcite Over Wide Ranges of Temperature, Pressure, and Ionic Strength With Mixed Electrolytes," *Rock Mechanics and Rock Engineering* 50 (2017): pp. 327–339.

92. C.A.J. Appelo, "Principles, Caveats and Improvements in Databases for Calculating Hydrogeochemical Reactions in Saline Waters From 0 to 200°C and 1 to 1000 atm," *Geochemical Speciation Codes and Databases* 55 (2015): pp. 62–71.
93. C.W. Blounot, F.W. Dickson, "The Solubility of Anhydrite ($CaSO_4$) in NaCl-H_2O from 100 to 450°C and 1 to 1000 Bars," *Geochimica et Cosmochimica Acta* 33 (1969): pp. 227–245.
94. W. Shi, A.T. Kan, C. Fan, et al., "Solubility of Barite Up to 250 °C and 1500 Bar in Up to 6 m NaCl Solution," *Industrial and Engineering Chemistry Research* 51 (2012): pp. 3119–3128.
95. W. Shi, A.T. Kan, N. Zhang, et al., "Dissolution of Calcite at Up to 250° C and 1450 Bar and the Presence of Mixed Salts," *Industrial and Engineering Chemistry Research* 52 (2013): pp. 2439–2448.
96. Z. Duan, D. Li, "Coupled Phase and Aqueous Species Equilibrium of the H_2O–CO_2–NaCl–$CaCO_3$ System From 0 to 250 °C, 1 to 1000 Bar With NaCl Concentrations Up to Saturation of Halite," *Geochimica et Cosmochimica Acta* 72 (2008): pp. 5128–5145.
97. F.W. Dickson, C.W. Blount, et al., "Use of Hydrothermal Solution Equipment to Determine the Solubility of Anhydrite in Water From 100 degrees C to 275 degrees C and From 1 Bar to 1000 Bars Pressure," *American Journal of Science* 261 (1963) 61–78.
98. G. Innorta, E. Rabbi, L. Tomadin, "The Gypsum-Anhydrite Equilibrium by Solubility Measurements," *Geochimica et Cosmochimica Acta* 44 (1980): pp. 1931–1936.
99. R.E. Hall, J.A. Robb, C.E. Coleman, "The Solubility of Calcium Sulfate at Boiler-Water Temperatures," *Journal of the American Chemical Society* 48 (1926): pp. 927–938.
100. W.L. Marshall, R. Slusher, E.V. Jones, "Aqueous Systems at high temperatures XIV. Solubility and thermodynamic relationships for $CaSO_4$ in NaCl-H_2O solutions from 40° to 200° C, 0 to 4 molal NaCl," *Journal of Chemical and Engineering Data* 9 (1964): pp. 187–191.
101. D. Freyer, W. Voigt, "The Measurement of Sulfate Mineral Solubilities in the Na-K-Ca-Cl-SO_4-H_2O System at Temperatures of 100, 150 and 200°C," *Geochimica et Cosmochimica Acta* 68 (2004): pp. 307–318.
102. Z. Li, G.P. Demopoulos, "Solubility of $CaSO_4$ Phases in aqueous HCl + $CaCl_2$ Solutions From 283 K to 353 K," *Journal of Chemical and Engineering Data* 50 (2005): pp. 1971–1982.
103. G. Azimi, V.G. Papangelakis, "The Solubility of Gypsum and Anhydrite in Simulated Laterite Pressure Acid Leach Solutions Up To 250°C," *Hydrometallurgy* 102 (2010): pp. 1–13.
104. G. Deng, A.T. Kan, Z. Dai, et al., "Impact of High Calcium Concentration on Sulfate Scale Prediction at High Temperature From 120° C To 220° C," SPE-190744-MS, SPE International Oilfield Scale Conference and Exhibition, held June 20–21, 2018 (Aberdeen, UK).
105. E.L. Rowan, M.A. Engle, T.F. Kraemer, et al., "Geochemical and Isotopic Evolution of Water Produced From Middle Devonian Marcellus Shale Gas Wells, Appalachian Basin, Pennsylvania," *AAPG Bulletin* 99 (2015): pp. 181–206.
106. C.J. Johnston, M.M. Jordan, "Prevention of Inorganic Scale in ESPs—A Holistic Review of Scale Inhibitor Deployment Options," SPE-209487-MS, SPE International Oilfield Scale Conference and Exhibition, held May 25–26, 2022 (Aberdeen, UK).
107. M.M. Jordan, L. Sutherland, C. Johnston, "Developments in Squeeze Treatments Design—Why Not Apply 24 Month Duration Squeeze to Offshore Production Wells?," SPE-200701-MS, PE International Oilfield Scale Conference and Exhibition, held June 24–25, 2020 (Virtual).
108. O. Vazquez, G. Ross, M.M. Jordan, et al., "Automatic Optimization of Oilfield-Scale-Inhibitor Squeeze Treatments Delivered By Diving-Support Vessel," *SPE Journal* 24 (2019): pp. 60–70.
109. A.T. Kan, Z. Dai, M.B. Tomson, "The State of the Art in Scale Inhibitor Squeeze Treatment," *Petroleum Science* 17 (2020): pp. 1579–1601.
110. J.E. Oddo, M.B. Tomson, "Elevated Temperature-Pressure Flow Simulator," Patent US5370799, 1994.

111. C. Yan, A.T. Kan, F. Zhang, et al., "Systematic Study of Barite Nucleation and Inhibition With Various Polymeric Scale Inhibitors By Novel Laser Apparatus," *SPE Journal* 20 (2015): pp. 642–651.
112. Z. Dai, F. Zhang, Y. Zhao, et al., "Predicting Calcite Crystallization and Inhibition Kinetics at Up to 175 °C Using a New Theoretical Inhibition Model," *Crystal Growth & Design* 20 (2020): pp. 8090–8100.
113. C. Dai, Z. Dai, S. Paudyal, et al., "New Kinetic Turbidity Test Method and Prediction Model for Calcite Inhibition," SPE-204398-MS, SPE International Conference on Oilfield Chemistry, held December 6–7, 2021 (The Woodlands, TX, USA).
114. N. Zhang, A. Kan, W. Shi, et al., "A Novel Approach to the Study of $CaCO_3$ Precipitation Kinetics on Carbon Steel Pipe," SPE-155263-MS, SPE International Conference on Oilfield Scale, held May 30–31, 2012 (Aberdeen, UK).
115. C. Fan, A.T. Kan, P. Zhang, et al., "Barite Nucleation and Inhibition at 0 to 200 °C With and Without Thermodynamic Hydrate Inhibitors," *SPE Journal* 16 (2011): pp. 440–450.
116. M.B. Tomson, G. Fu, M.A. Watson, et al., "Mechanisms of Mineral Scale Inhibition," *SPE Production & Facilities* 18 (2003): pp. 192–199.
117. S. He, A.T. Kan, M.B. Tomson, "Mathematical Inhibitor Model for Barium Sulfate Scale Control," *Langmuir* 12 (1996): pp. 1901–1905.
118. N. Bhandari, A.T. Kan, F. Zhang, et al., "The Effect of Pressure and TDS on Barite Scaling Kinetics," SPE-173790-MS, SPE International Symposium on Oilfield Chemistry, held April 13–15, 2015 (The Woodlands, TX, USA).
119. F. Zhang, Z. Dai, C. Yan, et al., "Barite-Scaling Risk and Inhibition at High Temperature," *SPE Journal* 22 (2017): pp. 69–79.
120. C. Dai, Z. Dai, Y. Zhao, et al., "Prediction Models of Barite Crystallization and Inhibition Kinetics: Applications for Oil and Gas Industry," *Sustainability* 13 (2021): p. 8533.
121. P. Zhang, Z. Zhang, Y. Liu, et al., "Investigation of the Impact of Ferrous Species on the Performance of Common Oilfield Scale Inhibitors for Mineral Scale Control," *Journal of Petroleum Science and Engineering* 172 (2019): pp. 288–296.
122. S.J. Friedfeld, S. He, M.B. Tomson, "The Temperature and Ionic Strength Dependence of the Solubility Product Constant of Ferrous Phosphonate," *Langmuir* 14 (1998): pp. 3698–3703.
123. S. He, A.T. Kan, M.B. Tomson, "Inhibition of Calcium Carbonate Precipitation in NaCl Brines From 25 to 90 °C," *Applied Geochemistry* 14 (1999): pp. 17–25.
124. Z. Dai, Y. Zhao, S. Paudyal, et al., "Gypsum Scale Formation and Inhibition Kinetics With Implications in Membrane System," *Water Research* 225 (2022): p. 119166.
125. S. Paudyal, S. Mateen, C. Dai, et al., "Developments on Calcium Sulfate Scaling Prediction and Control in Oil and Gas Production," SPE-200671-MS, SPE International Oilfield Scale Conference and Exhibition, held June 24–25, 2020 (Virtual).
126. L.M. Frostman, A.T. Kan, M.B. Tomson, "Mechanistic Aspects of Calcium Phosphonates Precipitation," in *Calcium Phosphates in Biological and Industrial Systems*, ed. Z. Amjad (Boston: Springer, 1998), pp. 493–506.
127. J. Xiao, A. Kan, M. Tomson, "Prediction of $BaSO_4$ Precipitation in the Presence and Absence of a Polymeric Inhibitor: Phosphino-Polycarboxylic," *Langmuir* 17 (2001): pp. 4668–4673.
128. G. Deng, A.T. Kan, F. Zhang, et al., "Novel Laser-Hydrothermal Apparatus for Nucleation and Inhibition Study of Scale Minerals At Temperatures Up To 250 °C," SPE-193556-MS, SPE International Conference on Oilfield Chemistry, held April 8–9, 2019 (Galveston, TX, USA).
129. W. Wang, A.T. Kan, M.B. Tomson, "A Novel and Comprehensive Study of Polymeric and Traditional Phosphonate Inhibitors for High-Temperature Scale Control," *SPE Journal* 18 (2013): pp. 575–582.

130. F. Zhang, Z. Dai, F. Yan, et al., "Acid/Base and Metal Complex Solution Chemistry of Sulfonated Polyacrylate Copolymer Versus Temperature and Ionic Strength," *Applied Geochemistry* 76 (2017): pp. 1–8.
131. C. Yan, A.T. Kan, W. Wang, et al., "Synthesis and Sorption Study of AlOOH Nanoparticle Cross-Linked Polymeric Scale Inhibitors and Their Squeeze Performance in Porous Media," SPE-164086-MS, SPE International Symposium on Oilfield Chemistry, held April 8–10, 2013 (The Woodlands, TX, USA).
132. X. Yao, X. Wang, S. Ko, et al., "Fate and Transport of Sulfonated Polymeric Inhibitors in the Reservoir: Studied by Column Experiments With Calcite Media," SPE-213787-MS, SPE International Conference on Oilfield Chemistry, held June 28–29, 2023 (The Woodlands, TX, USA).
133. S. Ko, X. Wang, A.T. Kan, et al., "Growth Inhibition and Deposition Prevention of Sulfide Scales Using Dispersants," *Journal of Petroleum Science and Engineering* 197 (2021): p. 108107.
134. S. Ko, Y. Zhao, X. Wang, et al., "Deposit Prevention of Mineral Scales Using a Universal Dispersant of Carboxymethyl Cellulose," *SPE Journal* 28 (2022): pp. 1–11.
135. X. Wang, Z. Dai, C. Dai, et al., "Development of Novel Iron Sulfide Control Model and Understanding of Iron Sulfide Dispersant Mechanism," NACE-2021-16883, NACE Annual Conference, held April 19–30, 2021 (Virtual).

Scale Prediction: Thermodynamics and Kinetics

Zhaoyi Dai[1], Amy T. Kan[2], Mason B. Tomson[2], Shu Jiang[3], Huaming Yang[4], and Fulong Ning[5]

[1]State Key Laboratory of Biogeology and Environmental Geology, School of Earth Sciences, China University of Geosciences, Wuhan, Hubei, China
[2]Brine Chemistry Consortium, Rice University, Houston, TX, USA
[3]Key Laboratory of Tectonics and Petroleum Resources of Ministry of Education, Faculty of Earth Resources, China University of Geosciences, Wuhan, Hubei, China
[4]Engineering Research Center of Nano-Geomaterials of Ministry of Education, China University of Geosciences, Wuhan, Hubei, China
[5]Faculty of Engineering, China University of Geosciences, Wuhan, Hubei, China

2.1 Introduction

As stated in Chapter 1, mineral scale formation can lead to significant economic losses, environmental issues, and safety threats throughout all processes of oil and gas production, including drilling, completion, production, separation, storage, transportation, refinery, and so on.[1] To optimize the mitigation and management of these mineral scale problems, engineers need to answer the following questions in a routine manner: if any mineral scale will form; if yes, how fast and where it will form; and how much scale inhibitor is needed to protect the operations from scale issues. Such engineering questions directly relate to the three key scientific challenges that will be discussed in this chapter, that is, (1) the thermodynamic model of the water-mineral-gas-oil system, (2) the kinetic model and mechanism of mineral scale formation, and (3) the kinetic model and mechanism of mineral scale inhibition.

The mineral scale risk prediction (*i.e.*, whether a mineral will form) is related to the mineral solubility prediction at operating conditions, which requires an accurate thermodynamic model of the water-mineral-ion system. In the disciplines of environmental engineering, hydrology, aquatic chemistry, marine science and engineering, and so on, various thermodynamic software models have been developed and widely used but mostly under surface or shallow conditions. For example, quite a few thermodynamic software models, such as VisualMinteq,[2] PHREEQC,[3] Geochemist's Workbench, have been developed to simulate the thermodynamic status of the water-mineral-ion system. However, most of these software models are limited to being used at temperatures below 100°C, at pressures no higher than the water vapor pressure (*i.e.*, 1 atm at 100°C), and in relatively dilute solutions (*e.g.*, mostly under 0.1 m). On the contrary, the operating conditions in the oil and gas industry can extend over very wide ranges, for example, temperature of 0–250°C or higher, pressure of 1–1,500 bars or higher, ionic strength up to halite saturation salinity (*e.g.*, 6.15 m at 25°C and 1 atm).[1, 4] In recent years, great efforts have been invested to extend the applicable ranges of temperature, pressure, and salinity in the presence of mixed electrolytes of various thermodynamic models.

To predict when and where a scale mineral will form, one needs to understand the kinetics and mechanisms of mineral crystallization in the aqueous phase and mineral deposition onto facility surfaces. After decades of efforts, some empirical models have been developed to predict the induction time (*i.e.*, the time elapsed between the creation of supersaturation and the detection of minerals in the aqueous phase) under wide ranges of conditions.[5–10] Similar to other empirical models, these models cannot be reliably extrapolated to conditions that have not been experimentally tested. Many studies have also focused on the mechanisms of each mineral scaling stage (*e.g.*, nucleation, crystal growth, aggregation, ripening, and deposition). To apply these mechanistic models to crystallization kinetics prediction, more efforts are required to develop quantitative models for each crystallization stage and couple them together as a complete crystallization model. Moreover, it has long been noticed that mineral crystallization in the aqueous phase does not necessarily lead to mineral scale formation on the facility surfaces (*i.e.*, scale deposition). The crystallized minerals can flow out of the system without doing any damage if no deposition happens. As the last critical step of mineral scale formation, however, few studies have focused on the deposition processes. The efforts to accurately predict the deposition rate will provide another critical criterion for scale control.

Scale inhibitors have been widely used as the most economic method for mineral scale control.[11, 12] Though billions of dollars are spent on scale inhibitors every year, the inhibition mechanisms have not yet been elucidated.[13] Great efforts have been made to develop empirical models to predict the efficiencies of various scale inhibitors.[5–7, 9, 10] Such empirical models have been widely used for minimum inhibitor concentration (MIC) predictions in oil and gas production. To better guide the development of new inhibitors and extrapolate model predictions to wider conditions, various mechanistic models have also been developed.[14–17] Great efforts are still needed to deepen the understanding of the inhibition process and improve the prediction accuracy.

This chapter will introduce the fundamental basics and current research status of three challenges mentioned above. More importantly, this chapter will go through these topics from an engineering perspective and focus on the practical applications of these research achievements in routine oil and gas operations.

2.2 Thermodynamic Models

The process of mineral crystallization and dissolution can be described by the following reaction as an example:

$$M_gX_h \Leftrightarrow gM^{n+} + hX^{n-}, \qquad K_{sp} = \{M^{n+}\}^g\{X^{n-}\}^h \tag{2.1}$$

where M^{n+} and X^{n-} represent the cation and anion of mineral MX; g and h are the stoichiometric coefficients; { } denotes the activity of the cation and anion; K_{sp} is the solubility product of the mineral MX and is the equilibrium constant of the dissolution reaction. Under equilibrium conditions, the activity product of the cation and anion equals the solubility product of the mineral when at equilibrium. The K_{sp} value varies with temperature and pressure but is independent of solution compositions (*i.e.*, ion concentrations, ionic strength). Different minerals can have K_{sp} values with several magnitudes of differences. Considering such large differences in K_{sp} values and cation to anion molar ratios, instead of using ion concentrations, saturation index (SI, base 10 logarithm of the saturation ration, SR) has been introduced to represent the mineral saturation status as follows:

$$\begin{aligned} SI &= \log_{10} SR = \log_{10}\frac{\{M^{n+}\}^g\{X^{n-}\}^h}{K_{sp}} \\ &= \log_{10}\frac{[M^{n+}]^g\gamma_M^g[X^{n-}]^h\gamma_X^h}{K_{sp}}, \quad \begin{cases} >0, & \text{supersaturated} \\ =0, & \text{equilibrium} \\ <0, & \text{undersaturated} \end{cases} \end{aligned} \tag{2.2}$$

where [ ] denotes the molality concentration of the ions (*i.e.*, mol/kg H_2O); γ represents the activity coefficient of each ion. As is presented in Equation (2.2), its SI value will be positive (*i.e.*, SR > 1 and $\{M^{n+}\}\{X^{n-}\} > K_{sp}$) when the aqueous solution is supersaturated in regard to a certain mineral, equal to 0 when at equilibrium, and negative when undersaturated, respectively. Note that measurement of concentrations as mg/L is proportional to molarity, M = (mg/L)/(mg/mole), whereas molality, m, can be obtained using the solution TDS and density: $m = M/$[desnity(kg/L) – TDS(kg/L)]. Theoretically, using Equation (2.2), engineers can answer the question that if a certain mineral has the potential to form scale at certain conditions. Though SI has been widely used for scale risk prediction, the following points need to be noticed to properly use SI values in real practice.

First, SI is a thermodynamic property rather than a kinetic one. A positive SI only implies that the mineral has the potential to form scale, given an infinite amount of time, but it cannot predict when, how fast or how severe the scale will happen (*i.e.*, scale kinetics). Though, in general, for the same mineral, the larger the SI or the higher the temperature usually represent larger scaling risks, the exact scaling kinetic prediction can be achieved using the scaling kinetics model which will be discussed in later sections. As a practical alternative, some researchers have proposed a series of threshold SI values for different minerals under different conditions depending on if the mineral scale can be observed in a realistically short time with or without scale inhibitors based on the empirical observations.[18–20] As listed in Table 2.1, four-scale risk regions have been used to represent the severity of the scale. In Scale Region 1, when the SI values are negative, no scale can form; in Scale Region 2, scales can form only when mineral solids are present in the system; in Scale Region 3, scales will form but can be inhibited via adding certain amounts of inhibitors; in Scale Region 4, scales will form and cannot be inhibited, even if substantial amounts of inhibitors are added. Different minerals have different sets of threshold SI values at different temperatures. Generally, the higher the temperature, the larger the corresponding threshold SI values (Table 2.1). Such empirical threshold SI values are not based on strict derivation but can provide a rough but convenient guideline for field operations.

Table 2.1. The threshold SI values of four different scale risk regions for calcite and barite.

Scale Regions	Calcite	Barite	
	25°C	25°C	120°C
Region 1, scale cannot form	(-∞, 0.00)	(-∞, 0.00)	(-∞, 0.00)
Region 2, scale forms only with seed	(0.00, 0.40)	(0.00, 1.30)	(0.00, 0.30)
Region 3, scale that can be inhibited	(0.40, 2.30)	(1.30, 3.00)	(0.30, 1.50)
Region 4, scale that cannot be inhibited	(2.30, + ∞)	(3.00, + ∞)	(1.50, + ∞)

Second, the SI value is not a static property and is strongly dependent on the operating conditions (*e.g.*, T, P, flow rates of water, oil, and gas). In real situations rather than the ideal laboratory testing environment, the operating conditions can vary significantly. The engineers need to keep in mind under what conditions the SI value is predicted, or how the SI value changes along the work flow. For example, in Table 2.2 are the listed SI values of calcite predicted for a typical produced water sample[21] under different temperature and pressure conditions using ScaleSoftPitzer (SSP) software, an Excel-based thermodynamic model. This produced water sample was collected from a well in East Texas. The SI value of calcite drops from 1.02 at bottom conditions (*i.e.*, 340°F, 7,000 psi) to 0.66 at a condition close to the wellhead (*i.e.*, 162°F, 1,133 psi), and then increases back to 0.94 at wellhead conditions (*i.e.*, 140°F, 400 psi). Such an unmonotonous change in SI value shows the complexity of scale risk prediction and the importance of measuring the operating conditions along the flow path accurately.

Table 2.2. The composition, flow rates, and calculated calcite SI values of the produced water sample collected from the Guerra Well in South Texas.

Species	Concentration (mg/L)	Flow Rates		
Na^+	19,872	Gas	Mcf/d	8,500
K^+	500	Oil	Barrel/d	1,000
Mg^{2+}	54	Water	Barrel/d	100
Ca^{2+}	6,500			
Sr^{2+}	700			
Ba^{2+}	550			
Fe^{2+}	12	**SI Predictions**		
Zn^{2+}	10	T (°F)	P (psi)	SI of calcite
Pb^{2+}	1	140	400	0.94
Cl^-	43,000	162	1,133	0.70
SO_4^{2-}	28.5	184	1,867	0.66
F^-	1	207	2,600	0.68
Br^-	10	229	3,333	0.72
Silica	10	251	4,067	0.77
Total alkalinity	281	273	4,800	0.83
CO_2%	0.28	296	5,533	0.89
H_2S_{aq}	4.32	318	6,267	0.96
pH, measured	7.62	340	7,000	1.02

Third, the SI values calculated using different thermodynamic software can be different from each other since they may use different K_{sp} values and different ion activity coefficient prediction models. Thus, the engineers need to check the valid ranges of each thermodynamic software and pick the best ones. It is also recommended that SI values cannot be compared directly against each other if they are predicted using different thermodynamic models. A general review of different thermodynamic models is discussed below.

2.2.1 Common Thermodynamic Models

In the past century, numerous models have been developed to predict the activity coefficients of aqueous species under different conditions based on the statistical thermodynamics. For example, in Figure 2.1 are plotted the activity coefficient of Ca^{2+} changing with ionic strength (*i.e.*, $I = \Sigma\ (m_i z_i^2)/2$) in $CaCl_2$-H_2O solutions at 25°C and 1 atm is calculated using different thermodynamic models. These models show similar predictions when I is less than 0.01 m (*i.e.*, mol/kg H_2O), but deviate significantly at higher I, especially above 0.1 m. The reason why most thermodynamic models are nearly identical at low I is because they all follow the Debye-Hückel (D-H) limiting law in the low I range. The shaded areas in Figure 2.1 represent the normal ionic strength ranges of solutions occurring in different natural or industrial conditions. It is obvious that except for drinking water or ground water, more accurate thermodynamic models are needed for aqueous solutions in other situations, for example, the oil and gas industry. Starting with the D-H limiting law, several of the most important models will be introduced in detail in this section.

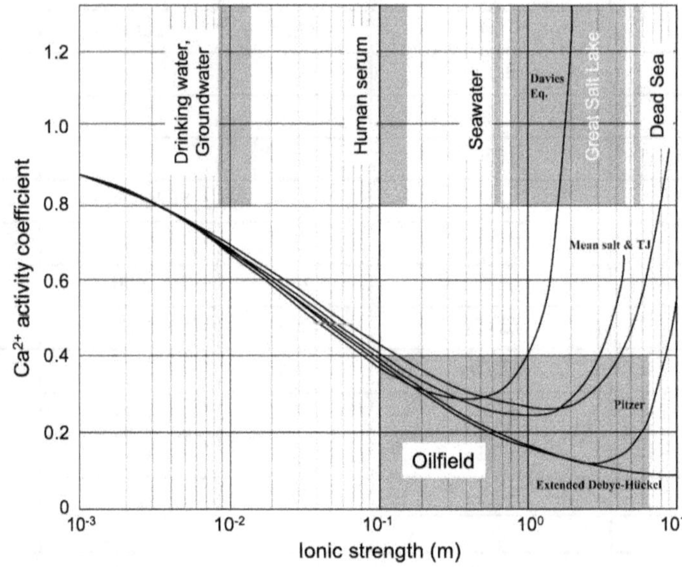

Figure 2.1. Activity of Ca^{2+} in $CaCl_2$-H_2O solutions changing with ionic strength at 25 °C and 1 atm with shaded areas showing the normal ionic strength ranges of various water types in different disciplines.

2.2.1.1 D-H Limiting Law

D-H limiting law or the D-H theory is developed based on the Boltzmann's law and the Poisson equation.[22] It is derived based on the following assumptions: (1) ions are treated as point charges and ion size has no impact on activity coefficients; (2) the solvent is treated as a dielectric continuum; (3) only long-range Coulombic electrostatic interactions are considered; (4) ions with the same sign charge do not interact with each other. By approximating the charged ion radial distribution function with the first two terms, ion the activity coefficient γ_i is calculated as:

$$\log_{10} \gamma_i = -A_\gamma z_i^2 I^{1/2} \tag{2.3}$$

$$A_\gamma = \frac{1}{\ln(10)} [2\pi N_{Av} \rho_w]^{1/2} \left[\frac{e^2}{4\pi\varepsilon k_B T} \right]^{3/2} \tag{2.4}$$

where z_i is the ion charge of interest; A_γ is the D-H limiting law slope for the activity coefficient; N_{Av} stands for Avogadro's number; e is the electron charges; ρ_w is the solvent (*i.e.*, water) density in kg/m^3 (*e.g.*, ρ_w = 997.05kg/m^3 at 25°C and 1 atm for water); and ε is the dielectric constant of the solvent and is usually calculated as the product of relative dielectric constant (*i.e.*, ε_r = 78.38 for pure water at 25°C and 1 atm) and the absolute dielectric constant of the vacuum (*i.e.*, ε_0 = 8.85 x 10^{-12} F/m), respectively, and both change with temperature and pressure; k_B is the Boltzmann's constant; the denominator ln(10) is for transition between Napierian logarithm (*i.e.*, natural logarithm, base *e*, ln) and Briggsian logarithm (*i.e.* scientific logarithm, base 10, $\log_{10}$); I (ionic strength) indicates the charge specific impacts of all solutes on individual ion interactions.[23]

A clarification has to be made here to explain the differences between the equations listed here and those listed in different literature. The main causes of such differences are: (1) different logarithm operators used in Equation (2.3) can lead to a correction factor of ln(10) or 2.3026; (2) D-H limiting law slope for activity coefficient (A_γ) is usually three times D-H limiting law slope for osmotic coefficient (A_φ) (*i.e.*, A_γ = 3 A_φ), when the same logarithm operator is used; (3) all parameters used in Equation (2.4) should be expressed in SI units; (4) different definitions of dielectric constants may lead to a 4π factor difference. When following the equations above and using the parameters in the correct units, the value of A_γ can be calculated to be 0.51 kg$^{1/2}$·mol$^{-1/2}$ at 25°C and 1 atm. Actually, Helgeson and Kirkham (1974) have developed a table for different D-H parameters (including A_γ) at different temperature and pressure conditions.[24]

The D-H theory has significant values for both theoretical developments and practical applications.[25] The assumptions that D-H law is derived on are valid under very dilute conditions (I < 0.01 m), and thus, it works perfectly for these dilute solutions (*e.g.*, ground water or drinking water as plotted in Figure 2.1). More importantly, it accurately describes the ion interactions due to long-range electrostatic interactions, which are an integral part that cannot be ignored in other thermodynamic models. This is also the reason why most thermodynamic models give almost identical predictions in the low ionic strength range as shown in Figure 2.1.

2.2.1.2 The Extended D-H Equation and Its Modifications

As the solute gets more concentrated and the ionic strength increases to above 0.01 m, ions get closer to each other, and they cannot be treated as point charges. Thus, assumption (1) above used in the D-H theory is not valid, and their size effect needs to be considered. For this reason, the extended D-H equation was developed to include the impact of ion sizes as follows:

$$\log_{10}\gamma_i = -A_\gamma z_i^2 \frac{I^{1/2}}{1+Ba_i \cdot I^{1/2}} \tag{2.5}$$

where a_i is the empirical parameter representing the distance of the closest approach between two interacting ions taking into account ion size and ranges from 3 Å to 9 Å and is independent of I; B is dependent on solvent dielectric constant and temperature and can be calculated by:

$$B = \left(\frac{8\pi N_{Av} \rho_w e^2}{4\pi\varepsilon k_B T} \right)^{1/2} \tag{2.6}$$

Equation (2.6) shows that B term is also a function of temperature and pressure. The value of the B term can be calculated to be 3.28×10^9 $kg^{1/2} \cdot m^{-1}$ at 25°C and 1 atm, and its value at other conditions can be found in Helgeson and Kirkham (1974) for convenience.[24] When Ba_i reduces to 1 or 1.5, the extended D-H equation becomes Güntelberg's Equation[26] and Scatchard's Equation,[27] respectively. The extended D-H equation and these modified equations work quite well at I up to 0.1 m. Its incorporation of ion size (*i.e.*, a_i) involves more ion-specific properties. Actually, the physical meaning of a_i is inadequate to represent a physically meaningful true distance, and it also contains other short-range effects.[28]

When the ionic strength exceeds 0.1 m, even closer distances between solute ions will lead to the occurrence of stronger short-range ion interactions between both likely and unlikely charged ions. Such short-range ion interactions violate the assumptions (3) and (4) used in the D-H theory derivation. In addition, as suggested by Robinson and Stokes (2002), the activity coefficient predicted by the extended D-H equation does not consider the ion-solvent solvation impacts.[28] When the ionic strength gets larger, solvent molecules will have more chances to interact with or be captured by solute ions (*e.g.*, solvation) leading to fewer free solvent molecules available in the solution. As a result, solutes have more chances to interact with solvent molecules and other solute molecules (*i.e.*, larger activities). Therefore, it is expected that more significant deviations will occur when using the extended D-H equation or the above modified equations at $I > 0.1$ m.

To compensate for such solute-solute short-range interactions and potential solvent-solute interaction effects, further modifications have been made in the following form to add a term that increases with I at higher I conditions:

$$\log_{10} \gamma_i = -A_\gamma z_i^2 \frac{I^{1/2}}{1 + Ba_i \cdot I^{1/2}} + b_\gamma I \tag{2.7}$$

where b_γ is the empirical parameter representing the characteristic of the solute species. For example, when Ba_i reduces to 1 and b_γ represents a species-specific interaction coefficient, Equation (2.7) becomes the Guggenheim Equation.[29] It becomes the famous Davies Equation when Ba_i reduces to 1 and b_γ reduces to $0.2A_\gamma z_i^2$ [30] or $0.3A_\gamma z_i^2$. With this modification, Davies Equation can work up to 0.5 m I for most solutions. Furthermore, when b_γ is another empirical ion size parameter, it becomes the Truesdell-Jones (TJ) Equation.[31] While it deserves notice that, the a_i values for each ion are different from those in the extended D-H equation (*e.g.*, a_i for H^+ is 9 Å and 4.78 Å in the extended D-H equation and TJ equation, respectively) though the formula is the same. With this modification, the TJ equation can work up to 2.0 m I for some solutions.

A side-by-side comparison between the extended D-H equation and its modified forms is listed in Table 2.3. As introduced above, starting from D-H theory, which accurately describes the long-range solute-solute columbic interactions, more modifications have been made to include the effects of ion size, solute-solute short-range interactions, and potential solute-solvent interactions. Most of these modifications are empirical and not ion-specific (except for the size parameter and the b_i term used in TJ equation). When solution I gets even larger (*e.g.*, >0.1 m or more), further modifications that can represent the ion-specific properties need to be included.

Table 2.3. The information of each term in Equation (2.7) and the valid ionic strength range of different D-H theory-based models.

Models	Ba_i	b_γ	Valid Ionic Strength (I) Range
D-H theory	0	0	<0.01 m
Güntelberg equation	1.0	0	<0.1 m
Scatchard equation	1.5	0	<0.1 m
Extended D-H	Ba_i	0	<0.1 m
Guggenheim equation	1	b_γ	<0.1 m
Davies equation	1	$0.2\,A_\gamma z_i^2$ or $0.3\,A_\gamma z_i^2$	<0.5 m
TJ equation	Ba_i (with different a_i)	b_i (size parameter)	<2.0 m for some solutions

2.2.1.3 Robinson-Stokes equation

Instead of adding on empirical terms to account for potential solutesolvent interactions, Robinson and Stokes (1955) considered the Gibbs free energy change in the solvation process in the solution.[32] They assume that a total of *h* moles of solvent interacts with *v* moles of electrolytes in the solution, which is unchanged even at infinite dilution. It ignores the "nonelectrolyte" contributions such as the mixing heat of solvated species with solvent. By combining the extended D-H equation, they derived the Robinson-Stokes (R-S) equation for mean ion activity coefficient of electrolyte $M_{v+}X_{v-}$ as below:

$$\log_{10}\gamma_{\pm} = -A_{\gamma}\left|z_{+}z_{-}\right|\frac{I^{1/2}}{1+Ba\cdot I^{1/2}} - \frac{h}{v}\log_{10}a_{w} - \log_{10}[1+10^{-3}\mathrm{MW}(v-h)m] \qquad (2.8)$$

$$\gamma_{\pm} = (\gamma_{\mathrm{M}}^{v_{+}}\gamma_{\mathrm{X}}^{v_{-}})^{1/(v_{+}+v_{-})} \qquad (2.9)$$

$$z_{+}v_{+} + z_{-}v_{-} = 0 \qquad (2.10)$$

where a_w is water activity; v_+ and v_- are the number of cation and anion atoms in the mineral molecule, respectively. The R-S equation has two parameters (*i.e.*, the size parameter *a* and the coordination parameter *h*) to be fitted. *a* values are not identical to those used in the previous models, though they are also derived from the extended D-H equation. *h* values represent the solvation state of the electrolytes, which means if *h* is larger, more water molecules are captured by the electrolytes. Based on the fitted *a* and *h*, the R-S equation can be used to accurately predict the activity coefficient of 1-1 electrolyte up to 4–5 M, and of 2-1 electrolyte up to 1.5–1.8 M. According to the fitted results, cations with larger crystallographic radii will hold fewer water molecules (*i.e.*, smaller *h*), which is consistent with other experimental results. For example, in NaCl and KCl solutions, *h* for sodium (3.5) is larger than for potassium (1.9) while the sodium ion has smaller radii of bare ions. However, it also shows inconsistency for anion species which is in a reverse order. It deserves notice that the h value is only meaningful for electrolytes but not for individual ions. Thus, the R-S equation is not suitable for mixed electrolyte solutions in which it is difficult to tell which electrolyte an individual ion belongs to. In addition, it ignores the "nonelectrolyte" effects, and the h values are not consistent with experiments to represent the number of hydrated water molecules especially when the ionic strength changes considerably. For solutions with mixed electrolytes, an ion-specific rather than an electrolyte-specific model is needed.

2.2.1.4 Meissner Model

Different from these models adjusted with extended D-H equation, Kusik and Meissner (1978) developed a thoroughly different model to predict the activity coefficients for electrolyte $M_{\upsilon+}X_{\upsilon-}$ which avoids of the inaccuracies of extended D-H equation at higher concentrations.[33] In their model, the reduced activity coefficient γ_r^0 is defined as follows:

$$\gamma_r^0 = \gamma_\pm^{1/\nu_+\nu_-} = [1 + B(1 + 0.1 \cdot I)^{q_t} - B]\gamma^* \tag{2.11}$$

where

$$B = 0.75 - 0.065 q_t \tag{2.12}$$

$$\log_{10} \gamma^* = -\frac{0.5107 \cdot I^{1/2}}{1 + C \cdot I^{1/2}} \tag{2.13}$$

$$C = 1 + 0.55 q_t \, e^{-0.023 \cdot I^3} \tag{2.14}$$

In this model, only one empirical electrolyte-specific parameter q_t needs to be defined at different temperature, and its temperature is:

$$\frac{q_t - q_{25}}{TC - 25} = a q_{25} + b^* \tag{2.15}$$

where TC is temperature in Celsius. It is reported that a, b^* are the same for most electrolytes with $a = -0.005$ and $b^* = 0.0085$, except for sulfuric acid with a, b^* values of −0.0079 and −0.0029, respectively. Equation (2.15) can only work up to 120°C. By replacing the constant, 0.5107 in Equation (2.13) with the A_γ term in Equation (2.4) can extend the valid range of temperature up to 300°C.[34]

For solutions with mixed electrolytes, the reduced activity coefficient for a specific electrolyte (*i.e.*, M_iX_j) is calculated by a weighted function:

$$\begin{aligned} \log_{10} \gamma_{r,MX} = {} & \frac{z_M}{z_M + z_X} \sum_{a=1}^{N_a} \left(\frac{(z_M + z_a)^2}{2 z_M z_a} \cdot \frac{m_a z_a^2}{2} \log_{10} \gamma_{r,Ma}^0 \right) / I \\ & + \frac{z_X}{z_M + z_X} \sum_{c=1}^{N_c} \left(\frac{(z_c + z_X)^2}{2 z_c z_X} \cdot \frac{m_c z_c^2}{2} \log_{10} \gamma_{r,cX}^0 \right) / I \end{aligned} \tag{2.16}$$

where z_i is the charge of ion i; m_i is the molality of ion i; *a* denotes for anions and c denotes for cations; N_a and N_c stand for the number of total anions and cations, respectively.

This model can be easily applied to different systems over wide ranges of *I* based on readily available data and parameters. Via the weighting equation (Equation [2.16], this model overcomes the disadvantage of the R-S model which is not suitable for mixed electrolyte solutions. However, the error can be as much as 20% for various mixed electrolyte solutions, and even larger at higher temperatures. It has also been pointed out that there are significant deviations for electrolytes such as zinc and cadmium chlorides, bromides, and iodides, sulfuric acid, and thorium nitrate. Moreover, for more complex mixed electrolytes, the weighting equations cannot accurately represent the complex ion interactions in mixed electrolyte solutions in which the chemical environment is totally different from that in single-electrolyte solutions, especially at higher temperatures.

2.2.1.5 Specific Ion Interaction Theory

When solute concentration increases even higher (>2.0 m) with mixed electrolytes, short-range solute-solute interactions are more significant and cannot be ignored. More importantly, such short-range interactions are species-specific and cannot be simply represented by universal constants. To resolve this issue, the Bronsted-Guggenheim-Scatchard-specific ion interaction (SIT) model is developed based on the extended D-H equation considering the ion-specific interactions as follows[27, 29, 35]:

$$\ln \gamma_{MX} = -\ln(10) A_\gamma |z_M z_X| \frac{I^{1/2}}{1+I^{1/2}} + \frac{2\nu_M}{\nu_M+\nu_X}\sum_a m_a \beta_{Ma} + \frac{2\nu_X}{\nu_M+\nu_X}\sum_c m_c \beta_{cX} \tag{2.17}$$

where in the first term of the extended D-H part, B^*a_i term is arbitrarily picked to be 1. The short-range ion-specific interactions are represented by the term β which is different for different ion interactions. The inclusion of such an ion-specific term improves the accuracy of the SIT model at higher solute concentrations. In addition, the SIT model also used a weighted function similar to that used in the Meissner model. However, it assumes that only oppositely charged ions have short-range interactions, and it ignores the interactions between the likely charged ions or the ones with neutral species. In addition, the β term is assumed to be independent of ionic strength in the SIT model. Such assumptions are reasonable when *I* is under 3.5 m, but when *I* is larger, they lead to large deviations especially for solutions with mixed electrolytes.

2.2.1.6 Pitzer Theory

Considering the shortcomings of the SIT model, Pitzer and his associate researchers expanded the SIT model by incorporating not only binary interactions but also ternary interactions, even including neutral species. At ionic strength over 3.5 m, or even higher, not only the binary interactions between anion and cation pairs but also binary interactions between pairs of ions of the same charge, and the ternary interactions among three ions among which two have the same sign charge, are all considered. The excess Gibbs free energy of the system is represented by the Pitzer equations as follows[25]:

$$\frac{G^{ex}}{\mathrm{RT}\cdot w_s} = -\frac{4A_\phi \cdot I}{b}\ln\left(1+b\cdot I^{1/2}\right) + 2\sum_{c=1}^{N_c}\sum_{a=1}^{N_a} m_c m_a\left(B_{ca} + ZC_{ca}/2\right)$$

$$+\sum_{c=1}^{N_c-1}\sum_{c'=c+1}^{N_c} m_c m_{c'}\left(2\Phi_{cc'} + \sum_{a=1}^{N_a} m_a \psi_{cc'a}\right)$$

$$+\sum_{a=1}^{N_a-1}\sum_{a'=a+1}^{N_a} m_a m_{a'}\left(2\Phi_{aa'} + \sum_{c=1}^{N_c} m_c \psi_{aa'c}\right) \tag{2.18}$$

$$+2\sum_{n=1}^{N_n}\sum_{c=1}^{N_c} m_n m_c \lambda_{nc} + 2\sum_{n=1}^{N_n}\sum_{a=1}^{N_a} m_n m_a \lambda_{na} + \sum_{n=1}^{N_n}\sum_{c=1}^{N_c}\sum_{a=1}^{N_a} m_n m_c m_a \xi_{nca}$$

$$A_\phi = \frac{\ln(10)}{3}A_\gamma = \frac{1}{3}\left[2\pi N_{\mathrm{Av}}\rho_{\mathrm{w}}\right]^{1/2}\left[\frac{e^2}{4\pi\varepsilon k_{\mathrm{B}}T}\right]^{3/2} \tag{2.19}$$

$$B_{ca} = \beta_{ca}^{(0)} + \beta_{ca}^{(1)} g\left(\alpha_{ca}\cdot I^{1/2}\right) + \beta_{ca}^{(2)} g\left(12\cdot I^{1/2}\right) \tag{2.20}$$

$$C_{ca} = \frac{C_{ca}^{\phi}}{2\left|z_c z_a\right|^{1/2}} \tag{2.21}$$

where m is the molality of species in mol/kg H_2O; z denotes the charge of an individual species; $Z = \sum_i m_i |z_i|$ is the total molality of charges in the system; b is arbitrarily chosen as 1.2 $(\mathrm{kg/mol})^{1/2}$ for all electrolytes; A^{φ} is the D-H limiting slope for the osmotic coefficient; B, C, Φ, ψ, λ, ξ are ion interaction terms; c, a, n in the subscripts represent cation, anion, and neutral species, respectively; function $g(x) = 2\left[1-(1+x)e^{-x}\right]/x^2$ is the ionic strength dependent function; α equals 2.0 $(\mathrm{kg/mol})^{1/2}$ for 1-1, 1-2, 2-1 ion interactions, 1.4 $(\mathrm{kg/mol})^{1/2}$ for 2-2 ion interactions, and α equals to $2.0 - 0.00181\times(T(\mathrm{K}) - 298.15)$ for Mg/Ca/Ba/Sr-Cl interactions with temperature dependence suggested by Holmes and his co-authors.[36, 37] The osmotic coefficients and activity coefficients of cations (M) and anions (X) are calculated as follows:

$$\phi = 1 + \frac{2}{\sum_i m_i}\left\{\begin{aligned} & -\frac{A_\phi \cdot I^{3/2}}{1+b\cdot I^{1/2}} + \sum_{c=1}^{N_c}\sum_{a=1}^{N_a} m_c m_a\left(B_{ca}^{\phi} + ZC_{ca}\right) \\ & +\sum_{c=1}^{N_c-1}\sum_{c'=c+1}^{N_c} m_c m_{c'}\left(\Phi_{cc'}^{\phi} + \sum_{a=1}^{N_a} m_a \psi_{cc'a}\right) \\ & +\sum_{a=1}^{N_a-1}\sum_{a'=a+1}^{N_a} m_a m_{a'}\left(\Phi_{aa'}^{\phi} + \sum_{c=1}^{N_c} m_c \psi_{aa'c}\right) \\ & +\sum_{n=1}^{N_n}\sum_{c=1}^{N_c} m_n m_c \lambda_{nc} + \sum_{n=1}^{N_n}\sum_{a=1}^{N_a} m_n m_a \lambda_{na} + \sum_{n=1}^{N_n}\sum_{c=1}^{N_c}\sum_{a=1}^{N_a} m_n m_c m_a \xi_{nca} \end{aligned}\right\}, \tag{2.22}$$

$$
\begin{aligned}
\ln \gamma_M = {} & z_M^2 F + \sum_{a=1}^{N_a} m_a \left(2B_{Ma} + ZC_{Ma}\right) + \sum_{c=1}^{N_c} m_c \left(2\Phi_{Mc} + \sum_{a=1}^{N_a} m_a \psi_{Mca}\right) \\
& + \sum_{a=1}^{N_a-1} \sum_{a'=a+1}^{N_a} m_a m_{a'} \psi_{aa'M} + |z_M| \sum_{c=1}^{N_c} \sum_{a=1}^{N_a} m_c m_a C_{ca} \\
& + \sum_{n=1}^{N_n} m_n \left(2\lambda_{nM}\right) + \sum_{n=1}^{N_n} \sum_{a=1}^{N_a} m_c m_a \xi_{nMa}
\end{aligned}
, \quad (2.23)
$$

$$
\begin{aligned}
\ln \gamma_X = {} & z_X^2 F + \sum_{c=1}^{N_c} m_c \left(2B_{cX} + ZC_{cX}\right) + \sum_{a=1}^{N_a} m_a \left(2\Phi_{aX} + \sum_{c=1}^{N_c} m_c \psi_{caX}\right) \\
& + \sum_{c=1}^{N_c-1} \sum_{c'=c+1}^{N_c} m_c m_{c'} \psi_{cc'X} + |z_X| \sum_{c=1}^{N_c} \sum_{a=1}^{N_a} m_c m_a C_{ca} \\
& + \sum_{n=1}^{N_n} m_n \left(2\lambda_{nX}\right) + \sum_{n=1}^{N_n} \sum_{c=1}^{N_c} m_c m_a \xi_{ncX}
\end{aligned}
, \quad (2.24)
$$

The F term includes two parts. The first part is the extended D-H equation term representing long-range interactions, and the second part is the terms with derivatives of virial coefficients with respect to ionic strength:

$$
\begin{aligned}
F = {} & -A_\phi \left\{ \frac{IS^{1/2}}{1 + b \cdot IS^{1/2}} + \frac{2}{b} \ln\left(1 + b \cdot IS^{1/2}\right) \right\} \\
& + \sum_{c=1}^{N_c} \sum_{a=1}^{N_a} m_c m_a B'_{ca} + \sum_{c=1}^{N_c-1} \sum_{c'=c+1}^{N_c} m_c m_{c'} \Phi'_{cc'} + \sum_{a=1}^{N_a-1} \sum_{a'=a+1}^{N_a} m_a m_{a'} \Phi'_{aa'}
\end{aligned}
. \quad (2.25)
$$

B, Φ, and λ terms are components of the second virial coefficients in which stand for binary interactions. Among them, B term is for binary interactions of unlikely charged ions; Φ term stands for likely charged electrolytes binary interactions; λ represents the binary interaction between neutral species and charged ions. These terms are theoretically a function of ionic strength, but for simplicity λ is set to be constant which has little impact. The Φ term contains two contributions and equals to $\theta_{ij} + {}^E\theta_{ij}$. θ_{ij} is the ion-specific contribution independent of ionic strength, while ${}^E\theta_{ij}$ is only dependent on charges (*i.e.*, z_i, z_j) and ionic strength C, ψ, and ξ terms are components of the third virial coefficients standing for ternary interactions. Among them, the C term represents the interactions between opposite charged ions in single-electrolyte solutions independent of ionic strength. ψ represents the ternary ion interactions with mixed electrolytes. ξ stands for interactions involving one neutral species, one cation and one anion.

In these, the ion-specific interaction terms, $\beta^{(0)}$, $\beta^{(2)}$, C^φ, Φ, ψ, λ, and ξ, are suggested to be only as a function of both temperature and pressure or constants. Empirically, the temperature and pressure dependences of these virial coefficients (PZ) are described by 12 or more parameters with T in K and P in atm:

$$
\begin{aligned}
PZ(T,P) = {} & z_1 + \frac{z_2}{T} + z_3 T + z_4 \ln T + \left(z_5 + \frac{z_6}{T} + z_7 T + z_8 \ln T \right) P \\
& + \left(z_9 + \frac{z_{10}}{T} + z_{11} T + z_{12} \ln T \right) P^2
\end{aligned}
. \quad (2.26)
$$

These temperature- and pressure-dependent parameters (*i.e.*, z_i) of these virial coefficients are usually fitted based on the different types of thermodynamic data (*e.g.*, mineral solubilities, solution density, reaction enthalpy).[36, 38, 39]

The Pitzer theory has been one of the most accurate and most widely used thermodynamic models for aquatic systems, especially for industrial applications where high salinity and mixed electrolytes occur commonly.[4, 38, 40–48] It has been widely adopted in many academic and industrial software, for example, PHREEQC,[49] ScaleSoftPitzer,[4, 19, 38, 41] PHRQPITZ,[50] SOLMINEQ.88.[51] However, when using the Pitzer theory, several limitations of various models need to be noted[46]:

1. The valid temperature and pressure ranges of certain virial coefficients might be limited. For example, Weare and his group developed a Pitzer theory-based thermodynamic model (*i.e.*, H-M-W model) for the H-Na-K-Ca-OH-Cl-HSO_4-SO_4-H_2O system with related virial coefficients valid up to 250°C, while the pressure dependence of many related virial coefficients (*e.g.*, Ca-Cl) are only valid up to saturated vapor pressure. Insufficient experimental data are one of the major causes and are still challenging to be resolved in the coming years.

2. The virial coefficients fitted from different studies might be incompatible with each other. For example, Shi *et al.* (2013) found that when using Ca^{2+}-SO_4^{2-} virial coefficients derived in H-M-W model, the predicted barite solubility in the presence of high Ca^{2+} ions will deviate from the experimental measurements.47

3. Some studies fitted virial coefficient based on different assumptions or using different types of modified thermodynamic models. For example, the H-M-W model explicitly treated the calcium sulfate ion pair as individual species with an association constant, which deviates from the assumption in Pitzer theory that the 2-2 ion pair is implicitly treated as strong ion interactions by a $\beta^{(2)}$ term2[5, 52]; Duan and Li treated Ca-HCO_3^- ion pair as an individual species and assigned virial coefficients to represent its interaction with other species[53]; Archer expanded the ternary ion interaction C term with two more I dependent terms to more accurately describe NaCl-H_2O system interactions over wider temperature and pressure ranges[54]; Wang *et al.* added a cation-anion interactions term (denoted as D) to fit the properties of $MgCl_2$ at from 250 K to 600 K and up to 1,000 bars.[55] As a result, significant errors can occur if using the virial coefficients are derived under different assumptions or different theoretical frameworks.

As a result, after decades of efforts from various scientists, it is still challenging to develop a comprehensive Pitzer theory-based thermodynamic model that works for complex multielectrolyte systems over wide ranges of temperature, pressure, and ionic strength. Integrating models developed for different systems under different conditions by putting their virial coefficients together can lead to significant unexpected errors. To avoid such errors, two possible ways are recommended. First, we can use the thermodynamic model specifically designed for the targeted system. For example, one can use Li and Duan's model for the H_2O-CO_2-NaCl-$CaCO_3$-$CaSO_4$ system valid from 0 to 250°C, 1 to 1,000 bar with NaCl concentrations up to halite saturation[56] but not for other systems with other species.

Second, after realizing the above incompatible issues among different models, scientists have tried to develop more comprehensive models by fitting all unknown virial coefficients simultaneously for different systems. For example, Dai and his collaborators have developed a thermodynamic model for the Na-K-Mg-Ca-Ba-Sr-Fe-Si-Cl-SO_4-CO_3-HCO_3-CO_2-H_2S-H_2O-oil system with conditions of temperature up to 250°C, pressure up to 1,500 bars, and ionic strength up to halite saturation salinity via incorporating different types of thermodynamic data (*e.g.*, mineral solubility, solution density) for different systems under different conditions.[4, 38, 41, 57–59] This model has been used as the thermodynamic engine for the SSP software, which will be introduced in detail later.

2.2.2 Common Thermodynamic Software

Based on the thermodynamic models mentioned above, many software has been developed and adopted in different disciplines. Table 2.4 briefly summarizes the basic information of several commonly used academic and industrial thermodynamic software followed by a more detailed introduction in this section.

Table 2.4. Basic information of some commonly used thermodynamic software.

Software	If Free	Thermodynamic Models	T, P, I Ranges	PVT Simulations
Visual MINTEQ	Yes	Davies equation Extended D-H equation SIT	*T*: mainly at 25°C, but with some temperature dependence using enthalpy *P*: 1 atm *I*: vary with selected models	No
PHREEQC	Yes	SIT Pitzer theory	*T*: up to 100°C or higher *P*: up to vapor pressure *I*: up to halite saturation	No
SOLMINEQ.88	Yes	Pitzer theory	*T*: up to 300°C *P*: up to 1,000 bars *I*: up to halite saturation	No
ScaleSim (Multi-Scale)	No	Pitzer theory	*T*: up to 250°C or higher *P*: vapor pressure *I*: up to halite saturation	Yes
ScaleSoftPitzer	No	Pitzer Theory	*T*: up to 250°C or higher *P*: up to 1,500 bars *I*: up to halite saturation	Yes

2.2.2.1 Visual MINTEQ

Visual MINTEQ is a software focused on chemical equilibrium calculation mainly for natural waters. It was originally built based on MINTEQA2 software by USEPA and is now maintained by Jon Petter Gustafsson at KTH since 2000. It is a free software, and the users can add more species that are not included. It is convenient and easy to use especially for natural waters at mild temperature and pressure conditions. However, errors can occur at high temperature and pressure especially when mixed electrolytes are present.

2.2.2.2 PHREEQC

PHREEQC (pH-REdox-Equilibrium written in C programming language) is developed and maintained by United States Geological Survey (USGS) with many different versions.[3] It is probably the most widely used aquatic chemistry model used by scientists and engineers from different disciplines. It uses SIT or Pitzer theory as the thermodynamic model. It provides different databases that include equilibrium constants, different group of species and their virial coefficients collected from various literature sources. The valid ranges of temperature, pressure, and ionic strength are much larger compared with Visual MINTEQ. It also gives users the flexibility to add more species that are not included in the database. However, its pressure dependence is limited to vapor pressure since it does not include pressure terms in the virial coefficients. In addition, as mentioned above in the section of "Pitzer theory" and as suggested in their introduction, the selection of the thermodynamic data and virial coefficients from different sources can be incompatible and need careful attention.

2.2.2.3 SOLMINEQ.88

SOLMINEQ.88 is a more general and comprehensive thermodynamic model that includes 260 inorganic species and 220 mineral species with wide temperature (0–350°C) and pressure (1–1,000 bars) ranges.[51] With the inclusion of pressure dependences, SOLMINEQ.88 has much wider applications in subsurface systems (*e.g.*, oil and gas, geothermal) than PHREEQC. However, it is still limited by the compatible issues of these virial coefficients from different literature sources.

2.2.2.4 MultiScale

Many companies have developed their own scale prediction software. ScaleSim, originally branded as MultiScale, is a software developed by Baard Kaasa for Statoil (now Equinor)[44] and now available from Scale Consult AS. Comparing with other geochemical models, this model includes the multiphase flow of oil and gas via P-V-T simulations and thus has wider applications in the oil and gas industry. More discussions about the P-V-T simulations are included in Chapter 3. In ScaleSim, the pressure effects are corrected in the equilibrium constants in a semiempirical way, instead of in the virial coefficients. Its pressure dependence has been verified for the common scale minerals in NaCl-dominated brines, but further check is recommended at high-pressure ranges in systems with high concentrations of multiple electrolytes.

2.2.2.5 ScaleSoftPitzer

ScaleSoftPitzer (SSP) is a software developed for scale prediction in the oil and gas industry by the Brine Chemistry Consortium at Rice University. It was also developed based on the Pitzer theory. As discussed above, instead of relying on virial coefficients from different literature sources without testing their compatibility, SSP simultaneously fitted a group of unknown virial coefficients based on a comprehensive thermodynamic database. The ranges of conditions for the experimental data in the database have been listed in Table 2.2 of Chapter 1. As listed in the last two columns of that table, most of the predicted SI values of these solubility experiments are very close to 0 (the theoretical value) with a small standard error of the SI mean (*i.e.*, <0.016). Such simultaneous fitting based on this comprehensive thermodynamic database assures the compatibility of these virial coefficients and significantly expands the valid ranges of temperature, pressure, I, and types of electrolytes. Considering the presence of oil, water, and gas phases, SSP also incorporated an equation of state (EoS) for multiphase flash calculations. These efforts in the thermodynamic model development equip SSP to be one of the most widely used and so-called standard oil and gas scale prediction software in the industry.[60–62]

2.2.2.6 Other Software

In the market, there are many other thermodynamic software available in addition to those introduced above, for example, French Creek, ScaleChem, Geochemist's Workbench, to mention a few. Most of them are also using the Pitzer theory or SIT. These models are usually designed for different purposes (*e.g.*, oil and gas, geothermal, cooling water system) or built with other software modules (*e.g.*, fate and transport model, multiphase flow simulator). It is recommended that the users pay special attention to the applicable ranges of each software before using it for scale prediction.

2.2.3 Semiempirical Saturation Indices

The above thermodynamic software can give accurate SI predictions for different minerals at various conditions if accurate water composition (and related gas and oil compositions) and operating conditions are provided. In the field, on the other hand, a more convenient index is usually preferred to tell if a mineral of interest is supersaturated or not. In the past decades, many semiempirical indices have been developed and used in different systems. Most of them are developed for calcite, one of the most commonly occurred scale minerals in the oil and gas industry. Following the same idea, Oddo and Tomson (1994) have also developed such semiempirical saturation indices for sulfate minerals (*i.e.*, barium sulfate, calcium sulfate, and strontium sulfate). Details are discussed below.

2.2.3.1 Langelier Saturation Index

Langelier saturation index (LSI) was first developed to reflect the corrosion risk of municipal water, since the formation of calcite can preclude corrosion. Actually, for calcite system, following Equation (2.2), the SI can be calculated as follows:

$$\mathrm{SI}(\text{calcite}) = \log_{10}\frac{\{\mathrm{Ca}^{2+}\}\{\mathrm{CO_3}^{2-}\}}{K_{\mathrm{sp,calcite}}} = \log_{10}\frac{\{\mathrm{Ca}^{2+}\}\{\mathrm{HCO_3}^{-}\}K_{\mathrm{a2}}}{K_{\mathrm{sp,calcite}}\{\mathrm{H}^{+}\}} \tag{2.27}$$

where the second equal sign was derived by replacing the $\{CO_3^{2-}\}$ term based on the second disassociation constant of carbonic acid $K_{a2} = \{CO_3^{2-}\}\{H^+\}/\{HCO_3^-\}$. When at equilibrium (*i.e.*, no calcite precipitation or dissolution), Equation (2.27) reduces to:

$$0 = \log_{10}\left[Ca^{2+}\right] + \log_{10}[Alk] - pK_{a2} + pK_{sp,calcite} + pH + \log_{10}\gamma_{Ca^{2+}} + \log_{10}\gamma_{HCO_3^-} \tag{2.28}$$

where [Alk] is the measured alkalinity of the solution and usually equals to the concentration of $[HCO_3^-]$ for most municipal water samples assuming no organic acids are present.[63] If replacing the pH at SI = 0 to be pH_s, and combing the $pK_{sp,calcite}$ term with the activity coefficients together as $pK_{c,calcite}$, the conditional solubility product of calcite, we can get:

$$pH_s = pK_{a2} - pK_{c,calcite} - \log_{10}\left[Ca^{2+}\right] - \log_{10}[Alk] \tag{2.29}$$

Therefore, the LSI can be calculated as below[64]:

$$LSI = pH - pH_s \quad \begin{cases} > 0, & \text{supersaturated} \\ = 0, & \text{equilibrium} \\ < 0, & \text{undersaturated} \end{cases} \tag{2.30}$$

When using LSI to evaluate calcite scale potential, one needs to find the $pK_{c,calcite}$ value and pK_{a2} values at operating conditions,[65] and then measure the concentrations of calcium ion and alkalinity. However, the accuracy of the LSI is limited, and erroneous predictions can occur especially for field samples.[20] One reason is that theoretically the $pK_{c,calcite}$ term is a compound value of $pK_{sp,calcite}$ and ion activity coefficients, all of which vary with temperature, pressure, and solution compositions. While the $pK_{c,calcite}$ term for LSI is usually empirically achieved, its valid ranges are limited by the experimental data that it was derived based on. Secondly, the $[Ca^{2+}]$ term refers to the free calcium ion concentration, rather than the total calcium ion concentration that is commonly measured. The misuse of the $[Ca^{2+}]$ concentration can lead to significant errors, especially when ion pairs (*e.g.*, $CaHCO_3^+$, $CaSO_4^0$) can play a major role. Third, the [Alk] is used to represent $[HCO_3^-]$, because [Alk] can be easily achieved via titrating the sample with acid to pH 4.3.[63] While the presence of organic acids (*e.g.*, acetic acid) can also contribute the [Alk] and leads to an overestimation of the alkalinity contributed by HCO_3^-. Fourth, pH is selected to represent the saturation status, because it can be conveniently measured in the field. However, the pH of the field in situ samples is difficult to be accurately determined due to the probe contamination by organics and the degassing process when samples are exposed to atmospheric conditions. Also, one needs to correct for the junction potential as TDS increases.

As a result, to achieve reliable prediction results, LSI is limited to be used in water samples with pH between 6.5 and 9.5, $I < 0.02$ M (about 4,000 mg/L) and below 90°C at atmospheric pressure conditions.

2.2.3.2 Larson-Buswell Saturation Index

Instead of using an empirical constant $pK_{c,calcite}$ to present the impacts of I, Larson and Buswell (1942) used a different pH_s value as follows:

$$pH_s = pK_{a2} - pK_{sp,calcite} - \log_{10}\left[Ca^{2+}\right] - \log_{10}\left[Alk\right] + 9.30 + 2.5 \cdot \frac{I^{1/2}}{1 + 5.3 \cdot I^{1/2} + 5.5 \cdot I} \quad (2.31)$$

where the last two added-on terms are used to represent the impacts of ionic strength on ion activity coefficients. As expected, this saturation index has the same limitations as the LSI discussed above.

2.2.3.3. SOLMINEQ.88

SOLMINEQ.88 is a more general and comprehensive thermodynamic model that includes 260 inorganic species and 220 mineral species with wide temperature (0–350°C) and pressure (1 to 1000 bars) ranges.[51] With the inclusion of pressure dependences, SOLMINEQ.88 has much wider applications in subsurface systems (*e.g.*, oil and gas, geothermal) than PHREEQC. However, it is still limited by the compatible issues of these virial coefficients from different literature sources.

2.2.3.4 Stiff-Davis Stability Index

To extend the applicable ranges of the LSI, Stiff and Davis replaced the $pK_{a2} - pK_{c,c}$ term in Equation (2.29) with an empirical constant K_{SSI} valid up to 90°C and I of 3.6 m based on more experimental data as follows[66]:

$$SSI = pH - p\left[Ca^{2+}\right] - p\left[Alk\right] - K_{SSI} \quad \begin{cases} > 0, & \text{supersaturated} \\ = 0, & \text{equilibrium} \\ < 0, & \text{undersaturated} \end{cases} \quad (2.32)$$

where SSI is the Stiff-Davis stability index. The empirical constant (K_{SSI}) is a function of temperature and I, and the values can be found in literature.[65, 66] However, similar to the LSI, SSI can only be applied to relatively simple systems without pressure dependence. The potential errors in the determinations of [Ca^{2+}], carbonate [Alk], and in situ pH still exist.

2.2.3.5 Ryznar Stability Index

Though in different forms, the above saturation indices are derived based on the SI calculation (*i.e.*, Equation [2.2]). However, as discussed above, a water sample with a positive SI (*i.e.*, a possible saturation index discussed above) does not necessarily precipitate within a certain time period. It also does not indicate how much calcite will deposit. Since calcite deposition layer can slow down or prevent further corrosion, Ryznar proposed a new stability index (RSI) as follows to indicate whether a calcite-supersaturated solution can form an anticorrosion layer of calcite[67]:

$$RSI = 2pH_s - pH \quad \begin{cases} < 6.5, & \text{scale dominated} \\ > 6.5, & \text{corrosion dominated, with very limited scale} \end{cases} \tag{2.33}$$

where the pH_s term follows the form of Equation (2.31). It deserves notice that RSI is used when the Larson-Buswell saturation index is positive.

2.2.3.6 Feitler Critical pH

As discussed above, saturation index is a thermodynamic term rather than a kinetic term. When the SI or the above semiempirical saturation or stability indices are positive, calcite scale may not always occur. Empirically, Feitler proposed that the pH should be 1.7 or 2.0 higher than the pH_s to observe scale immediately.[18] According to Equations (2.27) and (2.28), such empirical critical pH is equivalent to the threshold SI values listed in Table 2.1, though accurate pH measurement might not always be easy to achieve in the field.

2.2.3.7 Oddo-Tomson index

Though improvements have been made since the development of LSI, the semiempirical saturation indices introduced above can give vastly different results and erroneous predictions especially for samples with mixed electrolytes at higher temperature and pressure conditions. To overcome the limitations of the above semiempirical indices, Oddo and Tomson proposed a more sophisticated saturation index for calcite over much wider ranges of T, P and I.[68] Instead of relying on pH measurements which could have large uncertainties, the pH prediction and the saturation index calculation are calculated based on CO_2 gas percentage and alkalinity measurements. It also considers the impacts of hydrocarbons on CO_2 fugacity which makes it more valuable and popular in the oil and gas industry (Table 2.5).Following the same idea, the semiempirical saturation indices for sulfate minerals have also been developed as shown in Table 2.6. Instead of using the total ion concentration of cations, the free ion concentration has been calculated by considering the formation of ion pairs. In addition, it also considers the impacts of T, P, and I on solubility products and ion pair association constants.

Oddo-Tomson Index has been one of the most accurate semiempirical saturation indices being used in the oil and gas industry. However, with the Pitzer theory-based software well developed, it is recommended to use those thermodynamic models rather than the semiempirical saturation indices.

Table 2.5. The Oddo-Tomson index for calcite scale

Without gas phase

$$SI(calcite) = \log_{10} \frac{[Ca^{2+}]\cdot[HCO_3^-]^2}{C_{CO_2}} + 3.63 + 8.68\cdot10^{-3}\cdot T + 8.55\cdot10^{-6}\cdot T^2 - 6.56\cdot10^{-5}\cdot P - 3.42\cdot I^{1/2} + 1.373\cdot I$$

$$pH = \log_{10} \frac{[HCO_3^-]}{C_{CO_2}} + 6.39 - 1.198\cdot10^{-3}\cdot T + 7.94\cdot10^{-6}\cdot T^2 - 3.53\cdot10^{-5}\cdot P - 1.067\cdot I^{1/2} + 0.599\cdot I$$

$$C_{CO_2} = 7289.3 \cdot \frac{n_{CO_2}}{BWPD + BOPD}$$

With gas phase

$$SI(calcite) = \log_{10} \frac{[Ca^{2+}]\cdot[HCO_3^-]^2}{P\cdot y_{CO_2}\cdot\varphi_{CO_2}} + 5.85 + 15.19\cdot10^{-3}\cdot T - 1.64\cdot10^{-6}\cdot T^2 - 5.27\cdot10^{-5}\cdot P - 3.334\cdot I^{1/2} + 1.431\cdot I$$

$$pH = \log_{10} \frac{[HCO_3^-]}{P\cdot y_{CO_2}\cdot\varphi_{CO_2}} + 8.60 + 5.31\cdot10^{-3}\cdot T - 2.253\cdot10^{-6}\cdot T^2 - 2.237\cdot10^{-5}\cdot P - 0.99\cdot I^{1/2} + 0.658\cdot I$$

$$\varphi_{CO_2} = \exp\left[\left(2.84\cdot10^{-4} - \frac{0.255}{T + 460}\right)\cdot P\right]$$

Using measured pH

$$SI(calcite) = pH + \log_{10}[[Ca^{2+}]\cdot[HCO_3^-]] - 2.76 + 9.88\cdot10^{-3}\cdot T + 0.61\cdot10^{-6}\cdot T^2 - 3.03\cdot10^{-5}\cdot P - 2.348\cdot I^{1/2} + 0.77\cdot I$$

[ ] = measured ion concentration (M); P = total absolute pressure (psia); T = temperature (°F); I = ionic strength (M); φ_{CO_2} = fugacity coefficient of CO_2 gas; y_{CO_2} = molar volume fraction of CO_2 in produced gas at specific T and P; $y^T_{CO_2}$ = molar volume fraction of CO_2 in the gas at surface, or the volume fraction of CO_2 considering gas, oil and brine; BWPD = barrels of water produced per day; BOPD = barrels of oil produced per day; MMSCFD = million standard cubic feet per day.

Table 2.6. The Oddo-Tomson indices for sulfate scales

$$SI(barite) = \log_{10}([Ba^{2+}]_{free}\cdot[SO_4^{2-}]_{free}) + 10.03 - 4.8\cdot10^{-3}\cdot T + 11.4\cdot10^{-6}\cdot T^2 - 4.8\cdot10^{-5}\cdot P - 2.62\cdot I^{1/2} + 0.89\cdot I - 2.0\cdot10^{-3}\cdot T\cdot I^{1/2}$$

$$SI(celestite) = \log_{10}([Sr^{2+}]_{free}\cdot[SO_4^{2-}]_{free}) + 6.11 + 2.0\cdot10^{-3}\cdot T + 6.4\cdot10^{-6}\cdot T^2 - 4.6\cdot10^{-5}\cdot P - 1.89\cdot I^{1/2} + 0.67\cdot I - 1.9\cdot10^{-3}\cdot T\cdot I^{1/2}$$

$$SI(gypsum) = \log_{10}([Ca^{2+}]_{free}\cdot[SO_4^{2-}]_{free}) + 3.47 + 1.8\cdot10^{-3}\cdot T + 2.5\cdot10^{-6}\cdot T^2 - 5.9\cdot10^{-5}\cdot P - 1.13\cdot I^{1/2} + 0.37\cdot I - 2.0\cdot10^{-3}\cdot T\cdot I^{1/2}$$

$$SI(hemihydrate) = \log_{10}([Ca^{2+}]_{free}\cdot[SO_4^{2-}]_{free}) + 4.04 - 1.9\cdot10^{-3}\cdot T + 11.9\cdot10^{-6}\cdot T^2 - 6.9\cdot10^{-5}\cdot P - 1.66\cdot I^{1/2} + 0.49\cdot I - 0.66\cdot10^{-3}\cdot T\cdot I^{1/2}$$

$$SI(anhydrite) = \log_{10}([Ca^{2+}]_{free}\cdot[SO_4^{2-}]_{free}) + 2.52 + 9.98\cdot10^{-3}\cdot T - 0.97\cdot10^{-6}\cdot T^2 - 3.07\cdot10^{-5}\cdot P - 1.09\cdot I^{1/2} + 0.50\cdot I - 3.3\cdot10^{-3}\cdot T\cdot I^{1/2}$$

Free sulfate ions: $$[SO_4^{2-}]_{free} = \frac{(b^2 + 4\,K_{sc}\cdot[SO_4^{2-}])^{1/2} - b}{2K_{st}}$$ $$b = 1 + K_{sc}\cdot(\Sigma[Me^{2+}] - [SO_4^{2-}])$$

Free divalent cations: $$[Me^{2+}]_{free} = \frac{[Me^{2+}]}{1 + K_{st}\cdot[SO_4^{2-}]_{free}}$$ $$\Sigma[Me^{2+}] = [Ca^{2+}] + [Ba^{2+}] + [Sr^{2+}] + [Mg^{2+}]$$

Ion pair association constant: $$\log_{10}K_{sc} = 1.86 + 4.5\cdot10^{-3}\cdot T - 1.2\cdot10^{-6}\cdot T^2 + 10.7\cdot10^{-5}\cdot P - 2.38\cdot I^{1/2} + 0.58\cdot I - 1.3\cdot10^{-3}\cdot T\cdot I^{1/2}$$

$[\,]_{free}$ = free ion concentration (M); [ ] = measured ion concentration (M); P = total absolute pressure (psia); T = temperature (°F); I = ionic strength (M); Me^{2+} = divalent cation (Ba^{2+}, Sr^{2+}, Ca^{2+}); K_{sc} = stability constants of Me^{2+} ion and sulfate ion forming ion complexes.

2.2.4 Challenges in Scale Predictions

After decades of efforts, the current thermodynamic models have been significantly improved in terms of applicable types of aqueous species and valid ranges of temperature, pressure, and ionic strength. Though the Pitzer theory is developed based on the solid thermodynamic theory, the dependencies of temperature, pressure, and ionic strength on the virial coefficients for different species-specific interactions, that is, Equation (2.26), do not have a sound theoretical basis. As a result, extrapolating the models to the conditions out of the experimental data range can lead to significant errors. In another way, the experimental work of measuring different types of thermodynamic data at different conditions dictates the accuracy and applicable conditions of the thermodynamic model. For example, the ranges of T, P, I, and ion concentrations tabulated in Table 2.2 of Chapter 1 are the ranges of the experimental data that SSP was developed based on and also the applicable ranges of SSP. Focusing on the mineral scales in the oil and gas industry, more experimental and model development efforts are still required:

1. Sulfide and silicate mineral solubilities at all interested conditions,
2. Impacts of ionic strength and calcium ion concentrations (>0.5 m) at high temperature (>150°C) and high pressure (>500 bars) conditions,
3. Impacts of various ions on mineral composition and mineral solubility (*e.g.*, the formation of solid solutions of $Ca_x(Mg, Fe)_{1-x}CO_3$, $Ba_xSr_{1-x}SO_4$),
4. The presence of different mineral phases (*e.g.*, calcite, aragonite, vaterite and amorphous calcium carbonate for calcium carbonate mineral),
5. The impacts of organic components (*e.g.*, fracturing chemicals, scale inhibitors, hydrocarbons) on mineral solubilities,
6. The internal consistency of all virial coefficients when trying to include all aqueous species in a comprehensive thermodynamic model.

In addition to the development of the thermodynamic model, reliable sample collection and analysis are also key to accurate scale risk predictions. Though no standard procedures have been developed, some general recommendations can help improve the accuracy of the analysis:

1. Scale formation should be avoided during sample preparation and storage. Acid and scale inhibitors can be added to avoid scale formation.
2. pH measurements should be performed at in situ conditions if possible. However, one should notice that pH measurement in the field may not be accurate. It is recommended to calculate pH using the measured alkalinity and CO_2 gas composition rather than measure it.
3. The impacts of organic acids and potential scale formation need to be paid attention to. A recommended protocol has been developed to measure the amount of alkalinity and organic acids simultaneously.[69]

2.3 Kinetics of Scale Formation

Mineral scaling is a ubiquitous process that happens in various industrial, geologic, and natural processes, though it may be called by different names, for example, scale in the oil and gas industry, mineral fouling in the membrane system of water treatment, scale in heat transfer systems, lithiasis in pancreas, to mention a few. Though the conditions in different scenarios vary a lot, the mineral scaling processes are similar. Using the thermodynamic models introduced above, the scaling risks can be predicted. However, to accurately predict when, where and how much scale will form, it is critical to decipher the scaling mechanisms and develop quantitative models to accurately predict mineral scaling kinetics.

Generally, scale formation starts with the generation of supersaturated solution. A supersaturated solution can be generated from incompatible water source mixing, condition change, or solute concentration. For example, the high-sulfate water (*e.g.*, seawater) is usually injected into oil wells and mixed with reservoir water with high-barium content; during oil and gas production, when brine flows from the reservoir to the wellhead, a significant drop of temperature and pressure can lead to the generation of supersaturation; in reverse osmosis membrane systems, the dirty side gets more concentrated and supersaturated when pure water passes through the membrane.

Figure 2.2 plots the general mineral scaling processes (*i.e.*, mineral crystallization and surface deposition) after the creation of supersaturation, typical theoretical models developed for certain scaling stages, and a typical solution turbidity curve changing with time. This section will introduce the basic scaling processes, key definitions, and typical theoretical models for mineral scaling, without diving deep into the mathematical derivations.

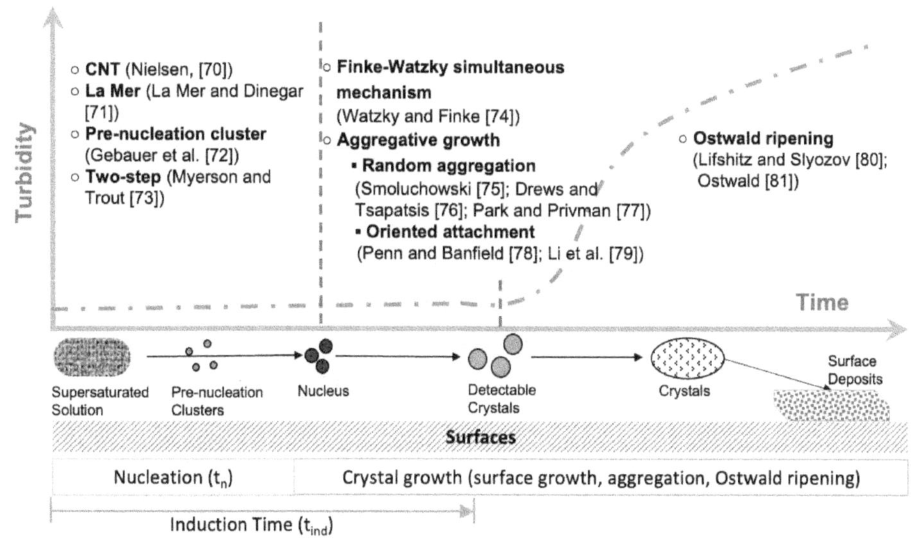

Figure 2.2. The processes of mineral scaling (crystallization and surface deposition), typical theoretical models developed for different stages,[70-81] and the change of solution turbidity as crystallization proceeds.

Theoretically, there should not be a distinct line between the nucleation and the crystal growth stages as is plotted in Figure 2.2, and they can coexist in the crystallization processes. Watzky and Finke have successfully incorporated nucleation and surface growth.[69] However, it is also believed that the nucleation rate drops significantly after the first period of crystallization, since it is strongly dependent on the mineral saturation index (or matrix ion concentration) which will drop once nucleation starts.[71, 82] As a result, scientists tend to divide the crystallization process into one period that nucleation dominates (*i.e.*, nucleation period) and another period that crystal growth dominates (*i.e.*, crystal growth period) as shown in Figure 2.2.

In the crystal growth stage, matrix ions keep coming to the surface of the existing nucleus or small crystals and form new crystal units (*i.e.*, surface growth in Figure 2.3). In addition to such surface growth, crystal particles can interact with each other. When multiple crystal particles coexist, smaller ones tend to dissolve while the larger ones become larger so that the whole system can achieve a smaller surface energy.[80, 81] This phenomenon is named as "Ostwald ripening" following the first proposer (*i.e.*, Ostwald ripening in Figure 2.3). At the same time, particles tend to randomly move around and aggregate together to become larger particles (*i.e.*, aggregation in Figures 2.2 and 2.3).[75, 83]

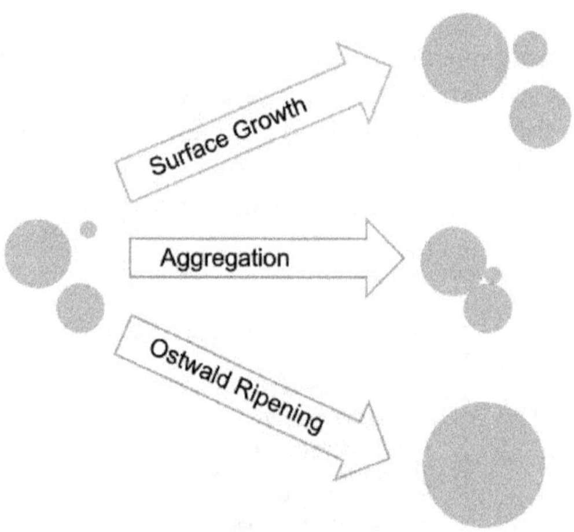

Figure 2.3. The processes of surface growth, aggregation, and Ostwald ripening.

In addition to the crystallization process in the bulk solution, crystal formation on the reactor surface is critical especially for various industrial applications.[84–87] For example, much less damage may occur if scale minerals flow out the subsurface system without depositing on the surfaces of downhole facilities. However, the mineral surface deposition process and mechanisms are still under debate.[84–88]

2.3.1 Nucleation

After the generation of a supersaturated solution, the solution will undergo nucleation process (Figure 2.2). To put it in a simple way, nucleation is the process that matrix ions get together and form a nucleus from the supersaturated solution. Mullin defines "primary nucleation" as the nucleation process that is not induced by crystals, while "secondary nucleation" is the one induced by crystals.[74] In the primary nucleation category, "homogeneous nucleation" and "heterogeneous nucleation" are distinguished depending on if nucleation happens in the solution or on foreign surfaces, respectively. To quantitatively describe the nucleation process, the classical nucleation theory has been proposed via evaluating the change of bulk energy and surface energy.[70, 83] In supersaturated solutions, the formation of nuclei in the nucleation process leads to a decrease in bulk Gibbs free energy (ΔG_b) and an increase in surface energy (ΔG_σ) due to the formation of interfaces between nucleus and bulk solutions. If a nucleus with a radius of r forms, the total nucleation Gibbs free energy (ΔG) change can be calculated as follows if n moles of matrix ions form the crystal:

$$\Delta G = \Delta G_b + \Delta G_\sigma = n\Delta G_r + \Delta G_\sigma = \frac{4\pi r^3}{3V_m}\Delta G_r + 4\pi r^2\sigma \tag{2.34}$$

where σ is the effective interfacial energy between the nucleus and the solution; V_m is the molar volume of the mineral; n is the number of moles of matrix ions that are needed to form a nucleus with radius r (*i.e.*, $n = 4\pi r^3/3V_m$); ΔG_r is the free bulk Gibbs free energy change per mole of the mineral nucleates. As shown in Figure 2.4, when a nucleus grows and the radius becomes larger, the total nucleation Gibbs free energy increases first and then decreases, since the bulk Gibbs free energy has a stronger radius dependence. The classical nucleation theory (CNT) assumes that when the particle size is larger than the critical radius (r_c), the decrease of bulk Gibbs free energy can cover the increase of surface interfacial energy, and the nucleus can spontaneously grow larger[70, 83]:

$$\left(\frac{d\Delta G}{dr}\right)\Big|_{r=r_c} = 0 \tag{2.35}$$

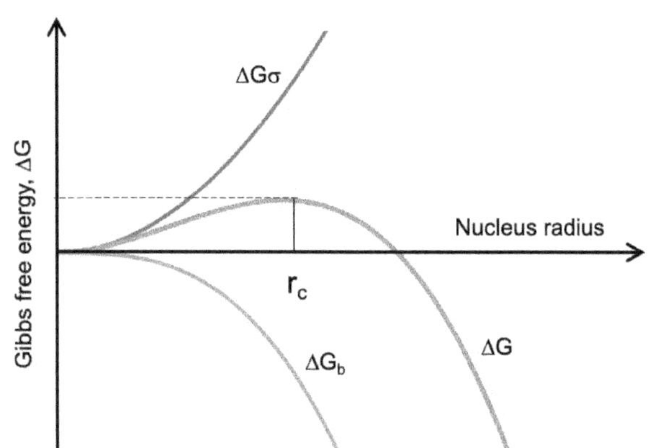

Figure 2.4. The change of total nucleation Gibbs free energy (ΔG), bulk Gibbs free energy (ΔG_b), and interfacial energy (ΔG_σ) with nucleus radius.

Then the critical radius (r_c) and critical nucleation Gibbs free energy (ΔG_c) can be calculated as functions of saturation index (SI), defined in Equation (2.2), molar volume, and interfacial energy:

$$r_c = \frac{2V_m\sigma}{RT \cdot \ln(10)SI} \tag{2.36}$$

$$\Delta G_c = \frac{4\pi\sigma r_c^2}{3} = \frac{16\pi\sigma^3 V_m^2}{3\left(RT \cdot \ln(10)SI\right)^2} \tag{2.37}$$

The nucleation rate (J), defined as the number of concentration of critical nuclei (#/m^3) formed during the nucleation stage (*i.e.*, n_0/t_n), can be calculated as:

$$J = \frac{n_0}{t_n} = A_n e^{-\frac{\Delta G_c}{RT}} = A_n e^{-\frac{16\pi\sigma^3 (V_m)^2}{3(RT)^3(\ln(10)\cdot I)^2}} \tag{2.38}$$

where A_n is the pre-exponential rate constant ($s^{-1}\cdot m^{-3}$) for nucleation. CNT successfully establishes the relationship between nucleation kinetics (J) and microscopic information (n_0, t_n) with the physical properties of the built minerals (σ, V_m). It is developed based on the solid mathematical derivations and has been widely adopted in crystallization kinetics predictions in various disciplines.[8, 89, 90]

Before the formation of critical nucleus (larger than which nucleus will grow larger spontaneously), as suggested in the classical nucleation theory (CNT), the stable prenucleation clusters have also been observed.[72, 91] Colfen and his group members observed that the measured free Ca^{2+} ion concentration is always lower than the dosed amount, and the "missing" part is believed to be present as stable prenucleation clusters in both under- and supersaturated solutions.[72] The cryo-TEM images taken by Pouget *et al.* further confirmed the formation of prenucleation clusters at the beginning of template-directed calcium carbonate formation.[91] Therefore, it is proposed that the stable prenucleation clusters are observed not only in supersaturated solutions before the formation of critical nucleus but also in undersaturated solutions.[72, 92, 93] The stable prenucleation cluster is a new starting point of crystal growth and can work as the precursor of the amorphous phase that usually occurs.

It is implied in CNT that the particle size dictates whether the clusters are stable critical nuclei, ignoring the differences in local densities or lattice structures.[94] It also ignores the much longer time it may take for possible molecule reorganization in the clusters or nucleus especially for complex molecules (*e.g.*, proteins).[94] Therefore, Myerson and his group proposed that the solute molecules form a sufficient size cluster following with reorganization into ordered structures rather than directly form a structured nucleus as suggested in CNT.[73, 94, 95]

Though these nonclassical nucleation mechanisms have provided more possible nucleation pathways, they failed to provide quantitative simulation models and cannot be incorporated with other mechanistic models to comprehensively describe the whole crystallization process.

2.3.2 Surface Growth

Surface growth of a crystal is the process of the growth unit (ions or molecules) becoming part of the nucleus. The growth units will transport from the bulk solutions to the nucleus surface following diffusion or convection. Then they move to the correct lattice position (*e.g.*, kink site) following adsorption, surface migration or diffusion, dehydration, kink site integration, and so on.[96] The orders of these reactions may vary from first order (*e.g.*, diffusion and convection) to second (*e.g.*, surface spiral step growth) or even higher orders. The relative speed of these reactions involved in the two processes may vary with the solution compositions, mineral properties, temperature, pressure, and so on. When diffusion or convection in the transport process is fast, the surface concentration of solute will be identical to the bulk concentration, and the rate limiting step is the surface reaction and vice versa. Therefore, it is hard to determine the superficial order of crystal growth and the prominent reactions in surface growth. In previous studies, the precipitation reaction order has been determined with different values, for example, 1.6th-order reaction was assumed for potash alum, and second-order reaction for potassium sulfate and barium sulfate.[83, 97, 98] While the dissolution reaction was usually treated as first-order reaction[83, 97, 98] or second-order reaction.[99–101] As with previous studies,[83, 102] for the primary nucleus, the surface growth can be described by the following equation, if first-order kinetics is assumed:

$$dV_1 = 4\pi r_1^2 dr_1 = \frac{k_r \cdot 4\pi r_1^2 MW \cdot dt}{\rho_p}\left(C_0 - \frac{4\pi \rho_p n_0 r_1^3}{3MW} - C_\infty\right) \tag{2.39}$$

where V_1 is the volume of the primary nucleus (formed with critical radius size at the nucleation time) changing with time (m^3); k_r is the first-order surface reaction rate ($m^4 \cdot mol^{-1} \cdot s^{-1}$); r_1 is the radius of primary nucleus increasing with time (*m*); MW is the molecular weight (*e.g.*, 0.233 $kg \cdot mol^{-1}$ for barite); ρ_p is the density of the particle, which is assumed to be the bulk density (*e.g.*, 4.48×10^3 $kg \cdot m^{-3}$ for barite); C_0 is the initial concentration ($mol \cdot m^{-3}$); C_∞ is the equilibrium concentration ($mol \cdot m^{-3}$). As a result, we can derive:

$$\frac{dr_1}{dt} = k_r\left(C_0 - \frac{4\pi \rho_p n_0 r_1^3}{3\text{MW}} - C_\infty\right)\frac{\text{MW}}{\rho_p} \tag{2.40}$$

As a boundary condition, at the end of the nucleation period (*i.e.*, $t = t_n$), the radius of primary nucleus equals to the critical radius r_c (Equation (2.36) according to CNT (*i.e.*, $r_1 |_{t=t_n} = r_c$).

2.3.3 Aggregation

When multiple particles are present in the solution, particles with the same or different sizes will aggregate with each other. The aggregation kinetics can be described by the second-order Smoluchowski theory with the following equation[75]:

$$\frac{dn_k}{dt} = \frac{1}{2}\sum_{i+j=k} \beta_{ij} n_i n_j - n_k \sum_{i=1}^{\infty} \beta_{ik} n_i \tag{2.41}$$

where subscripts i, j, and k represent the number of primary nuclei in the large particles; $n_{[]}$ represents the corresponding number concentration of the aggregating particles; β_{ij} is the second-order aggregation rate between particles with i and j primary nuclei inside. The first term is the formation rate of large particles from smaller particles, and the second term stands for the depletion rate of the particles through forming even larger particles. Since most of the particle sizes were nanoscale, Brownian motion was the dominant aggregation pathway. The size-dependent Brownian motion aggregation rate was calculated by

$$\beta_{ij} = \frac{2k_BT}{3\mu}\frac{\left(r_i + r_j\right)}{r_ir_j} \tag{2.42}$$

where μ is water dynamic viscosity (0.000891 kg·m^{-1}·s^{-1} at 25°C); r_i is the radius of particle i. It is commonly assumed that the aggregating particles are of equal size[76, 77, 102, 103] with the aggregation rate of $8k_BT/3\mu$. Based on this assumption, the total number concentrations of all the particles (n_p) is:

$$n_p = \frac{n_0}{1 + t/\tau} \tag{2.43}$$

and the number concentration of the particles aggregated from k primary particles (n_k) is:

$$n_k = \frac{n_0\left(t/\tau\right)^{k-1}}{\left(1 + t/\tau\right)^{k+1}} \tag{2.44}$$

where $\tau = 1/k_an_0$ is the characteristic time of aggregation, and k_a($k_a = 4k_BT/3\mu$) is the characteristic aggregation rate. After aggregation, the primary particles can be stabilized due to particle-particle short-range interactions, especially when the particle size is small.[104] Assuming that the primary particles have the same surface growth rate before and after aggregation:

$$V_k(t) = kV_1(t) \tag{2.45}$$

where V_k is the volume of the large particles aggregated from k primary nucleus; V_1 is the volume of the primary nucleus changing with time as shown in Equation (2.39). During aggregation, the structures of the aggregated particles can be very complex.[105] If assuming that the aggregated particle has the same scattering efficiency as a spherical particle with the same volume, their effective radius (*i.e.*, r_k) for turbidity calculation can be derived as

$$r_k = k^{1/3}r_1 \tag{2.46}$$

Oriented attachment has been observed to play an important role in defect and crystal morphology formations.[79, 106–108] Penn has derived a kinetic model for oriented attachment by assuming a rapid achievement of equilibrium between the association and dissociation of particles and the irreversible oriented attachment.[107] He derived a second-order reaction in which the reaction rate is related to the stability constants of the unstable intermediate particle complex.

2.3.4 Ostwald Ripening

Ostwald ripening describes the phenomenon that smaller particles sacrifice themselves to make the larger particles larger.[81] This phenomenon commonly occurs in an aqueous solution where particle size distribution is relatively wide.[109–111] The solubility of particles (C(r)) decreases with particle size following Ostwald-Freundlich (or Gibbs-Thomson) equation:

$$C(r) = C_{\infty} e^{\frac{2V_m \sigma}{\nu RTr}} \tag{2.47}$$

where ν is the number of ions in the mineral molecule; C_{∞} is the bulk solubility when radius is very large. Basically, particle solubility is greater when the size is smaller. As a result, the large particles tend to grow larger with the sacrifice of smaller particles, and the average particle size increases with time as the ripening goes on. The proposed Lifshitz-Slyozov-Wagner model derived the rate law for the average particle size change[80, 112]:

$$\bar{r}_t^3 - \bar{r}_0^3 = kt \tag{2.48}$$

where $\bar{r}_t$ and $\bar{r}_0$ are the average particle size at time t and 0, respectively; k is the reaction rate.

By fitting the particle size distribution at the early stages of crystallization collected by in situ TEM, Woehl *et al.* found that the crystal growth is dominated by random aggregation following Smoluchowski theory, rather than Ostwald ripening using Lifshitz-Slyozov-Wagner model.[103] Wang *et al.* suggested that CNT, aggregation, and Ostwald ripening occur in sequence, possibly with temporary overlaps between two consecutive regimes.[90] It implies that Ostwald ripening is more possible at much later stages of crystallization which matches Mullin's argument that Ostwald ripening is more likely to happen when particle size distribution is wide, and the supersaturation ratio is small.[83]

2.3.5 Induction Period

In experimental testing or field applications, the property changes of the system due to mineral crystallization are not always observed instantly after the generation of supersaturation. Instead, there is usually a certain time point before which no signal change is detected, while after which the signal change becomes visible. Thus, to be convenient and practical, the induction time (*i.e.*, length of the induction period, t_i) is defined as the time elapsed between the creation of supersaturation and the detection of a new phase.[89] Though it is a practical definition, it is experimentally convenient and is of great industrial engineering value, since it usually corresponds to the time when scale starts to cause noticeable problems. As a result, the induction time has been very widely used to represent the kinetics of mineral crystallization.[8, 70, 95, 113–115]

Generally, the induction period can be divided into three stages: (1) the relaxation stage (t_r) for the achievement of quasi steady-state distribution of molecular clusters, (2) the nucleation stage (t_n) for nuclei formation, and (3) the early crystal growth stage when the nuclei grow from a critical radius to a detectable size (t_g).[8, 83, 89, 96, 116] In low-viscosity solutions, the relaxation stage is usually ignored since it's believed to be much shorter than the nucleation stage,[89] thus the induction time can be approximated to be the sum of t_n and t_g (*i.e.*, $t_i = t_n + t_g$).

The definition of the induction period is experimentally oriented. From this practical definition perspective, Dai *et al.* developed a two-stage crystallization model to simulate the change of nucleus size, particle number concentration, solution turbidity, and induction time when monitoring the reaction via solution turbidity.[117] By definition, solution turbidity can be calculated as

$$\text{Turbidity} = -\frac{1}{l} \cdot \ln\left(\frac{I_t}{I_0}\right) = \sum_i n_i \pi r_i^2 Q_{s,i} \tag{2.49}$$

where l is the light path (m); I_t and I_0 are the transmitted and incident light energy, respectively; $Q_{s,i}$ is the scattering efficiency factor dependent on particle size and wavelength (λ_m). Dai *et al.* developed an eighth-order polynomial function to fit the scattering efficiency factor of the Mie light theory simulated by Monte Carlo methods within this size range ($0 < 2\pi r_i/\lambda_m < 7.85$, $0 < r_i < 500$ nm.)[118, 119]:

$$Q_{s,i} = \sum_{j=1}^{9} p_j \left(2\pi r_i / \lambda_m\right)^{j-1} \tag{2.50}$$

where the parameter p_j is the fitted parameters for the polynomial curve which can be found in Table 2.1 of the original article.[117] It deserves notice that to avoid multiple scattering, only the turbidity data less than 10 were used in the regression.[120] The Dai's two-stage crystallization model assumes that in the first stage (*i.e.*, nucleation stage), nuclei with radius r_c and number concentration n_0 will generate at a constant nucleation rate following the CNT. In the second stage, these nuclei undergo surface growth and aggregation after a period of nucleation (t_n). Via substituting Equations (2.50), (2.44), and (2.46) into (2.49) step by step, the turbidity can be calculated as[117]

$$\begin{aligned}\text{Turbidity} &= \sum_i n_i \pi r_i^2 \left(\sum_{j=1}^{9} p_j (2\pi r_i / \lambda_m)^{j-1}\right) \\ &= \sum_i^{\infty} \frac{n_0 (t/\tau)^{i-1}}{(1+t/\tau)^{i+1}} \left(\sum_{j=1}^{9} \frac{p_j}{\lambda_m^{j-1}} 2^{j-1} \pi^j r_1^{j+1} i^{(j+1)/3}\right) \\ &= \sum_{j=1}^{9} \frac{p_j}{\lambda_m^{j-1}} 2^{j-1} \pi^j r_1^{j+1} \frac{n_0}{(1+t/\tau)(t/\tau)} Li_{-(j+1)/3}\left(\frac{t/\tau}{(1+t/\tau)}\right)\end{aligned} \tag{2.51}$$

where, the polylogarithm function $Li_s(h)$ of order s and independent variable h is defined as $Li_s(h) = \sum_{i=1}^{\infty} h^i / i^s$ changing with time can be simulated by Equation (2.40) numerically via Runge-Kutta method. Equation (2.51) bridges the measured solution turbidity with the changes of particle size and number concentration with time. Via fitting the model with the turbidity changes of barite supersaturated solutions of different SI values, Dai's two-stage crystallization model can accurately predict the number and concentration of nuclei, particle size, surface tension, surface growth rate and induction time compared with experimental observations or literature data. For example, in Figure 2.5(a) is plotted the measured and simulated solution turbidity changing with time for barite supersaturated solution with SI of 1.57 using a green laser with wavelength of 532 nm. The induction period (blue arrow) is picked when the solution turbidity increases to a value of 0.05 m^{-1} due to barite scale, which matches well with experimental observations. Using a light path of 0.01 m in this study, the induction time is picked at the time when $I_t = 0.99 \times I_0$ (*i.e.*, 1% transmitted light intensity change). Moreover, the logarithm of induction time is observed to be linearly proportional to $1/SI^2$ ($\log_{10} t_{i,pred} = 7.21/SI^2 + 0.11$ with $R^2 = 0.997$ for barite crystallization at 25°C), even though no such linear relationship is explicitly or implicitly included in the derivation. Interestingly, such a linear relationship has also been derived mathematically from another perspective by Nielsen which will be discussed below.[82]

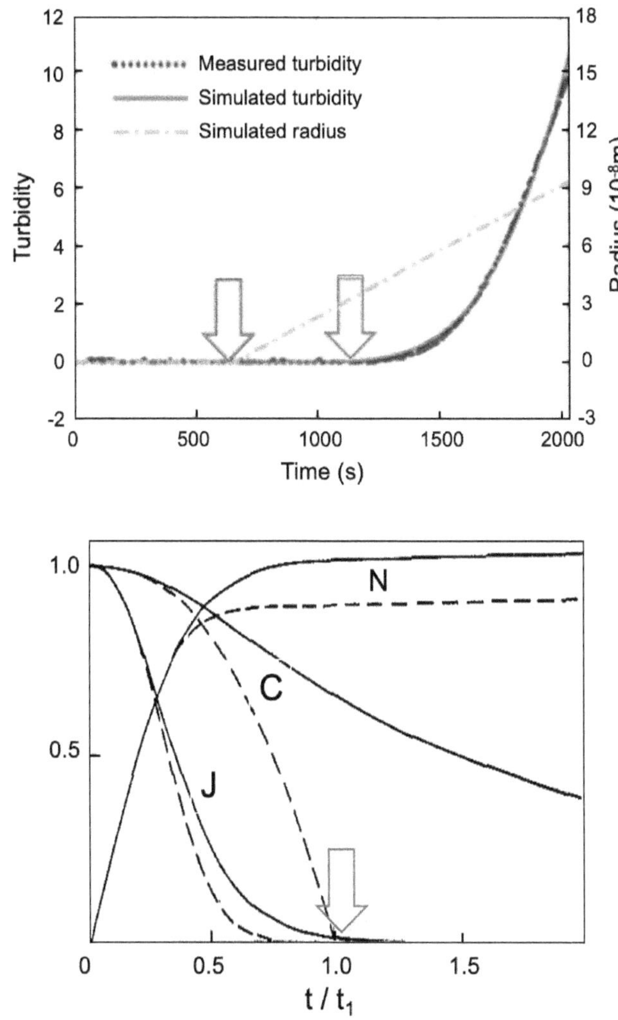

Figure 2.5. (a). The measured and simulated solution turbidity and the simulated primary particle radius change with time for barite crystallization process (SI = 1.57). (b). The actual (solid lines) and approximated (dashed lines) variations of matrix ion concentration (C), nucleation rate (J), and particle number concentration (N) changing with time during nucleation process. The blue arrows show the induction time picked in each paper.

Dai's two-stage crystallization model simulates the temporal change of nuclei size, number concentration, and solution turbidity. Via integrating the Mie light theory, it can also accurately predict the induction time by setting a threshold signal change of the transmitted light intensity (*i.e.*, 1% change in this case), strictly following the practical definition of induction time. More importantly, when the detection limit of the apparatus changes (*e.g.*, due to wavelength change or detector sensitivity change), this model can also give accurate induction time predictions. However, in real applications, it has been limited to predicting the induction time measured by solution turbidity. In addition, this model predicts the induction time in an implicit rather than explicit format that can be easily adopted.

Alternatively, Nielsen and others have made efforts to quantitively predict the induction time from another perspective based on theoretical derivations (t_1 in Figure 2.5[b] indicated by a blue arrow).[82, 89, 121] Nielsen assumed that nucleation happens continuously with a decreasing nucleation rate when more nuclei form and grow larger only via diffusion growth (*i.e.*, $dr/dt = D(V_m/N_{av})\left(C(t)-C_{eq}\right)/r \cong D(V_m/N_{av})C(t)/r$) for sparingly soluble minerals without considering aggregative growth. While simulating the temporal changes of the matrix ion concentration (*C*), nucleation rate (*J*), and particle number concentration (*N*), Nielsen used an iterative integration method and achieved an approximated result of the matrix ion concentration: $C(t)=C(0)\left[1-(t/t_1)^{5/2}\right]$ (*i.e.*, dashed line of C in Figure 2.5[b]). Based on this *C*(*t*) approximation, J and N can be simulated as shown as dashed lines in Figure 2.5[b]). Though this approximation of *C* differs a lot from the exact simulation (*i.e.*, solid line of *C* in Figure 2.5[b]), the simulations of *J* and *N* match well with the exact numerical simulation results (*i.e.*, solid lines of *J* and *N* in Figure 2.5[b]) with less than 5% error. Thus, the approximated *J* and *N* have been used to approximate their exact values which can only be simulated numerically. When the approximated nucleation rate (*i.e.*, *J*) and ion concentration (*i.e.*, *C(t)*) become zero (*i.e.*, $t = t_1$), the nuclei number concentration reaches a plateau and the crystallization process turns from nucleation dominated to crystal growth dominated (Figure 2.5[b]). Such transition time (t_1) has been treated as the induction time.[82, 89, 121] From Nielsen's theoretical induction time model, this theoretical derivation of induction time (t_1) can be calculated as follows[82]:

$$t_i = t_1 = \left(\frac{225}{512\pi^2}\right)^{1/5}\left(D^3C_0vA_n^2\right)^{-1/5}\exp\left(\frac{2\Delta G_c N_{av}}{5RT}\right)$$
$$= \left(\frac{225}{512\pi^2}\right)^{1/5}\left(D^5C_0v^{-7/3}\right)^{-1/5}\exp\left(\frac{2\beta\left(V_m^{2/3}\sigma\right)^3 N_{av}}{5(RT)^3\left(\ln(10)SI\right)^2}\right), \quad (2.52)$$

where $v = V_m/N_{av}$ is the volume of each molecule; $\beta = 16\pi/3$ for sphere particles. It also assumes that pre-exponential term A_n equals $DV^{-5/3}$ when D is the effective diffusion coefficient ($m^2 \cdot s^{-1}$) which is a compound value that includes the impacts of ion diffusion in solution, ion convection, ion diffusion in the adsorption layer of the crystal surface, and other processes on the crystal surface.[82] Based on Nielsen's theoretical induction time model, Söhnel and Mullin further considered other surface growth mechanisms (*e.g.*, polynuclear growth, screw-dislocation growth) and discarded the assumption of $C(t) - C_{eq} \cong C(t)$ adopted for sparingly soluble minerals.[89] As an example, for polynuclear growth the induction time can be calculated with:

$$t_i = t_1 = \left(\frac{3}{2\pi}\right)^{1/4}\left(\frac{V_m^{5/3}}{N_{av}^{8/3}D^4C_{eq}}\right)^{1/4}\left(\frac{C_0/C_{eq}}{\left(C_0/C_{eq}-1\right)^2}\right)^{1/4}\exp\left[\frac{\beta\left(V_m^{2/3}\sigma\right)^3 N_{av}}{4\left(RT\right)^3\left(\ln\left(10\right)SI\right)^2}+\frac{\beta'\left(V_m^{2/3}\sigma\right)^2 N_{av}^{2/3}}{4\left(RT\right)^2\left(\ln\left(10\right)SI\right)}\right], \quad (2.53)$$

where, β and β′ are the shape factors (*i.e.*, for spheres, β = 16π/3, β′ = π).

Nielsen's theoretical induction time model provides a different definition for induction time. However, such a theoretical definition of induction time is more of a mathematical approximation rather than a definition with strict physical meaning. The time when the nucleation rate approaches zero (t_1) does not necessarily represent the time when the nuclei or crystals can be detected (t_i) by a certain apparatus. The difference between these two time points can vary with experimental conditions, monitoring apparatus, and other factors. More importantly, in Nielsen's theoretical induction time model, the detection limit of the monitoring apparatus has no impact on the induction time, which is against experimental observations. For example, Yan *et al.* found that the induction time change monitored using a red laser (635 nm wavelength) is longer (*i.e.*, less sensitive) than that using a green laser (532 nm wavelength), as expected.[122] To a certain extent, it is reasonable to believe that t_1 and t_i have certain relationships since Dai's two-stage crystallization model and Nielsen's theoretical model both give a similar linear relationship between the logarithm of induction time and $1/SI^2$. However, more efforts are still needed to elucidate such relationships.

On the other hand, Nielsen's theoretical induction time models successfully bridged the gap between the induction time, which can be experimentally measured, and the key mineral properties (*e.g.*, V_m, σ, *D*, SI) based on the strict theoretical derivations. As a result, they have been used to simulate the crystallization processes of various minerals including nickel ammonium sulfate,[121, 123] barite,[16, 82, 124] calcite,[15, 89] gypsum[125] and celestite,[17] to mention a few. Such induction time prediction models have been widely used for scale risk evaluation, inhibitor efficiency determination, and scale management in the oil and gas industries. However, since such induction time data are measured using different apparatus with different detection limits, the fitted mineral properties (*e.g.*, σ, *D*) should be compared with each other in a relative way, and the potential deviations from their true values should not be surprised.

Since the induction time is widely used for field applications, a series of semiempirical models have been developed to predict the induction time of various minerals, including calcite,[5, 6] barite,[5, 8] celestite,[7, 9] and barium-strontium-sulfate solid solutions.[10] In these studies, the logarithm of induction time ($\log_{10}(t_i^0)$) is observed to be linearly correlated with $1/SI^2$ or $1/SI$, and with T^{-1}, which roughly follows the relationship derived from CNT (*i.e.*, Equation [2.52]). Within limited ranges of temperature and SI, a unified semiempirical equation has been proposed as follows:

$$\log_{10}\left(t_i^0\right) = \alpha_0 + \alpha_1/SI + \alpha_2/T\left(K\right) + \alpha_3/\left(SI\cdot T\left(K\right)\right) \quad (2.54)$$

where the α values are the empirical coefficients fitted from experimental data. Additional terms may be added to cover more complicated conditions. For example, in Table 2.7 are listed the coefficients of Equation (2.54) for barite,[5] calcite,[5] and celestite[9] based on measured induction time data from various sources.

Table 2.7. Coefficients in semiempirical equation for nucleation time prediction (Equation [2.54])

Scale Mineral	α_0	α_1	α_2	α_3	Condition Ranges
Barite$^{(A)}$	1.5232	-10.8784	-895.6683	5,476.992	*T*: 4°C-250°C SI: 0.4-2.8
Calcite	0.86	-7.69	-563.29	3,978.03	*T*: 25°C-90°C SI: 1.22-2.20 pH: 7.16-7.38
Celestite	-1.713	-3.411	0	2,646.1	*T*: 25°C-90°C SI: 0.70-2.60

$^{(A)}$Additional terms in Equation (2.54) for barite: $+0.8286[Ca^{2+}] + 0.225 \times \log_{10}([Ba^{2+}]/[SO_4^{2-}])$.

It also deserves notice that due to the potential occurrence of heterogeneous nucleation under certain conditions for various minerals, $\log_{10}(t_i^0)$ may not always show a linear relationship with SI^{-2} as is shown in Figure 2.6 for different minerals. Generally, they were treated as two separate nucleation regions (*i.e.*, homogeneous and heterogeneous, respectively).[7,89] Instead of two separate regions, Dai *et al.* (2022) proposed a sigmoid type function to quantify the gradual surface tension change of gypsum at different SI conditions as follows:

$$\sigma_{gypsum} = 4.54\times10^{-4}\times\left(1-0.00176\cdot T(K)\right)\times\left[1+\frac{79.19}{1+exp\left(-7.45\cdot(SI-0.125)\right)}\right] mJ/m^2 \quad (2.55)$$

Such surface tension change can accurately capture the induction time change within such a transition region from 25°C to 90°C.

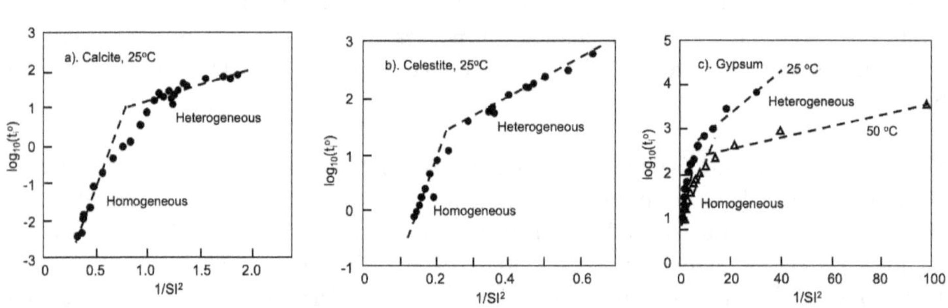

Figure 2.6. The relationship of $\log_{10}(t_i^0)$ with $1/SI^2$ for calcite, celestite, and (c) gypsum.

2.3.6 Surface Deposition

In many industrial processes, mineral formations on the facility surfaces can cause much more severe damage than those in aqueous solutions. For example, mineral scales formed inside the production tubing can lead to much less trouble if they can flow out of the production system when remaining in the fluids, rather than attached to the downhole facility surfaces. Though the mineral formations on the surfaces are usually called "deposition", it does not necessarily imply that mineral solids deposit onto the surfaces after they form in the aqueous solutions or the other way around. Actually, the deposition pathways are still under debate.

Vazirian *et al.* found that the surface deposition is more dominated by heterogeneous nucleation and crystal growth rather than the sticking of pre-existing crystals from the solution,[126] which is similar to other observations.[127, 128] In addition, many experiments show that the first detection of the heterogeneous nucleation is independent of surface material, but rather matches well with the induction time measured in batch crystallization experiments. For example, Cheong *et al.* observed no distinct difference between deposition mass gain within 1 h on six different materials[129]; Vazirian *et al.* only observed weak or no impacts of the surface roughness and surface energy of five different coating materials on calcite adhesion and deposition[126]; Mavredaki and Neville to measure the first occurrence time of calcite deposition used a QCM-D apparatus (quartz crystal microbalance—with dissipation) at different saturation levels, and they matched well with the induction time of calcite batch crystallization in a glass vial measured by a turbidity meter[130]; Dai *et al.* found that the measured gypsum induction time in batch testing matched well with those detected via QCM-D by Huang *et al.*[125, 131]

Based on such evidence, Dai *et al.* proposed a two-phase mineral deposition model as illustrated in Figure 2.7. During the first stage of deposition on a fresh surface, the very first mineral deposition layer forms in the stagnant laminar sublayer adjacent to the surface following the same batch crystallization kinetics.[85–87] In Phase 2, the flow induced crystal growth happens after the formation of the first mineral deposition layer. In many industrial applications, the facility surfaces may have already been fully or partially covered by a mineral scale layer if they are not freshly replaced. In addition, the formation of the first covered scale layer and the transition from Phase 1 to Phase 2 can be impacted by various factors, including but not limited to, surface roughness, surface material, flow regimes, and other fluid components. Currently, this two-phase deposition model can quantitatively predict the first occurrence of the scale deposition and the scale accumulation rate on facility surfaces but not the phase transition time.

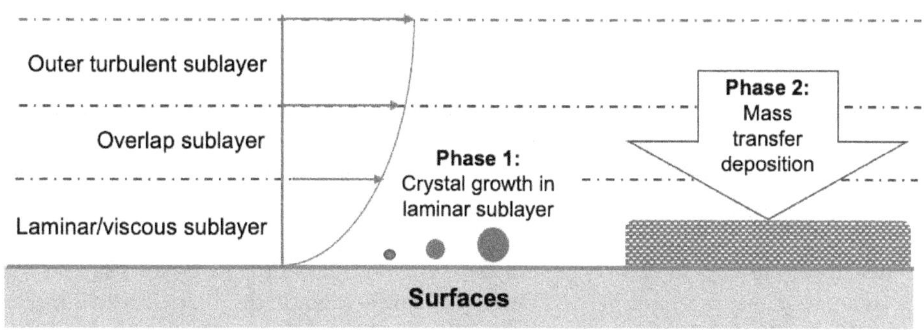

Figure 2.7. The illustration of the two-phase mineral deposition process.

This two-phase deposition model has been tested among different minerals (*e.g.*, calcite, barite, celestite), different flow regimes (*e.g.*, laminar and turbulent conditions), and under real oilfield operation conditions.[85–87, 125] For example, assuming the Guerra Well in South Texas is producing water (with compositions listed in Table 2.2) at a flow rate of 120 bbl/d, gas at 8.50 MMSCF/d, and oil at 50 bbls/d through a tubing with a 2-in ID, 0.25-mm roughness, and a length of 14,000 ft, a linear deposition rate is predicted to be 0.212 in/year of calcite and 0.045 in/y of barite (0.26 in/y in total) at the very bottom of the well.[85] It matches well with the observed deposition rate of about 0.2 in after half a year, given large uncertainties may exist in field operations.[21]

The quantification of scale deposition rate provides another perspective on evaluating scaling rates. Traditionally, the induction time (Section 2.3.1.5) is widely used and tested as the protection time (or time without scale formation) in the field. However, from the deposition perspective, the induction time only reflects the occurrence time of scale in the viscous sublayer in the first deposition phase. If some scale layer already exists, it is more critical to check the deposition rate that can be predicted by the deposition model in the second deposition phase. In another way, a short induction time for batch crystallization does not necessarily mean a fast deposition rate nor a more severe scale risk, which match well with the field observations of a worldwide scale rate evaluation program.[132]

2.3.7 Challenges in Scale Rate Predictions

The mechanisms and kinetics of mineral scaling are being constantly investigated by numerous research groups. However, two main challenges lead to the current large gap between scientific research and industrial applications.

First, the field-operating conditions (*e.g.*, T, P, I, mixed electrolytes) are inherently complicated and current laboratory studies barely cover these conditions. For example, very few studies have worked on the impacts of mixed electrolytes on the solid compositions and the kinetics of mineral scaling; few studies have worked on the mineral scaling kinetics at high temperature and high pressure conditions; most studies work on a static environment with constant T, P, and laminar flow regimes, while on the contrary, most field applications are present with dynamically changing T and P under turbulent flow regimes; the heterogeneity of reservoir properties (*e.g.*, rock porosity, permeability, facture network) can also potentially impact the scale rate predictions.

Second, scale management in the field needs the support of accurate predictions of scaling kinetics, which requires quantitative models. Unfortunately, many theories (*e.g.*, the non-classical nucleation theory) proposed a new scaling path but without a quantitative model to describe the process in a mathematical way. The lack of quantitative models limits the incorporation of such new theories into comprehensive scaling models.

To achieve accurate scale kinetics predictions, significant efforts are needed to elucidate the mineral scaling mechanisms and develop accurate quantitative models with considerations of complicated field operating conditions (*e.g.*, T, P, solution compositions, and flow regimes).

2.4 Scale Inhibition Mechanisms and Models

Scale inhibitors have been widely used for scale damage mitigation and scale formation control due to their low cost and high efficiency.[11, 12] The global scale inhibitor market is estimated to be more than 4.7 billion dollars per year.[13] Compared with the chelators (*e.g.*, EDTA, ethylenediaminetetraacetic acid), they are usually called threshold scale inhibitors since their effective dosage is much lower than the matrix ion concentrations of the scale minerals. In addition, scale inhibitors are commonly used to slow down the crystallization process rather than completely stop it. Though hundreds or even thousands of scale inhibitor products are available in the market, the active ingredients are not as abundant. In Section 1.5 of Chapter 1, a detailed discussion about the commonly used scale inhibitors has been included. Despite their wide usage, the inhibition mechanisms are still under debate and available quantitative models are very limited. The most commonly accepted inhibition mechanisms are discussed below, and Figure 2.8 briefly presents their comparisons.

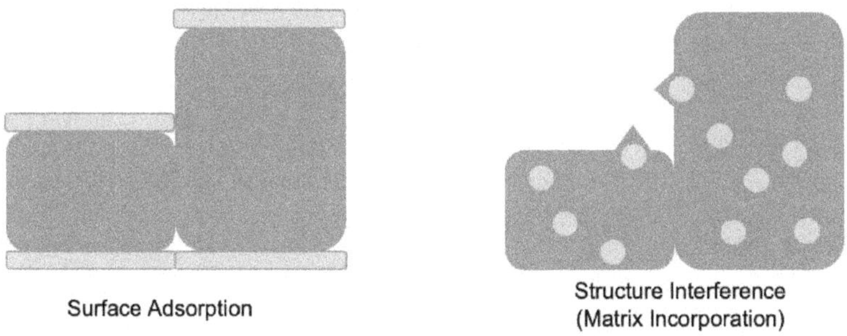

Figure 2.8. The comparison between surface adsorption and structure interference or matrix incorporation of inhibitors. The yellow parts represent the inhibitors and the blue parts represent the crystals.

2.4.1 Inhibition via Adsorption

Gebauer *et al.* found that the inhibitor (PAA, polyacrylic acid) can slow down nucleation process via adsorbing onto the prenucleation clusters of calcite.[133] They also suggest that such adsorption preferences may vary with inhibitors and nucleus structures.

In addition to the nucleation stage, Liu and Nancollas studied the barite seeded growth and dissolution in the presence of *N*,*N*,*N*′,*N*-triethylenediaminetetra(methylene phosphonic acid) (TENTMP), sodium tetrarnetaphosphate, and sodium tripolyphosphate at 25°C.[98] They suggest that the slowdown of growth rate and the initial surge of inhibitor concentration during dissolution are due to the adsorption of inhibitors onto seed surface or incorporation into the crystals. Tomson *et al.* measured the inhibition kinetics of barite in the presence of various P-containing inhibitors. Based on the adsorption/desorption tests, the hydrophobic repulsion of a macroneutral molecule (or complex) from solution is proposed to be the primary driving force for adsorption rather than a specific inhibitor-surface interaction.[134] The minimum inhibitor dosage matches well with field observations. To some extent, these explanations can be incorporated into the Burton-Cabrera-Frank (BCF) spiral growth mechanism.[135] The adsorption of inhibitors onto the kink sites in BCF theory can deactivate them and slow down the crystal growth or dissolution rates.[136, 137] Such adsorption can stop the propagation of kink sites without increasing the solubility or altering the effective surface tension.[138]

Following the adsorption mechanism, Zhao and Dai developed a mechanistic inhibition model and successfully applied to barite,[14] gypsum,[125] and celestite.[17] The adsorption of scale inhibitors onto the surfaces of mineral nuclei and crystal particles can be described as follows:

$$S + \mathrm{Inh} \Leftrightarrow S\,\mathrm{Inh}\,, \qquad K_\mathrm{L} = \frac{[S\,\mathrm{Inh}]}{[S][\mathrm{Inh}]} \tag{2.56}$$

where S is the active site on the surface (m^{-2}); Inh is the scale inhibitors; [ ] represents the concentration of the surface sites and aqueous species (mol/L), K_L is the adsorption affinity constant (L/mol). Thus, we have:

$$q_\mathrm{e} = \frac{q_\mathrm{m} K_\mathrm{L}[\mathrm{Inh}]}{1 + K_\mathrm{L}[\mathrm{Inh}]} \tag{2.57}$$

where q_e is the effective adsorbate concentration on the surface (mol/m^2 solid surface); q_m is the maximum adsorbate concentration on the surface (mol/m^2 solid surface). The isotherm can also be described as

$$\theta = \frac{K_\mathrm{L}[\mathrm{Inh}]}{1 + K_\mathrm{L}[\mathrm{Inh}]} \tag{2.58}$$

where $\theta = q_e/q_m$ and is the effective coverage fraction of the surface-active sites. There are different active sites on the crystal surface (*e.g.*, kinks, steps, and terraces) corresponding to different adsorption energy and growth rates.[139, 140] The adsorption of inhibitors onto nucleus can change the surface tension of the nucleus. Using the surface tension of a binary mixture proposed by Eberhart[141]:

$$\sigma_\mathrm{w/\,inh} = (1 - x_\mathrm{inh})\sigma_\mathrm{X} + x_\mathrm{inh}\sigma_\mathrm{inh} \cong \sigma_\mathrm{X} + x_\mathrm{inh}\sigma_\mathrm{inh} = \sigma_\mathrm{X} + B\theta\sigma_\mathrm{inh} \tag{2.59}$$

where σ_X and $\sigma_{w/inh}$ are the surface tension of the mineral nucleus (*e.g.*, barite) without or with inhibitors, respectively; σ_{Inh} is the surface tension of the inhibitor; x_{inh} is the mole fraction of inhibitor proportional to the surface coverage θ (*i.e.*, $x_{inh} = B\theta$); B is the molar ratio of the active sites over mineral molecule on the surface. Therefore, the surface tension in the presence of scale inhibitors can be described as

$$\sigma_\mathrm{w/\,inh} = \sigma_\mathrm{X} + \frac{(B\sigma_\mathrm{inh})K_\mathrm{L}[Inh]}{1 + K_\mathrm{L}[\mathrm{Inh}]} \tag{2.60}$$

During the induction period, the total adsorbed amount of inhibitor is small compared with the initial total inhibitor concentration. Then the equilibrium inhibitor concentration (*i.e.*, [Inh]) can be assumed to be the same as the initial total inhibitor concentration. The induction time of mineral crystallization in the presence of inhibitors can then be calculated using the new crystallization and inhibition model by substituting the mineral surface tension in Equation (2.52) with the modified surface tension (Equation [2.60]).

This model can predict the induction time of barite, celestite, and gypsum with or without the presence of various inhibitors. However, it is difficult to evaluate the validation of this model since surface tension measurement is difficult for nucleus. On the other hand, the K_L value between DETPMP (diethylene triamine-penta(methylene phosphonic) acid) and barite fitted from this model using induction time data is 0.88×10^6 L/mol, which is slightly larger than that measured by Liu *et al.*'s from a direct adsorption experiment.[142] Such a good match may indicate the validity of the adsorption assumption.

2.4.2 Inhibition via Structure Interference and Matrix Incorporation

In addition to the surface adsorption, the adsorbed inhibitors may possibly change the matrix structure or become part of the crystal as crystallization continues. In fact, Tomson proposed that the scale inhibitor can compete with matrix ions to form complex with nuclei or clusters.[143] Gebauer *et al.* also suggested the adsorption of inhibitors onto prenucleation clusters will influence the local structure of nucleated particles.[133] The structure distortion or matrix incorporation by inhibitors may lead to an increase of solubility and a decrease of supersaturation, which finally slow down the rate of crystallization.[144, 145]

Following the assumption of structure interference and matrix incorporation, Dai developed a mechanistic model for barite[16] and calcite.[15] This model assumes that inhibitors delay the scaling process via modifying the effective molar volume and effective SI values of the mineral. Using barite as an example, the effect of added inhibitors, Inh, on the nucleation process is modeled by assuming that a trace amount of inhibitor coprecipitates with the barite nuclei:

$$\mathrm{BaSO_4} \cdot \mathrm{Inh}_{r_{\mathrm{barite}}} \Leftrightarrow \mathrm{Ba^{2+}} + \mathrm{SO_4^{\,2-}} + r_{\mathrm{barite}} \cdot \mathrm{Inh}, \tag{2.61}$$

where r_{barite} is the molar concentration ratio of inhibitor over lattice ions with lower concentrations. For threshold inhibitors, r is expected to be much less than 1. The value of r essentially represents the probability of adding a lattice ion versus an inhibitor ion in each step of the nucleation process.

The Gibbs free energy change of this reaction (*i.e.*, $\Delta G^0_{\mathrm{r,new,barite}}$) can be calculated as

$$\Delta G^0_{r,\mathrm{new,barite}} = \Delta G^0_{\mathrm{Ba^{2+}}} + \Delta G^0_{\mathrm{SO_4^{2-}}} + r_{\mathrm{barite}} \Delta G^0_{\mathrm{Inh}} - \Delta G^0_{\mathrm{mix,barite}} \tag{2.62}$$

where $\Delta G^0_{\mathrm{mix,barite}}$ is the Gibbs free energy of $\mathrm{BaSO_4} \cdot \mathrm{Inh}_{r_{\mathrm{barite}}}$; $\Delta G^0_{[\,]i}$ represent the formation Gibbs free energies of each component. $\Delta G^0_{\mathrm{mix,barite}}$ can be calculated as

$$\Delta G^0_{\mathrm{mix,barite}} = \Delta G^0_{\mathrm{barite}} + r_{\mathrm{barite}} \Delta G^0_{\mathrm{inh}} + \Delta G^{\mathrm{ex}}_{\mathrm{barite}} \tag{2.63}$$

where $\Delta G^0_{\text{barite}}$ is the Gibbs free energy of barite; $\Delta G^{\text{ex}}_{\text{barite}}$ is the excess Gibbs free energy due to the nonideal mixture between barite and scale inhibitors. Similar to Margules's regular solution theory, when the concentration ratio between inhibitor concentration and lattice ions is small (*i.e.*, r_{barite} is small), the excess molar Gibbs free energy can be calculated as follows[25]:

$$\Delta G^{\text{ex}}_{\text{barite}} = wx_1x_2 \cong wr_{\text{barite}} \tag{2.64}$$

where w is a constant at a given temperature and pressure ($\text{J}\cdot\text{mol}^{-1}$); x is the mole fraction of different components of the mixture. By substituting Equations (2.63) and (2.64) into Equation (2.62), the following equation is obtained.

$$\Delta G^0_{r,\text{new,barite}} = \Delta G^0_{r,\text{barite}}\left(1 + r_{\text{barite}}a_{\text{G}}\right) \tag{2.65}$$

where $a_{\text{G}} = -w/\Delta G^0_{r,\text{barite}}$ is the relative Gibbs energy change due to the presence of a scale inhibitor. As a result, we can achieve

$$\text{SI}^{\text{new}}_{\text{barite}} = \text{SI}_{\text{barite}} + r_{\text{barite}}\log_{10}[\text{Inh}] - (r_{\text{barite}}a_{\text{G}})\log_{10}K_{\text{sp,barite}} \tag{2.66}$$

Assuming that inhibitor molecule forms a mixture with the barite lattice ions, the barite crystal lattice volume can be impacted as follows:

$$V^{\text{new}}_{\text{m,barite}} = V_{\text{m,barite}}\left(1 + r_{\text{barite}}a_V\right) \tag{2.67}$$

where a_V represents the relative molar volume change due to inhibitor addition. The induction time of mineral crystallization in the presence of inhibitors can then be calculated using the new crystallization and inhibition model by substituting the mineral surface tension in Equation (2.52) with the modified SI (Equation [2.66]) and modified molar volume (Equation [2.67]). This model has been successfully applied to barite and calcite in the presence of various inhibitors. On the other hand, the physical meaning of the a_G and a_V terms is unclear and cannot be readily measured experimentally.

2.4.3 Inhibition Model for Deposition Rate

In the deposition process of scale formation, scale inhibitors can also significantly slow down the kinetics. In the first deposition phase (Section 2.3.1.6), the inhibition mechanism is the same as that in the bulk crystallization process and the inhibition efficiencies can be described by the models introduced above. In the second deposition phase, a Langmuir-type kink site adsorption model has been proposed assuming that the adsorption of inhibitors on the kink site can block the kink sites and delay the reaction rate proportionally.[87]

$$\theta = \frac{K\,\text{ink}_m - K\,\text{ink}}{K\,\text{ink}_m} = \frac{k_{r,C_{\text{inh}}=0} - k_{r,C_{\text{inh}}}}{k_{r,C_{\text{inh}}=0}} = \frac{K_{\text{eq}}C_{\text{inh}}}{1 + K_{\text{eq}}C_{\text{inh}}}, \tag{2.68}$$

where θ is the fractional occupancy of active adsorption site and equals the relative change of the mass transfer coefficient of deposition; $K\,\text{ink}_m$ is the maximum adsorption site number; K ink is the unoccupied adsorption site number; K_{eq} is the associated equilibrium constant (ppm^{-1}); C_{inh} is the inhibitor concentration; $k_{r,C_{inh}=0}$ is the mass transfer coefficient without inhibitors; $k_{r,C_{inh}}$ is the mass transfer coefficient with inhibitors. However, only very limited studies have worked on the inhibition mechanisms of surface deposition. More efforts are needed in this area.

2.5 Monitoring Methods for Scaling Processes

Mineral scale is driven by the supersaturation. As crystallization proceeds, the supersaturation gradually drops. Monitoring the scaling process is critical for both scientific research and industrial applications. Via monitoring different system properties, various methods have been developed to monitor the scaling processes. In Table 2.8 summarized the monitoring methods most of which will also be discussed in detail below.

Table 2.8. Different monitoring methods for scaling processes.

Property	Apparatus	Temporal Resolution	Signal Resolution
Ion concentration	ICP-OES/MS IC Other methods NACE method	Low	ppb or lower (depending on methods)
TDS/conductivity	Conductivity meter	Real time	ppm or larger
Active ion concentration	Ion selective electrode	Real time	ppm, depending on meter sensitivity
pH	pH meter	Real time	0.01 pH or better
Turbidity	Laser Turbidimeter	Real time	1% signal change; about 10 nm particle size (depending on wavelength)
Partial pressure	DSL	Real time	Depending on pressure transducer
Microscopic property	TEM/HR-TEM Cryo-TEM In situ liquid TEM AFM Microfluidics	Low Low Real time Real time Real time	nm or less nm or less ~10 nm nm or less μm or less
Mass	QCM-D	Real time	0.1-1.0 ng/cm^2

DSL = dynamic scale loop; HR-TEM = high-resolution transmission electron microscopy; AFM = atomic force microscopy; QCM-D = quartz crystal microbalance with dissipation monitoring; ppb = part per billion; ppm = part per million; nm = nanometer; mm = micrometer; ng = nanogram.

2.5.1 Ion Concentration Methods

Ion concentrations have been one of the most commonly monitored properties of the mineral scaling system. Generally, a small portion of the solution is extracted from the system and prepared for ion concentrations using ICP-OES/MS, IC, or other methods that can be found in standard methods.[146] It is convenient and straightforward to monitor matrix ion concentrations. However, it is difficult to monitor the scaling processes in real time. In addition, the relatively large relative error of ion concentration measurements (*e.g.*, as large as 5% or more) limits the sensitivity of this method.

Two NACE (National Association of Corrosion Engineers) international standards (TM0197-2010 and TM0374-2007) have been developed to test the scale inhibitor performance for barium/strontium sulfates and calcium sulfate/carbonate minerals in static bottles, respectively, via monitoring matrix ion concentration changes.[147, 148] In this method, supersaturated solutions are put in glass bottles and kept in an oven at certain testing temperatures without stirring for a designated testing period from 1 h to 24 h (24 h being widely used) with or without scale inhibitors. The inhibitor efficiency is evaluated via measuring the cation concentration changes at the end of the experiment. The larger cation concentration indicates higher inhibition efficiency. This method is easy and economical to run, but it cannot simulate the dynamic flow conditions in the field, and its testing efficiency is limited by the oven space and the length of the testing period.

Alternatively, when the solution is mainly composed of the scaling ions, mineral scaling can also lead to the drop of TDS of the solution, which can be monitored in real time via conductivity. However, its sensitivity is still limited.

Instead of measuring the total ion concentration, Gebauer *et al.* utilized a calcium-ion-selective electrode (ISE) to monitor the active calcium ion concentration in real time during the calcium carbonate precipitation process.[72] The reliability and detection limit depend on the sensitivity of the meter and the experimental skills. Similarly, the precipitation of carbonate minerals can lead to the drop of solution pH (*i.e.*, $-\log_{10}\{H^+\}$, negative logarithm of active hydrogen ion concentration), which can also be easily monitored in real time. However, most pH and specific ion electrodes cannot tolerate high temperature or pressure.

2.5.2 Scattering Methods

The formation of nuclei and larger crystals can lead to the increase of light scattering, decrease of transmitted light energy, and the increase of solution turbidity (defined in Equation [2.49]). Different methods or standards use different apparatus and monitor light signal changes at different angles. By convention, the ones monitoring the transmitted light (*i.e.*, at 180°) are called turbidimetry, while the ones monitoring the scattered light (*i.e.*, at 90°) are called nephelometry (Figure 2.9). Actually, the difference between the incident light and the transmitted light is the sum of all scattered light intensity at all angles. Both U.S. EPA (United States Environmental Protection Agency) and ISO (International Organization for Standardization) have proposed the turbidity measurement standards (EPA 180.1 and ISO

7027-1:2016, respectively).[149, 150] Both standards use the nephelometry measurement, but EPA standard (NTU, Nephelometric Turbidity Units) uses a white light source, while ISO standard (FNU, Formazin Nephelometric Units) uses an infrared light source. According to the standard, a suspension of formazin reacted from 1.25 mg/L hydrazine sulfate and 12.5 mg/L hexamethylenetetramine in water has a turbidity of 1 NTU and 1 FNU, though the meter readings might be different. At standard conditions, turbidimeter has been used to monitor solution turbidity to measure the induction time of mineral scaling.[7, 151]

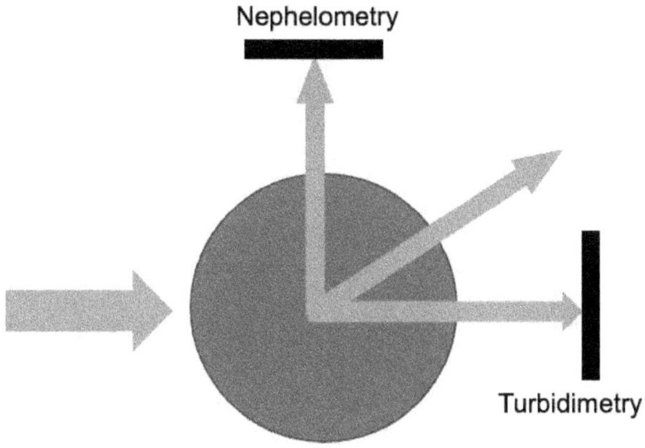

Figure 2.9. The difference between nephelometry and turbidimetry.

To achieve better temperature and pressure tolerance with more flexibility and real time measurement, a single wavelength laser apparatus has also been adopted, and the turbidity is measured via monitoring the transmitted light intensity (*i.e.*, turbidimetry) for induction time measurement.[122, 124, 152–154] The light source can be adjusted to a shorter wavelength to achieve better sensitivity. For example, Lu *et al.* used ultraviolet-visible (UV-Vis) spectrophotometer at UV wavelength range and called it a KTT (kinetic turbidity test) method.[153] The reactor can be customized to tolerate much higher temperature and pressure. For example, Deng *et al.* developed a high-pressure vessel with side glasses that can tolerate temperature up to 250°C and pressure 100 bars or higher.[152] This turbidity method has been more popularly adopted for scaling kinetics analysis than traditional NACE inhibitor efficiency testing method due to its convenient setup, high efficiency, larger temperature and pressure tolerances, real time signal recording, and dynamic flow conditions. Though not precisely quantified, the 1% transmitted light intensity change picked usually corresponds to the particle size of 20 nm (roughly 1/20 of the wavelength) for 532-nm laser.[117] Compared with the critical nucleus size of less than 1 nm, turbidity measurements do not have enough spatial resolution. To achieve better spatial resolutions, Jun *et al.* adopted the small-angle x-ray scattering techniques to achieve in situ monitoring of homo- and heterogeneous nucleation and crystal growth from critical radius[155] at subnanometer level.

2.5.3 Partial Pressure Methods

Oddo and Tomson patented a dynamic scale loop (DSL) testing method for scale kinetics monitoring under dynamic flow conditions.[156] Two high-pressure pumps are used to pump cationic and anionic solutions into a capillary tubing. The mixture of the cationic and anionic solutions is supersaturated in regard to a certain scaling mineral at the testing temperature and pressure. A pressure transducer is installed to monitor the pressure change between the inlet and outlet of the capillary tubing in real time. When mineral scale forms in the capillary tubing, the partial pressure increases due to the gradual shrinkage of the tubing diameter. The DSL testing method has been widely used in the oil and gas industry to test scaling risks and scale inhibitor performances due to its flow dynamics and high-temperature and high-pressure tolerances. Several companies have made this DSL apparatus commercially available and easier to operate, but with relatively higher costs.

2.5.4 Microscopic Methods

Transmission electron microscopy (TEM) is a microscopic technique that can achieve a much finer spatial resolution (nm or angstrom level) due to the smaller de Broglie wavelength of electrons. TEM has been widely used in different disciplines owing to its high spatial resolution. However, the sample preparation procedures limited its temporal resolution, and thus limited its applications in the real-time monitoring of the scaling process. As a semireal-time method, cryo-TEM can freeze the sample and pause the scaling process, though the sample preparation is even more complicated. Dai *et al.* utilized cryo-TEM and monitored the particle size change at induction time.[117] In the past decades, the development of *in situ* TEM shed light on the scaling process monitoring with both high-temporal and high-spatial resolutions.[157, 158] The supersaturated solution sample can be sealed in a closed sample holder for TEM imaging in real time with high-spatial resolution.

In addition, *in situ* atomic force microscopy (AFM) has also been adopted to monitor the heterogeneous nucleation or crystal growth processes with high-spatial resolution in real time by many scientists.[159–162] However, it cannot monitor the scaling processes in the bulk solutions.

2.5.5 Mass Methods

In batch crystallization experiments, it is difficult or unnecessary to measure the mass of precipitated solids in the solution. The development of quartz crystal microbalance with dissipation (QCM-D) significantly improves the detection limits of the deposition mass to as low as 0.1 ng/cm^2.[163] The basic mechanism of QCM is the correlation between the sensor mass and its resonance frequency, which can be described by the Sauerbrey relation (*i.e.*, $\Delta m = -C \cdot \Delta f/n$, where Δm and Δf are the change of sensor mass and frequency; C is the mass sensitivity constant; n is the harmonic number). Generally, the sensor is made of a piezoelectric material (*e.g.*, quartz) of which its electrical and mechanical state are coupled. To mimic different surfaces, the sensor can be coated with different materials (*e.g.*, stainless steel, gold). It has been widely used in various disciplines and is a powerful tool for mineral deposition mechanism analysis.[88, 131, 164, 165]

2.6 Conclusions

Scale prediction is the basis for mineral scale management in various industrial disciplines. Numerous scientists and engineers have been attracted to this area and have made enormous progresses in the past decades. For scale risks and scale rate predictions, both fundamental theories and practical software models have been developed. These models have been widely adopted in the oil and gas industry as guidelines of scale management. Despite such great success in the past, some outlooks are listed below in addition to the specific challenges discussed above:

- Scale prediction needs to be more integrated with field operations. Scale risks and scale rate predictions can help optimize the design of fracturing fluids, shut-in periods, production schedule, downhole material selection, artificial lifting methods, to mention a few. On the other hand, field operations can significantly impact scale management, for example, scale inhibitor type (*e.g.*, solid, liquid) selection, inhibitor injection point, and scale deposition rate (impacted by flow rates).
- In the big data era, scale management decisions can be made via scale risk and scale rate predictions based on much larger database from the field or even basin level. With the support of such large database, scale rate and risk predictions can be achieved from much more dimensions, for example, time, reservoir rock compositions, depth (or formations), area, basin, fracturing fluid design, shut-in period, to mention a few. Few studies have made progress in this direction,[166, 167] but more efforts are needed in building a comprehensive database, for example, USGS produced water database,[168] and integrating scale prediction software models.
- Considering the difficulties in simulating field-operating conditions in the laboratory, better sampling and testing methods are needed. In field sampling, lots of interferences need to be properly handled. For example, pH meter contamination; acid gas (CO_2 and H_2S) degassing; potential flashing, especially for gas wells; impacts of organic acids on alkalinity determination; and scale inhibition during sample transportation and storage. In laboratory scale rate testing, the significant differences in flow rate, Reynolds number, and shear can lead to the deviations between lab testing results and field observations. Thus, there is a great need to develop better testing apparatus, before which the engineers are recommended to perform testing and scale rate evaluations on the same apparatus.
- Scale prediction should consider other flow assurance problems. For example, corrosion can be the iron source of siderite or iron sulfide; sand accumulation can aggravate the damage that mineral scale brings about; the formation of wax or asphaltene may lead to more complicated scale formation mechanisms. Such interdiscipline research is needed.
- Since acid gas (*i.e.*, CO_2, H_2S) can significantly impact carbonate and sulfide mineral scaling and the pH of the produced water, it is recommended that flash calculations using different EoS are needed to more accurately determine the phase distributions of acid gas, water and hydrocarbons. Such discussions will be covered in Chapter 3.

Nomenclature

AFM	Atomic force microscopy
bbl	barrels of oil used in the oil and gas industry
BCF	Burton-Cabrera-Frank
BOPD	Barrels of oil produced per day
BWPD	Barrels of water produced per day
CNT	Classical nucleation theory
D-H	Debye-Hückel
DLS	Dynamic light scattering
DSL	Dynamic scale loop
DETPMP	Diethylene triamine-penta(methylene phosphonic) acid
EDTA	Ethylenediaminetetraacetic acid
FNU	Formazin nephelometric units
HTHP	High temperature and high pressure
I	Ionic strength
IC	Ion chromatography
ICP-OES/MS	Induced coupled plasma-optical emission spectroscopy/mass spectroscopy
ISE	Ion selective electrode
ISO	International Organization for Standardization
KTT	Kinetic turbidity test
LSI	Langelier saturation index
m	Molality concentration (=moles of dissolved ions in one kilogram of solvent)
M	Molarity concentration (=moles of dissolved ions in one liter of solution)
MIC	Minimum inhibitor concentration
NACE	National Association of Corrosion Engineers
NTU	Nephelometric turbidity unit
PAA	Polyacrylic acid
QCM-D	Quartz crystal microbalance—with dissipation
R-S equation	Robinson-Stokes equation
RSI	Ryznar stability index
SCF	Standard cubic feet
SI	Saturation index, the base 10 logarithm of the saturation ratio (SR)
SIT	Specific ion interaction
SR	Saturation ratio
SSI	Stiff-Davis stability index
SSP	ScaleSoftPitzer
TDS	Total dissolved solids
TEM	Transmission electron microscopy
TENTMP	N, N, N', N-triethylenediaminetetra (methylene phosphonic acid)

TJ equation	Truesdell-Jones equation
US EPA	United States Environmental Protection Agency
USGS	United States Geological Survey
UV-Vis	Ultraviolet-visible

References

1. M.A. Kelland, *Production Chemicals for the Oil and Gas Industry* (Boca Raton ,CRC Press, 2014). pp.51-88
2. J.P. Gustafsson, *Visual MINTEQ 3.0 User Guide* (Stockholm, Sweden: KTH Dep. Land Water Recourses, 2011).pp.3-39
3. D.L. Parkhurst, C.A.J. Appelo, *Description of Input and Examples for PHREEQC Version 3: A Computer Program for Speciation, Batch-Reaction, One-Dimensional Transport, and Inverse Geochemical Calculations* (Reston, Virginia:US Geological Survey, 2013).
4. Z. Dai, A.T. Kan, W. Shi, et al., "Solubility Measurements and Predictions of Gypsum, Anhydrite, and Calcite Over Wide Ranges of Temperature, Pressure, and Ionic Strength With Mixed Electrolytes," *Rock Mechanics and Rock Engineering* 50 (2017): pp. 327–339.
5. C. Dai, Z. Dai, F. Zhang, et al., "A Unified Experimental Method and Model for Predicting Scale Inhibition," SPE-193586-MS, SPE International Conference on Oilfield Chemistry, held April 8–9, 2019 (Galveston, Texas, USA:SPE).
6. S. He, A.T. Kan, M.B. Tomson, "Inhibition of Calcium Carbonate Precipitation in NaCl Brines From 25 to 90 °C," *Applied Geochemistry* 14 (1999): pp. 17–25.
7. S. He, J.E. Oddo, M.B. Tomson, "The Nucleation Kinetics of Strontium Sulfate in NaCl Solutions Up To 6 m and 90 C With or Without Inhibitors," *Journal of Colloid and Interface Science* 174 (1995): pp. 327–335.
8. S. He, J.E. Oddo, M.B. Tomson, "The Nucleation Kinetics of Barium Sulfate in NaCl Solutions Up To 6 m and 90 °C," *Journal of Colloid and Interface Science* 174 (1995): pp. 319–326.
9. Y. Zhao, Z. Dai, C. Dai, et al., "Semiempirical Model for Predicting Celestite Scale Formation and Inhibition in Oilfield Conditions," *Energy and Fuels* 35 (2021): pp. 11285–11294.
10. Y. Zhao, Z. Dai, C. Dai, et al., "A Semiempirical Model for Barium-Strontium-Sulfate Solid Solution Scale Crystallization and Inhibition Kinetics at Oilfield Conditions," *SPE Journal* 26 (2021): pp. 1–14.
11. S. He, A.T. Kan, M.B. Tomson, "Inhibition of Mineral Scale Precipitation By Polymers," in *Water Soluble Polymers Springer*, Z. Amjad, Springer, Boston, MA. (2002), pp. 163–171.
12. O. Vetter, "An Evaluation of Scale Inhibitors," *Journal of Petroleum Technology* 24 (1972): pp. 997–1006.
13. Data Bridge Market Research, *Scale Inhibitors Market–Global Industry Trends and Forecast to 2026* (2019).
14. Z. Dai, Y. Zhao, X. Wang, et al., "Novel Barite Crystallization and Inhibition Model Based on Surface Adsorption," *SPE Journal* 2 (2023): pp. 1818–1828.
15. Z. Dai, F. Zhang, Y. Zhao, et al., "Predicting Calcite Crystallization and Inhibition Kinetics at Up To 175 °C Using a New Theoretical Inhibition Model," *Crystal Growth & Design* 20 (2020): pp. 8090–8100.
16. Z. Dai, F. Zhang, F. Yan, et al., "A New Theoretical Model for the Induction Time and Scale Inhibitor Dosage Predictions for Calcite and Barite Over Wide Ranges of Temperature, Inhibition Concentration and Saturation Index," SPE-184547-MS, SPE International Conference on Oilfield Chemistry,held April 3–5, 2017 (Montgomery, TX, USA: SPE).

17. Y. Zhao, Z. Dai, X. Wang, et al., "A New Mechanistic Model for Mineral Crystallization and Inhibition Kinetics and Its Application to Celestite," *SPE Journal* 27 (2022): pp. 1553–1561.
18. H. Feitler, "Scale Meter: A New Method for Determining the Critical pH of Scaling," in *Journal of Materials Engineering and Performance* 11.1(United States), 1972.pp.31-35
19. A.T. Kan, M.B. Tomson, "Scale Prediction for Oil and Gas Production," *SPE Journal* 17 (2012): pp. 362–378.
20. M.B. Tomson, J.E. Oddo, *Calcite Scale Handbook: Measurement, Prediction, and Control* (Chicago: Gas Research Institute, 1998).pp.1-85
21. M.J. McKenzie, S. He, C. Hinrichsen, et al., "Controlling Barium Sulfate Scale Deposition by Inhibitor Squeeze at the Guerra Well in South Texas: A Case Study," American Petroleum Institute, Natural Occurring Radioactive Materials Conference (Houston, TX, USA:Proc, 1995): pp. 17–18.
22. P. Debye, E. Hückel, "De la theorie des electrolytes I. Abaissement du point de congelation et phenomenes associes," *Physikalishce Zeitschrift* 24 (1923): pp. 185–206.
23. D. Langmuir, *Aqueous Environmental Geochemistry* (New Jersey: Prentice Hall, 1997).pp. 586-586
24. H.C. Helgeson, D.H. Kirkham, "Theoretical Prediction of the Thermodynamic Behavior of Aqueous Electrolytes at High Pressures and Temperatures: I. Summary of the Thermodynamic/Electrostatic Properties of the Solvent," *American Journal of Science* 274 (1974): pp. 1089–1198.
25. K.S. Pitzer, *Thermodynamics* (New York: McGraw–Hill, Inc., 1995).pp.1-285
26. E. Güntelberg, "Untersuchungen Über Ioneninteraktion," *Zeitschrift für Physikalische Chemie* 123 (1926): pp. 199–247.
27. G. Scatchard, "Concentrated Solutions of Strong Electrolytes," *Chemical Reviews* 19 (1936): pp. 309–327.
28. R.A. Robinson, R.H. Stokes, *Electrolyte Solutions* (New York: Dover Publications, 2002).
29. E. Guggenheim, J. Turgeon, "Specific Interaction of Ions," *Transactions of the Faraday Society* 51 (1955): pp. 747–761.
30. C.W. Davies, "The Extent of Dissociation of Salts in Water. Part VIII. An Equation for the Mean Ionic Activity Coefficient of an Electrolyte in Water, and a Revision of the Dissociation Constants of Some Sulphates," *Journal of the Chemical Society (Resumed)* 397 (1938): pp. 2093–2098.
31. A.H. Truesdell, B.F. Jones, "WATEQ, A Computer Program for Calculating Chemical Equilibria of Natural Waters," *Journal of Research of the U.S. Geological Survey* 2 (1974): pp. 233–248.
32. R.A. Robinson, R.H. Stokes, *Electrolytil Solutions: The Measurement and Interpretation of Conductance, Chemical Potential and Diffusion in Solutions of Simple Electrolytes* (London: Butterworths, 1955).
33. C. Kusik, H. Meissner, "Electrolyte Activity Coefficients in Inorganic Processing," *Proceedings of AIChE Chemical Engineering Progress Symposium Series* 74 (1978): pp. 14–20.
34. P. Cohen, *The ASME Handbook on Water Technology for Thermal Power Systems* (New York: American Society of Mechanical Engineers, 1989).
35. J.N. Brønsted, "Studies on Solubility. IV. The Principle of the Specific Interaction of Ions," *Journal of the American Chemical Society* 44 (1922): pp. 877–898.
36. H.F. Holmes, R.H. Busey, J.M. Simonson, et al., "$CaCl_{2(aq)}$ at Elevated Temperatures. Enthalpies of Dilution, Isopiestic Molalities, and Thermodynamic Properties," *Journal of Chemical Thermodynamics* 26 (1994): pp. 271–298.
37. H.F. Holmes, R.E. Mesmer, "Aqueous Solutions of the Alkaline-Earth Metal Chlorides at Elevated Temperatures. Isopiestic Molalities and Thermodynamic Properties," *Journal of Chemical Thermodynamics* 28 (1996): pp. 1325–1358.

38. Z. Dai, A.T. Kan, W. Shi, et al., "Calcite and Barite Solubility Measurements in Mixed Electrolyte Solutions and the Development of a Comprehensive Model for Water-Mineral-Gas equilibrium of the Na-K-Mg-Ca-Ba-Sr-Cl-SO_4-CO_3-HCO_3-CO_2(aq)-H_2O System at Up To 250 °C and 1,500 Bars," *Industrial and Engineering Chemistry Research* 56 (2017): pp. 6548–6561.
39. K.S. Pitzer, G. Mayorga, "Thermodynamics of Electrolytes. II. Activity and Osmotic Coefficients for Strong Electrolytes With One or Both Ions Univalent," *Journal of Physical Chemistry* 77 (1973): pp. 2300–2308.
40. C. Christov, N. Moller, "A Chemical Equilibrium Model of Solution Behavior and Solubility in the H-Na-K-Ca-OH-Cl-HSO_4-SO_4-H_2O System to High Concentration and Temperature," *Geochimica et Cosmochimica Acta* 68 (2004): pp. 3717–3739.
41. Z. Dai, A. Kan, F. Zhang, et al., "A Thermodynamic Model for the Solubility Prediction of Barite, Calcite, Gypsum, and Anhydrite, and the Association Constant Estimation of $CaSO4^{(0)}$ Ion Pair up to 250 °C and 22000 psi," *Journal of Chemical and Engineering Data* 60 (2014): pp. 766–774.
42. Z. Duan, R. Sun, "An Improved Model Calculating CO_2 Solubility in Pure Water and Aqueous NaCl Solutions From 273 to 533 K and From 0 to 2000 Bar," *Chemical Geology* 193 (2003): pp. 257–271.
43. C.E. Harvie, N. Møller, J.H. Weare, "The Prediction of Mineral Solubilities in Natural Waters: The Na-K-Mg-Ca-H-Cl-SO_4-OH-HCO_3-CO_3-CO_2-H_2O System to High Ionic Strengths at 25 °C," *Geochimica et Cosmochimica Acta* 48 (1984): pp. 723–751.
44. B. Kaasa, "Prediction of pH, Mineral Precipitation and Multiphase Equilibria During Oil Recovery," (Ph.D. diss., Norwegian University of Science and Technology, 1998).
45. C. Monnin, "An Ion Interaction Model for the Volumetric Properties of Natural Waters: Density of the Solution and Partial Molal Volumes of Electrolytes to High Concentrations at 25 °C," *Geochimica et Cosmochimica Acta* 53 (1989): pp. 1177–1188.
46. D. Rowland, E. Königsberger, G. Hefter, et al., "Aqueous Electrolyte Solution Modelling: Some Limitations of the Pitzer Equations," *Applied Geochemistry* 55 (2015): pp. 170–183.
47. W. Shi, A.T. Kan, N. Zhang, et al., "Dissolution of Calcite at Up To 250° C and 1450 Bar and the Presence of Mixed Salts," *Industrial and Engineering Chemistry Research* 52 (2013): pp. 2439–2448.
48. W. Shi, A.T. Kan, C.F. Fan, et al., "Solubility of Barite Up To 250 °C and 1500 Bar in Up To 6 m NaCl Solution," *Industrial and Engineering Chemistry Research* 51 (2012): pp. 3119–3128.
49. C.A.J. Appelo, "Principles, Caveats and Improvements in Databases for Calculating Hydrogeochemical Reactions in Saline Waters From 0 to 200 °C and 1 to 1000 atm," *Applied Geochemistry* 55 (2015): pp. 62–71.
50. L.N. Plummer, D. Parkhurst, G. Fleming, et al., *A Computer Program Incorporating Pitzer's Equations for Calculation of Geochemical Reactions in Brines, Water-Resources Investigations Report* (US Geological Survey, 1988).pp.1-38
51. Y.K. Kharaka, W.D. Gunter, P.K. Aggarwal, et al., *SOLMINEQ. 88: A Computer Program for Geochemical Modeling of Water-Rock Interactions* (Department of the Interior, US Geological Survey, 1989).pp.7-140
52. K.S. Pitzer, G. Mayorga, "Thermodynamics of Electrolytes. III. Activity and Osmotic Coefficients for 2–2 Electrolytes," *Journal of Solution Chemistry* 3 (1974): pp. 539–546.
53. Z. Duan, D. Li, "Coupled Phase and Aqueous Species Equilibrium of the H_2O–CO_2–NaCl–$CaCO_3$ System From 0 to 250 °C, 1 to 1000 Bar with NaCl Concentrations up to Saturation of Halite," *Geochimica et Cosmochimica Acta* 72 (2008): pp. 5128–5145.
54. D.G. Archer, "Thermodynamic Properties of the NaCl + H_2O system II. Thermodynamic Properties of $NaCl_{(aq)}$, NaCl·$2H_2O$(Cr), and Phase-Equilibria," *Journal of Physical and Chemical Reference Data* 21 (1992): pp. 793–829.

55. P.M. Wang, K.S. Pitzer, J.M. Simonson, "Thermodynamic Properties of Aqueous Magnesium Chloride Solutions From 250 to 600 K and to 100 MPa," *Journal of Physical and Chemical Reference Data* 27 (1998): pp. 971–991.
56. J. Li, Z. Duan, "A Thermodynamic Model for the Prediction of Phase Equilibria and Speciation in the H_2O–CO_2–NaCl–$CaCO_3$–$CaSO_4$ System From 0 to 250 °C, 1 to 1000 Bar With NaCl Concentrations Up To Halite Saturation," *Geochimica et Cosmochimica Acta* 75 (2011): p. 4351.
57. Z. Dai, "Thermodynamics and Kinetics of Mineral Crystallization and Inhibition," (PhD thesis, Rice University, 2017).
58. A.T. Kan, Z. Dai, F. Zhang, et al., "Scale Prediction and Control at Ultra HTHP," SPE-173803-MS, SPE International Symposium on Oilfield Chemistry, held April 13–15, 2015 (The Woodlands, TX, USA: SPE).
59. X. Wang, Z. Dai, S. Ko, et al., "Iron Sulfide Solubility Measurement and Modeling Over Wide Ranges of Temperatures, Ionic Strength, and pH," *SPE Journal* 27 (2022): pp. 1–12.
60. M. Barber, S. Heath, "A New Approach to Testing Scale Inhibitors in Mild Scaling Brines—Are Dynamic Scale Loop Tests Needed?,"SPE-193580-MS, SPE International Conference on Oilfield Chemistry, held April 8–9 2019 (Galveston, TX, USA: SPE.).
61. K. Ho, T. Chen, P. Chen, et al., "Development of Novel Test Methodology To Understand the Mechanisms of Halite Inhibition and Environmentally Acceptable Halite Scale Inhibitors for High Temperature Application," SPE-169803-MS, SPE International Oilfield Scale Conference and Exhibition, held May 14–15, 2014 (Aberdeen, UK: SPE).
62. M.H. Van Spankeren, M.A. Hernandez, "Autonomous Corrosion and Scale Management in Electric Submersible Pump Wells," SPE-207732-MS, Abu Dhabi International Petroleum Exhibition & Conference, held November 15–18, 2021 (Abu Dhabi, UAE: SPE).
63. W. Stumm, J. Morgan, *Aquatic Chemistry, Chemical Equilibra and Rates in Natural Waters*, 3rd ed. (New Jersey: John Wiley & Sons, 1996).pp.148-240
64. W.F. Langelier, "The Analytical Control of Anti-Corrosion Water Treatment," *JOURNAL AWWA* 28 (1936): pp. 1500–1521.
65. J.C. Cowan, *Water-Formed Scale Deposits* (Houston: Gulf Publishing Co., 1976).
66. H.A. Stiff Jr., L.E. Davis, "A Method for Predicting the Tendency of Oil Field Waters to Deposit Calcium Carbonate," *Journal of Petroleum Technology* 4 (1952): pp. 213–216.
67. J.W. Ryznar, "A New Index for Determining Amount of Calcium Carbonate Scale Formed By a Water," *JOURNAL AWWA* 36 (1944): pp. 472–483.
68. J.E. Oddo, M.B. Tomson, "Why Scale Forms and How to Predict It," *SPE Production & Facilities* 9 (1994): pp. 47–54.
69. M.B. Tomson, A.T. Kan, G. Fu, et al., "Measurement of Total Alkalinity and Carboxylic Acid and Their Relation To Scaling and Corrosion," *SPE Journal* 11 (2006): pp. 103–110.
70. A.E. Nielsen, *Kinetics of Precipitation* (Pergamon Press Oxford, 1964).
71. V.K. La Mer, R.H. Dinegar, "Theory, Production and Mechanism of Formation of Monodispersed Hydrosols," *Journal of the American Chemical Society* 72 (1950): pp. 4847–4854.
72. D. Gebauer, A. Volkel, H. Colfen, "Stable Prenucleation Calcium Carbonate Clusters," *Science* 322 (2008): pp. 1819–1822.
73. A.S. Myerson, B.L. Trout, "Nucleation From Solution," *Science* 341 (2013): pp. 855–856.
74. M.A. Watzky, R.G. Finke, "Nanocluster Size-Control and "Magic Number" Investigations. Experimental Tests of the "Living-metal polymer" Concept and of Mechanism-Based Size-Control Predictions Leading To the Syntheses Of Iridium (0) Nanoclusters Centering About Four Sequential Magic Numbers," *Chemistry of Materials* 9 (1997): pp. 3083–3095.
75. M. Smoluchowski, "Drei Vortrage Uber Diffusion, Brownsche Bewegung und Koagulation Von Kolloidteilchen," *Zeitschrift für Physik* 17 (1916): pp. 557–585.

76. T.O. Drews, M. Tsapatsis, "Model of the Evolution of Nanoparticles to Crystals via an Aggregative Growth Mechanism," *Microporous and Mesoporous Materials* 101 (2007): pp. 97–107.
77. J. Park, V. Privman, "Growth of Monodispersed Colloids By Aggregation of Nucleating Subunits," *Recent Research Developments in Statistical Physics.*, (2000).
78. R.L. Penn, J.F. Banfield, "Imperfect Oriented Attachment: Dislocation Generation in Defect-Free Nanocrystals," *Science* 281 (1998): pp. 969–971.
79. D. Li, M.H. Nielsen, J.R. Lee, et al., "Direction-Specific Interactions Control Crystal Growth By Oriented Attachment," *Science* 336 (2012): pp. 1014–1018.
80. I.M. Lifshitz, V.V. Slyozov, "The Kinetics of Precipitation From Supersaturated Solid Solutions," *Journal of Physics and Chemistry of Solids* 19 (1961): pp. 35–50.
81. W. Ostwald, *Lehrbuch der Allgemeinen Chemie*, vol. 2, part 1. (Leipzig, Germany: Engelmann, 1896).
82. A.E. Nielsen, "Nucleation and Growth of Crystals at High Supersaturation," *Kristall und Technik* 4 (1969): pp. 17–38.
83. J.W. Mullin, *Crystallization* (Oxford: Butterworth-Heinemann, 2001).
84. O. Bello, "Calcium Carbonate Scale Deposition Kinetics on Stainless Steel Surfaces," (PhD thesis, University of Leeds, 2017).
85. Z. Dai, Y. Lu, A. Kan, et al., "A Mechanistic Software Platform for Mineral Surface Deposition and Inhibition Prediction Under Different Flow Conditions," *Desalination* 509 (2021): p. 115071.
86. A.Y.-T. Lu, K. Harouaka, S. Paudyal, et al., "Kinetics of Barium Sulfate Deposition and Crystallization Process in the Flowing Tube," *Industrial and Engineering Chemistry Research* 59 (2020): pp. 7299–7309.
87. A.Y.-T. Lu, W. Shi, J. Wang, et al., "The Mechanism of Barium Sulfate Deposition Inhibition and the Prediction of Inhibitor Dosage," *Journal of Chemical and Engineering Data* 64 (2019): pp. 4968–4976.
88. K. Zuo, X. Zhang, X. Huang, et al., "Ultrahigh Resistance of Hexagonal Boron Nitride to Mineral Scale Formation," *Nature Communications* 13 (2022): p. 4523.
89. O. Söhnel, J.W. Mullin, "Interpretation of Crystallization Induction Periods," *Journal of Colloid and Interface Science* 123 (1988): pp. 43–50.
90. F. Wang, V.N. Richards, S.P. Shields, et al., "Kinetics and Mechanisms of Aggregative Nanocrystal Growth," *Chemistry of Materials* 26 (2014): pp. 5–21.
91. E.M. Pouget, P.H.H. Bomans, J.A.C.M. Goos, et al., "The Initial Stages of Template-Controlled $CaCO_3$ Formation Revealed by Cryo-TEM," *Science* 323 (2009): pp. 1455–1458.
92. D. Gebauer, M. Kellermeier, J.D. Gale, et al., "Pre-Nucleation Clusters as Solute Precursors in Crystallisation," *Chemical Society Reviews* 43 (2014): pp. 2348–2371.
93. D. Gebauer, H. Cölfen, "Prenucleation Clusters and Non-Classical Nucleation," *Nano Today* 6 (2011): pp. 564–584.
94. D. Erdemir, A.Y. Lee, A.S. Myerson, "Nucleation of Crystals From Solution: Classical and Two-Step Models," *Accounts of Chemical Research* 42 (2009): pp. 621–629.
95. D. Knezic, J. Zaccaro, A.S. Myerson, "Nucleation Induction Time in Levitated Droplets," *Journal of Physical Chemistry B* 108 (2004): pp. 10672–10677.
96. A.E. Nielsen, "Mechanisms and Rate Laws in Electrolyte Crystal-Growth From Aqueous-Solution," *ACS Symposium Series* 323 (1986): pp. 600–614.
97. A.G. Christy, A. Putnis, "The Kinetics of Barite Dissolution and Precipitation in Water and Sodium Chloride Brines at 44–85°C," *Geochimica et Cosmochimica Acta* 57 (1993): pp. 2161–2168.
98. S.-T. Liu, G.H. Nancollas, "The Crystal Growth and Dissolution of Barium Sulfate in the Presence of Additives," *Journal of Colloid and Interface Science* 52 (1975): pp. 582–592.

99. C.H. Bovington, A.L. Jones, "Tracer Study of the Kinetics of Dissolution of Barium Sulphate," *Transactions of the Faraday Society* 66 (1970): pp. 764–768.
100. J.R. Campbell, G.H. Nancollas, "Crystallization and Dissolution of Strontium Sulfate in Aqueous Solution," *Journal of Physical Chemistry* 73 (1969): pp. 1735–1740.
101. P.M. Dove, C.A. Czank, "Crystal Chemical Controls on the Dissolution Kinetics of the Isostructural Sulfates: Celestite, Anglesite, and Barite," *Geochimica et Cosmochimica Acta* 59 (1995): pp. 1907–1915.
102. T.M. Davis, T.O. Drews, H. Ramanan, et al., "Mechanistic Principles of Nanoparticle Evolution to Zeolite Crystals," *Nature Materials* 5 (2006): pp. 400–408.
103. T.J. Woehl, C. Park, J.E. Evans, et al., "Direct Observation of Aggregative Nanoparticle Growth: Kinetic Modeling of the Size Distribution and Growth Rate," *Nano Letters* 14 (2014): pp. 373–378.
104. E.M. Hotze, T. Phenrat, G.V. Lowry, "Nanoparticle Aggregation: Challenges to Understanding Transport and Reactivity in the Environment," *Journal of Environmental Quality* 39 (2010): pp. 1909–1924.
105. M. Elimelech, J. Gregory, X. Jia, *Particle Deposition and Aggregation: Measurement, Modelling and Simulation* (Oxford: Butterworth-Heinemann, 2013).
106. H.G. Liao, K. Niu, H. Zheng, "Observation of Growth of Metal Nanoparticles," *Chemical Communications* 49 (2013) 11720–11727.
107. R.L. Penn, "Kinetics of Oriented Aggregation," *Journal of Physical Chemistry B* 108 (2004): pp. 12707–12712.
108. J.M. Yuk, J. Park, P. Ercius, et al., "High-Resolution EM of Colloidal Nanocrystal Growth Using Graphene Liquid Cells," *Science* 336 (2012): pp. 61–64.
109. J. Chai, X. Liao, L.R. Giam, et al., "Nanoreactors for Studying Single Nanoparticle Coarsening," *Journal of the American Chemical Society* 134 (2012): pp. 158–161.
110. Y. Hu, Q. Li, B. Lee, et al., "Aluminum Affects Heterogeneous Fe(III) (Hydr)oxide Nucleation, Growth, and Ostwald Ripening," *Environmental Science and Technology* 48 (2014): pp. 299–306.
111. F. Huang, H. Zhang, J.F. Banfield, "Two-Stage Crystal-Growth Kinetics Observed During Hydrothermal Coarsening of Nanocrystalline ZnS," *Nano Letters* 3 (2003): pp. 373–378.
112. C. Wagner, "Theorie der alterung von niederschlägen durch umlösen (Ostwald reifung)," *Zeitschrift für Elektrochemie, Berichte der Bunsengesellschaft für physikalische Chemie* 65 (1961): pp. 581–591.
113. E. Akyol, M. Öner, E. Barouda, et al., "Systematic structural determinants of the effects of tetraphosphonates on gypsum crystallization," *Crystal Growth & Design* 9 (2009): pp. 5145–5154.
114. E. Stamatakis, A. Stubos, J. Palyvos, et al., "An Improved Predictive Correlation for the Induction Time of CaCO3 Scale Formation During Flow in Porous Media," *Journal of Colloid and Interface Science* 286 (2005): pp. 7–13.
115. M. Van der Leeden, D. Kashchiev, G. Van Rosmalen, "Precipitation of Barium Sulfate: Induction Time and the Effect of an Additive on Nucleation and Growth," *Journal of Colloid and Interface Science* 152 (1992): pp. 338–350.
116. A.E. Nielsen, "Electrolyte Erystal-Growth Mechanisms," *Journal of Crystal Growth* 67 (1984): pp. 289–310.
117. Z. Dai, F. Zhang, A.T. Kan, et al., "Two-Stage Model Reveals Barite Crystallization Kinetics from Solution Turbidity," *Industrial and Engineering Chemistry Research* 58 (2019): pp. 10864–10874.
118. T.A. Bashkatova, A.N. Bashkatov, V.I. Kochubey, et al., "Light-Scattering Properties for Spherical and Cylindrical Particles: A Simple Approximation Derived from Mie Calculations," Saratov Fall Meeting 2000: Optical Technologies in Biophysics and Medicine II held 2001, SPIE. pp. 247–259.

119. C.F. Bohren, D.R. Huffman, *Absorption and Scattering of Light by Small Particles* (John Wiley & Sons, 2008).pp.57-81,158-165
120. H.C. Hulst, H. Van De Hulst, *Light Scattering by Small Particles* (Chelmsford: Courier Corporation, 1957).pp.1470-1471
121. J.W. Mullin, H.M. Ang, "Nucleation Characteristics of Aqueous Nickel Ammonium Sulphate Solutions," *Faraday Discussions of the Chemical Society* 61 (1976): pp. 141–148.
122. C. Yan, A.T. Kan, F. Zhang, et al., "Systematic Study of Barite Nucleation and Inhibition With Various Polymeric Scale Inhibitors By Novel Laser Apparatus," *SPE Journal* 20 (2015): pp. 642–651.
123. O. Söhnel, J. Mullin, "Kinetics of Precipitation of Nickel Ammonium Sulphate Hexahydrate From Aqueous Solutions," *Kristall und Technik* 14 (1979): pp. 217–228.
124. Z. Dai, F. Zhang, N. Bhandari, et al., "Development and Application of a New Theoretical Model for Additive Impacts on Mineral Crystallization," *Crystal Growth & Design* 17 (2017): pp. 4006–4014.
125. Z. Dai, Y. Zhao, S. Paudyal, et al., "Gypsum Scale Formation and Inhibition Kinetics With Implications in Membrane System," *Water Research* 225 (2022): p. 119166.
126. M.M. Vazirian, T.V.J. Charpentier, M. de Oliveira Penna, et al., "Surface inorganic scale formation in oil and gas industry: As adhesion and deposition processes," *Journal of Petroleum Science and Engineering* 137 (2016): pp. 22–32.
127. A.C. Lasaga, *Kinetic Theory in The Earth Sciences* (New Jersey:Princeton University Press, 2014). pp.528-550
128. Z. Amjad, K.D. Demadis, *Mineral Scales and Deposits: Scientific and Technological Approaches* (Amsterdam:Elsevier, 2015).pp.36-61
129. W.C. Cheong, P.H. Gaskell, A. Neville, "Substrate Effect on Surface Adhesion/Crystallisation Of Calcium Carbonate," *Journal of Crystal Growth* 363 (2013): pp. 7–21.
130. E. Mavredaki, A. Neville, "Prediction and Evaluation of Calcium Carbonate Deposition at Surfaces," SPE-169796-MS, SPE International Oilfield Scale Conference and Exhibition, held May 14–15, 2014 (Aberdeen, UK: SPE).
131. X. Huang, C. Li, K. Zuo, et al., "Predominant Effect Of Material Surface Hydrophobicity On Gypsum Scale Formation," *Environmental Science and Technology* 54 (2020): pp. 15395–15404.
132. K. Ramstad, K. Sandengen, A.F. Mitchell, et al., "Correlating Calcium Carbonate Scale Risk With Field Experience Data," SPE-200679-MS, SPE International Oilfield Scale Conference and Exhibition, held June 24–25, 2020 (SPE.Virtual).
133. D. Gebauer, H. Cölfen, A. Verch, et al., "The Multiple Roles of Additives in $CaCO_3$ Crystallization: A Quantitative Case Study," *Advanced Materials* 21 (2009): pp. 435–439.
134. M. Tomson, G. Fu, M. Watson, et al., "Mechanisms of Mineral Scale Inhibition," *SPE Production & Facilities* 18 (2003): pp. 192–199.
135. W. Burton, N. Cabrera, F. Frank, "The Growth of Crystals and the Equilibrium Structure of Their Surfaces," *Philosophical Transactions of the Royal Society A: Mathematical, Physical and Engineering Sciences* 243 (1951): pp. 299–358.
136. G. Bliznakow, "Crystal habit and adsorption of cosolutes," *Fortsch Min.* 36 (1958): p. 149.
137. A. Chernov, "The Spiral Growth of Crystals," *Physics-Uspekhi* 4 (1961): pp. 116–148.
138. J.J. De Yoreo, P.G. Vekilov, "Principles of Crystal Nucleation and Growth," *Reviews in Mineralogy and Geochemistry* 54 (2003): pp. 57–93.
139. J.W. Zhang, G.H. Nancollas, "Mechanisms of Growth and Dissolution of Sparingly Soluble Salts," *Reviews in Mineralogy and Geochemistry* 23 (1990): pp. 365–396.
140. A. Zieba, G. Sethuraman, F. Perez, et al., "Influence of Organic Phosphonates on Hydroxyapatite Crystal Growth Kinetics," *Langmuir* 12 (1996): pp. 2853–2858.

141. J.G. Eberhart, "The Surface Tension of Binary Liquid Mixtures," *Journal of Physical Chemistry* 70 (1966): pp. 1183–1186.

142. Y. Liu, Z. Dai, A.T. Kan, et al., "Investigation of Sorptive Interaction Between Phosphonate Inhibitor and Barium Sulfate for Oilfield Scale Control," *Journal of Petroleum Science and Engineering* 208 (2022): p. 109425.

143. M.B. Tomson, "Effect of Precipitation Inhibitors on Calcium Carbonate Scale Formation," *Journal of Crystal Growth* 62 (1983): pp. 106–112.

144. K. Davis, P. Dove, J. De Yoreo, "Resolving the Controversial Role of Mg2+ in Calcite Biomineral Formation," *Science* 290 (2000): pp. 1134–1137.

145. W. Van Enckevort, A. Van den Berg, "Impurity Blocking of Crystal Growth: A Monte Carlo Study," *Journal of Crystal Growth* 183 (1998): pp. 441–455.

146. R.B. Baird, A.D. Eaton, E.W. Rice, *Standard Methods for the Examination of Water and Wastewater*, 23rd ed. (Washington DC: American Public Health Association, 2017).pp.3-33

147. NACE International, *Standard test Method: Laboratory Screening Test To Determine the Ability of Scale Inhibitors To Prevent the Precipitation of Barium Sulfate and/or Strontium Sulfate From Solution (For Oil and Gas Production Systems)* (Houston: NACE International, 2002).

148. NACE International, *Standard Test Method: Laboratory Screening Tests To Determine the Ability of Scale Inhibitors To Prevent the Precipitation of Calcium Sulfate and Calcium Carbonate From Solution (For Oil and Gas Production Systems)* (Houston: NACE International, 2001).

149. International Organization for Standardization, *ISO 7027-1:2016: Water Quality–Determination of Turbidity* (Switzerland: IOS,2016).pp.2-6

150. J.W. O'Dell, *Method 180.1: Determination of Turbidity by Nephelometry* (Washington DC: USEPA, 1996).pp.2-7

151. J. Xiao, A. Kan, M. Tomson, "Prediction of $BaSO_4$ Precipitation in the Presence and Absence of a Polymeric Inhibitor: Phosphino-Polycarboxylic Acid," *Langmuir* 17 (2001): pp. 4668–4673.

152. G. Deng, A.T. Kan, Z. Dai, et al., "Impact of High Calcium Concentration on Sulfate Scale Prediction at High Temperature From 120 °C to 220 °C," SPE-190744-MS, SPE International Oilfield Scale Conference and Exhibition, held June 20–21, 2018 (Aberdeen, UK:SPE).

153. H. Lu, Z. Gu, T. Underwood, et al., "The Development of Novel Laboratory Test Method on Evaluation of Scale Inhibition and Dispersancy for Downstream Water Treatment Applications," NACE-2020-14491, Corrosion Annual Conference, held June 14–18, 2020 (NACE International.Virtual).

154. F. Zhang, C.J. Hinrichsen, A.T. Kan, et al., "Calcium Sulfate Scaling Risk and Inhibition for a Steamflood Project," *SPE Journal* 22 (2016): pp. 881–891.

155. Y.-S. Jun, B. Lee, G.A. Waychunas, "In Situ Observations of Nanoparticle Early Development Kinetics at Mineral–Water Interfaces," *Environmental Science and Technology* 44 (2010): pp. 8182–8189.

156. J.E. Oddo, M.B. Tomson, *Elevated Temperature-Pressure Flow Simulator, U.S. Patent No. 5,370,799.* (US.1994).

157. J.J. De Yoreo, "In-Situ Liquid Phase TEM Observations of Nucleation and Growth Processes," *Progress in Crystal Growth and Characterization of Materials* 62 (2016): pp. 69–88.

158. Z. Liu, Z. Zhang, Z. Wang, et al., "Shape-Preserving Amorphous-to-Crystalline Transformation of $CaCO_3$ Revealed By In Situ TEM," *Proceedings of the National Academy of Sciences* 117 (2020): pp. 3397–3404.

159. J.J. De Yoreo, G.A. Waychunas, Y.-S. Jun, et al., "In Situ Investigations of Carbonate Nucleation on Mineral and Organic Surfaces," *Reviews in Mineralogy and Geochemistry* 77 (2013): pp. 229–257.

160. T.A. Land, J.J. De Yoreo, J.D. Lee, "An In-Situ AFM Investigation of Canavalin Crystallization Kinetics," *Surface Science* 384 (1997): pp. 136–155.
161. A.J. Malkin, Y.G. Kuznetsov, A. McPherson, "In Situ Atomic Force Microscopy Studies of Surface Morphology, Growth Kinetics, Defect Structure and Dissolution in Macromolecular Crystallization," *Journal of Crystal Growth* 196 (1999): pp. 471–488.
162. L.E. Wasylenki, P.M. Dove, D.S. Wilson, et al., "Nanoscale Effects of Strontium on Calcite Growth: An In Situ AFM Study in the Absence of Vital Effects," *Geochimica et Cosmochimica Acta* 69 (2005): pp. 3017–3027.
163. A. Janshoff, C. Steinem, eds., *Piezoelectric Sensors* (Berlin: Springer Berlin Heidelberg, 2007). pp.117-119
164. N.H. Lin, Y. Cohen, "QCM Study of Mineral Surface Crystallization on Aromatic Polyamide Membrane Surfaces," *Journal of Membrane Science* 379 (2011): pp. 426–433.
165. I.G. Sedeva, R. Fetzer, D. Fornasiero, et al., "Adsorption of Modified Dextrins To a Hydrophobic Surface: QCM-D Studies, AFM Imaging, and Dynamic Contact Angle Measurements," *Journal of Colloid and Interface Science* 345 (2010): pp. 417–426.
166. Z. Dai, B. Ouyang, X. Wang, et al., "From Water Analysis to Scale and Corrosion Control Improvements: A Permian Basin Example," *Chemical Geology* 610 (2022): p. 121077.
167. Y. Kharaka, M.S. Lico, W.W. Carothers, "Predicted Corrosion and Scale-Formation Properties of Geopressured Geothermal Waters From the Northern Gulf of Mexico Basin," *Journal of Petroleum Technology* 32 (1980): pp. 319–324.
168. M.S. Blondes, K.D. Gans, M.A. Engle, et al., *U.S. Geological Survey National Produced Waters Geochemical Database v2.3* (Reston, VA: USGS,2019).

3 Role of PVT in Scale Modeling

Baard Kaasa[1]
[1]Scale Consult AS, Kongsberg, Norway

3.1 Introduction

Scale can form in any system containing salty water if the ion concentrations, pressure, and temperature are such that the system exceeds the solubility of one or several scale types. If there is no oil or gas present, the scale risk can be modeled by solving all chemical equilibria in the water phase. It is, of course, necessary to correct for nonideal behavior with activity coefficients and calculate all parameters at the correct pressure and temperature and to consider a gas or an oil phase and equilibrium between the phases.

For all systems where an oil and/or gas phase is present, it is necessary to include all phases in the modeling, because many of the important components will be distributed between the phases. Figure 3.1 shows a schematic illustration of a multiphase system. If we consider CO_2, it is necessary to calculate the phase equilibrium for CO_2 between the gas, oil and water, and simultaneously calculate how much CO_2 has dissociated to bicarbonate and carbonate, and if there is precipitation, how much CO_2 has precipitated, to solve the total mass balance for CO_2. Similar for H_2S, CH_4, and all other components. Some components will be mainly in the gas phase, some will be mainly in the oil phase, and some will be mainly in the water phase. For example, most of the heavy HCs will be in the oil phase, and it is safe to assume that ions will be in the water phase. But assuming that all ions will be in the water phase does not mean that the concentrations are constant and known, because the volume of water may of course change due to evaporation of water to the oil and gas phases.

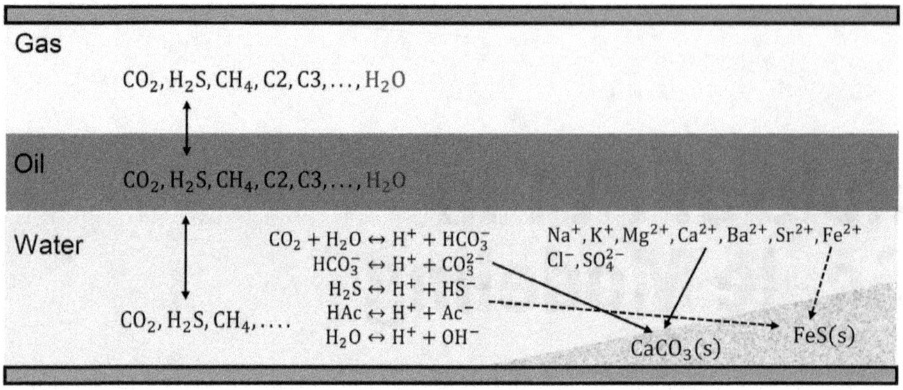

Figure 3.1. Components distribution between the phases potentially influence the chemical equilibria in the water phase, pH, and scale potentials.

To calculate the amount of CO_2 and H_2S in the aqueous phase in dissociated form, it is necessary to calculate pH simultaneously, and it is also necessary to include all organic acids and other possible acids/bases in the system, such as pH stabilization chemicals or H_2S scavengers.

To model the scale risk in oil and gas wells, it is therefore not sufficient to model the water phase only, it is necessary to include the oil and gas phases and model phase distribution simultaneously as all the equilibria in the water are calculated. Without calculating the phase distribution, for example, the phase distribution of CO_2, it is not possible to calculate pH in the water or the carbonate concentration and therefore scale potential for carbonate minerals. The same is also true for sulfides and hydroxides. It can be argued that sulfate minerals, such as barite ($BaSO_4$) or anhydrite ($CaSO_4$), are pH-independent and that it is not necessary to calculate phase distribution of CO_2 and H_2S to calculate scale potential for these types of minerals. However, for systems with low water cut or high-pressure high-temperature (HPHT) gas systems, water evaporation or condensation can change the ion concentrations in the aqueous phase and hence also the scale risk. Water evaporation due to pressure reduction is often one of the main reasons that halite (NaCl) forms in gas wells. To model water evaporation, it is necessary to include a PVT model and calculate phase distribution of water in the scale modeling.

This chapter will look further into the necessity of using a PVT model to model scale risk in multiphase systems, including the oil and gas phases, in scale modeling. The chapter will be divided into two main sections:

1. A theory section that will:
   - Explain the thermodynamic and chemical fundamentals of multiphase systems.
   - Explain what an equation of state (EoS) is, which parameters are needed and briefly how it is solved.

- Explain how oil and gas samples are taken, analyzed, and characterized to obtain the necessary input data to the EoS.

2. A simulation section: This section will show how varying PVT will influence the scale potential in different systems. The focus will be on typical oil systems. Scale modeling in gas systems will be discussed in Chapter 4.

Note that the theory part is meant as background information, and it is not necessary to read it to understand the second section. It is therefore possible to jump directly to Section 3.5 to read about how the PVT will influence scale simulations. However, if you are planning to perform scale modeling in systems containing oil or gas, it is recommended to read Sections 3.2–3.4 to better understand how the PVT model works and how it will be influenced by the input data given to the PVT model.

3.2 Multiphase Equilibria

3.2.1 Multiphase Systems

In most oil and gas wells, there are up to three phases present:

1. A water phase, also called the aqueous phase. The water phase contains all ions and, normally, most of the organic acids. It also contains dissolved gases/HCs such as CO_2, H_2S, CH_4, nitrogen, ethane, propane, and heavier HCs. The water phase is not always pure water (as solvent) but could also contain hydrate inhibitors such as methanol or MEG. When a hydrate inhibitor is present, it is normal to consider the water-methanol or water-MEG phase as the "solvent" and called as aqueous phase.
2. An oil phase: This is the liquid/heavy HC phase and contains most of the heavy HC components which typically mean heptane/hexane and heavier.
3. A gas phase: This is the gaseous HC phase and contains most of the light HC components.

All HCs will be distributed into all phases, and there will be heavy HCs in the gas phase and light/gaseous HCs in the liquid phase. The phase distribution varies with the change of pressure and temperature, and if the pressure is high, there will normally be only one HC phase containing all the HCs.

Very often, the number of phases will change when fluids flow from reservoir to surface. For example,

- The HC phase can be the liquid oil in the reservoir. When oil and water flow from the reservoir into the well and toward surface, pressure, and temperature change, and if/when the oil reaches the bubble point, a gas phase is formed, and the system becomes a three-phase flow.
- Alternatively, the reservoir may contain a gas phase. Water and gas can flow from the reservoir into the well, but as pressure decreases, a condensate phase can be formed (retrograde condensation), and the system becomes a three-phase system.

- A reservoir could also contain oil with a gas cap, and gas from the gas cap can come into the well together with the oil and water, meaning that there is a three-phase flow from the reservoir to the surface.
- In a HPHT well, gas and water (two-phase system) may flow from the reservoir and into the well, but due to high temperature and large pressure drop, the water phase may evaporate completely and disappear such that the system goes from a two-phase (gas + water) to a single gas phase system.
- When water is drained from a separator to the degasser, there will often be a large pressure drop, and gas flashes out of the water to form a gas phase.

These are all examples of how the number of phases can both increase and decrease when fluids flow from the reservoir to the surface or in a flowline or through a topside process.

3.2.2 Phase Equilibrium

When there are two or more phases present, equilibrium is defined as

- All phases have the same pressure and temperature.
- All components have the same "chemical energy" in all phases.

The first requirement is obvious; the second requires more explanation. The requirement that all components have the same chemical energy, or more precisely, *chemical potential*, means that it is not possible to "release" energy by moving a component from one phase to the other. For example, if CO_2 had higher "energy" in the gas phase than in the water phase, it would move from the gas phase to the water phase and thereby release energy. For example, if the pressure increases, the components in the gas phase will get higher energy and will dissolve into the other phases. Nature will always move in the direction of lowest energy, or more correctly, evenly distributed energy, and this is the same as saying that all components should have the same chemical energy/potential in all phases.

The chemical potential of a component in a mixture is given by

$$\mu_i(T,P) = \mu_i^\circ\left(T,P^\circ\right) + RT\ln\frac{f_i}{P^\circ} \tag{3.1}$$

The chemical potential in a mixture at P and T equals a reference chemical potential (normally a pure component as an ideal gas at a reference pressure) plus a term containing *fugacity* of the component in the mixture. The fugacity is defined as:

$$f_i = \varphi_i y_i P \tag{3.2}$$

which is the partial pressure multiplied with a *fugacity coefficient* to correct for nonideal behavior. Here, the partial pressure is expressed using mole fraction y which is normally used for a gas phase, but the definition of fugacity in the oil phase is identical, using the mole fraction in the oil phase, normally denoted as x. Equilibrium between phases is when all components have the same chemical potential in all phases. For a component i, we can write

$$\mu_i^{Oil} = \mu_i^{Gas} = \mu_i^{Aq} \tag{3.3}$$

We can now insert the definition of the chemical potential for each phase.

$$\mu_i^{\circ} + RT \ln \frac{f_i^{Oil}}{P^{\circ}} = \mu_i^{\circ} + RT \ln \frac{f_i^{Gas}}{P^{\circ}} = \mu_i^{\circ} + RT \ln \frac{f_i^{Aq}}{P^{\circ}} \tag{3.4}$$

The reference fugacity is the same for all phases and can be removed. Similarly, the RT term and reference pressure are also equal for all phases and can be removed. Finally, we can also remove the logarithm and the final equilibrium expression becomes that all components have the same fugacity in all phases.

$$f_i^{Oil} = f_i^{Gas} = f_i^{Aq} \tag{3.5}$$

If we look at the oil-gas equilibrium and insert the definition of fugacity, we get

$$\varphi_i^{Oil} x_i P = \varphi_i^{Gas} y_i P \tag{3.6}$$

This can be further simplified by dividing by pressure and we obtain a definition for equilibrium between oil and gas.

$$\varphi_i^{Oil} x_i = \varphi_i^{Gas} y_i \tag{3.7}$$

We can now define an equilibrium constant, often referred to as "K value" for component i between oil and gas.

$$K_i = \frac{y_i}{x_i} = \frac{\varphi_i^{Oil}}{\varphi_i^{Gas}} \tag{3.8}$$

The ratio of the mole fractions for a component in the gas phase vs. the oil phase equals the inverse ratio of the fugacity coefficients, and this is often used in calculation of phase equilibrium. To calculate phase equilibrium, it is possible to use models that predict the K values for the components, but a more common approach is to use an EoS to calculate the fugacity coefficients in the oil and gas phases, and then calculate the K values as the ratio between them and solve the necessary set of equations to calculate phase equilibrium (see Section 3.3.5 below on flash calculations).

3.3 Equation of State

An EoS is an equation that can describe the properties of a fluid (gas or liquid) for a given composition at a given set of conditions. The "conditions" normally mean at a given pressure and temperature, but it is also possible to define, for example, temperature and volume or temperature and enthalpy. There are many EoS models, but it is outside the scope of this chapter to give a detailed description of all of them. In this chapter, we will only explain the ideal gas law and briefly explain the Soave-Redlich-Kwong (SRK) equation which is one of the most widely used EoS models.

3.3.1 Ideal Gas

An ideal gas is defined as a gas where:

- The "particles" do not have volume.
- The "particles" do not interact with each other (no attractive or repulsive forces).
- The "particles" collide perfectly elastic.

With these assumptions, it is possible to derive a theoretical law describing the relation between pressure, volume, and temperature for a gas, and it is called the ideal gas law:

$$PV = nRT \tag{3.9}$$

where P is the pressure, V is the volume, n is number of moles, R is the gas constant, and T is the temperature in Kelvin. The value of the gas constant R depends on the units selected for pressure and temperature. If pressure is in Pascal (Pa) and volume is in cubic meters (m^3), the gas constant is 8.314 J/K×mole. The ideal gas law is accurate for gases at low pressure (below ca. 10–20 bar), and it is also better at higher temperatures. The ideal gas law can, however, not predict properties of a liquid (because it can only predict behavior of a gas), and it will never predict a two-phase solution. Therefore, it should not be used to describe a real oil-gas system.

3.3.2 Real Gases

In real gases at higher pressure, there will be a deviation from ideal behavior, and to correct for this, one can introduce the *compressibility*,

$$PV = ZnRT \tag{3.10}$$

The challenge is to know what compressibility is. There are several models for calculation of compressibility, for example, some virial equations express Z using several exponential terms. A more common approach for oil/gas systems is, however, to use a cubic EoS.

3.3.3 Cubic Equation of State

The most common approach today is to use a *cubic EoS*. There have been several EoSs proposed over the years; one of the most commonly used is the SRK EoS,[1] often called SRK. Another very popular EoS is the one developed by Peng and Robinson,[2] often called PR. More information about various equations of state is available in books edited by Pedersen[3] and Michelsen.[4] These equations attempt to correct the assumptions that particles have no volume by introducing a b-factor that corrects the volume and an a-factor to represent interaction (attractive or repulsive forces) between the molecules. The SRK equation is given by

$$P = \frac{RT}{v-b} - \frac{a}{v(v+b)} \tag{3.11}$$

The PR equation is very similar; the only difference is that the denominator in the second term is $v(v + b) + b(v - b)$. The reason that these equations are called cubic is that if one tries to solve them with respect to volume for a given P and T, the equation will be in the form $dv^3 + ev^2 + fv + g = 0$, where the constants d-g are expressions containing P, R, T, a, and b, and since this is a third-order equation, the equation is called cubic. This also means that for a given set of P and T (below the critical point), there will be three solutions to the equation.

The values of a and b can be determined from the critical point and for the SRK equation, they are given by

$$b = 0.08664\frac{RT_c}{P_c} \quad \text{and} \quad a = \alpha(T)0.42748\frac{R^2T_c^2}{P_c} \tag{3.12}$$

$$\alpha(T) = \left[a + m(\omega)\left(1 - \sqrt{\frac{T}{T_c}}\right)\right]^2 \quad \text{with} \quad m(\omega) = 0.480 + 1.57\omega - 0.175\omega^2 \tag{3.13}$$

The only three parameters needed for a component to solve the SRK equation are: critical pressure (P_c), critical temperature (T_c), and the acentric factor (ω). These parameters are well known for all defined components such as CO_2, H_2S, CH_4, ethane, propane, and so on.

For mixtures like natural gases and oils, it is necessary to introduce mixing rules. There are many different mixing rules that have been proposed over the years. The most common mixing rules are:

$$b_{mix} = \sum_i z_i b_i \quad \text{and} \quad a_{mix} = \sum_i \sum_j z_i z_j \sqrt{a_i a_j}\left(1 - k_{ij}\right) \tag{3.14}$$

where k_{ij} is the empirical binary interaction coefficient between the components i and j and must be determined from the experimental data for the various binary systems.

3.3.4 Solving the Equation of State

To solve the EoS, one has to first calculate a and b parameters, and then solve the EoS. This is not a trivial task, because it is a cubic equation, and under certain conditions (below the critical point), it will have three solutions, giving three possible roots (volumes), where the lower root corresponds to a liquid solution and the highest root corresponds to a vapor solution. To determine which one is correct, it is necessary to compare the Gibbs energy of the solutions, and the one with the lowest Gibbs energy is the correct solution. There are several methods to solve the EoS, but it is beyond the scope of this chapter to explain all the details. A detailed description of a simple and general method to solve the EoS is given by Michelsen.[4]

To calculate Gibbs energy, it is necessary to calculate the fugacity coefficients of all components. They can be found from

$$RT\ln\phi_i = -\int_\infty^V \left(\left(\frac{\partial P}{\partial n_i}\right)_{T,V,n_j} - \frac{RT}{V}\right)dV - RT\ln Z \tag{3.15}$$

where P represents the EoS, for example, the SRK equation. Solving this integral is not trivial and a detailed explanation is beyond the scope of this chapter. A general and systematic approach for solving fugacity coefficients and other thermodynamic properties from equations of state in general is given by Michelsen.[4]

3.3.5 Flash Calculations

It is, however, not sufficient to solve the EoS for the total composition, because HC systems will often have both oil and gas. It is then necessary to perform what is called a flash calculation. A flash calculation is performed to

- Determine whether the given fluid composition will be single phase or split into two phases (oil or gas) or perhaps several phases.
- Calculate compositions and properties of the phases.

When there are two phases, we can set up a material balance for each component:

$$n_i^{Gas} + n_i^{Oil} = n_i^{Total} \tag{3.16}$$

or

$$y_i N^{Gas} + x_i N^{Oil} = z_i N^{Total} \tag{3.17}$$

Dividing by total number of moles gives

$$\beta y_i + (1-\beta) x_i = z_i \quad \text{where } \beta = \frac{N^{Gas}}{N^{Total}} \tag{3.18}$$

If we do not introduce the gas-oil equilibrium constant defined in, we can express the mole fraction in the oil and gas phases as

$$x_i = \frac{z_i}{1-\beta+\beta K_i} \quad \text{and} \quad y_i = \frac{K_i z_i}{1-\beta+\beta K_i} \tag{3.19}$$

Further, we know that the sum of mole fractions must equal unity for each phase, we can therefore write

$$\sum_i (y_i - x_i) = 0 \tag{3.20}$$

Inserting the expressions for the respective mole fractions gives the Newton-Rapson formula:

$$\sum_i \frac{z_i (K_i - 1)}{1-\beta+\beta K_i} = 0 \tag{3.21}$$

If we have an estimate for the K values, the gas fraction can be found by solving Equation (3.21) as $F(\beta) = 0$ numerically, and then find the phase compositions from Equation (3.19). These phase compositions are then used to calculate fugacity coefficients and then update the K values with Equation (3.8). A good starting estimate for K values is Wilson's approximation:

$$\ln K_i = \ln\left(\frac{P_{ci}}{P}\right) + 5.373(1+\omega_i)\left(1-\frac{T_{ci}}{T}\right) \tag{3.22}$$

where P_{ci}, T_{ci}, and ω_i are the critical pressure, critical temperature, and acentric factor described in Section 3.3.3. For a typical HC mixture, the above iterative sequence will converge rapidly if the conditions are sufficiently far away from the critical point. If the conditions are such that the fluid corresponds to a single-phase oil or gas, the iteration will converge to either $\beta \leq 0$ corresponding to a single liquid phase solution or to $\beta \geq 1$ corresponding to a single gas phase solution.

The above iteration is, however, sensitive to the initial estimate, and there is a risk that it converges to a single-phase solution even if there is a stable two-phase solution if the initially estimated K values are incorrect. The above iteration method will also be very slow close to the critical point. There are many improvements of the above procedure that can be implemented to improve speed and stability, and the book by Michelsen[4] is recommended for further reading.

3.4 Multiphase Scale Prediction Models

A scale prediction model needs to include a minimum of three phases: oil, gas, and water. Solid phases are technically also separate phases. They are normally regarded as pure solid phases which can be treated differently and not as part of the multiphase equilibrium model. Other phases, such as hydrate, wax and asphaltenes, can, in theory, also be included because formation of these phases will change the composition of the other phases.

Chapter 2 describes the aqueous models for the calculation of mineral solubility and nonideal behavior of the species in the water phase where, for example, the Pitzer model is used to calculate activity coefficients together with aqueous equilibria and equilibrium constants to calculate concentrations of various species and scale potential for different salts. Sections 3.2 and 3.3 describe how oil and gas phases can be modeled using an EoS. Ideally, having one single model that could model all phases would make modeling of multiphase systems easier. Unfortunately, no such model exists yet or at least no good models.

An EoS such as the SRK equation described in Section 3.3.3 will not be able to model a water phase very accurately, even if the water is pure water. However, there are modified EoS models that can model water and aqueous phases. For example, if the Huron Vidal mixing rule is used, the SRK equation can model the aqueous phase quite accurately, but only if it is without any ions. An even better model is the cubic plus association models (CPA), where an association term that can model hydrogen bonding is added to a standard EoS equation. The CPA models can accurately model the aqueous phase, even if it is a mixture of water and other polar components, for example, alcohols. The CPA equation can also, to some extent, be used in water containing dissolved salts, but it cannot model the solubility of scale and all the aqueous equilibria needed in scale modeling.

Several attempts have been made over the years to find a single model that can describe both the oil and the gas phases and the aqueous phase with all the aqueous equilibria and mineral solubilities. In a recent review article, Kontogerogis *et al.*[5] reviewed electrolyte EoSs and concluded that although some of the models give good results for phase equilibrium in three-phase systems even when the water contains various ions, they are not able to model mineral solubility accurately, especially at elevated temperatures.

An alternative approach is therefore to combine an aqueous model with equations needed for modeling pH, mineral solubility, and scale potential in the water phase with an oil-gas phase equilibrium model. This can be done by combining an EoS for the oil/gas phases with an aqueous model such as the Pitzer model or the UNIversal QUAsi-Chemical (UNIQUAC) model, as illustrated in Figure 3.2. Equilibrium between the HC phase (oil/gas) and the aqueous phase is calculated using Henry's law for solubility of HCs in water and Rault's law for solubility of water in the HC phase. An iterative procedure with successive solutions of the EoS for the HC phase, the aqueous model for the aqueous phase and then updating equilibrium between the phases must be repeated until convergence.

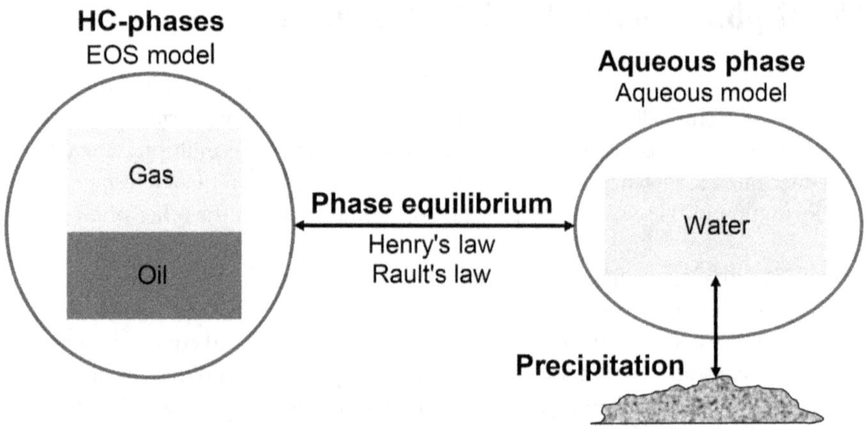

Figure 3.2. Multiphase modeling where an EoS model is combined with an aqueous model to calculate scale potential in multiphase systems.

In Section 3.2.2, it was defined that equilibrium between phases requires that fugacity for all components be the same in all phases, see Equation (3.5). This is also true if a system is modeled, as illustrated in Figure 3.2, with a combination of different models. The reason for introducing Henry's and Rault's law in the phase equilibrium calculations between the aqueous phase and the HC phase is that we will select a different reference state for the aqueous components in the aqueous phase than for the components in the HC phase. Rault's law will be used for the solvent (water), and the reference state is the pure solvent at its vapor pressure. Henry's law will be used for dissolved gases, and the reference state is the infinite dilution at the vapor pressure of the solvent.

If we simplify Rault's law and apply it to an ideal system, we could express it as follows: The partial pressure in the gas (or oil) phase equals the vapor pressure of the solvent (water). In a real system, we need to correct for nonideal behavior in the gas (or oil) phase and introduce fugacity. In the water phase, it is necessary to correct for presence of dissolved components, because pressure can be higher than the vapor pressure. The full Rault's law is given by

$$\varphi_{H_2O} y_{H_2O} P = a_{H_2O} \varphi_{H_sO}^{sat} P_{H_sO}^{sat} \exp\left[\int_{P^{sat}}^{P} \left(\frac{\bar{V}_{H_2O}}{RT}\right) dP\right] \tag{3.23}$$

where the left side is the fugacity of water in the gas phase. If there is no gas phase, this can be replaced by fugacity in the oil phase, since fugacity should be the same in both gas and water. a_{H_2O} is the water activity and corrects for the presence of other components and that the mole fraction of water may be smaller than 1. The water activity is calculated from the osmotic coefficient which again is calculated from a model describing nonideal behavior in the water phase, for example, the Pitzer model or a UNIQUAC model or similar. $\varphi_{H_sO}^{sat} P_{H_sO}^{sat}$ is the vapor pressure of water at the system temperature corrected with the fugacity coefficient of water vapor at its vapor pressure. The fugacity coefficient can be calculated with an EoS. The exponential term is the Poynting correction that corrects the vapor pressure to system pressure, where $\bar{V}_{H_2O}$ is the partial molar volume of water. With Rault's law, it is possible to calculate how much water there will be in the HC phase as a function of pressure and temperature, composition of the HC phase, and composition of the water phase.

Henry's constant is defined as the limiting value of the fugacity of a component divided by its concentration in the aqueous phase when the concentration goes toward zero. By integrating system pressure, we can derive the following equilibrium expression for components between the HC phase and the water phase:

$$\varphi_i y_i P = \gamma_i m_i H_{ij} \exp\left[\int_{P_j^s}^{P} \left(\frac{\bar{V}_i^{\infty}}{RT}\right) dP\right] \tag{3.24}$$

The left side is the fugacity of a component, for example, CO_2, in the gas (or oil) phase. The right side contains the activity of the same component in the water phase $a_i = \gamma_i m_i$, the Henry's law constant and the Pointing correction where $\bar{V}_i^{\infty}$ is the partial molar volume of the component, for example CO_2, in water at infinite dilution. Henry's law constant and the partial molar volume are found from experimental data, and the activity coefficient is calculated from, for example, the Pitzer model. When fugacity of a component is calculated for the gas/oil phase, the aqueous concentration can be calculated. More details about Henry's law are given by Carroll.[6]

3.5 Input Data to PVT Models

To model scale risk in a multiphase system, it is necessary to:

- Calculate phase distribution of CO_2, H_2S, and other HCs to find concentrations in the aqueous phase.

- Calculate water evaporation/condensation to find how much of the water in the well stream will be liquid water and can keep ions/salts in solution.
- Calculate phase distribution of organic acids, because, especially in gas wells, there can be significant amounts of organic acids in the gas phase.

To do this, it is necessary to calculate full phase distribution of all components in the well stream. This, of course, also means that it is necessary to calculate *which phases* are present. To do this, it is necessary to have a PVT model for calculation of phases and phase distribution coupled with the aqueous scale model to calculate all the equilibria, pH, and scale potentials in the aqueous phase.

The user of a scale simulation software does not have to know all the details about the flash calculation, how the EoS is solved and so on, but the resulting phase compositions, GOR, bubble point, and other properties of the oil and gas phases will of course influence the calculated scale potentials. This means that the results will depend on which PVT model is used, the input data given to the PVT model, and how this is done by the user. This means that the user of scale simulation software should at least have some basic knowledge of how the PVT model works and how the input data will influence the PVT model.

This chapter will give a brief introduction to:

- How PVT samples are taken
- How PVT samples are analyzed
- How a PVT report may look like and what data can be found
- How a fluid input is characterized before it is used in PVT calculations

3.5.1 PVT Sampling

PVT sampling means taking a sample of the oil and/or gas somewhere in the system. Oil companies will normally take samples during the drilling, completion, cleanup, or testing phases, but samples may also be taken when a well is in production. Examples of samples are:

- Samples of the reservoir fluid taken in the reservoir during drilling/completion.
- Surface samples of produced oil and gas taken from a test separator during well clean-up.
- Surface samples of produced oil and gas taken from test separator when the well is in production.

When a reservoir sample is taken, the sample contains the whole reservoir fluid and can be analyzed directly, and the result will be the composition of the reservoir fluid. If surface samples are taken as separate oil and gas samples, it is also necessary to measure the flow rates of oil and gas, so that oil and gas can be recombined in the correct ratio.

When a reservoir sample or oil sample from a separator is taken, it is important to transfer the fluid/oil to the sampling cylinder as a single phase without flashing, because if it flashes, there is a high risk that the oil and gas will not get into the sampling cylinder in the correct ratio. Oil samples are therefore taken on special cylinders, as the one shown in Figure 3.3. They are prefilled with either pressurized glycol or argon on one side of a moving piston. The cylinder is then connected to the sampling point, for example, the oil phase on a test separator, and the glycol/argon is then slowly bled out, so that the oil is filled into the cylinder at constant pressure. The cylinder is then closed and sent to the laboratory for analysis.

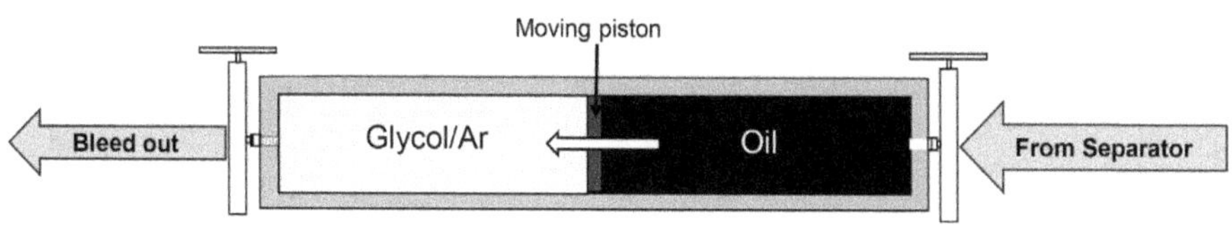

Figure 3.3. Schematic drawing of oil sample cylinder.

Gas samples are normally simply filled directly into a pre-evacuated gas sample bottle.

3.5.2 PVT Analysis and Reports

The two main analytical techniques used for analysis of oil and gas samples are:

- Gas chromatography (GC)
- True boiling point (TBP) distillation

If the sample is a reservoir fluid or a separator oil sample, it will first be flashed to ambient conditions to get a stabilized gas and a stabilized oil, as shown in Figure 3.4. The stabilized oil and gas are often referred to as stock tank oil and gas. The stabilized oil and gas can then be analyzed by GC. A GC analysis means that a small volume of gas or liquid is injected into a GC column together with a carrier gas. Different components will travel at different rates depending on molecular structure and will therefore be separated. The different components are then detected at the outlet and can be identified and quantified. A GC analysis is well suited for the gas sample but less suitable for oil samples due to the very large number of components. Oil samples can therefore be analyzed by a TBP analysis, where the oil is separated into boiling point cuts, and each cut is further analyzed by measuring weight%, molecular weight, and density. For example, when a PVT report states "C6," it does not mean hexane, but the boiling point cut that has a boiling point in a range close to hexane.

An example of a PVT report for an oil sample taken from a separator is given in Table 3.1. The oil sample has first been flashed as illustrated in Figure 3.4. The compositions of the resulting gas and oil phase were measured and reported as "stock tank gas" and "stock tank oil" in Table 3.1. It is common to report both weight% and mole%, where Table 3.1 has been simplified to show only mole%. The first components, nitrogen, carbon dioxide, and so on, up to normal-pentane, are pure components. The components C6–C9 are *pseudocom-*

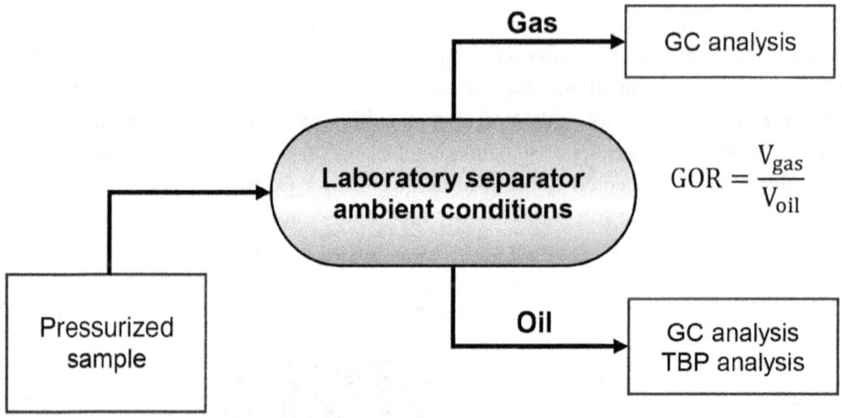

Figure 3.4. Analysis of a pressurized reservoir fluid sample.

ponents and are boiling point groups with boiling points in the range of hexane, heptane, octane, and so on. The last component on the list, in this case C10+, is "the rest" and represents all heavier components. The number of components that are reported in a PVT report can vary. Table 3.1 stops at C10+, but some PVT reports stop at C7+ or go to C12+ or C36+. Some PVT reports also split up the C6–C9 fractions into paraffinic, naphthenic, and aromatic components.

When the oil sample was flashed to ambient conditions, as shown in Figure 3.4, GOR and oil density were measured. It is then possible to mathematically recombine the gas and oil in correct ratios and *calculate* the recombined separator oil composition. If the sample had been a reservoir fluid sample, recombination of the gas and oil would yield the recombined reservoir fluid.

Table 3.1 shows the result of the analysis of a separator oil sample taken during a well test. Presumably, a gas sample was also taken, and this will be reported separately. Gas and oil rates measured at the test separator can then be used to recombine the separator gas and the recombined separator liquid to obtain the composition of the well stream.

Table 3.1. Example of PVT analysis report of a separator oil sample.

Component	Stock Tank Gas			Stock Tank Oil			Recombined Separator Liquid		
	Mole (%)	M_W (g/mol)	Density (kg/m³)	Mole (%)	MW (g/mol)	Density (kg/m³)	Mole (%)	M_W (g/mol)	Density (kg/m³)
Nitrogen	0.18						0.08		
Carbon dioxide	3.99						1.67		
Methane	49.01			0.08			20.55		
Ethane	17.36			0.4			7.5		
Propane	15.75			2.08			7.8		
iso-Butane	2.73			1			1.72		
n-Butane	6.08			3.49			4.57		
iso-Pentane	1.51			2.5			2.09		
n-Pentane	1.61			3.28			2.58		
C6	0.93	83.9	670.6	5.63	84.4	668.7	3.66	84.4	668.9
C7	0.64	89.6	748.5	9.2	91.8	737.6	5.62	91.7	738.1
C8	0.19	105.1	758.2	10.23	105.1	760.3	6.03	105.1	760.3
C9	0.02	120	764.1	6.72	120	769.7	3.92	120	769.7
C10+	0	156	-	55.39	261	874	32.21	261.3	874.1
Sum	100			100			100		

3.5.3 PVT Input and Fluid Characterization

A PVT report gives the *composition* of the fluid; to be able to use this with an EoS, it is necessary to know critical temperatures, critical pressures, and acentric factors for all components. For the pure components like nitrogen, CO_2, methane, and so on, this is straightforward, because these parameters are well known and listed. For the pseudocomponents C6–C9, these parameters are not known but can be calculated using correlations, for example, those suggested by Pedersen,[3] where the molecular weights and densities are used to calculate the critical parameters of the C6–C9 components.

$$T_c = c_1\rho + c_2 \ln M_w + c_3 M_w + \frac{c_4}{M_w} \tag{3.25}$$

$$\ln P_c = d_1 + d_2\rho^{d_5} + \frac{d_3}{M_w} + \frac{d_4}{M_w^2} \tag{3.26}$$

$$m = e_1 + e_2 M_w + e_3\rho + e_4 M_w^2 \tag{3.27}$$

where the parameters c_1–c_5, d_1–d_5, and e_1–e_4 are parameters determined from experimental PVT data, see Pedersen[3] for a detailed description. The molecular weights and densities that are entered for components C6–C9 will therefore have direct influence on how these components are modeled by the EoS and the properties of the oil and gas in the PVT model.

The C10+ fraction also requires further characterization to be used in the EoS model. A typical characterization routine is the one that is described by Pedersen[3] and consists of three steps:

1. Split C10+ up in a certain number of components, for example, C10–C80, where 10–80 represents carbon numbers.
2. Calculate properties of all the components using the correlations in Equations (3.25–3.27).
3. Lump components together to form pseudocomponent groups.

To split up C10+, it is normal to assume that there is a linear relationship between logarithm of concentration and carbon number:

$$\ln z_i = A + B \cdot CN_i \tag{3.28}$$

where CN_i is the carbon number. A and B must be determined such that sum of mole fractions equals the mole fraction of the C10+ fraction, and the average molecular weight must equal the molecular weight of the C10+ fraction:

$$\sum z_i = z_{+fraction} \quad \text{and} \quad \frac{\sum z_i Mw_i}{z_{+fraction}} = Mw_{+fraction} \tag{3.29}$$

When A and B are determined, the mole fractions can be calculated and the properties can be calculated using the correlations in Equations (3.25–3.27).

The final step in the C10+ characterization procedure is to lump the components together to reduce number of components in the EoS model. A typical lumping procedure is to collect C10–C80 components into, for example, nine pseudocomponent groups where critical temperature, critical pressure, and acentric factors for the groups are calculated as weight average.

An example of a fully characterized fluid is given in Table 3.2. In this example, the C10+ fraction had a molecular weight of 280 g/mol and a density of 872 kg/m^3, before it was characterized and grouped into nine pseudocomponents with carbon numbers ranging from 10 to 80. This characterized fluid can now be used with an EoS to calculate number of phases, phase compositions, and phase properties. The critical parameters for N_2-nC5 are well known and tabulated. The critical parameters for all the other components have been calculated from correlations using the measured and reported molecular weights and densities. The final properties of the HC phase will therefore depend on these data, and it is important that the user of scale simulation tools know how to extract the correct data from

PVT reports and correctly enter them into the software, because variations in the HC fluid properties will influence the scale modeling.

Table 3.2. Example of characterized fluid.

Component	Mol (%)	M_w (g/mole)	Density (kg/m^3)	T_c (°C)	P_c (bar)	Acentric Factor
Nitrogen	1.614	28.01	0	-147.0	33.94	0.04
Carbon dioxide	1.843	44.01	0	31.1	73.76	0.23
Methane	22.089	16.04	0	-82.6	46.00	0.01
Ethane	7.155	30.07	0	32.3	48.84	0.10
Propane	7.882	44.10	0	96.7	42.46	0.15
iso-Butane	1.64	58.12	0	135.0	36.48	0.18
n-Butane	4.936	58.12	0	152.1	38.00	0.19
iso-Pentane	1.991	72.15	0	187.3	33.84	0.23
n-Pentane	2.863	72.15	0	196.5	33.74	0.25
C6	3.657	84.80	667.3	234.3	29.69	0.30
C7	5.325	92.90	727	255.6	32.46	0.46
C8	5.44	106.80	747.5	279.6	28.55	0.50
C9	3.759	121.30	764.6	301.7	25.52	0.54
C10-C12	7.167	146.51	788.05	336.1	21.95	0.61
C13-C15	5.446	189.39	816.86	384.3	18.41	0.73
C16-C18	4.139	235.78	839.97	428.9	16.33	0.84
C19-C21	3.145	275.48	859.34	463.7	15.38	0.93
C22-C24	2.39	317.21	875.99	497.6	14.71	1.02
C25-C29	2.778	370.99	894.66	538.7	14.16	1.12
C30-C35	2.023	447.28	916.55	592.5	13.71	1.24
C36-C44	1.551	547.56	940.49	658.7	13.46	1.34
C45-C80	1.168	752.68	978.94	789.9	13.45	1.28

3.6 Scale Modeling in Oil and Gas Wells

To model scale in oil and gas wells, the oil/gas phases must be included to model phase distribution of CO_2, H_2S, water, and other components properly. In addition to a water analysis report, it is also necessary to have a PVT report that gives the composition of the reservoir fluid and the flow rates of water and oil/gas. Instead of defining reservoir fluid composition, it is also possible to enter the composition of produced oil and gas together with respective flow rates. When the user has entered all necessary data, including flow rates, pressure, and temperature of interest, the scale prediction software can be used to calculate:

- The phases are present.
- Composition of all phases.
- Detailed ionic composition of the aqueous phase, including dissociation of CO_2, H_2S, and organic acids.
- Calculate the properties of all phases, such as densities, fugacity coefficients, compressibility, pH, volumes, and so on.
- Calculate scale potentials.
- Calculate theoretical mass of scale that can form.

To illustrate how the oil and gas phases and the PVT model will influence scale simulations, the characterized reservoir fluid given in Table 3.2 will be used together with the formation water analysis in Table 3.3. The reservoir conditions are 409 bar and 125°C. The reservoir fluid in Table 3.2 is an undersaturated oil with a bubble point of 121.5 bar at reservoir temperature and GOR = 84.7 Sm^3/Sm^3. Normal wellhead conditions are 60 bar and 109°C. The water rate is 75 m^3/d, and the oil rate is 1,325 Sm^3/d, which gives a water cut of 5.4%.

Table 3.3. Formation water composition.

Ion	(mg/L)
Sodium (Na^+)	50 600
Potassium (K^+)	1 940
Magnesium (Mg^{2+})	1 124
Calcium (Ca^{2+})	1 232
Barium (Ba^{2+})	200
Strontium (Sr^{2+})	744
Iron (Fe^{2+})	27
Chloride (Cl^-)	85 621
Sulfate (SO_4^{2-})	14
Total alkalinity as HCO_3^-	600
Acetic acid	42

Before running simulations, it is necessary to perform a couple of fluid tuning steps:

- **Water-saturate the PVT**. PVT reports never include water; they only report the *dry* composition, like the one in Table 3.2. However, all reservoir fluids will contain some water, and the amount depends on pressure, temperature, and type of fluid. For example, a gas in a HPHT reservoir can easily contain 1%–3% water. A heavy oil in a low-temperature reservoir will contain much less. Before a simulation can be started, the reservoir fluid must be saturated with water at reservoir conditions using the formation water composition.
- **Gas-saturate the water**. A formation water analysis like the one in Table 3.3 normally does not report the concentration of dissolved gas. It is possible to take pressurized water samples and measure flash gas, but this is rare and not always very accurate. Therefore, it has to perform a calculation and saturate the formation water with gases like CO_2, H_2S, CH_4, and so on at reservoir conditions using the PVT model.

- **Salt-saturate the water.** If the water being analyzed is a formation water, it should not be supersaturated with any minerals. Formation waters have been in the reservoir for a very long time, and supersaturated minerals should have precipitated and equilibrated. If the formation water analysis results in any supersaturated salts, the ion concentrations should be adjusted. It is also a common assumption that most formation waters are saturated with calcite because most reservoir rocks contain at least some calcite. A standard approach is therefore to fix the CO_2 pressure (or fugacity) from the PVT analysis, and then tune the alkalinity of the water to SR($CaCO_3$) = 1 at initial reservoir conditions.

If the above tuning steps are performed for the PVT composition in Table 3.2 and the formation water in Table 3.3, the following results are obtained:

- Water in reservoir fluid: 0.94 mole% corresponding to 1.8 m^3/d when the oil rate is 1,325 Sm^3/d.
- Alkalinity is adjusted to 632 mg/L to make water calcite-saturated; no salts are supersaturated.
- Water is saturated with 2226 mg/L CO_2 and 688 mg/L CH_4 to make it in equilibrium with the reservoir fluid.

Figure 3.5 shows the calculated SR($CaCO_3$) and pH when water and oil are produced from reservoir conditions (409 bar, 125°C) to wellhead conditions (60 bar, 109°C). SR($CaCO_3$) starts at 1.0 (saturation) in the reservoir, increases slowly to about 120 bars, and then increases rapidly to 2.2 at the wellhead. pH shows the same trend. The reason for the sudden change around 120 bar is that the oil reaches the bubble point, and a gas phase is formed. This results in a sudden reduction in CO_2 concentration in the aqueous phase and when pH increases, the scale potential for $CaCO_3$ increases.

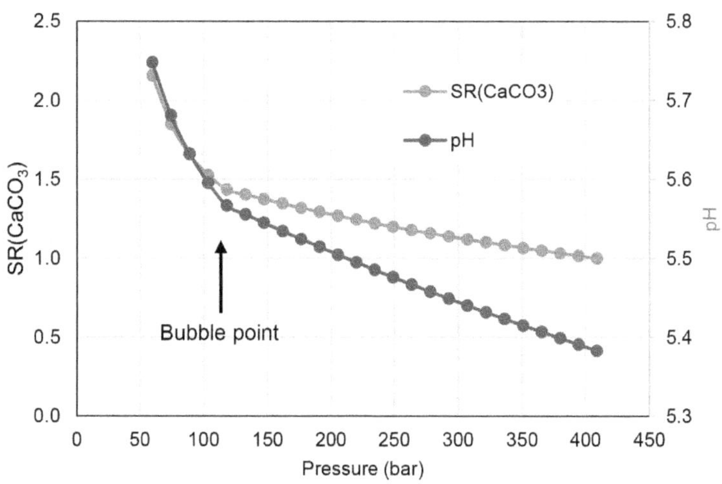

Figure 3.5. Variation in SR($CaCO_3$) and pH with pressure.

Figure 3.5 shows that the SR($CaCO_3$) increases when pressure decreases. This is expected because CO_2 goes out of the water phase when pressure decreases, and pH will increase. However, there is a large difference in the changes of pH and SR($CaCO_3$), when there is a gas phase and when there is oil only. In the above case, the GOR of the oil is only 84 Sm^3/Sm^3, so the amount of gas formed is very low. What will happen if the GOR of the oil was much higher?

3.6.1 Effect of GOR

It is not possible to vary GOR without changing the fluid composition, but one method to change GOR while still having the same gas and oil compositions is to flash the reservoir fluid at a pressure below the bubble point and recombine the gas and oil phases in a different ratio with more gas. The new reservoir fluid will have a different overall composition with more light components, but the gas phase that forms when it flashes will be very similar and have almost the same CO_2 concentration. The GOR was increased to 200 Sm^3/Sm^3 and 650 Sm^3/Sm^3, and simulations were run from reservoir to wellhead and the resulting SR($CaCO_3$) is shown in Figure 3.6.

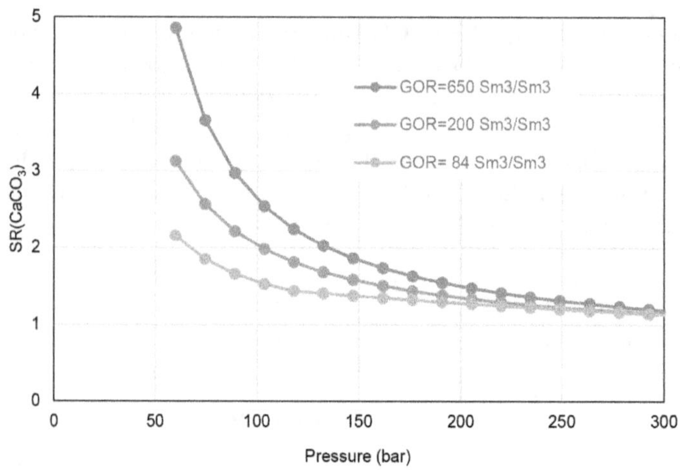

Figure 3.6. Variation in SR($CaCO_3$) with pressure for fluids with varying GOR.

When GOR increases, the system becomes more pressure-sensitive and SR($CaCO_3$) increases much more when pressure decreases, even if the CO_2 concentration in the flashed gas phase is almost the same in all three cases.

3.6.2 Effect of Water Cut

The water cut in an oil well will normally increase throughout its lifetime. Many oil wells can start to produce oil with virtually no water, and at some point, there is water breakthrough and the water cut starts to increase. Wells with low water cut will be more sensitive to changes in pressure and temperature, because there will be more CO_2 flashing and potentially more water evaporation or condensation. Figure 3.7 shows examples of how SR($CaCO_3$) can vary with water cut. The example above with GOR = 650 Sm^3/Sm^3 was used and the water cut was varied from 1% to 20%. When water cut is 10%–20% or higher, there is very little effect of water cut, because water evaporation will not have any significant effect on the ion concentrations in the water. However, when the water cut is lower, evaporation of water will have a huge impact on the scale potential. For the WCUT = 1% case, SR($CaCO_3$) goes to 31 (outside the graph) at wellhead conditions. If the water cut had been lower, there would be a risk of complete evaporation of the water phase.

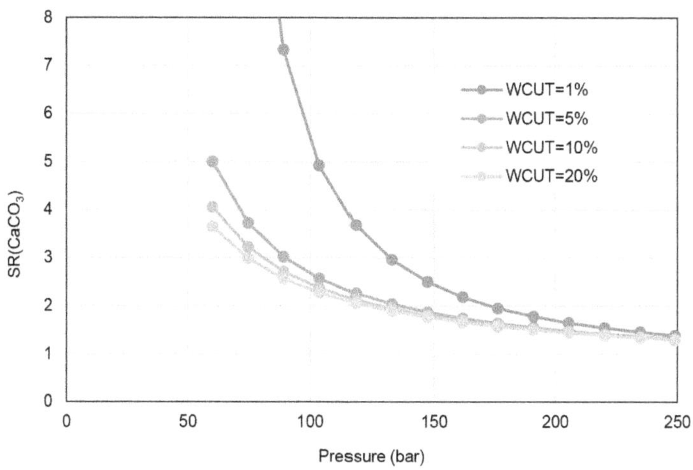

Figure 3.7. Effect of varying water cut on scale potential for $CaCO_3$.

3.6.3 The Effect of Temperature and Water Evaporation/Condensation

The above simulations were run from reservoir (409 bar, 125°C) to wellhead at 60 bar, 109°C. If the temperature profile had been changed, the result would have been different. The simulations with 1% WCUT were repeated, but with wellhead temperatures of 100°C and 85°C. This means that the temperature decreases faster in the well. The resulting scale potential for $CaCO_3$ is shown in Figure 3.8. When temperature decreases more, SR(-$CaCO_3$) will increase slower. In the case where wellhead temperature is 85°C, the scale potential will decrease in the beginning before the pressure effect becomes so strong that it starts to increase again. Most of these variations are related to water evaporation or condensation. In the case of a wellhead temperature of 85°C, the temperature decreases so fast that water condenses in the lower parts of the well, but then evaporates again when pressure goes below 140 bar.

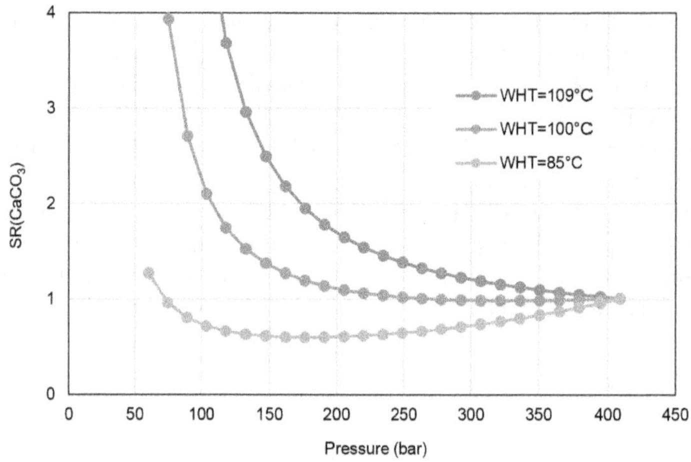

Figure 3.8. Effect of decreasing wellhead temperature. Graphs show how scale potential varies from reservoir (409 bar, 125°C) to wellhead at 60 bar and three different temperatures.

3.6.4 Effect of Lift Gas Injection

Some wells are produced with gas lift. Gas lift means that gas is injected into the well stream down in the well. Injecting gas will reduce the overall density of the fluid column in the well and help to lift the liquids out. The lift gas is normally taken from the gas treatment process on the host and can also come from a different host. If the lift gas comes from the same host as the well produces, the lift gas will probably have a similar composition as the gas in the well, and injection lift gas will not change the composition in the well. Otherwise, the lift gas could have a completely different composition. For example, if a field produces from some oil wells and some gas wells, the majority of the gas produced from the host will probably come from the gas wells, while lift gas is only needed in the oil wells. In this situation, the lift gas injected to the oil wells may be gas from the gas wells, and it may have a very different composition. For example, if the lift gas has a higher or lower CO_2 concentration than the "natural" gas at the lift gas injection point, this will change the overall partial pressure of CO_2 in the well and, therefore, also change the pH and scale potential for all carbonates and sulfides. If the lift gas has been dried, it will take up water when it is injected into the well, and if the water cut is low, this means that some of the liquid water that keeps ions in solution evaporates and all ion concentrations increase. In extreme cases, this can lead to halite formation and even complete evaporation.

The example from the previous chapters with oil composition from Table 3.2 was used. This oil has a CO_2 concentration of 1.843 mole%, while the released gas contains about 3.5 mole% CO_2. A lift gas containing 1.48% CO_2 at a rate of 90 kSm^3/d was injected in the well at a location corresponding to about 200 bar. Figure 3.9 shows that SR($CaCO_3$) varies with pressure without and with lift gas injection. Injecting lift gas gives a significant increase in scale potential for $CaCO_3$, partly because CO_2 concentration is lower and partly because the well stream contains more gas. As discussed in Section 3.6.1, a higher gas rate tends to give higher scale potentials for $CaCO_3$, because more CO_2 will flash out of the water.

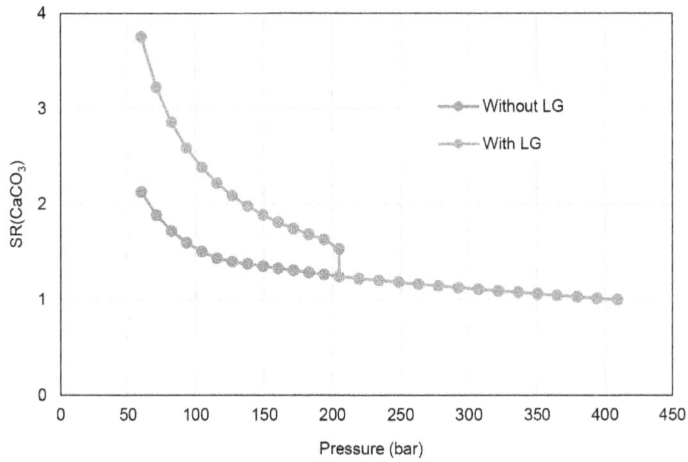

Figure 3.9. Effect of injecting a lift gas with low CO_2 concentration.

3.7 Scale Modeling in Topside Processes

Scale can also form in a topside process, for example, in separators, heaters, pumps, regulation valves, and so on. Consider the example of a topside process in Figure 3.10. Only separators, heaters, coolers, and scrubbers are shown in Figure 3.10. Locations with a potential scale risk are:

- Manifold where well streams are mixed and fed to an inlet separator. Mixing well streams can lead to scale formation if the well streams have different compositions.
- At the oil outlet from the inlet separator, there is a level regulation valve where pressure is decreased to second-stage pressure. There is some carryover water following the oil, and when there is a pressure drop, CO_2 will flash out and the scale potential for carbonates and sulfides will increase.
- After the second-stage separator, there is a heater with a further pressure drop to the third-stage separator. Carryover water following the oil from the second-stage separator will then experience a pressure drop, an increase in temperature, and a gas phase formed when the oil flashes. All factors will result in CO_2 and H_2S flashing out of the water, pos-

sible water evaporation, lower solubility of carbonates, and so on. All factors contribute to increasing the scale risk for carbonates and sulfides in the third-stage separator.

- From the various scrubbers, there could be condensate and condensed water formed when the gas streams are cooled. These streams are drained down back into the process, for example, as shown in Figure 3.10 (pumps may be needed). If these condensate streams are large, they can influence the scale potential in the locations where they are drained. For example, the amount of condensed water coming from the recompression scrubbers can be large enough to give a significant dilution of the water in the third-stage separator.
- Depressurization of the waters from the separators down to the produced water (PW) degasser will flash out dissolved gas and cause the increase of pH of water and scale potential for carbonates and sulfides.
- When there is an oil heater in the process, there is a risk of water evaporation in the heater, which can lead to formation of all types of scale and salt. A typical example is that the separation efficiency in the second-stage separator is improved by the addition of a new demulsifier or lower production rates giving a longer residence time. If the carryover rate decreases, there may not be enough water present to keep the oil and the flash gas in the heater water-saturated, and scale and salt can form. There are several examples of halite formation in oil heaters caused by water evaporation.

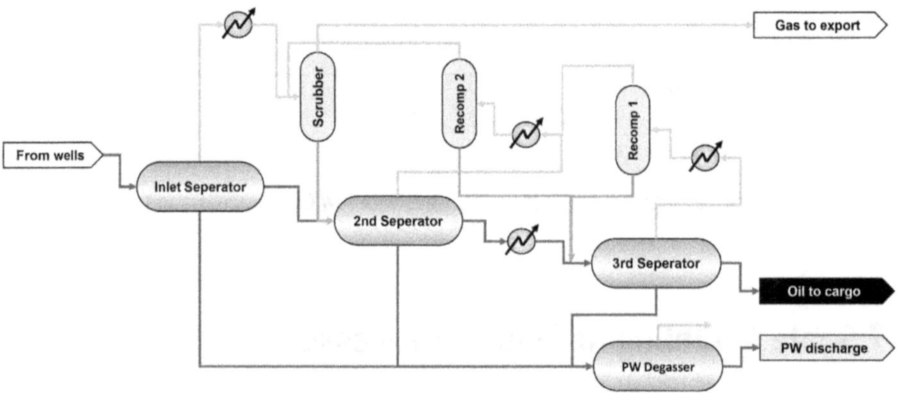

Figure 3.10. Schematic drawing of topside process.

Modeling a topside process, as shown in Figure 3.10, requires that the scale modeling software has a full PVT model integrated with the capabilities of modeling a process with recirculating streams.

3.8 Summary

Scale and salt precipitation can occur when saline waters are exposed to changes in pressure and temperature, when different waters are mixed or when the water evaporates. The main focus when modeling scale formation is therefore naturally on the aqueous phase and the various ions in the water. However, some of the common scale types, such as calcite

($CaCO_3$), siderite ($FeCO_3$), iron sulfide (FeS), all other carbonates, sulfides, and hydroxides, also depend on pH and concentration of acidic gases (CO_2 and H_2S) in the water phase. In multiphase systems, when oil and/or gas are present, the concentrations of CO_2 and H_2S in the water phase are changed when fluids flow from reservoir to surface and through the topside process, depending on their distribution among the phases, which are affected by the change of temperature and pressure.

To model the phase distribution of CO_2 and H_2S, it is necessary to combine the aqueous model with a PVT model that can simulate the behavior of the oil and gas phases. One of the most commonly used models PVT models are EoS-based models, for example the SRK model or the PR model. These models can simulate the behavior of most oil/gas systems, and predict bubble and dew points, GOR/GCR, phase compositions, phase densities and perhaps most important, fugacity coefficients of components. By combining the PVT model with an aqueous model, phase equilibrium and phase distribution can be modeled.

This chapter clearly illustrates the importance of combining a PVT model in the scale modeling with support of case studies under various scenarios. Variations in GOR of the oil or water cut have large impact on pH, and therefore, on scale potential for $CaCO_3$ and other carbonates/sulfides. The presence of oil/gas combined with water cut and variations in pressure and temperature can also lead to water evaporation or condensation, which could have a major impact on the scale potential in the water phase.

The purpose of this chapter was to explain the necessity to include a PVT model when simulating scale risk in oil and gas wells and explain how it can be modeled. The book edited by Pedersen is recommended for further reading about PVT models, and the book edited by Michelsen is recommended for those who want more insight into the mathematics of using EoS models and calculating phase properties.

Nomenclature

CPA	Cubic plus association—extension of an EoS to include associating components such as water
EoS	Equation of state
GC	Gas chromatography
GCR	Gas condensate ratio, volume of gas divided by volume of condensate (same as GOR)
GOR	Gas oil ratio, volume gas divided by volume of oil
H_{ij}	Henry's law constant for component *i* in solvent *j*
HPHT	High pressure high temperature
K_i	Equilibrium constant, in this context ratio between mole fractions in gas and oil phase
MEG	Monoethylene glycol
M_W	Molecular weight
N, n, n_i	Number of moles, either total or for component *i*
$P^{sat}_{H_2O}$	Saturation pressure/vapor pressure of water
$P, P°$	Pressure of system and reference pressure (normally 1 bar)

P_c	Critical pressure
PR	Peng-Robinson equation of state
PVT	Pressure, volume, temperature
PW	Produced water
R	Gas constant (=8.314 J/K×mole)
Sm^3	Standard cubic meters
SR	Saturation ratio
SRK	Soave Redlich Kwong equation of state
T_c	Critical temperature
TBP	True boiling point
UNIQUAC	UNIversal QUAsi-Chemical
V, v	Volume, total volume or molar volume
$\overline{V}_{H_2O}$	Partial molar volume of water
WHT	Wellhead temperature
WUCT°	Water cut–vol% water in the liquid phase (water + oil)
x, y	Mole fraction, x used for oil phase, y used for gas phase
z, z_+	Mole fraction for total phase or mole fraction of ± fraction.
Z	Compressibility factor
a_{H_2O}	Activity of water in water phase
β	Gas fraction. Number of moles in gas phase relative to total in gas and oil
f_i	Fugacity of component i
φ_i	Fugacity coefficient of component i in either oil or gas phase
$\varphi^{sat}_{H_2O}$	Fugacity coefficient for water in water vapor at saturation pressure
γ_i	Activity coefficient of component i in water phase
ρ	Density
μ_i, μ_i°	Chemical potential of component i in current state or in reference state.
ω	Acentric factor, used in EoS

References

1. G. Soave. Equilibrium Constants From a Modified Redlich-Kwong Equation of State. *Chemical Engineering Science* 27 (1972): pp. 1197–1203.
2. D.Y. Peng, D.B. Robinson, A New Two-Constant Equation of State. *Industrial and Engineering Chemistry* 15 (1976): pp. 59–64.
3. K.S. Pedersen, P.L. Christensen, *Phase Behavior of Petroleum Reservoir Fluids* (London: Taylor & Francis, 2007).
4. M.L. Michelsen, J.M. Mollerup, *Thermodynamic Models: Fundamentals Computational Aspects*, 2nd ed. (Tie-Line Publications, 2007).
5. G.M. Kontogeorgis, A. Schlaikjer, M.D. Olsen, et al., A Review of Electrolyte Equations of State With Emphasis On Those Based On Cubic and Cubic-Plus-Association (CPA) Models. *International Journal of Thermophysics* 43 (2022): p. 54.
6. J.J. Carroll, What is Henry's law. *Chemical Engineering Progress* 87 (1991): pp. 48–52.

4 Scale Modeling for Gas Wells

Baard Kaasa[1]
[1]Scale Consult AS, Kongsberg, Norway

4.1 Introduction

Modeling scale risk in gas wells is, in principle, similar to predicting scale risk in oil wells; the chemical and thermodynamic laws are the same. But because gas and oil (liquid) behave differently, the factors that influence the scale risk when gas and water are flowing from reservoir to wellhead will often be different than those for an oil well. Gas wells will typically have a lower water cut, or more precisely, less water compared to gas volume, than oil wells, and this makes the scale potential in gas wells more sensitive to water evaporation and condensation. The pressure and temperature profiles in a gas well will also be different compared to an oil well mainly due to the lower density and lower heat capacity of the gas compared to the liquid oil.

When predicting scale risk in gas wells, it is often more important to focus on temperature variations along the well path, because this will influence not only the solubility of scale minerals but also if there is water evaporation or condensation. For gas wells with a large drawdown or high pressure drop over the choke, the Joule-Thomson (J-T) effect can cause changes in temperature, which again has a large effect on mineral solubility and water evaporation and condensation.

In this chapter, we will look at these effects and explain how they all influence the scale risk and how to include them correctly in the scale modeling for gas wells.

4.2 Gas Reservoirs

Before we look at scale modeling in gas wells, it is important to understand what a "gas well" is and what makes it different from an "oil well." One obvious difference is that gas wells produce predominantly gas, and if they produce a hydrocarbon liquid phase, it is normally a liquid called condensate. To illustrate this, consider the phase envelope in Figure 4.1. A phase envelope is a graphic illustration showing if a given composition is liquid (oil), gas, or two phase (gas and liquid). If the reservoir is located at a pressure above the phase envelope, it is either single-phase oil or single-phase gas. If the reservoir conditions are to the *left* of the critical point, we would call the fluid oil. Reducing pressure will make the fluid cross the *bubble point line*, and a gas phase will be formed. If the reservoir is located to the right of the critical point, the reservoir is gas or gas condensate.

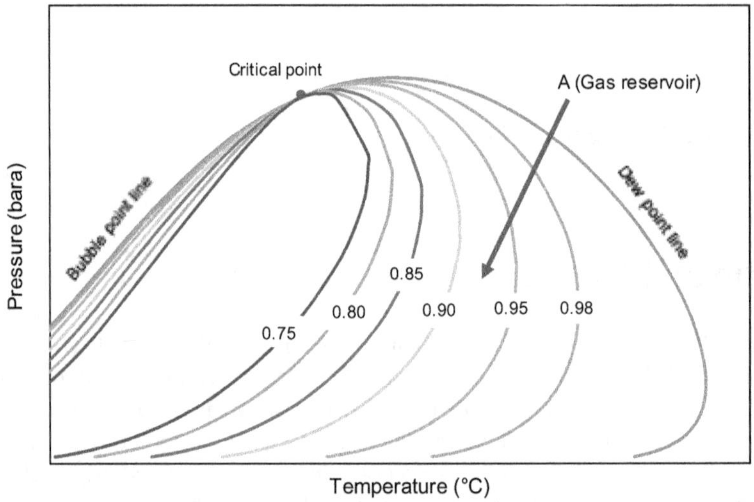

Figure 4.1. Phase envelope for gas reservoirs.

In a gas condensate reservoir, marked as A in Figure 4.1, there will be single-phase gas in the reservoir. When pressure decreases, one will hit the *dewpoint line*, and a condensate phase will condense out of the gas. When pressure continues to decrease, the amount of liquid will increase. It is not intuitive that the amount of liquid *increases* when pressure is decreased, and the behavior is called *retrograde* condensation. The amount of liquid will increase as indicated by the liquid fraction lines inside the phase envelope until a certain pressure is reached, at which point the condensate will start to re-evaporate. The amount of condensate at surface conditions depends on where the surface conditions are located compared to the phase envelope.

In a dry gas reservoir, the reservoir conditions are located to the right of the whole phase envelope such that when pressure decreases, one will not touch or pass through the phase envelope, and there will not be any liquid production or at least very little.

Gas condensate reservoirs are normally produced by pressure depletion. This means that reservoir pressure will decrease, and when the reservoir pressure hits the dew point line, there will be condensation. This means that condensate will drop out of the gas in the whole reservoir. Because gas has a much lower viscosity than condensate, it is more mobile than condensate and will flow easier toward the production well(s). The result is that some of the condensate is left in the reservoir, the produced gas is leaner, and the gas condensate ratio (GCR) will increase. If the reservoir pressure decreases enough, the condensate will later re-evaporate.

4.3 Thermodynamics

Before we look at how scale risk is modeled in gas wells, it is necessary to look into some of the thermodynamic effects that will have a significant influence on the sale risk in gas wells:

- Water evaporation and condensation. If some of the water phase evaporates when gas and water flow from the reservoir, all ion concentrations in the water will increase, leading to an increase in all scale potentials. In some gas wells, water evaporation is the main reason for scale formation. In extreme situation, the whole water phase can evaporate, and all salts will precipitate.
- Temperature variation during inflow and in the well are often much larger in a gas well compared to an oil well.

This chapter aims to explain why water will evaporate and/or condense when gas is produced from the reservoir to the surface and how and why temperature will vary.

4.3.1 Water Evaporation and Condensation

Most reservoirs contain water, both as an aquifer below the gas zones and as residual water in the gas zone. The water saturation, S_w, in the gas zone will of course vary with reservoir conditions and height above the aquifer, but there will be a liquid water phase present, and the reservoir gas will be saturated with water.[1] The water concentration in the gas will depend on several factors, and the most important of which are pressure and temperature, but gas composition and the salinity of the water will also have some influence. To calculate the water equilibrium between the phases and the water concentration in the gas phase, we can use Rault's law. When there is equilibrium between the gas phase and liquid water in the reservoir, the fugacity of water will be equal in the gas phase and the water phase.

$$f_{H_2O}^{v} = f_{H_2O}^{aq} \tag{4.1}$$

On the left side, we can insert the definition for fugacity in a gas. On the right side, we can insert the pure solvent reference state that relates the fugacity of water (in a mixture) to the fugacity of pure water. $f_{H_2O}^{aq} = a_{H_2O} f_{H_2O}^{0}$ and the expression are then integrated from the vapor pressure of water to the pressure of interest. The full Rault's law then becomes

$$\varphi_{H_2O}^{v} y_{H_2O} P = a_{H_2O} \varphi_{H_2O}^{sat} P_{H_2O}^{sat} \exp\left[\int_{P^{sat}}^{P} \left(\frac{\bar{V}_{H_2O}}{RT} \right) dP \right] \tag{4.2}$$

The left side of Equation (4.2) expresses the fugacity of water in the gas phase, but it can equally well be in condensate by replacing the fugacity coefficient of water in the gas phase with the one for the condensate phase and using the mole fraction of water in the condensate phase. The right side includes the water activity, the fugacity of pure water at its vapor pressure and the pressure effect on the water fugacity which is the exponential term often referred to as the Poynting correction. The water activity can be calculated from the osmotic coefficient of water which can be calculated using, for example, the Pitzer model.[2] The full Rault's law is complex, and it is not intuitive to see how water content varies with pressure, temperature, and other factors. If we consider moderate changes in pressure and temperature, the variation in the fugacity coefficients and water activity will be moderate. The variation in the Poynting correction is also small for moderate pressure changes. To better see how water content in the gas phase varies with pressure and temperature, we can therefore rewrite and simplify Equation (4.2) to

$$y_{H_2O} \propto \frac{P_{H_2O}^{sat}}{P} \tag{4.3}$$

Equation (4.3) shows that the water content in the gas will be proportional to the vapor pressure of water and inversely proportional to the pressure. The vapor pressure of water, and most other liquids, is often expressed as an exponential function of temperature, for example, the Antoine equation or something similar. Equation (4.3) therefore shows that

- Water content in a gas will increase exponentially when temperature increases.
- Water content in a gas will decrease inversely proportional to increasing pressure.

By using Rault's law, water concentration in a natural gas can be calculated as a function of pressure and temperature. Figure 4.2(a) shows how the water content in a natural gas varies with temperature and how it increases exponentially with increasing temperature. As expected, the water content will be lower with higher pressure. Figure 4.2(b) shows how the water content in natural gas varies with pressure and shows that the water content decreases inversely proportional to pressure as predicted by Equation (4.3).

The water concentration in the gas will also depend on the composition of the water phase through water activity and the composition of the gas phase through the fugacity coefficient. When water has high salinity, the water activity will be lower than 1, meaning that the vapor pressure is lower, and there will be less water in the gas phase. The effect is not very strong but can give some reduction of water concentration in the gas. For example, if the salinity of the water is 170,000 mg/L, the water activity will be about 0.9, meaning that the vapor pressure is 10% lower, and there will be 10% less water in the gas compared to, if the water were pure water. This is small compared to the effect of temperature/pressure, but it can be significant in some cases. The composition of natural gas has very little effect on the fugacity coefficient of water and, hence, very little effect on the water concentration in the gas.

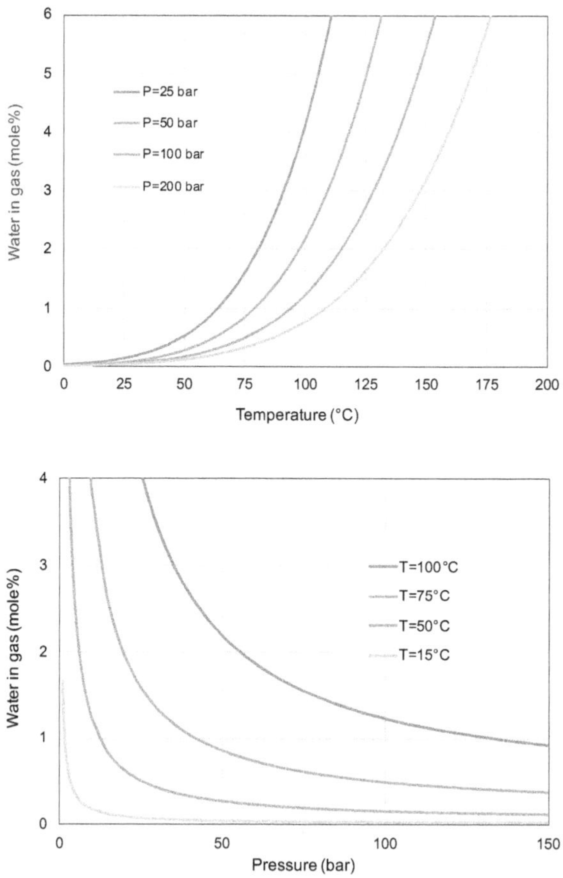

Figure 4.2. Variations of water content in natural gas with (a) temperature and (b) pressure.

When gas and water flow from the reservoir to the surface, both pressure and temperature will decrease.

- Lower pressure will increase the water content in gas and evaporation.
- Lower temperature will result in lower water content in gas and condensation.

The total effect, whether there will be evaporation or condensation, will therefore depend on how pressure and temperature vary relative to each other. To see if there will be net evaporation or condensation when gas and water flow from the reservoir to the surface, a plot showing the iso-evaporation line is practical. The iso-evaporation line is the combination of pressure and temperature where there is no net evaporation of water. An example of an iso-evaporation line is shown in Figure 4.3. This shows a natural gas system with reservoir conditions of 400 bar and 135°C, marked with the green dot. If temperature follows the blue line when pressure decreases, the water content in the gas (and condensate) will be constant, and there will be no net condensation or evaporation. The ion concentrations will not change due to evaporation. It does not mean that the scale potentials do not change, because there will be changes in the solubility of the minerals, equilibria for CO_2, H_2S, and organic acids, and pH will change. But there will not be any effect of evaporation/condensation.

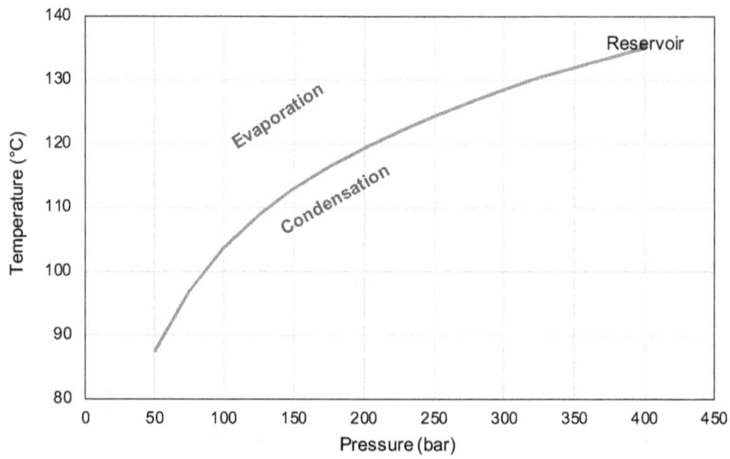

Figure 4.3. Iso-evaporation line showing how temperature should vary with pressure from reservoir for no net evaporation or condensation.

4.3.2 Temperature Change During Inflow to the Well

To produce gas, it is necessary to have a pressure gradient from the reservoir to the well. This pressure difference is called the drawdown and is the driving force for flowing gas, condensate, and water from the reservoir into the well. The magnitude of the drawdown can vary a lot depending on reservoir permeability and the layout of the well, and it can range from a few bars to several hundred bars.

When a gas expands, it normally cools down, and the effect is known as the J-T effect. For example, when gas is depressurized over a valve, the temperature will normally drop. To quantify how much temperature changes when pressure changes, we can use the J-T (or η) coefficient, which is given by

$$\eta = \left(\frac{\partial T}{\partial P}\right)_{H,x} \approx \frac{\Delta T}{\Delta P} \tag{4.4}$$

which tells how much temperature will change for a given pressure change. In the International System of Units (SI) units, the J-T coefficient is given in K/bar. The value of the J-T coefficient can be calculated from thermodynamic models, such as an equation of state. For example, a typical natural gas mixture at 200 bar and 100°C will have a J-T coefficient of 0.147 K/bar. This means that if there is a pressure drop of 50 bar ($\Delta P = -50$ bar), the temperature change will be −7.35 K (= −50 bar·0.147 K/bar), meaning that the temperature will drop by 7.35 K from 100°C down to 92.65°C. This is, however, a simplification, because the JT coefficient changes with pressure and temperature. If we look at the same natural gas after pressure has been reduced to 150 bar and temperature has dropped to 92.65°C, the J-T coefficient has also changed to 0.2235 K/bar. Because the J-T coefficient varies with pressure and temperature, it is necessary to perform a sort of numerical integration/iteration over the pressure-temperature (P-T) change, where the pressure drop is divided up in many small steps and calculate an updated J-T coefficient at each state. Even if this is done, it will still not be correct for many systems, because the J-T coefficient can only be used in *single-phase* systems. During inflow from a reservoir to a well, there may also be an oil/condensate phase and a water phase. Liquid phases will normally have negative J-T coefficients and will therefore heat up when pressure decreases. In addition, there will be evaporation/condensation of both water and hydrocarbons and changes in the phase equilibria which are not captured by the J-T coefficient. To overcome this problem, one can instead look at inflow from the reservoir to the well as an irreversible expansion with no exchange of heat with the surroundings, which makes the inflow process an isenthalpic process as suggested by Vassenden et al.[3] The assumption of no exchange of heat with the surroundings is often a reasonable assumption, because the internal energy of the flowing gas will be much greater than the heat flux to/from the surrounding reservoir. With these assumptions, the inflow from the reservoir to the well can be simulated as an isenthalpic process, meaning that the enthalpy of the fluids (gas, oil, condensate, and water) in the reservoir is equal to the enthalpy of the same fluid in the well. If there is any evaporation, condensation or changes in the phase equilibria when pressure and temperature change, this will be captured by calculating the enthalpy.

To calculate the temperature change or temperature profile during inflow from the reservoir to the well, a simulation must first be done at reservoir pressure and temperature to calculate the enthalpy of the inflowing fluids. Simulations can then be done at lower pressure, and it is necessary to iterate to find the temperature that gives the same enthalpy as in the reservoir. The result will show how temperature changes with pressure during inflow, and how all the equilibria, pH, and scale potentials vary with pressure and temperature during inflow.

The J-T coefficient is normally positive, meaning that when a gas experiences a pressure drop, there will be a temperature reduction. This is, however, only the case for systems up to a certain pressure. When pressure exceeds about 450–550 bar, the J-T coefficient of most natural gas mixtures becomes negative, and that gas will heat up when pressure decreases. If the reservoir pressure is, for example, 700 bar, and gas is produced with a drawdown of, for example, 100 bar, it will heat up several degrees during inflow to the well. This temperature increase has been observed during PLT-logging of gas wells in production.[4] When the production logging tool (PLT) passed through the inflowing zone, the measured temperature was 1.9°C higher than expected based on static temperature. The expected temperature increase based on simulations was 2.7°C. It is suggested that the difference is due to some heat exchange with the surroundings.

Figure 4.4 shows an example of how temperature will change, if there is a 100-bar pressure drop from varying initial pressures. The initial reservoir temperature is set at 125°C. The inflowing fluids are a mixture of 1 MSm^3/d gas and 25 m^3/d water. If the reservoir pressure is 800 bar, the temperature will increase by 2°C during inflow. If the reservoir pressure is 700 bar, the increase is only 1°C. If the reservoir pressure is 600 bar, there will be almost no change in temperature with a 100-bar pressure reduction. If the reservoir pressure is 500 bar, the temperature will decrease by 2°C. If the reservoir pressure is lower, the temperature reduction caused by the 100-bar drawdown is larger.

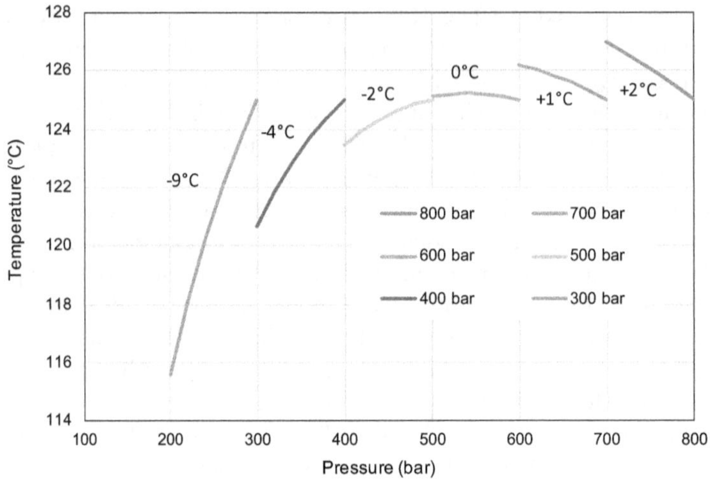

Figure 4.4. Example of temperature change caused by 100-bar drawdown from various start pressures, with gas rate of 1 MSm^3/d and water rate of 25 m^3/d.

A temperature change of a few degrees up or down will often have a significant impact on how much water evaporates, and this will again have a large effect on the scale potential for all scale minerals and also halite.

In reality, the temperature may not change as much as predicted by the J-T effect, or more precisely, by assuming isenthalpic flow. There will be some exchange of heat with the surrounding reservoir, and there is also a time effect. If a well is closed in, we can assume that the temperature equals the static geothermal gradient. When the well is started, and the conditions are such that the gas will cool down, it will take time to cool down the well and the surrounding rocks where the gas is flowing through. It will therefore take some time before the full JT cooling/heating temperature is reached. The assumption of no heat exchange with the surroundings can also be uncertain. The heat conductivity of reservoir rock is generally low, but the well completion may also have a large influence. For example, if we imagine a cylinder around the well, a vertical well with a short perforated interwall will have a much smaller surface area toward the surrounding reservoir than a long horizontal, and it will therefore exchange much less heat with the surroundings.

A reduction in temperature due to the J-T effect will, in most cases, have *lower* scale potentials. A conservative approach may therefore be to calculate the theoretical temperature reduction, but in the evaluation of scale risk, also include scale risk, if the temperature change is only, for example, 50% of the theoretical change.

4.4 Scale Modeling for Gas Wells

As explained earlier, modeling scale risk in gas wells is in principle not different from modeling scale risk in oil wells. But because gas wells often have a low water cut, evaporation/

condensation will often be more important than it is for oil wells. Also, because the gas volume is often large compared to the water volume, the scale potential for the carbonate scale will be more sensitive to pressure changes and CO_2 evaporation.

4.4.1 Input Data and Fluid Tuning

To illustrate scale predictions in a well, we will use the formation water (FW) composition given in Table 4.1 and the reservoir gas composition in Table 4.2.

Table 4.1: Formation water composition.

Component	Concentration (mg/L)
Sodium (Na^+)	92,230
Potassium (K^+)	1,680
Magnesium (Mg^{2+})	690
Calcium (Ca^{2+})	3,710
Barium (Ba^{2+})	250
Strontium (Sr^{2+})	265
Iron (Fe^{2+})	4.5
Chloride (Cl^-)	152,455
Sulphate (SO_4^{2-})	6
Total alkalinity (as HCO_3^-)	320
Acetic acid (HAc)	184

Table 4.2: Reservoir gas composition.

Component	Mole%
Nitrogen (N_2)	0.681
Carbon dioxide (CO_2)	1.73
Methane (CH_4)	82.221
Ethane	8.782
Propane	3.247
i-Butane	0.551
n-Butane	0.945
i-Pentane	0.291
n-Pentane	0.314
C6	0.317
C7	0.362
C8	0.331
C9	0.125
C10+	0.103
M_w_C10+ (g/mole)	190
Density (kg/m^3)	810

For simplicity, the gas rate will be set to 1 MSm^3/d, and the FW will be varied. Reservoir conditions are 420 bar and 137°C. Before simulations can commence, it is necessary to do all the standard fluid checks and tuning steps, the same as the tuning and checks performed for oil systems:

- Verify that water is electroneutral and adjust Na^+ or Cl^- if not.
- Add water to the reservoir gas such that it becomes water-saturated at reservoir conditions. This should be done using water with the correct salinity. If the FW contains organic acids, the reservoir gas should also be saturated with organic acids. This step is normally necessary, because PVT reports normally only state the dry gas composition.
- Add dissolved gas (CO_2, H_2S, CH_4, *etc.*) to the water phase such that it is in equilibrium with the reservoir gas. This step is necessary because most water analyses only report ion concentrations. Amounts of dissolved gases in the water are normally not reported.
- If the water is a FW analysis, check that it is not supersaturated with any minerals; if it is, adjust ion concentrations.
- If there is calcite in the reservoir rock, and most reservoirs contain some calcite, the water composition should be adjusted to make the water saturated with calcite. This is normally done by adjusting the alkalinity, because this is often the most uncertain parameter.

In the present example, the gas will contain 1.195 mole% water at reservoir conditions when it is in equilibrium with the water. If the gas rate is 1 MSm^3/d, this corresponds to 9.16 m^3/d water. It is essential to add this water into the gas before simulation. If not, the gas phase will "steal" this volume of water from the added FW. For example, the FW rate is 15 m^3/d, two-third of it will immediately evaporate to saturate the gas, and that will increase all ions concentrations in the remaining water a lot and give far too high-scale potentials.

If the water in Table 4.1 and the gas in Table 4.2 (after it has been water-saturated) are used in a simulation at reservoir conditions (420 bar, 137°C), the results shown in Table 4.3 are obtained. Simulations show that the pH of the FW is 5.03, and the water is undersaturated with calcite having a SR($CaCO_3$) of 0.83 (SI of −0.08). No other minerals are supersaturated. SR(NaCl) is 0.29, which is far from saturation, but still relatively high. A FW should normally be saturated with calcite and by tuning the alkalinity from 320 mg/L to 362 mg/L (as bicarbonate), the water will become saturated with calcite. The after-tuning results are also shown in Table 4.3. Increasing the alkalinity to make the water saturated with calcite also increased the pH a little. Increasing alkalinity has no effect on pH-independent minerals like $BaSO_4$, NaCl, and others.

Table 4.3: Simulation results at reservoir conditions (420 bar, 137°C) using water from Table 4.1 and gas from Table 4.2 after it has been water-saturated.

Parameter	Before Tuning	After Alkalinity Tuning
pH	5.03	5.07
SR($CaCO_3$)	0.83	1.00
SR($BaSO_4$)	0.35	0.35
SR(NaCl)	0.29	0.29

After equilibrating water and gas and making sure that the water is not supersaturated, scale simulations can be run.

4.4.2 Scale Prediction During Well Inflow

The well is produced with a high drawdown, and the flowing well bottom hole pressure is 300 bar, giving a drawdown of 120 bar. If it is assumed that the temperature is *constant* during inflow, scale potentials can be simulated as a function of the pressure reduction during inflow, and the result shown in Figure 4.5 is obtained. As expected, SR($CaCO_3$) increases when pressure decreases. The main reason for this is that CO_2 flashes out of the water, pH increases and SR($CaCO_3$) increases. But Figure 4.5 also shows that the effect is much larger when the water rate is low, and the reason for this is water evaporation. When pressure decreases, the gas will take up more water as shown in Equation (4.3); and when the water rate is lower, the effect of evaporating a certain volume of water is relatively larger, and the scale potential increases more.

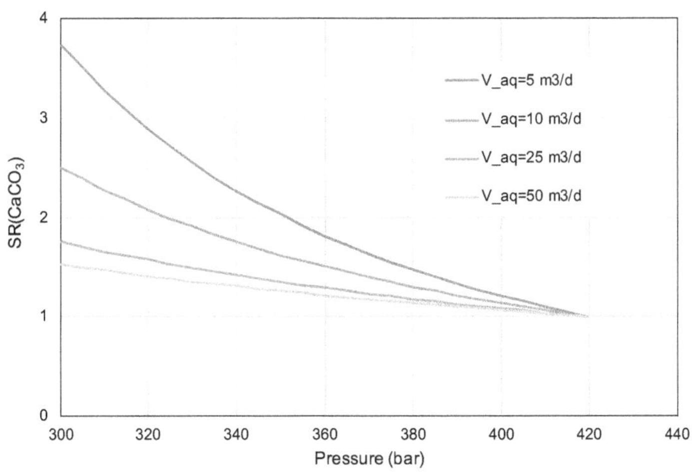

Figure 4.5. Calculated SR($CaCO_3$) versus pressure during well inflow from a reservoir at 420 bar, 137°C assuming constant temperature. Simulations are done with a gas rate of 1 MSm^3/d and varying formation water rates.

Holding the temperature constant during inflow is, however, not correct, because when a gas is expanded, the temperature can change as explained in Section 4.3.2. In this case, when pressure decreases from 420 bar, it is expected that the temperature will decrease. If it is assumed that the inflow to the well is isenthalpic flow, the temperature profile can be calculated, and the results for different water rates are shown in Figure 4.6. When the water rate is low, the temperature is determined almost only by the temperature change in the gas phase. When the water rate increases, the high heat capacity of the water will hold the temperature a little higher, but the difference is not very large. Simulations show that temperature is expected to decrease by approximately 5°C when pressure decreases from

420 to 300 bar. This temperature reduction will reduce water evaporation and will also affect all the other equilibria in the water. Figure 4.6 also shows the iso-evaporation line, which shows how temperature must change so that there is no net evaporation of water. The iso-evaporation line is lower than the expected temperature profile, so there will still be some water evaporation during inflow.

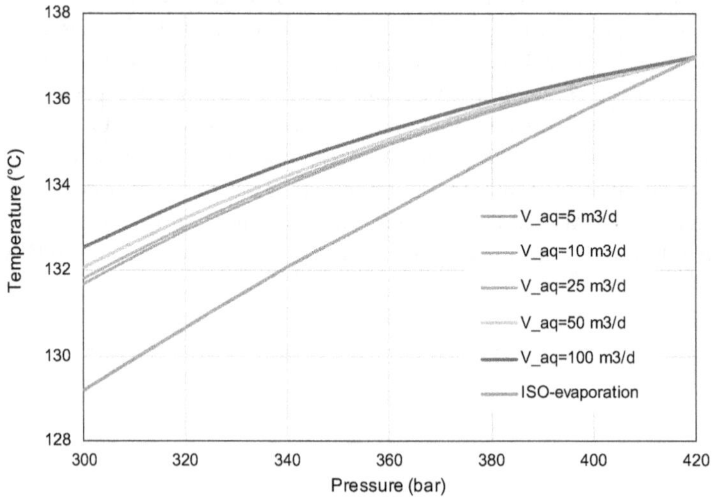

Figure 4.6. Calculated temperature during well inflow assuming isenthalpic flow. Reservoir conditions are 420 bar, 137°C. Simulations are done with a gas rate of 1 MSm³/d and varying formation water rates.

Figure 4.7 shows the simulated variation in SR($CaCO_3$) using the isenthalpic temperature profile during inflow. When the reduction in temperature is included, the increase in scale potential is much smaller, and there is much less difference between simulations with high and low water rates, because there is less water evaporation. SR($CaCO_3$) will increase to levels very close to or above the critical limit where $CaCO_3$ formation is expected given the high temperature, but the scale potential is much lower and scale formation will be much easier to prevent by, for example, applying scale inhibitors via a squeeze treatment. Compared to the scale potential prediction in Figure 4.5 where the temperature was held constant, it is clear that a reduction in temperature will significantly reduce the scale risk during well inflow.

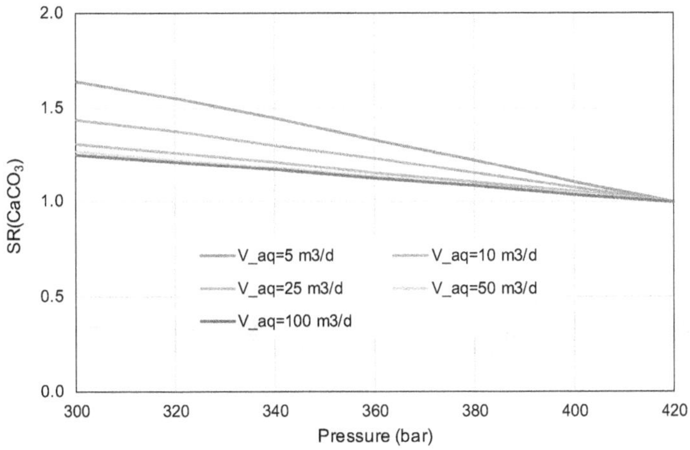

Figure 4.7. Calculated SR($CaCO_3$) versus pressure during inflow from a reservoir at 420 bar, 137°C. Simulations are done assuming isenthalpic flow with calculated temperatures from Figure 4.6. Gas rate is 1 MSm^3/d, and simulations are performed for different water rates.

4.4.3 Temperature Profile from Bottom Hole to Wellhead

How pressure and temperature vary from bottom hole to surface varies a lot and is influenced by several factors such as density of produced fluids, depth, flowrate (friction), geothermal gradient, well insulation, and so on. The pressure drop is mainly determined by the weight of the fluid column and friction loss. This means that the pressure difference between the well bottom and the wellhead will be much lower for gas wells than for oil wells, because the gas column is much lighter than an oil column. And wells with a high water cut will have a larger pressure difference between the well bottom and wellhead. Gas wells will, in general, have a larger temperature reduction when gas flows from the bottom hole to the surface. According to Vassenden,[3] the flow in the well can be assumed to be a reversible adiabatic process which means that it is a process with constant entropy. If one simulates the change in temperature with pressure for an isentropic flow, one will see that the temperature change for an oil well with a high liquid rate is much smaller than for a gas well with mainly gas. The reason for this is partly that liquids have much higher heat capacity, but also because liquids in general have a negative J-T coefficient and will therefore heat up when pressure decreases, while gases will in general cool down. Sandengen[5] has demonstrated this by comparing simulated temperature profiles in a typical oil well and a typical gas well. The cases in the oil well with water cut varying from 10% to –99%, the reduction in temperature from bottom to wellhead was typically 5–10°C, depending on how the simulation was performed. For the gas well with 10 m^3/d water and 1 MSm^3/d gas, the temperature reduction from bottom to wellhead was 20–40°C, depending on bottom hole pressure and wellhead pressure.

To illustrate this, simulations for an oil well and a gas well with characteristics given in Table 4.4 have been performed. One set of simulations simply assumes that there is constant entropy ($dS = 0$). The other simulation set is performed with OLGA,[6] a flow simulator that also includes friction loss and other flow parameters. The resulting temperature profiles are shown in Figure 4.8. In both cases, the well bottom temperature is 100°C. The full lines are for the oil well, and the stippled lines are for the gas well. The red lines show the simplified approach with constant entropy without considering things like friction loss, and so on, while the blue lines are with a full flow simulator. It is clear from the simulations that the temperature gradient in a gas well is much steeper than in an oil well, and the temperature drops much faster.

Table 4.4: Fluids used for well temperature simulations.

Parameter	Oil well	Gas well
Oil/gas rate	1,000 Sm^3/d	1 MSm^3/d
GOR	100 Sm^3/Sm^3	3,032 Sm^3/Sm^3
Water rate	100 m^3/d	30 m^3/d
Well bottom pressure	125 bar	60 bar
Reservoir temperature	100°C	100°C

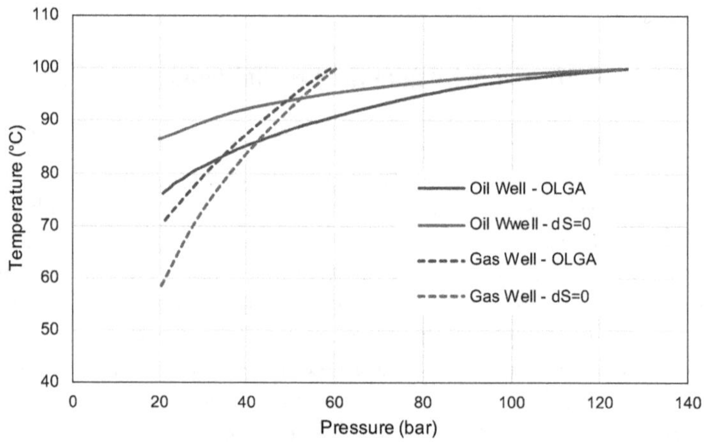

Figure 4.8. Temperature profiles in an oil well and in a gas well.

The difference in pressure/temperature change between oil and gas wells will have a direct impact on the scale potential of several minerals in the well. For wells with low water cut, there will be much less water evaporation in gas wells, because the reduction in pressure is smaller, and the temperature decreases more. Both factors will cause less water evaporation, and in many cases, they will actually cause water condensation. This means that the produced water (PW) will be diluted with condensed water, and scale potentials for all minerals will decrease. This is in contrast to oil wells where there will normally be water evaporation due to the larger pressure drop. The effect of water evaporation or condensation will, of course, only be important for wells with low water cut.

For carbonates and sulfides, the scale potential will normally *increase* with a reduction in pressure, because CO_2 and H_2S flash out of the water, and this gives an increase in pH. This is the main reason scale potentials for carbonates and sulfides increase from bottom to wellhead in oil wells. A reduction in temperature, however, will *reduce* the scale potential for carbonates and sulfides, because the solubility of CO_2 and H_2S in water *increases* with decreasing temperature, and more CO_2 and H_2S in the water will keep pH lower. The total effect of water evaporation/condensation and temperature variation can therefore result in very different variations in scale potential for carbonates/sulfides along the well path. Figure 4.9 shows how SR($CaCO_3$) will vary from downhole to surface for the oil-well and gas-well cases given in Table 4.4 with the constant entropy temperature profiles shown in Figure 4.8. The scale potential in the oil well will increase, while in the gas well, it will decrease.

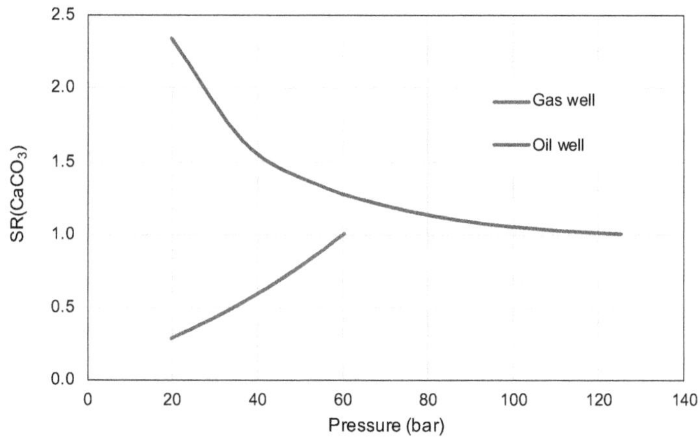

Figure 4.9. Example of variation in SR($CaCO_3$) with pressure in an oil well and a gas well.

There may, of course, be situations where the variation in SR with pressure is different; it depends on fluid composition, fluid gas oil ratio (GOR), water cut, reservoir, and surface pressure and temperatures, well layout, if there is a gas lift, if there is an electric submersible pump, and so on. But in general, the pressure/temperature profiles in gas wells are often such that the increase in scale risk for carbonates is low, or even decreasing, while for oil wells, it will normally increase.

For $BaSO_4$, the situation is the opposite. The solubility of $BaSO_4$ is not very pressure-sensitive, but it has a maximum around 100°C. When the temperature decreases below 100°C, the solubility will decrease, and the scale potential will increase. But the scale potential for $BaSO_4$ is normally very low in most FWs; it is normally a result of mixing FW with injected seawater, and gas wells will not have seawater injection for pressure support.

To perform scale simulations along a well path, it is necessary to calculate the P-T profile along the well. There are several commercially available tools for this type of calculation, for example, OLGA[6] or Prosper.[7]

4.5 Halite Formation in Gas Wells

Halite (NaCl) is not a typical scale mineral, because the solubility in water is very high, but it is found quite frequently, and when it forms, it is normally due to one of the following reasons[8]:

- The FW is saturated, or close to saturated, with NaCl in the reservoir and halite forms, because solubility decreases as temperature and pressure decrease on the way from reservoir to surface.
- Water evaporation results in a very high concentration and eventually NaCl precipitation.

A combination of the two is, of course, also possible, because it is more likely to form halite due to water evaporation, if the salinity is high in the first place.

Figure 4.10 shows how the solubility of NaCl in water varies with temperature. As expected, solubility increases with temperature. At temperatures that are typical for reservoirs/wells, the solubility of NaCl is roughly between 6 mol/kg H_2O and 7 mol/kg H_2O. This corresponds to 350,000–400,000 mg/L NaCl in the water. Most FWs contain less NaCl, but there are also many reservoirs where the FW is saturated with NaCl or close to it. If, for example, the reservoir temperature is 110°C, the NaCl solubility is about 6.7 mol/kg. The Na-Cl stoichiometry does not necessarily have to be 1:1, but the NaCl concentration will be in this order. If the water cools down to, for example, 50°C at surface concentration, the solubility will decrease to about 6.2 mol/kg. The difference is 0.5 mol/kg which corresponds to 29,000 mg/L or 29 kg/m^3. This means that the reduction in temperature as the water flows from reservoir to surface will produce about 29 kg/m^3, and if the well produces a few 100 m^3/d of water, there will be tons of NaCl every day.

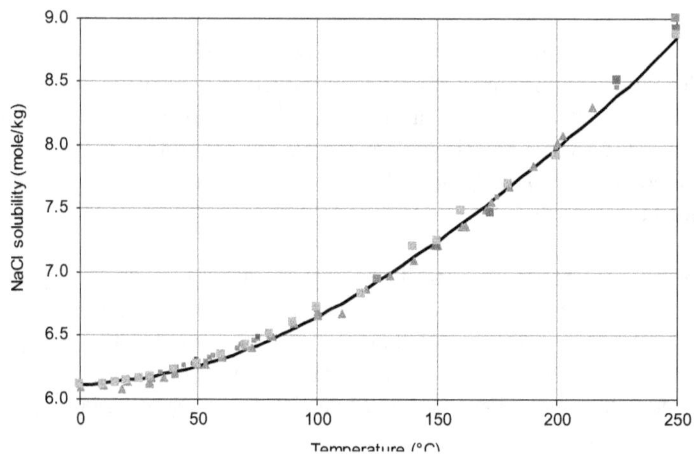

Figure 4.10. Solubility of NaCl in water versus temperature, model simulation and various literature data.

The solubility of halite depends on pressure too, but the effect is weak compared to temperature, as shown in Figure 4.11. For example, it is necessary to have a pressure reduction of approximately 500 bar to get a reduction in solubility of 0.1 mol/L.

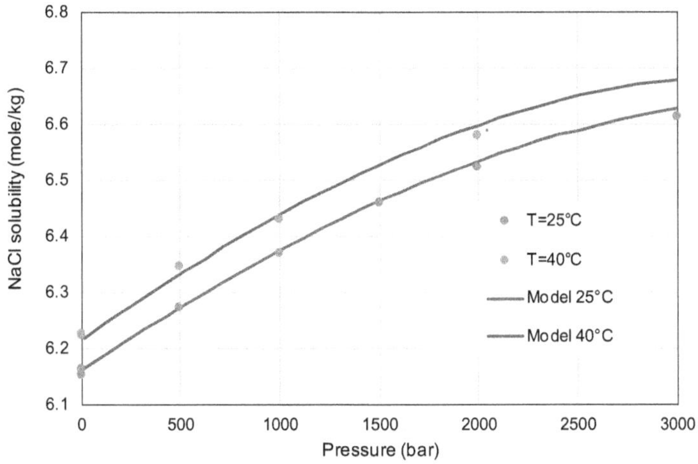

Figure 4.11. Solubility of NaCl in water versus pressure, model simulation and literature data.

Even if solubility is more dependent on temperature, the two effects will add up. When water flows together with oil or gas from the reservoir to the surface, both pressure and temperature will decrease, and both changes will lower halite solubility and increase the risk of halite precipitation.

The other main reason why halite forms is water evaporation. When water evaporates, all ion concentrations increase, and even if the FW was far below halite saturation in the reservoir, it can easily become saturated when water evaporates.[8, 9] The typical example where halite is formed due to water evaporation is in a (high-pressure high-temperature) HPHT gas well with a large drawdown and low water cut. As shown in Section 4.3.1, the water content in the gas is inversely proportional to the pressure. This means that as pressure drops when gas flows from the reservoir and into the well, the gas will absorb water. When some of the water evaporates, all ion concentrations become higher. This will, of course, also mean that the scale potential of all minerals will increase, and there can also be the formation of $CaCO_3$, $BaSO_4$, $CaSO_4$, $FeCO_3$, FeS, PbS, and so on, simply because all ion concentrations increase. The main difference between all the "normal" scale minerals and halite is that the "normal" scale types typically form scale masses in the mg/L range, while halite can form salt masses in the g/l range, that is, 1,000 times more.

Let us consider an example with reservoir conditions of 650 bar, 142°C, and the water with salinity that gives SR(NaCl) = 0.66 in the reservoir. The gas is a typical natural gas, and the gas rate is 1 MSm^3/d. The FW rate will be varied. Saturating the gas with water at reservoir conditions gives a water concentration in the reservoir gas of 0.96 mol% and with a gas rate of 1 MSm^3/d, this corresponds to 7.3 m^3/d water dissolved in the gas. This means that there can be up to 7.3 m^3/d of condensed water in this well. It has been assumed that the flowing well pressure is 530 bar, giving a drawdown of 120 bar. It is further assumed that the inflow can be simulated as an isenthalpic process ($\Delta H = 0$), and the temperature was calculated as a function of pressure. However, because pressure is in the range of 500–600 bar, the J-T coefficient is close to zero (it is actually negative at 650 bar and turning positive at ca. 550 bar, so the change in temperature is marginal in this case. This means that water evaporation is driven by pressure reduction only.

Figure 4.12 shows how SR(NaCl) varies with pressure during inflow to the well for different water rates. The gas rate is the same in all cases, and the gas phase is saturated with water at reservoir conditions. When pressure decreases, the gas will take up water according to Equation (4.3), and this will have a larger effect on water composition when there is less water present. Figure 4.13 shows how much water is taken up by the gas when pressure decreases. It may be surprising to see that gas will take up *more* water when the water rate is high. The reason for this is that when water is evaporated, salinity will increase, and this will reduce the vapor pressure of water, because the water activity will be lower (see Rault's law Equation [4.2]). Therefore, when the water rate is lower, salinity increases more, water activity is lower, and less water evaporates. When the water becomes NaCl-saturated, the water composition does not change much; it stays at saturation, and the amount of evaporated water will be the same. This is seen when comparing results for 1 m^3/d and 2 m^3/d. When the water rate is even lower, the system is remarkably close to complete evaporation and that is why the line for 0.5 m^3/d deviates. If the water rate is 25 m^3/d, evaporation of up to ~0.9 m^3/d will not change the salinity very much and SR(NaCl) only increases slightly. But when the water rate is lower, the effect is larger, and if the water rate is 5 m^3/d, NaCl will become supersaturated in the well. When the water rate is lower, NaCl will become supersaturated at higher pressures, (earlier during inflow) and the final SR will be higher.

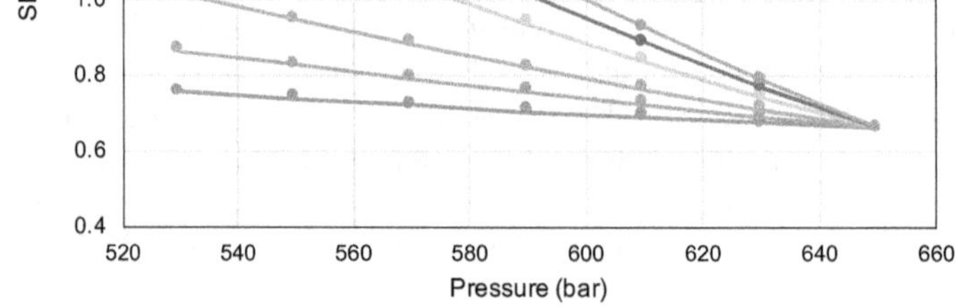

Figure 4.12. SR(NaCl) as function of pressure during inflow to well for different water rates.

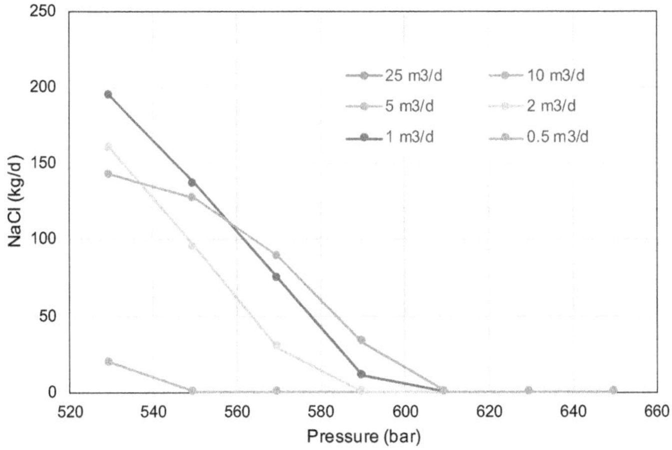

Figure 4.13. Water evaporation during inflow as function of pressure for different water rates.

Figure 4.14 shows how much NaCl can form as a function of pressure for different water rates. Note that the line shows how much can precipitate at the given pressure, if there has not been any previous precipitation. Normally, one would expect that the mass of scale/salt is proportional to the water rate. But for halite precipitation when it is driven by water evaporation, low water rates can give *more* salt than high water rates. The figure shows that there is no precipitation before the water rate is 5 m^3/d or lower. Then, because the effect of evaporation is stronger at low water rates, more salt precipitates when the water rate is lower. A maximum is reached when the water rate is about 1 m^3/d. At lower water rates, the system is very close to complete evaporation meaning that everything precipitates, and the mass of NaCl will, of course, start to decrease, if there is lower water production. But it should be noted that with only 0.5 m^3/d FW, there can be up to 142 kg/d of NaCl. For comparison, the simulation also shows that other scale types can form, for example, $CaCO_3$, but the mass of CO_2 that can form if the water rate is only 0.5 m^3/d is less than 100 g/d.

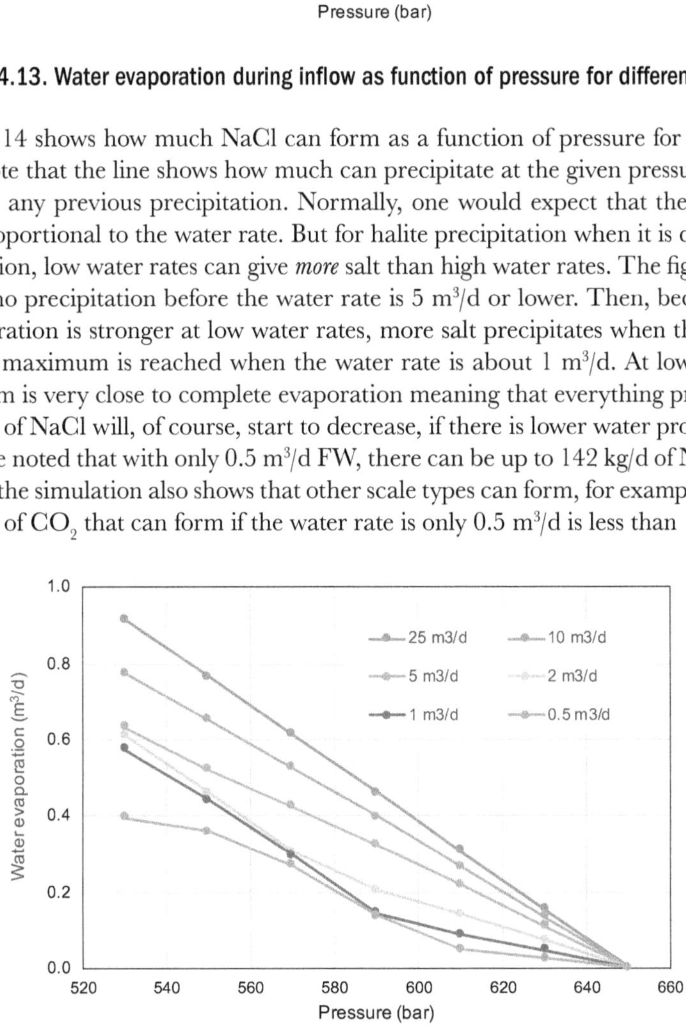

Figure 4.14. NaCl precipitation as function of pressure during inflow to well for different water rates.

4.6 Gas Well Water Sampling and Estimation of Formation Water Composition

If gas wells produce water, the operators will normally try to take PW samples and analyze the ionic composition. The results can often be very surprising and show that the PW composition is very different from the expected FW composition.[1] The explanation for this is very often that sampling/surface conditions are such that water has condensed from the gas and therefore diluted the PW, resulting in much lower ion concentrations than expected. Alternatively, the conditions are such that more water has evaporated which, of course, gives much higher concentrations.

An example is shown in Figure 4.15 where the measured chloride concentration (Cl^-) in PW samples from a North Sea field has been plotted over time. The red line shows the expected chloride concentration in the PW. For the first 2 years, the PW was almost fresh water with chloride concentrations often below 100 mg/L. Then, the chloride concentration increased rapidly in most of the water samples. All the other ions also showed similar variations in ion concentration, but it was uncertain, if there were one or several sources for salty water.

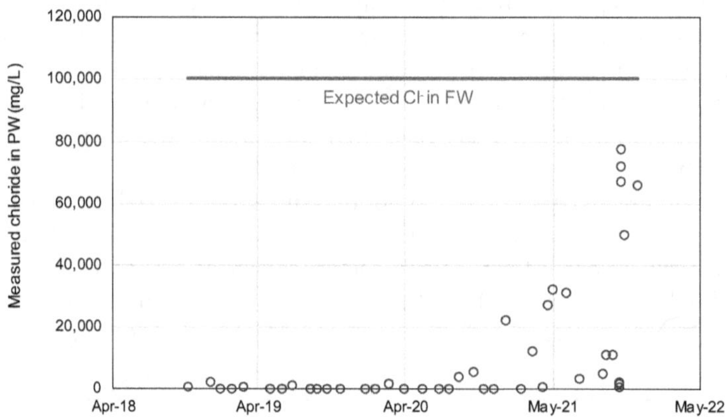

Figure 4.15. Chloride concentrations in produced water from a gas well.

It was suspected that the reason for this was the varying content of condensed water that diluted all ions. A simple method to verify this is to look at ion concentration ratios. Figure 4.16 shows a plot of various ion ratios, plotted versus the total dissolved solids (TDS) of the samples. Some of the ion concentrations have been multiplied by a factor to scale the ratios so that they fit into the same plot. It is clear from this graph that as soon as the TDS increases, which means that the FW rate has increased, the ion ratios stabilize, and this shows that the PW contains the same FW, but that it has been diluted with condensed water. As expected, there will be more scatter when the TDS is low, but most of the ratios at TDS < 10,000 mg/L show approximately the same ion ratio as at higher TDS. If, however, a plot like this shows that the ion ratios vary, it would mean that the well is producing different types of water, for example, if there is water production from several zones having different water compositions. A possible risk when plotting ion/chloride ratios, as shown in Figure 4.16, is that if there is an error in the chloride concentration, all ratios become wrong. This may be the reason all ratios are a bit low for the samples taken when TDS was about 90,000 mg/L. A safer method can be to plot ion concentrations vs. TDS or also look at ratios between other ions.

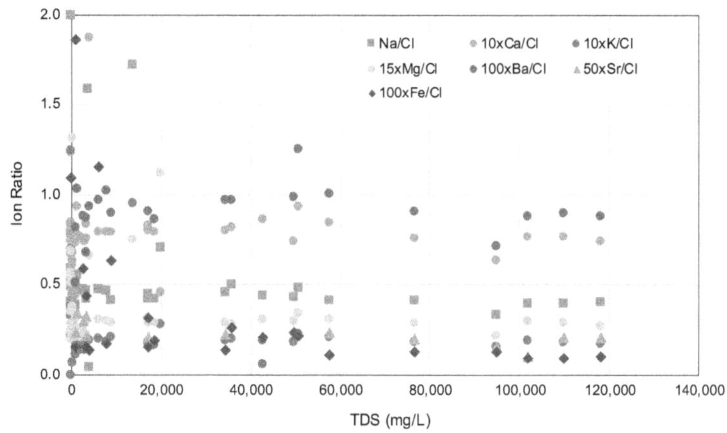

Figure 4.16. Ion ratios in produced water samples.

When the PW is diluted with condensed water, topside water samples will not show the FW composition, because all ion concentrations will be diluted. This means that even if all the ratios between the ions are correct and the same as in the FW, we do not know the actual concentrations in the reservoir. To estimate the FW concentrations, it is necessary to estimate the degree of dilution by condensed water. A possible calculation sequence is:

1. Calculate the water concentration in the reservoir gas at reservoir conditions using current reservoir pressure and temperature and an estimated FW composition.
2. Use the measured gas rate and conditions at the water sampling point to calculate how much of the water that was dissolved in the reservoir gas has condensed.
3. From the total water rate, subtract the condensed water rate calculated in step 2, and adjust the measured ion concentrations accordingly. This gives a good estimate of the water composition. If this composition is very different from the composition used to saturate the gas in step 1, repeat the three steps.

An alternative approach is possible with some of the scale simulation programs. The user can then

- Enter the reservoir gas composition and measure gas rate.
- Enter the PW composition and measure water rate.
- Let the program *add* water to make the gas/condensate water-saturated at sampling conditions.

- Run a calculation at reservoir conditions (back calculate to reservoir). Some of the water will then be evaporated to saturate the gas at reservoir conditions, and the remaining water will now have higher ion concentrations which will be a good estimate for the FW composition.
- Tune the water to reservoir conditions by adjusting some ion concentrations if any minerals are supersaturated and alkalinity to make the water calcite saturated.

A detailed description of water sampling from both gas and oil wells is given by McCartney.[1]

4.7 Summary

Scale modeling in gas wells is, in theory, not different from scale modeling in oil wells, because all chemical reactions, equilibria, and scale formation reactions will be controlled by the same chemical and thermodynamic laws. But because gas wells normally have low water rates compared to gas rates, and because pressure and temperature vary differently, some parameters are more important for gas wells than for oil wells.

To classify a reservoir as a gas reservoir, the reservoir temperature (and normally also pressure) must be higher than the critical point of the reservoir fluid. When the gas is produced, pressure and temperature will decrease, and the gas may cross the dew point line of the phase envelope, and condensate is formed. How much condensate is formed can vary a lot, but a typical gas condensate fluid will have GOR (or GCR) ranging from perhaps 2,000 to 10,000 Sm^3/Sm^3 and higher. In rare cases, the reservoir conditions are located to the right of the phase envelope, and the gas wells can produce "dry" gas without condensate.

The water production rate from a gas well is normally lower than from an oil well and the gas-to-water ratio will be much higher in gas wells than in oil wells. Because gas can take up more water than oil, gas wells will be more sensitive to water evaporation or condensation than typical oil wells. It is therefore very important to include water evaporation/condensation and how this changes the water composition in the scale modeling of gas wells to be able to simulate scale risk correctly in gas wells. This is, of course, also very important for many oil wells, but often more important for gas wells. To be able to model scale risk in gas wells, it is therefore necessary to use a scale simulation tool that has a full PVT model that also includes water evaporation and condensation.

The main factors influencing water evaporation and condensation are pressure and temperature. A reduction in pressure will result in evaporation, while a reduction in temperature will cause condensation. Whether there will be evaporation or condensation when gas flows from the reservoir, into the well and further to the surface, will therefore depend on how pressure and temperature change *relative* to each other. In many cases, there will be water evaporation during inflow from the reservoir to the well but condensation when the gas flows up the well toward the surface. It is therefore very important to know the P-T profile in the well. The P-T profile of a well can be calculated by considering both thermodynamics and hydrodynamics with tools like OLGA or Prosper.

When gas expands, it will normally cool down. This is known as the J-T effect. This is true for most gases and gas mixtures but only up to a certain pressure. Above ~450–550 bar, the J-T coefficient is negative, and the gas will heat up when depressurized. Therefore, if the reservoir is, for example, 700 bar, gas will head up when pressure decreases during inflow from the reservoir to the well. If the reservoir pressure is lower than 450–550 bar, for example, only 300 bar, the gas will cool down when flowing into the well. When gas and water (and condensate) flow together, there is multiphase flow, and the J-T coefficient is not valid for a multiphase system. Instead, inflow from the reservoir to the well can be assumed to be an irreversible adiabatic process, which means that it is an isenthalpic process and that the enthalpy is constant during inflow. By calculating the enthalpy of the fluid at reservoir conditions, the temperature in the well when pressure has decreased can be found as the temperature giving the same enthalpy as in the reservoir. From the assumption that inflow to the well is isenthalpic, the inflow temperature profile can be estimated.

Similarly, flow from the bottom of the well to the surface is close to a reversible adiabatic process and can be modeled as an isentropic expansion. This is, however, not always accurate, because there will also be friction loss in the well, and in a lengthy well, there will also be heat flux through the tubing/casing. But because of the differences in fluid properties between gas and oil wells, the temperature will decrease much faster for a gas well compared to an oil well. This means that even if a pressure reduction normally results in an increase in scale potential for $CaCO_3$ and other carbonates/sulfides, the reduction in temperature will cause both water condensation and a higher solubility of CO_2 in the water such that the scale potential for $CaCO_3$ increases much slower in gas wells than in typical for oil wells. The scale potential for carbonates may even decrease. For $BaSO_4$, the solubility decreases with temperature (at least below 100°C), and a reduction in temperature will therefore increase the scale potential.

Halite (NaCl) has very high solubility and is not a typical scale mineral. Still, it can sometimes form in both oil and gas wells, and there are two typical situations when halite forms. The first is when the FW is saturated or close to saturated with halite in the reservoir. When pressure and temperature decrease, both changes cause lower solubility and halite precipitation will occur, if the system goes below the solubility limit. The other reason is halite formation due to water evaporation. The typical case where water evaporation is the main driving mechanism for halite formation is in a HPHT gas well with a low water cut and high drawdown. The reduction in pressure will result in water evaporation. If the pressure is lower than the J-T inversion point, there may be some cooling that compensates for the pressure drop and reduces evaporation, but if the pressure is high, there will be heating during inflow to the well, which will cause more evaporation and increase the risk of halite formation.

PW from gas wells may not always have the same water composition, because it has been diluted with condensed water from the gas. Some gas wells will produce condensed water only for a long period of time before it is possible to detect a gradual increase in salinity. Eventually, if the formation water rate becomes high enough, the PW will be similar to the FW. A good way to check if diluted water samples come from the same formation is by logging the ion ratios and checking if they are stable. It is possible to back-calculate the PW composition to FW by estimating the rate of condensed water, subtracting this from the total water rate and correcting all ion concentrations accordingly. Some scale simulation packages can do this automatically, and the user only has to enter gas composition and PW composition together with conditions at the sampling point and reservoir conditions, and the PW is back-calculated to the reservoir.

Nomenclature

a_{H_2O}	Activity of water in water phase
d	Day
FW	Formation water
GCR	Gas condensate ratio, volume gas divided by volume of condensate
GOR	Gas oil ratio, volume gas divided by volume of oil
HPHT	High pressure, high temperature
J-T	Joule-Thomson
MSm^3	Thousand standard cubic meter
P	Pressure
$P^{sat}_{H_2O}$	Saturation pressure or vapor pressure of water
PLT	Production logging tool
PVT	Pressure-volume-temperature
PW	Produced water
R	Gas constant
SI	Saturation index
Sm^3	Standard cubic meter
SR	Saturation ratio
T	Temperature (absolute temperature)
TDS	Total dissolved solids
$\overline{V}_{H_2O}$	Molar volume of water
x, y	Mole fraction, *x* used for oil phase, *y* used for gas phase
η	Joule-Thomson (J-T) coefficient
φ_i	Fugacity coefficient of component *i* in either oil or gas phase
$\varphi^{sat}_{H_2O}$	Fugacity coefficient for water in water vapor at saturation pressure

References

1. R.A. McCarney, T. Østvold, "Mass Transfer of H_2O Between Petroleum and Water: Implications for Oilfield Water Sample Quality," *Applied Geochemistry* 20 (2005): pp. 1518–1532.
2. K.S. Pitzer, "Thermodynamics of Electrolytes. I. Theoretical Basis and General Equations," *Journal of Physical Chemistry* 77 (1973): pp. 268–277.
3. F. Vassenden, Ø. Gustavsen, F.M. Nielsen, et al., "Why Didn't All the Wells at Smørbukk Scale in?," SPE-94578-MS, SPE International Symposium on Oilfield Scale, held May 11–12, 2005 (Aberdeen, UK).
4. R.A. McCarney, S. Duppenbecker, R. Cone, "Constraining the Conditions of Scale Deposition in a Gas Condensate Well: A Case Study," SPE-169804-MS, International Oilfield Scale conference and Exhibition, held May 14–15, 2014 (Aberdeen, UK).
5. K. Sandengen, K. Ramstad, S. Høiland, et al., "Should a New Well be Equipped With Downhole Chemical Injection Lines," SPE-209491-MS, SPE International Oilfield Scale Conference and Exhibition, May 25–26, 2022 (Aberdeen, UK).
6. Schlumberger, "OLGA Dynamic Multiphase Flow Simulator," 2022, https://www.software.slb.com/products/olga
7. Petroleum Experts, "Prosper, Multiphase Well and Pipeline Nodal Analysis," 2023, https://www.petex.com/products/ipm-suite/prosper/
8. R. Jasinski, D. Frigo, "The Modeling and Prediction of Halite Scale," 3rd International Conference Solving Oilfield Scaling, held January 22–23, 1997 (Aberdeen, UK).
9. P.J. Schuler, E.J. Daniels, L. Burton, et al., "Modeling of Scale Deposition in Gas Wells With Very Saline Produced Water," NACE-00118, Annual Corrosion Conference, held March 26–31, 2000 (Orlando, Florida, USA).

References

5 Scale Inhibitor and Evaluation

Chunfang Fan[1]
[1]SLB, Houston, Texas, USA

5.1 Introduction

Since the deposition of sparingly soluble inorganic minerals has been found to cause operation issues in oil and gas production, tremendous efforts have been made to mitigate scaling issues. In the early days after scaling issues were recognized, acids, such as inhibited hydrochloric acid, were suggested and used as an effective remediation method to remove acid-soluble scales, such as carbonate scales. However, due to the corrosive nature of acids and the requirement of large quantities to achieve desired results, it is not efficient and feasible to use acids to prevent the deposition of acid-soluble scales. During that time, in seeking a better way to prevent scale formation in the first place, many types of chemicals, such as phosphates, were suggested as possible inhibitors to accomplish the goal.[1–3] Among those chemicals, polyphosphates became the most widely used scale inhibitors because of the small quantities required and their corrosion control properties, while some of the other chemistries suggested are corrosive and/or require large quantities to prevent scale formation, which makes them not ideal chemicals as scale preventives.[2] In the late 1930s, a concept called "Threshold Theory" was introduced by Hatch, Rice, and Partridge.[4–6] They studied the inhibition of calcium carbonate using hexametaphosphate with specific emphasis on the peculiar ability of as little as 1 or 2 parts per million (ppm) to inhibit the precipitation of calcium carbonate, and they concluded that the threshold level of hexametaphosphate increases stabilization for a condition of supersaturation, and the deposition of calcium carbonate nuclei formed is inhibited by the adsorption of hexametaphosphate on it, which thus prevents the growth of the nuclei beyond colloidal dimensions.[4–6] In 1940, Reitemeier and Buehrer's study supported this "Threshold Theory," and their findings suggested glassy polyphosphate at a very low concentration level in the milligram-per-liter range could effectively inhibit calcium carbonate formation as the results of the adsorption of polyphosphate on potential scale-forming crystal faces restricting calcium carbonate crystallization.[7, 8] Since then, many different chemistries, such as phosphonates, phosphate esters, phosphino polymers, polyacrylates, polymaleates, and polysulfonates have been found to effectively retard or delay scale formation in threshold dosages (typically at the level of a few milligrams per liter). With these findings, scale inhibitors, commonly referred to those threshold inhibitors, are widely used in oilfield and industrial water treatment to prevent scale formation and deposition. In contrast to threshold scale inhibitors, chelating agents, such as ethylenediaminetetracetic acid, diethylenetriaminepentacetic acid, and glutamic acid diacetic acid, which can be used to inhibit scale formation as well, often need high dosages to form water-soluble metal ions and chelant complexes in which one molecule of a chelant coordinates with one or several molecules of metal ions. The scale inhibition mechanism of chelating agents is different from that of threshold scale inhibitors. Chelating agents prevent metal ions from entering into the reactions with anionic ions in solutions by forming stable bonds with metal ions in which the ion activities of metal ions are reduced due to the formation of ion complexes. From a thermodynamic point of view, a supersaturated solution could become undersaturated as long as there are enough molecules of chelating agents in the solution to prevent metal ions from participating in the reaction with anionic ions. As a result, in contrast to threshold scale inhibitors, such chelating agents used in scale control are often called thermodynamic scale inhibitors. Due to the much higher treatment rates of these chelating agents compared with threshold scale inhibitors, they are not commonly used as oilfield scale inhibitors.

The evaluation and selection of oilfield scale inhibitors for a given application in earlier days were often done with a limited laboratory investigation through static precipitation tests and/or dynamic flow tests in an autoclave with a weighted tubular coupon, where scales could be deposited, observed, and measured. Because of the difficulties of simulating oilfield conditions in such laboratory tests in the earlier days, a scale inhibitor that is selected based on such laboratory tests could be found to be ineffective after being deployed into an actual oil well. As a result, field trials of the selected scale inhibitors need to be conducted to assess the scale inhibitors' inhibition performance. In recent decades, the dynamic tube blocking test has been gradually accepted as an industrial standard testing method to evaluate a scale inhibitor's inhibition performance, as it could better simulate field conditions than a static jar test, especially for a high-temperature and high-pressure environment. In this chapter, the common oilfield scale inhibitor chemistries mainly for controlling carbonate and sulfates and the evaluation of scale inhibitors are discussed in detail.

5.2 Scale Inhibitor Chemistries

Since the "Threshold Theory" was introduced in the late 1930s, enormous progress has been made in developing scale inhibitor chemistries. From polyphosphate to various phosphonates, phosphate esters, phosphino polymers, polyacrylates, polymaleates, and polysulfonates, a long list of chemistries has been found, developed, and applied in field applications with the understanding of the mechanisms of scale formation and inhibition. Some reviews of scale inhibitors have been published[9–12] over the last 30 years. In this section, the scale inhibitor chemistries reviewed and discussed are the "Threshold Inhibitors" commonly used in oil field scale control.

5.2.1 Polyphosphates

Polyphosphates (Figure 5.1), mainly sodium tripolyphosphate and sodium hexametaphosphate (Figure 5.2), could be considered the first type of threshold inhibitor used for oilfield scale inhibition.[2] This type of chemistry has long been found to be able to inhibit the precipitation of calcium carbonate at low dosages and can also be used in corrosion inhibition. However, due to its limitations in thermal stability and compatibility for oilfield scale control, this type of chemistry is mainly used in municipal and industrial water systems, such as a boiler system at a low-calcium concentration.[13]

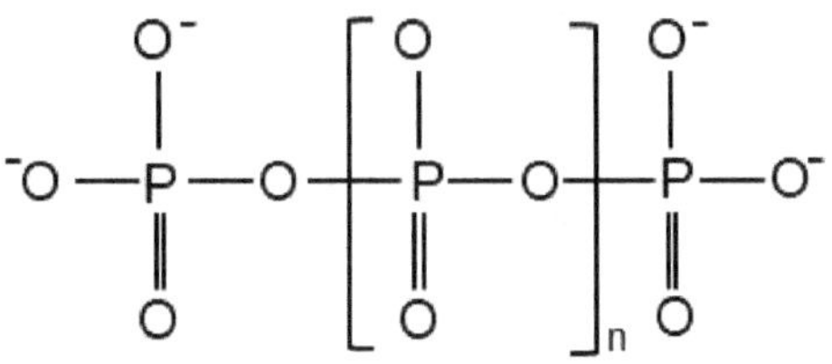

Figure 5.1. Polyphosphate.

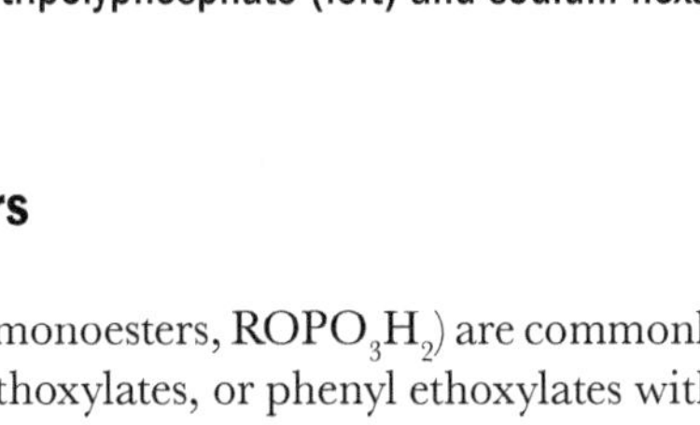

Figure 5.2. Sodium tripolyphosphate (left) and sodium hexametaphosphate (right).

5.2.2 Phosphate Esters

Phosphate esters (mainly monoesters, $ROPO_3H_2$) are commonly produced by the reaction of either alcohols, alcohol ethoxylates, or phenyl ethoxylates with polyphosphoric acid. Phosphate esters are highly versatile anionic surfactants. Their unique range of properties makes them widely used in emulsion polymerization, textile auxiliaries, maintenance chemicals, metal finishing, and many other industry applications. The common type of phosphate ester used as a scale inhibitor is triethanolamine phosphate ester (Figure 5.3). Compared to other phosphonate types of scale inhibitors, phosphate esters are considered environment-friendly scale inhibitors. Phosphate esters play a crucial role in scale control within various industrial processes, particularly in water treatment systems. These compounds are adept at inhibiting the formation of scales, such as calcium carbonate and calcium sulfate. Phosphate esters are excellent scale inhibitors for preventing calcium carbonate and are generally compatible with high-calcium brines. Phosphate esters function by forming a protective layer on metal surfaces, preventing the deposition of scale-forming minerals. Their ability to sequester metal ions helps mitigate scale buildup, ensuring the smooth operation of industrial machinery. Despite their effectiveness, phosphate esters have limited thermal stability and are generally recommended to treat scale formation up to 95°C (203°F).[13]

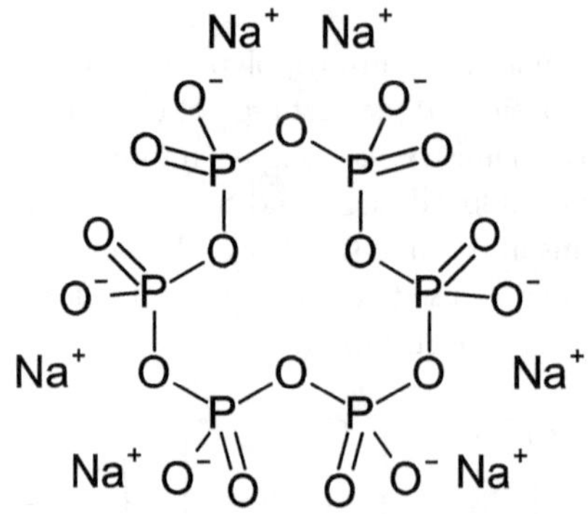

Figure 5.3. Triethanolamine phosphate ester.

5.2.3 Phosphonates

Phosphonates are the most common scale inhibitor chemistries used in oilfields for scale control due to several reasons. Phosphonates are generally low-cost scale inhibitors, and various chemistries of phosphonates show excellent inhibition performance against different types of scales, such as calcium carbonate, iron carbonate, calcium sulfate, and barium sulfate. Additionally, phosphonates can be easily detected by either measuring elemental phosphorous using inductively coupled plasma optical emission spectroscopy (ICP-OES) or measuring phosphate ions using a colorimetric method, such as the Hach method. Phosphonates in general have a better adsorption profile on rock/mineral surfaces than polymeric scale inhibitors, which makes them more suitable for downhole scale squeeze applications. Figure 5.4 shows the molecular structures of several most commonly used phosphonates.

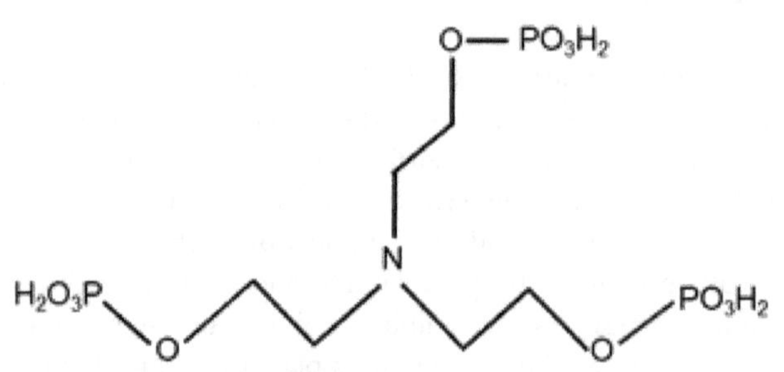

Figure 5.4. Molecular structures of the common phosphonate scale inhibitors.

Among these phosphonates, bis-hexamethylene triamine-penta(methylene phosphonic) acid (BHPMP) has an excellent inhibition performance against sulfates, especially barium sulfate, and it is also an effective inhibitor for calcium carbonate. Diethylene triamine-penta(methylene phosphonic) acid (DETPMP) is an excellent scale inhibitor against both carbonates and sulfates, and this chemistry is not only widely used in continuous injection applications but also in downhole scale squeeze applications. It is probably the most commonly used phosphonate scale inhibitor in the oil industry.[13]

Nitrilotris(methylene phosphonic) acid (NTMP), or amino tris(methylene phosphonic) acid (ATMP), has a simpler molecular structure compared to BHPMP or DETPMP. NTMP is not the most effective inhibitor for controlling both carbonates and sulfates, and it is commonly used to inhibit calcium carbonate scale only.

Hydroxyethyl amino bis(methylene phosphonic) acid is another excellent calcium carbonate scale inhibitor. It has superior calcium tolerance, which makes it one of the top candidate scale inhibitor chemistries selected for any high-calcium brines. 1-Hydroxyethylidene-1,1-diphosphonic acid (HEDP) and 2-phosphono-1,2,4-butanetricarboxylic acid (PBTC) are the other two common phosphonate scale inhibitors. Unlike the other phosphonates, HEDP and PBTC are not widely used in oilfield scale control but more in industrial water treatments, boiler applications, and liquid dishwashing and detergents.

Hexamethylene diamine-tetra(methylene phosphonic) acid (HDTMP) is a highly effective scale inhibitor widely utilized in water treatment processes and applications such as cooling water systems, boilers, and oil extraction processes. Its stability under varying conditions, low toxicity and efficient scale inhibition, makes HDTMP a key player in safeguarding industrial equipment and enhancing the longevity of water-based systems. Recognized for its exceptional ability to control the formation of calcium sulfate and barium sulfate scale deposits, in particular calcium sulfate, HDTMP is often used to prevent calcium sulfate precipitation and subsequent scaling issues in oilfields. P,P′-[[[2-[(2-hydroxyethyl) (phosphonomethyl) amino]ethyl]imino]bis(methylene)]bis[phosphonic acid], also commonly known as amineothylethanolamine phosphonate, has very high-calcium tolerance with excellent scale inhibition effects on both calcium carbonate and barium sulfate.

5.2.4 Phosphorus-Based Polymers

Since the phosphorus-containing function groups show good scale inhibition effectiveness and having phosphorus elements in the backbones of polymers makes measuring polymers' residual concentrations much easier compared to those polymeric scale inhibitors that do not contain any phosphorus groups, various phosphorus-containing polymer inhibitors have been developed and used extensively in oilfield scale control. Phosphino-polycarboxylic acid (PPCA) is one of the typical phosphorus-containing polymers, which comprises a single phosphino group bridging two polyacrylic or polymaleic chains (Figure 5.5). PPCA has high-calcium ion tolerance and is a great barium sulfate scale inhibitor. PPCA is used in downhole scale squeeze applications. However, it has less adsorption than phosphonates.

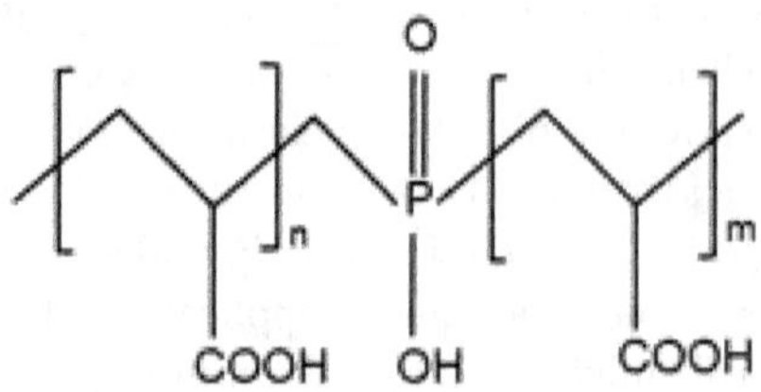

Figure 5.5. Molecular structure of phosphino-polycarboxylic acid.

N-phosphonomethylated amino-2-hydroxypropylene polymer (PMPA) is another type of phosphorus-containing polymer, which has a molecular weight of between 300 and 5,000 Dalton (Figure 5.6). PMPA has very good inhibition efficiency against barium sulfate and is also good for squeeze applications.[14–17]

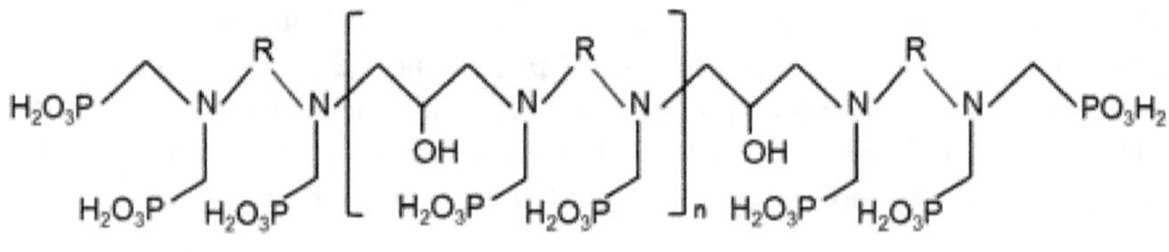

Figure 5.6. Molecular structure of N-phosphonomethylated amino-2-hydroxypropylene polymer.

Polyamino polyether methylene phosphonic acid (PAPEMP), as shown in Figure 5.7, is one of the most common phosphorus-containing polymers widely used in the oil field and industrial water treatments. PAPEMP is commonly produced by the reactions of polypropylene glycol diamine with phosphorous acid and formaldehyde. PAPEMP is very effective in preventing calcium carbonate precipitation at high supersaturation and high pH.[12, 18–20] PAPEMP has very high-calcium and high-iron tolerance, which means it could still have excellent inhibition efficiency against calcium carbonate, when a brine contains very high-calcium ion concentrations and/or high-iron ion concentrations when compared to other phosphonates or polymers.[21–23] PAPEMP is also shown to have an excellent inhibition performance against gypsum precipitation[24] and silica scales.[25] There are other types of phosphorus-based polymers, such as phosphino telomers, polyphosphinates, maleic acid–sodium q-styrenesulfonate copolymer, or phosphonate-end-capped phosphino polymers, which have been synthesized and used in scale prevention.[12, 13]

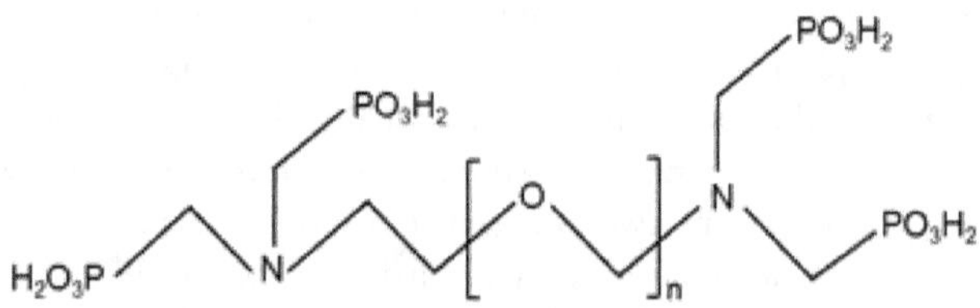

Figure 5.7. Molecular structure of polyamino polyether methylene phosphonic acid.

5.2.5 Sulfonated Polymers

Sulfonated polymeric scale inhibitors are another common group of chemistries showing excellent scale inhibition effectiveness and thermal stability at extremely high temperatures (>175°C). There are many commercially available sulfonated polymeric scale inhibitors, such as polyvinyl sulfonate (PVS), sulfonated polycarboxylate (SPCA), and maleic acid–sodium q-styrenesulfonate copolymer (Figure 5.8). Most of those sulfonated polymers have a polyvinyl backbone. The most common monomers used to make sulfonated polymeric scale inhibitors are acrylamido(methyl)propylsulfonic acid (AMPS), allyloxy-2-hydroxypropyl sulfonic acid, styrene sulfonic acid (SSA), maleic anhydride (MA), and vinyl sulfonic acid (VSA; Figure 5.8). PVS is commonly used for preventing sulfate scales, while it also shows good inhibition performance on carbonate scales.[26–28] Several studies[27, 29] have shown that PVS has better inhibition performance at low-pH conditions compared to other polymeric scale inhibitors containing carboxylate function groups, such as carboxymethyl inulin (CMI) and SPCA. That is because PVS has a lower pK_a value than scale inhibitor chemistries with phosphonic or carboxylic function groups and has lower stability constants with Ca or Mg ions.[13, 30] PVS does not have a high-iron tolerance[31] and has a shorter squeeze life due to poor adsorption compared to phosphonates.[32] Sorbie and Laing[30] suggested that PVS has excellent inhibition performance at low temperatures (4°C–5°C). Similar to PVS, SPCA has shown its inhibition efficiency at lower temperatures is better than that at higher temperatures[29] and has a poor squeeze life compared to phosphonates as well.[32] Both PVS and SPCA have very good thermal stability, while PVS is more thermally stable than SPCA. This is because the amount of sulfonation in SPCA is much less than in PVS or other sulfonated polymers. One study[33] has shown that PVS inhibition performance against the barite scale is completely unaffected by exposure to 200°C for 7 days, and the Fourier transform infrared (FTIR) data show that the spectra of PVS after 7 days of exposure to 200°C are identical to the ones of the untreated sample, while the spectra of SPCA after just 2 or 4 h of exposure to 200°C show significant changes compared to the untreated sample.[33] Such changes are most evident in the carbonyl absorption band, which has shifted slightly after thermal exposure, indicating an isomerization.[33]

Maleic anhydride and allyl sulfonic acid copolymers are the other sulfonated polymers that have been known to have excellent scale inhibition on sulfates, especially barium sulfate.[34–36] Maleic anhydride and allyl sulfonic acid copolymers or maleic acid and acrylic acid copolymers also have very good inhibition performance on calcium carbonate.[37, 38] There are some tagged copolymers/terpolymers were developed by polymerized monomers, such as acrylic acid, VSA or its salts, vinyl phosphoric acid or its salts, maleic anhydride, AMPS, and hydroxyphosphonoacetic acid.[39–41] These tagged copolymers/terpolymers have shown to be more suitable for squeeze applications, as they could be detected by using phosphorus element concentration or fluorescence wavelength.

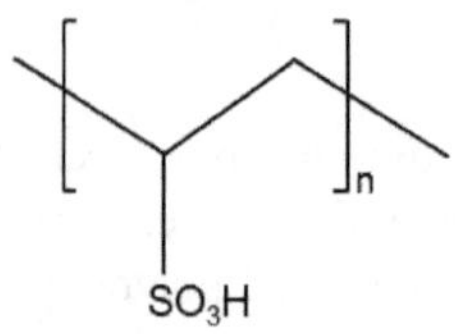

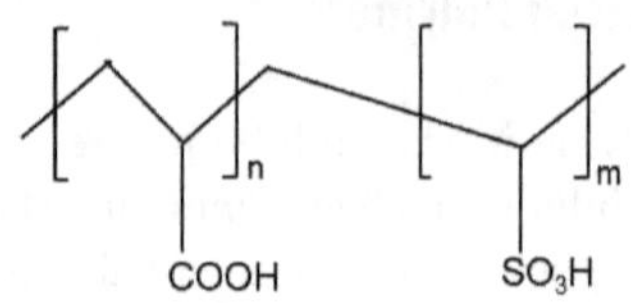

Polyvinyl sulfonate, PVS

Sulfonated polycarboxylate, SPCA

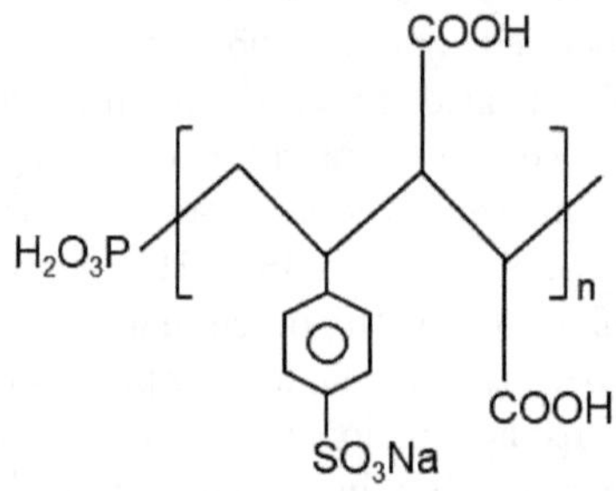

Maleic acid–sodium q-styrene sulfonate copolymer

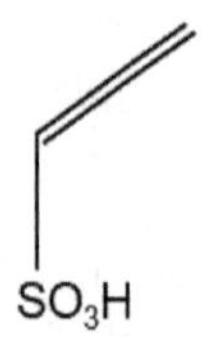

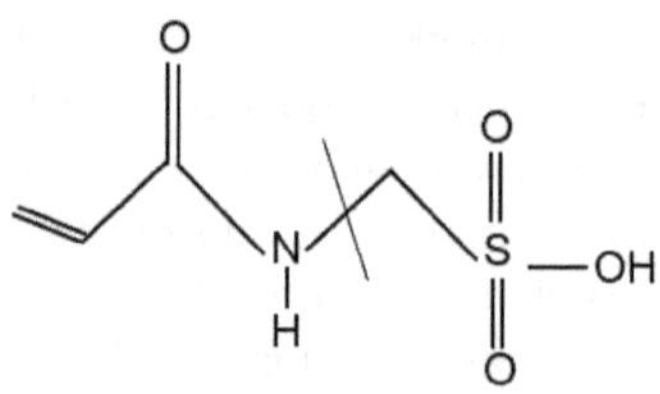

Vinyl sulfonic acid, VSA

Acrylamido(methyl)propylsulfonic acid, AMPS

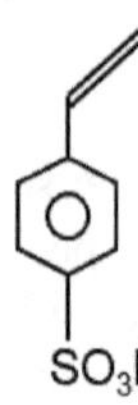

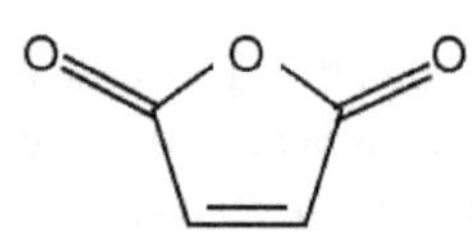

Styrene sulfonic acid, SSA

Maleic anhydride, MA

Figure 5.8. Molecular structures of common sulfonated polymers and monomers.

5.2.6 Biodegradable Polymeric Scale Inhibitors

As the increasingly strict environmental regulations on oil field chemicals, biodegradable scale inhibitors have been developed to meet regulatory requirements for applications in environmentally sensitive areas, such as the North Sea. For example, the Oslo and Paris Commission (OSPAR) has been implementing the Harmonized Offshore Chemical Notification Format as the standard environmental reporting and declaration format for chemical products discharged to the Northeast Atlantic Area and North Sea areas, which has been widely used as the standard reporting requirements in other areas as well.[42, 43] The Harmonized Mandatory Control System is used by the OSPAR to protect the marine environment by identifying those chemicals used in offshore oil and gas operations with the potential to cause an adverse environmental impact and restricting their use and discharge to the sea. Typically, color systems are used to distinguish the levels of bioaccumulation, biodegradability, and toxicity of a chemical product, such as the one used in the Norwegian area. For example, Figure 5.9 shows the four-color system that is used to indicate whether a product can be discharged based on these properties. For yellow banding, the following requirements must be met[44]:

Norwegian Pre-Screening Color System
Black = Not allowed to be discharged under any circumstance
Red = Only discharged under special conditions
Yellow = Can generally be discharged by the offshore industry
Green = No risk to the aquatic environment Poses little or no risk to the environment (PLONOR)

Figure 5.9. Norwegian HOCNF prescreening color system.

- Toxicity EC50 or LC50 > 10 mg/L
- Log Pow < 3
- Biodegradation > 20%

There are several commercially available biodegradable scale inhibitor chemistries commonly used in oilfield applications including polyaspartates, polytartaric acid, or their derivatives, CMI. Polyaspartates are highly biodegradable scale inhibitors produced by polymerizations of aspartic acid monomers, and they are fairly good corrosion inhibitors as well.[45–47] In the polymerization process, aspartic acid could form a linear thermal polycondensation polymer known as polysuccinimide through heating with or without a catalyst.[45] The polysuccinimide going through a hydrolysis reaction could form a random copolymer of α and β aspartate groups (III).[45] The structure of polyaspartates contains both α and β aspartate groups, and it is complex because of the certain degree of branching regardless of the manufacturing procedure.[13] As shown in Figure 5.10, α aspartate groups contain pendant carboxylic acid groups having one carbon atom further away from the peptide backbone than the β aspartate groups. The inhibition efficiency of polyaspartates against different types of scales is related to their molecular weight. There is an article suggesting polyaspartates in the range of 3,000–4,000 Dalton of MW are most effective against calcium carbonate and barium sulfate, whereas the optimum molecular weight of polyaspartates is around 1,000–2,000 Dalton of MW for controlling calcium sulfate.[45] Although polyaspartates have been shown to have a much poorer adsorption profile on rock surfaces than commonly used phosphonates based on the core flood data,[48, 49] they have been used in squeeze applications in environmentally sensitive areas showing a shorter squeeze life than a phosphonate.[48] Some lab data show that iron ions could impair polyaspartates' inhibition performance, and their tolerance to iron ions is limited.[50, 51] Polyaspartates also have limited tolerance to calcium ions and are thermally stable after aging at 110°C, but their performance is reduced after aging at 130°C.[52]

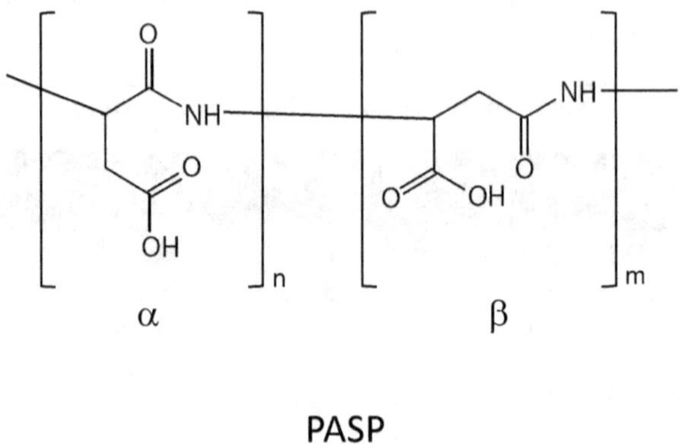

Figure 5.10. Molecular structure of polyaspartic acid.

Polytartaric acid (PTA) and its derivatives are another type of biodegradable polymer used as scale and corrosion inhibitors. PTA contains a mixture of monomers, dimers, trimers, and so on, of tartaric acid, and its molecular structure is shown in Figure 5.11. It is reported that *n* is less than 4 for optimum corrosion inhibition, and *n* is higher than 4 for best scale inhibition performance.[53]

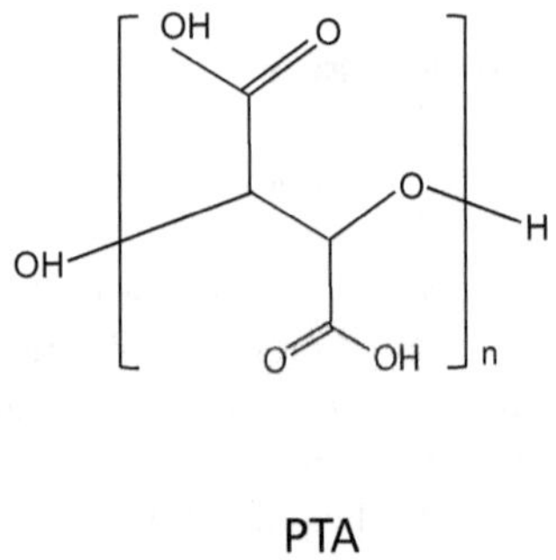

Figure 5.11. Molecular structure of polytartartic acid.

CMI is another commercialized biodegradable polymeric scale inhibitor commonly used in environmentally sensitive areas (Figure 5.12). CMI is a polysaccharide-based polycarboxylate with good carbonate scale inhibition performance, whereas its inhibition performance against sulfate is poor.[54–57] Similar to polyaspartates, some core flood data have suggested CMI has a much poorer adsorption profile on rock surfaces than commonly used phosphonates,[58, 59] which indicates the polymer needs to be optimized or used with a scale squeeze aid agent in squeeze applications. More information on biodegradable/environment-friendly scale inhibitors can be found in the following Chapter 18 of green scale inhibitors.

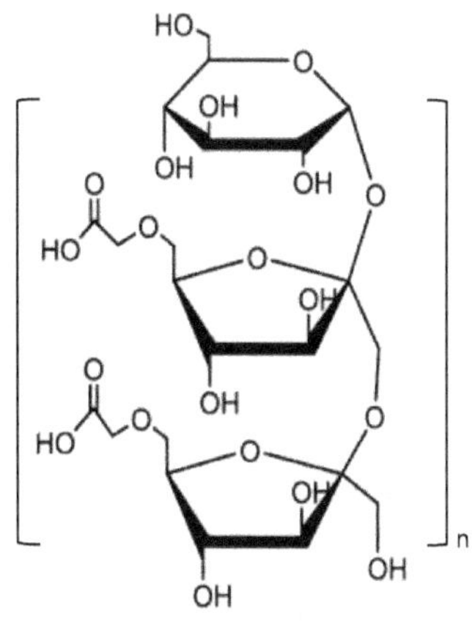

Figure 5.12. Molecular structure of carboxymethyl inulin.

5.3 Laboratory Evaluation of Scale Inhibitors

Before deploying scale inhibitors to a field application, multiple lab evaluation tests need to be done to ensure the successful field application of scale inhibitors. As discussed in the introduction part of this chapter, in earlier days, the evaluation and selection of oilfield scale inhibitors for a given application were often done with a very limited laboratory investigation mostly through static precipitation tests and/or dynamic flow precipitation tests in an autoclave with a weighted tubular coupon. The difficulties of simulating oilfield conditions in these laboratory tests often cause a selected scale inhibitor to be found to be ineffective after being deployed into an actual oil well. With the development/improvement of the evaluation methods and equipment in the past several decades, although field trials of the selected scale inhibitors are still being conducted to assess the scale inhibitors' inhibition performance, the success rate of selected scale inhibitors for field applications based on laboratory evaluations is much higher than in earlier days. There are several aspects of laboratory evaluations involving checking the inhibition performance of scale inhibitors, compatibility of scale inhibitors with field brines or synthetic brines made based on brine analyses, compatibility of scale inhibitors with metals and nonmetallic materials, compatibility of scale inhibitors with solvents and other chemicals in use, the thermal stability of scale inhibitors, deep water and/or umbilical injection qualification tests, emulsion and foaming tendency tests, product stability and cleanliness tests, environmental toxicity and bioaccumulation tests, and adsorption/desorption tests. Depending on the type of applications and field conditions, not all laboratory evaluations need to be conducted before selecting a scale inhibitor for a field application.

5.3.1 Inhibition Performance Evaluation of Scale Inhibitors

To assess a scale inhibitor against a scale, such as carbonates or sulfates, there are two methods widely used in oilfield evaluations. One is called the static jar test, and the other is the dynamic tube blocking test. Static jar tests have been used since earlier days to check scale inhibitors' inhibition efficiency, since it is a relatively simple and effective method to evaluate scale inhibitors' inhibition performance. In 1974, NACE International published a standard method for laboratory screening tests to determine the ability of scale inhibitors to prevent the precipitation of calcium sulfate and calcium carbonate from solution for oil and gas productions. Later on, it was reviewed and revised several times with the most recent version as NACE Standard TM0347-2015. There is another standard method for laboratory screening tests to determine the ability of scale inhibitors to prevent the precipitation of barium sulfate and/or strontium sulfate from solution for oil and gas productions that NACE published initially in 1997 and was revised several times. The most recent version is NACE TM0197-2022. Those two standard methods are accepted by the oil and gas industry and are widely used for evaluating scale inhibitors with or without slight modifications based on field conditions and applications. In these methods, generally two testing brines are prepared based on a specific field brine composition that simulates the field scaling brine, so mixing these two brines in equal volumes results in a composite brine with the ion concentrations equivalent to the specific field brine composition chosen for assessing scale inhibitors' inhibition performance. One brine is often called cationic brine or Brine A, which shall contain all of the polyvalent cations, such as Ca^{2+}, Ba^{2+}, Sr^{2+}, and so on, and other ions such as Na^{+}, K^{+}, Cl^{-}, and so on. The other brine is often called anionic brine or Brine B, which shall contain all of the scaling anions, such as SO_4^{2-} or bicarbonate ions, and the other ions, such as Na^{+}, K^{+}, Cl^{-}, and so on. Before starting a test, the pH of the prepared brines shall be adjusted based on field conditions. Both brines shall be preheated to the desired temperature before mixing. The test should have at least one "blank" sample, which does not contain any scale inhibitors, one "control" sample, which only contains scaling cationic brine mixed with Brine C (typically a NaCl solution to maintain equivalent ionic strength to that of Brine B/anionic brine), and several other samples of the mixture of cationic brine and anionic brine with different dosages of a scale inhibitor added. Static jar tests often last for a few hours a day. For a simplified static jar test, only visual observations are conducted, which are typically made right after mixing, 1 h, 2 h, 4 h, and so on. For example, Figure 5.13 shows a picture of sample bottles after 24 h of testing. The visual observations suggest the sample bottle with 30 ppm of the selected scale inhibitor is clear, while the blank sample and the sample with either 5 ppm or 15 ppm of the same scale inhibitor show precipitation. This indicates that 30 ppm of the selected scale inhibitor could inhibit scale formation for this brine at the condition tested.

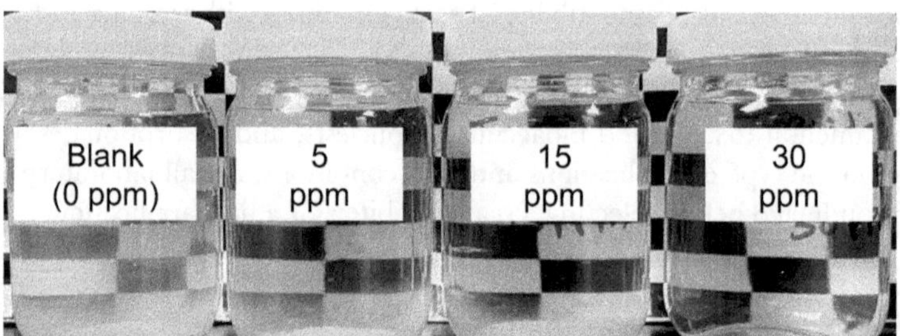

Figure 5.13. Picture of a typical static jar test of a scale inhibitor.

Besides visual observations, samples of the mixed brine could be taken from each sample bottle right after mixing, 1 h, 2 h, 4 h, and so on, and the sample analysis is often done by using an ICP-OES to measure scaling cationic ions' concentration. Other analytical methods, such as a wet chemistry titration method or atomic absorption spectroscopy, could be used to measure scaling ion concentrations. The ion concentration results of the brine samples are used to calculate the inhibition percentages of the tested scale inhibitor using the following equation.

$$\%\,\text{Inhibition}\ MSO_4\ (\text{or}\ MCaCO_3) = \frac{[M]_{\text{Sample}} - [M]_{\text{Blank}}}{[M]_{\text{Control}} - [M]_{\text{Blank}}} \times 100, \tag{5.1}$$

where $[M]_{\text{sample}}$ is the concentration of solution ion M in the inhibited solution after incubation. $[M]_{\text{blank}}$ is the concentration of soluble ion M in the blank sample after incubation. $[M]_{\text{control}}$ is the concentration of soluble ion M in the control after incubation.

Figure 5.14 shows an example of the inhibition percentage results of a static jar test conducted at 60°C for a phosphonate-based inhibitor.

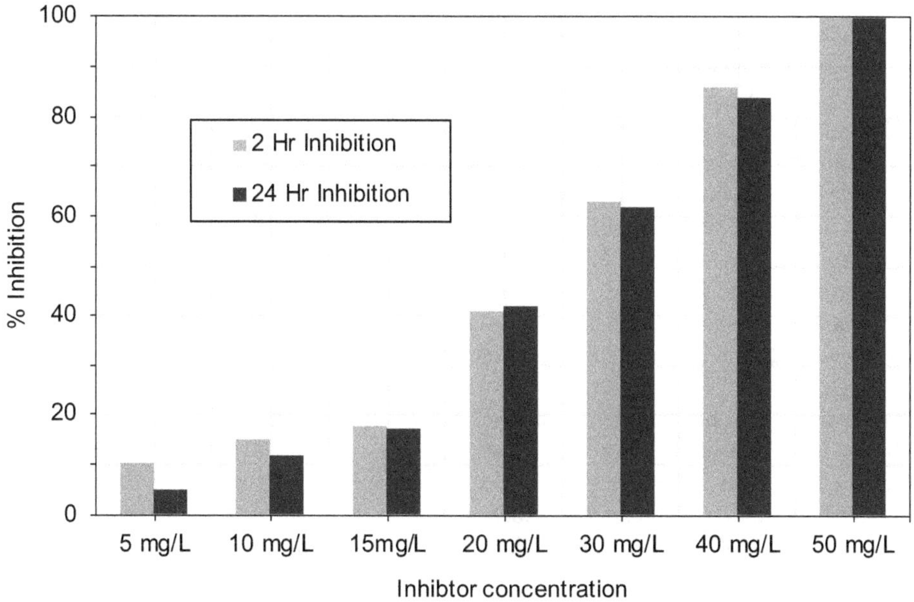

Figure 5.14. An example of barite inhibition percentage result of a static jar test.

Although the static jar test is a simple and effective way to screen scale inhibitors, there are some disadvantages of this method. Firstly, there are limitations on the testing temperature and pressure. The testing pressure is ambient pressure, which may not match field conditions. The testing temperature is typically below 100°C, when using regular glass bottles. While high-pressure rating glass bottles or tubes are used, the testing temperature could be up to 120–150°C depending on the maximum pressure rating of the bottles/tubes in use, but the testing pressure is not adjustable and is only related to the water vapor pressure at the testing temperature. Secondly, because the test is conducted in static jars and there is no fresh brine flowing into the testing jars, the pH of the brine in the jars will not be maintained constantly. It will change over time, especially when carbonate scales precipitate, which could affect the carbonate inhibition results of a scale inhibitor. Thirdly, when a test is performed at high temperatures (>80°C), evaporation could significantly affect the ion concentration results of the samples taken during the test, resulting in incorrect inhibition percentages calculated. Fourthly, due to analytical errors coming from sampling, dilution, and the instrument used, for some brines, the calculated inhibition percentages could not make any sense. For example, for a brine with a very high-calcium ion concentration, like 10,000 mg/L, the change in the calcium ion concentration due to the precipitation of calcium carbonate could be smaller than the errors in measuring the calcium ion concentration, since a large factor of dilution of the taken samples has to be made, so the samples could be measured by ICP-OES. If that is the case, considering the 10%–20% analytical error coming from ICP-OES, even a 1 mg/L difference in the reading of a diluted sample could result in a 100 mg/L difference when the dilution factor is 100. Such huge errors in the measured ion concentrations could be larger than the concentration difference between the *blank* sample and the *control* sample, which will make the calculated inhibition percentages meaningless according to Equation (5.1). The other issue of this method is related to the grade of the reagents used to make synthetic brines. In general, analytical reagent grade or ACS grade will be suitable for use in static jar tests. However, in some cases, impurities, mainly sulfates, from an analytical reagent grade or ACS grade reagent could cause precipitation to be seen when making a cationic brine that contains a very high concentration of barium. For example, for a field brine with high barium and sodium ions, when making synthetic brines to simulate the field brine used in the test, the sulfate impurity coming from the NaCl reagent could cause barium sulfate formation in cationic brine. If that is the case, an ultrapure-grade reagent needs to be used. Considering these disadvantages of the static jar test, this method is often recommended for screening scale inhibitors for a field with a low-temperature (<95°C) application, especially for sulfate inhibition since pH drifting will not significantly change the testing results for carbonates.

The dynamic tube-blocking test is another commonly used method to evaluate scale inhibitors' inhibition performance. The tests are often done with a commercially available instrument called dynamic scale loop (DSL). Compared to the static jar test, the dynamic tube-blocking test could be done at much higher temperatures and pressures. Depending on the capability of commercially available DSLs, the testing temperature could be up to 200°C, and the testing pressure could be up to 3,000 psi or higher. Figure 5.15 shows a typical setup of a dynamic scale loop, and there could be some differences between those commercially available DSLs.

Similar to the static jar test, two synthetic brines, a cationic brine and an anionic brine, will need to be prepared based on the field brine composition before the test. A scale inhibitor stock solution is made by adding the desired amount of a scale inhibitor into the anionic brine. Depending on the setup of a DSL (two-pump system or three-pump system) in use, the scale inhibitor stock solution is mixed with the anionic brine either by a third pump or by a flow control device to mix two solutions at a ratio, so the final mixture with the cationic brine contains the scale inhibitor at the desired testing concentration. In a typical DSL test,

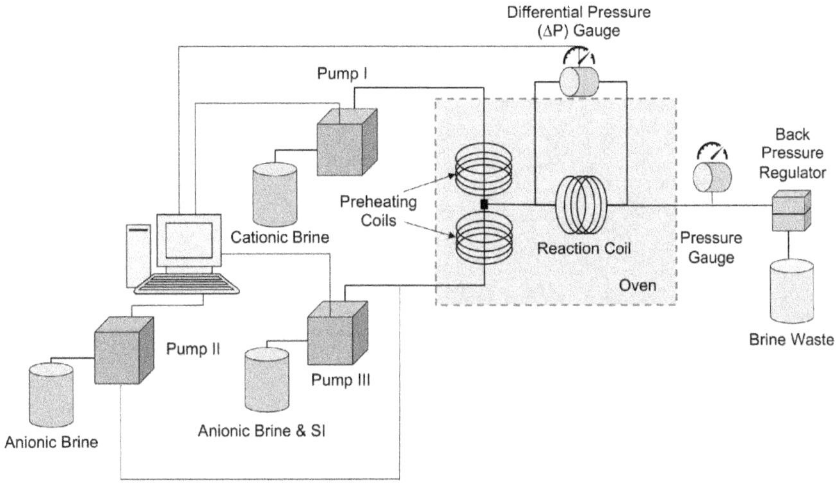

Figure 5.15. Schematic drawing of a dynamic scale loop setup.

both cationic brine and anionic brine with or without a scale inhibitor are pumped separately at 5 ml/min into the preheating coils, which are located inside an oven that has been set up at a specific temperature. The two brines mix at a 50:50 ratio at the entry point to a capillary coil (reaction coil) forming a scaling brine simulating the field brine composition. A typical coil size is 0.5–1.0 mm internal diameter and 1–3 m in length. The change of the differential pressure (ΔP) across the coil as a result of scale buildup in the coil is monitored through a differential pressure gauge. At the end of the flow pass, a back pressure regulator is used to control the system pressure at a specific testing pressure condition. In a test without any scale inhibitor added (blank run), the time taken for the differential pressure to reach a certain level (typically 2–5 psi higher than the baseline) is recorded as the blank scaling time. Then, a DSL test with a scale inhibitor at the desired dosage level is conducted to evaluate the scale inhibitor's inhibition performance by monitoring the differential pressure change across the reaction coil.

The efficacy of the inhibitor is judged by how much longer it takes for the differential pressure across the coil to increase by more than 0.5 or 1 psi compared to the baseline differential pressure. The industrially recognized passing criteria are that it needs to last a minimum of 1 h or at least three times the blank scaling time, whichever is longer should be used. Figure 5.16 shows typical DSL testing results. In this example, the blank DSL testing result indicates the blank scaling time is ~40 min. Therefore, the passing criteria are the differential pressure needs to be less than 1 psi above the baseline ΔP after 120 min (~3 times of blank scaling time) testing run. The scale inhibition testing results show that at a 5-ppm level of a scale inhibitor, the differential pressure increased more than 1 psi from the ΔP baseline after ~90 min of testing, whereas at 10 ppm, the change of differential pressure is less than 1 psi after 120 min, and at 15 ppm, there is no noticeable change in the differential pressure during the whole test. Though the differential pressure kept increasing above 1 psi from the ΔP baseline after ~133 min, this scale inhibitor is considered as passing at 10 ppm, and its minimum inhibitor concentration (MIC) is determined as 10 ppm.

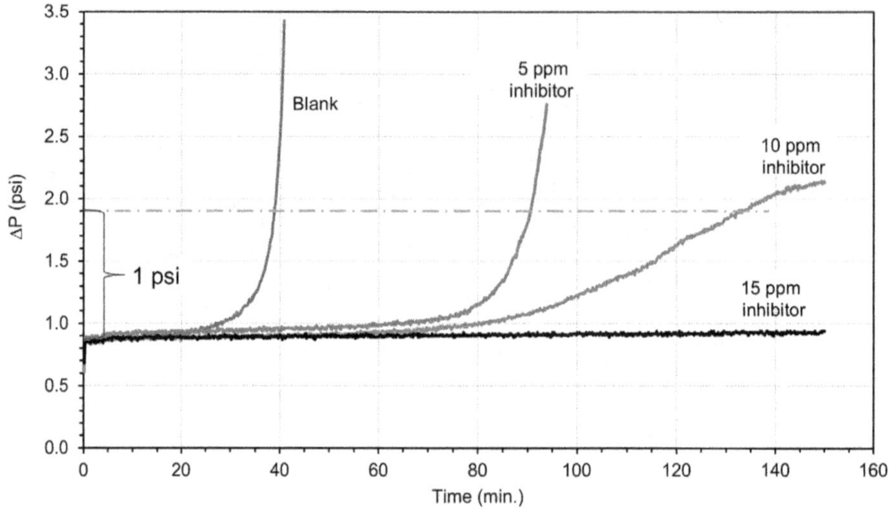

Figure 5.16. Typical dynamic scale loop testing example results.

In general, since the dynamic tube-blocking test could simulate field scaling conditions better than the static jar test as it is a pressurized dynamic flow system, it is often the first choice for evaluating scale inhibitors for field applications. However, the dynamic tube-blocking test has some disadvantages as well. For example, for a brine with a low-scaling tendency (e.g., a saturation index close to 1), the blank scaling time of the brine could be very long or is not able to be determined. This could also happen for a brine with a low- to moderate-scaling tendency, while the testing temperature is low, which indicates slow kinetics in crystal growth, precipitation, and attachment. When the blank scaling time is several hours long, it is difficult to conduct a scale inhibition test three times the blank scaling time. There are some modifications/adjustments in testing conditions and setups that may help solve the issue. A coil with a smaller ID and longer length could be used. The flow rates of the cationic and anionic brines could be reduced to increase the residence time of the mixed brine in the reaction coil. The testing temperature may need to be adjusted to a higher level, or the testing brines could be modified by adding more scaling ions.

5.3.2 Brine, Solvents, and Chemical Compatibility Evaluation

To assess whether a scale inhibitor could be deployed for a field application, a brine compatibility test is often needed. This test method is used to confirm that there are no adverse effects, when mixing the selected scale inhibitor with a field brine. This test is often done with a synthetic brine made in a laboratory based on the field brine water composition. The anionic scaling ions are removed to prevent any precipitation caused by scaling when preparing the testing brine. The scale inhibitor-brine compatibility test is often done at a temperature of no more than 110–120°C for a 24-h duration, with observations taken at initial, 1 h, 2 h, 4 h, and 24 h for any evidence of incompatibility such as precipitation, flocculation, or destabilization. A scale inhibitor is typically assessed at several different concentrations, such as from 50 ppm up to 200,000 ppm (20%), as shown in Figure 5.17.

Figure 5.17. Typical scale inhibitor-brine compatibility test.

Similar compatibility tests are often performed for a scale inhibitor with solvents, such as monoethylene glycol and methanol, and other production chemicals in use, such as corrosion inhibitors, wax inhibitors, and asphaltene inhibitors. The tests are typically conducted at an elevated temperature and ambient pressure for 1 day or several days. In the test, the scale inhibitor is mixed at the following ratios: 99:1, 90:10, 75:25, 50:50, 25:75, 10:90, and 1:99. Visual observations are made periodically during the tests to check the scale inhibitor's compatibility with the solvents and chemicals tested. Typical ranking criteria are: (1) clear with no precipitation, (2) slight haze, (3) hazy with precipitation; (4) cloudy with heavy precipitation, (5) clear with precipitation, (6) phase separation and/or emulsion, and (7) exothermal reaction.

5.3.3 Material Compatibility Evaluation

Scale inhibitors used in oilfield applications, especially for continuous injection, need to be evaluated with the materials they could contact in the production system to minimize the corrosion risk associated with the deployment of scale inhibitors since the injection of some scale inhibitors could cause an increasing downhole corrosion rate.[60] A scale inhibitor shall be fully compatible with all the components that it contacts in the chemical injection system under all the field application scenarios. With that, the laboratory evaluation of a scale inhibitor shall consider several aspects including but not limited to field temperature and pressure, high-shear environment, neat/dilution form of a product, extended contact time, with or without other production chemicals, such as corrosion inhibitors, gas and brine compositions, crude oil, and so on. Material compatibility tests are conducted with corrosion coupons made with various types of materials including metals and nonmetallic materials. Depending on testing conditions, material compatibility tests could be done with regular glass bottles, high-pressure glass tubes, aging cells, and autoclaves. Coupon immersion testing with glass bottles/high-pressure tubes under static conditions is the most commonly used method to assess a scale inhibitor's material compatibility at a temperature up to 120–130°C and an ambient pressure condition. If the testing temperature and pressure are higher than that, aging cells or autoclaves are needed. In a typical coupon immersion test, corrosion coupons are cleaned, weighed, and measured in dimensions before being put into the testing bottles containing a scale inhibitor in neat or diluted form. For metallic coupons, the testing duration is typically 7 days, 14 days, or 28 days, whereas for nonmetallic coupons, the typical testing duration is 28 days. The material compatibility of the tested scale inhibitor is evaluated based on coupon weight loss and pitting profile analysis for metals and based on changes in coupon weight, hardness, and volume/dimensions for elastomers and thermoplastics. The changes in the testing fluids containing the scale inhibitor should be analyzed. The corrosion rates of a metallic coupon expressed as mpy (milli-inches per year) can be calculated using the equation below.

$$\text{Corrosion rate}\,(\text{mpy}) = \frac{M \times 3.45 \times 10^6}{S \times T \times D}, \tag{5.2}$$

where M = mass loss (g), S = initial exposed area of coupon (cm^3), T = exposure time (h), and D = density of metal coupon (g/cm^3).

The corrosion rate in mpy can be converted to millimeter per year (mm/y) by the following conversation equation.

$$\text{Corrosion rate}\,(\text{mm/y}) = \text{Corrosion rate}\,(\text{mpy}) \times 0.0254. \tag{5.3}$$

For nonmetallic coupons, the changes in weight, hardness, and dimension generally should be less than 10%, and the coupons should have no cracking, blistering, or physical deformation.

5.3.4 Thermal Stability Testing

Scale inhibitors, depending on their chemistries, have different thermal stability temperatures due to degradation at high temperatures.[61–63] The possible mechanisms of such degradation include both hydrolysis and free radical attack.[61] Since the degradation of a scale inhibitor often means the loss or partial loss of its inhibition efficiency, this could pose a great risk to scaling control and flow assurance for oil and gas production, especially for high-temperature reservoirs. Before deploying a scale inhibitor to the field application, its thermal stability needs to be tested to make sure it will be thermally stable up to the maximum temperature it could encounter in such a field application. Generally, the tests are performed using neat scale inhibitor chemicals with aging cells. In a typical thermal stability test, samples of scale inhibitors are put in glass tubes, and then the filled glass tubes are put into thermal aging cells. The aging cells are pressurized and depressurized with nitrogen gas several times to remove oxygen. The pressure of the aging cells needs to be higher than the boiling point of the water at the testing temperature. If the tested product mainly contains solvents other than water, their boiling points need to be considered to choose a proper pressure. The duration of the test is typically 7 days or shorter/longer depending on the field application conditions. Physical properties of the testing scale inhibitor chemicals, such as appearance, pH, and density, should be measured before and after the test. FTIR analysis, viscosity, and particle size distribution of a sample before and after the test need to be conducted as well. Additionally, the inhibition performance of an aged sample needs to be compared with that of the fresh sample. Similar to the thermal stability test, scale inhibitors are often tested for their stability under long-term storage at a given temperature, typically 60°C as the maximum storage temperature and −40°C as the lowest storage temperature. The duration of such tests could be 1–6 months. The testing products should not have any major changes in appearance, viscosity, pH, density, particle size distribution, and/or solvent concentrations and not have any solid precipitation, phase separation, or loss of inhibition performance.

5.3.5 Emulsion Tendency and Foaming Tendency Testing

Scale inhibitors, depending on their chemistries and solvent packages, may cause unwanted emulsion or foaming in the production system at their dosage levels or at much higher dosage levels at the injection/mixing point. To avoid any emulsion or foaming risk associated with the deployment of a scale inhibitor, tests to determine whether a scale inhibitor is likely to promote emulsion formation and/or foam formation when deployed into an oilfield production system are needed.

For the emulsion tendency test, a scale inhibitor is typically tested at various dosages (*e.g.* ½ recommended dosage, recommended dosage, five times recommended dosage, and 10 times recommended dosage) at 25%, 50%, and 75% water cut with field crude and a synthetic brine. All the samples with different mixing ratios of oil and water with or without a scale inhibitor should be placed in a water bath to be heated to the desired testing temperature. Once the samples reach the desired testing temperature, depending on the severity of the agitation determined by the field conditions, the samples could be agitated either by hand shaking or using a blender/mixer. Observations typically should be made at the initial, 1 min, 5 min, 10 min, 20 min, 30 min, and 60 min to check the speed of separation, interface, percentage of water drop, water quality, and brightness of the crude layer.

Foaming tendency tests are typically performed on an aqueous phase involving brine and a scale inhibitor only, but other testing fluids, such as crude oil, could be included as well to better simulate the system conditions. The test is often conducted at ambient temperature but can be done at an elevated temperature using a water bath. The foaming tendency test is performed with a foamer column/graduated measuring cylinder and a flowmeter-and-pressure-controlled compressed air/nitrogen gas source. In a typical foaming tendency test, 50 mL or 100 mL of testing fluid containing various dosages (e.g., 1× recommended dosage, 5× recommended dosage, and 10× recommended dosage) of a scale inhibitor is put into a graduated glass measuring cylinder. Using a water bath to preheat the fluid if the testing temperature is at an elevated temperature. Then, turn on the compressed air/nitrogen gas source to sparge the testing fluid for about 1 min or until stable foam forms and record the volume of the foam formed by reading from the level of the liquid surface in the cylinder every 15–20 s for 5 min. After that, switch off the gas source and record the foam collapse time.

5.3.6 Subsea/Deepwater Umbilical Application Qualification Testing

Continuous injection of scale inhibitors is one of the major deployment methods to apply a scale inhibitor to prevent and control scale formation in oilfields. Downhole scale inhibitor continuous injection (DHCI) through a chemical injection system, such as capillary tubing, has been widely used in the oil and gas industry. Especially for offshore applications, production chemicals, including scale inhibitors, are injected through several miles-long umbilical lines running from a platform to wellheads sitting on the seafloor. Downhole scale inhibitor continuous injection has several advantages, such as improved control over inhibitor dose, providing constant delivery of scale inhibitors without interruption of production, and being automated and remotely controlled. Furthermore, DHCI can be used on wells that are not applicable for scale inhibitor downhole squeeze treatments due to a high-scaling tendency of the water or the risk of formation damage for a sensitive reservoir. Before a scale inhibitor can be applied through a downhole chemical continuous injection system, it needs to pass a series of rigorous qualification tests for any chemicals to be delivered through a capillary injection line.[64–68] These tests not only include thermal stability tests, material compatibility tests, and brine/solvent compatibility, which have been discussed in the above sessions, but also include other tests, such as rheology test, pressurized flow loop test, and gunking test. The typical subsea/deepwater downhole umbilical injection qualification tests are summarized in Table 5.1.

Table 5.1: Summary of subsea/deepwater downhole umbilical application qualification testing.

Test Type	Description
Rheology test	Measure chemicals' viscosities at field operating temperatures and pressures
Static stability testing	Check chemicals' long-term static stability test at desired temperatures
Accelerated stability by refrigerated centrifuge test	Check chemicals stability under shear and at seabed temperature, ~4°C
Pressurized flow loop test	Check chemical stability under field conditions in a pressurized flow loop to simulate a sub-sea umbilical environment
Metal materials compatibility	Compatibility tests with various metals
Elastomers/plastics materials compatibility	Compatibility tests with various elastomers and plastics
U-tube gunking test	Examine the effect of solvent evaporation which may result in blockages and solids formation
Umbilical flushing solvent compatibility	Compatibility tests with various solvents, such as methanol, MEG, and EGMBE
Seawater compatibility	Compatibility tests with seawater
Chemical cleanliness	To minimize the risk of umbilical blockage, there are industry standards on production chemical cleanliness, such as NAS 1638 Class 6 or SAE AS 4059, Class 8B
Hydrate tendency	The hydrate formation tendency theoretical evaluation

5.3.7 Core-flood Test

Downhole scale inhibitor squeeze application is a common deployment method to apply a scale inhibitor into a reservoir to protect a well at the downhole and its production tubing from scale deposition, and formation damage caused by scale precipitation at the reservoir. In a typical squeeze treatment, a solution containing the selected scale inhibitor is injected into the well, and an overflush solution is injected after the injection of the scale inhibitor solution to push the scale inhibitor further into the near-wellbore formation rock pore spaces. Then, the well is shut in for several hours to allow the scale inhibitor to react with rock/mineral surfaces and the reservoir fluid inside the pore spaces, so some of the scale inhibitors could be retained in the reservoir and slowly flow back with the produced water after the well is switched back to production. Once the scale inhibitor concentration in the produced water is below the MIC level, it means the squeeze life has been reached, and the well should be squeezed again. Since a longer squeeze lifetime generally means fewer squeeze treatments are needed and the overall downhole scale squeeze treatment cost will be less as well, selecting a scale inhibitor that could provide a longer protection time is desired. The retention of a scale inhibitor in formation rocks is affected by various mechanisms. There are two major mechanisms, the precipitation mechanism and the adsorption mechanism.[69] The precipitation mechanism suggests scale inhibitors could react with divalent ions in reservoir fluids, such as calcium ions, to form precipitates in the pores of the formation rock, and then slowly release back to the reservoir fluid during production.[69, 70] The adsorption mechanism proposes that scale inhibitors could be adsorbed on the rock/mineral surface due to electrostatic van der Waals forces[70], and then released back to the water phase through desorption. Both mechanisms involve understanding the chemistry of the rock-scale inhibitor-brine interactions, which is important for selecting the best scale inhibitor chemistry for a specific squeeze application and helps in the squeeze design.[71, 72]

To study and understand rock-scale inhibitor-brine interactions, core flooding tests are the most commonly used laboratory testing methods. Core flood tests can help to understand the mechanism of scale inhibitor transport and retention of aqueous and nonaqueous scale inhibitor solutions in real field systems, and the results of core flood tests are the keys to determining the sensitivities to inhibitor retention, inhibitor squeeze lifetimes, and possible formation damage in field reservoir formations. The adsorption/desorption data obtained through a core flood test for a scale inhibitor could be used for squeeze design modeling to derive a correct adsorption isotherm for a specific field. There are several commercially available core flood apparatuses, each with slightly different functions and capabilities. Figure 5.18 is a sketch of a typical core flood apparatus.

In core flood tests, real field core samples are always recommended to be used. Considering that getting a real field core sample is not always possible, a reference core sample with a similar composition as the field core sample can be used to study the interaction of scale inhibitors with core materials as well. For example, Berea sandstone core samples are often used in core flood tests as a reference core material, as they provide a representative distribution of common oilfield sandstone mineral assemblages. Besides using a core sample, sand-packing columns using silica sand or crushed field formation rock samples are also commonly used in core flood tests to study scale inhibitors' adsorption and desorption properties.

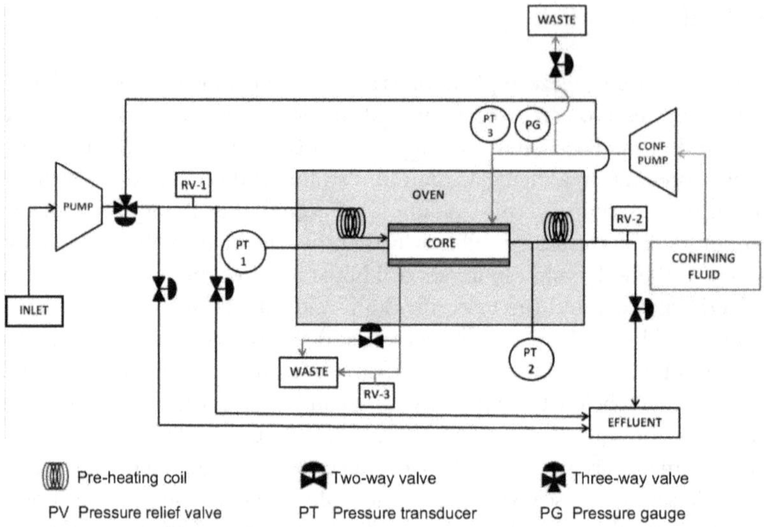

Figure 5.18. Schematic drawing of a core flood apparatus.

Depending on the field application conditions, core flood tests could be conducted with or without an oil phase. Table 5.2 shows outlines of the stages of a typical scale inhibitor core flood test involving an oil phase. There could be some slight differences in core flood testing procedures reported,[69, 70, 73–75] but the results from those core flood tests include mainly scale inhibitor return profile, brine/oil permeabilities before and after chemical injection, injection pressure profiles, pH profiles of effluents, and elemental analysis of effluents.

Table 5.2: Outlines of the stages of a typical SI core flood test.

Testing Stage	Description
1. Ambient stage	Prepare a core sample including core analysis and taking a photograph; mount the core and use solvents to clean the core; measure ambient brine permeability; heat the core to the reservoir temperature.
2. Pretreatment stage	Measure brine permeability and pore volume (PV) at 100% water saturation; inject oil and measure oil permeability; inject brine and measure brine permeability and PV in the presence of oil residual; inject oil and measure oil permeability
3. Scale inhibitor injection stage	Preflush the core; inject scale inhibitor solution (typically containing 10%–20% scale inhibitor in brine) and collect effluent samples; shut in
4. Flowback stage	Inject oil and measure oil permeability; collect effluent samples; elemental analysis of the effluent samples; formation damage analysis
5. Post-treatment stage	Injection brine and measure brine permeability; collect effluent samples for SI residual analysis until the SI concentration in the effluent samples is below the MIC or a desired level; elemental analysis of the effluent samples; inject oil and measure oil permeability
6. Final stage	Cool the system; clean the core with solvents; release the core holder pressure; remove the core sample and conduct a core analysis

Figure 5.19 shows a typical scale inhibitor return profile from a core flood test. Scale inhibitor return data from a core flood test are crucial in determining how long a squeeze life could be, and they are important data for squeeze design modeling. Since the accuracy of scale inhibitor residual analysis is very important to get correct data for squeeze design and for operators to know when the concentration of scale inhibitor in the produced fluid of a well is below the MIC, whether scale inhibitors' concentrations could be easily and reliably analyzed often needs to be considered when choosing a scale inhibitor for core flood test and actual field squeeze applications. Depending on the chemistry of a scale inhibitor, there are different methods available. For example, any scale inhibitors containing phosphorus, such as phosphonates, phosphate esters, and P-tagged polymers/copolymers, could be measured by ICP-OES based on elemental phosphorus readings. Polymeric scale inhibitors could be measured using high-performance liquid chromatography[76] or a hyamine turbidity method.[77] Additionally, there are some scale inhibitors with fluorescent-tagged polymeric chemistries that emit light/fluoresces at a unique wavelength which could be detected by traditional fluorescence techniques.[78, 79] More detailed information on scale inhibitor analysis can be found in Chapter 12.

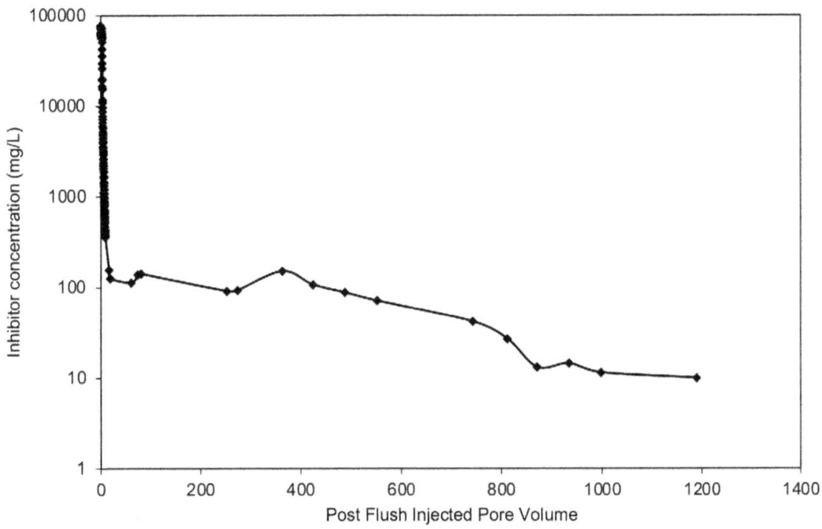

Figure 5.19. Scale inhibitor returns profile of a typical core flood test.

Permeability evaluations are often carried out to compare oil/brine permeability before and after a chemical injection. The mechanisms of reductions in permeability could involve a water block, solid precipitation, emulsion formation, fines mobilization, and so on.[73] Any reductions in oil/brine permeability after the chemical injection would imply a certain degree of pathway blockage, which may indicate a potential risk of formation damage if applying the chemical in a specific field squeeze application. The injection pressure profiles could indicate whether there is any injection problem in the application of the chemical. To understand scale inhibitor and rock/mineral interactions, elemental analysis of effluents and pH profile of the effluents are often conducted. The data obtained from a core flood test could provide a comprehensive study of a scale inhibitor squeeze treatment for a specific well, significantly increase the successful rate of scale inhibitor squeeze applications by choosing a proper product, provide good data for the initial squeeze design, and lower the risk of formation damage. The details of squeeze treatment are discussed in the following two chapters.

5.4 Final Remarks

Understanding scale inhibitor chemistries, their inhibition mechanisms, and inhibition performances against various scale types plays an important role in selecting a proper product for a specific scale control application and helps better manage scale-related issues for various industrial applications. Since the scale inhibitor chemistry information available in the public domain is limited, and there are quite a lot of scale inhibitor chemistries that cannot be found in published literature, because they are proprietary information, only the most common scale inhibitors were reviewed in this chapter. Additionally, common scale inhibitors used for exotic scales, such as sulfide scales, halite scales, and silicates, are not discussed in the chapter. These are covered in the later chapters on exotic scales. The scale inhibitor evaluation tests discussed in this chapter are based on the author's experience and various testing procedures found in the published articles. For any specific tests, due to the differences in testing equipment in use and testing conditions, such testing procedures could be different from those of the different oil field service companies, chemical suppliers, and oilfield operators, especially when no standard testing procedures exist. The discussion on those testing methods is intended to provide an overview of various scale inhibitor evaluation tests and their general testing procedures.

As reviewed in this chapter, there has been massive progress and improvement made in both scale inhibitor chemistries and their evaluation methods compared to the past several decades. However, challenges still exist in this area, and certainly more work needs to be done to address those challenges and improve scale control efficiency for the oil and gas industry. There are some outlooks listed below that may shed some light on this:

Almost every industry has sustainability and low-carbon strategies set to create a sustainable and balanced planet for current and future generations. Many industries have set up goals to achieve net-zero emissions by 2050 through the energy transition. The oil and gas industry has an important role to play in the global energy transition by reducing emissions, decarbonizing energy production, and lowering carbon footprints. There is a need to develop new scale inhibitor chemistries that have a low-carbon footprint and to evaluate the carbon footprint of existing scale inhibitor chemistries. Currently, there is very limited data available in this area. Certainly, more work is needed.

The effectiveness of the currently available biodegradable scale inhibitors under certain conditions is still not as good as those commonly used scale inhibitors with limited biodegradable properties. Although significant progress has been made in finding "green scale inhibitors," the challenge remains in this area. More discussion on this topic can be found in Chapter 18.

Although some scale inhibitor chemistries, such as sulfonated polymers like PVS and SPCA, are thermally stable and show good inhibition performance under high-temperature conditions, these types of scale inhibitor chemistries may not work under extremely high temperatures, such as 200–250°C, which is often seen in geothermal wells. More efforts are still needed to develop thermally stable and effective scale inhibitor chemistries under very high-temperature conditions. The other challenge is about the compatibility between scale inhibitors and high-calcium and/or high-iron brines. There are only a few scale inhibitor chemistries that have a high-calcium tolerance, and most scale inhibitors' inhibition performance will drop when iron ions are present. The improvement of existing chemistries and the development of new chemistries are still needed.

Effectively controlling the exotic scales, such as sulfide scales, halite scales, and silicates, is still very challenging as only limited chemistries work to a certain degree, which is not as effective as controlling the precipitation of carbonate or sulfate scales. Although this chapter did not cover specific scale inhibitor chemistries in this area, it is clear that this will remain one of the main challenges for scale control in the oil and gas industry. More discussions can be found in the following chapters.

Testing scale inhibitors' inhibition performance under harsh conditions, such as high temperature, high pressure, or in the presence of H_2S is still very difficult with current testing equipment and methods. For example, a typical DSL can only be used to conduct inhibition tests at conditions below 200°C in temperature and 3,000–5,000 psi in pressure with no presence of H_2S gas, unless modification of the equipment has been done. To better simulate field conditions for testing scale inhibitors under harsh field conditions, improvements to the current testing apparatus and testing methods are needed.

Nomenclature

ACS	American Chemical Society
AEEAP	Amineothylethanolamine phosphonate
AMPS	Acrylamido(methyl)propylsulfonic acid
ATMP	Aminotri(methylene) phosphonic acid
BHPMP	Bis-hexamethylene triamine-penta(methylene phosphonic) acid
CMI	Carboxymethyl inulin
DHCI	Downhole scale inhibitor continuous injection
DSL	Dynamic scale loop
DTPA	Diethylenetriaminepentacetic acid
DETPMP	Diethylene triamine-penta(methylene phosphonic) acid
EDTA	Ethylenediaminetetracetic acid
EGMBE	Ethylene Glycol Monobutyl Ether
FTIR	Fourier transform infrared
GLDA	Glutamic acid diacetic acid
HDTMP	Hexamethylene diamine-tetra(methylene phosphonic) acid

HEDP	1-Hydroxyethylidene-1,1-diphosphonic acid
HEBMP	Hydroxyethyl amino bis(methylene phosphonic) acid
HMCS	Harmonized mandatory control system
HPA	Hydroxyphosphonoacetic acid
HPLC	High performance liquid chromatography
HOCNF	Harmonized offshore chemical notification format
ICP-OES	Inductively coupled plasma optical emission spectroscopy
ID	Internal diameter
MEG	Monoethylene glycol
MIC	Minimum inhibitor concentration
MW	Molecular weight
NACE	National Association of Corrosion Engineers
NTMP	Nitrilo tris(methylene phosphonic) acid
OSPAR	Oslo and Paris Commission
PAPEMP	Polyamino polyether methylene phosphonic acid
PASP	Polyaspartic acid
PBTC	2-Phosphono-1,2,4-butanetricarboxylic acid
PLONOR	Poses little or no risk to the environment
PPCA	Phosphinopolycarboxylic acid
ppm	Parts per million
PMPA	N-phosphonomethylated amino-2-hydroxypropylene polymer
PTA	Polytartaric acid
PV	Pore volume
SSA	Styrene sulfonic acid
TEAPE	Triethanolamine phosphate ester
VSA	Vinyl sulfonic acid
ΔP	Change of the differential pressure

References

1. E.J. Burcik, "The Inhibition of Gypsum Precipitation by Sodium Polyphosphates," *Producers Monthly* 19 (1954): pp. 42–44.
2. A.B. Featherston, R.G. Mirham, A.B. Waters, "Minimization of Scale Deposits in Oil Wells by Placement of Phosphates in Producing Zones," *Journal of Petroleum Technology* 11 (1959): pp. 29–32.
3. C.F. Smith, T.J. Nolan III, P.L. Crenshaw, "Removal and Inhibition of Calcium Sulfate Scale in Waterflood Projects," *Journal of Petroleum Technology* 11 (1968): pp. 1249–1256.
4. G.B. Hatch, O. Rice, "Surface-Active Properties of Hexametaphosphate," *Industrial & Engineering Chemistry* 31 (1939): pp. 51–57.
5. O. Rice, E.P. Partridge, "Threshold Treatment Elimination of Calcium Carbonate Deposits from Industrial Waters," *Industrial & Engineering Chemistry* 31 (1939): pp. 58–63.
6. O. Rice, G.B. Hatch, "Threshold Treatment of Municipal Water Supplies," *Journal of American Water Works Association* 31 (1939): pp. 1171–1185.

7. R.F. Reitemeier, T.F. Buehrer, "The Inhibiting Action of Minute Amounts of Sodium Hexametaphosphate on the Precipitation of Calcium Carbonate from Ammoniacal Solutions. I," *Journal of Physical Chemistry*. 44 (1940): pp. 535–551.
8. R.F. Reitemeier, T.F. Buehrer, "The Inhibiting Action of Minute Amounts of Sodium Hexametaphosphate on the Precipitation of Calcium Carbonate from Ammoniacal Solutions. II, *Journal of Physical Chemistry*. 44 (1940): pp. 552–574.
9. J.S. Gill, "Development of Scale Inhibitors," NACE-96229, NACE Annual Corrosion Conference, held March 24–29, 1996 (Denver, CO, USA).
10. Z. Amjad, "Performance of Polymers as Precipitation Inhibitors for Calcium Phosphonate," *Tenside Surfactants Detergents* 34 (1997):102–107.
11. M. Mpelwa, S. Tang, "State of the Art of Synthetic Threshold Scale Inhibitors for Mineral Scaling in the Petroleum Industry: A Review," *Petroleum Science* 16 (2019):830–849.
12. Y. Liu, P. Zhang, "Review of Phosphorus-Based Polymers for Mineral Scale and Corrosion Control in Oilfield," *Polymers* 14 (2022): p. 2673.
13. M.A. Kelland, *Production Chemicals for the Oil and Gas Industry,* 2nd ed. (Boca Raton: CRC Press, 2014).
14. G.E. Jackson, K. Mclaughlin, N. Poynton, et al., "Scale Inhibition by Squeeze Application of a Phosphonomethylated Polyamine Inhibitor," International Patent Application WO 97/21905, 1997.
15. J.L. Przybylinski, R.H. Nguyen, J.W. Ruggeri, "Field Test of a Unique Phosphonate Scale Inhibitor for Extended-Life Squeeze Treatment in Sandstone," SPE-50703-MS, SPE International Symposium on Oilfield Chemistry, held February 16–19, 1999 (Houston, Texas, USA).
16. M.A. Singleton, J.A. Collins, N. Poynton, et al., "Developments in Phosphonomethylated Polyamine (PMPA) Scale Inhibitor Chemistry for Severe $BaSO_4$ Scaling Conditions," SPE-60216-MS, SPE International Symposium on Oilfield Scale, held January 26–27, 2000 (Aberdeen, UK).
17. T.L. Herrera, M. Guzmann, K. Neubecker, et al., "Process and Polymer for Preventing Ba/Sr Scale with Incorporated Detectable Phosphorus Functionality," International Patent Application WO 2008/095945, 2008.
18. M.L. Re, J.S. Gill, "The Effect of Polyamino Polyether Methylene Phosphonate On the Crystallization Kinetics of Calcium Carbonate: A Constant Composition study," NACE-96157, NACE Annual Corrosion Conference, held March 24–29, 1996 (Denver, CO, USA).
19. J.S. Gill, J.R. Parsons, P.H. Wrede, "A New Cooling Water Treatment to Reduce Blow Down," NACE-97482, NACE Annual Corrosion Conference, held March 24–29, 1996 (Denver, CO, USA).
20. N.S. Sherwood, "Inhibitor Performance in Process Water Containing Ammonia," NACE-98572, NACE Annual Corrosion Conference, held March 22–27, 1998 (San Diego, CA, USA).
21. Z. Amjad, A. Kweik, "Formation and Control of Iron-Based Deposits in Industrial Water Systems," NACE-2016-7591, NACE Annual Corrosion Conference, held March 6–10, 2016 (Vancouver, Canada).
22. M. Hamdi, A. Kweik, A. Almakrani, et al., "Extending Phosphonates Performance in Simulated Cooling Waters," NACE-2018-11054, NACE Annual Corrosion Conference, held April 15–19, 2018 (Phoenix, AZ, USA).
23. D. Shen, R. Perkins, D. Schielke, et al., "Method for Preventing Scale Formation in the Presence of Dissolved Iron," U.S. Patent 8236734 B1, 2012.
24. S.O. Adegoke, O.A. Falode, P.C. Nwankwo, "Understanding Oilfield Scale Deposition and Inhibition Mechanisms for Optimum Management: A Review," SPE-207133-MS, SPE Nigeria Annual International Conference and Exhibition, held August 2–4, 2021 (Lagos, Nigeria).
25. J. S. Gill, "Silica Scale control," NACE-98226, NACE Annual Corrosion Conference, held March 22–27, 1998 (San Diego, CA, USA).
26. J. J. Wylde, G. D. M. Williams, F. Careil, et al., "Deep Downhole Chemical Injection on BP-Operated Miller: Experience and Learning," *SPE Production & Operations* v21 (2006): pp. 188–193.

27. R. Pirri, C. Hurtevent, P. Leconte, "New Scale Inhibitor for Harsch Field Conditions," SPE-60218-MS, SPE International Symposium on Oilfield Scale, held January 26–27, 2000 (Aberdeen, UK).
28. O. Karazincir, S. Thach, W. Wei, et al., "Scale Formation Prevention During ASP Flooding," SPE-141410-MS, SPE International Symposium on Oilfield Chemistry, held April 11–13, 2011 (The Woodlands, TX, USA).
29. C. Yan, A. T. Kan, F. Zhang, et al., "Systematic Study of Barite Nucleation and Inhibition With Various Polymeric Scale Inhibitors by Novel Laser Apparatus," *SPE Journal* v20 (2015): pp. 642–651.
30. K. S. Sorbie, N. Laing, "How Scale Inhibitors Work: Mechanisms of Selected Barium Sulphate Scale Inhibitors Across a Wide Temperature Range," SPE-87470-MS, SPE International Symposium on Oilfield Scale, held May 26–27, 2004 (Aberdeen, UK).
31. Z. Zhang, F. Zhang, Q. Wang, et al., "Ferrous Iron Impact on Phosphonate and Polymeric Scale Inhibitors," SPE-173770-MS, SPE International Symposium on Oilfield Chemistry, held April 13–15, 2015 (The Woodlands, TX, USA).
32. X. Yao, X. Wang, S. Ko, et al., "Fate and Transport of Sulfonated Polymeric Inhibitors in the Reservoir: Studied by Column Experiments With Calcite Media," SPE-213787-MS, SPE International Symposium on Oilfield Chemistry, held June 28–29, 2023 (The Woodlands, TX, USA).
33. J. R. Griffin, J. R. Johnstone, T. E. Cotter, et al., "Factors Affecting the Stability of Scale Inhibitors Used for Capillary Injection in High Temperature Wells," SPE-164128-MS, SPE International Symposium on Oilfield Chemistry, held April 8–10, 2013 (The Woodlands, TX, USA).
34. B. Cook, "Scale Inhibition," Patent EP0297049 A1, 1988.
35. Z. Amjad, "Kinetic and Morphological Investigation of Calcium Sulfate Dihydrate (Gypsum) Scale Formation on Heat Exchanger Surfaces in the Presence of Inhibitors," NACE-2017-9020, NACE Annual Corrosion Conference, held March 26–30, 2017 (New Orleans, LA, USA).
36. K.K. Kommanapalli, P. Lyot, J. R. Sunkara, et al., "Synthesis and Characterization of Maleic Acid and Sodium Methallyl Disulfonate New Copolymer: Application as a Barium Sulfate Scale Inhibitor," *Journal of Petroleum Exploration and Production Technology* 9 (2019): pp. 223–232.
37. P. Bainbridge, V. Rowan, "Interpolymers for Barium Sulfate Inhibition," U.S. Patent 4937002, 1990.
38. S.T. Chen, C.W. Vaughan, "Method for Controlling Scale Deposition in Aqueous Systems Using Allyl Sulfonate/Maleic Anhydride Polymers," U.S. Patent 5000856, 1991.
39. L. Moore, L Clapp, "Tagged Scale Inhibitor Compositions and Methods of Inhibiting Scale," U.S. Patent 0032093 A1, 2012.
40. L. Liu, J. Zhao, Y. Zuo, et al., "Synthesis of AA-MA-HPA Polymer and Its Inhibition for BaSO4 Scale," *Fine Chem* 21 (2004): pp. 58–60.
41. M. Zhou, Y. Gu, R. Yi, et al., Synthesis and Property Study of Ter-Copolymer P(MA-AMPS-HPA) Scale Inhibitor," *Journal of Polymer Research*. 27 (2020): p. 294.
42. H.J. Beck, K. Salte, B.T. Brun, "A Common Database for Ecotoxicological Documentation of Chemicals," SPE-35823, SPE Health, Safety and Environment in Oil and Gas Exploration and Production Conference, held June 9–12, 1996 (New Orleans, LA, USA).
43. M. Thatcher, G. Payne, "Impact of the OSPAR Decision on the Harmonised Mandatory Control System on the Offshore Chemical Supply Industry," *Special Publication-Royal Society of Chemistry* 280 (2002): pp. 21–30.
44. W. Dickinson, R. Griffin, L. Sanders, et al., "Development and Performance of Biodegradable Antiscalants for Oilfield Applications," Offshore Technology Conference, held May 2–5, 2011 (Houston, Texas, USA).
45. R.J. Ross, K. Low, J. E. Shannon, "Polyaspartate Scale Inhibitors—Biodegradable Alternatives to Polyacrylates," NACE-96162, NACE Annual Corrosion Conference, held March 24–29, 1996 (Denver, CO, USA).
46. W. Hater, "Environmental Compatible Scale Inhibitor for the Mining Industry," NACE-98213, NACE Annual Corrosion Conference, held March 22–27, 1998 (San Diego, CA, USA).

47. C. Estievenart, N. Kohler, F. Ropital, et al., "Mechanisms of Scale and Corrosion Inhibition by Polyaspartates," NACE-04536, NACE Annual Corrosion Conference, held March 28–April 1, 2004 (New Orleans, LA, USA).
48. P. Chen, T. Hagen, A. MacLean, et al., "Meeting the Challenges of Downhole Scale Inhibitor Selection for An Environmentally Sensitive Part of the Norwegian North Sea," SPE-74652-MS, SPE International Symposium on Oilfield Scale, held January 30–31, 2002 (Aberdeen, UK).
49. M.M. Jordan, C.J Johnston, L. Sutherland, "Impact of Reservoir Temperature on Scale Inhibitor Retention—The Challenge of Ultra-Low Temperature Sandstone Reservoirs," SPE-179889-MS, SPE International Oilfield Scale Conference and Exhibition, held May 11–12, 2016 (Aberdeen, UK).
50. M. Barber, W.Y. Mok, "The Impact of Ferrous Iron(II) on Scale Inhibitors Performance in the Field—Is it a Problem?," NACE-2021-16428, AMPP Annual Corrosion Conference, held April 19–30, 2021 (virtual).
51. H. Davies, M. Toole, R. Higgins, "Development and Field Application of an Iron Tolerant Scale Inhibitor for Use in Fracturing Completions and Production," SPE-190728-MS, SPE International Oilfield Scale Conference and Exhibition, held June 20–21, 2018 (Aberdeen, UK).
52. A. Fraser, J. Davidson, N. Feasey, et al., "Limitations of Common Oilfield Scale Inhibitor Chemistries," NACE-2016-7617, NACE Annual Corrosion Conference, held March 6–10, 2016 (Vancouver, Canada).
53. V. Jovancicevic, D. Hartwick, "Recent Developments in Environmentally-Safe Corrosion Inhibitors," NACE-96226, NACE Annual Corrosion Conference, held March 24–29, 1996 (Denver, CO, USA).
54. D.L Verraest, J.A. Peters, H. Van Bekkum, et al., "Carboxymethyl Inulin: A New Inhibitor for Calcium Carbonate Precipitation," *Journal of the American Oil Chemists' Society* 73 (1996): pp. 55–62.
55. D.L. Verraest, J.G. Batelaan, J.A. Peters, et al., "Carboxymethyl Inulin," U.S. Patent 5777090A, 1998.
56. H.C. Kuzee, H. Wilhelmus, C. Raaijmakers, "Method for Preventing Deposits in Oil Extraction," U.S. Patent 6613899 B1, 2003.
57. M.F. Mady, S. Abdel-Azeim, M.A. Kelland, "Investigation of the Antiscaling Performance of Phosphonated Chitosan for Upstream Petroleum Industry Application," *ACS Sustainable Chemistry & Engineering* 48 (2021): pp. 16494–16505.
58. N. Kohler, B. Bazin, A. Zaitoun, et al., "Green Inhibitors for Squeeze Treatments: A Promising Alternative," NACE-04537, NACE Annual Corrosion Conference, held March 28–April 1, 2004 (New Orleans, LA, USA).
59. S. Baraka-Lokmane, K.S. Sorbie, N. Poisson, et al., "Can Green Scale Inhibitors Replace Phosphonate Scale Inhibitors?: Carbonate Coreflooding Experiments," *Petroleum Science and Technology* 27 (2009): pp. 427–441.
60. A. Daminov, V. Ragulin, A. Voloshin, "Mechanism Formations of Corrosion Damage of Inter Equipment in Wells By Continuous Scale-Inhibitor Dosing Utilizing Surface Dosing Systems: Testing Scale and Corrosion Inhibitors," SPE-100476-MS, SPE International Oilfield Corrosion Symposium, held May 30, 2006 (Aberdeen, UK).
61. G.M. Graham, M.M. Jordan, G.C. Graham, et al., "The Implication of HP/HT Reservoir Conditions on the Selection and Application of Conventional Scale Inhibitors: Thermal Stability Studies," SPE-37274-MS, SPE International Symposium on Oilfield Chemistry, February 18–21, 1997 (Houston, TX, USA).
62. S.J. Dyer, G.M. Graham, K.S. Sorbie, "Factors Affecting the Thermal Stability of Conventional Scale Inhibitors for Application in High Pressure/High Temperature Reservoirs," SPE-50717-MS, SPE International Symposium on Oilfield Chemistry, held February 16–19, 1999 (Houston, TX, USA).
63. W. Wang, A.T. Kan, C. Yan, et al., "The Use of Inhibition Kinetics and NMR Spectroscopy in Thermal Stability Study of Scale Inhibitors," SPE-164047-MS, SPE International Symposium on Oilfield Chemistry, held April 8–10, 2013 (The Woodlands, TX, USA).

64. H. Guan, A. Jenkins, "Testing Equipment and Procedures Suitable to Evaluate Deepwater Scale and Corrosion Inhibitors," NACE-11345, NACE Annual Corrosion Conference, held March 13–17, 2011 (Houston, TX, USA).
65. J.H. Olsen, "Statoil Experiences and Consequences Related to Continuous Chemical Injection," SPE-146625, SPE Annual Technical Conference and Exhibition, held October 20–November 2, 2011 (Denver, CO, USA).
66. C. Stewart-Liddon, N.J. Goodwin, G.M. Graham, et al., "Qualification of Chemicals/Chemical Injection Systems for Downhole Continuous Chemical Injection," SPE-169782-MS, SPE International Oilfield Scale Conference and Exhibition, held May 14–15, 2014 (Aberdeen, Scotland, UK).
67. M. Grutters, J. Kuzniarek, S. Jackson, "Risk-Free Delivery of Production Chemicals to Deepwater Oil and Gas Fields," Offshore Technology Conference Asia, held November 2–6, 2020 (Kuala Lumpur, Malaysia).
68. D. Shen, J. Stewart, K. Chao, et al., "The Application of Scale Inhibitors for Downhole Capillary Injection Under Ultra High Temperature," SPE-213615-MS, Middle East Oil, Gas, and Geosciences Show, held February 19–21, 2023 (Manama, Bahrain).
69. M.M. Jordan, K.S. Sorbie, P. Chen, et al., "The Design of Polymer and Phosphonate Scale Inhibitor Precipitation Treatments and the Importance of Precipitate Solubility in Extending Squeeze Lifetime," SPE-37275-MS, SPE International Symposium on Oilfield Chemistry, held February 18–21, 1997 (Houston, TX, USA).
70. L. Sutherland, M. Jordan, "Enhancing Scale Inhibitor Squeeze Retention in Carbonate Reservoirs," SPE-190715-MS, SPE International Oilfield Scale Conference and Exhibition, held June 20–21 2018 (Aberdeen, UK).
71. A.T. Kan, G. Fu, M. Al-Thubaiti, et al., "A New Approach to Inhibitor Squeeze Design," SPE-80230-MS, SPE International Symposium on Oilfield Chemistry, held February 5–7, 2003 (Houston, USA).
72. M.B. Tomson, A.T. Kan, G. Fu, "Control of Inhibitor Squeeze Through Mechanistic Understanding of Inhibitor Chemistry," *SPE Journal* 13 (2008): pp. 325–336.
73. H. Montgomerie, P. Chen, T. Hagen, et al., "Development of a New Polymer Inhibitor Chemistry for Downhole Squeeze Applications," SPE-113926-MS, SPE International Oilfield Scale Conference, held May 28–29, 2008 (Aberdeen, UK).
74. H. Guan, H.C. Rohde, "The Mitigation of Formation Damage and the Application of a New Scale Squeeze Inhibitor," SPE-122026-MS, SPE European Formation Damage Conference, held May 27–29, 2009 (Scheveningen, The Netherlands).
75. T. Chen, P. Chen, H. Montgomerie, et al., "Scale Squeeze Treatments in Short Perforation and High Water Production ESP Wells—Application of Oilfield Scale Management Toolbox," IPTC-16844-MS, International Petroleum Technology Conference, held March 26–28, 2013 (Beijing, China).
76. J. Johnstone, B. Peltokoski, S. Toivonen, et al., "Application of a Novel Method for Real-Time Monitoring of Scale Control Products at the Site of Use," SPE International Symposium on Oilfield Chemistry, held April 13–15, 2015 (The Woodlands, TX, USA).
77. N.P. Chilcott, D.A. Phillips, M.G. Sanders, et al., "The Development and Application of an Accurate Assay Technique for Sulphonated Polyacrylate Co-Polymer Oilfield Scale Inhibitors," SPE-60194-MS, SPE International Oilfield Scale Conference, held January 26–27, 2000 (Aberdeen, UK).
78. Y. Liu, D. Lee, H. Lu, et al., "Development of fluorescence tagged scale inhibitors for squeeze applications in Gulf of Mexico," SPE-204349-MS, SPE International Symposium on Oilfield Chemistry, held December 6–7, 2021 (The Woodlands, TX, USA).
79. J. Russek, A. Leontieff, M. Barclay-Dunbar, et al., "Detection of Four Discretely Tagged Scale Inhibitors in a Single Produced Water," SPE-213842-MS, SPE International Symposium on Oilfield Chemistry, held June 28–29, 2023 (The Woodlands, TX, USA).

Squeeze Treatment: Theory, Design, and Optimization

Oscar Vazquez[1]
[1]School of Energy, Geoscience, Infrastructure and Society, Heriot-Watt University, Edinburgh, UK

6.1 Introduction

Scale inhibitors are characterized by the ability to inhibit scale at low concentrations, generally at a few ppm, known as threshold scale inhibition.[1–4] These chemicals have strong interactions with the scaling crystal surface by bonding between the ions on the crystal surface, as they are more energetically favorable than remain in solution.

There are several techniques to deploy scale inhibitors in the fields. The most common are continuous injection and squeeze treatment. Continuous injection, also known as downhole inhibitor injection, involves the injection of chemical downhole through a capillary injection line connected to the surface facilities. It is commonly used to protect against the precipitation of carbonate and sulfide scales, which precipitate mainly due to pressure drop. The highest-pressure reductions occur in the production lines. If injection lines are not available, scale inhibitors can also be deployed via the gas lift system.[5–7]

Squeeze treatment is the technique to deploy scale an inhibitor slug downhole by bullheading in a producer well, providing protection downstream from the near well bore area, that is, perforations, safety valves, and production lines. This technique is commonly used to prevent the deposition of sulphate scales, formed due to the mixing of incompatible brines such as seawater and formation brine. Sulphate scales may occur at any location, at injection wells, within the formation, in the near-well bore area, or/and production lines.

6.2 Squeeze Treatment Basics

In a squeeze treatment, a solution of scale inhibitor is injected into the well above the formation pressure and normally below fracture pressure. The chemical slug is pushed into the rock formation, then the well is shut in for some time to allow the chemical to be retained in the rock formation. When the well production is resumed, inhibitors will slowly be released into the produced brine. The well will be protected if the inhibitor return concentration is above a certain threshold level, known as minimum inhibitor concentration (MIC).

Squeeze treatment design consists of the following stages, as illustrated in Figure 6.1: (i) preflush; (ii) main treatment, where the chemical slug is injected, usually at a concentration ranging from 5% to 20%; (iii) overflush, a stage to push the chemical deeper into the formation, with the goal to expose the scale inhibitor (SI) to clean the rock; (iv) shut in, a period to allow the chemical to further be retained in the rock surface; and (v) the well is back in production, and the chemical is slowly released and produced back. Normally, the main treatment and overflush are injected as aqueous solutions, although in certain scenarios, diesel overflush may be deployed instead.[8]

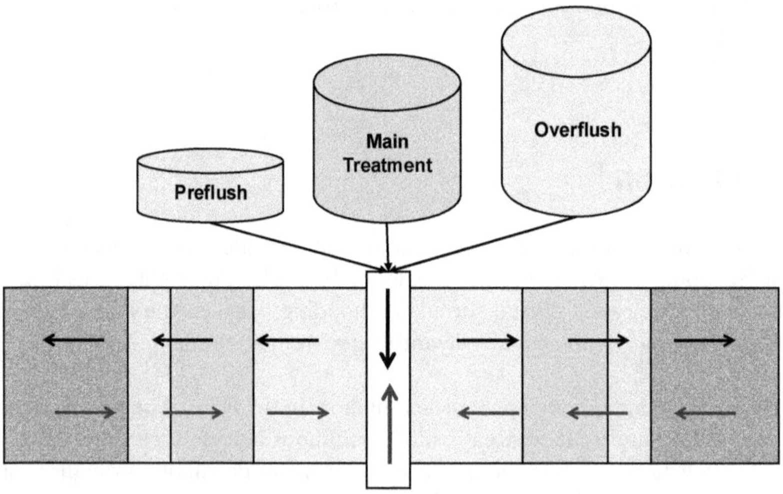

Figure 6.1. Scale inhibitor treatment.

The squeeze treatment lifetime is defined as the time at which the SI return concentration falls below the MIC, which may be reported as volume of water protected. A typical squeeze return concentration profile can be found in Figure 6.2.

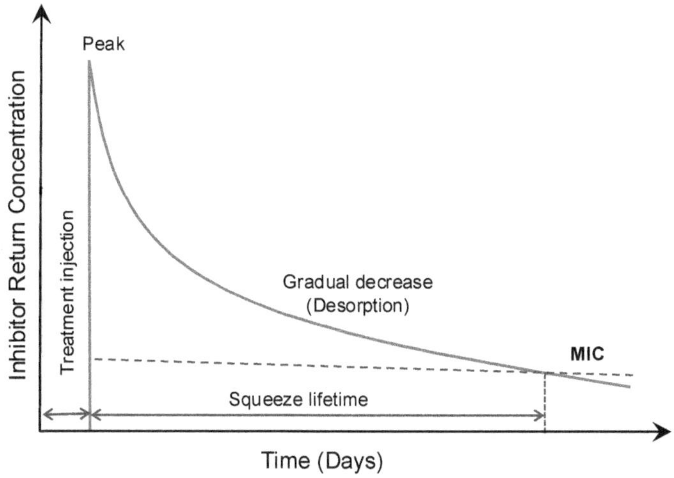

Figure 6.2. Scale inhibitor treatment return profile.

6.2.1 Preflush

Preflush is commonly injected as a spacer between the main chemical slug and wellbore fluids to clean the formation for enhancing retention and/or reducing the cleanup time. Preflush stage may consist of spiking with a low dosage of scale inhibitor in carrier brine. On other occasions, mutual solvents and surfactants have been used as additives. Surfactants alter the wettability of the formation, as they reduce the interfacial tension between the fluid phases as they adsorb on the rock surface.[9, 10] Surfactant preflush has been reported to extend the inhibitor retention in core flood experiments.[11, 12] Mutual solvents are soluble in water and oil, with the primary objective to minimize residual water saturation, reducing the risk of oil permeability impairment, due to water injection in the near wellbore area. Mutual solvent preflush in squeeze treatments in the Norwegian North Sea sector has been reported to minimize the formation damage.[13] This type of preflush was also used in the deepwater wells offshore Brazil.[14] In addition, it has been reported to double the treatment lifetime in a field trial, due to increased retention by reducing the solubility of Ca-inhibitor complexes.[15]

A survey of over 100 wells in Cedar Lake Unit, located in Gaines County, Texas, also indicated the beneficial impact of preflush additives.[16] The squeeze treatments were divided into three categories: no preflush, mutual solvent, and surfactant preflush. It was concluded that the well cleanup was most accelerated by mutual solvent preflush, followed by surfactant. In addition, mutual solvent was more effective than surfactant in extending squeeze treatment lifetime, on average 60% and 25% over no pretreatment, respectively.

6.2.2 Main Treatment

This stage of the treatment is where the main chemical slug is deployed in the well. Normally, a chemical pump is used to add the neat chemical into the main water injection line; both pumps need to be set at the appropriate flow rates for the required target injection concentration.[17] It is recommended to assess the compatibility with other additives, such as corrosion inhibitors, clay stabilizers, and so on, when deploying high concentrations of scale inhibitor.[18, 19] When the overflush volume is fixed, the squeeze lifetime is proportional to the main treatment volume. For a fixed mass of inhibitor, injection at a lower concentration with a larger volume increases squeeze lifetime. Operational costs may offset the benefits.[20] Therefore, a full optimization study should be performed.[21, 22]

6.2.3 Overflush

The objective of the overflush stage is to push the chemical deeper into the formation to achieve maximum retention. Generally, the larger the overflush volume, the longer the lifetime. However, in certain scenarios, it is not recommended to inject large volumes of water, such as in water-sensitive formations, pre-emptive treatments (performed at low water cuts), and/or where artificial lift is required, as water is denser than hydrocarbons, and will be more difficult to lift. In these circumstances, where large water overflush is not an option, the overflush might be split between diesel and water.[8, 23]

6.2.4 Inhibitor Retention Mechanism

The retention of scale inhibitors in the rock formation and the corresponding threshold concentration, known as MIC, will determine the efficacy of the squeeze treatment, in other words, how long the well will be protected, which may be reported as volume of water produced or production time. The retention level of scale inhibitors is governed by two mechanisms, adsorption and precipitation. Adsorption occurs at low concentration, and the coupled adsorption/precipitation occurs at higher concentration. The coupled adsorption/precipitation can be determined by the experiment shown in Figure 6.3. Adsorption/precipitation experiments have been reported for phosphonate and polymeric species.[24–31] It is generally concluded that the retention level depends on pH, temperature, and concentration of Ca^{2+} and Mg^{2+}.

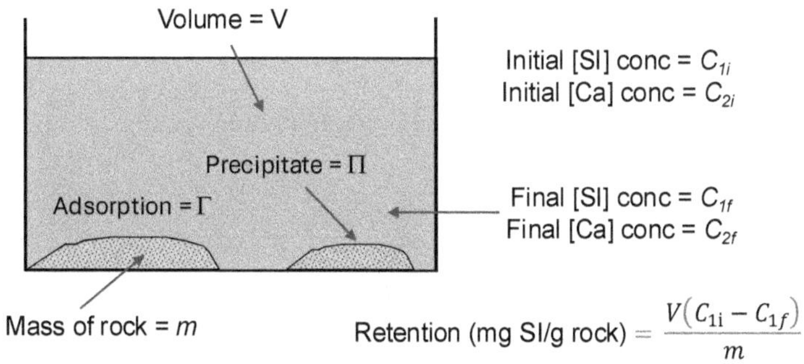

Figure 6.3. Coupled inhibitor retention by adsorption and precipitation.

6.2.5 Placement Considerations

Oil reservoirs are usually large and heterogeneous rock bodies, where wells are drilled through rock formations with different properties, such as permeability, porosity, and thickness. Scale inhibitors are bullheaded into the completion interval. Therefore, for successful treatment, it is imperative that the scale inhibitor is placed in the water-producing zones, or upstream, so at least the production line can be protected. Correct placement depends on two factors: on the one hand, identifying the offending production zones, and secondly, monitoring where the treatment package is placed. Production-logging-tool (PLT) provides a pressure log (or injection/production flow rate log), which may be used to identify the water production zones.[32] Reservoir simulation has been used to successfully design squeeze treatments.[33–36] In addition, another alternative has been proposed before, consisting in the injection of a tracer package as a part of the squeeze treatment design.[37] Lithium chloride (LiCl) was co-injected in the early stages of squeeze treatments to estimate the effectiveness of the diversion stage of a treatment in the Strathspey field in the North Sea.[36] Potassium chloride (KCl) has been used in the Norne field to estimate the layer flow rate profile in a producer to optimize scale squeeze treatments.[38]

Several analytical expressions to calculate the flows and relative penetrations of injected fluid in both linear and radial heterogeneous formations with no crossflow, assuming that all the layers were at the same pressure, have been proposed before.[39–41] Later, analytical expressions for multilayered systems in single- and two-phase flow with no crossflow have been applied to squeeze treatments, with the similar assumption that all the layers are at the same pressure.[42] In addition, the derivation of an analytical expression to simulate the partitioning of flow with variously pressured layers in radial coordinates assuming no cross flow was proposed before,[43] and a brief derivation is presented below.

The scenario is depicted in Figure 6.4, where the pressure at the wellbore for all layers is P_w. This assumption is generally valid, except for long horizontal wells, where well friction may have a significant impact. Secondly, each layer is assumed to be at a different pressure, P_i. Then, the pressure drop in each layer can be expressed as a function of the pressure drop in the first layer (ΔP) and the pressure difference between the first layer and the others ($\Delta_{1l} = P_i - P_1$), as shown below:

$$\Delta P_j = P_j - P_w = \left(P_1 - \delta_{1j}\right) - P_w = \Delta P + \delta_{1j} \tag{1}$$

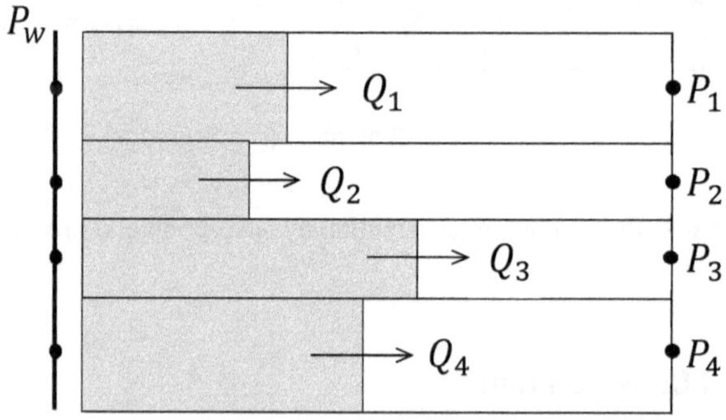

Figure 6.4. Partitioning of total flow in a multilayer system.

Applying Darcy's law in radial coordinates and assuming that the volume is conserved, the sum of the layer flow rates is equal to the total, $\sum_{l=1}^{N} Q_l = Q_T$, expressions describing the flow partitioning during injection and production are shown below:

$$Q_j = \frac{1}{E_j} \cdot \left[\frac{Q_T + \sum_{i=2}^{N} \frac{\delta_{1i}}{E_i}}{\sum_{i=1}^{N} \frac{1}{E_i}} - \delta_{1j} \right], \text{ injection where } P_w > P_i \tag{2}$$

$$Q_j = \frac{1}{E_j} \cdot \left[\frac{Q_T - \sum_{i=2}^{N} \frac{\delta_{1i}}{E_i}}{\sum_{i=1}^{N} \frac{1}{E_i}} + \delta_{1j} \right], \text{ production where } P_w < P_i \tag{3}$$

where $\quad \Delta_j = E_j \cdot Q_j,\ E_j = \dfrac{1}{2\pi h_j} \displaystyle\int_{R_{Well}}^{R_{Max}} \frac{1}{K_j \left(\frac{k_{ro}}{\mu_o} + \frac{k_{rw}}{\mu_w} \right)} \frac{dr}{r}.$

6.2.6 Treatment Monitoring

After a squeeze treatment has been deployed, it is paramount to monitor the return concentration. As mentioned before, the well will be protected, that is, scale deposition will be prevented, if the return concentration is above a certain threshold. Two aspects are of vital importance for accurate monitoring of the treatment, which are sample preservation and accurate analysis.

Therefore, the scale inhibitor return concentration should be monitored to identify when the well would need to be treated again, that is, where the concentration drops below MIC. Adequate sampling is critical to preserving the produced water samples when taken offshore, so that the samples are not compromised, preventing any misinterpretation when the samples are analyzed in the lab.[44, 45] Previous studies reported that significant precipitation can occur in nonpreserved produced water samples.[46, 47]

6.2.7 Treatment Design Optimization

Optimization is the process of selecting something as effective as possible, in this context, a treatment design. A squeeze treatment design is determined by several input parameters, such as the injection chemical concentration, the injection volume (main treatment) and the overflush volume, where a simulator is used to predict the lifetime, and thus to identify an optimum design. This goal could be achieved by varying each parameter, which is known as a sensitivity study. Several of these kinds of studies have been reported before.[20, 22, 48] Although the analysis of this type of study provides very valuable information, it can be extremely time-consuming, onerous, and error prone, as traditionally, it has been performed manually. This process can be automated by using optimization algorithms, which has many advantages, among others the possibility of the exploration of a wider search space.

There are a great variety of optimization algorithms or techniques, which can be classified as deterministic or stochastic.[49] Deterministic optimization is based on the computation of the objective function gradient to establish the search direction of the optimal solution, such as bisection method, Newton Raphson, and the steepest descent or gradient method.[50] Stochastic search algorithms are commonly used for scenarios where the functional relationship between input variables and misfit is not clear, that is, the objective function might not be differentiable or even discontinuous. This type of algorithm is classified as local and population search-based. The hill climber algorithm is a well-known local search algorithm, which has been successfully used for the automatic adsorption isotherm matching.[51] Population search algorithm has been extensively used in automatic reservoir history matching, due to its flexibility to accommodate complex real-life problems. Some of the most popular population-based algorithms are the genetic algorithm,[52] differential evolution[53] and particle swarm optimization.[54]

To find the optimum squeeze treatment design, the operational constraints need to be considered; a brief introduction will be discussed in Section 6.4. Then, it is necessary to use a suitable optimization technique for the case under study. Two main techniques have been proposed before, fixed and global squeeze treatment lifetime, which will be described briefly below.

6.3 Treatment Modeling

In this section, the elements necessary to simulate a squeeze treatment will be described, which may be used to predict the lifetime of the treatment, based on the operating parameters, rock formation, and inhibitor properties. This simulation tool has been used to design squeeze treatments and to identify optimum designs. Two processes need to be described mathematically: on the one hand, the transport of the chemical through the porous media, and on the other, the interaction of the chemical with the rock surface, which will be described as an adsorption process.

Commonly, scale squeeze treatments are injected as aqueous solutions, as mineral scale precipitation is associated with the aqueous phase, and also the transport and subsequent retention are processes occurring in the aqueous phase. However, there are certain situations where an aqueous-based treatment might not be desirable, generally due to concerns of water formation damage, hydrate formation, and/or gas lifting limitations. In these circumstances, squeeze treatments become nonaqueous, where some of the stages of the treatment are nonaqueous, such as the chemical injection carrier phase, preflush, and/or overflush. In the next sections, the governing equations for the transport and adsorption for single aqueous phase and nonaqueous two phases will be shown.

In terms of squeeze modeling, particularly regarding scale inhibitor retention, there are two main schools of thought. Heriot Watt University's approach is based on adsorption isotherms to determine the relationship between the SI concentration in the aqueous phase and the mass of SI adsorbed or retained on the rock surface, which are commonly described by Langmuir or Freundlich functions.[42, 55] While Rice University's approach describes the scale inhibitor retention as precipitation and subsequent dissolution of the Ca-inhibitor complexes.[24, 25] In this manuscript, the Heriot Watt approach will be adopted.

6.3.1 Scale Inhibitor Transport and Adsorption

To simulate the transport and subsequent adsorption of the scale inhibitor in the rock formation, the advection-dispersion-reaction equation will be used. The advection and dispersion will simulate the transport of the chemical in the porous media, and the reaction is the adsorption process. The single-phase transport equation with adsorption has the following form:

$$\frac{\partial C}{\partial t} = -v\frac{\partial C}{\partial x} - \frac{\rho}{\phi}\frac{\partial \Gamma}{\partial t}, \tag{4}$$

where Γ = actual adsorbed level of inhibitor, C = mobile concentration, D = dispersion coefficient (L^2/T), v = superficial velocity (L/T) = $Q/A\phi$, Q = water injection rate (L^3/T), A = cross-sectional area (L^2), ρ = rock density (M/L^3), and ϕ = porosity.

To simulate nonaqueous squeeze treatments, where one or more stages are nonaqueous, the governing equations of two-phase flow and two-phase transport in porous media will be presented. To expand the single-phase transport equation to two phases, the fractional flow must be introduced. Considering that water and oil phases are incompressible, the continuity equation for each phase (water or oil) has the following form:

$$\frac{\partial S_p}{\partial t} = -\frac{Q_t}{A\phi}\frac{\partial f_p}{\partial x} \tag{5}$$

This equation is based on the fractional flow theory and mass conservation principle. It specifies the rate of change of phase saturation for a given change in time for a block of pore volume $A\phi\delta x$, due to the volume of the phase entering the block by advection, where f_p is the fractional flow equation, which defines the fraction of the flowing stream, phase p, either aqueous or nonaqueous.[56–59]

To derive an expression describing the water fraction of flowing fluid, the following physical assumptions proposed before[60] need to be considered. First, the simultaneous flow of oil and water is an immiscible displacement. Second, the displacement is assumed to be incompressible, that is, all the pressure-dependent variables, viscosities, and densities are ignored. This kind of displacement will occur if the volume is conserved.

$$Q_t = Q_o + Q_w = Q_I, \tag{6}$$

where Q_t = total flow rate, Q_o = oil flow rate, Q_w = water flow rate, and Q_I = water injection rate.

Then the fractional flow of water (f_w) can be described by the expression below:

$$f_w = \frac{Q_w}{Q_w + Q_o} = \frac{Q_w}{Q_t} \tag{7}$$

Next, the displacement of oil or water is assumed to take place under diffuse flow conditions, that is, the fluid saturation distributions are to be uniform with respect to the thickness of the reservoir, where the simultaneous flow of oil and water can be used using the relative permeabilities. Applying two-phase Darcy law for oil and water in a linear horizontal block, the one-dimensional equations for simultaneous flow, the expressions below are obtained:

$$Q_o = -\frac{Kk_{ro}A}{\mu_o}\frac{\partial P_o}{\partial x}, \tag{8}$$

$$Q_w = -\frac{Kk_{rw}A}{\mu_w}\frac{\partial P_w}{\partial x}, \tag{9}$$

where K = absolute permeability, k_{rw} = water relative permeability, and k_{ro} = oil relative permeability.

By subtracting the equations above and expressing the oil flow rate as, $Q_o = Q_t - Q_w$, the following expression is obtained, where $\frac{\partial P_c}{\partial x} = \frac{\partial P_o}{\partial r} - \frac{\partial P_w}{\partial r}$ is the capillary pressure gradient in the flow direction.

$$Q_w\left(\frac{\mu_w}{Kk_{rw}} + \frac{\mu_o}{Kk_{ro}}\right) = \frac{Q_t\mu_o}{Kk_{ro}} + A\frac{\partial P_c}{\partial x} \tag{10}$$

The water fraction of flowing fluid is described by the following expression:

$$f_w = \frac{Q_w}{Q_w + Q_o} = \frac{Q_w}{Q_t} \tag{11}$$

After substituting the above equations in the water fractional flow equations for the displacement of oil and water in one dimension, it can be written as follows in hybrid units, more details are in[60]:

$$f_w = \frac{1 + \frac{Kk_{ro}A}{Q_t \cdot \mu_o} \cdot \frac{\partial P_c}{\partial x}}{1 + \frac{\mu_w}{\mu_o} \cdot \frac{k_{ro}}{k_{rw}}} \tag{12}$$

$$f_o = 1 - f_w \tag{13}$$

Assuming high flow rates which commonly occur in the near wellbore area, it is reasonable to neglect the capillary pressure gradient. Thus, the water fractional flow is reduced to the expression below:

$$f_w = \frac{1}{1 + \frac{\mu_w}{\mu_o} \cdot \frac{k_{ro}}{k_{rw}}} \tag{14}$$

For constant temperature and the water and oil viscosities, the immiscible displacement is strictly a function of the water saturation through the relative permeabilities and the characteristic shape of the fractional flow, as depicted in Figure 6.5.

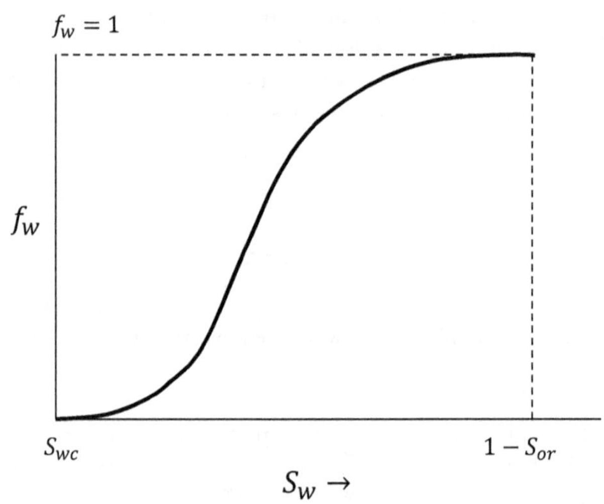

Figure 6.5. Water fractional flow.

The one-dimensional two-phase flow component transport and adsorption equation is shown below. It describes the isothermal ideal miscible displacement, where all the components and phases, mix in all proportions without an interface forming between them. The dispersion of solute within the phase is neglected, which is a reasonable assumption due to the high flow velocities occurring in the near wellbore area. The solute concentration units are mass of solute per unit volume of phase. The form of this equation is described below:

$$\frac{\partial\left(S_p C_p\right)}{\partial t}=-\frac{Q_t}{A\phi}\frac{\partial\left(f_p C_p\right)}{\partial x}-\frac{\rho}{\phi}\frac{\partial\Gamma}{\partial t}, \tag{15}$$

where C_p = is the component concentration in phase p, either aqueous or nonaqueous, Γ = actual adsorbed level of inhibitor, described by an adsorption isotherm.

6.3.2 Inhibitor Adsorption Characteristics

The concentration of the retained species has been commonly described by an adsorption isotherm, despite the retention mechanism being either adsorption and/or precipitation.[42, 55] There are various empirical expressions, such as Langmuir, Freundlich, and Brunauer-Emmett-Teller (BET). The Langmuir and BET isotherms are derived theoretically, based on the concept of monolayer and multilayer adsorption, whereas Freundlich is an empirical relationship. However, the Freundlich isotherm form may be derived from the Langmuir equation by considering repulsive interactions between adsorbed solute particles and the rock surface heterogeneities.[61]

The BET expression has the following formula for adsorption from a liquid solution:

$$\Gamma=\frac{bKC}{\left(C_S-C\right)\left[1+\left(K-1\right)\left(\frac{C}{C_S}\right)\right]}, \tag{16}$$

where b = value of x/m, when monolayer has been completed (mg/mg); x = mass of adsorbate (mg); m = mass of dry adsorbent (mg); C_s = concentration of adsorbate at saturation of all layers; and K = constant related to energy of interaction between adsorbent and adsorbate.

Langmuir expression is described as below, with a and b are constants:

$$\Gamma=\frac{abC}{\left(1+aC\right)} \tag{17}$$

Freundlich expression is described below, with k and n are constants:

$$\Gamma=kC^n \tag{18}$$

The Freundlich isotherm has been used more often to describe field and laboratory data.[20, 43]

The adsorption process of scale inhibitors on the rock surface may be instantaneous or time-dependent. Instantaneous or equilibrium adsorption corresponds to the situation where the adsorption reaction is fast enough relative to the flow velocity to reach equilibrium; in other words, the residence time is long enough. On the other hand, there are situations where the adsorption process is slow enough compared to the flow velocity, so equilibrium is not achieved. If the process is not at equilibrium, therefore it is time-dependent, and the rate to equilibrium needs to be described. The first-order expression below was proposed before[55, 62, 63]:

$$\frac{\partial \Gamma}{\partial t} = k_{ad} \cdot \left(\Gamma_{eq}(C) - \Gamma \right) \tag{19}$$

where k_{ad} = the kinetic desorption rate, Γ_{eq} = the equilibrium adsorption level at a given mobile concentration (C), and Γ = the actual adsorption.

6.3.3 Field Treatment Design Lifecycle

Once scale precipitation has been identified as a flow assurance problem, the first step is to identify an appropriate chemical inhibitor, as shown in Figure 6.6. During the production period, it might be necessary to treat the well several times. If the well has not been treated before, normally the first step is the best-in-class study, where suitable scale inhibitors are identified, normally one or two. Then, the next step is to perform a core flood experiment, which has two objectives: to evaluate the retention level where a dynamic adsorption isotherm is routinely derived[64] and to monitor treatment-induced formation damage.

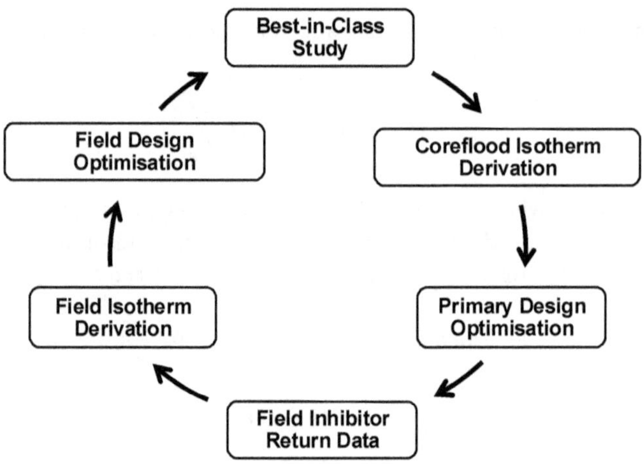

Figure 6.6. Squeeze treatment design lifecycle.

Once the dynamic adsorption isotherm is derived, an initial primary treatment design optimization can be performed. At this stage, it is the best descriptor of the retention, if the well has not been treated before or a new chemical is going to be injected. Alternatively, if core flooding is not available, it is possible to use an analogue from a sister well or sister reservoir, with similar formation characteristics and the same chemical. Once the well has been treated, and the return concentration profile is available. It is common that the core flood isotherm does not achieve good predictions, for a variety of reasons, such as assumptions on inhibitor placement, changes in lithology, a core flood experiment performed at a single flow rate, and so on.[48] Therefore, it is necessary to derive a field isotherm, which generally predicts with a significant degree of accuracy[48, 51] and is suitable to perform a field design optimization.[21, 22] This step should be repeated during the well lifetime, as generally, the well will need to be treated periodically.

The lifecycle will start again, even if the well has been treated before, as it is common that a new chemical will be deployed during well production, due to change of service contracts, well conditions, or, on occasions, environmental concerns, where greener chemistries should be employed. In these circumstances, the cycle is restarted with the best-in-class step.

6.3.4 Core Flood Isotherm Derivation

Deriving an adsorption isotherm from core flood experiments is a common practice in designing squeeze treatments.[62] A core flood experiment consists in flooding a chemical slug in a core for several pore volumes, then the core is postflush, and the chemical effluent concentration is monitored (Figure 6.7).

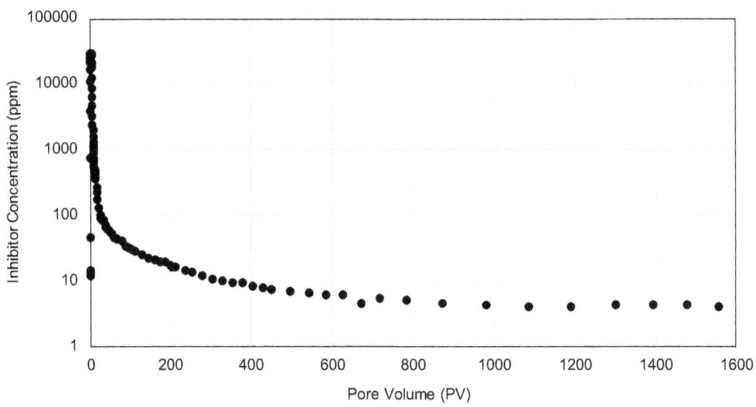

Figure 6.7. Core flood concentration effluent profile.

As the scale inhibitor is transported through the core, it will be retained; and the chemical slug will be retarded against an inert tracer. The effluent concentration profile can be used to estimate the retention of scale inhibitor, which then will be used to derive an adsorption isotherm that describes the inhibitor-rock interaction (Figure 6.8), which relates the solution concentration, and the mass of inhibitor adsorbed per unit of volume of the rock.

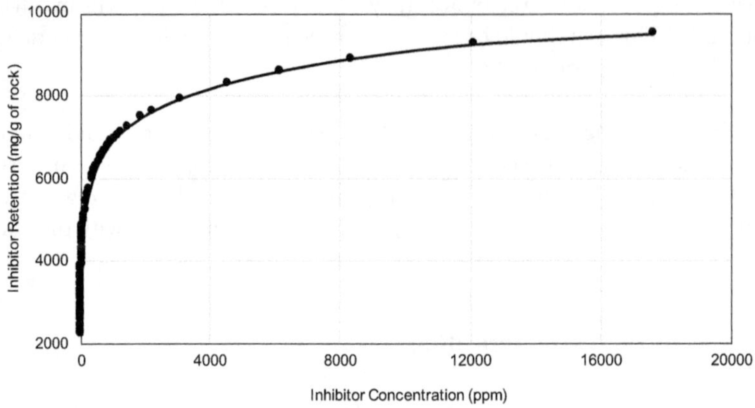

Figure 6.8. Derived adsorption isotherm.

Assuming that the adsorption process is in equilibrium, the single-phase transport equation can be rearranged as follows:

$$\frac{\partial C}{\partial t} = -V_C \frac{\partial C}{\partial x} = -\frac{v}{1+\frac{(1-\phi)}{\phi}\frac{\partial \Gamma}{\partial C}}\frac{\partial C}{\partial x}, \tag{20}$$

where V_C = velocity at which concentration C returns to the well, v = fluid velocity, ϕ the porosity, and $\frac{\partial \Gamma}{\partial C}$ = gradient of the isotherm at concentration C.

Therefore, V_C, which describes how retarded the inhibitor is vs. an inert tracer, can be used to estimate the level of adsorption. A steep gradient reflects a low velocity; therefore, a steep gradient at concentrations close to MIC will result in inhibitor concentrations close to MIC taking a long time to return to the well, which is a desirable characteristic of scale inhibitors. Once the isotherm is derived, it is common to validate the derived isotherm vs. the effluent concentration profile, as shown in Figure 6.9.

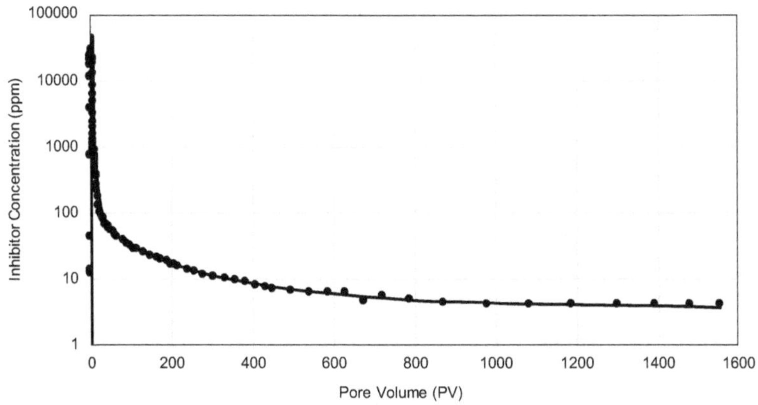

Figure 6.9. Validation of the derived adsorption isotherm.

6.4 Operational Constraints

As any other field operation, squeeze treatment designs are generally constrained by logistics, operational, and/or reservoir conditions. In the design stage of squeeze treatments, this needs to be addressed. Generally, operational constraints consist of the volume of neat chemical volume, the injection concentration, and the total volume of water injected.

Logistical challenges depend on the location of the well to be treated, particularly, offshore or onshore. In offshore locations, there are two scenarios. On one hand, treatments delivered from the platform, where the chemicals are normally delivered in tanks, and due to space limitation on deck, there will be a limitation of the volume of chemical. On the other hand, for subsea wells, the treatment might be deployed from a diving support vessel (DSV), where the volume of chemical is restricted by the storage capacity on the vessel.[65] Scale inhibitor concentration might be constrained by two aspects. First, chemical pumps are used to add neat chemical into the main water injection line to provide the required injection concentration. If the chemical pump is not strong enough or the water injection rate is high, there might be some limitation on the injection concentration. The second aspect to consider is that due to incompatibility with the formation brine, which may result in the precipitation of complexes of divalent ions, it may lead to formation damage. Thus, a limitation in the injection concentration might be due to the risk of formation damage. Finally, the volume of water injected may be of concern in water-sensitive formations, and/or if the well has lifting constraints. In addition, the treatment should be performed as quickly as possible to minimize deferred oil and cleanup time.

6.5 Squeeze Treatment Lifetime

6.5.1 Fixed Target Lifetime

In offshore fields, where treatments are part of a campaign, treatments are delivered by a DSV, such as in the North Sea. This type of operation is commonly scheduled for the summer period, the squeeze treatment needs to achieve a specific lifetime. In the North Sea, it will be annually, that is, 12 or 24 months.[65, 66] A methodology was presented before,[21] based on the gradient method, to find the squeeze designs for a target lifetime. A squeeze treatment lifetime increases continuously as a function of main treatment volume, inhibitor injection concentration, and overflush.[48] This methodology might be used to calculate the "iso-lifetime" curves for a fixed target treatment lifetime for a range of chemical concentration, a function of main treatment, and overflush volume, which describes all the treatment designs for a given target lifetime, see Figure 6.10. This type of graphical description presents a general overview of all the designs, which will provide the target squeeze lifetime, then it is a very valuable tool to identify an optimum design.

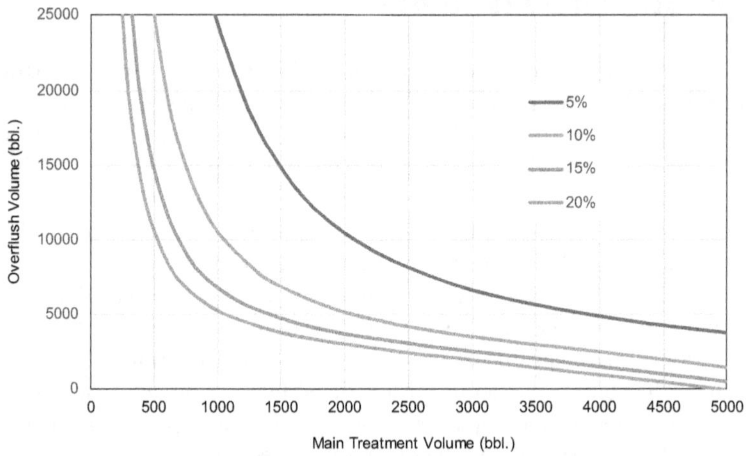

Figure 6.10. Squeeze treatment "iso-lifetime" curves with fixed target lifetime for several inhibitor concentrations.

6.5.2 Global Treatment Optimization

There are certain scenarios where the squeeze lifetime is not critical as in the previous case, where it is not strictly necessary to achieve a certain treatment lifetime target, such as platform wells with easy access, or onshore wells. In these cases, the optimum design might not be the one with the longest lifetime, as it might not be the most cost-effective or/and nor the most efficient, since the other objectives should be considered:

- Squeeze lifetime
- Total chemical volume
- Total injected water volume

The treatment lifetime is the most important objective. However, the volume of chemical and the total volume of water injected are also important. In offshore locations, the chemical cost is a small proportion of the total expense.

However, the injection concentration might be of concern, and there might be limitation of deck space. In onshore locations, the chemical cost might be an important contribution to the total treatment expense. The water-injected volume might be of paramount importance in offshore wells, but it might not be as important as in onshore wells. Therefore, not only should the treatment lifetime be considered but also the total SI and total water volume injected. Therefore, the problem becomes multiobjective, where there might not be a single design which honors each objective. Thus, objectives are said to be in conflict, and therefore, several Pareto optimal treatment designs might exist. A design will be part of the Pareto front, if it is not dominated by any other design, that is, if there is not any other design capable of improving all the objectives.

The methodology proposed was based on a well-known stochastic algorithm, PSO,[22, 65] which is a popular algorithm from the category of population-based search algorithms, which works by simulating the social behavior of bird flock or fish school seeking food.[54] The methodology assumes one objective in the optimization exercise, particularly, squeeze lifetime, since the priority is to identify a design that is capable of providing a particular lifetime. However, since other objectives need to be considered, the Pareto front of the suggested designs by the algorithm is calculated, to identify the most effective design, in terms of the operations management criteria. The Pareto front considering the treatment lifetime vs. total chemical volume injected is shown in Figure 6.11, and treatment lifetime vs. total chemical volume injected (Figure 6.12), which provides a general overview of squeeze treatment design. The methodology may be used to identify the most efficient designs, in terms of cost, chemical volume, and injected water volume.

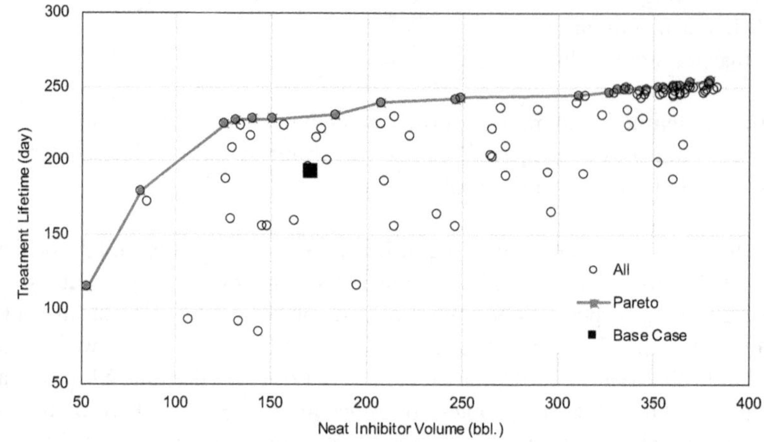

Figure 6.11. Pareto front considering treatment lifetime vs. volume of scale inhibitor.

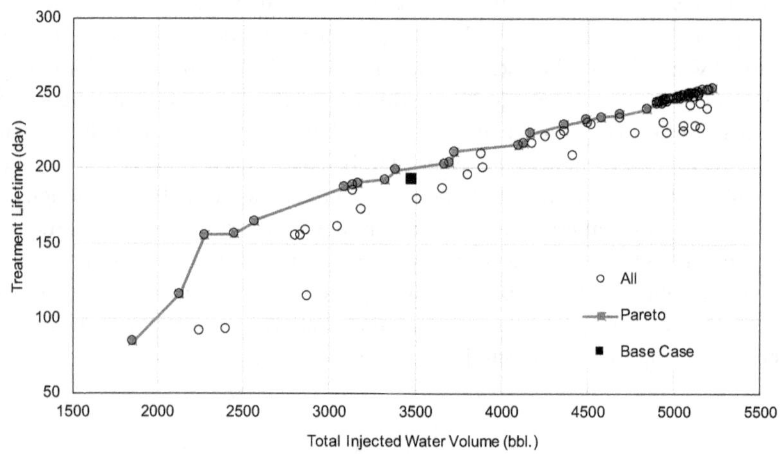

Figure 6.12. Pareto front considering treatment lifetime vs. total injected water volume.

6.6 Nonaqueous Squeeze Treatments

When an aqueous treatment is not suitable or has undesirable side effects, such as treatments at low water cut or in water-sensitive formations, a nonaqueous scale inhibitor delivery system should be deployed. Despite the name, most of these treatments still contain some water, as scale inhibitors only truly dissolve in water due to their hydrophilic nature. The nonaqueous squeeze inhibitor can be classified as below.[67]

6.6.1 Oil-Soluble

These materials contain less than 8% water content. They essentially consist of a solvent, a scale inhibitor, and an oil-solubilizing agent (OSA). The OSA typically consists of a low-molecular-weight base, designed to interact with the scale inhibitor to incorporate it into the carrier solvent typically a common hydrocarbon, such as diesel or kerosene.[67, 68] In the presence of water, the inhibitor strongly partitions in the water phase.

6.6.2 Invert/Reverse Emulsions

This system is a temporarily stable mixture of water, hydrocarbon (diesel or kerosene), and surfactant, with dispersed water droplets of size between 0.2 μm and 40 μm, containing the scale inhibitor.[69] The breakdown of the emulsion in the presence of water delivers the chemical. Other attempts to improve the delivery of the emulsions and prevent formation damage have been proposed.[70–72]

6.6.3 Amphiphilic Solvent Systems

These materials consist of an amphiphilic solvent, a liquid which has affinity for both water and oil, and a high concentration of conventional aqueous scale inhibitor solution that may be dissolved in the amphiphilic solvent. The successful deployment of these kinds of treatments have been reported in Milne Point, North Slope Alaska. The treatment design consisted of a nonionic amphiphile preflush, a 100% oil-soluble scale inhibitor as main treatment, and finally dead crude oil as overflush for low-water-cut wells to improve well cleanup time and extend squeeze life due to enhanced retention.[73] Also, it has been reported that an amphiphilic solvent-based system for the inhibitor application was developed with the ability to formulate a clay stabilizer within the package and applied in the Asgard Field of Norway.[74]

6.6.4 Microemulsions

Microemulsions are thermodynamically stable mixtures typically consisting of an oil phase, scale inhibitor, and surfactant. The droplet size is typically less than 50 nm, making these materials optically transparent. Microemulsions can be potentially used as carrier deployment and squeeze treatment lifetime enhancer, due to the increase of surface area by the displacement of organic material by the surfactant.[75]

6.6.5 Water-Free Materials

These custom-made scale inhibitors do not contain any water in the manufactured formulation. The formulations dissolve in oil and convert into active water-soluble species in the producing formation after contact with water.[67, 76]

6.6.6 Part-Aqueous Systems

The term "part-aqueous" has been coined to refer to squeeze treatments that include both aqueous and nonaqueous phases. Usually, the treatment design preflush and/or the overflush are nonaqueous. The overflush generally constitutes the largest volumetric proportion of the treatment. There have been numerous field examples deploying this type of treatment, such as injection of mutual solvent preflush,[13, 77, 78] split overflush between diesel and water,[79] and the use of diesel for entire overflush in squeeze treatments, in high-value wells producing at low water cut, less than 10% of water produced of the total produced.[14, 80]

6.7 Combined Treatments

Squeeze treatments may include the deployment of other chemicals. Two techniques have been proposed before. One is matrix stimulation treatments to enhance production or remove completion/production-related damage. Another one is a tracer package to evaluate layer flow rate profile along the completion interval, which is a significantly cheaper alternative to running PLTs to evaluate treatment placement.

6.7.1 Squimulation

The term "squimulation" was coined to describe the simultaneous squeeze and stimulation treatments.[81] This technique consists of the displacement of bullhead stimulation treatment (HCl and HCl/HF) to be displaced deep into the formation using the squeeze treatment stages. Four successful field treatments in terms of well stimulation/skin reduction and squeeze treatment lifetime have been reported.[81]

6.7.2 Squeeze Treatment and Tracer Program

A successful squeeze design will prevent scale deposition, resulting in maintaining well productivity, no evidence of scale deposition, such as reduction in scaling ion concentrations, no scale crystals detected by the environmental scanning electron microscope analysis of filtered samples, and so on.[37] On occasion, an undesirable situation may occur, where the return scale inhibitor concentration is above the concentration threshold or MIC, and the well shows sign of scale deposition. This may be due to inadequate placement of chemical in the offending zones, particularly in long horizontal wells or/and in wells with pronounced crossflow. To evaluate any placement issues, it might be necessary to run a PLT, which may result in a prohibitively expensive operation, due to the extra expenses and the deferred oil. An alternative is to include a tracer package as part of the squeeze treatment, where a tracer package in the overflush stage, as proposed before.[37] LiCl was co-injected in the early stages of squeeze treatments to estimate the effectiveness of the diversion stage of a treatment in the Strathspey field in the North Sea.[36] KCl has been used in the Norne field to estimate the layer flow rate profile in a producer to optimize the scale squeeze treatments.[38]

6.8 Summary

Squeeze treatment is one of the most common techniques to prevent mineral oilfield scale deposition in oil production wells. This type of treatment consists of the injection of special type of chemicals, capable of inhibiting scale deposition at low concentrations. A critical aspect for a successful squeeze treatment is the retention on the reservoir rock, where the chemical is slowly released in the producing brine. The well will be protected if the return chemical concentration is above a certain threshold, known as MIC.

The lifecycle of squeeze treatment was introduced in this chapter, where treatment design steps are described, including best-in-class study, isotherm core flooding, primary treatment optimization, and field design optimization. To design and to have the ability to optimize treatments, it is necessary to have a mathematical framework. In terms of design optimization, two techniques were presented for applying in different scenarios. In addition, the nonconventional squeeze treatments, which include nonaqueous treatments and combine treatments, were described.

Nomenclature

ADR	Advection-dispersion-reaction
bbl	Barrels
BET	Brunauer-Emmett-Teller
DHCI	Downhole inhibitor injection
DSV	Diving support vessel
ESEM	Environmental scanning electron microscope
MIC	Minimum inhibitor concentration
OSA	Oil-solubilizing agent
PLT	Production-logging-tool
ppm	Parts per million
PSO	Particle swarm optimization
PV	Pore volume
SI	Scale inhibitor

References

1. P.J. Breen, B.N. Diel, H.H. Downs, "Correlation of Scale Inhibitor Structure With Adsorption Thermodynamics and Performance in Inhibition of Barium Sulfate in Low-pH Environments," SPE-20688-MS, SPE Annual Technical Conference and Exhibition (1990).
2. S.-T. Liu, G.H. Nancollas, "The Crystal Growth and Dissolution of Barium Sulfate in the Presence of Additives," *Journal of Colloid and Interface Science* 52 (1975): pp. 582–592.
3. G.M. Graham, L.S. Boak, K.S. Sorbie, "The Influence of Formation Calcium and Magnesium on the Effectiveness of Generically Different Barium Sulphate Oilfield Scale Inhibitors," *SPE Production & Facilities* 18 (2003): pp. 28–44.

4. M.C. van der Leeden, G.M. van Rosmalen, "Inhibition of Barium Sulfate Deposition By Polycarboxylates of Various Molecular Structures," *SPE Production Engineering* 5 (1990): pp. 70–76.
5. G. Poggesi, C. Hurtevent, J.L. Brazy, "Scale Inhibitor Injection Via the Gas Lift System in High Temperature Block 3 Fields in Angola," SPE-68301-MS, International Symposium on Oilfield Scale, held January 30–31, 2001 (Aberdeen, UK).
6. M.A. Daas, T.S. Golczynski, A.A. Magalhães, et al., "Downhole Chemical Injection Through Gas Lift Lines: Options and Consequences," SPE-142951-MS, Brasil Offshore, held June 14–17, 2011 (Macaé, Brazil).
7. N. Fleming, J.A. Stokkan, A.M. Mathisen, et al., "Maintaining Well Productivity Through Deployment of a Gas Lift Scale Inhibitor: Laboratory and Field Challenges," SPE-80374-MS, International Symposium on Oilfield Scale, held January 29–30, 2003 (Aberdeen, UK).
8. O. Vazquez, P. Herrero, E.J. Mackay, et al., "Non-Aqueous vs Aqueous Overflush Scale Inhibitor Squeeze Treatment in an Oilfield Offshore Norway," *Journal of Petroleum Science and Engineering* 138 (2016): pp. 1–10.
9. J.J. Hinkel, J.E. Brown, B.R. Gadiyar, et al., "New Environmentally Friendly Surfactant Enhances Well Cleanup," SPE-82214-MS, SPE European Formation Damage Conference, held May 13–14, 2003 (The Hague, Netherlands).
10. R. Gdanski, "Modeling the Impact of Capillary Pressure Reduction By Surfactants," SPE-106062-MS, International Symposium on Oilfield Chemistry, held February 28–March 2, 2007 (Houston, Texas, USA).
11. M.M. Jordan, K.S. Sorbie, P. Griffin, et al., "The Application of Surfactants to Improve Scale Inhibitor Squeeze Lifetime in Mixed-Wet Reservoirs: A Coreflood Study Using Oil-Reconditioned Core From a Large North Sea field," 6th Tekna Oilfield Chemistry Symposium, held March 23–26, 1995 (Geilo, Norway).
12. B. Ghosh, X. Li, M.Y. Alklih, "Preflush Design for Oil-Wet Carbonate Formations: Key to Enhance Scale-Inhibitor-Squeeze Lifetime," *SPE Journal* 21 (2016): pp. 776–785.
13. T. Tjomsland, O. Vikane, M. Zettlitzer, et al., "Veslefrikk scale control strategy and economic implications: revisited 7 years later—Did we improve?," SPE-114086-MS, SPE International Oilfield Scale Conference, held May 28–29, 2008 (Aberdeen, UK).
14. P. Bogaert, M. Berredo, C. Toschi, et al., "Scale Inhibitor Squeeze Treatments Deployed from an FPSO in Deepwater Subsea Fields in the Campos Basin," *SPE Production & Operations* 22 (2007): pp. 24–27.
15. I.R. Collins, L.G. Cowie, M. Nicol, et al., "Field Application of a Scale Inhibitor Squeeze Enhancing Additive," *SPE Production & Facilities* 14 (1999): pp. 21–29.
16. J.L. Przybylinski, J.W. Ruggeri, W.D. Wakley, et al., "Optimization of Scale Inhibitor Squeeze Treatments in the Cedar Lake Unit," NACE-96183, NACE Annual Conference, held March 24–29, 1996 (Denver, Colorado, USA).
17. V. Azari, O. Vazquez, E. Mackay, et al., "Squeeze Design Optimization By Considering Operational Constraints, Numerical Simulation and Mathematical Modelling," SPE-200687-MS, SPE International Oilfield Scale Conference and Exhibition, held June 24–25, 2020 (Aberdeen, UK).
18. R. Stalker, I.R. Collins, G.M. Graham, "The Impact of Chemical Incompatibilities in Commingled Fluids on the Efficiency of a Produced Water Reinjection System: A North Sea Example," SPE-80257-MS, SPE International Symposium on Oilfield Chemistry, held February 5–7, 2003 (Houston, Texas, USA).
19. G.M. Graham, D.M. Frigo, I R. McCracken, et al., "The Influence of Corrosion Inhibitor/Scale Inhibitor Interference on the Selection of Chemical Treatments Under Harsh (HP/HT/HS) Reservoir Conditions," SPE-68330-MS, SPE International Symposium on Oilfield Scale, held January 30–31, 2001 (Aberdeen, UK).
20. S.A. Hong, P.J. Shuler, "Mathematical Model for the Scale-Inhibitor Squeeze Process," *SPE Production Engineering* 3 (1988): pp. 597–607.

21. V. Azari, O. Vazquez, E.J. Mackay, et al., "Gradient Descent Algorithm to Optimize the Offshore Scale Squeeze Treatments," *Journal of Petroleum Science and Engineering* 208 (2022): pp. 109469.
22. O. Vazquez, I. Fursov, E.J. Mackay, "Automatic Optimization of Oilfield Scale Inhibitor Squeeze Treatment Designs," *Journal of Petroleum Science and Engineering* 147 (2016): pp. 302–307.
23. O. Vazquez, T. van Ommen, P. Chen, et al., "Modelling a series of nonaqueous field-scale inhibitor squeeze treatments in the Heidrun field," *SPE Production and Operations* 26 (2011): pp. 98–110.
24. A.T. Kan, G. Fu, M.B. Tomson, "Prediction of Scale Inhibitor Squeeze and Return in Calcite-Bearing Formation," SPE-93265-MS, SPE International Symposium on Oilfield Chemistry, held February 2–4, 2005 (The Woodlands, Texas, USA).
25. M.B. Tomson, A.T. Kan, G. Fu, et al., "Mechanistic Understanding of Rock/Phosphonate Interactions and Effect of Metal Ions on Inhibitor Retention," *SPE Journal* 13 (2008): pp. 325–336.
26. M. Kahrwad, K.S. Sorbie, L.S. Boak, "Coupled Adsorption/Precipitation of Scale Inhibitors: Experimental Results and Modelling," *SPE Production & Operations* 24 (2009): pp. 481–491.
27. A. Valiakhmetova, K.S. Sorbie, L.S. Boak, et al., "Solubility and Inhibition Efficiency of Phosphonate Scale Inhibitor-Calcium-Magnesium Complexes for Application in a Precipitation-Squeeze Treatment," *SPE Production and Operations* 32 (2017): pp. 343–350.
28. K. Jarrahian, K.S. Sorbie, M.A. Singleton, et al., "The Effect of pH and Mineralogy on the Retention of Polymeric Scale Inhibitors on Carbonate Rocks for Application in Squeeze Treatments," *SPE Production and Operations* 34 (2019): pp. 344–360.
29. K. Jarrahian, M. Singleton, L. Boak, et al., "Surface Chemistry of Phosphonate Scale Inhibitor Retention Mechanisms in Carbonate Reservoirs," *Crystal Growth Design* 20 (2020): pp. 5356–5372.
30. K. Jarrahian, K. Sorbie, M. Singleton, et al., "Building a Fundamental Understanding of Scale-Inhibitor Retention in Carbonate Formations," *SPE Production & Operations* 35 (2020): pp. 85–97.
31. K. Jarrahian, L.S. Boak, A.J. Graham, et al., "Experimental Investigation of the Interaction Between a Phosphate Ester Scale Inhibitor and Carbonate Rocks for Application in Squeeze Treatments," *Energy & Fuels* 33 (2019): pp. 4089–4103.
32. M.M. Jordan, "The Effect of Production-Logging-Tool Data on Scale-Squeeze Lifetime and Management of Scale Risk in Norwegian Subsea Production Wells: A Case Study," *SPE Production and Operations* 35 (2020): pp. 49–67.
33. E.J. Mackay, M.M. Jordan, O. Ishkov, et al., "Reservoir Simulation, Ion Reactions, and Near-Wellbore Modelling to Aid Scale Management in a Subsea Gulf of Mexico field," *SPE Production & Operations* 29 (2014): pp. 172–182.
34. E.J. Mackay, M.M. Jordan, N.D. Feasey, et al., "Integrated risk analysis for scale management in deepwater developments," *SPE Production & Facilities* 20 (2005): pp. 138–154.
35. E.J. Mackay, A.P. Matharu, K.S. Sorbie, et al., "Modeling Scale-Inhibitor Treatments in Horizontal Wells: Application to the Alba Field," *SPE Production & Facilities* 15 (2000): pp. 107–114.
36. M.M. Jordan, M.C. Edgerton, E.J. Mackay, "Application of Computer Simulation Techniques and Solid Divertor to Improve Inhibitor Squeeze Treatments in Horizontal Wells," SPE-50713-MS, SPE International Symposium on Oilfield Chemistry, held February 16–19, 1999 (Houston, Texas, USA).
37. O. Vazquez, I. Giannakouras, E.J. Mackay, "Simulation of Squeeze Treatment/Tracer Programme Designs," SPE-190752-MS, SPE International Oilfield Scale Conference and Exhibition, held June 20–21, 2018 (Aberdeen, UK).
38. O. Vazquez, E. Mackay, T. Tjomsland, et al., "Use of Tracers to Evaluate and Optimize Scale-Squeeze-Treatment Design in the Norne Field," *SPE Production & Operations* 29 (2014): pp. 5–13.

39. R.S. Seright, "Impact of Dispersion on Gel Placement for Profile Control," *SPE Reservoir Engineering* 6 (1991): pp. 343–352.
40. R.S. Seright, "Effect of Rheology on Gel Placement," *SPE Reservoir Engineering* 6 (1991): pp. 212–218.
41. R.S. Seright, "Placement of Gels to Modify Injection Profiles," SPE-17332-MS, SPE Enhanced Oil Recovery Symposium, held April 16–21, 1988 (Tulsa, Oklahoma, USA).
42. O. Vazquez, E.J. Mackay, K.S. Sorbie, "Development of a Non-Aqueous Scale Inhibitor Squeeze Simulator," SPE-100521-MS, SPE International Oilfield Scale Symposium, held May 31–June 1, 2006 (Aberdeen, UK).
43. O. Vazquez, E.J. Mackay, K.S. Sorbie, "Modelling the Placement of Scale Squeeze Treatments in Heterogeneous Formations With Pressurised Layers," SPE-121856-MS, 8th European Formation Damage Conference, held May 27–29, 2009 (Scheveningen, The Netherlands).
44. K. Ramstad, H.C. Rohde, T. Tydal, et al., "Scale Squeeze Evaluation Through Improved Sample Preservation, Inhibitor Detection, and Minimum Inhibitor Concentration Monitoring," *SPE Production and Operations* 24 (2009): pp. 530–542.
45. C.J. Johnston, M.M. Jordan, "The Procedures that Time Forgot: Sample Preservation, Back to Basics," SPE-190734-MS, SPE International Oilfield Scale Conference, held June 20–21, 2018 (Aberdeen, UK).
46. E.A.M. Myhr, K. Ramstad, T. Tydal, et al., "Preservation of Produced Water Samples for Better Scale Management and Extended Squeeze Lives," 12th Tekna Oil Field Chemistry Symposium, held April 1–4, 2001 (Geilo, Norway).
47. E.A.M. Myhr, K. Ramstad, T. Tydal, et al., "Preservation of Produced Water Samples for Better Scale Management—An Update Including Non-Squeezed Wells," 13rd Tekna Oil Field Chemistry Symposium, held March 17–20, 2002 (Geilo, Norway).
48. E.J. Mackay, M.M. Jordan, "SQUEEZE Modelling: Treatment Design and Case Histories," SPE-82227-MS, SPE European Formation Damage Conference, held May 13–14, 2003 (The Hague, Netherlands).
49. G.C. Onwubolu, B.V. Babu. *New Optimization Techniques in Engineering* (Berlin: Springer, 2004): pp. 1–11.
50. J.C. Spall. *Introduction to Stochastic Search and Optimization: Estimation, Simulation, and Control* (New York: Wiley-Interscience, 2003): pp. 126–147..
51. O. Vazquez, D. Corne, E.J. Mackay, et al., "Automatic Isotherm Derivation from Field Data for Oilfield Scale-Inhibitor Squeeze Treatments," *SPE Journal* 18 (2013): pp. 563–574.
52. D.E. Goldberg. *Genetic Algorithms in Search Optimization and Machine Learning* (Boston: Addison Wesley, 1989) pp. 59–88.
53. R. Storn, K. Price, "Differential Evolution—A Simple and Efficient Adaptive Scheme for Global Optimization Over Continuous Spaces," *Journal of Global Optimization* 23 (1995): pp. 1–12.
54. J. Kennedy, R. Eberhart, "Particle Swarm Optimization," Proceedings of ICNN'95, IEEE International Conference on Neural Networks, held November 27–December 1, 1995 (Perth, Australia).
55. O. Vazquez, E.J. Mackay, K.S. Sorbie, "A Two-Phase Near-Wellbore Simulator to Model Non-Aqueous Scale Inhibitor Squeeze Treatments," *Journal of Petroleum Science and Engineering* 82–83 (2012): pp. 90–99.
56. L.W. Lake. *Enhanced Oil Recovery* (Hoboken: Prentice-Hall Inc., 1989): pp. 131–171.
57. K.S. Sorbie. *Polymer Improved Oil Recovery* (Berlin: Springer, 1991)): pp. 208-245.
58. G.A. Pope, "Application of Fractional Flow Theory to Enhanced Oil Recovery," *Society of Petroleum Engineers of AIME Journal* 20 (1980): pp. 191–205.
59. S.E. Buckley, M.C. Leverett, "Mechanism of Fluid Displacement in Sands," *Transactions of the AIME* 146 (1942): pp. 107–116.
60. L.P. Dake. *Fundamentals of Reservoir Engineering* (Amsterdam: Elsevier, 1998): pp. 131-139.

61. L.K. Wang, Y. Hung, N.K. Shammas. *Advanced Physicochemical Treatment Processes* (Totowa: Human Press, 2006): pp. 123-153.
62. K.S. Sorbie, R.M.S. Wat, A.C. Todd, "Interpretation and Theoretical Modeling of Scale-Inhibitor/Tracer Corefloods," *SPE Production Engineering* 7 (1992): pp. 307–312.
63. K.S. Sorbie, P.A.V. Johnson, S. Hubbard, et al., "Non-Equilibrium Effects in the Adsorption of Polyacrylamide Onto Sandstone," *In Situ* 13 (1989): pp. 121–163.
64. V. Azari, O. Vazquez, E.J. Mackay, et al., "Gradient Descent Algorithm to Optimize the Offshore Scale Squeeze Treatments," *Journal of Petroleum Science and Engineering* 208 (2022): pp. 109469.
65. O. Vazquez, G. Ross, M.M. Jordan, et al., "Automatic Optimization of Oilfield-Scale-Inhibitor Squeeze Treatments Delivered By Diving-Support Vessel," *SPE Journal* 24 (2019): pp. 60–70.
66. M.M. Jordan, L. Sutherland, C. Johnston, "Developments in Squeeze Treatments Design—Why Not Apply 24 Month Duration Squeeze to Offshore Production Wells," SPE-200701-MS, SPE International Oilfield Scale Conference and Exhibition, held 24–25 June, 2020 (Virtual).
67. A.F. Miles, O. Vikane, D.S. Healey, et al., "Field Experiences Using 'Oil Soluble' Non-Aqueous Scale Inhibitor Delivery Systems," SPE-87431-MS, SPE International Symposium on Oilfield Scale, held May 26–27, 2004 (Aberdeen, UK).
68. J.M. Reizer, M.G. Rudel, C.D. Sitz, et al., "Scale Inhibitors," U.S. Patent 6379612, 2002.
69. E.D. Kaufman, R.W. Dexter, "Delivery of Polymeric Antiprecipitants in Oil Wells Employing an Oil Soluble Carrier System," U.S. Patent 4518511, 1985.
70. I.R. Collins, M.M. Jordan, S.E. Taylor, "The Development and Application of a Novel Scale Inhibitor of Deployment System," *SPE Production and Facilities* 17 (2002): pp. 221–228.
71. R.N. Smith, T.A. Lawless, H.M. Bourne, et al., "Planning and Execution of a Field Trial Utilising New Invert Emulsion Squeeze Technology," SPE-60210-MS, SPE International Symposium on Oilfield Scale, held January 26–27, 2000 (Aberdeen, UK).
72. T.A. Lawless, R.N. Smith, "Experimental and Theoretical Investigation into the New Invert Emulsion Scale Squeeze Technology," SPE-50590-MS, SPE European Petroleum Conference, held October 20–22, 1998 (The Hague, Netherlands).
73. M.M. Jordan, C.J. Graff, K.N. Cooper, "Development and Deployment of a Scale Squeeze Enhancer and Oil-Soluble Scale Inhibitor to Avoid Deferred Oil Production Losses During Squeezing Low-Water Cut Wells, North Slope, Alaska," SPE-58725-MS, SPE International Symposium on Formation Damage Control, held February 23–24, 2000 (Lafayette, Louisiana, USA).
74. P. Chen, et al., "The Challenge of Squeezing a Water Sensitive HP/HT Reservoir—Lab and Field Experiences With a Novel Non-Aqueous Inhibitor/Squeeze Enhancer Package," SPE-87435-MS, SPE International Symposium on Oilfield Scale, held May 26–27, 2004 (Aberdeen, UK).
75. I.R. Collins, J.A. Hewartson, "Extending Squeeze Lifetimes Using Miscible Displacement," SPE-74650-MS, International Symposium on Oilfield Scale, held January 30–31, 2002 (Aberdeen, UK).
76. A.F. Miles, H.M. Bourne, R.G. Smith, et al., "Development of a Novel Water in Oil Microemulsion Based Scale Inhibitor Delivery System," SPE-80390-MS, Symposium on Oilfield Scale, held January 29–30, 2003 (Aberdeen, UK).
77. N. Poynton, R. Tidswell, J. Steele, et al., "Squeezing Aqueous Based Scale Inhibitors Into a Water Sensitive Reservoir-Development of a Squeeze Strategy," SPE-60219-MS, International Symposium on Oilfield Scale, held January 26–27, 2000 (Aberdeen, UK).
78. P. Bogaert, M.C. Berredo, C. Toschi, et al., "Managing Formation-Damage Risk From Scale-Inhibitor Squeeze Treatments in Deepwater, Subsea Fields in the Campos Basin," *SPE Production & Operations* 23 (2008): pp. 468–477.
79. K.U. Raju, M.H. Al-Buali, T. Al-Saadoun, et al., "Scale Inhibitor Squeeze Treatments Based on Modelling Studies," SPE-130280-MS, SPE International Conference on Oilfield Scale, held May 26–27, 2010 (Aberdeen, UK).

80. M.M. Jordan, K. Sjuraether, I.R. Colllns, et al., "Life Cycle Management of Scale Control Within Subsea Fields and Its Impact on Flow Assurance, Gulf of Mexico and the North Sea Basin," SPE-71557-MS, SPE Annual Technical Conference and Exhibition, held September 30–October 3, 2001 (New Orleans, Louisiana, USA).
81. D. Patterson, M. Kendrick, W. Williams, et al., ""Squimulation"—Simultaneous Well-Stimulation and Scale-Squeeze Treatments in Deepwater West Africa," *SPE Production & Operations* 28 (2013): pp. 55–66.

7 Advanced Squeeze Treatment

Ping Chen[1] and Tao Chen[2]
[1]CNPC R&D (DIFC) Company Ltd., Dubai, United Arab Emirates
[2]EXPEC Advanced Research Center, Saudi Aramco, Dhahran, Saudi Arabia

7.1 Introduction

For a squeeze treatment, a scale inhibitor is usually injected at a concentration between 5% and 20% into the formation through a producing well after a preflush. After overflush and shut in, well production is resumed. The scale inhibitor is expected to flow back to the surface above or at a required minimum concentration to prevent scale precipitation in the well and the near well bore area. The squeeze life normally lasts from 6 to 24 months, depending on the inhibitor retention and release properties in the formation and the water cut/production rate in the well. Obviously, a successful squeeze treatment design should last as long as possible and protect the volume of produced water as much as possible for economic reasons. The three main properties required for scale inhibitors in the most general case are as follows:

- Threshold inhibition: These scale inhibitors must inhibit scale formation at very low concentrations, typically between 1- and 20-ppm inhibitor concentrations.
- Long squeeze life: These scale inhibitors must show long return profiles from the reservoir at levels higher than the required threshold level or minimum inhibitor concentration (MIC).
- No injectivity problem during fluid placement and no formation damage after treatment.

Challenges are always presented in the fields in order to meet these requirements. These are often seen in the following respects:

- Squeezed polymer inhibitor residual analysis at the low-MIC level in the produced brine.
- Longer scale inhibitor treatment life is required.
- Inhibitor placement in a horizontal well.
- Treating water-sensitive and clay particle-rich sandstone formation.
- Inhibitor placement in a low-pressure reservoir.

To meet and overcome these challenges, new squeeze treatment technologies have been developed in recent years. These new technologies are summarized under the title of advanced squeeze treatment, and the general descriptions are presented in this chapter. They are outlined in the following sections:

- Tagged polymer inhibitor for the downhole squeeze treatments.
- Special additives developed to extend the scale inhibitor treatment life.
- Enhanced squeeze treatments through surface charge alternation and inhibitor surface precipitation.
- Foamed scale inhibitor for low-pressure reservoirs.
- Gelled scale inhibitor placement for a horizontal well.
- Nonaqueous scale inhibitor to treat low-pressure reservoirs and water-sensitive formations.

The latest development on nanoparticle-based squeeze technology will be discussed in Chapter 18.

7.2 Tagged Polymer Scale Inhibitor

The retention of a scale inhibitor within formation rocks is fundamentally governed by the affinity nature of the inhibitor molecules for a given minerology. This is particularly important in order to select an inhibitor to meet the treatment life target. Moreover, not only to keep a good formation retention/adsorption property, but the selected inhibitor also needs to show an excellent desorption property to give a good release profile to the formation brine. It is well known that phosphonate-based inhibitors have good rock adsorption characteristics, whereas most common polymeric structures have poor rock adsorption properties. In addition, phosphonate inhibitor can be analyzed by an inductively coupled plasma (ICP) instrument at the MIC level, down to 1 ppm. The ICP can detect the phosphorus at the ppb level to ensure an accurate measurement of the phosphonate scale inhibitor. However, polymeric inhibitors are often the ultimate choice for regions with strict environmental rules or for certain scaling types. Polymer inhibitors are often shown better properties in terms of high-temperature thermal stability, where a reservoir presents a very high temperature, for example, the temperature is above 180°C. Many efforts have been made to generate new polymeric scale inhibitors with strong interactions with rock minerals, which would provide an extended well treatment lifetime, when applied in squeeze jobs.

Once a polymer scale inhibitor is deployed into the formations, it is important and critical to measure the inhibitor concentration accurately, since inhibitor residual analysis results will be part of the factors deciding when the next squeeze treatment should be carried out. It is particularly important for the inhibitor analysis to be properly performed at inhibitor MIC levels, which often stand between 1 ppm and 20 ppm. In the past years, it was found that it was difficult to do a proper analysis of some of the scale inhibitors, due to the interference from brine salinity and other chemicals in the brine. The residual polymer inhibitors were found to be even more difficult to be analyzed accurately.

Since the ICP method is a common way used to determine the phosphorus content for phosphonate inhibitor concentrations, tremendous efforts have been made to develop a more effective inhibitor with a P-tagging or label system which can be used to accurately determine inhibitor concentration at ppm levels. An end-capped polymer based on a phosphorus head—vinylidene diphosphonic acid and the sulfonate copolymer backbone (Figure 7.1)—was reported to have increased adsorption onto typical formation rocks, improved thermal stability (up to at least 200°C) and biodegradability, and exceptional scale inhibition performance against barium sulphate and calcium carbonate scales.[1–3] These end-capped polymers have >20% biodegradability in the OECD306 28-day biodegradation test. By incorporating the novel end-capped technology, squeeze lifetimes were significantly extended compared to the regular treatment using conventional polymeric scale inhibitors.

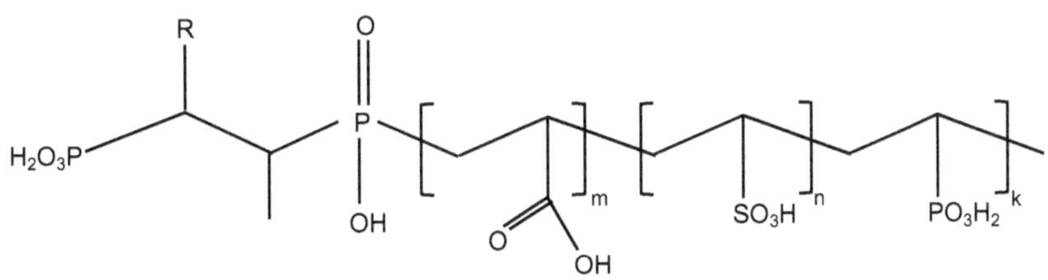

R = H or preferably PO_3H_2; m, n, and k can be zero or any number

Figure 7.1. Molecular structure of end-capped scale inhibitor. *R* = *H* or preferably PO_3H_2; *m*, *n*, and *k* can be zero or any number.

To demonstrate that the end-capped P tagged polymer could offer a good squeeze treatment life, the static beaker tests were carried out.[2] The sand adsorption data determined for the two polymers showed that, compared with the adsorption profile of a leading conventional scale control polymer, the end-capped polymers were adsorbed more strongly than the conventional polymer. At all three pH values tested (pH 3, 4, and 5), the inhibitor adsorption level is almost doubled in comparison with the conventional polymer inhibitor.

Diallyldimethyl ammonium chloride (DADMAC) embodied polymers with P tags were reported for squeeze application through a series of publications and patents.[4–9] The P-tagged polymer inhibitor was found particularly useful, when applied as a squeeze treatment. Figure 7.2 shows the typical structure of the polymer scale inhibitor. Like other polymer inhibitors widely used in the oil industry, the monomer containing carboxylic function group was utilized in the synthesis of the polymer. However, while the conventional carboxylic function group is included in the structure, a quaternary amine monomer was also introduced to the polymer. The quaternary amine is positively charged. The introduction of this positive charged monomer amine would offer a favorite property for the inhibitor to be adsorbed to the surfaces of the negatively charged sandstone formations. This is particularly useful for the inhibitor to be applied as a squeeze scale inhibitor in downhole applications. It is thought that by "fixing" the polymers to the surface of the formation (*i.e.*, by electrostatic attraction), the inhibitor adsorption can be enhanced. The amount of the positively charged polymer amine monomer had to be optimized during the polymerization in order to allow the adsorbed inhibitors to be released into the formation brine. This concept of the introduction of a positive charged monomer to the scale inhibitor with phosphorus tagged has been further extended recently. Based on the chemical structure, R1, R2, or R3 on the polymer can be replaced by a sulphonated group to make a new type of inhibitor. The new polymer scale inhibitor can be prepared by a conventional polymerization procedure known as solution polymerization.

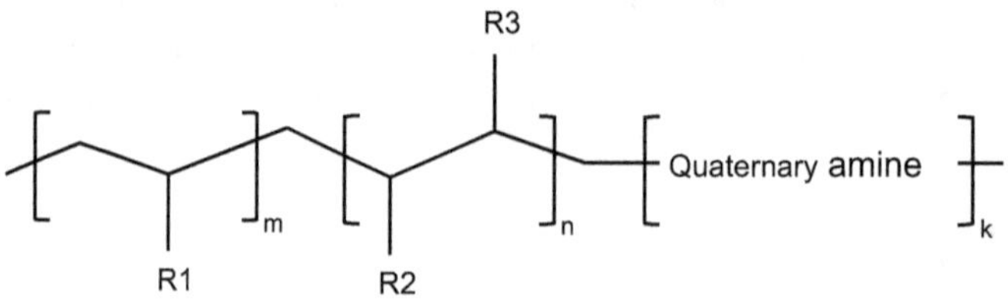

R1, R2, R3 = carboxylic function group; m, n, and k can be zero or any number

Figure 7.2. Molecular structure of co-polymer scale inhibitor containing quaternary amine. R1, R2, and R3 = carboxylic function group.

The resultant polymers showed great brine compatibility and scale inhibition performance in inhibitor testing and ideal long inhibitor return profiles with little permeability impairment in core flood experiments and have been applied many times in North sea regions.[4–9] While showing good performance data through conventional scale laboratory screening tests, the inhibitor chemistry displays good environmental properties. The field applications of the polymer from the fields showed an excellent inhibitor squeeze life while no negative impact on the formation permeability was seen, as shown in Figure 7.3. Residual inhibitor returns from the treatments have shown that the wells have been protected from scale damage for more than 200 days, and the wells are still protected at the time of writing the article, indicating the potential for a long treatment life for the developed polymer inhibitor.

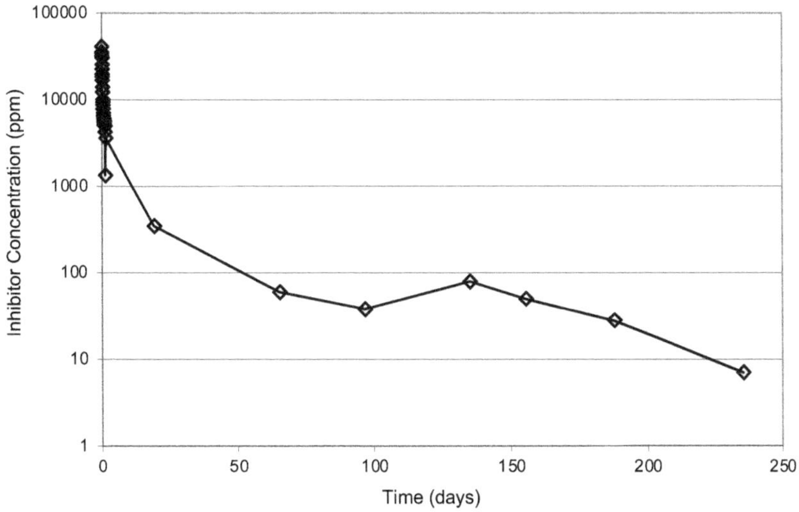

Figure 7.3. Inhibitor returns from a treated well.

A new method for the synthesis of phosphorus-tagged polymeric scale inhibitors was reported in 2010.[10] The inhibitor molecules contain a greater level of phosphorus which is achieved by polymerizing novel phosphonate monomers. This results in highly detectable, environmental-friendly molecules with enhanced oilfield reservoir adsorption characteristics. These phosphorus-tagged polymers are readily detected by ICP techniques, offering advantages over conventional phosphonate and polymeric scale inhibitors. The exact structure of the inhibitor molecules was not published. However, the further field application results showed that the phosphorus-tagged polymeric scale inhibitor had improved treatment life.[11] The inhibitor was successfully field deployed in two carbonate formations and one sandstone formation. No problems were encountered during treatment, and all wells were brought back online without incident. In all instances, this new polymer inhibitor showed an improvement over the incumbent copolymer product on carbonate fields and the phosphorus functionalized poly maleic acrylic copolymer on sandstone fields.

A fluorescent-tagged polymeric scale squeeze inhibitor was reported by Poynton *et al.*[12] The inhibitor can be readily detected at low concentrations by a spectrofluorometric procedure, including sample pretreatment and cleanup. Under this development, this chemical can be compared with the phosphorus-tagged scale inhibitor. As a result, two tagged inhibitors are available, and the advantage can be seen as squeezing two subsea completed wells with essentially the same squeeze inhibitor but being able to detect the residual inhibitor concentration from each well independently of each other. To detect the fluorescent-tagged inhibitor produced from the individual wells in the mixed comingled fluids, a precleaning method was developed. Different contaminants may give a background fluorescence or quench the fluorescence originating from the fluorescence moiety attached to the scale inhibiting polymer. The use of solid phase extraction cartridges clean technique was considered. After evaluation of several different cartridges, those with a suitable stationary phase and the right conditioning and eluting procedure were identified. They enabled the cleanup of all interfering contaminants in the samples of real produced waters. The produced water samples were precleaned with cartridges prior to the fluorescent measurement. The cartridges were used to separate the produced water sample into desired and undesired components. The interfering impurities in the produced water sample are retained on the stationary phase, while the tagged scale inhibitor passes through it. This precleaning procedure eliminates any background fluorescence or quenching contaminants in these produced waters, and thus a calibration curve can be achieved accurately.[12]

7.3 Special Additives Developed to Extend Scale Inhibitor Treatment Life

Apart from changing chemical structure and adding special function groups favorable to inhibitor adsorption on the reservoir rock surfaces, other efforts have also been made to increase inhibitor retention and hence to extend inhibitor squeeze life. In the following sections, typical technologies, such as precipitation squeeze, mutual solvent preflush, and so on, are discussed and outlined.

7.3.1 Precipitation Squeeze

The retention mechanism for scale inhibitors on rock matrix can be categorized into adsorption and precipitation types. Forced precipitation is a step further than "natural" precipitation squeeze, as it intentionally introduces foreign chemicals or alters fluid conditions to promote the formation of less soluble inhibitor-metal complexes. They are sometimes referred to as "pseudoscale" precipitations that are considered better "fixed" within rock pores during the inhibitor injection, and then slowly dissolve back into the water when the production starts, and the fluid pH drops.[13]

Calcium salt is a common treatment applied to force scale inhibitor precipitation during squeeze treatment. Most scale inhibitors, including both phosphonates and polymers, are somewhat incompatible with high concentrations of calcium, especially at elevated reservoir temperatures. If the formation water has low total dissolved solids levels, injection of additional Ca-rich fluids can effectively produce the pseudo-inhibitor precipitation in the near-well bore zone, obtaining longer retention than the inhibitor alone.[14]

Extensive experimental core floods were carried out in order to compare the inhibitor return performances between designed inhibitor adsorption and precipitation with various inhibitors,[15] using test procedures illustrated in Figure 7.4. For the adsorption floods, the inhibitor solutions were made in the synthetic seawater at room temperature. The solution bottles were then placed in a preheated oven at 70°C. The samples were checked, and no precipitation reaction occurred. For the precipitation floods, the inhibitor solutions at the given inhibitor concentration were made with the addition of $CaCl_2$ at the fixed pH and room temperature. The sample bottles were then placed in an oven, where the temperature was set at 70°C. With the temperature increase, precipitations were observed, where $CaCl_2$ was added.

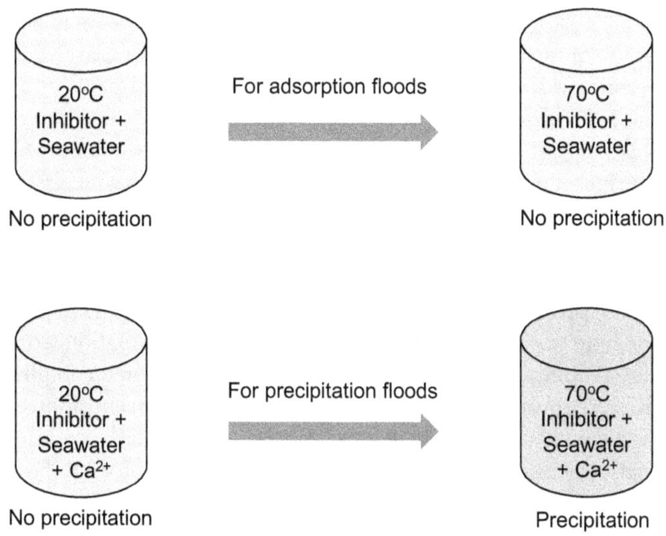

Figure 7.4. Inhibitor tests for the comparison between with and without precipitation.

Figure 7.5 shows that the inhibitor returns with and without the designed precipitations. Three inhibitors were tested: triethylene-tetramine hexmethanephonic acid, polyphosphinocarboxylic acid (PPCA), and diethylenetriaminepenta phosphonic acid (DETPMP). The results demonstrated that at the forced inhibitor precipitation with the addition of Ca^{2+}, the inhibitor returns are better than that of adsorption floods with the same type of scale inhibitor tested, suggesting an improved inhibitor treatment life. However, PPCA, as a polymer, has a lower performance of inhibitor return than that of the other two phosphonate scale inhibitors, regardless of precipitation or adsorption retention mechanism.

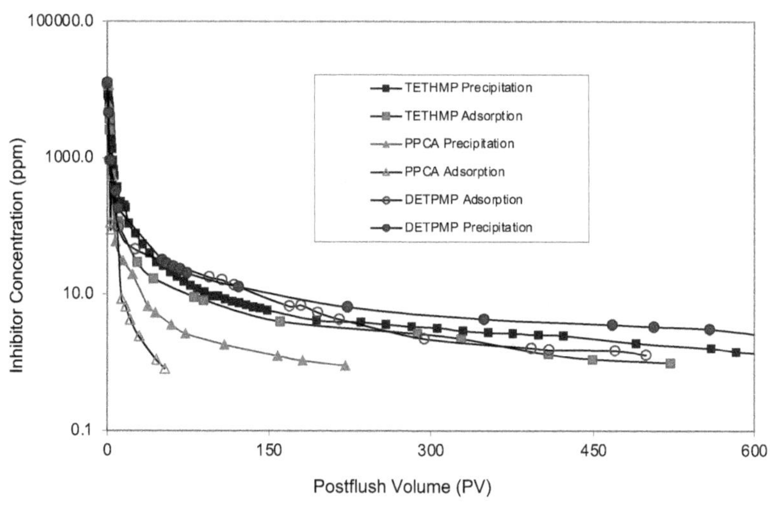

Figure 7.5. Inhibitor returns for precipitation and adsorption core floods.

When a high concentration of phosphonates is in contact with a cation, such as Ca^{2+}, the initial precipitate is most likely amorphous, due to the random nature of the phase separation of a macroneutral molecule. This initial precipitate transforms into a well-defined crystalline phase after dissolving excess phosphonate out of the amorphous phase during the inhibitor flow back. It was proposed that the formation of the amorphous precipitate contributes to the initial retention of the scale inhibitor following a squeeze treatment, while the dissolution of the crystalline phase precipitate controls the long-term inhibitor return concentration.[16]

The solubility of the precipitated scale inhibitor calcium complex is a key parameter in controlling inhibitor returns. Based on the laboratory tests and simulation work, Jordan *et al.* reported that the solubility of the phosphonate/calcium precipitated complex is sensitive to the produced brine chemistry, principally the divalent cation concentrations of calcium and magnesium. High concentrations of calcium suppress the solubility resulting in a low-concentration return curve of inhibitor over an extended period of time. For the tests conducted on the penta-phosphonate, it is clear that the solubility also declines with increased temperature. The change in test temperature extends the lower concentration region of the squeeze for both adsorption and precipitation processes. The solubility of a single formulation of precipitated scale inhibitor may vary significantly over the production life of a well as the formation water composition changes with increasing sea water breakthrough. As a result, a precipitation formulation should be optimized to cope with the varying brine chemistry (seawater ratio) and increasing water volumes.[17]

Significant progress has been made in recent years to understand the inhibitor precipitation/retention mechanism in carbonate formation. The carbonate formation is much more chemically reactive than the sandstone minerals. An extensive study on the inhibitor retention mechanism via adsorption/precipitation on carbonate formation was reported by Jarrahian *et al.*[18] Three commercially available scale inhibitors were selected in their study. They are DETPMP, PPCA, and a sulfonated polyacrylic acid copolymer (VS-Co). The main findings are: (i) for three inhibitors, regions of both pure adsorption at lower scale inhibitor concentrations and coupled adsorption/precipitation retention mechanisms at higher inhibitor concentrations were observed, with the dominant mechanism depending on scale inhibitor chemistry, temperature, and mineralogy. For most scale inhibitors, precipitation was the dominant component in the carbonate system, except for VS-Co, where adsorption was much more prominent. The highest retention was for DETPMP, because of the higher P content and greater reactivity of DETPMPs with Ca^{2+}, resulting in more precipitation; (ii) the region of precipitation was determined by scale inhibitor type, the levels of Ca^{2+}, and the final pH. Both Ca^{2+} concentration and pH determined the amount of precipitate for a given scale inhibitor type, because pH governs the state of dissociation of the scale inhibitor (more dissociation at higher pH values leads to more Ca^{2+} chelation and more precipitate) and the actual Ca^{2+} concentration in solution.

Zinc ion is believed to perform similarly to calcium in the forced precipitation squeeze. Inclusion of zinc ion into a squeeze main pill has shown a significant increase in inhibitor retention in laboratory studies, but no field trial has been reported yet. According to the Irving-Williams order of stability of bivalent transition metal complexes ($Ca^{2+} < Fe^{2+} << Zn^{2+}$), the complexation of zinc and phosphonate ligands is much more stable than that of calcium and inhibitors. The stable complexes may facilitate rock adsorption but draw concerns regarding whether the Zn-free inhibitor molecules can be available for scale inhibition when released back from the formation. Zinc may also trigger strict regulation issues in certain geological regions.[19]

Thermally activated pH modifier (TAPM) is another technique that raises the pH of the inhibitor pill in situ in the reservoir, instead of the calcium levels, to promote inhibitor precipitation. TAPM is the urea or alkyl urea derivatives which will decompose to release alkaline to raise the pH of the treating fluids at elevated temperatures. The TAPM approach is more applicable to polymeric scale inhibitors that cannot be easily forced precipitated through a calcium-inducing methodology. Squeeze jobs incorporating the TAPM technology have been executed in dozens of wells in the North Sea caulk formation, significant increases in treatment lifetime and overall savings in chemical cost were reported.[20, 21] Since urea derives only break down above 80°C, it cannot be used in low-temperature wells. A modified approach was developed using enzyme breakers to release alkaline at temperature as low as 40°C and has been successfully used in the same oilfield.[22]

7.3.2 Mutual Solvents as Preflush

In 2000, it was reported that the squeeze in the initial treatment on the well showed significant damage even though reasonable care was taken to limit the water loading placed in the reservoir. The possible damage mechanisms were highlighted by the author, as described below.[23]

- Ingress of a small volume of methanol, used for hydrate inhibition, in the near well bore
- Water blocks in lower water cut zones
- Insufficient drawdown available to assist in cleanup

Core flood experiments showed that water ingress into the formation could cause significant changes in oil permeability. The treatment strategy was then to focus on using a mutual solvent-based preflush. The field treatment results showed that with the mutual solvent preflush, coupled with diesel over flush, well stimulation was achieved, and the scale inhibitor squeeze life was extended. It was believed that the increased oil production was attributed to the stimulation, by the mutual solvent-based preflush. The water removing properties of the preflush could also help to nullify any damage, due to the presence of water initially retained within the formation. The extended squeeze life resulted from the preflush cleaning of the formation which led to a cleaner reservoir rock surface for the inhibitor to adsorb onto. The further study by Vazquez *et al.*[24] showed a surfactant preflush to enhance scale inhibitor placement in both aqueous and nonaqueous squeeze treatments. They concluded that the improvement resulted from the reduction of residual water saturations, due to the application of the mutual solvent package. The impact of the mutual solvent on scale inhibitor retention was due to the increased rock surface area available for adsorption.

The mechanism related to the changes of rock surface wettability from oil wet to water wet is well described by Ghosh *et al.*[25] Due to the adsorption of the surfactant, the rock surfaces became more water wet, and hence increased the inhibitor attention. The field data showed the effects of surfactant composition in preflush on reservoir wettability and scale inhibitor squeeze lifetime. Using 4% of ethylene glycol monobutyl ether (EGMBE) solution as a benchmark, a series of core floods were carried out, the anionic surfactants showed the most beneficial improvement on oil-wet carbonate reservoirs. However, anionic surfactant may not be the right choice for carbonate formation, because of its higher adsorption and competition with scale inhibitor molecules. Mild alkali seems to be very effective in favor of inhibitor adsorption.[26, 27] The anionic surfactant used in the tests was the modified carboxylated C10/16 alkylpolyglucoside (APG) with one of the –OH groups replaced by a –COOH group.[27] The structures of EGMBE and APG are shown in Figures 7.6 and 7.7, respectively.

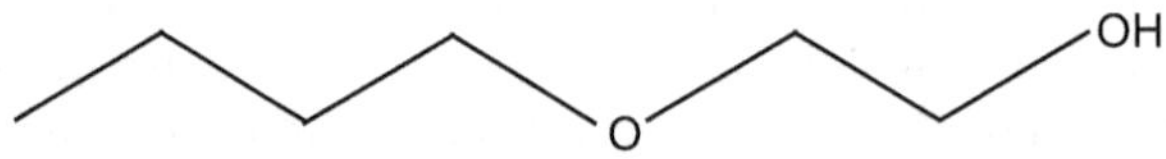

Figure 7.6. General chemical structure of ethylene glycol monobutyl ether.

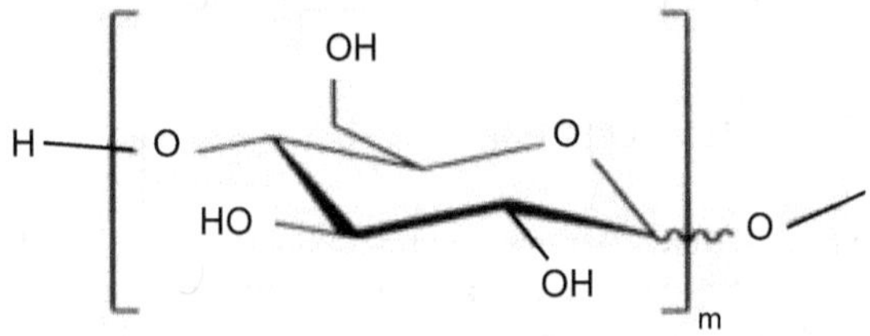

Figure 7.7. General chemical structure of alkylpolyglucoside.

7.3.3 Kaolinite

The mineralogy of sandstone core was found to be a key factor in the selection of inhibitors for field application. Based on the results from an extensive series of core flooding experiments with phosphonate and PPCA inhibitors, it was concluded that the presence of clay minerals, principally kaolinite, can significantly enhance inhibitor adsorption, leading to an increased squeeze treatment life.[28] Then, it was proposed to inject small kaolinite particles (average size of 2 µm) into clean and high-permeability sandstones for improving squeeze treatment. Combined with the fixation agent and scale inhibitor, the injected kaolinite microcrystals could mechanically alter the near wellbore mineralogy and surface property characteristics of sandstone reservoirs. This concept was validated with core flood experiments with kaolinite included in the preflush solution.[29]

7.4 Enhanced Squeeze Treatments through the Surface Charge Alternation and Inhibitor Surface Precipitation

To achieve the desired squeeze life, it is essential to maximize inhibitor retention without damaging the formation. Scale inhibitors can be retained in the reservoir by a number of mechanisms, including adsorption and precipitation. The retention mechanism would be dictated by the inhibitor type and concentration, brine chemistry, reservoir temperature, and rock mineralogy. At lower pH, the weakly acidic inhibitors tend to be more protonated/associated and, thus, can be retained on the rock surface through hydrogen bonding and Van der Waals forces. At higher pH, the scale inhibitors are more dissociated, and then can be retained on the rock surface by weak Van der Waals forces and static electric forces, due to metal ion bridging.

As discussed in the previous section, enhanced inhibitor adsorption could be achieved through a surface-binding process with the presence of calcium ions.[13] The calcium ions (Ca^{2+}) may modify either the electrochemical properties of the solid surface, or they may complex with the inhibitor to aid adsorption. This Ca^{2+}-enhanced adsorption process has been referred to as surface precipitation, in differentiating to forced precipitation mechanism discussed in Section 3.1. It was also reported that inhibitor retention could be enhanced by the ester cross-linking of polycarboxylic scale inhibitors, leading to an improved inhibitor return curve.[30]

7.4.1 Enhanced Inhibitor Retention via Bridging

Several chemical additives have been identified to provide an effective method to improve inhibitor squeeze life. The experimental test method and field results in the applications of these additives have been presented in recent publications.[31–40]

Enhanced inhibitor adsorption involving a polymer was reported in recent year.[32] The technology is based on the modification of the rock surface charge to create a more positively charged and less repulsive surface to the negative-charged scale inhibitors. The additive contained the desired functionality to adsorb onto the more negatively charged rock surface. It is believed that the adsorbed additives present a less negatively charged surface than the rock itself, altering the double layer, and thereby the interaction forces between chemicals and the mineral surface with which the scale inhibitor compounds can interact. The inhibitor was expected to be adsorbed on the additive coated rock surfaces through static electric forces, hydrogen bonding, and Van der Waals force. The pretreatment of the reservoir zone with the additive or a combination of the squeeze chemical and the additive is therefore expected to enhance scale inhibitor adsorption through a bridging-type mechanism.

The typical additive used as the bridging agent is a quaternary amine polymer, which is a positively charged polymer. The inhibitor used is a phosphonate. The laboratory test methodology to confirm the concept involves beaker tests and core floods. In the beaker tests, the crushed core was preconditioned with the bridging agent at the given concentration and pH value. The test results are presented in Table 7.1. The test results demonstrated that inhibitor adsorption increased significantly in the presence of an additive quaternary amine polymer. At pH 3, the presence of additive effectively doubled the amount of inhibitor adsorbed. At pH 5 and 7, the presence of the additive increased the adsorption of inhibitor by approximately 60%. The data indicated that the adsorption of inhibitor was enhanced through the application of the selected quaternary amine. It is postulated that the adsorption of the additive leads to a less negative rock surface and a less repulsive surface to the negatively charged scale inhibitor. Thus, the subsequent application of the scale inhibitor to the core material is enhanced by bridging to the additive already retained on the rock surface.[32]

Table 7.1. Inhibitor adsorption comparison with and without bridging agent.

Sample No.	Preconditioning pH = 5	pH (10% Inhibitor in Seawater)	Inhibitor Adsorption (mg/g)
1	6% NaCl	3	25.3
2	10% additive in 6% NaCl	3	56.1
3	6% NaCl	5	39.1
4	10% additive in 6% NaCl	5	63.8
5	6% NaCl	7	38.8
6	10% additive in 6% NaCl	7	62.3

The sandstone core floods were also performed to confirm the effect of the bridging agent on extending the inhibitor treatment life with the same quaternary amine and scale inhibitor. The oil permeability was checked to ensure that there was no potential permeability reduction, while the chemicals were introduced.

The field trials on a selected well were undertaken with phosphonate inhibitors, and the bridging agent was only used in the second treatment for comparison purpose. The first squeeze treatments (Treatment I) without bridging agent were carried out using the following design:

- Seawater preflush: 10 M^3
- Main pill: 140 M^3 (10% scale inhibitor product in seawater)
- Overflush: 100 M^3
- Tubing volumes: 47 M^3
- Shut in: 15 h

The second treatment (Treatment II) design with the bridging agent was carried out as follows:

- Seawater preflush: 10 M^3
- Preflush 2: 50 M^3 (10% bridging agent in 6% NaCl)
- Main pill: 140 M^3 (5% scale inhibitor product in seawater).
- Overflush: 100 M^3
- Tubing volumes: 47 M^3
- Shut in: 15 h

The phosphonate scale inhibitor returns concentration as a function of the produced water volumes from two squeeze treatments are presented in Figure 7.8, for comparison purposes.

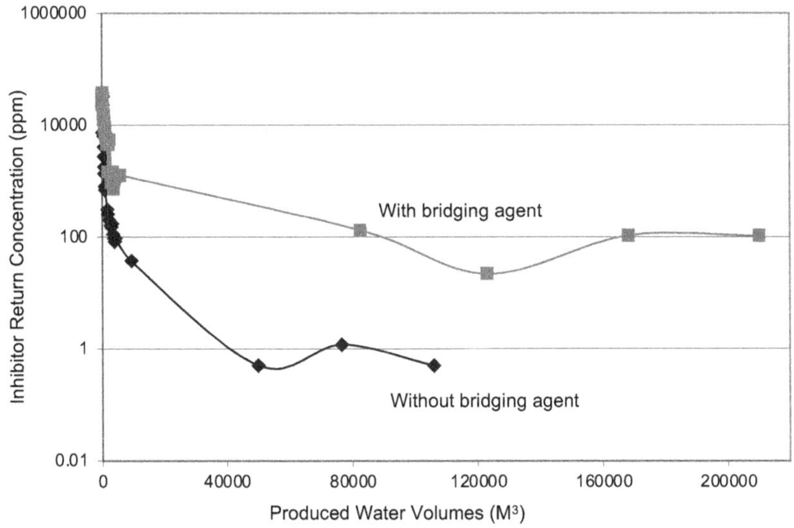

Figure 7.8. Scale inhibitor return comparison between with and without bridging agent.

The phosphonate deployed in Treatment I dropped below the laboratory determined MIC at approximately 10,000 M^3 cumulative water production. In Treatment II, the bridging agent was deployed as part of a pretreatment to enhance scale inhibitor adsorption. After the pretreatment stage, the inhibitor was squeezed. This treatment, using the same volume but half the inhibitor concentration used in the first treatment, had protected the production of over 220,000 m^3 of brine. The ability to protect significant volumes of brine offers significant economic benefits to the field asset in terms of increased well availability, lower total chemical costs, improved operational logistics as well as simplifying the job execution.

7.4.2 Enhance Inhibitor Retention via Surface Precipitation

In addition to the surface charge alternation for enhanced inhibitor adsorption, an additional function was identified when a polymer scale inhibitor was introduced in the inhibitor squeeze treatment. With the same quaternary amine bridging agent mentioned in the above section, it was found that under a given pH, temperature, and concentration, the quaternary amine polymer reacted with the polymer inhibitor and formed gel-like precipitates. The precipitates could dissolve, when the conditions, such as solution pH, chemical concentration, and temperature were changed. Further investigations showed that the dissolving process of the gel-like precipitates could be a hydrolysis process in which the inhibitor dissociated from the additive.[33] Like the process of calcium/scale inhibitor precipitation, the precipitation interaction between the bridging agent (quaternary amine polymer) and scale inhibitor can significantly increase inhibitor retention, and hence extend the inhibitor treatment life. In other words, under a controlled manner with given pH, temperature, and chemical concentrations, the preflushed bridging agent can react with the subsequently injected scale inhibitor, leading to a precipitation reaction that increases the inhibitor retention. Unlike the conventional calcium/inhibitor precipitation approach, the additive (bridging agent) can be preinjected and rapidly adsorbed on a rock surface. The subsequently injected inhibitor will react with the additive on the rock surfaces. This avoids bulk precipitations, where a permeability reduction is often observed in the calcium/inhibitor precipitation squeeze process. This phenomenon was termed as "surface precipitation."

The static beaker tests were carried out to demonstrate that enhanced inhibitor retention could be achieved through the interaction/precipitation between the selected scale inhibitor and additive. The crushed rock materials were preconditioned with 6% NaCl brine or 10% additive at pH 3. The inhibitor adsorption/precipitation were performed with the addition of 10,000 ppm of polymer inhibitor at pH 3 and 5, respectively. The precipitation was observed and a scanning electron microscope (SEM) picture was taken for the precipitated polymer inhibitor with quaternary amine polymer (Figure 7.9).

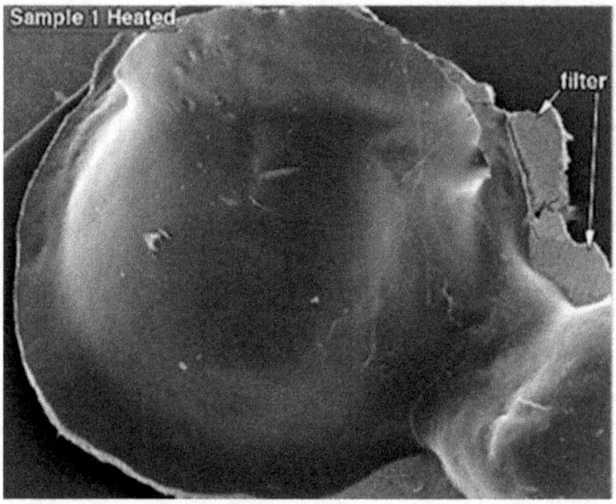

Figure 7.9. Scanning electron microscopic image of precipitated polymer inhibitor with quaternary amine polymer.

Specially designed core floods were conducted to assess the potential formation damage related to surface precipitation. The procedure involved oil permeability evaluations before and after the chemical applications with surface precipitation after both bridging agent and scale inhibitor were injected into the core plug. Table 7.2 lists the determined crude oil permeability recorded in both forward and reverse flow directions at the stages of before and after chemical applications. It can be seen that the average oil permeability was recovered at >100%, after the injections of the bridging agent quaternary amine and polymer inhibitor. The initial oil permeability at 98°C was recorded at 155 mD in the forward direction and 161 mD in the reverse directions. After the quaternary amine and polymer inhibitor injection, the flow back permeability was determined at 185 mD in the forward direction and 184 mD in the reverse directions. The enhanced oil permeability was likely caused by a possible wettability change as a result of surface precipitation due to the reaction between quaternary amine and polymer inhibitor.

Table 7.2. Permeability results before and after chemical application.

Flood No.	Injection Direction	Before Chemical Injection (mD)	After Chemical Injection (mD)	Recovery %	Average Recovery %
1	Forward	155	185	120	117
1	Reverse	161	184	114	

Figure 7.10 shows the inhibitor returns from experimental core floods. As seen from the figure, the inhibitor return with the surface precipitation is kept at much higher concentrations than the inhibitor return without the surface precipitation, indicating an improved inhibitor retention and release profile.

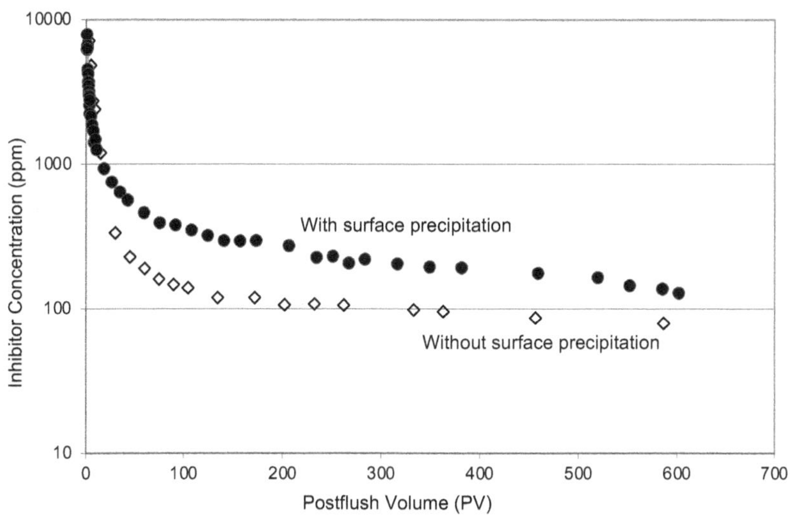

Figure 7.10. Inhibitor core flood return comparison between with and without surface precipitation.

A number of squeeze treatments using quaternary amine and polymer inhibitor had been successfully conducted in a North Sea field. The typical treatment design is summarized below:

- Preflus 1: 10 M^3 seawater
- Preflush 2: 60 M^3 10% additive in 6% NaCl.
- Main pill: 220 M^3 inhibitor solution (10% inhibitor in seawater)
- Overflush: 100 M^3 seawater
- Tubing volumes: 48 M^3 seawater
- Shut in period: 12 h

After the treatment, the wells were brought back without any problem. Also, no process problems were reported during the back production. Using quaternary amine as a preflush agent, the squeeze jobs provided protection for more than 200,000 m^3 of brine produced in each treatment.

Through both experimental test results and field treatments, it is believed that the quaternary amine worked in both enhanced adsorption and precipitation mechanisms. The quaternary amine worked as a bridging agent to modify the surface charge in favor of inhibitor adsorption. Also, the quaternary amine can react with the selected scale inhibitors to form precipitates under certain conditions. This precipitation interaction can significantly increase inhibitor retention for an inhibitor squeeze treatment. The slow release of inhibitor into the produced brine through the hydrolysis/dissolution process can enhance the inhibitor treatment life.

One of the field-proven bridging agents used as a squeeze life enhancer is polydiallyldimethyl ammonium chloride (polyDADMAC). The chemical structure of the polymer is shown in Figure 7.11. The satisfactory results of prolonged squeeze lifetime were achieved through the applications in some North Sea fields.[36] Different bridging agents based on nonpolymeric dia-quaternary amine were also reported to have similar enhancement on squeeze life.[38] The best performances were achieved by placing the bridging agent/squeeze life enhancer into preflush or postflush at optimized concentrations of 2.5%–10%.[38, 39] For phosphate ester and phosphonate inhibitor chemistries, the agent incorporated into the postflush stage had more significant impacts on the squeeze lifetime enhancement than that into the preflush stage. Under laboratory settings, a 2.5% agent in the postflush was equivalent to the 10% in the preflush. The possible mechanism is that the additional wrapping formed by the bridging chemical in postflush restrains the release of the inhibitor that have been adsorbed on rock surfaces.[39]

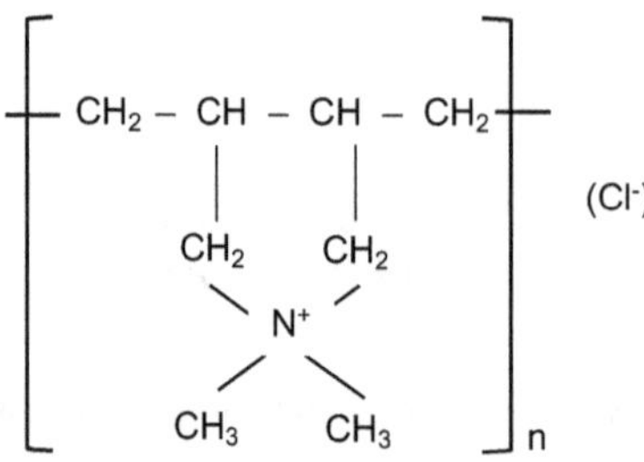

Figure 7.11. Molecular structure of polydiallyldimethyl ammonium chloride.

7.5 Foamed Scale Inhibitor Squeeze Treatment

Foam has been used in the oil industry for many years and for many purposes. As a diverting agent for acids, foam slugs became an accepted treatment method in the mid-1980s. It was found useful as a successful diverting agent in sandstone acidizing for almost any type of completion or production.[41] Foam has also been widely used in carbonate matrix acidizing.[42, 43]

The placement of a scale inhibitor using a foam application with nitrogen gas is particularly beneficial for low-pressure, low-permeability, and low-water-producing formations. The reduced volume of brine compared with a traditional squeeze application significantly reduces the potential risk of formation damage and of killing the wells. This advantage becomes critical to the squeeze treatment targeting highly water-sensitive formations. Additionally, using nitrogen gas aids with return flow of the well on restart following a squeeze application. The technologies would involve using nitrogen gas as a carrier, surfactant as a foamed agent and scale inhibitor solution. The schematic view of the technology is shown in Figure 7.12. The concept is to blend the nitrogen, the small volumes of foam agent, and scale inhibitor solutions at the well head. As a result, the foam is formed, and the scale inhibitor is carried into the reservoir by the foam with the injection pumps.

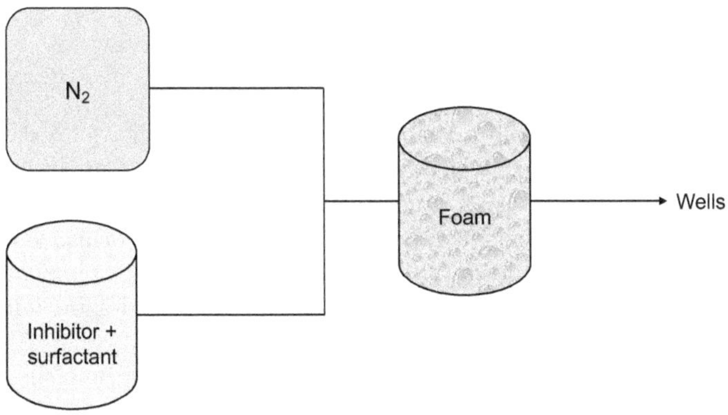

Figure 7.12. Schematic diagram of foamed scale inhibitor.

The first successful foamed scale inhibitor squeeze treatment was performed in Statoil's Norne field.[44] The well had scale inhibitor/dissolver placement challenges from its horizontal section with low reservoir pressure. The conventional water-based scale inhibitor injection had difficulty placing the scale inhibitor in the target zones, and the well had no gas lift facility installed. The placement of a large amount of water-based fluid would cause difficulty in lifting the well after it shuts in. After carefully planned laboratory qualification works, including inhibitor efficiency, compatibility with both foam and formation water, foam function tests, and foamed scale inhibitor/dissolver core floods, the foamed scale inhibitor/dissolver treatments were performed. Because of the optimized inhibitor/dissolver placement by the foamed chemical placement method, the well went back to normal production without any difficulty. The well was maintained at good productivity, and a longer scale inhibitor return life than the conventional water-based scale inhibitor squeeze treatment was achieved.

Foamed scale inhibitor was also applied in squeeze treatment to gas wells with severe sulfate scale problem.[45] Due to low reservoir pressure, the introduction of a large amount of scale inhibitor fluid to the depleted gas wells would cause formation damage. After a careful scale control strategy study and laboratory development tests, the foamed scale inhibitor package was selected. The field results showed that the scale inhibitor was successfully squeezed into a tight reservoir (<10 mD) by the foamed technologies without any damage to the reservoir.

7.5.1 Laboratory Test to Qualify the Foamed Scale Inhibitor for Downhole Squeeze Treatments

Apart from the common squeeze inhibitor qualification tests, such as inhibitor compatibility, inhibitor dynamic loop test, and inhibitor static beaker efficiency, the additional special tests were required to qualify a foamed scale inhibitor package for the downhole applications. These tests include: (i) inhibitor compatibility with foam, (ii) foam function with the addition of the scale inhibitor, and (iii) foamed scale inhibitor core flood for the foamed scale inhibitor injection ability and inhibitor return property.

7.5.1.1 Scale Inhibitor Compatibility with the Foam Agent

The selected scale inhibitor and foaming agent are aged together for 7 days at the given temperature, and then their equivalent performance was assessed using the dynamic tube blocking test method. Performance in terms of the MIC of the aged scale inhibitor, with and without the aged foaming agent, was compared to the performance of the nonaged scale inhibitor test results. Figure 7.13 shows the tests obtained at a temperature of 135°C. The inhibitor MIC is the same for both cases with and without aging with the foaming agent. Similar results were observed in tests using the static bottle test method, indicating the proposed scale inhibitor is compatible with the foaming agent. Both chemicals are thermally stable and compatible after being aged at a given temperature for seven days.

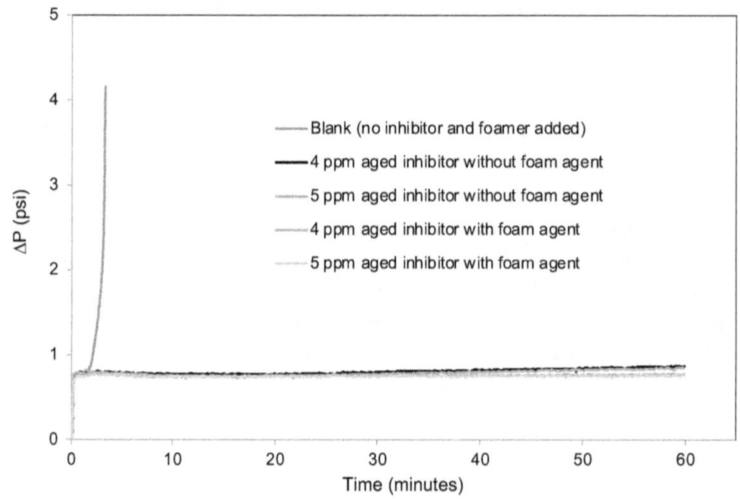

Figure 7.13. Scale loop test results of the aged and unaged scale inhibitor with and without mixing with the foaming agent (test temperature: 135°C).

7.5.1.2 Foam Half-Life and Foam Friction Loop Tests

The foam half-life is the time that stable foam under static conditions is separated to 50% of its original volume. It is one of the most important parameters to ensure a successful treatment. The prepared foam with the addition of the scale inhibitor needs to be alive during the squeeze treatment. The measurements of both half-life and viscosity under dynamic conditions are necessary steps before the application of the foamed scale inhibitor package. Some test results are listed in Tables 7.3 and 7.4 as an example. For the half-life of the foam tests, the fluids were assembled by making 2% KCl water using 20 g of KCl salt in each liter of fresh water, adding 50 mL (5%) of scale inhibitor, and adding the desired amount of foaming agent. The initial pH of 2% KCl with the scale inhibitor was 5.8, and the final pH of the fluid was a maximum of 5.86 with the foam agent included. Table 7.3 shows that at 1.0% and 1.5%, good foam height was achieved, and the foam agent generated sufficient half-life time at ambient temperature, as shown in Table 7.3.

Table 7.3. Results of the screening test and maximum foam and half-life determination.

% Foam agent	0.25	0.50	1.00	1.50
Max foam volume	425 mL	600 mL	750 mL	900 mL
Foam Stability	**Regained Fluid Volume from Foam Drainage**			
Time to 25 mL	0.3 min	1.2 min	2.4 min	3.5 min
50 mL	0.4	1.7	5.1	6.5
75 mL	1.1	2.1	7.0	9.0
100 mL	2.1	4.3	10.4	17.5
125 mL	3.4	6.5	12.4	-
150 mL	5.0	11.1	-	-
Half-life time (min)	2.1	4.3	7.0	9.0

Dynamic foam testing was completed with a foam rheometer. Multiple friction loop tests were conducted at foam quality levels of 50% and 70%. The test at 70% foam quality with 1.0% foam agent demonstrated good viscosity at 10 s^{-1} and 85°C and had fair 10 s^{-1} shear qualities at the upper temperatures of 100°C and 140°C.

Table 7.4 shows an overview of measured foam viscosity as a function of shear rate and temperature. At each temperature, the viscosity difference between the shear range is high, which is a desired ratio for diverting the flow from a high- to low-permeability zone. Although the low shear rate exhibited instability, the foam was stable in the site glass. Furthermore, it is anticipated that the formation is between 100°C and 140°C, cooling the formation during placement. It is recommended to use 1.0% foam agent and 70 foam quality during the foamed scale squeeze operation with 5% scale inhibitor.

Table 7.4. Overview of measured foam viscosity (cP) as a function of shear rate and temperature.

Shear Rate (s^{-1})	Temperature		
	85°C	100°C	140°C
511	35.2	30.2	17.0
300	44.2	37.2	19.2
170	58.1	47.6	24.0

7.5.1.3 Scale Inhibitor Core Flood

Core flood test was conducted to examine the potential formation damage issues (reduced flow back gas permeability) associated with applying the scale inhibitor treatments in producer wells from the reservoir formations. The primary purpose of the core flood is to determine: (i) if there is any significant potential formation damage by the injection of the foamed scale inhibitor into the core, and (ii) if the injected scale inhibitor provides a good retention/release profile for long-term well protection from the scale damage. The injectivity data confirmed that stable pretreatment differential pressures were recorded for both gas and brine. During the primary treatment, elevated differential pressure and a visual assessment of the effluent indicated that the viscous foam was successfully produced. In post-treatment, stable gas flow was achieved. The permeability recovery measurements demonstrate that 74% gas recovery was achieved immediately following the treatment. During the primary chemical application, some retardation of the scale inhibitor against the inert lithium tracer was observed, suggesting retention of the chemical within the core matrix. Additionally, initial post-flush analysis indicates additional chemical retention during shut in. The return profile suggests the potential for scale protection over a long production period.

7.5.2 Field Trial Results

When the treatment design stage is reached, two parameters are important factors for ensuring a successful treatment: foam quality (the percentage of gas present) and foam half-life. With the selected foam quality and foam half-life, the foamed scale inhibitor at the given concentration can be injected into the reservoir as foam during the pumping state. During well shut in, the foam is broken, and the scale inhibitor will be adsorbed on the rock surfaces. When the well is back to production, the scale inhibitor will slowly be released into the produced brine to protect the near wellbore and well tubing from scale deposition damage. A treatment design is listed below as an example:

- Preflush: 10 m^3 diesel
- Main pill I: 20 m^3 13% scale inhibitor in seawater
- Main pill II: 53 m^3 13% scale inhibitor in seawater + N_2 with 1%–2% foaming agent (foam quality 50%)
- Main pill III: 25 m^3 13% scale inhibitor in seawater
- Overflush: 138 m^3 seawater + N^2 with 1%–2% foaming agent (foam quality 50%)
- Tubing volumes: 4 m^3 50% MEG + 50% seawater
- Shut in period: 19 h

The volumes for the foamed stages (main pill II, overflush) are volumes after N_2 has expanded downhole. The foamed fluids were injected successfully. The well pressure during the shut in period was as high as 80 bar allowing easier production start. The inhibitor return results are presented in Figure 7.14.

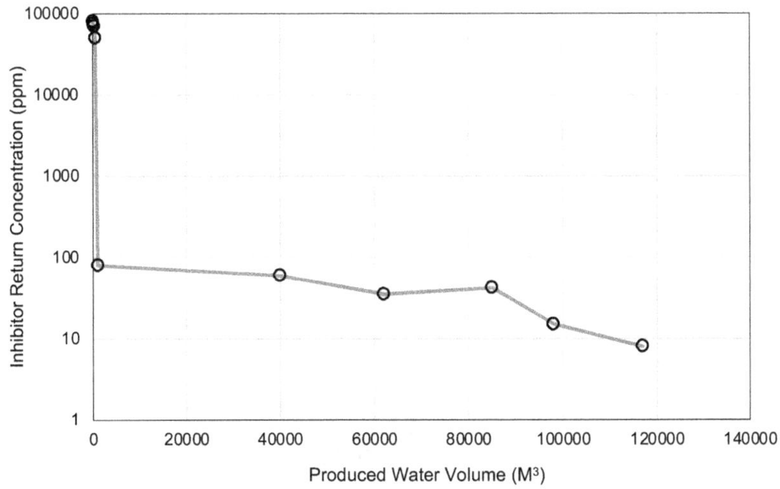

Figure 7.14. Return curve for back production of the foamed scale inhibitor treatment.

7.5.3 Summary

The foamed scale inhibitor technology for the downhole squeeze treatment involves using N_2 gas as a carrier, surfactant as a foaming agent, and a scale inhibitor solution by blending the N_2 and small volumes of the foaming agent and scale inhibitor solution at the well head. A foam is formed, and the scale inhibitor is carried into the reservoir by the foam with the injection pumps. Advantages of applying the foamed scale inhibitor include:

- Suitable over a range of pressure, temperature, and reservoir permeability
- Contains no solid particles
- Enhances treatment flow back
- Treats low-pressure reservoirs
- Optimal inhibitor placement (at well formation layers with large permeability/pressure contrast)
- Eliminates concern of diverter agent cleanup

The placement of a scale inhibitor using a foam application with N_2 gas is particularly beneficial for low-pressure, low-permeability, and low-water-producing formations. The reduced volume of brine compared with a traditional squeeze application significantly reduces the potential risk of formation damage and of killing the wells. This advantage becomes extremely beneficial to the squeeze treatment targeting highly water-sensitive formations.

Additionally, using nitrogen gas aids with return flow of the well on restart following a squeeze application.

7.6 Gelled Scale Inhibitor Squeeze Treatment

The success of a scale inhibitor treatment depends on its placement efficiency. The scale inhibitor should be placed so that all water-producing intervals could accept a sufficient quantity of the total treatment volume. If significant permeability or pressure variations are present in the interval to be treated using the bull-heading method, treatment fluid will enter the zones with the higher permeability and lower pressure leaving little fluid to treat the other zones, which can potentially be the water-producing zones. The challenge is even greater in long horizontal wells with significant permeability and pressure contrast.[46]

Diversion technologies were utilized in order to deliver the scale inhibitor pills to the target zones through the long open interval in horizontal wells. The advantages and disadvantages of these technologies are summarized in Table 7.5.

Table 7.5. Summary of diversion technologies.

Methods	Comments
Coil tubing	• Delivery chemical anywhere • Running cost • Time concern
Sliding sleeve	• Optimizes the completion of multistage wellbores by enabling highly accurate placement of stimulation treatments without intervention • Scale deposition can cause a problem
Wax beads	• Easy to operate • Temperature-sensitive • Wax beads did not melt if not properly selected
Gelled scale inhibitor	• Can be pumped via bull-heading method • Easy to operate • Cost-effective • Successful case histories
Foamed scale inhibitor	• Can be pumped via bull-heading method • Easy to operate • Cost-effective • Successful case histories • N_2 gas consumption

Besides the technologies listed above, pumping at a high rate (without exceeding fracturing pressure) might be possible for diversion purpose.[47] Modeling work with dynamic simulation software showed that fully viscosified treatments gave a better distribution along the length of an open-hole horizontal well.[48]

After a careful review of the technologies in response to the challenge of scale inhibitor placement in a horizontal well, a joint study with the objective of improving the placement of treatment fluids was initiated by a major operator in the North Sea and two service companies.[46] As a result of this study, a fully viscosified gelled scale inhibitor system was developed. The system comprises a purified xanthan viscosifying agent, a standard scale inhibitor and a breaker to achieve controlled gel breaking down. The viscosified gelled scale inhibitor system was applied at both Norne and Heidrun fields with long horizontal wells, all with significant permeability variations and cross flow. The field trial results showed that the operations were successful, and the scale treatments protected the wells from scaling.

By using the technology, the gelled scale inhibitor was used as a diverter through the bull-heading method. The gel can block the high-permeability and low-pressure zones, when it is pumped to the downhole. The subsequent water-based scale inhibitor can then be diverted to the low-permeability and high-pressure zones. The water-based scale inhibitor pill could be injected first into the high-permeability zones as a treatment design option, and then followed by the gelled scale inhibitor pills. As a result, all zones with both low pressure/high permeability and high pressure/low permeability would be placed with the inhibitor pills. Figure 7.15 shows the changes in viscosity of the gelled scale inhibitor with time from the injection at wellhead to flow back. The viscosity of the gelled scale inhibitor is high at the beginning in order for the gelled polymer package to act as a diverter. The viscosity decreased with time and when the temperature went up. After shut in, the diversion function of the gel is completed, and the gel is broken for an easy flow back without damaging the formation.

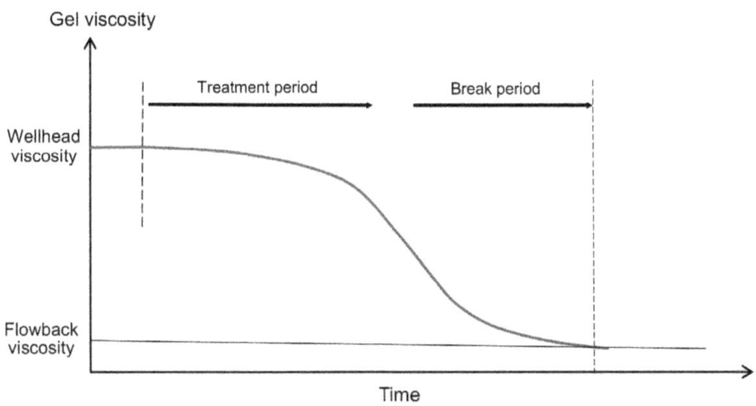

Figure 7.15. Viscosity curve changes with time from injection to flow back.

Another benefit of using the gelled scale inhibitor pill is its non-Newtonian shear thinning properties. The instantaneous sharp drop of rate in the initial period and the shear thinning effect could contribute to reducing the pressure buildup during injection.

7.6.1 Qualification of Gelled Scale Inhibitor

The purified xanthan was found to meet the qualification criteria and was first qualified for the gelled organic acid system. Xanthan is a shear-thinning polysaccharide that breaks thermally when heated above 115°C, depending on the salinity. Salinity increases the breaking temperature. A proper amount of gel breaker at the design condition is needed.

7.6.1.1 Polymer Gel/Scale Inhibitor Compatibility and Temperature Stability

The stability and compatibility tests could show whether the scale inhibitor is compatible with the gel. Scale inhibitor was mixed in the gel solution and aged at reservoir temperature. Performance tests were carried out for both the un-aged and the "thermal-aged" samples. Experimental results showed that the gel did not affect the performance of the chemical even under field conditions. In addition, static adsorption tests were done under the field conditions showing a good inhibitor adsorption level.

7.6.1.2 Gelled Scale Inhibitor-Breaking Characteristics and Shear Thinning Properties

To establish how long the gel mixtures had to be heated before breaking and becoming Newtonian, that is, behave like water, rheology measurements were used to determine the gelled scale inhibitor viscosity as a function of temperature and shear rate. The breaker concentration was optimized through the tests.

7.6.1.3 Gelled Scale Inhibitor by Core Flooding

The purposes of the designed core flood are to check the effect of the gel on the return permeability of the reservoir rock sample as well as the effect of the gel on the adsorption/return profile of the scale inhibitor. Some permeability test data are listed in Table 7.6. These results show that only a slight loss in recovery was observed, when the gel was used as fluid with the scale inhibitor instead of brine, with respective recovery levels of about 81%–85% and 83%–115% in the forward flow direction and of 83%–99% and 101%–104% in the reverse flow direction.

Table 7.6. Permeability data from core flood tests.

Core Flood No.	Flow Direction	Initial Oil	Flowback oil		Final oil	
		Permeability (mD)	Permeability (mD)	% Recovery	Permeability (mD)	% Recovery
Aqueous inhibitor	Forward	311	312	100	239	77
	Reverse	304	307	101	230	76
Gelled inhibitor	Forward	115	98	85	93	81
	Reverse	114	83	83	113	99
Gelled inhibitor	Forward	100	115	115	83	83
	Reverse	130	135	104	131	101

The inhibitor returns were compared with and without gel. No impairment was seen for the gelled scale inhibitor system under one of the field conditions.

7.6.2 Field Applications

The candidate well is a long horizontal oil producer with significant pressure and permeability contrast producing from about 700-m perforated zone. The treatment was done with pumping 100 M^3 of a brine-based scale inhibitor pill at a concentration of a 13%, and then followed by 13% gelled scale inhibitor as the diverter. The subsequent 13% brine-based scale inhibitor was injected and diverted. The overflush involved in pumping 358 M^3 of 1% brine-based scale inhibitor and 48.8 M^3 of gelled scale inhibitor at 1%.

The well was equipped with downhole pressure and temperature gauges. The bottomhole pressure (BHP) significantly increased when the gel reaches the formation, and the BHP flattens out when the gel is displaced into the formation. The increase in downhole injection pressure and the decrease in surface pumping pressure clearly illustrate the shear thinning property of the fluid. The increase in injection pressure also signaled the arrival of the gelled scale inhibitor and its self-diverting capability.

Figure 7.16 shows water analysis results prior and post treatment. After the treatment, the Ba^{2+} concentration in the produced water increased. This is a clear indication that $BaSO_4$ was forming close to or in the well bore prior to the treatment. After the treatment, the Ba^{2+} concentration stayed high for a long time. Cleanup of the well and subsequent well test showed no decrease in well productivity. The minimum inhibitor concentration dropped below the limit of 5 ppm only after a total of 500,000 m^3 of cumulative water production.

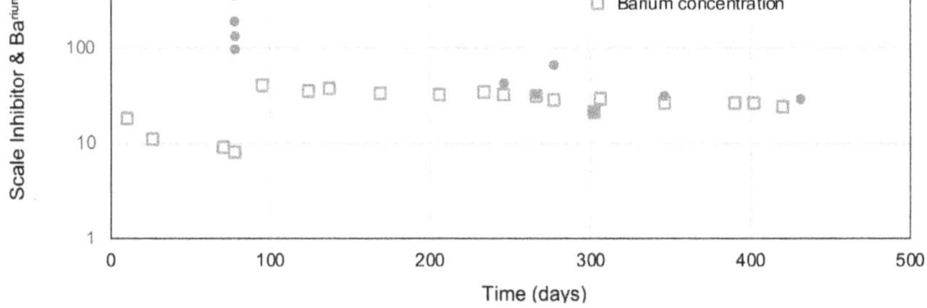

Figure 7.16. Produced water analysis (Ba^{2+}) and scale inhibitor return profiles.

7.6.3 Summary

Gelled scale inhibitor can be one of the suitable methods to treat horizontal wells with long-perforated intervals with permeability contrast. The gel can act as a diverter and also the inhibitor placement benefits from the shear thinning property of the gelled fluid. The field application data demonstrated that the scale protection and squeeze lifetime in the wells are as good as or better than the standard water-based treatments. Good quality rheology measurements are essential in understanding light viscosity systems regarding shear thinning and breaking mechanisms.

7.7 Nonaqueous Scale Inhibitors

A number of nonaqueous inhibitor treatments for downhole squeezes have been reported in the last 20 years.[49–62] The reported benefits of applying the nonaqueous products include minimizing water-related formation damage (water and or emulsion blocking, impact on relative permeability, clay mobilization and or swelling), rapid well cleanup, reducing the well hydrostatic head, and extending inhibitor squeeze life. These nonaqueous products have been described as oil-soluble, amphiphilic solvent solution, invert emulsion, and microemulsion systems. In this section, two case studies are reviewed to focus on the treatments on low-pressure reservoir and water-sensitive/fine-rich formation with the mutual solvent-based scale inhibitor package.

7.7.1 Mutual Solvent-Based Package for Low-Pressure Reservoir

The nonaqueous products in publications have been described as water-free materials, oil-soluble, amphiphilic solvent solution, invert emulsion, and microemulsion systems. However, very limited field data are available on these low-density nonaqueous scale inhibitor package for treating low-pressure reservoirs. The advantages of applying a low-density product are a rapid well cleanup and reducing the well hydrostatic head. More importantly, a large volume of brine-based scale inhibitor pill (higher density) could be avoided to prevent a likely formation damage, where the low-pressure reservoir could be difficult to resume normal production after the shut in.

7.7.1.1 Requirements for the Nonaqueous Scale Inhibitor Package

A number of issues such as product environmental requirements and product compatibility limited the choice of oil-soluble formulation, invert emulsion, and microemulsion systems. The research was focused on the system using the amphiphilic/mutual solvent solution technology. The mutual solvent system normally contains a certain amount of aqueous scale inhibitor and amphiphilic solvents. Some of the phosphonate and polymer inhibitors can be formulated with a popular and widely used amphiphilic solvent in industry. However, the density of these formulations is usually higher than 1. Furthermore, these polymer and phosphonate scale inhibitors have a limited biodegradation property and were not allowed to be used in some fields. Through the experimental work, it was found that a mutual solvent and a polymer inhibitor were fully compatible, while other widely used phosphonate and polymer scale inhibitors showed poor compatibility.[62]

The selected scale inhibitor for the squeeze treatments had a low MIC at 5 ppm and could offer a good squeeze life. The scale inhibitor was fully compatible with the brine produced in the field. The laboratory work was carried out to realize its full potential and for it to be fully accepted in the field. That is, the new package must be able to demonstrate its ability:

- Specific gravity < 0.94.
- Partitioning back to the water phase.
- Compatible with the brine and crude at bottom hole temperature.
- No damage to formation.

7.7.1.2 Laboratory Qualification Tests

A range of tests were designed to assess the suitability of the formulated mutual solvent scale inhibitor squeeze package for application in the field. The tests included fluid compatibility, partitioning tests, and compatibility between core materials and the formulated product packages. The product stability tests were performed on the scale inhibitor formulated with a mutual solvent. The product was evaluated at temperatures between −15°C and 150°C for visual evidence of instability over an extended test period. No change in physical appearance was observed across the full temperature spectrum evaluated.

The inhibitor partitioning tests were also carried out. The ability of the scale inhibitor in the formulation package to partition back into the aqueous phase is one of the most important properties. A high rate of mass transfer and a high extent of partitioning between the contacting water phase and itself are essential. The aim of this partitioning test was to determine how much of the scale inhibitor formulated in the low-density/nonaqueous package could be moved to the water phase, when mixtures of the low-density product, formation water, and crude oil were heated up to the reservoir temperature of 98°C. Table 7.7 shows the mixing ratios evaluated. Each test sample contains 20% synthetic formation water, with the remaining 80% being a mixture of low-density product and crude oil. Test sample bottles were placed in the preheated oven (98°C). A small amount of the water sample was taken to analyze the inhibitor concentration after predetermined test duration.

Table 7.7. Content of mixing samples for partitioning tests.

Sample No.	Product:Oil	% FW	% Product	% Oil
1	10:90	20	8	72
2	25:75	20	20	60
3	50:50	20	40	40
4	75:25	20	60	20
5	90:10	20	72	8
6	100:0	20	80	0

Figure 7.17 summarizes the percentages of scale inhibitor partitioned into the synthetic water with time. It can be seen that over 90% of the scale inhibitor was transferred into the water phase for the majority of samples after 4 h and 16 h. These data were confirmed with further tests using field waters. All results indicated that the scale inhibitor moves into the water phase rapidly and does not partition into the oil phase.

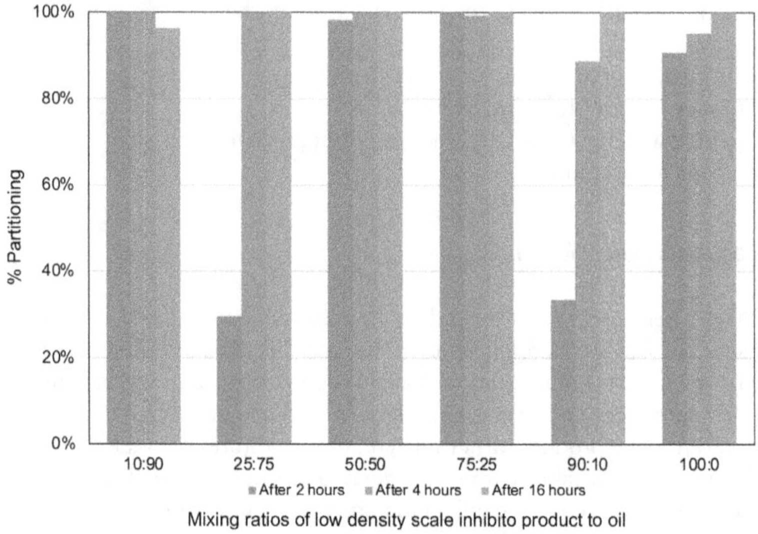

Figure 7.17. Partitioning of scale inhibitor from nonaqueous package to water phase, when mixed with crude oil at 98°C.

The formulation package was further evaluated for compatibility with reservoir rocks. Table 7.8 shows the determined crude oil permeability in forward/reverse flow directions at the initial and final stages. The initial oil permeability before the chemical injection was measured at 22 mD in the forward direction and 24 mD in the reverse direction. After the brine post flush, the flow back permeability was measured at 41 mD in the forward direction and 40 mD in the reverse direction. The oil permeability was increased by 186% in forward flow and 167% in reverse flow, indicating reduced water saturation and increased oil saturation.

Table 7.8. Summary of permeability to crude oil data measured from core flood works.

Stage	Fluid	Flow Direction	Permeability (mD)	% Recovery
Initial	Crude oil	Forward	22	
		Reverse	24	
Final	Crude oil	Forward	41	186
		Reverse	40	167

The inhibitor desorption behavior was also studied. By analyzing the inhibitor returns from the core flood works, the squeeze treatment life can be estimated. Furthermore, an inhibitor squeeze radial model can be built based on the core flood data, and the squeeze life for a given well can be predicted. The core flood results show that the scale inhibitor residual was ~20 ppm after 400-PV overflush, which is well above the MIC of 5 ppm, indicating a good adsorption/desorption property of the inhibitor. A long treatment life is anticipated.

7.7.1.3 Field Applications

Several precautions were taken in the treatment design to minimize potential formation damage caused by the inhibitor squeeze treatments. Diesel was selected as the preflush and overflush fluid. The volumes of squeezed chemical and fluid were carefully calculated, leaving a radial injection of approximately 4 m into the formation. Also, the shut in time was limited to 16 hours, and all fluids were filtered before being pumped into the well.

Field trials were carried out, and candidate wells were selected based on their bottom-hole pressure which is close to 250 bars. The mutual solvent scale inhibitor product was pumped without any significant pressure buildup in the treatments, and no flow back issues were recorded after shut in. The well was cleaned up rapidly to full production immediately following the squeeze treatments. More interestingly, an increase in oil production and a decrease in water cut were also observed after the treatment. The oil production rate was increased from 199 m^3/d to 263 m^3/d, and the water cut was reduced from 25% to 18%. The oil production increase lasted for more than 3 months. Scale inhibitor residual data also suggested a long treatment life.

7.7.2 Mutual Solvent-Based Package for Water-Sensitive/Fine-Rich Reservoir

One field in the North Sea produces from low-permeability formations which are water-sensitive and clay-rich. The reservoir also has a high temperature (165°C) and pressure (up to 475 bar). This limited the choice of chemistry that can be applied for the squeeze treatments. A squeeze treatment was carried out on one of the wells in the field using a water-based product initially. Significant formation damage was observed after the treatment. The possible damage mechanism was reported to be due to the relative permeability effect, fines mobilization, possible emulsion formation, and uncontrolled inhibitor precipitation.[63] After careful consideration of the available technologies, a treatment strategy using a non-aqueous scale inhibitor package was planned. After the full laboratory evaluations, the amphiphilic solvent-based system was applied for the inhibitor squeeze treatment in the field. The selection of the amphiphilic solvent system was based on the fact that the formation was water-sensitive and injection of the water product could result in formation damage. In addition, selecting an amphiphile-based system was the ability to formulate a clay stabilizer within the package. This was essential to minimize the risks of fines-related damage.

7.7.2.1 Laboratory Evaluations

A range of tests were designed to assess the suitability of the various components of the proposed nonaqueous scale inhibitor/clay stabilizer squeeze package for application in the field. The tests included thermal stability, fluid compatibility, inhibitor performance, and compatibility between the core materials and the formulated product packages.

The selected scale inhibitor had a track record of successful application for the control of carbonate scales in calcium-rich brines under HP/HT conditions in aqueous-based treatments, while the particular fines stabilizer agent derivative was being deployed. Thermal stability tests were performed on the scale inhibitor formulated with an amphiphilic solvent, and the clay stabilizer agent, formulated with an amphiphilic solvent. Both products were evaluated at temperatures between −15°C and 165°C for visual evidence of color changes, phase change, and precipitation over an extended test period. Neither of the formulated products underwent any phase or color changes, nor was any evidence of precipitation observed across the full temperature spectrum evaluated.

Compatibility tests were performed with field crude, synthetic formation water, inhibitor, and clay stabilizer. The initial tests were performed with ratios of crude oil and either inhibitor or clay stabilizer (10:90, 25:75, 50:50, 75:25, and 90:10) mixed with synthetic formation water to create 20%, 40%, 60%, and 80% brine-oil/product mixtures, respectively. The test results indicated that all formulations and combinations of brine and oil/inhibitor remained clear and stable after 24 hours. The test data also indicated that all formulations and combinations of brine and oil/clay stabilizer remained clear and stable after 24 hours.

Core plugs were analyzed using SEM after core flood tests. Clay particles stayed intact after treatment (Figure 7.18). No migration of the chlorite clay phase and other solid damaging mechanisms were observed.

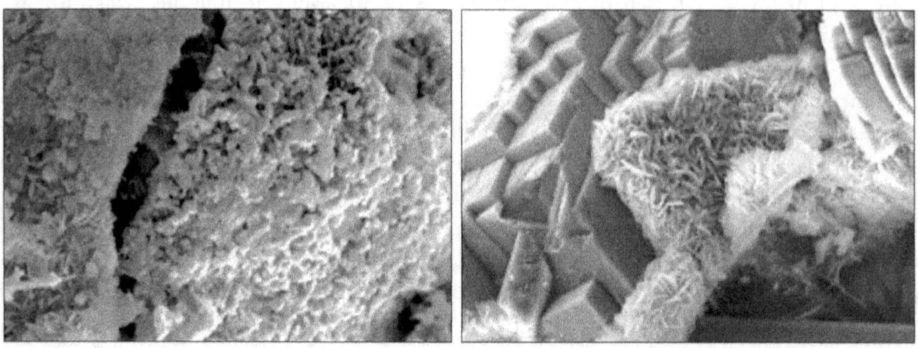

Figure 7.18. Scanning electron microscopic analysis with core plug after scale inhibitor flood.

7.7.2.2 Field Applications

The nonaqueous clay stabilizer and scale inhibitor were bullheaded using injection quality base oil as a preflush and overflush. Small volumes of squeezed chemical were used, leaving a radial injection of approximately 1 m`into formation. Also, the shut in time was limited to 6 hours, and all fluids were filtered before being pumped into the well. An analysis of the return to production following the squeeze on one of the treated wells, showed that (i) no flow back issues were recorded, and the well was cleaned up rapidly to full production immediately following the squeeze treatments; (ii) the production index (PI) was recorded at 13.2 m^3/bar·d following the squeeze. This compared well with a calculated pretreatment PI of 14.0 m^3/bar·d; and (iii) the squeeze treatment life reached the design target of 12 months, before the scale inhibitor concentration was below the MIC level.

7.7.3 Summary

Special chemical formulations and designs are often considered when setting up a downhole squeeze treatment strategy for treating low reservoir pressure, water-sensitive and clay-rich reservoirs. The case histories presented in this section showed that the amphiphile-based nonaqueous scale inhibitor squeeze package is able to meet the treatment challenges. This technology can minimize the risks of formation damage, while meeting all the physical requirements in terms of product stability, inhibitor partitioning, environmental properties, treatment life, and inhibition efficiency. In addition, the reported clay stabilizer could also be formulated with the amphiphile-based non-aqueous without any compatibility issue while successfully controlling the fines mobilization during the squeeze treatments.

7.8 Conclusions

Scale inhibitor downhole squeeze treatment is one of the most common methods used in oil fields to control the damages related to inorganic scale depositions. Scale inhibition can be achieved in reservoirs, where the conventional continuous injection method is limited in the tubing. Significant progress has been made in recent years to overcome the treatment challenges. In this chapter, the special downhole scale inhibitor treatment methods and chemicals applied are summarized.

Polymer scale inhibitors display advantages in treating high-temperature reservoirs. In addition, the polymer scale inhibitors offer a good biodegradation property, which is usually more environmental-friendly in comparison with some of the phosphonate scale inhibitor chemistries. Two typical polymer inhibitors presented in this chapter are the end-capped polymer and the polymer containing a DADMAC monomer. Both polymer inhibitors are tagged with phosphorus and can be easily detected by ICP. While these polymer scale inhibitors displayed strong performance in scale inhibition, they also showed excellent performance in rock retention and release properties. Hence, the designed treatment life targets have been achieved by the application of such polymeric scale inhibitors.

Many efforts were made to develop chemical additives to extend the scale inhibitor treatment life, including divalent cations (Ca^{2+}, Zn^{2+}) and kaolinite particles. The inhibitor precipitation with Ca^{2+} without damaging the formation became one of the choices, when squeeze treatment strategy was considered. The field results showed good squeeze treatment lives. The mineralogy of sandstone reservoirs was found to be significant, when inhibitors were selected for field application. Kaolinite had a clear effect on enhanced inhibitor adsorption, leading to an increased squeeze treatment life. Mutual solvents, like EGMBE, became one of the popular choices in treatment designs. It was often used in the preflush stage to change the rock surface wettability in favor of scale inhibitor adsorption.

Quaternary amine polymer (polyDADMAC) is one of the most suitable chemical additives for extending the treatment life. It is believed that the adsorbed additives present a less negatively charged surface than the rock itself, altering the double layer, and thereby the interaction forces between chemicals and the mineral surface with which the scale inhibitor compounds can interact. The pretreatment of the reservoir zone with the additive or a combination of the squeeze chemical and the additive is therefore expected to enhance scale inhibitor adsorption through a bridging-type mechanism. Furthermore, polyDADMAC can react with polymeric scale inhibitors, forming white gel-like precipitates. The precipitates could dissolve when the solution pH, chemical concentration, and temperature are changed. Similar to the process of calcium/scale inhibitor precipitation in improving an inhibitor squeeze life, the precipitation interaction between the quaternary amine polymer, and scale inhibitor can significantly increase the inhibitor retention, and hence extend the squeeze treatment life.

The foamed scale inhibitor technology for the downhole squeeze treatment involves using nitrogen gas as a carrier, surfactant as a foaming agent, and a scale inhibitor solution by blending the nitrogen and small volumes of the foaming agent and scale inhibitor solution at the well head. The scale inhibitor is carried into the reservoir by the foam. The placement of a scale inhibitor using a foam application with nitrogen gas is particularly beneficial for low-pressure, low-permeability, and low-water-producing formations. The reduced volume of brine compared with a traditional squeeze application significantly reduces the potential risk of formation damage.

Gelled scale inhibitor can be one of the suitable methods to treat horizontal wells with long-perforated intervals with permeability contrast. The gel can act as a diverter and also benefit the inhibitor placement, due to its shear thinning property. The field application results showed similar or even better squeeze life than the standard water-based treatments. Rheology measurements are essential during lab qualification studies in determining the shear thinning and breaking behaviors.

Special chemical formulations and designs are often considered in treating low reservoir pressure, water-sensitive, and clay-rich reservoirs. The amphiphile-based nonaqueous scale inhibitor squeeze package is a good choice to meet the treatment challenges. This technology can minimize the risks of formation damage, while meeting all the physical requirements in terms of product stability, inhibitor partitioning, environmental properties, treatment life, and inhibition efficiency. The clay stabilizer additive could also be formulated with the amphiphile-based nonaqueous package for controlling the fines mobilization during squeeze treatments.

Nomenclature

APG	Alkylpolyglucoside
BHP	Bottomhole pressure
CT	Computed tomography
DADMAC	Diallyldimethyl ammonium chloride
DETPMP	Diethylenetriaminepenta phosphonic acid
DTS	Distributed temperature sensing
EGMBE	Ethylene glycol monobutyl ether
HP/HT	High pressure/high temperature
ICP	Inductively coupled plasma
mD	Milli-Darcy
mL	Milliliter
M^3	Cubic meter (1 M^3 = 6.29 bbl)
MIC	Minimum inhibitor concentration
MEG	Monoethylene glycol
OECD	Organization for Economic Cooperation and Development
P	Phosphorus
PPCA	Polyphosphinocarboxylic acid
PV	Pore volume
PVS	Polyvinyl sulphonate
SEM	Scanning electron microscopy
TAPM	Thermally activated pH modifier
TDS	Total dissolved solids
TETHMP	Triethylene-tetramine hexmethanephonic acid
VDPA	Vinylidene diphosphonic acid
VS-Co	Sulfonated polyacrylic acid copolymer
ΔP	Differential pressure

References

1. K.P. Davis, S.D. Fidoe, G.P. Otter, et al., "Novel Scale Inhibitor Polymers With Enhanced Adsorption Properties," SPE-80381-MS, International Symposium on Oilfield Scale, held January 29–30, 2003 (Aberdeen, UK).
2. R. Gillian, H. Claire, N. Fleming, "Enhanced Scale Inhibitor Performance Through Utilization of End Capped Technology," NACE-04387, NACE Annual Conference, held March 28–April 1, 2004 (New Orleans, LA, USA).
3. G. Woodward, G.P. Otter, K.P. Davis, et al., "Biodegradable Polymers," International Patent Application WO/2004/056886, 2004.
4. P. Chen, T. Hagen, H. Montgomerie, et al., "A Scale Inhibitor Chemistry Developed for Downhole Squeeze Treatments in a Water Sensitive and HTHP Reservoir," Oil Field Chemistry Symposium, held March 9–12, 2008 (Geilo, Norway).

5. H. Montgomerie, P. Chen, T. Hagen, et al., "Development of a New Polymer Inhibitor Chemistry for Downhole Squeeze Applications," SPE-113926-MS, SPE International Oilfield Scale Conference, held May 28–29, 2008 (Aberdeen, UK).
6. P. Chen, H. Montgomerie, T.H. Hagen, et al., "Meeting the Challenges in Downhole Scale Control for High-Water-Production Horizontal Wells in an Environmentally Sensitive Part of the Norwegian North Sea," *SPE Production & Operations* 27 (2012): pp. 382–389.
7. H. Montgomerie, P. Chen, T. Hagen, et al., "Development of a New Polymer Inhibitor Chemistry for Downhole Squeeze Applications," *SPE Production & Operations* 24 (2009): pp. 459–564.
8. T. Chen, P. Chen, T. Hagen, et al., "Development of an Environmentally Friendly Polymer Squeeze Inhibitor for Short Perforation and High Water Production ESP Wells," NACE-2013-2661, NACE Annual Conference, held March 17–21, 2013 (Orlando, Florida, USA).
9. P. Chen, T. Hagen, H. Montgomerie, et al., "Scale Inhibiting Well Treatment," Patent AU2006274668A1, 2011.
10. M.J. Todd, C.J. Strachan, G. Moir, et al., "Development of the Next Generation of Phosphorus Tagged Polymeric Scale Inhibitors," SPE-130733-MS, SPE International Conference on Oilfield Scale, held May 26–27, 2010 (Aberdeen, UK).
11. M.J. Todd, A.R Thornton, J. Wylde, et al., "Phosphorus Functionalized Polymeric Scale Inhibitors, Further Developments and Field Deployment," SPE-154135-MS, SPE International Conference and Exhibition on Oilfield Scale, UK, held May 30–31, 2012 (Aberdeen, UK).
12. N. Poynton, A. Molliet, A. Leontieff, "Development of a New Tagged Polymeric Scale Inhibitor With Accurate Low Level Residual Inhibitor Detection, for Squeeze Applications," SPE-155187-MS, SPE International Conference and Exhibition on Oilfield Scale, held May 30–31, 2012 (Aberdeen, UK).
13. K.S. Sorbie., P. Jiang, M.D. Yuan, et al., "The Effect of pH, Calcium, and Temperature on the Adsorption of Phosphonate Inhibitor Onto Consolidated and Crushed Sandstone," SPE-26605-MS, SPE Annual Technical Conference and Exhibition, held October 3–6, 1993 (Houston, Texas, USA).
14. M. Tomson, A. Kan, G. Fu, et al., "Mechanistic Understanding of Rock/Phosphonate Interactions and Effect of Metal Ions on Inhibitor Retention," *SPE Journal* 13 (2008): pp. 325–336.
15. P. Chen, G.M. Graham, "Examination of the Influence Flow Rate on Inhibitor Return Concentrations in Non-Equilibrium Core Flooding of Generically Different Scale Inhibitors Applied IN Both Adsorption and Precipitation Treatments," 11th International Oil Field Chemical Symposium, held March 19–22, 2000 (Fagernes, Norway).
16. A. Kan, Z. Dai, M. Tomson, "The State of the Art in Scale Inhibitor Squeeze Treatment," *Petroleum Science* 17 (2020): pp. 1579–1601.
17. M. Jordan, K.S. Sorbie, P. Chen, et al., "The Design of Polymer and Phosphonate Scale Inhibitor Precipitation Treatments and Importance of Precipitates Solubility in Extending Squeeze Life," SPE-37275-MS, SPE International Symposium on Oilfield Chemistry, held February 18–21, 1997 (Houston, Texas, TX).
18. K. Jarrahian, K. Sorbie, M. Singleton, et al., "Building a Fundamental Understanding of Scale-Inhibitor Retention in Carbonate Formations," *SPE Production & Operations* 35 (2020): pp. 85–97.
19. A. Kan, T. Fu, D. Shen, et al., "Enhanced Inhibitor Treatments With the Addition of Transition Metal Ions," SPE-114060-MS, SPE International Oilfield Scale Conference, held May 28–29, 2008 (Aberdeen, UK).
20. H. Bourne, M. Booth, L. Steven, "Combining Innovative Technologies to Maximize Scale Squeeze Cost Reduction," SPE-50718-MS, SPE International Symposium on Oilfield Chemistry, held February 16–19, 1999 (Houston, Texas, USA).
21. M.J. Todd, G. Lamont, A.R. Thornton, "A New Precipitation Squeeze Alternative for Treating Harsh Barium Sulphate Scaling in a Highly Naturally Fractured North Sea Carbonate Reser-

voir," Trinidad and Tobago Energy Resources Conference, SPE-132901-MS, held June 27–30, 2010 (Port of Spain, Trinidad).
22. J.A. McRae, S.M. Heath, C. Strachan, et al., "Development of an Enzyme Activated, Low Temperature, Scale Inhibitor Precipitation Squeeze System," SPE-87441-MS, SPE International Symposium on Oilfield Scale, held May 26–27, 2004 (Aberdeen, UK).
23. N. Poynton, R. Tidswell, J. Steele, et al., "Squeezing Aqueous Based Scale Inhibitors Into a Water Sensitive Reservoir—Development of a Squeeze Strategy," SPE-60219-MS, International Symposium on Oilfield Scale, held January 26–27, 2000 (Aberdeen, UK).
24. O. Vazquez, E. Mackay, K. Sorbie, et al., "Modelling a Surfactant Preflush With Non-Aqueous and Aqueous Scale Inhibitor Squeeze Treatments," SPE-113212-MS, European Formation Damage Conference, held May 9–12, 2009 (Scheveningen, The Netherlands).
25. B. Ghosh, M.Y. Alklih, X. Li, "Pre-Flush Design for Extended Scale Inhibitor Squeeze Lifetime in Limestone Reservoir—Laboratory Investigations and Cost Benefit Analysis," OTC-26604-MS, Offshore Technology Conference Asia, held March 22–25, 2016 (Kuala Lumpur, Malaysia).
26. B. Ghosh, X. Li, "Effect of Surfactant Composition on Reservoir Wettability and Scale Inhibitor Squeeze Lifetime in Oil Wet Carbonate Reservoir," *Journal of Petroleum Science and Engineering* 108 (2013): pp. 250–258.
27. B. Ghosh, X. Li, M.Y. Alklih, "Preflush Design for Oil-Wet Carbonate Formations: Key to Enhance Scale-Inhibitor-Squeeze Lifetime," *SPE Journal* 21 (2016): pp. 776–785.
28. M.M. Jordan, K.S. Sorbie, P. Jiang, et al., "Mineralogical Controls on Inhibitor Adsorption/Desorption in Brent Group Sandstone and Their Importance in Predicting and Extending Field Squeeze Lifetimes," SPE-27607-MS, European Production Operations Conference and Exhibition, held March 15–17, 1994 (Aberdeen, UK).
29. N. Fleming, K. Ramstad, A.C. Nelson, et al., "Innovative Use of Kaolinite in Downhole Scale Management: Squeeze Life Enhancement and Water Shutoff," *SPE Production & Operations* 24 (2009): pp. 439–449.
30. O. Bache, S. Nilsson, "Ester Cross-Linking of Polycarboxylic Acid Scale Inhibitors As a Possible Means to Increase Inhibitor Squeeze Lifetime," SPE-60190-MS, International Symposium on Oilfield January 26–27, 2000 (Aberdeen, UK).
31. O.M. Selle, M.S. Wat, O. Vikane, et al., "A Way Beyond Scale Inhibitors—Extending Scale Inhibitor Squeeze Life Through Bridging," SPE-80377-MS, International Symposium on Oilfield Scale, held January 30–31, 2003 (Aberdeen, UK).
32. P. Chen, T. Hagen, H. Bourne, et al., "The Challenge of Squeezing a Water Sensitive HP/HT Reservoir—Lab and Field Experiences With a Novel Non-Aqueous Inhibitor/Squeeze Enhancer Package," SPE-87435-MS, International Symposium on Oilfield Scale, held May 27–28, 2004 (Aberdeen, UK).
33. P. Chen, H. Bourne, R. Thoraval, "Extending Squeeze Life With a Novel Scale Inhibitor Squeeze Additive," Paper #16, 15th Norwegian International Oilfield Chemical Symposium, held March 28–31, 2004 (Geilo, Norway).
34. P. Chen, T. Hagen, T. Haaland, et al., "Enhanced Inhibitor Retention Through Surface Precipitation and Molecular Interaction," Paper #10, 16th Norwegian International Oilfield Chemistry Symposium March 13–16, 2005 (Geilo, Norway).
35. P. Chen, T. Hagen, H. Montgomerie, et al., "Field Experiences in the Application of an Inhibitor/Additive Interaction Package to Extend an Inhibitor Squeeze Life," SPE-100466-MS, SPE International Oilfield Scale Symposium, held May 31–June 1, 2006 (Aberdeen, UK).
36. P. Chen, H. Montgomerie, S.M. Heath, et al., "Method of Increasing Scale Inhibitor Retention," U.S. Patent 20170002258A1, 2018.
37. C. Sitz, J. Hardy, J. Merryman, "A New Class of Non-Damaging, Environmentally Responsible Scale Squeeze Enhancement Aids," SPE-155254-MS, SPE International Conference on Oilfield Scale May 30–31, 2012 (Aberdeen, UK).

38. S. Heath, B. Juliussen, P. Chen, et al., "Novel Scale Squeeze Technology and Treatment Designs for Improving Scale Inhibitor Retention and Treatment Lifetimes—Use of Ionic Polymers in the Overflush," SPE-154935-MS, International Symposium on Oilfield Scale May 30–31, 2012 (Aberdeen, UK).
39. L. Sutherland, M. Jordan, "Enhancing Scale Inhibitor Squeeze Retention With Additives," SPE-179888-MS, SPE International Oilfield Scale Conference and Exhibition May 11–12, 2016 (Aberdeen, UK).
40. P. Chen, T. Al-Ghamdi, Z. Yue, et al., "Enhanced Scale Inhibitor Squeeze Treatment With a Newly Developed Chemical Additive," NACE-2021-16802, NACE Annual Conference April 19–30, 2021 (Virtual).
41. K.E. Thompson, R.D. Gdanski, "Laboratory Study Provides Guidelines for Diverting Acid With Foam," *SPE Production & Operations* 8 (1993): pp. 285–290.
42. F.F Chang, X. Qiu, H.A. Nasr-El-Din, "Chemical Diversion Techniques Used for Carbonate Matrix Acidizing: An Overview and Case Histories," SPE-106444-MS, SPE International Symposium on Oilfield Chemistry, held February 28–March 2, 2007 (Houston, TX, USA).
43. S.I. Kam, W.W. Frenier, S.N. Davies, et al., "Experimental Study of High-Temperature Foam for Acid Diversion," SPE-82266-MS, SPE European Formation Damage Conference May 13–14, 2003 (The Hauge, Netherlands).
44. O.M. Selle, O.K. Nygård, E. Storås, et al., "Foamed Scale Inhibitor and Scale Dissolver Treatments for Improved Placement in a Subsea Norne Well," SPE-155407-MS, SPE International Conference and Exhibition on Oilfield Scale May 30–31, 2012 (Aberdeen, UK).
45. P. Chen, P. Vijn, T. Hagen, et al., "Applications of a Foamed Scale Inhibitor Package in a High Temperature and Low Pressure/Low Permeability Gas Reservoir," MECCOCT18-12544, 17th Middle East Corrosion Conference, held September 30–October 3, 2018 (Manama, Bahrain).
46. O.M. Selle, M. Springer, I.H. Auflem, et al., "Gelled Scale Inhibitor Treatment for Improved Placement in Long Horizontal Wells at Norne and Heidrun Fields," SPE-112464-MS, SPE International Symposium and Exhibition on Formation Damage Control, held February 13–15, 2008 (Lafayette, LA, USA).
47. N.A. Menzies, E.J. Mackay, K.S. Sorbie, "Modeling of Gel Diverter Placement in Horizontal Wells," SPE-56742-MS, SPE Annual Technical Conference and Exhibition, held October 3–6, 1999 (Houston, TX, USA).
48. O.M. Selle, R.M. Wat, H. Nasvik, et al., "Gelled Organic Acid System for Improved $CaCO_3$ Removal in Horizontal Openhole Wells at the Heidrun Field," SPE-90359-MS, SPE Annual Technical Conference and Exhibition, held September 26–29, 2004 (Houston, TX, USA).
49. R.M. Wat, A.F. MacLean, L. Barber, et al., Oil Soluble Scale Inhibitor (OSI) Squeeze Treatment—The 1st North Sea Case History," Symposium on Oilfield Scale, held January 27–28, 1999 (Aberdeen, UK).
50. I.R. Collins, M.M. Jordan, S.E. Taylor, "The Development and Application of a Novel Scale Inhibitor for Deployment in Low Water Cut, Water Sensitive or Low Pressure Reservoirs," SPE-60192-MS, International Symposium on Oilfield Scale, held January 26–27, 2000 (Aberdeen, UK).
51. H.M. Bourne, A.S. Buckley, I. Spark, et al., "A Mechanistic Investigation of the Dunbar Scale Challenge—The Use of Cryogenic SEM to Determine Wettability Alteration," SPE Oilfield Scale Symposium, January 27–28, 1999 (Aberdeen, UK).
52. S.M. Heath, A. Thornton, C. Strachan, et al., "Development of an 'Environmentally Friendly' Non-Aqueous Scale Inhibitor for Harsh Scaling Environments," 14th International Oil Field Chemicals Symposium, held March 24–26, 2003 (Geilo, Norway).
53. G.M. Graham, A. Gyani, M.M. Jordan, et al., "Selection and Application of a Non-Damaging Scale Inhibitor Package for Pre-Emptive Squeeze in Mungo Production Wells," SPE-74665-MS, SPE Oilfield Scale Symposium, held January 30–31, 2000 (Aberdeen, UK).

54. M.M. Jordan, C. Graff, K. Cooper, "Development and Deployment of a Scale Squeeze Enhancer and Oil-Soluble Scale Inhibitor to Avoid Deferred Oil Production Losses During Squeezing Low-Water Cut Wells, North Slope, Alaska," SPE-58725-MS, SPE International Symposium on Formation Damage Control, held February 23–24, 2000 (Lafayette, LA, USA).
55. A.F. Miles, O. Vikane, D.S. Healey, et al., "Field Experiences Using 'Oil Soluble' Non-Aqueous Scale Inhibitor Delivery Systems," SPE-87431-MS, International Symposium on Oilfield Scale, held May 26–27, 2004 (Aberdeen, UK).
56. N.J. Jenvey, A.F. MacLean, A.F. Miles, et al., "The Application of Oil Soluble Scale Inhibitors Into the Texaco Galley Reservoir. A Comparison of Traditional Squeeze Techniques to Avoid Problems Associated With Wettability Modification in Low Water-Cut Wells," SPE-60197-MS, International Symposium on Oilfield Scale, held January 26–27, 2000 (Aberdeen, UK).
57. A.F. Miles, H.M. Bourne, R.G. Smith, et al., "Development of a Novel Water in Oil Microemulsion Based Scale Inhibitor Delivery System," SPE-80390-MS, International Symposium on Oilfield Scale, held January 29–30, 2003 (Aberdeen, UK).
58. C. Romero, B. Bazin, A. Zaitoun, et al., "Behavior of a Scale Inhibitor Water-In-Oil Emulsion in Porous Media," SPE-98275-MS, SPE international Symposium and Exhibition on Formation Damage Control, held February 15–17, 2006 (Lafayette, LA, USA).
59. H.A. Nasr-El-Din, H.A. Al-Saiari, A. Al-Ruwayily, et al., "Field Application of Emulsified Scale Inhibitor Treatment to Mitigate Calcium Carbonate Scale in Horizontal Wells," SPE-100456-MS, International Symposium on Oilfield Scale, held May 30–June 1, 2006 (Aberdeen, UK).
60. R.A. Shields, K.S. Sorbie, M.A. Singleton, et al., "Analysis of the Mechanism of Transport and Retention of Non-Aqueous Scale Inhibitor Treatments in Cores Using Novel Tracer Techniques," SPE-100518-MS, International Symposium on Oilfield Scale, held May 30–June 1, 2006 (Aberdeen, UK).
61. T. Tjomsland, J.O. Sæten, O. Vikane, et al., "Veslefrikk Scale Control Strategy and Economic Implications: Revisited 7 Years Later—Did we Improve?," SPE-114086-MS, SPE International Oilfield Scale Conference, held May 28–29, 2008 (Aberdeen, UK).
62. P. Chen, J. Bjørn, O. Vikane, et al., "Development of Non-Aqueous/Low Density Scale Inhibitor Package for Down-Hole Squeeze Treatments," SPE-130686-MS, SPE International Conference on Oilfield Scale, held May 26–27, 2010 (Aberdeen, UK).
63. T.I. Asheim, B. Kaasa, F. Nielsen, et al., "Inhibitor Squeeze Treatment for Preventive Carbonate Scale Control in a Subsea Completed Well on Smfbukk," SPE-60201-MS, SPE Oilfield Scale Symposium, held January 26–27, 2002 (Aberdeen, UK).

8 Solid Scale Inhibitor Treatment

Dong Shen[1]
[1]Baker Hughes, Sugar Land, TX, USA

8.1 Introduction

In oilfield applications, scale inhibitors are commonly squeezed onto the subsurface, where they precipitate or adsorb into the surface of formation rock, and then slowly release into the produced fluids to inhibit the formation of mineral scale during production; this is referred to as a "squeeze treatment."[1] A major shortcoming of conventional squeeze treatments is that most of the scale inhibitor is precipitated near the entrance of the formation as an acidic scale inhibitor is applied. The reaction of acid with carbonate and the formation of divalent cations (Ca^{2+}, Mg^{2+}) and inhibitor precipitates take place very quickly. As a result, only a limited reservoir protection distance can be achieved (near wellbore area). Adsorption squeezes are often performed in reservoirs where a neutralized inhibitor pill solution or a high-soluble scale inhibitor product is injected. Then, the injected inhibitor adsorbs onto the native rock and clay surfaces. Compared with precipitation squeeze, only very small amounts of inhibitor can be retained in the formation. Therefore, a large amount of the scale inhibitor tends to flow back quickly, and then decreases to a low value which is not enough to effectively inhibit scale formation. It is common to see a significant fraction of scale inhibitors (approximately one-third) return within only a few days at a very high concentration. Often, only one-third of the injected inhibitor may return in concentration range to efficiently inhibit scale formation, while another one-third of the injected inhibitor will not return at a concentration sufficient for scale control, referred to as being lost into the formation. The scale inhibitor return concentrations depend on the nature of reservoir (mineralogy of formation, temperature, formation water composition), the scale inhibitor chemistry, and also treatment design. Once the squeeze treatment is executed, there is not much that can be done to reduce the initial inhibitor spike and extend the scale inhibitor protection lifetime. To overcome this drawback, various solid scale inhibitors have been developed as an alternative treatment option to achieve a controlled scale inhibitor release rate.

8.2 Solid Scale Inhibitors for Hydraulic Fracturing

It was reported that more than 25,000 hydraulic proppant fracturing treatments were conducted per year to enhance oil and gas production.[2] Most treated wells may experience mineral scale issues as the wells start to produce. Although a certain amount of liquid scale inhibitor can also be added to the fracturing fluid to provide inhibitor protection, there may not be sufficient active adsorption sites available inside the fracturing fluid to assist in the binding of scale inhibitors to the rock or proppant surface. The adsorption/retention of liquid scale inhibitors may heavily depend on the amount of scale inhibitors leaking off into the formation rock matrix. As a result, there is little control over the scale inhibitor release rate, and most of the inhibitor may flush out at the very early stage of production.

The addition of scale inhibitors to fracturing fluid can combine two treatments into one operation, and thus possibly lead to a cost reduction and less operation time. In this approach, the solid inhibitor particles are mixed with the sand/proppant and can be placed evenly inside the full length of the fracture created by the fracturing fluid. To ensure that, it is well mixed with fracturing sand/proppants, the solid inhibitor is normally made to a similar size to the sand/proppants (10–100 mesh). At this particle size range, various approaches are available to encapsulate the solid inhibitor particles to control/adjust their release rate to meet different treatment requirements.

Webb and Collins compared the economic and technical features of liquid and solid scale inhibitor delivery methods for offshore-fractured and gravel-packed wells.[3] The conventional bullhead squeeze treatments had difficulties placing liquid scale inhibitor into water-producing zones, because of the unfavorable pressure regimes in the subsea field developed with advanced horizontal- and multilateral-type well completions. The cost of squeeze treatments using coiling tubing and packers, setting of bridge plugs, or flow diversion to improve the liquid inhibitor placements would increase significantly and could be 6–10 times more than a typical bullhead treatment. Moreover, the cost of deferred oil production was another major concern for operators as inhibitor squeeze treatment was applied. It was concluded that the application of solid inhibitors could reduce at least 25% production loss and create 2–12 million dollars of value for vertical wells and horizontal wells.

The following sections detail the two types of commercially available solid inhibitors which have been used in oilfield fracturing operations.

8.2.1 Encapsulated Bivalent Metal Salt Scale Inhibitor

Powell and McCabe first reported a solid inhibitor based on the calcium/magnesium polyphosphate precipitates for fracturing application.[4] This solid inhibitor had a CaO/MgO to

P_2O_5 ratio of 1:2.7 and an average particle size of 10–30 mesh. Continuous flow release tests were performed at various temperatures with a solid inhibitor loading of 1% by weight. At low temperatures (<100°F), an almost constant inhibitor release rate was recorded. However, the inhibitor release rate gradually increased with the increase in temperature. Two wells with a downhole temperature of 75°F were picked for the field trial. Based on the first 3 months of data on inhibitor residuals, a 50-week protection lifetime was expected.

However, the polyphosphate scale inhibitors tend to decompose at high temperature (>150°F), and thus lead to very limited field applications. A different type of solid scale inhibitor for use in fracturing treatment was further developed to treat the wells with a downhole temperature up to 225°F by this group.[5] This solid inhibitor was prepared by mixing phosphonate or polyacrylic acid scale inhibitor with alkaline earth oxides first to form solid particles. These particles were then grounded and sieved to the desired size (12–40 mesh). A fluidized bed process was applied to coat the particles with 20% semipermeable materials. Continuous flow release testing results indicated that the phosphonate-based solid inhibitor was released at approximately 4.5 ppm for 30 days with only 17.5% chemical depletion, whereas 36.8% inhibitor was released at a rate of 17.8 ppm for an uncoated solid inhibitor sample. One polyacrylic acid-based solid inhibitor showed a similar inhibitor release profile to the coated phosphonate sample. However, another polymer solid inhibitor was released too slowly (0.6 ppm), which was not enough to meet the minimum inhibitor concentration (MIC) requirements for most scale inhibition treatments. The encapsulated phosphonate solid inhibitor was applied in a well with a projected calcite scale issue. A total of 100 lb of solid inhibitor was added to the fracturing fluid with 1,000 sk of synthetic proppant. The phosphonate scale inhibitor residuals were monitored for 5 months. Field data showed that the inhibitor return concentrations dropped to ~1 ppm after 60 days (Figure 8.1), which was much faster than projection based on the lab test results and very low water production rate (<2 bbl of water per day).

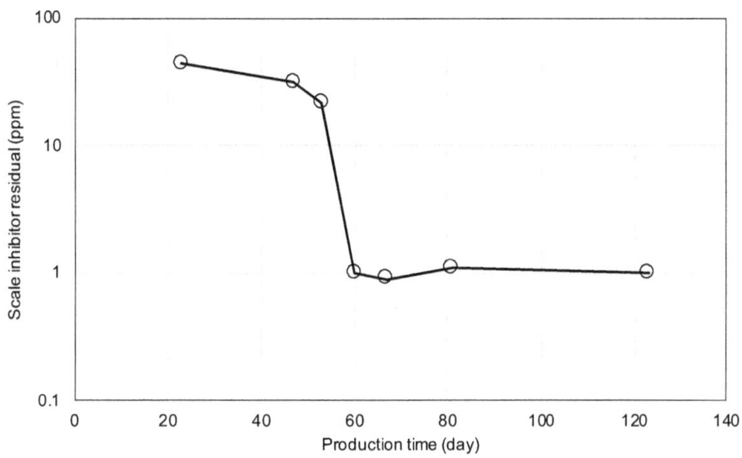

Figure 8.1. Inhibitor residuals for a treatment with phosphonate-based solid inhibitor.

8.2.2 Scale Inhibitor Impregnated Porous Proppants

Rather than encapsulated solid scale inhibitor salt with a polymeric coating, a solid inhibitor was introduced by infusing scale inhibitor into a porous proppant.[3, 6–9] To meet the API recommended guidelines for fracture and gravel pack proppant selection, extensive efforts have been made to develop the porous proppant substrate.[8] For example, spherical particles with a size up to 1 mm in diameter were made first by spraying dry a mixture of ground bauxite ($Al_2O_3 \cdot H_2O$) and mullite ($Al_6Si_2O_{13}$) slurry. These dried spherical particles were then sintered at high temperatures (from 870°C to 1,500°C) to obtain porous proppants with a porosity range of 10%–20%. The porous proppants were placed into a pressure vessel and vacuumed to about 0.1 mbar, to ensure no air or vapors remained in the pore space. Desired scale inhibitor solutions were then injected into the vessel under a vacuum to prompt the absorption or deposition into porous proppants for 1 h. After the vacuum was released, the wet porous proppants were dried in an oven to remove the moisture. This inhibitor impregnation and drying process could be repeated to increase the number of inhibitors retained inside the pore space of proppants. The impregnated porous proppant can be further coated with polymeric materials to control the inhibitor release rate.

It was suggested that the selected porous proppants should be crush resisted and have negligible impact on the fracture/gravel pack conductivity.[3] However, an inverse proportionality between porous proppant absorption capacity and its strength was observed. Noticeable proppant pack conductivity loss was observed, when 25% porous proppants were used to replace the regular proppant. The first field trial of this type of scale inhibitor was performed in the Prudhoe Bay field of Alaska.[10, 11] For this field trial, approximately 794 lb of active diethylene triamine-penta(methylene phosphonic) acid scale inhibitor was impregnated into 21,500 lb of 20–40 mesh porous proppant. A total of 21,338 lb of this solid inhibitor was pumped into a propped fracture at an average downhole pressure of 3,600 psi. The scale inhibitor return profile showed that the inhibitor residuals experienced a sharp drop in the early stages of production. The inhibitor return concentration was 50% below MIC (1.4 ppm) after 81 days of production, and then bounced back to 20 ppm 8 months later. This might be attributed to the limited water flowing through the impregnated porous proppant in the first couple months.

A new impregnated porous proppant was also developed for application in high-humidity environments, which is commonly experienced in offshore stimulation vessels.[7] Porous proppant strength, impregnated solid inhibitor compatibility with fracturing fluid, and scale inhibitor release characteristics were fully evaluated before being applied in two North Sea fields. Two different scale inhibitors were selected to make impregnated porous proppant, one was phosphonate, and the other was polymeric. The finished solid inhibitor products have a very low moisture content (<0.2%), which is beneficial for product storage properties in humid environments. Dynamic scale inhibitor release test results indicated that the polymeric-based solid scale inhibitor could be released at a relatively high concentration, and thus was suitable for wells with a high MIC expected (6 ppm). A total of 18,000 lb of solid inhibitor was applied in the treatment and 320,000 bbl of produced water (or 2.9 years) protection was expected. The field inhibitor monitoring data showed that polymeric inhibitor residuals decreased to at or below MIC after only 25,000-bbl water was produced, although the mass balance calculation on scale inhibitor depletion suggested that only 4.5% of impregnated scale inhibitor was released. For phosphonate-based solid inhibitor, the long-term inhibitor return concentrations were approximately 1 ppm, which was a good fit for some wells with a low MIC (0.6 ppm). A small fracturing treatment was conducted for this application by blending 8% phosphonate impregnated porous proppant into 50,000 lb of resin-coated lightweight proppant. Residual analysis showed that the phosphonate scale inhibitor was released at low dosages of 1–10 ppm.

8.2.3 Comparison Studies

Comparison studies between the aforementioned two types of solid scale inhibitors were conducted for an offshore field application. The wells were completed with a frac-pack for sand control, due to the reservoir's low geomechanically strength and other rock characteristics. One of the major concerns was the sulfate scale deposition behind a frac-pack screen and within proppant-filled perforations.[12] Conventional subsea scale inhibitor squeeze was very costly and had an uncertain treatment life, because of the limited industry experience in treating the frac-packed and multizoned wells. As a result, placing scale inhibitor throughout the frac-pack was proposed to prevent the formation of mineral scales, when unexpected early water breakthroughs occurred in the production well. Lab core flood testing results indicated that the scale inhibitor protection lifetime is short, if liquid scale inhibitors were added into the fracturing fluid, because the field core materials have insufficient active sites available to prompt the adsorption and retention of scale inhibitor. Therefore, solid inhibitors were selected as a viable alternative to enhance the inhibitor protection time.

Both solid inhibitors evaluated, an encapsulated Ca/Mg salt product and inhibitor impregnated porous proppants, showed superior compatibility when compared to the liquid scale inhibitors, due to their slow dissolving rates in produced water. Conductivity tests, which are used to determine whether the presence of solid scale inhibitor particles had any negative impact on the conductivity of frac-pack, indicated that the impregnated porous proppants had less crush resistance than the conventional proppant of equivalent size. This could be managed by adjusting the loading rate of solid inhibitor by limiting the quantity added into frac-pack. The encapsulated salt product could be crushed at high pressure. At a solid inhibitor loading rate of 2% or below, there was no significant conductivity loss observed at 4,000 psi, but approximately 50% conductivity loss was recorded as closure pressure was increased to 6,000 psi.

The encapsulated Ca/Mg salt solid inhibitor was used to treat one subsea well, by adding 3,000-lb inhibitor product into 150,000 lb of proppants. Water broke through slowly in the treated well, and 58,000 bbl of water was produced after 4 years. The collected water sample showed the scale inhibitor residuals had dropped to 3 ppm, which was less than the lab determined MIC of 5 ppm. However, no loss of production due to scale depositions was observed. The impregnated porous proppant product was used to treat two subsea wells in the same field. The scale inhibitor returned to a concentration of 2–3 ppm after 100,000 bbl of cumulated water production in the treated wells.

Field results from this study suggested that the application of solid inhibitors could provide effective scale control, during the early stage of water breakthrough, and thus reduce the need to perform the conventional squeeze treatment in wells with very low water cuts. The solid scale inhibitors could also provide sufficient lead time to fully investigate and carefully plan the squeeze job for long-term protection.

8.2.4 Latest Development on Impregnated Porous Proppants

The maximum quantity of inhibitor that could be impregnated into the above discussed porous proppants was only 12%, which was not suitable to provide long-term scale protection.[9] To overcome this limitation, a solid inhibitor product named scale inhibitor impregnated gravel (SIIG) was developed to increase the inhibitor content to 20% for gravel pack application. Gravel packing completion is a commonly used technique to control fines generation and sand mobilization in oil and gas wells. A screen is placed in the wellbore across the production area in a typical gravel pack completion operation. The spherical gravel pack proppants are circulated into the well with a gel fluid to fill the annular space between the screen and formation rock. The strength requirements for gravel pack proppant are normally not as high as those for frac-pack proppant. The permeability of SIIG was comparable with conventional gravel pack proppants before the impregnated inhibitor was depleted, when the closure pressure was below 6,000 psi. During the field trial, a flush scale inhibitor loss to the gravel pack fluid was observed, but the monitored scale inhibitor residuals still worked as planned. However, the liquid productivity index dropped significantly after the treated well started to produce. The scaling Ba^{2+} concentration in the produced water fell from 74 ppm to zero within hours, which indicated that this well experienced a severe barite ($BaSO_4$) scale problem. These observations suggested that the scaling issue happened inside the formation and was far behind the impregnated gravel pack.

To control the scale inhibitor initial flush release and ensure a steady release rate in the long run, semipermeable membranes were used to encapsulate the impregnated solid scale inhibitors.[13, 14] Lab testing results showed that a significant improvement in the inhibitor release profile was achieved, when certain semipermeable membrane coatings were applied. In field trials at the Unita Basin of Eastern Utah, various encapsulated inhibitor impregnated porous proppant loading rates were applied to optimize the treatment design. Three wells were treated with a 5% solid inhibitor incorporated with frac sand, and the other two wells were treated with 7.5% loading rate. Collected field data indicated that the scale inhibitor residuals dropped from 100 ppm to approximately 10 ppm, as the cumulated water production was close to 2,000 bbl. After 18,000 bbl of water were produced, the inhibitor flow back concentrations were further decreased to near the MIC level of 2 ppm, with still more than 95% of scale inhibitor remaining inside the frac-pack.

Normally, pore space can largely reduce the compressive strength of the proppant. As a result, the presence of porous proppant tends to reduce the frac-pack conductivity. With the improvement of the proppant manufacturing process, a porous proppant that can resist up to 20,000 psi of closure pressure was developed to make the infused and encapsulated solid inhibitor products suitable for high-pressure reservoir deployment in deepwater fields.[15] The new process allowed the porous proppant to be tailored to a desired porosity specification without losing compressive strength via optimizing the proppant microstructure and reducing imperfection. Conductivity testing results indicated that no noticeable conductivity loss was recorded when up to 20% of this high-strength porous proppant was mixed with regular high-strength proppant at a testing pressure of 12,000 psi. The long-term inhibitor release profile of this high-strength inhibitor-infused porous proppant was comparable to that of previously developed inhibitor-infused porous proppant. This product was initially applied in two high-pressure wells in Gulf of Mexico (GOM) at loading rates of 13% and 20% weight of proppants. Up to 2020, over 500 wells had been treated with this solid scale inhibitor product worldwide,[16] covering the Permian Basin, Uinta Basin, Colombia, North Sea, Middle East, and offshore Congo.[17, 18] The inhibitor release data showed that hundreds of thousands of barrels cumulated produced water could be protected in the treated wells.

8.3 Proppant-Sized Solid Scale Inhibitor

Proppant-sized solid scale inhibitor has been commercially available on the market since 2004,[19] which is larger than the encapsulated scale inhibitor bivalent metal salt and inhibitor-impregnated porous proppant. However, this proppant-sized solid scale inhibitor is probably the most commercially successful product in the industry, with more than 25,000 wells having been treated before the end of 2017. This product was made by absorbing the scale inhibitor chemicals into a water-soluble adsorbent with proppant-sized. Available adsorbents included activated carbon silica particulate, zeolite, and diatomaceous earth (which is preferred).[20, 21] Diatomaceous earth has been widely used in different industries as an adsorbent with various particle sizes available. It can absorb up to 160% wt/wt of water and 120% wt/wt of organic solvents. Therefore, diatomaceous earth can be used to absorb various production chemicals, including inhibitors for controlling scaling, corrosion, and paraffin and asphaltene accumulation, to make a product into a solid form. A solid scale inhibitor with more than 40% active content can be easily made through a single absorption and dry process. Commercially available solid scale inhibitor with proppant size normally contains 25% activity.[22] This ensures a sufficient scale inhibitor could be placed in the frac-pack, even if a relatively small amount of solid inhibitor is applied.

8.3.1 Case Studies

The application of this proppant-sized solid scale inhibitor was first reported by Szymczak and colleagues.[19] It was claimed that a prolonged scale inhibitor protection lifetime of up to 3 years could be achieved. The application of this solid inhibitor would save operation costs and encourage the operator to develop new fields that might be considered unprofitable, due to the frequency and cost of mineral scale remediation.

A special class of new wells was selected for the field trial. These wells were expected to produce water with scale problems and had a need for hydraulic fracturing stimulation. Lab tests confirmed that the addition of 1% or 2% of this solid scale inhibitor based on the weight of the proppant had negligible impact on fracturing fluid performance and frac-pack conductivity under downhole temperature and pressure. The solid inhibitor was transferred to a hopper and fed into the proppant at a consistent rate to ensure an even distribution throughout the frac-pack. The treated two wells are shallow (<2,000 ft) coalbed methane producers located on the border of Virginia and West Virginia. These wells typically have long, perforated producing intervals across different zones. Therefore, multiple stages of fracturing stimulation were applied for each well with a total 150,000 lb of proppant sand used, with a solid inhibitor loading rate of 0.5% weight of sand. To compare the performance of different scale inhibitor treatment techniques, an offset well in the same area was squeezed with a liquid scale inhibitor. The scale inhibitor return concentration comparison between the squeezed well and solid inhibitor treated well is plotted in Figure 8.2. It appeared that a large amount of scale inhibitor had been washed out in the early life of production after a liquid scale inhibitor squeeze job, whereas the solid scale inhibitor was released at a relatively steady dosage. After 2 years of production, both wells still released scale inhibitor at a level above MIC, and no scale deposition-related well failures were reported.

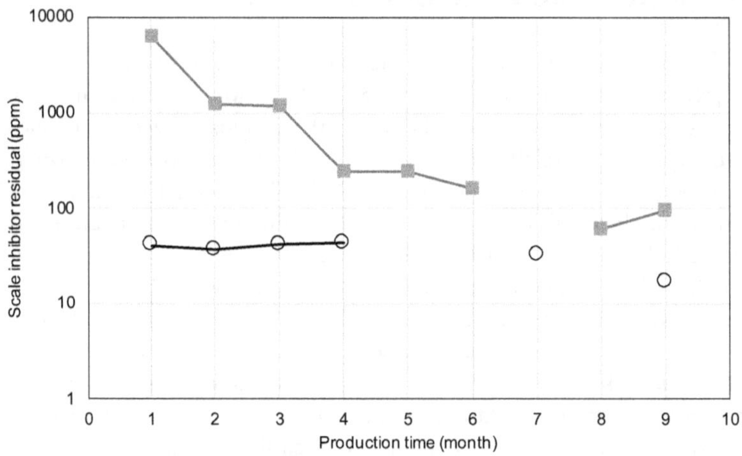

Figure 8.2. Scale inhibitor residuals comparison between liquid squeeze and solid inhibitor treated wells.

With the success of the trial test on solid inhibitor, the same treatment was deployed in over 500 wells over a short period of time. Gupta *et al.* provided a detailed discussion on the performance of the 24 treated wells from two regions.[22] They also discussed the impact of adding this proppant-sized solid inhibitor to the frac proppant pack conductivity. The API crush resistance testing (REF API 60) results indicated that the fine generated at 6,000 psi closure pressure was less than 10%, when 8% of this solid inhibitor was mixed with economy ceramic proppant (Table 8.1). Based on API 60, a crush fine of less than 10% is acceptable at a given stress.

Table 8.1: Crush test data for various amount of proppant-sized solid inhibitor added to economy ceramic proppant.

Closure Pressure (psi)	Solid Inhibitor Loading (%)	Fines Generated (%)
1,000	0	0.07
	2	0.2
	4	0.2
	8	NA
4,000	0	0.19
	2	1.15
	4	2.09
	8	5.11
6,000	0	1.77
	2	2.82
	4	4.77
	8	9.02

Conductivity tests were performed by using a mixture of the industry standard lightweight ceramic proppant with a solid inhibitor at a loading rate of 1% and 2% by weight, and the results are listed in Table 8.2. It appears that the presence of this solid inhibitor did not have a negative impact on the ceramic proppant pack conductivity, when up to 2% solid inhibitor was added. For some data points, the conductivity of the mixture of proppant and solid inhibitor was higher than the baseline data on proppant alone. This may be attributed to an acceptable experimental error. Based on the crushed and conductivity testing results, a conservative conclusion could be drawn as follows: there was no negative effect on the frac-pack conductivity, when 2% or below solid inhibitor was added, and the closure pressure was less than 6,000 psi. This was slightly better than the capsulated Ca/Mg inhibitor solid (2% for 4,000 psi) reported by Fitzgerald and Cowie.[12]

Table 8.2: Conductivity data for lightweight ceramic proppant and mixture with solid inhibitor at 250°F.

Closure Pressure (psi)	Solid Inhibitor Loading (%)	Conductivity (mD-ft)
4,000	0	5,183
	1	7,230
	2	5,236
5,000	0	4,405
	1	6,450
	2	4,782
6,000	0	3,810
	1	4,950
	2	3,560

To ensure compatibility with the fracturing fluid, 0.5%–2% loading of this solid scale inhibitor in the proppant was applied in field application. The total proppant used in each treated well was 100,000–1,000,000 lb. The solid inhibitor was placed at certain stages or at certain intervals in the larger horizontal completion wells.

Table 8.3 summarizes the treatment and production data for the 24 monitored wells. The downhole temperatures are 80°F for these in the Appalachia region (A1-A16) and 175°F for wells located in California (C1-C8). The scale inhibitor residuals produced in the Appalachian wells were relatively high (Table 8.3), because of the low water production and high solid inhibitor loading. The expected scale inhibitor protection time could be several years. The first two treated wells (A1 and A2) showed no sign of scale inhibitor return concentrations declining even after 3 years. After the 5 years mark was reached, over 2,000 wells had been treated with this proppant-sized solid inhibitor.[23] Wells A1 and A2 still produced scale inhibitor at a concentration above 10 ppm with approximately 8,000 bbl of cumulated water production. No scale-related failure was reported within this 5-year production time. All other treated wells were also producing with a sufficient inhibitor return concentration for scale control, except one coalbed methane well in Appalachia (Well A4). However, this well produced normally, and no indication of scale formation in the system was observed, suggesting the scale inhibitor residual was sufficient to prevent scale deposition under operation conditions.

The 16 coalbed methane wells had a very low water production rate (<10 bbl/d). Therefore, most of the inhibitor return concentrations were still well maintained at 10 ppm or above after 5 years of production. Based on the mass balance estimation, the solid scale inhibitors that remained in the formation should have been continuously produced at a residual above MIC for at least another 5 years for each well.

For wells in the California area, the scale inhibitor returned at a relatively low concentration over time due to the high water production and low solid inhibitor loading quantity. All treated wells have produced for approximately 4 years without any scale-related failures. Additional 2–3 years of production at inhibitor residual above MIC are expected at current water production rates. The scale inhibitor residuals were closely monitored for Well C8 in this area (Figure 8.3). Over the first 650 days of production (>180,000 bbl of water produced), the scale inhibitor residuals were well above the MIC level (1 ppm). After 45 months, the inhibitor residual in produced water was still 3.9 ppm. The success of solid inhibitor treatment was also demonstrated by the fact that no failures attributed to scale deposition were identified in all these treated wells.

Table 8.3: Summary of treatment and production data for wells located in California area.

Well ID	Solid Inhibitor Loading (lb)	Short term Production Duration (days)	Short term Inhibitor Residual (ppm)	Short term Water Produced (bbl)	Long term Production Duration (months)	Long term Inhibitor Residual (ppm)	Long term Water Produced (bbl)
A1	800	1,110	16	5,550	60	11	8,700
A2	800	1,109	31	5,545	60	20	7,630
A3	550	983	7.6	4,915	54	6.0	8,065
A4	800	983	4.0	4,915	54	<1.0	8,065
A5	550	982	6.0	4,910	54	5.0	8,060
A6	840	693	22.0	3,465	44	11.0	6,615
A7	1,520	687	54.0	3,435	44	20.0	6,585
A8	1,400	667	18.0	3,335	43	10.0	6,445
A9	1,375	659	15.0	3,295	43	15.0	6,385
A10	880	647	23.0	3,235	42	9.0	6,280
A11	1,100	626	51.0	3,130	41	3.0	6,100
A12	1,580	590	4.0	2,950	41	21.0	6,085
A13	1,225	587	26.0	2,935	38	15.0	5,740
A14	1,580	521	23.0	2,605	38	16.0	5,655
A15	1,750	501	23.0	2,505	37	12.0	5,560
A16	624	482	42.0	2,410	36	10.0	5,420
C1	200	924	2.7	115,500	54	2.0	NA
C2	200	924	2.5	115,500	54	2.0	133,464
C3	200	924	5.0	92,400	54	3.5	NA
C4	460	514	1.4	84,810	48	1.0	192,780
C5	970	709	3.7	88,625	47	3.2	NA
C6	495	649	8.1	210,925	45	6.0	NA
C7	450	649	7.4	194,700	45	4.8	NA
C8	450	650	8.3	181,720	45	3.9	NA

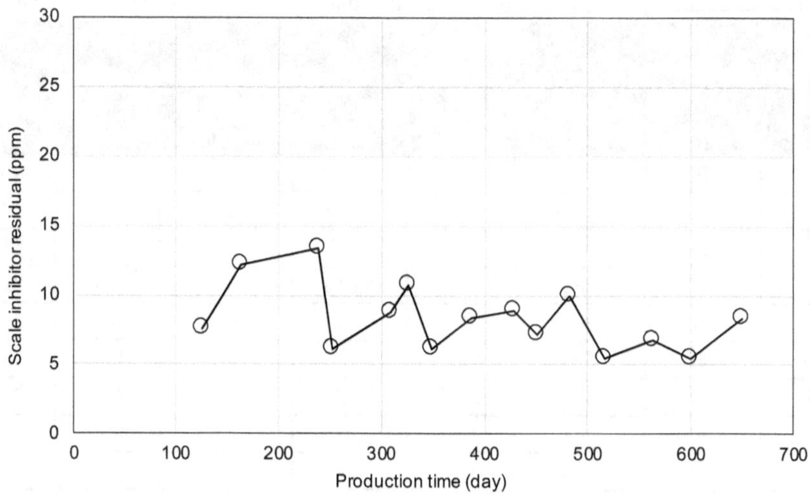

Figure 8.3. Scale inhibitor residuals after solid inhibitor treatment for Well C8.

8.3.2 Combined Application with Paraffin and Asphaltene Solid Inhibitors

Unlike scale inhibitors, paraffin and asphaltene inhibitors are formulated with organic solvents and are not soluble in water-based fracturing fluid. Thus, it is difficult to place such liquid inhibitors into formation during the hydraulic fracturing process. Such a challenge can be resolved with the use of absorbents like diatomaceous earth. As stated previously, these solid substrates can absorb paraffin and asphaltene inhibitors to turn them into solid form.

Lab testing results have shown that the addition of solid paraffin and asphaltene inhibitors had no adverse impact on fracturing fluid performance, and no noticeable characteristic differences between scale inhibitor solid particles and paraffin/asphaltene inhibitor solid particles were observed in the crush and conductivity tests conducted.[24]

There are several reported field case studies on the use of solid inhibitors to control mineral and organic deposit problems after hydraulic fracturing stimulation treatment. One well in South Texas was treated with both paraffin and scale solid inhibitors. After the treatment, the field observation results indicated that this well outperformed the offset wells in the same area. One deepwater well demanded scale, paraffin, and asphaltene inhibition treatment. It was desired to combine all three inhibitors to make a combination product that could be applied to the fracturing fluid at one time. This would not be possible, if three liquid inhibitor products were selected, but viable with solid inhibitors. Three different solid inhibitors were preblended and monitored to ensure the bulk density was suitable for shipping and settling, and treatment was carried out successfully. Long treatment lives were achieved in some cases. Gupta *et al.*[23] reported a production time of 29 months after a solid paraffin inhibitor treatment, with 40,000 bbl of cumulated oil production. For the asphaltene solid inhibitor, the one well was protected for over 6 months with 500,000 bbl of cumulated oil production.

8.3.3 Solid Inhibitor Loading Effect

Field practices have proved that the addition of 1%–2% solid scale inhibitor by weight of proppant would be sufficient to provide prolonged protection longevity. However, with the advances in hydraulic fracturing technology and the developments of unconventional shale oil and gas, it was not uncommon to have more than a million pounds of proppants pumped into a larger horizontal completion well. A fundamental understanding of the effect of solid inhibitor loading rates (from 0.05% to 10%) on inhibitor release behavior is critical to guide future job design and to develop a mathematical model for predicting long-term inhibitor return profile.[25]

In the lab study, the scale inhibitor release rate was determined by packing 20/40-mesh Ottawa sand with a predetermined amount of solid inhibitor into a 35-cm-long stainless steel column (ID = 1.08 cm). The pore volume (PV) of the sand pack is approximately 12 mL. The column was flushed with synthetic brine (0.025 mol/L $CaCl_2$, 0.015 mol/L $NaHCO_3$, 1 mol/L NaCl, sparged with 100% CO_2) at 60°C with a flow rate of 2 mL/min. The effluent solution was collected periodically and analyzed for phosphorus and Ca concentrations in each sample to obtain the inhibitor release curve. At relatively low solid inhibitor loading rates (from 0.05% to 1%), the scale inhibitor concentration in the return fluid gradually decreased, as synthetic brine was pumped through the column, as shown in Figure 8.4. As expected, the inhibitor return concentrations increased with the increase in solid inhibitor loading. This suggested that a longer inhibitor protection lifetime could be achieved by increasing solid inhibitor loading rates in the proppant pack, as the loading was less than 1%. A similar trend was observed as the solid inhibitor loading was further increased from 1% to 3% (Figure 8.5). Assuming that the MIC is 1 ppm, the scale inhibitor protection could last for approximately 1,800 PVs of return fluid, when a 2% solid scale inhibitor is loaded. However, only marginal increases in the scale inhibitor long-term return concentration were observed as the solid inhibitor loading was increased from 3% to 5%. As the solid inhibitor loaded at more than 5%, the further increase of loading rates had negligible effects on the inhibitor's overall return concentrations, as shown in Figure 8.6. The elution data indicated that the increased solid inhibitor loading rates as a percentage of the total proppant pack resulted in a significant increase in protection time but only a slight increase in inhibitor return concentration (Figures 8.5 and 8.6). Intuitively, it was expected that the inhibitor return concentration would increase in the effluent with little to no increase in treatment life as a result of the increase in solid inhibitor loading rate. This unexpected finding provided very useful information for fracturing treatment design.

In the field application, several solid inhibitor treatments were used in a producing field in California. The fracturing treatment job sizes were similar for all wells, but the amount of solid inhibitor in each well was different, based on the projected water production rate. Scale inhibitor residuals in the produced waters were monitored after treatment. The amounts of inhibitor produced were calculated based on the tracked cumulative water production, and a percentage of inhibitor depletion was also tracked. One well treated with only 200 lb of solid inhibitor was nearly 80% depleted after 4 years of production at an average water production of 67 bbl/d. Another well treated with 2,000 lb of solid inhibitor had only 17% of the inhibitor depleted after 3 years at an average water production of 185 bbl/d. By reviewing the inhibitor depletion data for the seven wells treated with 100, 200, 450, 1,500, 2,000, 2,050, and 2,250 lb of solid inhibitor respectively versus production time, a conclusion was drawn that increasing the amount of solid inhibitor in the fracturing treatment could significantly increase the scale inhibitor treatment life. Using a linear regression analysis, a longevity model was developed by the author to predict the inhibitor depletion rate at various solid inhibitor loading rates.

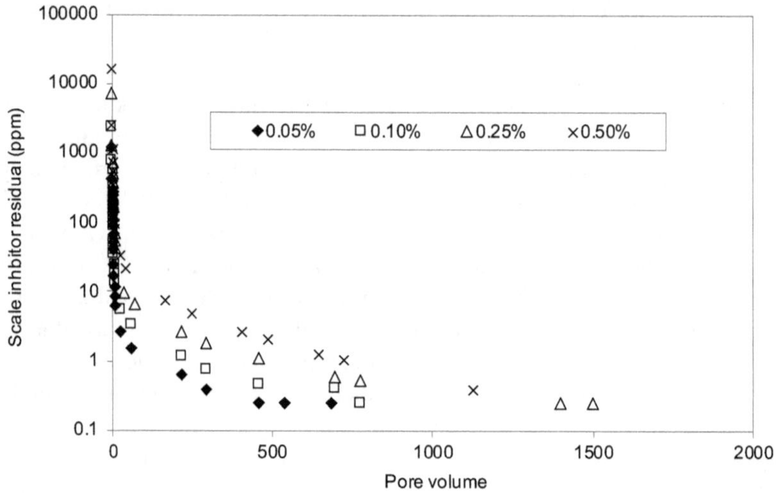

Figure 8.4. Inhibitors return concentrations at low solid inhibitor loading rates (<1%).

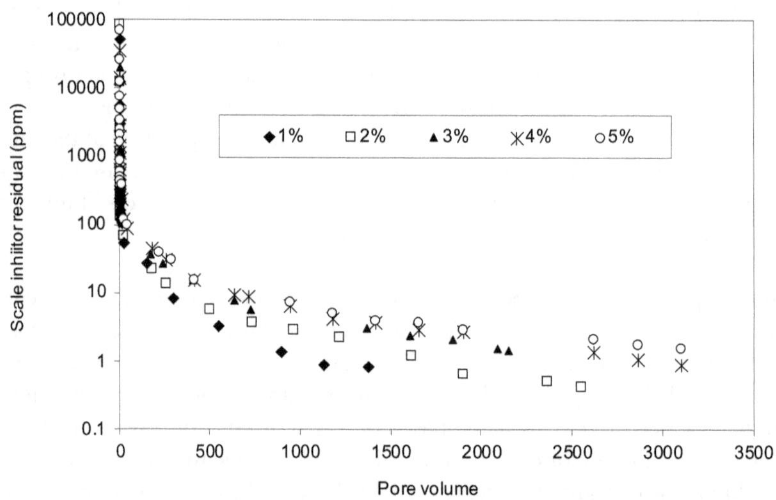

Figure 8.5. Inhibitors return concentrations at medium solid inhibitor loading rates (1%–5%).

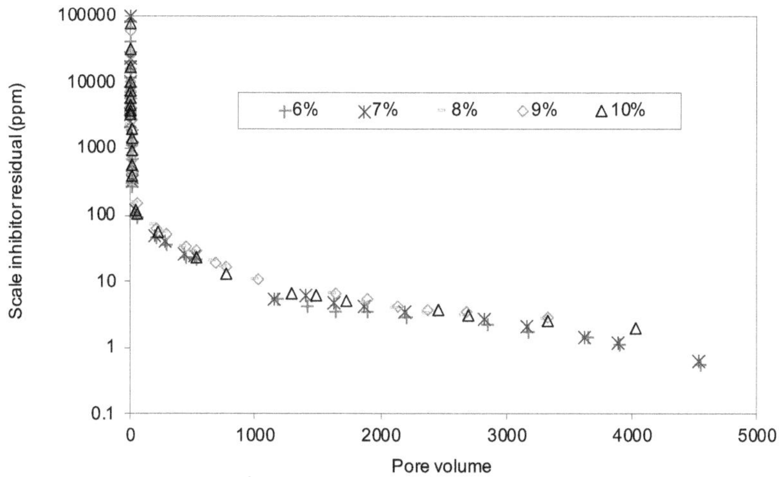

Figure 8.6. Inhibitors return concentrations at high solid inhibitor loading rates (>5%).

Based on this model, the solid inhibitor depletion curve could be calculated at each solid inhibitor loading. Figure 8.7 presents the results of this modeling in a plot of cumulated produced water volume (bbl) versus total pounds of solid inhibitor applied as 80% solid inhibitor was depleted. This curve could be used to predict the inhibitor protection lifetime, based on the amount of solid inhibitor applied in a fracturing treatment. It should be noted that the model is expected to be field-specific. Therefore, not all wells treated with the solid inhibitor have shown the same level of response.

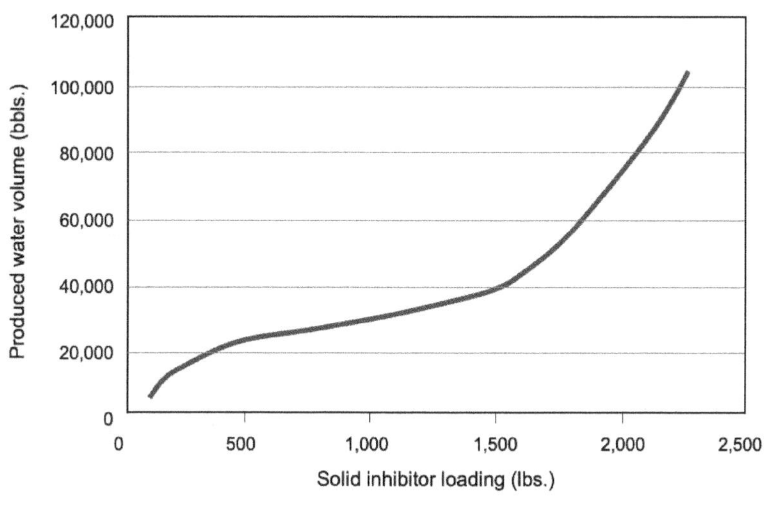

Figure 8.7. Produced water volume to 80% inhibitor depletion vs. amount of solid inhibitor applied.

8.3.4 Environment-Friendly Proppant-Sized Solid Inhibitor

The developments of horizontal drilling and high-volume hydraulic fracturing processes have stimulated the vast potential for oil and gas production within the shale plays throughout the world. Meanwhile, operators and fracturing service companies in the United States have been targeted by federal and state agencies due to increasing citizens' concern about the safety of water resources used in the fracturing process. Therefore, extensive efforts have been made to develop more environmentally acceptable fracturing fluid additives. As a result, a novel biodegradable solid scale inhibitor with an excellent ecotoxicity profile for fresh water was developed and deployed in the Bakken Formation in North Dakota.[26, 27] This new scale inhibitor is prepared by using natural monomers and could be used in food processing. The aquatic toxicity comparison between this new scale inhibitor and a conventional polyacrylate scale inhibitor is summarized in Table 8.4.

Table 8.4: Scale inhibitor aquatic toxicity comparison.

Organism	Parameter	Polyacrylate Inhibitor (ppm)	Environmentally Preferred New Inhibitor (ppm)
Fish	LCO	200	>10,000
Bacteria	EC10		>10,000
Daphnia	ECO		2,000
	EC50	240	5,500

EC10: the concentration that will have an effect on 10% of the population of test organisms.
ECO and EC50: none or 50% of the daphnia were unable to swim after exposure time of 24 h.
LCO: lethal concentration to fish.

Bakken Shale Formation production zones typically are at a depth of 8,000–11,000 ft, where temperatures can reach 240°F with downhole pressures of 5,500–6,000 psi. The water produced in this area generally has high TDS (>200,000 ppm) along with high levels of dissolved Fe^{2+} (200–300 ppm). The presence of such a high concentration of dissolved Fe^{2+} could have a severe adverse effect on the inhibitor performance of conventional phosphonate and polyacrylic scale inhibitors against the carbonate scale.[28, 29] This was further proved during a field practice performed by an operator in this area at the expense of 22 scale-related well failures.[30] Scale inhibitor performance tests indicated that this new chemical was an excellent calcium carbonate and barium sulfate scale inhibitor in the presence of dissolved Fe^{2+} and was compatible with high TDS brine. For a typical Bakken formation, produced water with calcite scale saturation index (SI) of 2.27, 4–5 ppm of this new scale inhibitor should be sufficient to control the formation of scale.

More than 500 wells in this area were treated with this new solid inhibitor after the scale problem in the first set of nine wells was successfully controlled. In most wells, 200 lb of new solid inhibitor per stage was applied to ensure a prolonged protection lifetime. Less than the designed amount of solid inhibitor could be deployed to some wells, depending on an evaluation of carbonate scaling risk. The statistics of a solid inhibitor used for a 460-well pool are summarized in Table 8.5. A total of 389 wells were treated with more than 2,000 lb of solid inhibitor per well. There were 71 of the 460 treated wells that had less than 2,000 lb of solid inhibitor added. Of these, 24 of the 71 wells that were fractured with a solid inhibitor loading less than 1,000 lb.

Table 8.5: Amount of solid inhibitor used in a 460-well pool.

Solid Inhibitor Used for Each Well (lb)	Wells Treated
<1,000	33
1,000-2,000	38
2,000-4,000	91
4,000-6,000	134
6,000-8,000	164

More than 500 monitored inhibitor residual data points were collected from 11 operators with 130 treated wells. Most of the residual data showed that the scale inhibitor flew back at a concentration above MIC (5 ppm), with 23 inhibitor residuals less than 5 ppm. Of these, 11 of the 23 inhibitor residuals less than 5 ppm were collected from a single operator. Over 90% of this operator's wells fractured with less than 1,100 lb of new solid inhibitor received per well. These results suggested that a treated well has a higher chance to produce scale inhibitor at 5 ppm or below after a couple months of production, when lesser amounts of solid inhibitor are deployed. Another operator deployed a large amount of the new solid scale inhibitor (>2,650 lb) per well. A total of 94 scale inhibitor residual data points showed no inhibitor return concentrations less than 5 ppm after 11 months of production.

The cumulative water and oil production were also monitored for 354 wells treated with this new solid inhibitor. Results can be summarized as below:

- 36 wells had produced over 100,000 bbl of cumulated water, and two of these 36 wells produced approximately 180,000 bbl water.
- 18 wells that produced over 200,000 bbl of oil. One well that was fractured with 7,200 lb of the new solid inhibitor has produced approximately 440,000 bbl of cumulated oil.
- 92 wells produced more than 100,000 bbl of oil in 2–3 years of production.

These results demonstrated that tremendous financial benefit could be achieved by using this new solid inhibitor even for wells producing high-scaling-risk water.

8.3.5 Application for Metal Cross-Linked Fluid

Typical scale inhibitors, such as phosphonates or polycarboxylates, are only compatible with borate-cross-linked fracturing fluids. When used with organometallic cross-linking agents, scale inhibitors tend to interfere with fracturing fluid stability and severely impact proppant transport capabilities.[31] This could be caused by the competitive chemical interactions between the scale inhibitors and the metal cross-linking agents.

Fracturing fluids typically contain a cross-linkable, viscosifying polymer and a cross-linking agent. Commonly used viscosifying polymers include guar, guar derivatives, and cellulosic derivatives. Derivatized guar and cellulose can be cross-linked by zirconium over a wide pH range (3.5–10.5) and exhibit tremendous fluid stability at high temperature.[32] This metal cross-linked fracturing fluid is preferred for a stimulation treatment where low-pH or low-quality water, as well as foaming or energizing with CO_2, is used for the stimulation job.

Various commercially available scale inhibitors were tested and proved to be incompatible with Zr-cross-linked fracture fluid at a concentration of 0.25 gptg (gallons per thousand gallons). Phosphonate scale inhibitors were identified as the most incompatible.[33] It was found that two polymer-based liquid scale inhibitors have minimal impact on fluid rheology for the selected Zr-cross-linked fluids. However, the amount of scale inhibitor that can be added without losing fluid stability was still low (well below 250 ppm). Along with the lack of control over the liquid inhibitor release rate during production, long-term inhibitor protection was also not expected.

Further investigation found that the interference between a conventional phosphonate-based scale inhibitor and metal cross-linkers can be mitigated, if a very small amount of liquid scale inhibitor is present in the fracturing fluid. Figure 8.8 illustrates the rheology profile for a carboxymethyl cellulose (CMC)-based zirconium (Zr)-cross-linked fluid with various low-phosphonate scale inhibitor loading rates. There was little or no adverse effect on fluid viscosity at a concentration of 40-ppm scale inhibitor, and a slight viscosity drop was observed, when the inhibitor concentration was increased to 80 ppm. At 120-ppm inhibitor concentration, the fluid viscosity declined from around 600 cP to 300 cP, representing ~50% decrease.

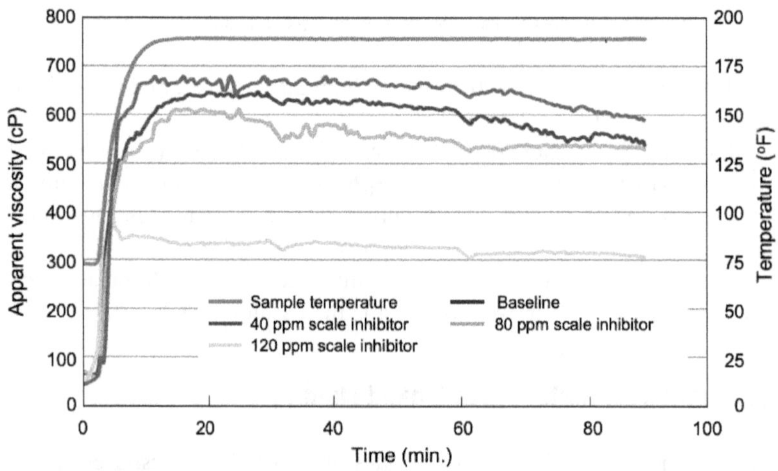

Figure 8.8. The effect of phosphonate scale inhibitor concentrations on the rheology profiles of Zr-cross-linked fluid at 190°F.

Static release tests showed that approximately 130 ppm of scale inhibitor was released at the solid inhibitor loading rate of 5 pptg, if there were no release control agents applied. This number was much higher than the minimum inhibitor concentration (80 ppm) required for a Zr-cross-linked fracture fluid (Table 8.6). However, the inhibitor's initial release rate dropped significantly to around 40 ppm, when release control agents were added. The inhibitor release rate was still below 80 ppm even with 10 pptg of solid inhibitor added. This result indicated that solid inhibitor with a controlled release rate may have little or no compatibility issues with Zr-cross-linked fracturing fluid at a loading rate of 10 pptg or less.

Table 8.6: Solid inhibitor static release tests.

Solid Inhibitor Loading (pptg)	Inhibitor Leaching Concentrations (ppm)	
	Solid Inhibitor	Solid Inhibitor with Controlled Release Rate
5	130	40
10	235	74
20	425	107
50	764	168

Figure 8.9 shows the rheology of one fracturing fluid exposed to a solid inhibitor with a controlled release rate compared to a control fluid at 190°F. As expected, the level of interaction with the cross-linked gel performance was negligible, when the composite was added. Moreover, no changes in gel breaking time were evident confirming that the solid scale inhibitor used was fully compatible with the fracturing fluid formulations.

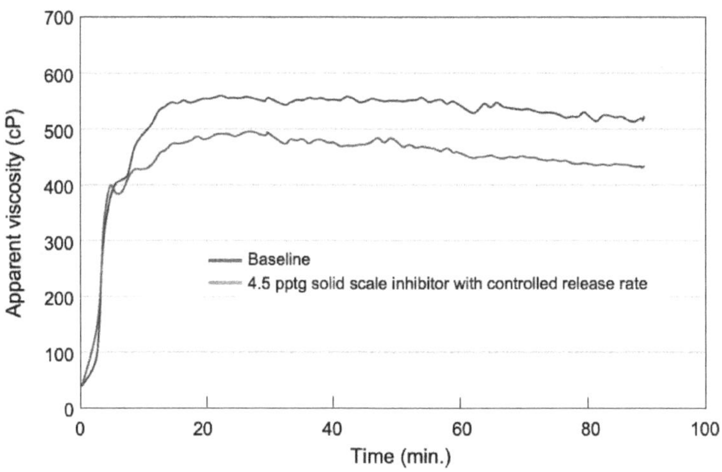

Figure 8.9. Fluid rheology comparison between fluids with and without solid scale inhibitor added.

One field in the North Central/Eastern Colorado region of the United States was experiencing calcite- and barite-related mineral deposition problems in wellbore and production tubing. The problem was sufficiently severe, as it caused the wells to be plugged during production. Hydraulic fracturing operations have been widely used in this area to stimulate the oil and gas production, with 30 stages per well being common. Therefore, adding scale inhibitor into the fracture fluid would be a realistic and cost-effective approach to controlling scale deposition for both short- and long-term production durations. However, in order to lower the gel damage caused by insoluble viscosifying polymer residues, Zr-cross-linked fracturing fluids were preferred in this area. As discussed in the previous section, adding a liquid scale inhibitor to Zr-cross-linked fluid may cause serious incompatibility problems or a limited protection lifetime. As an alternative approach, a new solid inhibitor with controlled release was developed and applied in this area to successfully mitigate fluid compatibility problems and provide a comprehensive mineral scale mitigation solution.

To date, more than 160 wells in this area have been treated with this new solid scale inhibitor. In the field, the solid additive was added into the fracturing fluid with a proportioning screw set at the designated rate. Each stage received a predetermined amount of solid inhibitor. The load rate based on water would be less than 5 pptg, depending on the volume of fracturing fluid to achieve the necessary amount per stage. The average solid inhibitor amount in these applications was 150 lb per stage. No fluid compatibility issues were reported for all the wells treated, though the fracturing operations were performed by different service companies with various metal cross-linked fluids.

The inhibitor return profile suggested that around 50-ppm scale inhibitor was washed out in the produced fluid during the first month of production, and then the inhibitor residuals gradually dropped to approximately 10 ppm after 3 months of production. Laboratory dynamic tube blocking tests suggested that 3 ppm of this solid scale inhibitor would be adequate to effectively prevent the formation of scale in this produced water. There are several wells which have been operating for more than 11 months, and the inhibitor residuals were kept in the range of 5–10 ppm. These results indicated that at least 6–12 months of inhibitor protection could be achieved at the selected solid inhibitor loading rate.

8.3.6 Application in Prepacked Screen and Rechargeable Annular Space

After the proppant-sized solid inhibitor was successfully used for several years in fracture treatment to provide prolonged scale protection time, it also earned a wider interest for applications beyond fracturing operation. Weirich discussed the possibility of using prepacked well screen to convey the proppant-sized solid scale inhibitor in a gravel packed well.[34] From the perspective of sand control, the introduction of solid inhibitor to the prepacked screen is necessary, as prepacked material should not produce resistance to fluid and should not hold fines. From the perspective of scale control, the prepacked screen needs to be designed to maximize the amount of solid scale inhibitor loaded to obtain long-term inhibitor protection. Lab small scale screen testing showed that the pressure response reading was far below the detectable limit, when carrier fluid was injected through a solid inhibitor prepacked screen. The screen inspection indicated that no plugging of the screen was detected. A field trial was performed in a well in the northeastern part of GOM. The prepacked screen with an enlarged space was successfully installed to allow the placement of 166 lb of solid scale inhibitor. The effectiveness of this treatment remains unclear, as only a very small amount of water was produced over the monitoring period.

Bhaduri discussed the possible application of proppant-sized solid inhibitor in annular space to mitigate the scale problem during production.[35] This work addressed the benefits of the placement of a proppant-sized solid inhibitor into the annular space between the production tubing and formation in a long horizontal open hole well. An appropriate inhibitor chemical was first selected to treat the produced water after a comprehensive scaling analysis was performed. A proppant-sized solid scale inhibitor was then prepared by absorbing the chemical onto diatomaceous earth particles, and multiple surface modifications were made to adjust the inhibitor release rate.

The solid inhibitor column elution tests show the scale inhibitor slowly released 30,000 PVs of return fluid at the end of the test. The predicted inhibitor protection would last for over 60,000 PVs, before the scale inhibitor was substantially depleted. Based on this result, approximately 330,000 bbl of cumulated produced water could be protected, if 3,000 lb of this solid inhibitor was placed in the annular space. The protection lifetime could be more than 5 years for a well producing 170 bbl of water per day.

Becker and Gupta claimed that the presence of a high-surface-area diatomaceous earth substrate could allow an inhibitor to be recharged after it was completely depleted.[36, 37] The recharge characteristic of a depleted solid inhibitor that remained in annular space was also evaluated. After being recharged with liquid scale inhibitor, a mixture of 50% diatomaceous earth and 50% Carbolite sand pack could protect at least 16,000 PVs of produced water, whereas the inhibitor protection lasted less than 2,000 PVs, when the diatomaceous earth was replaced by mineral calcite.

8.3.7 Ultra-High-Strength Proppant-Sized Solid Scale Inhibitor

The financial and operational benefits of establishing long-term scale treatment with a single operation have been well recognized. Several review articles have been presented to summarize the application of proppant-sized solid inhibitor to extend the treatment lifetime, enhance the production, and reduce the subsequent intervention costs.[22, 23, 27, 38] However, the strength of the solid inhibitors could limit the quantity of inhibitor chemicals added, especially for a deepwater well, where a high closure pressure is commonly applied. The proppant-sized solid inhibitors prepared with diatomaceous earth particles are not designed to resist high closure pressure. Therefore, a new type of proppant-sized solid inhibitor, which could stand up to 16,000 psi stress, was developed.[39, 40] This ultra-high-strength absorbent substrate can be prepared as follows:

- Mixing aluminum oxide hydrosol and an additive component selected from carbon (such as carbon black) or a high-molecular-weight natural organic material (such as wood flour and starch) which is insoluble in aqueous solutions up to a temperature of 50°C and carbon with a solution of hydrolyzable base to form a mixture.

- Introducing the mixture in dispersed form into a water-immiscible liquid having a temperature of about 60–100°C, whereby gel particles are formed.
- Aging the gel particles in the liquid at that temperature and subsequently in an aqueous base, such as an aqueous ammonia solution.
- Recovering the aged particles, and then calcining the recovered particles. During calcination, the additive component is removed. The calcined particles have a lower bulk density, when the additive component is present during calcinations compared to when the additive component is not present. Combustion of the additive component during calcinations of the hydrosol results in the formation of pores in the calcined aluminum oxide.

The prepared ultra-high-strength solid inhibitor substrate has a 30-mesh spherical shape (Figure 8.10) and a very high surface area and can absorb up to 20% of active inhibitor chemicals. The results of the crush test at 10,000 psi showed that only a very small fraction of the crushed sample (3.2%) was smaller than 60-mesh and could be described as fine. Conductivity testing results indicated that a slight conductivity drop was recorded as the closure pressure reached 12,000 psi (Figure 8.11).

The first field trial was performed with a GOM subsea well in 1,700 ft deep of water to prevent the barium sulfate scale. This well was fractured with a total of 14,881 lb of proppant including 10% loading of a high-strength solid inhibitor (1,488 lb) under 6,893 psi at 199°F.[41] This treated well was put in production right after the completion. All production data indicated the job was completed successfully. The initial oil production was 636 bbl/d.

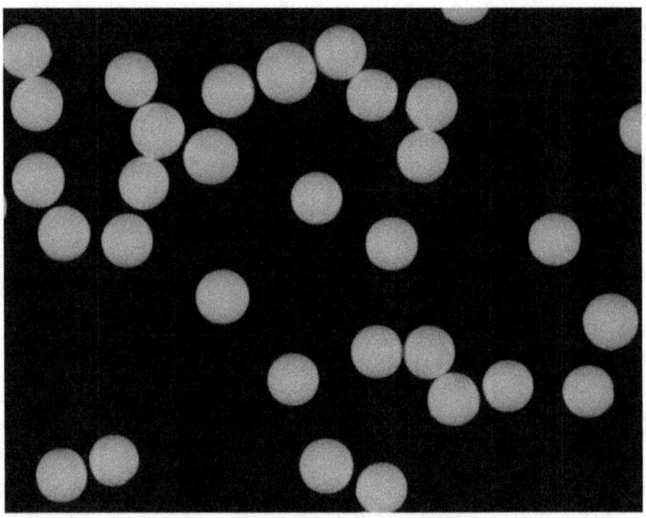

Figure 8.10. Photograph of 30-mesh ultra-high-strength solid inhibitor substrate.

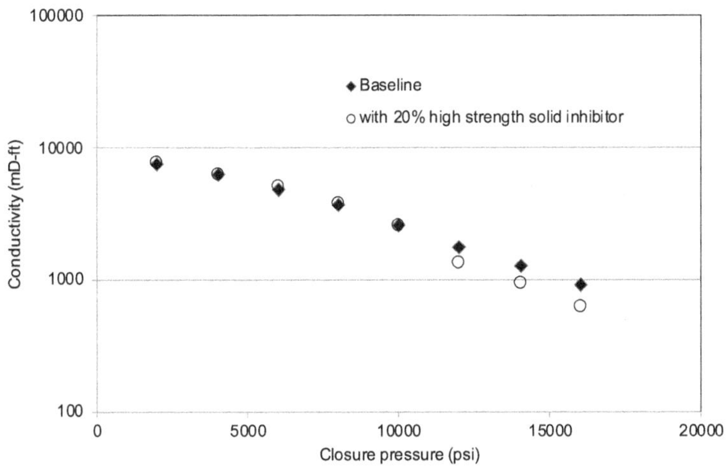

Figure 8.11. Conductivity of the addition of 30-mesh ultra-high-strength solid inhibitor substrate to high-strength bauxite proppant.

Shen reported the deployment of this high-strength proppant-sized scale inhibitor under very hash downhole conditions (420°F, 13,500 psi).[42, 43] A total of six wells located in Eastern offshore basin of India were treated to protect wellbore and production tubing from anhydrite deposition. Zr-cross-linked fluid was used to maintain the fracturing fluid's stability at such high downhole temperature and pressure. As discussed in Section 10.3.5, most scale inhibitors are not compatible with Zr-cross-linked fracturing fluid. Therefore, adding a liquid scale inhibitor was not a viable option to achieve long-term inhibitor protection. Various surface modifications were made to control the scale inhibitor flush release from the solid absorbent, and hence allowed the metal cross-linker fracturing fluid to take up to 480 pptg of this solid inhibitor. In the field application, a total of 200,000 lb of proppant was fractured for each well with an average operation pressure of 9,000 psi. The solid inhibitor was placed at a rate of 2% weight ratio to the proppant, which represented 4,000 lb loading per well. The wells produced gas with a very small amount of water after treatment, the overall inhibition performance would be evaluated as more water was produced from the reservoir.

In another field application performed in two GOM deepwater wells, the solid inhibitor residuals were continuously monitored for 17 months of production.[44] The high-strength solid scale inhibitor used was a phosphonate-based inhibitor. Lab dynamic release testing results showed that this solid inhibitor could generate a nearly constant long-term release rate profile after multiple release control techniques were applied. 8% solid inhibitor by weight loading in proppant pack could protect at least 10,000 pore volumes of elution fluid. Four zones in these two 6,500-ft wells were fractured with this high-strength solid inhibitor at a formation closure pressure of 7,500 psi. The solid inhibitor used was 13,000 lb per well, which represented a 7% weight ratio to the proppant. After being shut-in for several months, the treated wells started to produce. The scale inhibitor was released at 70–100 ppm in the first 3 months of production, and then gradually decreased to 20 ppm or below. After 17 months of production, the scale inhibitor still returned at a concentration well above the MIC level (6 ppm), and the cumulated water and oil production reached 1.1-MM bbl and 7.8-MM bbl, respectively.

8.4 Other Solid Inhibitors for Hydraulic Fracturing

Given that the combined application of solid inhibitor in fracturing fluid has brought significant benefits, extensive efforts have been made to develop various new solid inhibitors.

Based on a similar approach to preparing inhibitor solid in nanoscale by esterifying polymeric scale inhibitors with an alcohol-based chemical,[45–47] a proppant-sized cross-linked polymeric controlled release particle was developed by Paul and others.[48, 49] It was claimed that this cross-linked polymeric particle could continuously release scale inhibitor for months at high temperatures, and no additional surface coating techniques were required. After a more complicated spray drying and roll compaction process, the newly prepared cross-linked polymeric particle could resist 4,000-psi closure pressure. By varying the formulation and reaction conditions, three products were prepared for low-, medium-, and high-temperature application to overcome the temperate sensitivity for a single product.

Also, a new nonsubstrate solid inhibitor was developed for hydraulic fracturing application by Tang *et al.*[50] This solid inhibitor is capable of incorporating various active components, such as scale inhibitor or paraffin inhibitor, but does not need to infuse the inhibitor chemicals into a porous substrate. Lab data indicated that the solid scale inhibitor had a comparable long-term release profile to other low-strength solid inhibitor products.

In addition, a sand-type encapsulated spheres solid scale inhibitor was developed and put into field trial in the Permian Basin field. The product was specially formulated to provide long term inhibitor protection. It was prepared by infusing a phosphate scale inhibitor into a high-surface-area amorphous silica substrate, followed by an encapsulation technique to control the release rate.

8.5 Encapsulated Scale Inhibitor for Batch Treatment

Compared with the placement of solid inhibitor into formation via hydraulic fracturing process, the addition of release-controlled solid inhibitor directly into the wellbore or production tubing has a much longer treatment history. It was first reported by Sloat in 1960, when a controlled solubility phosphate scale inhibitor pellet was developed and deployed for long-term scale protection.[51] One year of protection with cumulate water production of 18,000 bbl was achieved in a treated well. Several field applications have been reported worldwide, from Meddle East,[52, 53] North Sea,[54] to South America.[55]

A common solid scale inhibitor used inside wellbore includes liquid scale inhibitor and weight agent surrounded by encapsulating polymer for release control (Figure 8.12). In most cases, the weighted solid scale inhibitor can be pumped into the wellhead and allowed to settle down to the rathole by gravity after being mixed with a carrier fluid. The rathole is the dead space between the bottom of the well and below the perforation and pump. To reduce the risk of blocking the perforation area, only 80% of the free rathole space was allowed for solid inhibitor placement (Figure 8.13). The 20% "head space" was left to allow the active scale inhibitor to leach out to its optimum level before encountering the produced fluids. The produced fluids from production interval reaching wellbore could lead to turbulence at the top of the "head space." As a result, an inhibitor concertation gradient could be established, because the inhibitor concertation at the top of rathole was diluted by the produced fluid. The dilution of the inhibitor on the top upsets the inhibitor dissolution equilibrium and encourages the capsules to release more inhibitor to regain the equilibrium. Hence, a continuously releasing scale inhibitor into the produced water was achieved.

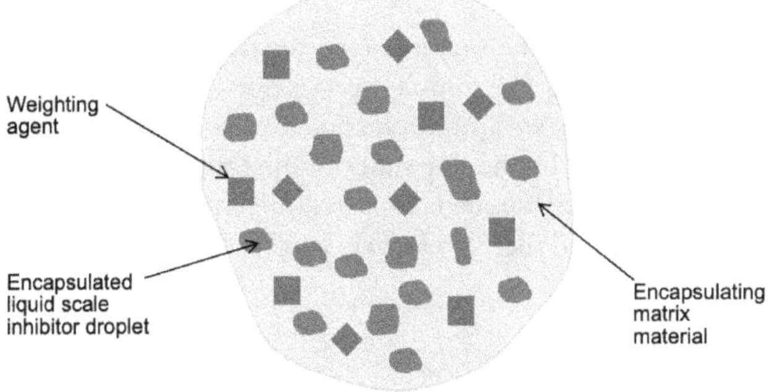

Figure 8.12. Illustration of encapsulated liquid inhibitor structure.

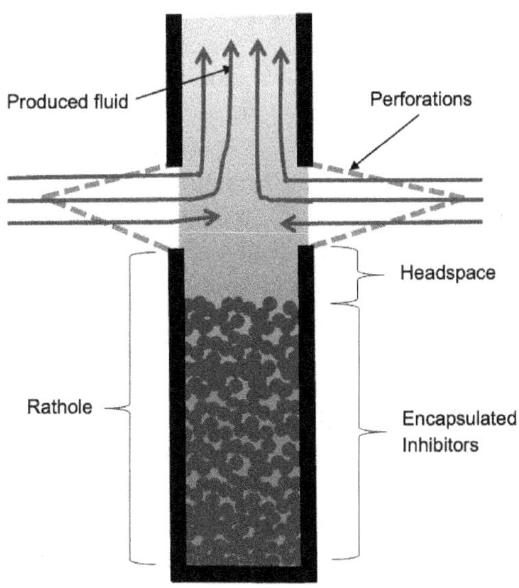

Figure 8.13. Deployment of encapsulated liquid inhibitor product in well rathole.

Besides the placement of encapsulated product into the rathole, the industry also attempted to pack this encapsulated inhibitor into a basket to hang right below the electric submersible pump inside the production tubing to ensure the scale inhibitor could slowly release into the target area to prevent mineral scale-related pump failure. However, no field application case has been published yet.

Gupta and Shen reported the development of shaped compressed solid scale inhibitor pellets which could be used to treat a deviated well such as a horizontal well or well with multiple deviations like an "S" shaped completion.[56, 57] Such solid inhibitor was prepared by compressing a mixture of proppant-sized solid scale inhibitor with a binder to the desired shape (sphere is preferred). These shaped pellets can be dropped directly into the well from the wellhead, since they easily pass through the restrictions in the wellbore and flow into low points of the horizontal well or past obstruction in a deviated well.

Adding encapsulated inhibitor to the wellbore for scale control does no damage to the formation. However, this may lead to the inhibitor protection only being limited to production tubing and around perforations, with no protection available inside the formation. Moreover, the job design is restricted by the free space of the rathole, which may lead to insufficient solid inhibitor loading being applied. This treatment process is most suitable for wells with a low-scaling tendency, and then a relatively low MIC is required for scale control.

8.6 Final Remarks

The financial and operational benefits of placing solid scale inhibitor particles into formation through a hydraulic fracturing process have been well established. Instead of two separate treatments, fracturing and inhibitor squeeze can be combined into one process to save operation time and cost. Also, the solid inhibitor can be evenly placed throughout the frac-pack to ensure inhibitor protection is available in all the producing zones. The continuous liquid scale inhibitor injection treatment through a capillary string can only protect production tubing and annular. Although squeeze treatment can push the liquid scale inhibitor into the formation near the wellbore area, conventional bullhead pumping cannot guarantee the placement of liquid scale inhibitor into water-producing zones, because of the unfavorable pressure regimes in the long-reach-horizontal-, multilateral-, and multistage-type well. A squeeze treatment equipped with coiling tubing and packers, setting of bridge plugs, or flow diversion to improve the liquid inhibitor placements significantly increase the operation difficulty and cost.

The solid inhibitor release rate can be well controlled via surface modification or encapsulation to achieve a prolonged treatment lifetime, when compared to liquid squeeze treatment. There may be some arguments regarding this point, because long-term scale inhibitor protection does not only depend on the inhibitor release rate but also on the quantity of inhibitor that can be spent. For a large, land-based, multistage horizontal well, the hydraulic fracture job size can easily reach over million pounds of proppant with millions of gallons of fracturing fluid. Meanwhile, the water production rate is just in the low or medium range (less than 500 bbl/d), when the treated wells start to produce. This will ensure that a sufficient solid scale inhibitor could be placed into the formation without compromising the conductivity and permeability of the proppant pack. In a typical field application, 200–400 lb of loaded solid inhibitor per stage, which is less than 10 pptg in the fluid, is a common practice. By adjusting the solid inhibitor loading, it is not impossible to achieve complete protection during the lifetime of a production well. However, for offshore application in typical GOM completions, the fracturing job size is relatively small, and the total proppant loading is generally less than 200,000 lb. As the offshore well starts to produce, the water production can easily reach several thousand barrels per day. Though various efforts have been made to increase the active inhibitor loading in an offshore fracturing operation, the treatment longevity would still be restricted by the frac-pack size. A subsequent continuous injection or squeeze treatment is often required. Nevertheless, the placement of solid scale inhibitor in a subsea well can still provide sufficient protection during the early stages of production. This could allow the operators to have enough time to evaluate the producing well's scaling issue and conclude a thorough treatment strategy.

Since a well may have a scaling problem after the hydraulic fracturing stimulation, the benefit of adding scale inhibitor into the fluid is undoubtable. The remaining question is whether to add a liquid or solid inhibitor. In some applications, solid inhibitor should be chosen for prolonged protection. For example, as a metal cross-linked fracturing fluid was deployed, flush release-controlled solid inhibitor is a much better option, due to the incompatibility of liquid inhibitor with the fracturing fluid at high loading. Besides, only very few scale inhibitors are compatible with the cationic friction reducer in a slickwater treatments,[58, 59] and thus certain types of encapsulated solid scale inhibitor might be selected to address a specific scaling circumstance, where available liquid inhibitors could not offer effective inhibition.

However, in many cases, liquid scale inhibitor can compete with solid inhibitor, because it is less expensive. Although the liquid scale inhibitor placed in frac-pack generally releases quickly compared with solid inhibitor, this can be offset by increasing the liquid inhibitor loading rate in some applications. Therefore, one of the challenges to solid scale inhibitor application is to reduce costs without sacrificing the release control. Regardless, the balance between cost and release control must be considered during solid inhibitors' development.

The scale inhibitor release characteristics can be controlled through adjusting the intrinsic properties of the chemical, such as volubility, dissolution rate, or as a product of a chemical reaction initiated by a change of environmental temperature or pressure. Alternatively, some polymer materials, such as polyvinylidene dichloride or epoxy resin, can be used to coat around the chemical as a diffusion barrier to control the chemical release. Other substances, which tend to gradually dissolve or degrade over time, can also be applied as coating materials.

Nomenclature

bbl	Barrels
CMC	Carboxymethyl cellulose
DETPMP	Diethylene triamine-penta(methylene phosphonic) acid
ft	Feet
GOM	Gulf of Mexico
pptg	Pounds per thousand gallons
lb	Pounds
mD	Milli-Darcy
MIC	Minimum inhibitor concentration
SIIG	Scale inhibitor impregnated gravel
TDS	Total dissolved solids
wt	Weight
Zr	Zirconium

References

1. D. Shen, P. Zhang, A.T. Kan, et al., "Control Placement of Scale Inhibitors in the Formation With Stable Ca-DTPMP Nanoparticle Suspension and Its Transport in Porous Media," SPE-114063-MS, SPE International Oilfield Scale Symposium, held May 28–29, 2008 (Aberdeen, UK).
2. M. Norris, D. Perez, H.M. Bourne, et al., "Maintaining Fracture Performance Through Active Scale Control," SPE-68300-MS, SPE International Symposium on Oilfield Scale, held January 30–31, 2001 (Aberdeen, UK).
3. P.J.C. Webb, T.A. Nistad, I.R. Collins, "Economic and Technical Features of a Revolutionary Chemical Scale Inhibitor Delivery Method for Fractured and Gravel Packed Wells: Comparative Analysis of Onshore and Offshore Subsea Applications. SPE-39451-MS," SPE International Symposium on Formation Damage Control, held February 18–19, 1998 (Lafayette, LA, USA).
4. R.J. Powell, R.D. Gdanski, M.A. McCabe, "Controlled Release Scale Inhibitor for Use in Fracturing Treatment," SPE-28999-MS, SPE International Symposium on Oilfield Chemistry, held February 14–17, 1995 (San Antonio, TX, USA).
5. R.J. Powell, R.D. Gdanski, M.A. McCabe, "Encapsulated Scale Inhibitor for Use in Fracturing Treatments," SPE-30700-MS, SPE Annual Conference and Exhibition, held October 22–25, 1995 (Dallas, TX, USA).
6. I.R. Collins, "Scale Inhibitor Impregnated Particles—Field Applications," IBC Conference Solving Oilfield Scaling, held January 22–23, 1997 (Aberdeen, UK).
7. M. Norris, D. Perez, H.M. Bourne, et al., "Maintaining Fracture Performance Through Active Scale Control," SPE-68300-MS, International Symposium on Oilfield Scale, held January 30–31, 2001 (Aberdeen, UK).
8. H.M. Bourne, P.A. Read, "Well treatment," U.S. Patent 5964291, 1995.
9. O. Selle, F. Haavind, M.H. Haukland, "Downhole Scale Control on Heidrun Field Using Scale Inhibitor Impregnated Gravel," SPE-130788-MS, SPE International Conference on Oilfield Scale, held May 26–27, 2010 (Aberdeen, UK).

10. P.J.C. Webb, T.A. Nistad, B. Knapstad, et al., "Revolutionary New Chemical Delivery System for Fractured, Gravel Packed and Prepacked Screen Wells," SPE-38164-MS, SPE European Formation Damage Conference, held June 2–3, 1997 (Hague, The Netherland).
11. P.J.C. Webb, T.A. Nistad, B. Knapstad, et al., "Economic and Technical Advantage of Revolutionary New Chemical Delivery System for Fractured and Gravel Packed Wells," SPE-38548-MS, Offshore Europe Conference, held September 9–12, 1997 (Aberdeen, UK).
12. A.M. Fitzgerald, L.G. Cowie, "A History of Frac-Pack Scale-Inhibitor Deployment," SPE-112474-MS, SPE International Symposium on Formation Damage Control, held February 13–15, 2008 (Lafayette, LA, USA).
13. R.J. Duenckel, J.G. Leasure, T. Palisch, "Improvements in Downhole Chemical Delivery: Development of Multifunctional Proppants," SPE-168605-MS, SPE Hydraulic Fracturing Technology Conference, held February 4–6, 2014 (The Woodlands, TX, USA).
14. J.G. Leasure, R.J. Duenckel, J. Hebert, "Effective Scale Prevention Using Chemically Infused Proppant—A Uinta Basin Case History," SPE-173792-MS, SPE International Symposium on Oilfield Chemistry, held April 13–15, 2015 (The Woodlands, TX, USA).
15. J. Hehert, J. Leasure, "Pioneering a New Method to Prevent Scale in Deep Water, High Pressure Reservoirs—A Gulf of Mexico Case Study," SPE-179882-MS, SPE International Conference on Oilfield Scale, held May 11–12, 2016 (Aberdeen, UK).
16. J. Leasure, T. Lieng, B. Hicks, et al., "Multi-Year Scale Inhibition Using a Ceramic Proppant Based Particulate Chemical Carrier—Learning From the First Application Offshore Congo," SPE-200686-MS, SPE International Oilfield Scale Conference and Exhibition, held June 24–25, 2020 (Aberdeen, UK).
17. J. Leasure, M. Marotz, "Case Histories of Extended Scale Prevention From Single Treatment of Chemically Infused, Slow Release Media," SPE-199255-MS, SPE International Conference and Exhibition on Formation Damage Control, held February 19–21, 2020 (Lafayette, LA, USA).
18. T. Palisch, J. Leasure, "Proppant Delivered Scale Inhibition—Unconventional Case Histories," URTec-5656-MS, Unconventional Resources Technology Conference, held July 26–28, 2021 (Houston, TX, USA).
19. S. Szymczak, J.M. Brown, S. Noe, et al., "Long-Term Scale Inhibition Using a Solid Scale Inhibitor in a Fracture Fluid," SPE-102720-MS, SPE Annual Technical Conference and Exhibition, held September 24–27, 2006 (San Antonio, TX, USA).
20. D.V.S. Gupta, J.W. Kirk, "Method of Inhibiting or Controlling Formation of Inorganic Scales," U.S. Patent 7491682, 2005.
21. D.V.S. Gupta, J.W. Kirk, "Well Treating Compositions for Slow Release of Treatment Agents and Method Using the Same," U.S. Patent 7493955, 2004.
22. D.V.S. Gupta, J.M. Brown, S. Szymczak, "Multi-Year Scale Inhibition From a Solid Inhibitor Applied During Stimulation," SPE-115655-MS, SPE Annual Technical Conference and Exhibition, held September 21–24, 2008 (Denver, CO, USA).
23. D.V.S. Gupta, J.M. Brown, S. Szymczak, "A 5-Year Survey of Application and Results of Placing Solid Chemical Inhibitors in the Formation Via Hydraulic Fracturing," SPE-134414-MS, SPE Annual Technical Conference and Exhibition, held September 19–22, 2010 (Florence, Italy).
24. D.V.S. Gupta, J.M. Brown, S. Szymczak, "Solid Production Chemicals Added With the Frac for Scale, Paraffin and Asphaltene Inhibition," SPE-119393-MS, SPE Hydraulic Fracturing Technology Conference, held January 19–21, 2009 (The Woodlands, TX, USA).
25. J.M. Brown, D. Shen, D.V.S. Gupta, et al., "Laboratory and Field Studies of Long-Term Release Rates for a Solid Scale Inhibitor," SPE-140177-MS, SPE International Conference on Oilfield Scale, held April 11–13, 2011 (The Woodlands, TX, USA).
26. S. Szymczak, D. Shen, D.V.S. Gupta, et al., "Minimizing Environmental and Economic Risks With a Proppant-Sizes Solid Scale Inhibitor Additive in the Bakken Formation," SPE-159701-

MS, SPE Annual Technical Conference and Exhibition, held October 8–10, 2012 (San Antonio, TX, USA).

27. D. Shen, S. Szymczak, D.V.S. Gupta, et al., "A 4-Year Survey of the Application of an Environmentally Preferred Solid Inhibitor in North Dakota," SPE-170638-MS, SPE Annual Technical Conference and Exhibition, held October 27–29, 2014 (Amsterdam, The Netherlands).
28. G.M. Graham, R. Stalker, R. McIntosh, "The Impact of Dissolved Iron on the Performance of Scale Inhibitors Under Carbonate Scaling Conditions," SPE-80254-MS, International Symposium on Oilfield Chemistry, held February 5–7, 2003 (Houston, TX, USA).
29. D. Shen, D. Shcolnik, R. Perkins, et al., "Evaluation of Scale Inhibitors in Marcellus Waters Containing High Levels of Dissolved Iron," SPE-141145-MS, SPE international Symposium on Oilfield Chemistry, held April 11–13, 2011 (The Woodlands, TX, USA).
30. L.M. Cenegy, C.A. McAfee, L.J. Kalfayan, "Field Study of the Physical and Chemical Factors Affecting Downhole Scale Deposition in the North Dakota Bakken Formation," SPE-140977-MS, SPE international Symposium on Oilfield Chemistry, held April 11–13, 2011 (The Woodlands, TX, USA).
31. D. Shen, W.H. Steiner, F. Debenedictis, et al., "The Development of a Novel Solid Inhibitor Additive and Its Application," MECCFEB 16-7889, 17th Middle East Corrosion Conference, held September 30–October 3, 2016 (Manama, Bahrain).
32. P.C. Harris, A. Sabhapondit, "Chemistry Applied to Fracture Stimulation of Petroleum Wells," SPE-120029-MS, Middle East Oil and Gas Show, held March 15–18, 2009 (Manama, Bahrain).
33. Z. Yue, Q. Fu, N. Lang, et al., "Liquid Scale Inhibitors for Metallic-Crosslinked Gel Fracturing Systems," SPE-169806-MS, SPE International Oilfield Scale Conference and Exhibition, held May 14–15, 2014 (Aberdeen, UK).
34. J.B. Weirich, T.D. Monre, B.B. Beall, et al., "Field Application of Chemically Treated Substrate in Pre-Packed Well Screen," SPE-141054-MS, SPE Middle East Oil and Gas Show and Conference, held September 25–28, 2011 (Manama, Bahrain).
35. S. Bhaduri, D. Shen, D.V. Gupta, "Application of Solid Scale Inhibitor in Annular Space to Reduce Well Intervention Cost," SPE-189489-MS, SPE International Symposium on Formation Damage Control, held February 7–9, 2018 (Lafayette, LA, USA).
36. H.L. Becker, "Squeeze Process for Reactivation of Well Treatment Fluids Containing Diatomaceous Earth," U.S. Patent 7686081, 2008.
37. H.L. Becker, D.V.S. Gupta, "Squeeze Process for Reactivation of Well Treatment Fluids Containing a Water Insoluble Absorbent," U.S. Patent 8336624, 2010.
38. S. Fears, S. Bolton, N. Chollar, "Mitigation of Downhole Scale Deposition Leads to Production Enhancement for a South Texas Eagle Ford Shale Operator," SPE-173616-MS, SPE Production and Operations Symposium, held March 1–5, 2015 (Oklahoma City, OK, USA).
39. D.V.S. Gupta, D. Shen, S. Szymczak, "A Chemical Inhibitor-Infused, High-Strength Proppant Additive for Reservoir Flow Assurance Against Scale Deposition in Wells With High Intervention Costs," SPE-165078-MS, SPE formation Damage Conference and Exhibition, held June 5–7, 2013 (Noordwijk, The Netherlands).
40. D.V.S. Gupta, "Composites for Controlled Release of Well Treatment Agents," U.S. Patent 9574130, 2017.
41. S. Szymczak, D.V.S. Gupta, D. Shen, "Gulf of Mexico Frac-pack With a 10% Loading of an Intermediate Strength, Chemical Infused Proppant-Sized Particles Designed to Provide Long Term Inhibition for Barium Sulfate Scale," SPE-168615-MS, SPE Hydraulic Fracturing Technology Conference, held February 4–6, 2014 (The Woodlands, TX, USA).
42. D. Shen, S. Bhaduri, M. Vo, et al., "A Novel Solid Inhibitor for Anhydrite Scale Control Under Extreme Well Conditions," SPE-184525-MS, SPE International Symposium on Oilfield Chemistry, held April 3–5, 2017 (Montgomery, TX, USA).

43. D. Shen, S. Bhaduri, M. Vo, et al., "The Application of Solid Inhibitor for Calcium Sulfate Scale Control Under High Pressure and Temperature Conditions," SPE-183778-MS, SPE Middle East Oil & Gas Show and Conference, held March 6–9, 2017 (Manama, Bahrain).
44. S. Bhaduri, C. Lowe, D. Shen, "The Application of Solid Scale Inhibitor in Deepwater Completion to Reduce Well Intervention Costs," NACE-2018-10534, Corrosion Annual Conference, held April 15–19, 2018 (Phoenix, AZ, USA).
45. I.R. Collins, S.D. Duncum, M.M. Jordan, et al., "The Development of a Revolutionary Scale-Control Product for the Control Of Near-Well Bore Sulfate Scale Within Production Wells By the Treatment of Injection Seawater," SPE-100357-MS, SPE International Oilfield Scale Symposium, held May 31–June 1, 2006 (Aberdeen, UK).
46. J. Hen, "Method of Increasing Retention of Scale Inhibitors in Subterranean Formations," U.S. Patent 5089150, 1992.
47. Ø. Bache, S. Nilsson, "Ester Cross-Linking of Polycarboxylic Acid Scale Inhibitors As a Possible Means to Increase Inhibitor Squeeze Lifetime," SPE-60190-MS, International Symposium on Oilfield Scale, held January 26–27, 2000 (Aberdeen, UK).
48. F. Paul, K.A. Rodrigues, J.S. Thomaides, et al., "Particles Containing One or More Controlled Release Cross-Linked Active Agents," U.S. Patent Application 0338915, 2014.
49. F. Paul, G. Jeff, H. Kurt, et al., "Cross-Linked Polymeric Controlled Release Particles for Inorganic Scale Inhibitor," SPE-193557-MS, SPE International Conference on Oilfield Chemistry, held April 8–9, 2019 (Galveston, TX, USA).
50. C. Tang, R. Navarrete, M. Asadi, et al., "Novel Slow-Release Particle Technology for Use in Hydraulic Fracturing Applications," SPE-195920-MS, SPE Annual Technical Conference and Exhibition, held September 30–October 2, 2019 (Calgary, Canada).
51. B. Sloat, "Controlled Solubility Phosphates—Versatile Solution to Oilfield Scale Problems," *Journal of Petroleum Technology* 12 (1960): pp. 30–36.
52. J.S. Al-Thuwaini, B.J. Burr, "Encapsulated Scale Inhibitor Treatment," SPE-37790-MS, SPE Middle East Oil Show, held March 15–18, 1997 (Manama, Bahrain).
53. J.F. Hsu, A.K. Al-Zain, K.U. Raju, "Encapsulated Scale Inhibitor Treatments Experience in the Ghawar Field," SPE-60209-MS, SPE International Symposium on Oilfield Scale, held January 26–27, 2000 (Aberdeen, UK).
54. H.M. Bourne, S.M. Heath, S. Mckay, "Effective treatment of subsea wells with a solid inhibitor system," SPE-60207-MS, SPE International Symposium on Oilfield Scale, held January 26–27, 2000 (Aberdeen, UK).
55. D. Ruiz, O.F. Torres, M. Jackson, et al., "Application of Encapsulated Scale Inhibitors—Understanding the Field Success Factors," NACE-2017-9527, Corrosion Annual Conference, held March 26–30, 2017 (New Orleans, LA, USA).
56. J.G. Darby, D.V.S. Gupta, D. Shen, "Method of Using Shaped Compressed Pellets in Treating a Well," U.S. Patent 9010430, 2015.
57. D.V.S. Gupta, D. Shen, "Method of Using Shaped Compressed Pellets in Well Treatment Operations," U.S. Patent 9976070, 2018.
58. L. Ke, H. Sun, D. Shen, et al., "Dry Cationic Friction Reducers: New Alternative for High TDS Slickwater," SPE-204286-MS, SPE International Conference on Oilfield Chemistry, held December 6–7, 2021 (The Woodlands, TX).
59. D. Shen, H. Sun, "Scale Inhibitors are Compatible With Cationic or Amphoteric Friction Reducers," U.S. Patent 11479710, 2020.

Scale Control in EOR Processes

Qiongwei Li[1], Tao Chen[2], Xiaolin Wang[1], and Huiying Yuan[1]
[1]Changqing Oilfield Branch Company, PetroChina, Xian, Shanxi, China
[2]EXPEC Advanced Research Center, Saudi Aramco, Dhahran, Saudi Arabia

9.1 Introduction

Conventional oil production or recovery typically contains three main production or recovery stages: primary recovery, secondary recovery, and tertiary recovery.[1–3] Primary recovery is the first stage of crude oil extraction. The primary recovery uses the natural energy of the reservoir itself to push or an artificial lift such as an electric submersible pump or gas lift to pump crude oil to the surface. Generally, the amount of oil that can be extracted from oilfields by using primary recovery is less than 30% of the total reserved oil.

When the natural drive energy of the reservoir is depleted or too low for economic oil recovery, secondary or tertiary recovery techniques can be applied to enhance oil recovery and compensate for declining production. Secondary recovery involves the injection of fluids, usually aquifer water, seawater, or natural gas, into the injection wells to maintain or increase the reservoir pressure and improve the flow rates of oil in the oil production wells. Secondary recovery is often applied at an early stage in the production lifetime of an oilfield, especially when the natural pressure of the reservoir is low. The process mainly depends on physical displacement to recover additional oil. The amount of oil that can be extracted from oilfields by using primary and secondary recovery can reach 30%–50% of the total reserved oil. Both primary and secondary recovery (conventional recovery) target the mobile oil in the reservoir.[4]

After conventional secondary recovery, the residual oil in the reservoir remains in a discontinuous phase in the form of oil drops trapped by capillary forces. In order to prolong the stable production period of oil fields and achieve the highest input-output ratio, tertiary recovery, also called enhanced oil recovery (EOR), is implemented in the middle and late stages of production or when secondary recovery is no longer economic. EOR involves the use of different techniques to change the properties of the oil, the reservoir rock, or flow patterns within the reservoir. It aims to make the oil flow more easily to the well as well as increase reservoir pressure, which requires injecting into the reservoir a substance that will interact with the oil to alter its density or viscosity, reduce the adhesiveness of the rock to the oil (wettability), or plug high-permeability flow paths in the reservoir.[5] Different from primary or secondary recoveries where nature energy or physical displacement mechanisms are applied, EOR is usually induced by the addition of heat, chemical interaction between the injected fluid and the reservoir oil, mass transfer, and/or changing of oil properties in such a way that the process facilitates oil movement through the reservoir.[2] Moreover, EOR targets immobile oil, which cannot be produced due to capillary and viscous forces.[4] EOR recovery methods mainly include (1) chemical injection (such as surfactant flooding, polymer flooding, alkaline water flooding, foam flooding, chemical composite flooding, etc.); (2) gas injection (gas miscible flooding and nonmiscible flooding, including hydrocarbon, carbon dioxide, nitrogen, flue gas, gas-water alternating displacement, mixed-gas water displacement, etc.); and (3) thermal injection (which usually involves injecting steam, hot water, etc.). In recent years, the exploration of microbial flooding and in situ thermochemical/gasification recovery technology has also been carried out for special unconventional resources, such as shale oil and coal rock gas. The choice of recovery technique depends on the characteristics of the reservoir, the degree of depletion, and economic considerations.[2–4] EOR can extract >50% and up to 80% of crude oil from a reservoir.

The oil recovery stages are summarized and shown in Figure 9.1.

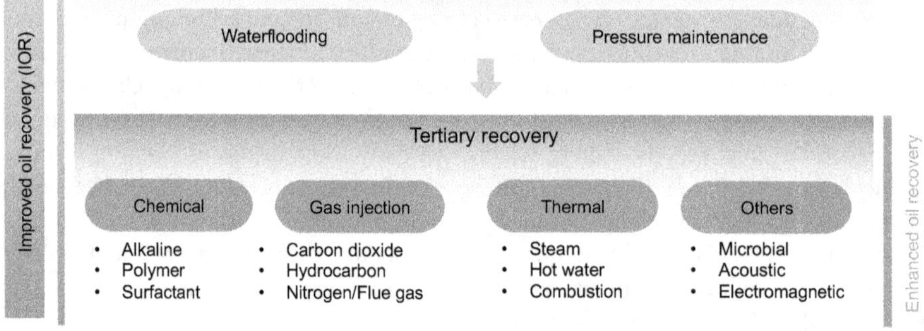

Figure 9.1. The oil recovery stages during oil exploration.

Energy demand across the globe is increasing. EOR is still one of the major recovery methods to improve recovery from a wide variety of fields to meet the rising energy demands globally. It is estimated that there are currently around 400 EOR operating projects globally, producing over 2 million barrels per day of crude oil using EOR, with a trend of increasing to over 4.5 million barrels per day of crude oil using EOR in 2040.[6–8]

EOR is usually induced by chemicals, heat, acid gas, and so on, which will cause changes in the reservoir fluid environment, such as pH, temperature, and fluid chemistry. These changes could trigger the deposition of some types of inorganic mineral scales sensitive to pH, temperature, and so on, for example, calcium carbonate, calcium sulfate, and silica scales. The formation of mineral scales during the EOR process could cause blockage in the near-wellbore regions, downhole completion tubulars, and surface facilities have a negative effect on the efficiency of EOR or even cause the failure of EOR treatment. This chapter aims to provide an overview of oilfield scale formation and mitigation during the EOR, including chemical, gas, and thermal EOR flooding. The mechanisms of inorganic scale deposition and the common mitigation strategies will be discussed in detail, especially during the alkalis-surfactants-polymers (ASP), CO_2, and steam-assisted gravity drainage (SAGD) flooding. The mitigation strategies cover scale removal, scale inhibition, and EOR design to inhibit or prevent inorganic scale deposition. Case studies were also used to provide a holistic understanding of inorganic scale deposition and inhibition in real-specific oilfields.

9.2 Scale Formation and Control in Chemical-Enhanced Oil Recovery Process

9.2.1 Chemical Injection-Enhanced Oil Recovery

Chemical injection-EOR is a process to improve crude oil recovery through the injection of various chemicals, such as surfactant, polymer, alkaline, and foam. The injection of various chemicals, usually as dilute solutions, is aimed at aiding mobility and reducing the surface tension of crude oil. The chemicals are injected into injection wells, and the production occurs in other nearby wells.

Alkaline flooding is a process to add alkaline or caustic solutions, such as sodium hydroxide or sodium carbonate, into injection waters (IWs), which will result in the production of soap that may lower the interfacial tension, reverse the rock wet ability, emulsify, and mobilize the oil to increase recovery.

Surfactant flooding consists of the injection of dilute solutions of surfactants, such as petroleum sulfonates or biosurfactants, to lower the oil-water interfacial tension or capillary pressure that impedes oil droplets from moving through a reservoir and improve the wettability of porous rocks, allowing better displacement of the oil by injected water. Cosurfactants, synergistic agents, and cosolvents are usually added to primary surfactants to improve the stability of the formulation. This reduces the residual oil saturation and improves the macroscopic efficiency of the process. Surfactants may be used in conjunction with polymers and hyperbranched polyglycerols to decrease the interfacial tension between the oil and water.[9]

Polymer flooding has been a long-standing EOR technology since the 1960s, with considerable interest and application since then.[10–15] It consists of mixing long-chain polymer molecules, typically by adding a high-molecular-weight polyacrylamide-based polymer or xanthan gum to the injection brine, to increase the water viscosity, thereby reducing the permeability of water to the porous rock and creating a more efficient and uniform front to displace oil from the reservoir. These polymers are often mechanically degraded to a lower molecular weight on passing through high-shear injection facilities and thermally hydrolyzed to a higher acrylic acid content during transport through the reservoir.[16–20] This method improves the vertical and areal sweep efficiency as a consequence of improving the water/oil mobility ratio and increasing the amount of oil recovered in formations.[21]

Combined with these chemical additives in chemical EOR, it can achieve the synergistic effect of surfactants, alkalis, and polymers to significantly improve the oil driving efficiency.

Binary flooding is mainly the surfactant-polymer composition of the oil repellent system. The use of the surface-active agent greatly reduces the interfacial tension between oil and water and the polymer's high viscoelasticity, and the fluidity control ability to play a role in emulsification at the same time improves the wave coefficient and the efficiency of the oil flooding in the reservoir conditions. Binary flooding has little harm to the reservoir, no dissolution of the reservoir, and little change in the pore structure. After the extracted water is treated, it can be recycled to formulate an oil-driven system.[22]

ASP ternary flooding is generally described as the application of low-salinity brine to displace oil in reservoirs with additional alkalinity, surfactant, and polymer.[23–25] It aims to use the synergistic effect of surfactants, alkalis, and polymers to significantly improve the oil-driving efficiency by reducing the interfacial tension while enlarging the wave volume, thus substantially increasing the oil recovery rate. In APS flooding, the polymer acts as a mobility control agent to increase the viscosity, reduce the mobility ratio, and hence allow a greater volumetric swept efficiency; the surfactant is responsible for reducing the interfacial tension between oil and water phases to a level that promotes the mobilization of trapped oil drops and improves the porous rock wettability, allowing water to flow faster displacing more oil; and the alkaline agent is used to decrease the adsorption of surfactant and polymer on the rock surface and activate indigenous surfactants present in acidic or heavy crude oils or is intended to react with the acids to generate in situ surfactant to overcome the surfactant depletion in the liquid phases due to retention. The alkali (Na_2CO_3 or NaOH) and surfactant act synergistically to widen the range of ultralow interfacial tension and strengthen interfacial activity and stability.[26]

ASP flooding has been proven to increase oil recovery in lab studies, pilot testing, and field deployments. It has been identified as one of the best EOR methods to further recover the residual oil in some regions. It has been broadly applied in the oilfields globally, achieving some important accomplishments that lay a solid foundation for the extension of this technique to practical application in oil fields.[27–39] Despite the promising recovery performance, ASP flooding always comes with risks and challenges. Two major challenges of ASP flooding are the formation of inorganic scales in the near wellbore and in the production system and the formation of tight microemulsions with surfactant breakthrough. These challenges cause injectivity and formation damage, which impacts flow assurance to the production facilities as well as impairing the crude quality.[27] The formation of inorganic scales and mitigation strategies will be discussed in detail in this chapter.

9.2.2 Mechanism of Scale Deposition

In the process of ASP flooding, the formation of inorganic scale could be a major issue in injection wells, production wells, and topside facilities. The typical scale deposits during ASP flooding include calcite (i.e., calcium carbonate, $CaCO_3$), magnesium carbonate ($MgCO_3$), silica (SiO_2), and even magnesium hydroxide ($Mg(OH)_2$) and calcium hydroxide ($Ca(OH)_2$).[24, 40, 41]

ASP flooding can intensify scale issues due to the addition of alkaline chemicals, which can tremendously increase the pH of the water to ~10–12 at surface conditions. When the alkaline agents are injected into the injection wells, various reactions with formation rock and formation water (FW) can take place and be affected by many factors such as the formation temperature, pressure, ions, and pH of the injection system. The high-pH alkaline chemicals could react with divalent cations (Ca^{2+} and Mg^{2+}) in the injection and FWs and primarily form pH-sensitive carbonate scales (i.e., calcium carbonate, $CaCO_3$) in both injection and production wells, along with hydroxide deposition (i.e., calcium hydroxide, $Ca(OH)_2$; and magnesium hydroxide, $Mg(OH)_2$), especially in the injection wells.[42] If the alkaline chemical is sodium carbonate, the carbonate scale issue will be more severe.[39, 40, 42, 43]

The precipitation of $Ca(OH)_2$ and $Mg(OH)_2$ is controlled by the equilibrium shown in Equation (9.1):

$$Ca_{(aq)}^{2+} / Mg_{(aq)}^{2+} + 2OH^- \leftrightarrow Ca(OH)_{2(s)} / Mg(OH)_{2(s)} \tag{9.1}$$

One of the most common mineral scales formed during ASP flooding is calcium carbonate ($CaCO_3$). The origin of the calcium carbonate scale is due to the chemical reaction of calcium ions and bicarbonate ions in the produced water, which may precipitate calcite ($CaCO_3$). The precipitation of $CaCO_3$ is controlled by equilibrium, as shown in Equation (9.2).

$$Ca_{(aq)}^{2+} + 2HCO_{3(aq)}^- \leftrightarrow 2OH^- \leftrightarrow CaCO_{3(s)} + CO_{2(g)} + H_2O_{(l)} \tag{9.2}$$

In the calcium carbonate scale formation system, the equilibrium among CO_3^{2-}, HCO_3^-, and H_2CO_3 is an important thermodynamic factor to drive calcium carbonate deposition.[44] The inorganic carbon equilibrium is shown in Equation (9.3). The pH change at the beginning of scale formation in the system will cause a change in equilibrium, as shown in Figure 9.2.

$$CO_{2(g)} + H_2O_{(aq)} \leftrightarrow H_2CO_{3(aq)} \leftrightarrow HCO_{3(aq)}^- + H_{(aq)}^+ \tag{9.3}$$

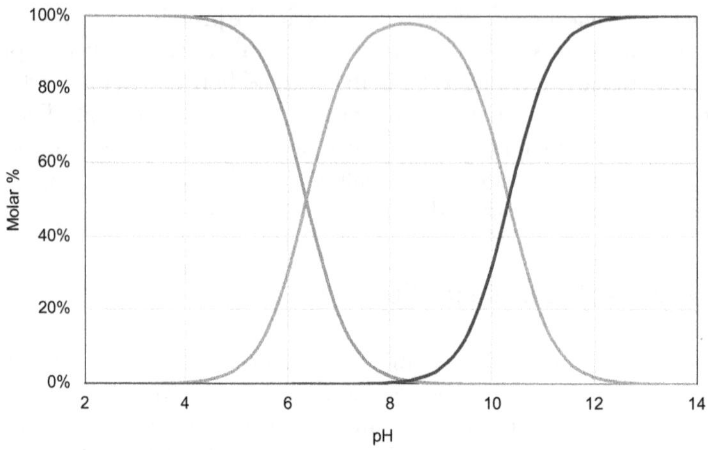

Figure 9.2. Percentage of CO_3^{2-}, HCO_3^-, and H_2CO_3 as a function of pH.

Due to the addition of alkaline chemicals, ASP flooding tremendously increases the pH of the flooding water to ~10–12 at surface conditions and increases the pH of the produced water, when the flooding water breakthroughs and is produced in the production wells. This increase in pH of the solutions will shift the reaction to the right and produce more bicarbonate ions, as shown in Equation (9.3), along with introduced carbonate ions if applying sodium carbonate in ASP flooding, which will increase the supersaturation ratio of calcium carbonate and consequently result in the deposition of calcium carbonate.

Silica scale deposition is another common mineral scale potentially deposited during ASP flooding, especially in sandstone formations. Silica is less soluble in water under ambient conditions, and its solubility in water depends on the pH, temperature, and concentration. The precipitation and dissolution of SiO_2 are controlled by equilibrium,[45] as shown in Equation (9.4):

$$SiO_{2(s)} + 2H_2O_{(aq)} \leftrightarrow H4SiO_{4(aq)} \tag{9.4}$$

The silica compounds are more soluble at high pH (>9) or low pH (<5), elevated temperatures, and low-salinity brines.[23] As the solubility of silica increases under high pH, the high-pH alkaline ASP flooding leads to the dissolution of quartz and clay minerals in the reservoir. The dissolved silica in the ASP flood close to the injection wells, and reservoir is transported to the production wells and could potentially form amorphous silica, magnesium silicate ($Mg_3Si_2O_5(OH)_4$), or magnesium/calcium silicate ($CaMgSi_2O_6$) scales in the production wells and topside facilities, when the production conditions change, such as reduced temperature, mixed with low-pH FW, and changes in water composition. If the FW contains ferrous iron (Fe^{2+}), it potentially forms iron-silicate (Fe3Si2O5(OH)4) scale.[40, 41, 46–50] If the formation rock contains feldspar, illite, kaolinite, and montmorillonite, alkali can also dissolve these components, generate Si and Al ions, and form aluminosilicate scales with the change of the formation conditions. The greater the concentrations of silicon, aluminum, calcium, and magnesium ions in the solution, the higher the reaction temperature and the higher the pH value (pH between 11 and 13.5 in ASP flooding conditions), the more serious the silica scaling.[51, 52]

Moreover, barium sulfate is a type of scale mainly due to the mixing of incompatible waters, one containing barium ions and the other containing sulfate ions, especially with the injection of sulfate-rich seawater into barium-rich FW.[51] This could happen during seawater flooding, followed by ASP flooding. As barium sulfate deposition is a major scale issue due to seawater flooding (a secondary recovery), this chapter will not consider barium sulfate scale deposition as a major concern in the ASP EOR flooding process.

9.2.3 Scale Control Method

Mineral scale deposition is one of the most serious problems encountered during ASP oil recovery. Scale precipitation occurs in the injection well and reservoir, but it is mainly a serious problem in near-wellbore regions, downhole completion tubing in production wells, and surface facilities. Severe $CaCO_3$ or silica scale precipitates can lead to a reduction in rock permeability, formation damage around wellbores, failure of downhole and surface equipment, flow restriction in wells, and ultimately loss of production.[53, 54] Effective scale control is critical in ASP flooding.

Both chemical and mechanical methods can be used to remove the silica and carbonate scales deposited during the application of ASP flooding.

Once silica scale forms, it can be remediated by chemical treatment, either using a complex high-pH fluid containing organic chelates or hazardous chemicals (e.g., hydrofluoric acid). Special cautions should be taken, when the hazardous hydrofluoric acid is applied. Alternatively, silica removal through pH manipulation by shifting the pH of the scale dissolver to above 9 with the addition of chelates can be one of the most common silica removal treatments. However, the high-pH caustic environment will potentially drive the deposition of more $CaCO_3$. Therefore, the remaining scale mitigation task is $CaCO_3$ removal.[55]

Acid carbonate scale removal is commonly applied to remove acid-soluble scale deposition, like calcium carbonate, iron carbonate, and so on. The acids could be hydrochloric acid (HCl), acetic acid (CH_3COOH), formic acid (HCOOH), and so on. HCl is more effective but has unacceptable issues of corrosion, especially under high-temperature applications. The carbonate scale dissolution using CH_3COOH or HCOOH is not as effective as HCl, with the issue of potential re-precipitation, when the environment changes, for example, with a reduced temperature.

Alternatively, chelating agents can be applied to remove $CaCO_3$ scales by chemically binding the calcium cation to form a soluble complex, for example, ethylenediaminetetraacetic acid (EDTA). The reaction of chelating calcium carbonate dissolution is shown in Equation (9.5). The kinetics of carbonate scale removal using chelating agents are not as quick as acid dissolution. However, the chelating agent complexes are stable in alkaline solutions, due to the presence of the chelate bonds.[56, 57]

$$CaCO_{3(s)} + Na_4EDTA_{(aq)} \leftrightarrow CaNa_2EDTA_{(aq)} + Na_2CO_{2(aq)} \quad (9.5)$$

Karazincir et al. studied the feasibility of using complexing agents and scale inhibitors to prevent carbonate scale in ASP flooding.[42] A high concentration of the complexing agent, 11,000-ppm EDTA, effectively inhibited $CaCO_3$ at a pH value of 10.5 in waters with 450-ppm bicarbonate alkalinity and 1,000-ppm Ca^{2+}/Mg^{2+} divalent cations. With the addition of 200 ppm to 500 ppm of phosphonate or polyvinyl sulfonate-based scale inhibitors, the required complexing agent EDTA concentration was reduced to 5,500 ppm. Karazincir also used complexing reagents, like methylglycinediacetic acid (MGDA), together with common scale inhibitors for calcite inhibition. Similar to EDTA, high concentrations of MGDA and scale inhibitors are requested to effectively inhibit $CaCO_3$ under the same conditions.[42] The organic complexing reagents are expensive and will enormously increase the operational cost of the scale treatment.

To reduce the concentration of chemicals and cost applied for $CaCO_3$ scale mitigation generated in ASP flooding, Noraliza et al. conducted research and found a combination of scale inhibitors and synergistic additives to achieve calcite inhibition at around 100 ppm in a scaling water at pH 9.5 with a calcite saturation index of 2.08.[54] Lo studied the $CaCO_3$ scale control, when ASP was injected into a reservoir containing low-salinity/low divalent ion ($Ca^{2+} < 100$ mg/L) FW. They found that penta-phosphonate-based inhibitors could provide effective control under these particular field conditions; however, when the previously injected seawater was also coproduced, scale control was only achieved at limited FW-seawater mixing ratios.[24]

Due to the overwhelming effect of high pH on $CaCO_3$ scale formation and inhibition during ASP flooding, the effect of EOR surfactants or polymers was seldom considered in the previous studies. Although the majority of the injected surfactant and polymer are consumed by reservoir rocks on the way from the injection wells to the production wells, they can still reach the production wells at concentrations much higher than the applied scale inhibitor dosage.

Liu et al. studied the interference of EOR polymers and conventional phosphonate and polymer-based scale inhibitors on $CaCO_3$ scale inhibition.[58] It was found that the EOR polymer can serve as a scale inhibitor in moderate calcite scaling brines by itself, although the required dosage was significantly higher than that of common scale inhibitors. There are strong synergistic effects between EOR polymer and phosphonate scale inhibitors on calcite control, which can significantly reduce scale inhibitor dosage for calcite control in ASP flooding, regardless of whether in severe or moderate calcite brines. The impact of EOR polymers on polymeric scale inhibitors varied depending on polymer types. There was weak synergism between EOR polymer and acrylic copolymer inhibitors, while antagonism was observed between EOR polymer and sulfonated copolymer inhibitors. The synergism is valid at a wide range of EOR polymer dosages, from 100 ppm to 1,672 ppm. The synergism effect also showed the dispersing effect of calcite crystals into small particles with an average size of 1,000 nm in diameter and suspended in the solution without precipitation.[58] Similarly, Beteta investigated the impact of polymer-hydrolyzed polyacrylamide (HPAM) on scale inhibitor diethylenetriaminepenta(methylenephosphonic acid (DETPMP) against $CaCO_3$ inhibition. The performance of DETPMP in a dynamic $CaCO_3$ system was improved in the presence of a low concentration of polymer HPAM with an approved synergistic inhibition effect. The $CaCO_3$ crystals formed with 2.5-ppm DETPMP and 2.5-ppm HPAM have a similar morphology to DETPMP, although more rounded, and some remaining rhombi $CaCO_3$ crystals formed. However, an amorphous solid was formed at high HPAM concentrations (500 ppm), which indicates the synergistic inhibition effect may be dependent on the HPAM concentration at the production well.[59] According to these studies, the impact of EOR polymers on scale inhibitor performance should be considered when selecting scale inhibitors for polymer flooding.

Wang also carried out comprehensive studies to investigate the effects of both the EOR polymer and surfactant on $CaCO_3$ scale formation and scale inhibitor performance.[29, 60] The studies showed that the effectiveness of common scale inhibitor chemistries, such as tri-phosphonate, penta-phosphonate, and polyacrylate, can be significantly deteriorated by the EOR chemicals, with reduced efficacies on the order of 10 times under the test conditions. It could be the EOR chemicals, in particular surfactant, adsorbing on the $CaCO_3$ surface, and then preventing scale inhibitor molecules from reaching and interacting with the active growth sites of $CaCO_3$ crystals. The EOR chemicals may also combine the scale inhibitors to form complexes, poison the scale inhibitors, and then reduce the scale inhibition. The interactions of the EOR chemicals with $CaCO_3$ scale and scale inhibitors are also exhibited by the significant changes in $CaCO_3$ morphology and polymorph from cubic calcite crystals to spheroid, flowering structure nanorod aragonite or vaterite crystals. According to these studies, the tested EOR surfactant and polymer combination could significantly negatively affect the inhibition performance of tested scale inhibitors. Although the actual field concentrations of the EOR chemicals could be much lower than the concentration applied in lab evaluation, the potential interferences of the EOR chemicals on scale inhibitor performance shall be investigated and understood to ensure effective scale control in the chemical EOR field applications.[60]

Scale inhibitor squeeze treatment is widely applied to protect scale formation in the near wellbore, downhole completion tubing, and surface facilities during chemical EOR flooding. Generally, the scale inhibitor is injected into the near wellbore of the production well and shut in to allow for adsorption, where the inhibitor slowly desorbs back into the flowing brine to inhibit scale formation after resuming production.[61] A key aspect of scale squeeze treatment is the adsorption and desorption of the inhibitor on the rock. Any modification to the rock surface may significantly impact scale inhibitor adsorption and desorption and the ability to prevent the formation of oilfield scales. With increasing activity in ASP-EOR floods, consideration must be given to the potential impacts of the produced EOR chemical packages on the scale squeeze treatment.[62]

Beteta et al. investigated the interaction between a scale inhibitor, DTPMP, and an EOR polymer, HPAM, with respect to adsorption and squeeze lifetime.[63] The study demonstrated the negative impacts of polymer-EOR on scale squeezes, the possible mechanisms, and the way to manage these risks. The EOR polymer HPAM can significantly reduce the squeeze lifetime of neutral-pH DTPMP (pH ~7.7) by competing with DTPMP to adsorb on the rock surface. However, the HPAM does not impair the adsorption of DETPMP, when a low-pH DETPMP (pH ~2.2) is applied in core flooding tests, where EOR polymers may be stripped from the rock surface by an acidic fluid and tend to precipitate in acidic conditions.

It concluded that the polymer-EOR could pose a significant risk to effective scale squeeze treatment with a significant reduction in scale squeeze lifetime, if an appropriately low-pH DETPMP inhibitor is not used.[63, 64]

Jordan and Mackay studied the effect of high-pH and low-salinity IW on the performance of scale inhibitor squeeze treatment using the core flood test method.[40] Two core flood tests were carried out with the injection of 10% of a conventional phosphonate scale inhibitor, aminotris(methylphosphonic acid) (ATMP), in a 3 wt% KCl solution at 50°C. Two postflush brines were tested: calcium-containing (36 ppm) FW (pH 7) and non-calcium-containing (0 ppm) IW (IW) with 1% Na_2CO_3 (pH 11). Both the FW and the IW have a similar salinity. Excellent retention for these types of phosphonates was published previously. The flooding with the 1% Na_2CO_3 low-salinity IW at pH 11 causes a very rapid decline in the inhibitor concentration. It appeared that the combination of non-calcium-containing and high-pH IW caused the inhibitor to desorb and dissolve more rapidly relative to the calcium-containing and lower pH FW. The impact of pH on phosphonate retention can be clearly observed, where increasing pH results in less adsorption and more rapid desorption of ATMP scale inhibitors from both quartz and the clay mineral kaolinite. In addition, the brine with no calcium could cause less retention of the inhibitor on the rock. Although it is very unlikely that the pH of IW will still be 11, and the IW contains zero calcium, due to ion exchange or dissolution reactions after flow through the reservoir, the effect of the pH and low calcium content of the water on the performance of the scale inhibitor squeeze treatment should be considered during the field application.[40]

Similar to calcium carbonate inhibitors, the selection and application of silica scale inhibitors in ASP-EOR flooding should consider the interference of the selected inhibitor with ASP-EOR chemicals, including polymers and surfactants, on the inhibition performance and adsorption and desorption characteristics for scale inhibitor squeeze treatment. The detailed silica and silicate scale inhibitor selection and evaluation are presented in Chapter 16.

Except for scale inhibition using effective scale inhibitors, engineering EOR flooding design is an alternative method to mitigate scales deposited during EOR flooding, such as carbonate and silica scales.

Al-Murayri et al. introduced a new overflush ASP-EOR design strategy to alleviate carbonate and silica scaling concerns in ASP-EOR flooding into a sandstone reservoir.[43] To mitigate the scaling in the production wells, the conventional approach is to inject a small pore volume (targeting a voidage replacement ratio of 1) of softened water preflushing prior to the injection of an ASP-EOR chemical slug. The softened water preflush will dilute the high-divalent ion-containing reservoir brine to reduce the risk of incompatibility issues (carbonate scale and hydroxide deposition), especially in injection wells with the high-pH ASP chemicals in injection wells and production wells when ASP slug breakthrough. However, significant carbonate scaling risks at the injection and production wells are still high with the application of such a small volume of preflush, especially for reservoir brines containing high concentrations of calcium ions. In order to reduce and overcome the scaling challenges during ASP-EOR flooding, an extended softened water preflush strategy (overflush) was developed with a large volume of softened water (overflush) injection before the injection of an ASP-EOR chemical slug. The large volume of softened water injection can precondition the reservoir to achieve optimal salinity requirements for the proposed ASP formulation and reduce divalent ion production to a much lower level to significantly mitigate the risk of carbonate scale in the injection wells and production wells when the ASP chemicals breakthrough. Moreover, the large volume of softened water injection will result in significant cooling of the reservoir. The temperature reduction due to overflushing will result in a reduction of dissolved quartz and subsequent silica scale deposition in the production wells. Simulations using various scenarios can be applied to identify the optimal overflush strategy to mitigate the risk of carbonate and silica scale formation at the producers. Therefore, a properly designed overflush can mitigate both carbonate and silica scales in ASP-EOR flooding. An additional consequence of overflushing is that the ASP formulation that was originally designed for a reservoir of high temperature had to be adjusted to a lower temperature as the original formulation could show poor oil recovery performance at a lower temperature. The formulation can be adjusted by varying the surfactant blend ratio, cosolvent concentration, and so on.

Due to the complexity of scale inhibition in EOR flooding, more research is definitely necessary in scale control in ASP flooding conditions, especially in the presence of EOR polymers and surfactants.

9.2.4 Case Studies in China

Daqing oilfields have been deploying ternary ASP flooding since 2010 to improve the crude oil recovery rate at the high water cut stage. When the ASP fluid is injected into the formation, it could react with the FW, interact with mineral rock, and cause the deposition of mineral scales. The scaling issue is worsening, with more than 70% of wells suffering scaling issues in CWD blocks in Daqing Oilfields, for example, pump stickiness, rod failure, tubing blockage, and so on. The scaling issues in the ASP-flooding wells result in a much higher workover operation cycle than in water-flooding or polymer-flooding wells, which seriously affects recovery performance and significantly increases operation costs. It has become one of the bottlenecks restricting the industrialization of the ASP-EOR flooding technology.[65, 66]

The main mineral components in the comprehensive analysis of downhole scale are amorphous silica, hexagonal spherical or common calcite, and so on. The secondary minerals are detrital quartz, feldspar and clay particles, and pyrite. In addition, there are minor trace minerals.

Generally, the composition of scales and formation mechanisms are various during the different stages of ASP flooding, which can be roughly divided into three stages[65, 66]:

1. Early stage of scaling. Scales mostly form at this stage. The growth and accumulation of scale are fast. Carbonate scale deposits first, followed by silica deposition.
2. Middle stage of scaling. The growth rate of the carbonate scale decreases, along with an increased growth rate of the silica scale. Generally, the growth rate of scales at the middle stage is relatively stable and less than at the early stage of scaling.
3. Late stages of scaling. The silica scale content increases. The rate of total scaling (carbonate and silica scales) further slows down.

An example of scaling deposition at each stage is shown in Table 9.1. The majority of scales form at the early stage with the appearance of a gray-brown, loose, flaky plate. The carbonate scale is major and deposits first, followed by silica deposition. The growth and accumulation of scale are fast at this stage, reaching 3–4 mm within 30 days. In the middle stage, the growth rate of the carbonate scale decreases and the growth rate of the silica scale increases, resulting in silica-dominant scale deposition at this stage. At the late stage of scaling, the rate of scaling further slows down, and the silica scale content reaches 70% or more.

Table 9.1. Composition of scale deposition at different stages (%).

Scaling Stage	Scaling Rate	Organic Component	$CaCO_3$	SiO_2	MgO	Al_2O_3	Fe_2O_3
Early	High	15.90	55.30	20.09	0.84	0.25	0.51
Middle	Stable	10.57	16.93	66.96	0.27	0.15	2.96
Late	Slow down	9.90	14.90	70.80	0.62	0.14	1.10

The scaling scenarios are complex. Various compositions of scales deposit in the area of blocks with different physical properties of the reservoirs. Figure 9.3 illustrates the composition of scale deposition in different areas at the early and middle stages. The scale deposited in the M1 area is always a $CaCO_3$-dominated scale from the early stage, where the content of the $CaCO_3$ scale is higher than that of the silica scale.[67] The S5 and LM areas have a $CaCO_3$-dominated scale at the early stage of scaling but a silica-dominated scale at the middle stage of scaling. The content of the $BaSO_4$ scale in the eastern II block of the S6 area and X1-2 area is relatively higher, with the content of $BaSO_4$ being about 20%.

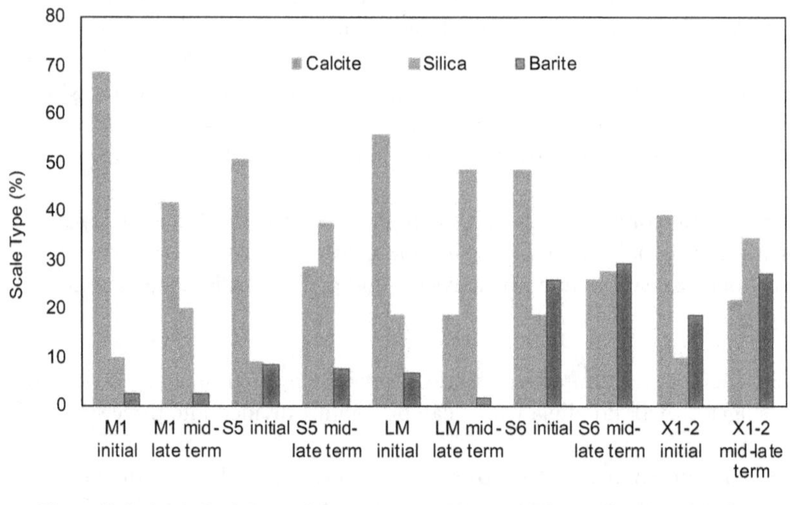

Figure 9.3. Analytical data of scale composition in different flooding blocks.

In the same block with the same reservoir physical properties, the scales were found at different times in the wells. For example, scale deposition was observed earlier in the center wells than in the side wells in the N5 block. The scale was first observed in the center wells in September 2007, compared to the side wells in June 2008. Until September 2008, 75% of center wells and 15.8% of side wells had scaling issues at different levels.

In addition, the distribution of scale deposited in the downhole completion tubing is not uniform. Scaling coupon sections were set in the tubular at different depths to monitor the distribution and changes in scaling. With the increase in depth, more severe scaling is monitored. Much less scale is observed at depths less than 500 m. The most serious scale was observed in the lower part of the pump and the pump barrel. It is assumed that the pressure gradient change and higher surface temperature in the pump make a significant contribution to scale deposition in the pump, especially for the pressure- and temperature-sensitive calcium carbonate scale. The proportion of carbonate scale increases and the proportion of silica scale decreases with depth at the early stage of the scaling. The proportion of carbonate scale decreases, and the proportion of silica scale increases in the middle stage of scaling.

Similar to Daqing oilfields, Shengli oilfields suffered scale issues in the area of L-Island Oilfield after deploying an ASP-flooding pilot test in 1998. A more serious scaling deposition was formed in the wellbore and pump, which significantly affected normal production. The scales deposited in the pump barrel are mainly composed of $CaCO_3$, accompanied by a small amount of SiO_2 (2.65–5.56 wt%).

Scale inhibitors are proposed and deployed to inhibit scale deposition in these ASP-flooded oilfields.[68]

More than 15 candidate scale inhibitors were evaluated at a concentration of 50 mg/L for calcium carbonate inhibition, including polyphosphine, polyacrylic acid, sodium polyacrylate, ethylenediamine tetrakis(methylenephosphonate), inorganic polyphosphonates, polycarboxylic acid copolymer, hydrolyzed polymaleic anhydride, acrylic acid copolymer, aminotrimethylenephosphonic acid, acrylic acid-polyacrylic copolymer, carboxylic copolymer, polyphosphonic acid, polyphosphine carboxylic acid, organophosphonic acid, and phosphine carboxylic acid copolymer. Organophosphonic acid and phosphine carboxylic copolymers were selected as suitable calcium carbonate scale inhibitors and provided the best inhibition performance with inhibition efficiencies of 85.9% and 81.9%, respectively.

As most carbonate scale inhibitors are not effective on SiO_2 inhibition, a series of combined calcium carbonate and silica scale inhibitors were formulated and evaluated. The combinations of organophosphonic acid, phosphine carboxylic copolymer, and a polymeric silica scale inhibitor provided more than 95% and 80% inhibition efficiency against calcium carbonate and silica, respectively. The scale inhibitors were deployed in more than hundreds of wells through squeeze treatment. 6–12 months after scale squeeze treatment, the concentration of scale inhibitor in the produced fluid is relatively constant, and the concentration of scaling ion Ca^{2+} increases and stays stable. The average pump workover cycle increases from less than 30 days to more than 200 days. In Shengli L-island Oilfields, an (hydroxymethyl)phosphonate scale inhibitor was developed and applied in ASP-flooding wells through squeeze treatment in eight wells. Before the use of scale inhibitors, acid treatment was required every 20 days to clean the scale deposition in the downhole completion tubing and pumps. The scale inhibitor treatment prolongs the acid treatment to about 100 days. All these indicate successful field scale inhibition.

9.3 Scale Formation and Control in Gas-Enhanced Oil Recovery

9.3.1 Gas Injection-Enhanced Oil Recovery

Gas injection, either in the immiscible or miscible mode (mainly involving CO_2), is one of the most commonly used EOR approaches to enhancing oil recovery. Gas injection accounts for about 60% of EOR production in the United States. It boosts recovery by 5%–15% of the original oil in place, depending on the miscibility of oil and CO_2.[69]

Miscible gas flooding is an EOR approach to introducing miscible gases into the reservoir, such as CO_2, natural gas, or nitrogen. The flooding fluid and crude oil reach complete mutual dissolution, and the interfacial tension between the two phases is zero. CO_2 (supercritical CO_2) is the most commonly used gas for miscible displacement, as it reduces the oil viscosity, which has a low minimum miscible pressure, has a better oil driving effect, and is less expensive than liquefied petroleum gas.[70] Moreover, by reducing capillary forces and promoting better microscopic displacement, CO_2 could potentially reach residual or non-swept oil that remained in the reservoir after conventional recovery.[71]

Generally, the miscible flooding maintains reservoir pressure and reduces the interfacial tension between oil and gas, which improves the displacement efficiency and results in the enhancement of oil recovery. However, the miscible CO_2 flooding results in reduced oil viscosity, where the less dense phase leads to poor sweep efficiency.[40]

Immiscible gas flooding is an alternative gas flooding approach to mitigate the negative effects of miscible flooding. In immiscible gas flooding, the injection gas and reservoir oil remain in two separate phases.

Foam flooding and water-alternating-gas (WAG) flooding are two major immiscible gas floodings. Foam flooding is injecting foam into a reservoir to improve the sweep efficiency of a driving fluid. The foam can be generated either in the reservoir pore space or at the surface before injection. Foam system has a certain blocking effect on the high-permeability pore throat and expands the gas wave volume. At the same time, due to the high interfacial activity of the foam, it can further improve oil driving efficiency. In WAG, a saline solution and CO_2 are injected into the oil well for larger recovery. WAG flooding is a technique applied to provide better mobility control, reducing the tendency of the injected CO_2 to override the hydrocarbon bearing formation due to its lower density by reducing its tendency to finger past the oil due to its higher mobility (lower viscosity), and promoting a more uniform macroscopic sweep.[71, 72]

CO_2-EOR is a proven technology for improving oil recovery in mature oilfields. It is also a means of carbon capture, utilization, and storage technology and supports a gas reutilization strategy to pave the path for a low-carbon future.[5] However, injection of slugs of CO_2

into reactive rocks under certain thermodynamic conditions can drive mineral dissolution and precipitation processes and cause severe calcium carbonate scaling problems in the production wells.[71]

Scales can potentially deposit in the downhole completion tubing and surface facilities during the CO_2-EOR, when the water breakthroughs in production wells.

9.3.2 Mechanism of Scale Deposition

Calcium carbonate scale is expected to deposit in the production wells, where CO_2 is applied to enhance oil recovery.

CO_2 is miscible with oil and also very soluble in brine, especially at high pressure in reservoirs. The brine is acidified, and the pH of the brine decreases when CO_2 dissolves in it, as shown in Equation (9.2). For example, in CO_2 miscible floods, the pH of the connate water is reduced by the initial dissolution of the gas, making the fluid more acidic at around pH 4.5.[73, 74]

Sufficient acidification of brine near the injection well could result in significant dissolution of carbonate minerals or calcite cements in the near-wellbore zone in both immiscible and miscible CO_2 flooding, as shown in Equation (9.6).

$$CaCO_{3(s)} + 2H_{(aq)}^{+} \leftrightarrow Ca_{(aq)}^{2+} + CO_{2(g)} + H_2O_{(aq)} \tag{9.6}$$

Due to the dissolution of carbonate formation, the pH of brine increases until it reaches around the original pH of brine before CO_2 injection. There is no expected calcium carbonate scale deposition during this stage. However, the mass of dissolved carbonate could reprecipitate out of solution in the production wells, when the pH of the brine increases due to pressure drop and CO_2 released from fluids (as shown in Equation [9.2]).[40, 72, 74]

The calcium carbonate dissolution close to injection wells and precipitation in production wells and surface facilities can be more severe, when the CO_2 concentration in injection gas and WAG ratio increases.[71]

9.3.3 Scale Control Method

Calcium carbonate scale deposition for CO_2 gas flooding is not as severe as in other water injection floodings, as the amount of water produced is minimal and only calcium carbonate scale occurs due to CO_2 evolving from the produced water in the production wells and surface facilities.[71] The scale deposited on the inner surface of the production system seriously affects the normal operation and safety of production wells. The scale mitigation strategies are different depending on the water quality, temperature, wellbore trajectory, and scaling conditions of each oilfield.[75–77]

The selection of a suitable scale mitigation strategy in CO_2 gas flooding can start with a scale risk analysis by assessing scaling tendency using reservoir simulation models coupled with scale prediction models to forecast reactive multiphase flow in heterogeneous reservoirs, wells, and surface facilities at the early stages of their development. During the production stage, the production data, water chemistry, downhole pressures and temperatures, gas production, and so on, are available to validate and match the model applied or scale risk analysis carried out at the early-stage development. This process involves preventive design, validation using field data, and review and decision-making to mitigate the scale deposition in CO_2 gas flooding.[78]

Rodrigues conducted a comprehensive investigation on the scale risk of CO_2 flooding injection and production systems (formation, wellbore, and surface systems), mastering the data and information on scale, analyzing the composition of scale samples and water samples of FW and production fluid according to the theories of ionization equilibrium, thermodynamics, crystalline dynamics, and fluid dynamics. Using a geochemistry model embedded in the reservoir simulation calculations is a powerful tool to understand the complex interactions of the injected fluid, as it travels throughout the porous medium, encountering FW, reactive rock, and hydrocarbon phases. The output of the simulations, such as ion and flow rate distributions, can be used to predict the amount and location of calcite scale deposition that could occur and support the scale management strategy. According to the simulations, the path from the production wellbore to the surface facilities, for example, manifolds, risers, valves, and surface equipment, is the most crucial location for calcite scale deposition, where significant pressure drops, CO_2 releases, and pH increases occur in the produced fluids. In addition, the simulation results show the total calcite scale risk of the WAG injection scheme lies in between two extreme cases: water flooding and gas flooding. The simulation results from reservoir simulation and scale prediction models provide crucial information to identify CO_2-WAG production strategies that minimize scale risk.[78]

The scale mitigation could be achieved through CO_2 gas flooding design. It is desirable to delay CO_2 breakthrough and ideally ensure CO_2 permanent geological storage, which will not only increase oil productivity and mitigate the carbon footprint but also prevent flow assurance hazards such as inorganic scale and corrosion.[71, 78] There are many factors that come into play when designing a CO_2-EOR flood. The displacement can be very complex, because of channeling caused by stratification, segregated flow caused by buoyancy effects, crossflow between strata, permeability changes caused by the dissolution or precipitation of rock, the trapping of oil and gas by mobile water, relative permeability and capillary pressure hysteresis, and so on. The optimal application of miscible flooding schemes in reservoir scenarios can only start to be grasped with reservoir simulation models that comprise at least a few of the many actual phenomena happening in the subsurface. Rodrigues simulated and evaluated the effect of CO_2 concentration in the injection gas stream on calcite scale risk at CO_2 concentrations of 20%, 50%, and 90%. The WAG ratio and slug size constants varied with the CO_2 concentration in the injected gas. As expected, calcite scale risk significantly rose with greater CO_2 injection concentration. The brine is more acidic with a higher CO_2 concentration and partial pressure, which induces more rock dissolution and higher calcium and bicarbonate ions released in the brine. Thus, greater calcite precipitation is expected in production wells and surface facilities, where the pressure is reduced, causing CO_2 to be released and pH to increase in brines. In addition, high concentrations of CO_2 are accompanied by a smaller amount of water throughout the production lifetime. The oil viscosity is lower in high-concentration CO_2 flooding, the relative permeability of oil increases, and there is even less water production. The reduced amount of water was capable of practically doubling the scale damage due to the greater availability of CO_2 in the system. So, the low concentration of CO_2 in a high amount of water can significantly reduce the scale risk in the production well.[71] However, it is in the best interest of operators to determine CO_2-WAG design parameters, such as CO_2 concentration, volume of injection, ratio of water and gas, and so on, that balance the scale risk, profitability of the project, CO_2 emissions, and so on.

Except CO_2 flooding design using reservoir simulation and scale prediction models, conventional scale inhibitor squeeze treatment is still one of the main approaches applied to protect production wells from carbonate scale deposition generated due to CO_2 gas flooding. The scale inhibitor is pumped into the near-wellbore region of the production well, adsorbs on the rock surface, and gradually desorbs and releases into the produced fluid to inhibit scale deposition. The near-wellbore region remains protected until the concentration of scale inhibitors produced falls below the minimum inhibitor concentration. The frequent squeeze treatments also depend on the volumes of water produced. Generally, the higher the volumes of water produced, the more frequent squeeze treatment is. To avoid scale precipitation downstream of the wellbore, continuous SI injection can also be applied through an umbilical in the perforation tubing or surface injection points.[78]

Calcium carbonate scale is the major type of scale deposited during CO_2 gas flooding. Along with the calcium carbonate scale, the barium sulfate ($BaSO_4$) scale could be intensified, if there was any $BaSO_4$ deposition before CO_2 gas flooding.[79] Although the low-pH brine can lead to decreased formation of calcium carbonate, it creates problems in the prevention of non-acid-soluble scales such as barium sulfate.[79–82] Ramsey et al. found the $BaSO_4$ scaling problem suddenly intensified in the CO_2 flooding pilot test.[81] The calcium carbonate scale crystals formed in CO_2 injection flooding can increase the $BaSO_4$ scaling rate by up to 15%. Moreover, $BaSO_4$ scale inhibitors could significantly lose inhibition performance at low pH conditions during CO_2 flooding due to the protonation of the scale inhibitor molecule, preventing its interaction with scale crystals, especially in sandstone reservoirs with a significant pH depression when the carbonic acid forms. Ramsey et al. studied the effect of the pH of the produced water on the $BaSO_4$ inhibition performance of five different scale inhibitors: polyacrylates, 1-hydroxyethylidenediphosphonic acid (HEDP), diethylene triamine pentamethane phosphonic acid (DETAPMPA), triethanolamine phosphate ester, and sodium hexametaphosphate (SHMP). The pH of the produced water was reduced from 7 to 4. Most of the tested barium sulfate scale inhibitors were affected by the pH and showed significantly reduced $BaSO_4$ inhibition performance, except for SHMP. At pH 7.0, all the tested inhibitors gave 80% $BaSO_4$ inhibition at 5 ppm under test conditions. At pH 4.0, DETAPMPA and HEDPA provided inhibition only at extremely high concentrations of 1,250 ppm and 750 ppm, respectively, and SHMP still gave 80% $BaSO_4$ inhibition at 7 ppm. However, SHMP is not widely applicable in oilfields due to its low thermal stability of about 66°C. It appeared to be a relationship between the pK_a of a scale inhibitor and its effectiveness at low pH: the lower the pK_a, the more effective the inhibitor. Thus, scale inhibitors with a low pK_a should be selected for testing before those with a higher pK_a.[81] Chesnut et al. evaluated a low-pK_a polyanionic-based scale inhibitor specific for barium sulfate scale prevention.[83] It appeared that the polyanionic-based scale inhibitor overcame the effect of pH and allowed a favorable interaction with barium sulfate crystals, although it did not work against the calcium carbonate scale. The relationship between the anionic sites of the inhibitor and the geometry of the crystal may play a significant role in determining the effectiveness of scale inhibitors. So, the $BaSO_4$ scale inhibitors to be applied in CO_2 gas flooding should be selected and evaluated under low pH conditions.

9.3.4 Case Studies

In China, Jilin Oilfield carried out WAG flooding pilot tests at the Hei 46 Station and Hei 59 Station to improve the crude recovery rate. When CO_2 is injected into an aquifer, the concentration of scaling ions increases due to the dissolution of carbonate formation during the fluid transport from injection wells to production wells. The scaling and CO_2 corrosion problems were much more severe in the wellbore and surface systems. The scale is mainly deposited at the lower part of the completion tubing in the sections below 1,000 m, 500–1,200 m below the pumping rods, interior of the pump barrel, pump inlet, and so on, mainly due to the sudden drop in pressure, especially in the inlet of the pump. Calcium carbonate scale accounts for 80%–95% of the total solid deposition. The mud and sand in the formation also accelerated the scale formation.

Scale inhibitors, including HEDP, phosphono-butane-1,2,4-tricarboxylic acid (PBTCA), ATMP, and so on, were evaluated with synthesized field water bubbling with CO_2. HEDP inhibitor showed the best calcium carbonate inhibition performance at 10 ppm, followed by PBTCA and ATMP. The pilot tests were carried out in the presence of 10-ppm HEDP using a water source from the field under CO_2 partial pressure of 16 MPa, 20 MPa, 24 MPa, and 28 MPa, respectively. The calcium ion concentration did not decrease before and after the test under low-CO_2 pressure test conditions, indicating effective calcium carbonate scale inhibition under the test conditions.

HEDP was successfully applied in the fields through scale inhibitor squeeze treatment and resulted in a relatively constant concentration of calcium ions in the production water, less scale deposition observed in the downhole tubing, pump and surface facilities, and less frequency of requested acid treatment for scale removal.

9.4 Scale Formation and Control in Thermal-Enhanced Oil Recovery

9.4.1 Thermal Injection-Enhanced Oil Recovery

Thermal oil recovery is an EOR method to introduce heat into a reservoir for the recovery of heavy or extra-heavy oil (API gravities less than 20) through oil mobility improvement by viscosity.[84] These kinds of heavy oils cannot flow toward producing wells unless their viscosity is reduced enough by heating. Thermal recovery can mainly be subdivided into two types: hot fluid injection, such as steam injection and hot waterflooding, and in situ combustion processes (fire flooding).

Steam flooding, cyclic steam drive, or SAGD is a thermal recovery method that involves injecting steam generated at the surface or in situ into the reservoir through injection wells near the base of a heavy oil reservoir. The steam injected into the reservoir can heat up the heavy crude oil, physically reduce crude viscosity, and generate an artificial driving force to sweep oil toward production wells. SAGD is a combined drilling technique and thermal in situ heavy oil recovery process used to extract heavy crude oil. A typical SAGD setup is a parallel pair of wells (the injection and production wells) that are drilled horizontally and spaced apart vertically, with the injection well on top.[55] It could be twice as efficient as cyclic steam stimulation, allowing for up to 80% production of bitumen in place.

Fire flooding is the process of continuously injecting compressed, high-oxygen-content air or other gas mixtures into a reservoir and burning some of the oil to drive the rest of the oil into production wells. Combustion generates heat inside the reservoir, physically reducing oil viscosity and vaporizing reservoir water to steam. At the same time, chemical changes, such as cracking, destruct carbon-carbon bonds to generate more mobile, lower molecular weight compounds. The steam, hot water, combustion gas, and a bank of distilled solvent all act to flood oil in front of the fire toward production wells. The average efficiency of fire oil flooding can reach up to 85%.

Thermal flooding could significantly improve the recovery rate of heavy crude oil. However, different types of temperature-sensitive scales, such as calcium carbonate and silica scale, could deposit due to the significant temperature increases with the introduction of thermal flooding.

9.4.2 Mechanism of Scale Deposition

Silica scales can form in oilfield production systems associated with steam flooding.[85–88] Silica deposits occur via a similar process to other inorganic scales, where saturated brine becomes supersaturated, due to changes in the environment the fluids are exposed to.

The precipitation and dissolution of SiO_2 are controlled by equilibrium, as shown in Equation (9.4).[45] The formation of silica is driven by certain pH and temperature ranges, with precipitation generally favored at $pH < 9$ and at lower temperatures.[23] Whereas solubility increases proportionally with temperature, solubility is virtually independent of pH at below 9 and increases proportionally above that. These silicate deposits can be rich in iron, calcium, magnesium, sodium, or aluminum,[85] and most commonly, these deposits are amorphous silica or colloidal silica.[87, 88]

The high temperature introduced during the thermal flooding can cause the dissolution of silica in the formation. The dissolved silica could reprecipitate out of solution in the production wells, when the temperature of the fluids reduces in the production wells and surface facilities. Moreover, the high temperature can cause the deposition of temperature-sensitive scales, for example, calcium carbonate and calcium sulfate.

Calcium carbonate scaling can also occur when the liquefied bitumen and water enter the production well and encounter a pressure drop. As the pressure decreases, carbon dioxide comes out of solution, and a bicarbonate-carbonate equilibrium shifts toward the latter and precipitates out calcite ($CaCO_3$) with a corresponding pH drop. The reduction of carbon dioxide partial pressure disturbs the equilibrium present at reservoir conditions, as given by Equation (9.2).

9.4.3 Scale Control Method

As with the calcium carbonate and silica scale removal introduced in the session of ASP flooding, both chemical and mechanical methods can be used to remove the silica, carbonate, and sulfate scales deposited during the application of thermal flooding.

Chemical treatment, either using a complex high-pH-fluid-containing organic chelates or hazardous chemicals (*e.g.*, hydrofluoric acid), or pH manipulation by shifting the pH of the scale dissolver to above 8.5 with the addition of chelates, can be applied to remove silica scale.[55] Acid-soluble calcium carbonate or iron carbonate scale can be removed using hydrochloric acid (HCl), acetic acid (CH_3COOH), formic acid (HCOOH), and so on. HCl is more effective but has unacceptable issues of corrosion, especially under high-temperature applications. Chelating agents can be applied to remove both $CaCO_3$ and $CaSO_4$ scales by chemically binding the calcium cation to form a soluble complex. The kinetics of carbonate scale removal using chelating agents are not as quick as acid dissolution. However, the chelating agent complexes are stable in alkaline solutions due to the presence of the chelate bonds. Moreover, EDTA presented no risk of corrosion to well equipment and excellent thermal stability up to 200°C, which makes it suitable for application in high-temperature environments.[56, 57]

It also should be noted that silica has susceptibility to coprecipitation with $CaCO_3$, as the $CaCO_3$ scale provides a crystalline matrix, where silica nuclei can form and grow. Therefore, the incomplete removal of $CaCO_3$ will indirectly exacerbate silica scaling.[55]

Similar to the scale mitigation using scale inhibitors discussed previously in ASP and CO_2 flooding, scale inhibition using conventional scale inhibitor squeeze treatment or continuous inhibitor injection can be applied to inhibit different types of scales deposited during the thermal flooding. The conventional scale inhibitors only prevent the carbonate or sulfate scales instead of the silica scale. In addition, as a higher operating temperature may degrade the treatment agent or its efficacy, the thermal stability of scale inhibitors and the compatibility between scale inhibitors and FW under high temperature should be evaluated to select effective scale inhibitors for scale mitigation to be applied in thermal floodings.

9.3.4 Case Studies

Shell carried out a SAGD operation at the Orion field in Alberta, Canada, in 1997. Bitumen viscosity is close to solid at virgin reservoirs and is not suitable for production. SAGD treatment can heat the fluid up to 200°C, significantly reduce the viscosity of bitumen, and make production possible. However, the buildup of scales due to the pressure drops during bitumen and water recovery leads to well equipment plugging up, increased differential pressure, and lower production.[55]

Initially, $CaCO_3$ deposition was hypothesized to be the main scale, referring to the similar problem encountered in other fields. Conventional acid treatment using 1 wt% HCl acid and CH_3COOH was applied successfully, with evidence of a large spike of calcium ions flowing back immediately after treatment. This resulted in more than 1,000 kg of $CaCO_3$ removal, pressure recovery, and increased oil production of ~ 10 m^3 of oil per day. However, $CaCO_3$ scale removal did not resolve the root cause of the problem, as DP increased relatively quickly and subsequent applications gave diminishing returns. It was assumed that acid-insoluble scales could be deposited along with $CaCO_3$ deposition.

The solid samples deposited at the wellhead were analyzed and confirmed to contain acid insoluble scale, silica, and minor $CaSO_4$ deposition. The remediation of silica deposits requires either a complex of high-pH fluids containing organic chelates, hazardous hydrofluoric acid, or mechanical removal. When selecting the method to remediate silica deposition, the higher the operating temperature, the potential for severe $CaCO_3$ deposition in high-pH fluids, and the thermal stability of chemicals under high temperature should be considered and balanced.

Silica removal through pH manipulation was selected and applied in the field to mitigate silica scale deposition. The use of alkaline boiler feed water and the reservoir's inherent high temperature assisted in the dissolution of silica scale.

As pH manipulation provided a caustic environment, which favors $CaCO_3$ deposition in high-pH fluids, especially at high-temperature SAGD treatment, $CaCO_3$ removal was planned and executed as well.

EDTA was applied to mitigate and remove both the $CaCO_3$ and $CaSO_4$ scales. The treatment was exceptionally successful, with evidence of reduced DP by over 2,000 kPa, improved SOR by 25% with analogous increases in oil production, and increased average retreatment to six months. It achieved high cost-effectiveness and an extremely low payout for scale removal treatment.[55]

9.5 Summary

EOR is a process that increases the amount of crude oil that can be extracted from an oil field through the application of specific chemicals, acidic gas, heat, and so on, which can cause changes in the reservoir fluid environment, such as pH, temperature, and fluid chemistry. These changes could trigger the deposition of some types of inorganic mineral scales sensitive to pH and temperature, for example, calcium carbonate, calcium sulfate, and silica scales. The formation of mineral scales during the EOR process could cause blockage in the near-wellbore regions, downhole completion tubulars, and surface facilities, have a negative effect on the efficiency of EOR, or even cause the failure of EOR treatment.

Different from conventional scale deposition, the scales deposited during the EOR are usually driven by an unconventionally high pH (up to 11 for ASP flooding), a low pH (low to ~4 during CO_2 flooding), or a high temperature (>200°C during thermal flooding). The scale mitigation strategies should be evaluated under these unconventional conditions. The effect of the pH on the performance of scale inhibitors, the thermal stability of the chemicals under high temperatures, the treatment of unconventional silica scales, and the interference of combined scales should be evaluated when selecting fixed-purpose scale inhibition strategies. Due to the complexity of scale formation and inhibition in EOR flooding, more research is definitely necessary to understand the mechanisms of scale formation and mitigation in EOR flooding.

Nomenclature

ASP	Alkalis surfactants polymers
ATMP	Aminotris(methylphosphonic acid)
DETAPMPA	Diethylene triamine pentamethane phosphonic acid
DETPMP	Diethylenetriaminepenta(methylenephosphonic acid)
EDTA	Ethylenediaminetetraacetic acid
EOR	Enhanced oil recovery
FW	Formation water
HEDP	1-Hydroxyethylidenediphosphonic acid
HPAM	Polymer-hydrolyzed polyacrylamide
HMP	(Hydroxymethyl)phosphonate
IW	Injection water
MGDA	Methylglycinediacetic acid
MIC	Minimum inhibitor concentration
PAA	Polyacrylates
PBTC	Phosphono-butane-1, 2, 4-tricarboxylic acid
SAGD	Steam-assisted gravity drainage
SHMP	Sodium hexametaphosphate
SI	Scale inhibitor
TEAPE	Triethanolamine phosphate ester

WAG	Water alternating gas
VRR	Voidage replacement ratio

References

1. A.H. Alagorni, Z.B. Yaacob, A.H. Nour, "An Overview of Oil Production Stages: Enhanced Oilrecovery Techniques and Nitrogen Injection," *International Journal of Environmental Science and Development* 6 (2015): pp. 693–701.
2. G.J. Stosur, J.G. Hite, N.F. Carnahan, et al., "The Alphabet Soup of IOR, EOR and AOR: Effective Communication Requires a Definition of Terms," SPE-84908-MS, SPE International Improved Oil Recovery Conference, held October 20–21, 2003 (Kuala Lumpur, Malaysia).
3. G. Kristopher, "Why it's Important to Know the Crude Oil Extraction Process," Market Realist, January 15, 2015.
4. S. Kokal, A. Al-Kaabi, *Enhanced Oil Recovery: Challenges & Opportunities* (World Petroleum Council: Official Publication, 2010), pp. 64–69.
5. Asian Development Bank, "Carbon Dioxide-Enhanced Oil Recovery in Indonesia: An Assessment of Its Role in a Carbon Capture and Storage Pathway," ADB Report, December, 2019.
6. C. McGlade, G. Sondak, M. Han, "Whatever Happened to Enhanced Oil Recovery?," IEA Newsroom, November 28, 2018.
7. P. Teigland, J. Kleppe, "EOR Survey in the North Sea," SPE-99546-MS, SPE/DOE Symposium on Improved Oil Recovery, held April 22–26, 2006 (Tulsa, Oklahoma, USA).
8. J.R. Wilkinson, G.F. Teletzke, K.C. King, "Opportunities and Challenges for Enhanced Recovery in the Middle East," SPE-101679-MS, Abu Dhabi International Petroleum Exhibition and Conference, held November 5–8, 2006 (Abu Dhabi, UAE).
9. O. Massarweh, A.S. Abushaikha, "The Use of Surfactants in Enhanced Oil Recovery: A Review of Recent Advances," *Energy Reports* 6 (2020): pp. 3150–3178.
10. D.J. Pye, "Improved Secondary Recovery by Control of Water Mobility," *Journal of Petroleum Technology* 16 (1964): pp. 911–916.
11. B.B. Sandiford, "Laboratory and Field Studies of Water Floods Using Polymer Solutions to Increase Oil Recoveries," *Journal of Petroleum Technology* 16 (1964): pp. 917–922.
12. D.C. Standnes, I. Skjevrak, "Literature Review of Implemented Polymer Field Projects," *Journal of Petroleum Science and Engineering* 122 (2014): pp. 761–775.
13. A. Poulsen, G.M. Shook, A. Jackson, et al., "Results of the UK Captain Field Interwell EOR Pilot," SPE-190175-MS, SPE Improved Oil Recovery Conference, held April 14–18, 2018 (Tulsa, Oklahoma, USA).
14. N. Mehta, G. Kapadia, P.V. Selvam, "Challenges in Full Field Polymer Injection in Mangala Field," SPE-179807-MS, SPE EOR Conference at Oil and Gas West Asia, held March 21–23, 2016 (Muscat, Oman).
15. J. Juri, A. Ruiz, F. Schein, et al., "Grimbeek Successful Polymer Pilot Extends to 80 Injectors in Factory-Mode Development at CGSJ Basin," 20th European Symposium on Improved Oil Recovery, held April 8–11, 2019 (Pau, France).
16. A.Z. Abidin, T. Puspasari, W.A. Nugroho, "Polymers for Enhanced Oil Recovery Technology. *Procedia Chemistry* 4 (2012): pp. 11–16.
17. A. Beteta, L. Nurmi, L. Rosati, et al., "Polymer Chemical Structure and Its Impact on EOR Performance," SPE-200441-MS, SPE Improved Oil Recovery Conference, held August 31–September 4, 2020 (Virtual).
18. T. Divers, N. Gaillard, S. Bataille, et al., "Successful Polymer Selection for CEOR: Brine Hardness and Mechanical Degradation Considerations," SPE-185418-MS, SPE Oil and Gas India Conference and Exhibition, held April 4–6, 2017 (Mumbai, India).

19. M. Han, A. AlSofi, A. Fuseni, et al., "Development of Chemical EOR Formulations for a High Temperature and High Salinity Carbonate," IPTC-17084-MS, International Petroleum Technology Conference, held March 26–28, 2013 (Beijing, China).
20. J.J. Sheng, "Status of surfactant EOR technology, *Petroleum* 1 (2015): pp. 97–105.
21. J.J. Sheng, *Modern Chemical Enhanced Oil Recovery: Theory and Practice* (Boston: Elsevier, 2011).
22. L. Ma, X. Liu, Q. Yang, "Study on the Adaptability of Binary Flooding in a Certain Oil Field, *Open Physics* 19 (2021): pp. 447–453.
23. Q. Jia, B. Zhou, R. Zhang, et al., "Development and Application of a Silicate Scale Inhibitor for ASP Flooding Production Scale," SPE-74675-MS, International Symposium on Oilfield Scale, held January 30–31, 2002 (Aberdeen, UK).
24. S.W. Lo, G.T. Shahin, G. Graham, et al., "Scale Control and Inhibition Evaluation of an Alkaline Surfactant Polymer Flood," SPE-141551-MS, SPE International Symposium on Oilfield Chemistry, held April 11–13, 2011 (Woodlands, TX, USA).
25. W.M. Stoll, H. Shureqi, S.A. Al-Harthy, et al., "Alkali/Surfactant/Polymer Flood; From Laboratory to the Field," SPE-129164-MS, SPE EOR Conference Oil and Gas West Asia, held April 11–13, 2010 (Muscat, Oman).
26. V. Alvarado, E. Manrique, *Enhanced Oil Recovery–Field Planning and Development Strategies* (Houston: Gulf Professional Publication, 2010).
27. A. Noraliza, N. Borhan, B.M. Ibrahim, et al., "Managing Micro-Emulsion and Scale During ASP Flooding for North Sabah field EOR," OTC-24984-MS, Offshore Technology Conference-Asia, held March 25–28, 2014 (Kuala Lumpur, Malaysia).
28. H. Guo, R. Ma, D. Kong, "Success and Lessons Learned From ASP Flooding Field Tests in China," SPE-186931-MS, SPE/IATMIAsia Pacific Oil & Gas Conference and Exhibition, held October 17–19, 2017 (Jakarta, Indonesia).
29. Q. Wang, W. Al-Nasser, F. Al-Dawood, et al., "Inhibition of Calcium Carbonate Scale in the Presence of EOR Chemicals," SPE-184534-MS, SPE International Conference on Oilfield Chemistry, held April 3–5, 2017 (Montgomery, Texas, USA).
30. S.R. Clark, M.J. Pitts, S.M. Smith, et al., "Design and Application of an Alkaline-Surfactant-Polymer Recovery System to the West Kiehl Field," *SPE Advanced Technology Series* 1 (1993):172–179.
31. J. Vargo, J. Turner, B. Vergnani, et al., "Alkaline-Surfactant-Polymer Flooding of the Cambridge Minnelusa Field," SPE-55633-MS, SPE Rocky Mountain Regional Meeting, held May 15–18, 1999 (Gillette, Wyoming, USA).
32. M. Baviere, P. Glenat, V. Plazanet, et al., "Improvement of the Efficiency/Cost Ratio of Chemical EOR Processes by Using Surfactants, Polymers, and Alkalis in Combination," *SPE Reservoir Engineering,* 10 (1995): pp. 187–193.
33. X.M. Yang, G. Liao, P. Han, et al., "An Extendedfield Test Study on Alkaline-Surfactant-Polymer Flooding in Beiyi-Duanxi of Daqing Oilfield," SPE-80532-MS, SPE AsiaPacific Oil and Gas Conference and Exhibition, held September 9–11, 2003 (Jakarta, Indonisia).
34. H. Li, G. Liao, P. Han, et al., "Alkline/Surfactant/Polymer (ASP) Commercial Flooding Test in Central Xing Area of Daqing Oilfield," SPE International Improved Oil Recovery Conference in Asia Pacific, held October 20–21, 2003 (Kuala Lumpur, Malaysia).
35. S. Al-Mutairi, S. Koka, "EOR Potential in the Middle East: Current and Future Trends," SPE-143287-MS, SPE EUROPEC/EAGE Annual Conference and Exhibition, held May 23–26, 2011 (Vienna, Austria).
36. M. Han, A. Fuseni, B. Zahrani, et al., "Laboratory Study on Polymers for Chemical Flooding In Carbonate Reservoirs," SPE-169724-MS, SPE EOR Conference at Oil and Gas West Asia, SPE EOR Conference at Oil and Gas West Asia, held March 31–April 2, 2014 (Muscat, Oman).
37. S. Gao, Q. Gao, "Recent Progress and Evaluation of ASP Flooding for EOR in Daqing Oilfield," SPE-127714-MS, SPE EOR Conference at Oil & Gas West Asia, held April 11–13, 2010 (Muscat, Oman).

38. N. Ghadami, N. Zakaria, A. Manap, et al., "Robust Chemical EOR Modelling From Coreflood to Full Field Scale In A Brown Field, Offshore," SPE-182376-MS, SPE Asia Pacific Oil & Gas Conference and Exhibition, held October 25–27, 2016 (Perth, Australia).
39. Y. Zhu, "Current Developments and Remaining Challenges of Chemical Flooding EOR Techniques in China," SPE-174566-MS, SPE Asia Pacific Enhanced Oil Recovery Conference, held August 11–13, 2015 (Kuala Lumpur, Malaysia).
40. M. Jordan, E. Mackay, "Predicting and Managing Inorganic Scale Associated with Produced Water From EOR Projects," SPE-177144-MS, SPE Latin American and Caribbean Petroleum Engineering Conference, held November 18–20, 2015 (Quito, Ecuador).
41. J. Sonne, S. Kerr, K. Miner, "Application of Silicate Scale Inhibitors for ASP Flooded Oilfields: A Novel Approach to Testing and Delivery," SPE-154332-MS, SPE International Conference on Oilfield Scale, held May 30–31, 2012 (Aberdeen, UK).
42. O. Karazincir, S. Thach, W. Wei, et al., "Scale Formation Prevention During ASP Flooding," SPE-141410-MS, SPE International Symposium on Oilfield Chemistry, held April 11–13, 2011 (The Woodlands, Texas, USA).
43. M.T. Al-Murayri, D.S. Kamal, M.Y. Khan, et al., "Alleviation of Scaling Concerns and Controlling Of Chemical EOR Under-Performance Risk for a Giant Sandstone Reservoir in North Kuwait," SPE-197249-MS, Abu Dhabi International Petroleum Exhibition & Conference, held November 11–14, 2019 (Abu Dhabi, UAE).
44. T. Chen, "New Insights into the Mechanisms of Calcium Carbonate Mineral Scale Formation and Inhibition," (Thesis, Heriot-Watt University, 2005).
45. J.D. Rimstidt, "Rate Equations for Sodiumcatalyzed Quartz Dissolution," *Geochimica et Cosmochimica Acta* 167 (2015) 195–204.
46. T.C. Campbell, P.H. Krumrine, "Laboratory Studies on Alkaline Waterflooding," SPE-8328-MS, SPE Annual Technical Conference and Exhibition, held September 23–26, 1979 (Las Vegas, Nevada, USA).
47. J. Qing, B. Zhou, R. Zhang, et al., "Development and Application of a Silicate Scale Inhibitor for ASP Flooding Production Scale," SPE-74675-MS, International Symposium on Oilfield Scale, held January 30–31, 2002 (Aberdeen, UK).
48. J. Sonne, K. Miner, S. Kerr, "Potential for Inhibitor Squeeze Application for Silicate Scale Control in ASP Flood," SPE-154331-MS, SPE EOR Conference at Oil and Gas West Asia, held April 16–18, 2012 (Muscat, Oman).
49. J. Southwick, "Solubility of Silica in Alkaline Solutions: Implications for Alkaline Flooding," *SPE Journal* 25 (1985): pp. 857–864.
50. A. Pandey, N. Koduru, M. Stanley, "Results of ASP Pilot in Mangala Field: A Success Story," SPE-179700-MS, SPE Improved Oil Recovery Conference, held April 11–13, 2016 (Tulsa, Oklahoma, USA).
51. G. Jing, S. Tang, X. Li, et al., "The scaling conditions for ASP flooding oilfield in the fourth Plant of Daqing oilfield," *Journal of Petroleum Exploration and Production Technology* 3 (2013), 175–178.
52. M. Jordan, I.R. Collins, E. Mackay, "Low Sulfate Seawater Injection for Barium Sulfate Scale Control: A Life-of-Field Solution to a Complex Challenge," *SPE Production & Operations* 23 (2008) 192–209.
53. M.M. Vazirian, F.B. Alvim, M de Oliveira Penna, et al., "Scale Deposition in the Oil and Gas Industry: From a Systematic Experimental Scale to Real-Time Field Data," NACE-2016-7632, CORROSION 2016, held March 6–10, 2016 (Vancouver, British Columbia, Canada).
54. A. Noraliza, I.K. Salleh, J.M. Ibrahim, et al., "Development of Improved Chemical Formulations for Scale Control for Alkaline Surfactant Polymer Flood in a Seawater Flooded Reservoir," SPE-179883-MS, SPE International Oilfield Scale Conference and Exhibition, held May 11–12, 2016 (Aberdeen, UK).

55. T. Cheung, M. Scheck, "Novel scale Remediation for Steam Assisted Gravity Drainage (SAGD) Operations," SPE-164048-MS, SPE International Symposium on Oilfield Chemistry, held April 8–10, 2013 (The Woodlands, Texas, USA).
56. M. Cikes, B. Vranješevic, M. Tomic, et al., "A successful Treatment of Formation Damage Cause by High-Density Brine," *SPE Production Engineer* 5 (1990): pp. 175–179.
57. R.J. Motekaitis, X.B. Cox, P. Taylor, et al., "Thermal degradation of EDTA chelates in aqueous solution," *Canadian Journal of Chemistry* 60 (1982) 1207–1213.
58. Y. Liu, R. Rebecca, D. Shen, "How Does EOR Polymer Impact Scale Control During ASP Flooding?," SPE-204350-MS, SPE International Conference on Oilfield Chemistry, held December 6–7, 2021 (The Woodlands, Texas, USA).
59. A. Beteta, L. Boak, K. McIver, et al., "Mechanistic Understanding of the Impact of EOR Polymer on the Inhibition Mechanism and Performance of Phosphonate Scale Inhibitors," SPE-204383-MS, SPE International Conference on Oilfield Chemistry, held December 6–7, 2021 (The Woodlands, Texas, USA).
60. Q. Wang, W. Al-Nasser, T. Chen, et al., "Calcium Carbonate Scale Inhibition: Effects of EOR Chemicals," NACE-2018-10546, CORROSION 2018, held April 15–19, 2018 (Phoenix, Arizona, USA).
61. E.J. Mackay, M.M. Jordan, "SQUEEZE Modelling: Treatment Design and Case Histories," SPE-82227-MS, SPE European Formation Damage Conference, held May 13–14, 2003 (The Hague, Netherlands).
62. K.O. Meyers, H.L. Skillman, "The Chemistry and Design of Scale Inhibitor Squeeze Treatments," SPE-13550-MS, SPE Oilfield and Geothermal Chemistry Symposium, held April 9–11, 1985 (Phoenix, Arizona, USA).
63. A. Beteta, K. McIver, O. Vazquez, et al., "The Impact of EOR Polymers on Scale Inhibitor Adsorption," SPE-200692-MS, SPE International Oilfield Scale Conference and Exhibition, held June 24–25, 2020 (Virtual).
64. C. Silva, A. Beteta, K. McIver, et al., "The Impact of EOR Polymers on the Adsorption of Phosphonate Scale Inhibitors," SPE-213820-MS, SPE International Conference on Oilfield Chemistry, held June 28–29, 2023 (The Woodlands, Texas, USA).
65. X.J. Wang, C.Y. Xie, Q.G. Wang, "Analytical Study on the Composition of Scaling Materials in Ternary Composite Flooding Oil Recovery Wells," *Oilfield Chemistry* 20 (2003): pp. 307–330.
66. X.J. Wang, Q.G. Wang, "Research and Application of scale formation and Scale Remover in Oil Recovery Wells of Ternary Composite Flooding Deposits," *Oilfield Chemistry* 20 (2003): pp. 1–3.
67. X.Q. Cao, X.G. Sui, "Awareness of Some Problems of Ternary Composite Flooding in North Area I Break West," *Daqing Petroleum Geology and Development* 20 (2001): pp. 111–114.
68. J.C. Cheng, J.Z. Wu, K. Luo, et al., *ASP Flooding Technology* (Beijing: Petroleum Industry Press, 2019).
69. Department of Energy, *Enhanced Oil Recovery* (Washington, DC: Office of Fossil Energy, 2019).
70. A. Cavanagh, P. Ringrose, "Improving Oil Recovery and Enabling CCS: A Comparison of Offshore Gas-Recycling in Europe to CCUS in North America," *Energy Procedia* 63 (2014): pp. 7677–7684.
71. H. Rodrigues, E. Mackay, A. Daniel, "Impact of WAG Design on Calcite Scaling Risk in Coupled CO_2-EOR and Storage Projects in Carbonate Reservoirs," OTC-29823-MS, SPE Reservoir Simulation Conference, held April 10–11, 2019 (Galveston, Texas, USA).
72. P.D. Torrey, *A Review of some Important Water-Injection Projects, API-51-286* (New York, USA: American Petroleum Institute, January 1951).
73. D.C. Silva, K.S. Sorbie, E.J. Mackay, "Thermodynamic Model Development Relevant to Calcium Carbonate Scale Mitigation in CO_2 Water Alternating Gas Schemes," IBP-1927-14, Rio Oil & Gas Expo and Conference, held September 15–18, 2014 (Rio de Janeiro, Brazil).

74. E.J. Mackay, A.P. Martins de Souza, "Modelling of CO_2 and Seawater Injection in Carbonate Reservoirs to Evaluate Inorganic Scaling Risk," SPE-169766-MS, SPE International Oilfield Scale Conference and Exhibition, held May 14–15, 2014 (Aberdeen, UK).
75. Y.H. Chen, W.D. Pan, S.L. Sun, "Study on the Effect of CO_2 Throughput on Reservoir Scaling Trend," *Petrochemical Corrosion and Protection* 20 (2003): pp. 21–23.
76. B.N. Ryzhenko, "Genesis of Dawsonite Mineralization: Thermodynamic Analysis and Alternatives," *Geochemistry International* 44 (2006): pp. 835–840.
77. M.S. Kapelke, E.P. Caballero, "Prevention of CaCO3 Precipitation From Calcium Chloride Kill Fluid in CO2-Laden Formations," SPE-12752-MS, SPE California Regional Meeting, held April 11–13, 1984 (Long Beach, California, USA).
78. H. Rodrigues, E. Mackay, D. Arnold, et al., "Economic Optimization and Calcite Scale Management of CO_2-EOR in Carbonate Reservoirs," SPE-200678-MS, SPE International Oilfield Scale Conference and Exhibition, held June 24–25, 2020 (Virtual).
79. A.J. Ellis, "The Solubility of Calcite in Carbon Dioxide Solutions," *American Journal of Science* 257 (1959): pp. 259–266.
80. E.R. McCartney, A.E. Alexander, "The Effect of Additives Upon the Process of Crystallization–I. Crystallization of Calcium Sulfate," *Journal Colloid Science* 13 (1958): pp. 383–396.
81. J.E. Ramsey, L.M. Cenegy, "A Laboratory Evaluation of Barium Sulfate Scale Inhibitors at Low pH for Use in Carbon Dioxide EOR Floods," SPE-14407-MS, SPE Annual Technical Conference and Exhibition, held September 22–26, 1985 (Las Vegas, Nevada, USA).
82. D.W. Griffiths, S.D. Roberts, S.T. Lin, "Inhibition of Calcium Sulfate Dihydrate Crystal Growth by Phosphonic Acids—Influence of Inhibitor Structure and Solution pH," SPE-7862-MS, SPE Oilfield and Geothermal Chemistry Symposium, held January 22–24, 1979 (Houston, Texas, USA).
83. G.R. Chesnut, G.D. Chappell, D.H. Emmons, "The Development of Scale Inhibitors and Inhibitor Evaluation Techniques for Carbon Dioxide EOR Floods," SPE-16260-MS, SPE International Symposium on Oilfield Chemistry, held February 4–6, 1987 (San Antonio, Texas, USA).
84. A. Khalilnezhad, H. Rezvani, A. Abdi, et al., "Chapter 8—Hybrid Thermal Chemical EOR Methods," in *Thermal Methods Enhanced Oil Recovery Series*, eds. A. Hemmati-Sarapardeh, A. Alamatsaz, M. Dong, Z. Li Editor (Amsterdam: Elsevier, 2023), pp. 269–314.
85. D.L. Gallup, "Alumium Silicate Scale Formation and Inhibition: Scale Characterization and Laboratory Experiments," *Geothermic* 26 (1997): pp. 483–499.
86. Z.I. Khatib, E.E. Olson, M.C. Place, "Effect of High Silica Content on Scale Deposition and Pipe-Wall Loss in Oilfield Steam Generation," *SPE Production Engineering* 7 (1992): pp. 357–362.
87. R.W. Bowman, L.C. Gramms, R.R. Craycraft, "High-Silica Water in Steamflood Operations," *SPE Production and Facilities* 15 (2000): pp. 123–125.
88. P. Pedenaud, C. Goulay, F. Potier, "Silica Scale Inhibitor for Steam Generation in OTSG Boiler," SPE-86934-MS, SPE International Thermal Operations and Heavy Oil Symposium and Western Regional Meeting, held March 16–18, 2004 (Bakersfield, California, USA).

10 Scale Prevention in Shale Production

Zhiwei (David) Yue[1]
[1]Halliburton, Houston, Texas, USA

10.1 Introduction

Shale is a type of sedimentary rock that primarily consists of clay minerals like kaolinite and smectite, with additional minor components such as quartz, calcite, iron oxides, and other species. It exhibits fissility, low permeability, and contains organic solid matter or kerogen that is often considered a precursor to oil and natural gas. Shale gas unanimously refers to natural gas trapped within shale formations. However, the term "shale oil" caused some confusion within academia and industry. It should literally indicate the crudes produced from the shale formation, but sometimes can be used to specify the synthetic oil converted from oil-bearing shales through pyrolysis or other thermal processes.[1] To eliminate ambiguity, the World Energy Council and other resources suggest using the term "tight oil" to describe natural oil extracted from shale rock with low permeability.[2, 3] Though tight oil can be produced both from the shale formation and tight sandstone reservoirs, the process involves the same horizontal drilling and hydraulic fracturing (HF) technologies, therefore, posing similar scale challenges toward the operation. Throughout this chapter, unless specified otherwise, the terms "tight oil," "shale gas," as well as "unconventional" are used to discuss the oil and gas production in shale plays.

The mechanisms of mineral solid precipitation in water are universally driven by two factors: thermodynamics and kinetics. Major influencing parameters in oilfield systems include the inherent chemistry, temperature (*T*), and pressure (*P*) of the brine, the *T*/*P* profile for the well and reservoir, brine evaporation, water commingling process, water residence time in the system, brine compatibility with other treating fluids (*e.g.*, frac fluids, acids, scale inhibitors, and other chemicals), injection water and rock interaction, preexisting crystalline seed, fluid turbulence, and sulfides from reservoir souring.[4, 5] From these aspects, mineral scaling in shale gas and oil operations shares similarities with scaling issues in conventional fields but still poses some unique challenges due to the characteristics of the shale formations and the specific processes involved in hydrocarbon extraction.

10.2 Challenges for Mineral Scaling in Shale Production

10.2.1 Varieties of Shale Formation Characteristics

Formations can contain minerals that are prone to scaling, such as calcite, dolomite, siderite, magnetite, pyrite, and barite. Numerous investigations such as Na-Cl-Br systematics, stable and radiogenic isotopes, and geochemical modeling have demonstrated that the interactions between reservoir rocks and water imprint distinct chemical characteristics onto the chemistries of the connate or formation water over extended maturation periods.[6] The geochemical processes contributing to the final connate water chemistry include but are not limited to mineral dissolution and precipitation, ion exchange between rock and water, adsorption, rock-fluid equilibrium, spatial variation, and so on. For instance, brines in western Pennsylvania, originating from Salina Group seawater evaporation and undergoing interactions with the Ridgeley Sandstone aquifer, continued to evolve by displacing and mixing with connate waters. This interaction yielded a stabilized water chemistry characterized by high salinity since mid-Paleozoic time.[7] Consequently, a comprehensive understanding of the formation's mineralogy is crucial for anticipating and managing scaling challenges for each shale plays. Unfortunately, the mineralogy and reservoir conditions vary significantly among shale basins.[6–9] Figure 10.1 illustrates the locations and the recent production data of the major onshore shale plays in the United States. The Marcellus region is the most prolific for shale gas, while the Permian is the predominant producer of tight oil and the second largest producer of shale gas.

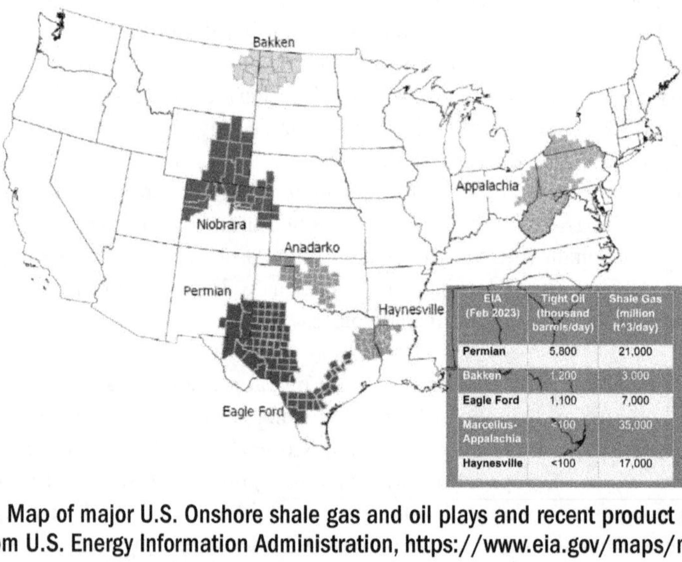

EIA (Feb 2023)	Tight Oil (thousand barrels/day)	Shale Gas (million ft^3/day)
Permian	5,800	21,000
Bakken	1,200	3,000
Eagle Ford	1,100	7,000
Marcellus-Appalachia	<100	35,000
Haynesville	<100	17,000

Figure 10.1. Map of major U.S. Onshore shale gas and oil plays and recent product rates. Reproduced from U.S. Energy Information Administration, https://www.eia.gov/maps/maps.htm.

The Permian Shale, located in West Texas and southeastern New Mexico, is one of the most productive and active basins in the United States. It is abundant in both shale oil and gas, ranking as the No. 1 oil producer and No. 2 gas producer among all U.S. onshore plays. The basin is characterized by multiple sub-basins, including Delaware Basin, Midland Basin, and Central Basin Platform. Shale reservoir temperatures are typically in the moderate range of 180–250°F. Operators in the regions tend to utilize the abundant brackish groundwater or reused produced water for fractured wells.[10]

The Marcellus Shale, primarily situated in the Appalachian Basin, spans parts of Pennsylvania, West Virginia, Ohio, and New York. Shale reservoir temperatures in this region are relatively low, typically in the range of 120–180°F. Wells in the Marcellus produce high rates of shale gas with minimal produced water. However, they also consume the highest volume of water in the fracturing process for each well, approximately 9 million gallons per well compared to the national average of 5 million per well. Notably, almost all the produced water with high salinity is reused in hydraulic fracturing operations in the region.[7]

The Utica Shale, situated a few thousand feet below the Marcellus formation, is frequently discussed alongside Marcellus and exhibits notable similarities in mineralogy. It is recognized as one of the newest discoveries in shale gas reservoirs.[6]

The Bakken Shale, primarily situated in North Dakota and Montana and spanning most of the Williston Basin, comprises three layers. The upper and lower members consist of organic-rich black shales deposited in a deep anoxic environment. The middle unit, approximately 85-ft thick, is mainly composed of fine-grained, often dolomitic siltstone deposited in a shallow matrix, and represents the most produced zone. The Bakken formation is abundant in both shale oil and natural gas. Notably, the shale exhibits high water sensitivity, potentially leading to swelling and damage to productive fracture systems. Moreover, the formation rocks contain reactive pyrite minerals, which, upon exposure to the hydrochloric acid commonly applied as the fracturing spearhead fluids, may generate soluble iron species in the produced water. The Bakken region faces technical challenges due to the elevated levels of total dissolved solids (TDS), calcium, salinity, and iron, posing difficulties in utilizing recycled produced water for fracturing operations, especially in a water-scarce environment.[9–11]

The Eagle Ford Shale, sprawling over South Texas, is abundant in both shale oil and natural gas, featuring liquid-rich production for crude oil, condensate, and natural gas liquids. The most productive section of the Eagle Ford shale, approximately 50- to 400-ft thick, is positioned between the Austin Chalk at the top and the Buda Limestone at the bottom. Notably, the region benefited from numerous conventional wells drilled both above and below the shale layer, leading to reduced capital investment for unconventional wells, particularly from around 2008 onward.[12]

The Haynesville Shale, primarily situated in northwest Louisiana, east Texas, and southwestern Arkansas, is known for its substantial reserves of dry gas. Positioned at depths exceeding 10,000 ft below the surface, it stands out as one of the deepest shale plays in North American Land. With an average thickness ranging from 200 ft to 300 ft, the shale operates under elevated temperatures, typically in the range of 300–350°F. Notably, the Bossier Shale lies just above the Haynesville Shale and beneath the top Cotton Valley sandstone formation, leading to their occasional reference together as the Haynesville-Bossier Shale.[10]

The Barnett Shale, spreading within the Bend Arch-Fort Worth Basin in Northeast Texas, previously held fame as the largest dry gas producer among the U.S. plays until its production began to decline around 2012. A significant portion of the formation is located beneath the urban area of the Dallas-Fort Worth metroplex, imposing some challenges for future field development. The region continues to be dominated by shale gas production, with recent activity focusing on the liquid-rich segments of the play. Wells within the Barnett Shale typically reach depths in the range of 6,000–8,000 ft.[13]

The Niobrara Shale, situated within the Denver-Julesburg Basin (DJ Basin), extends across Colorado, Wyoming, Nebraska, and Kansas. The basin encompasses various submembers such as the Niobrara Formation, Niobrara A/B/C Benches, Codell Sandstone, and the Parkman Formation in the Powder River Basin. The shale manifests mineralogical variations, with certain segments exhibiting a notable abundance of carbonate minerals in addition to prevalent clay minerals. The Niobrara Shale has garnered recognition for its dual production of oil and natural gas.[10]

The mentioned shale plays have been vital in making the United States a major oil and gas producer. Benefiting from advanced technologies like hydraulic fracturing and horizontal drilling, we can efficiently extract a lot of energy from these shale formations. However, the unique geographic features of these plays bring challenges with the water produced during extraction. Dealing with the chemistry and scaling of this water adds complexity to developing and improving shale operations in these areas.

10.2.2 Intensified Rock-Fluid Interactions during Shale Development

The influence of subsurface fluid interactions with the formation rocks on brine composition is widely acknowledged. This geochemical phenomenon may undergo intensification in the stimulation of shale reservoirs. In conventional vertical drilling, depicted in the left panel of Figure 10.2, the rock-fluid equilibrium can be disrupted aggressively only near the wellbore "pay zone," which is a relatively small region in comparison to the entire volume of reservoirs. This zone is constrained by the thickness of the oil-bearing shale formation, typically ranging from tens to several hundred feet. In the production stage, the bulk brine within the reservoir is anticipated to experience some additional imbalance and interactions with the surrounding rock. The process is typically related to pressure drop or reservoir drawdown. Though with potentials to scale deposition or even formation damage, it is relatively slow in time span like months to years and concentrates locally at certain regions, like near wellbore. However, the widespread adoption of horizontal drilling and hydraulic fracturing in shale development has introduced significant changes. In contemporary U.S. plays, wells often incorporate horizontal laterals extending for considerable distances (miles), fractured across multiple stages. The right panel in Figure 10.2 illustrates a schematic of multistage dual-lateral technology applied to a shale well, exponentially expanding the pay zone and enabling economic production rates. Nonetheless, this process can also rapidly reshuffle the connate water within a short time frame, on the order of hours or days, leading to substantial rock-fluid interactions that are meant to establish new equilibriums. When managed inappropriately, some of these interactions can result in unforeseen mineral scale depositions, in terms of scale types, severity, locations, and adverse impacts on well production.[14]

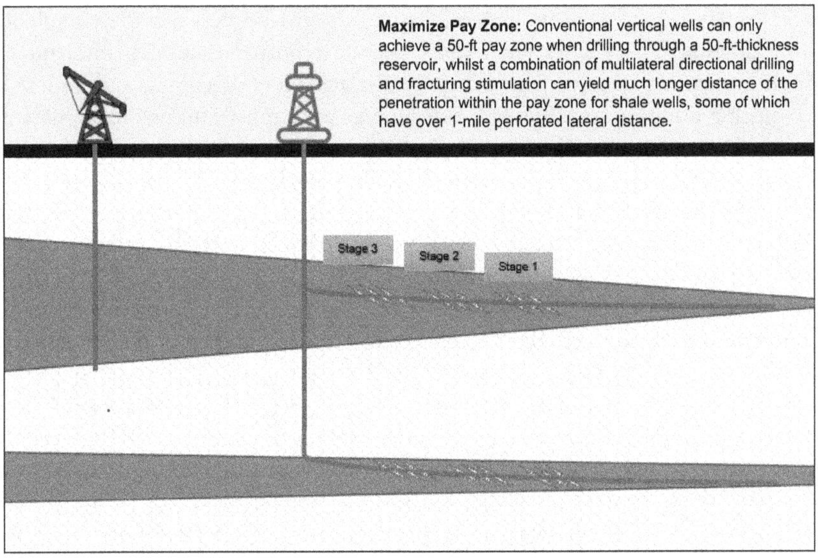

Figure 10.2. Vertical (left) and horizontal (right) wells. The pump jack represents a conventional well on the production stage, whereas the drilling rig shows an unconventional well utilizing stimulation technologies.

Components in fracturing fluids can also promote interactions with the rock matrix. Khan *et al.*[15] summarized the fracturing process into four steps: spearhead acid to clean the wellbore with pH≤0.3, neutral pad fluid to open the formation/fractures, a slightly alkaline proppant step to transport and place proppants (pH 7.5–10, depending on fracture fluids such as slickwater or gels), and the final flush step to clean excess solids. The dramatic swing of pH from a strong acidic to an alkaline environment occurs both temporally over treating stages and spatially through the lateral into the reservoir, without giving the brine systems sufficient time for gradual equilibration. The acid gradient can cause some coupled mineral dissolution, precipitation, transportation, adsorption, and secondary scaling. Calcium and magnesium can be dissolved from acid-sensitive formations, such as chalk and dolomite from fracture faces; relocated during pressured shut-in; and reprecipitated in high-alkaline and low-pressure zones. Iron can be released from siderite in acid treatment or from pyrite and chlorite under the co-influence of acidic and oxidative fluids, forming iron carbonates or iron sulfides in sour reservoirs.[16] Controversially, barium in shale and completion fluids was also reported to be susceptible to acid matrix stimulation under shale conditions, thus it can migrate to secondary locations for reprecipitation in hydraulic fracturing systems.[17]

Enriched strontium in the formation water of some plays like Permian can coprecipitate with high levels of barium in the flowback water postfracturing.[15] To have a comprehensive scale management program, interactions between the fracturing fluids and the target shale should be placed into consideration.

The extended residence time for fluid within the reservoir is another factor contributing to intensified rock-fluid interactions in unconventional wells. This fluid residence time, sometimes referred to as shut-in, generally spans the period between the completion and put-on-production (POP) dates for a well. Aramco reported that horizontal multilateral wells with maximum reservoir contact and extreme reservoir contact fall under the category requiring additional scaling risk assessments and more cautious scale management programs.[18] Researchers at Chevron developed a conceptual model to minimize scale formation and its impacts at the subsurface for hydraulically fractured wells. One of the key processes involves analyzing rock-fluid interactions during the shut-in period. They observed that the prolonged residence time led to noticeable impacts on scaling ions and scale inhibitor performance, requiring significantly higher doses of scale inhibitors to provide adequate protection during the extended incubation time. As a corrective measure, they proposed that the 24-hour standard scale inhibition performance in the laboratory is applicable for conventional wells but may be insufficient to address the needs of unconventional fields. Therefore, testing periods extending into weeks should be considered.[5, 14, 19] It is noteworthy to highlight that analogous concerns become even more pronounced in the development of offshore unconventional fields. The duration of shut-in periods is susceptible to various factors beyond the design to maximize the performance for individual wells. Emergency events such as hurricanes or rig fires can force the cessation of well production for durations spanning from days to weeks. Additionally, operational constraints related to the field infrastructure may contribute to prolonged shut-in periods. Recent case studies have revealed instances, where wells already fractured with seawater-based fluids in earlier phases were designed to undergo shut-ins ranging from 6 weeks to 6 months, until newer wells and the necessary field infrastructure were all ready for field production in the Gulf of Mexico. Given the utilization of seawater-based fracturing fluids to treat the barium-rich formation water, stringent criteria were complied with to ensure the efficacy of barite inhibition under reservoir conditions. Laboratory incubation periods for scale inhibition reportedly extended up to 6 months to comprehensively address scale-related concerns arising from those unconventional activities.[20–22]

10.2.3 Sensitivity to Scaling for Low-Permeability Shale Formation

The accumulation of scale within the pore throats of rock formations has the potential to induce formation damage, thereby significantly compromising the inherent productivity of reservoirs. In unconventional reservoirs, the susceptibility to scale deposition within the formation is heightened due to the low permeability characteristic of shale. Shale rocks feature a fine-grained structure with pore sizes ranging from 0.1 μm to 0.005 μm, and their permeabilities are typically reported at submilliDarcy or even micro-Darcy (mD) levels. In contrast, conventional reservoir rocks with high productivity, either in sandstone or chalk matrix, possess permeabilities on the order of hundreds of millidarcy and pore throats exceeding 10 μm.[1, 5, 23] As shales ordinarily have insufficient permeability to allow significant hydrocarbon flow to the wellbore, the development of shale fields relies heavily on hydraulic fracturing. This technology involves the creation of extensive artificial fractures in rocks through high-pressure water injection, followed by the maintenance of these fracture openings using small proppants like sand.

While the fracturing process substantially enhances productivity for unconventional wells by increasing rock-fluid contacting surfaces, it does not alter the mineralogy or physical properties of the bulk shale rock, leaving it susceptible to blockage from various substances in the fluids, including mineral scale solids. Recent simulation studies focused on a Marcellus shale formation revealed that unwanted scale deposition could lead to a loss of fracture conductivity of up to 10%. The study indicated that moderate plugging of the fracture face might occur during the early production stage, resulting in a reduction in the peak production rate. Subsequently, more severe plugging onto the fracture tip could occur during later stages of production. The primary factors influencing the overall scaling risk were deemed to be the local formation water chemistry, the composition of the HF fluid, and the injection rate.[24, 25] To secure the unconventional well performance and the return on capital employed, it is critical to have a thorough understanding of the scale potentials to be encountered and have an effective scale control program in place before the execution of the treatment.

10.2.4. Complexity of Water Management in Unconventional Operation

Controlling scale involves approaches intertwined with effective water management. Essentially, as the water consumption or production, the variety of water sources utilized, and the diversification in water chemistry increase, so do the scale issues in daily operations. Unconventional production seems to present an ideal scenario for these challenges.

10.2.4.1 Elevated Water Footprint

Hydraulic fracturing liberates gas and oil from shale formations using large volumes of fluids, approximately 99% of which is water. With the rapid growth of unconventional development and the widespread adoption of hydraulic fracturing operations, the demand for HF water usage has substantially increased since 2010. For example, the annual total HF water usage increased nearly linearly from 140 million·m^3 in 2011 to 400 million·m^3 in 2014. It then declined to 320 million·m^3 in 2015, possibly due to a drop in crude oil prices during that period.[10] In terms of average HF use per well (as shown in Figure 10.3), the national average increased from 9,900 m^3 per well (CMPW) in 2011 to 20,000 CMPW in 2015, presumably due to longer laterals used to increase the contact area with the shale formation. Among the major plays, Marcellus used the highest CMPW (34,000 in 2015) to fracture each well, Eagle Ford ranked second, and Barnett required the least CMPW (11,000 in 2015). While most regions had the CMPW nearly double from 2011 to 2015, Permian was observed as the fastest growing region with a CMPW increase of 510%, whereas Barnett was the only region that experienced a 30% decrease in CMPW during this period.

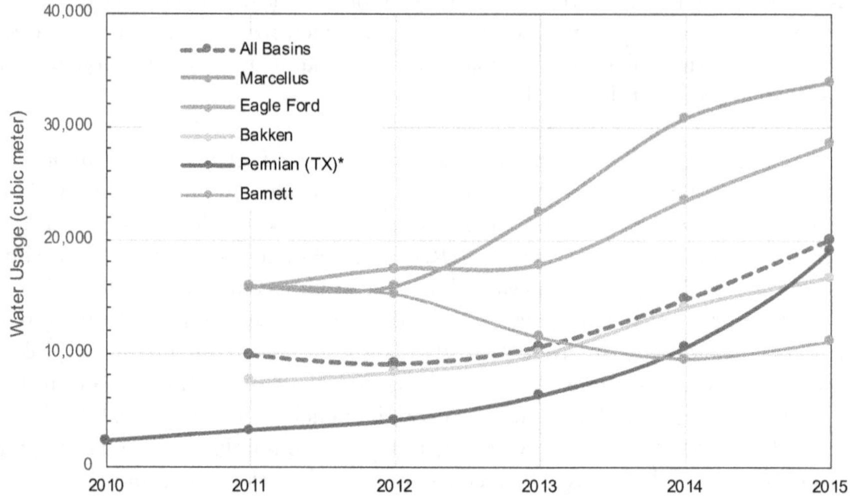

Figure 10.3. Average water used in major U.S. basins. Reproduced with data from.[10, 26]

Similar to conventional wells, fractured wells also produce water after completion. Typically, during the early stage, which lasts for approximately 2–4 weeks, the fluid from the well is considered more of a recovery of the injection and drilling fluids and is sometimes referred to as "flowback" water. Flowback is reported to recover 10%–40% of the fractured fluids, with the remaining portion continuing to flow out gradually throughout the lifetime of the well. Water generated after the flowback period is termed classical "produced water." The definition distinction between flowback and produced water (FP water) can be subjective. Consequently, the concept of combined FP water has been applied in water management studies for unconventional wells. Kondash et al. illustrated a year-by-year comparison of FP water production rates with gas or oil production rates for the top U.S. plays from 2011 to 2016. As shown in Figure 10.4, the folded lines represent the average FP water (in m^3) per well—CMPW for the first year after the well is fractured, whereas the bars depict the average gas or oil produced per well for each play during the same period.[27]

In shale gas plays, Haynesville and Marcellus exhibited not only the highest but also most rapid growth in single-well gas production, while their produced FP water remained low and relatively stable. Permian showed the opposite pattern, with each fractured well displaying moderate increases in gas production rates but producing the greatest amount of FP water, experiencing the fastest upsurge (350% increase). In the scenario of tight oil, all four major regions demonstrated substantial progression in single-well oil production. However, Eagle Ford outpaced the rest with a 560% increase in FP water production for each well. Permian secured the second-place position with a 290% growth rate, but once again ranked first for the absolute volume of FP water produced for each well. Representative wells from Bakken produced a significant volume of FP water, but the annual increase rate was moderate (140%). Niobrara showed both low production rates and relatively small growth for the FP water produced from each oil well during the same period.[27]

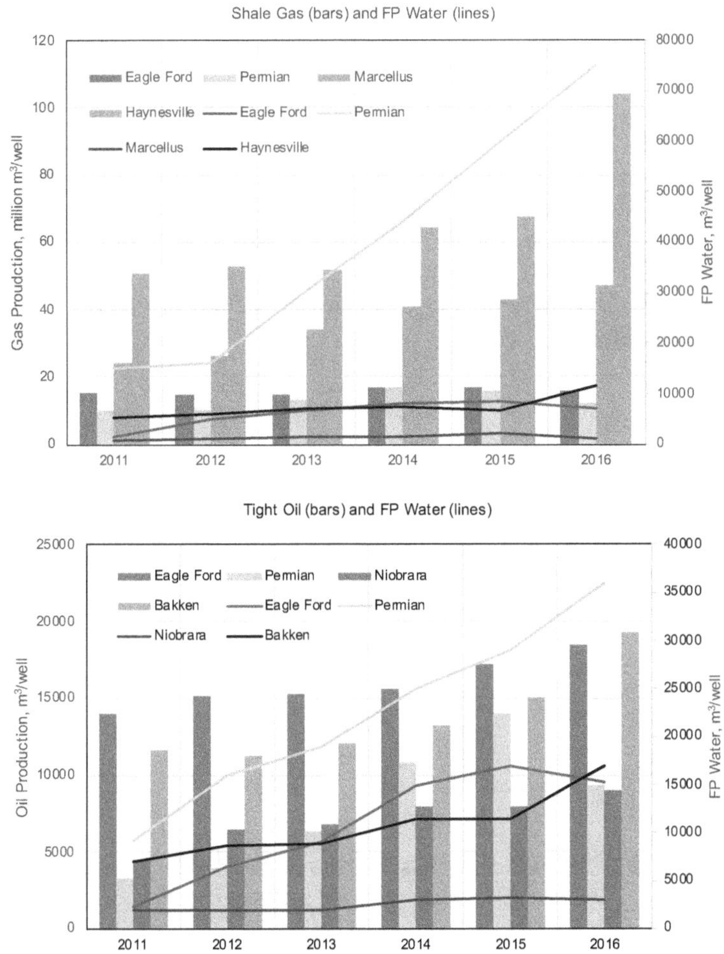

Figure 10.4. Per-well shale gas, tight oil, and FP water production in major U.S. shale plays. Produced with data from.[27]

The tremendous water requirements for fracturing injection and the water returned from these operations contribute to elevated water footprints in unconventional fields. An increase in the water footprint for a specific play doesn't necessarily correlate to an affirmative index of increased scaling risk. This correlation depends on various factors, including well types, the geology of the formation, and operators' fracturing design practices. Nevertheless, it can be considered a reflection that emphasizes the need for more proactive scale management, especially in Permian and Eagle Ford.

10.2.4.2 Water-Commingling Scenario

Commingling different water sources can introduce unexpected scaling risks. In a conventional well, formation water is produced from the reservoir through the wellbore, pipes, and finally to the wellhead. The water starts typically from an equilibrium and could initiate self-scaling due to the gradients in temperature and pressure generated during this production process. If the scaling is prevented before the water chemistry is sampled and analyzed from the wellhead, the overall scaling risk can be fairly assessed through common scale modeling software.[4] The solid lines in Figure 10.5 depict representative Bakken formation water with high TDS and calcium, starting at equilibrium with saturation index (SI) near zero for calcite and barite in the reservoir and ending with a slight scaling tendency for calcite (SI = 0.3) after reaching the surface. Unless in specific scenarios such as enhanced oil recoveries, water injection fields, scale squeeze treatments, or interlinking multiple producing zones, water commingling typically occurs at surface facilities, enabling scale management to be handled spatially and over time.

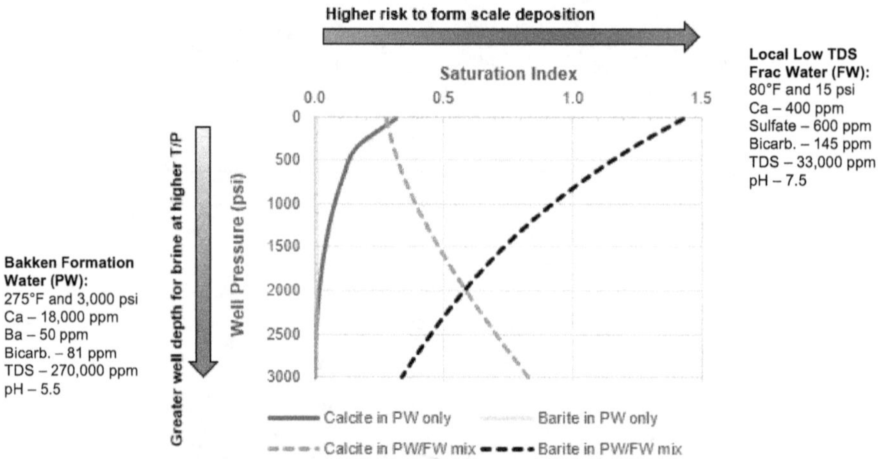

Figure 10.5. Schematics of self-scaling for produced water in conventional wells and water-commingling scales in fracturing.

However, water-commingling is an inherent challenge in every design of fracturing operation. In the same Bakken case, when typical local creek water with a low-scaling tendency is injected into the same formation and contacts the connate water, assuming a mixing ratio of 50/50, the scaling scenario is expected to alter significantly, leading to severer downhole calcite scaling (SI = 0.8) and additional barite from the wellhead to downhole locations. The situation becomes more intricate considering that the mixing ratio of injection water and formation water is a variable, and the injection water chemistry can be revised due to the rock-water interaction discussed in the previous section. Furthermore, in many cases, the laterals of unconventional wells can spread so widely under the surface that their formation water chemistries can be distinct from each other, even when sharing the same well location. Researchers at Chevron reported a case study for shale development in the Delaware Basin, in which unexpected barite scale plugged the production due to unforeseen Ba-rich water invasion from another formation zone. However, the neighboring well from the same pad did not experience the barite issue at all.[18] This is not considered an isolated incident in shale plays. Given that formation water chemistry is typically unavailable or inaccurate for unconventional wells before being drilled, the scale control strategy is often initiated inconsiderately only with the low-risk injection water, representing significant biases from the actual scaling scenarios. As a proactive measure for scale prevention, "the best time for scale inhibitor placement is at the frac stage" for hydraulically fractured horizontal wells.[18] And this principle should be applied for most fracturing jobs, even utilizing "fresh water" for the injection.

10.2.4.3 Uncertainty in Flowback Water Chemistry

With water-commingling, the presence of other components from drilling and fracturing fluids, and recent rock-fluid interactions, either the injection water or even the formation water can sufficiently represent the subterranean fluid for hydraulic fracturing. The flowback water results from all these mechanisms. Thus, it can be considered a realistic matrix to provide a better understanding of the scaling potentials in a fracturing job. Unfortunately, there are still a couple of drawbacks to this strategy. First, unless for refracturing, flowback water for a specific well is typically not available until the target well is completed and POP. Planning scale control for a specific well based on its flowback water would tend to focus on remediation rather than a preventative measure. In practice, flowback water from neighboring wells or the same field can be referred, at the cost of encountering unexpected scales. Second, as discussed previously, the large quantity of the injection water brings a "shock" to the connate fluids, without providing sufficient period for the fractured shale to rebalance. Thus, the water chemistry can still be evolving even after the flowback stage starts.

Figure 10.6 illustrates a flowback scenario when the local freshwater was used as the injection source in the Permian Basin. The major ions and the potential mass of precipitation at "–10 days" represent the initial injection water, with values that were extremely low. Time at zero indicates the POP date, when the well started to produce after a shut-in of 10 days. As seen in the graph, the calcium concentration increased rapidly from nearly 0 ppm to 4,000 ppm within 3 days and peaked at 6,000 ppm at about 50 days after POP, then gradually declined to 2,000 ppm by 90 days after POP. Alkalinity increased initially as well but appeared to decline continuously, whereas sulfate showed more fluctuations—similar to alkalinity but with an increase after 60 days post POP. The temporal evolution of major scaling ions in the flowback water delineates the dynamic scaling risks associated with the fractured well at various stages. In this particular case, a combination of calcite and anhydrite scales is anticipated in the early stages, yet a rapid transition to a dominant anhydrite scale is projected. Consequently, an effective scale management program should prioritize considerations for the scale inhibitor performance over sulfate scales.

Numerical flow models have been developed to improve the understanding of the evolving composition of brine within the fractured shale. These models incorporate inputs such as reservoir properties, fracture characteristics, and the composition of injection fluids and flowback waters. Consequently, they can generate outputs including in situ formation water chemistries, source apportionment for scaling ions in the flowback, and more precise predictions of reservoir scaling.[9] However, the information derived from these models remains relevant for root-cause scaling analysis in wells already completed. Looking ahead, models capable of predicting the composition of flowback or in situ formation water based on the injection water, geological parameters, and engineering design will emerge as powerful tools for preventative scale management in unconventional plays.

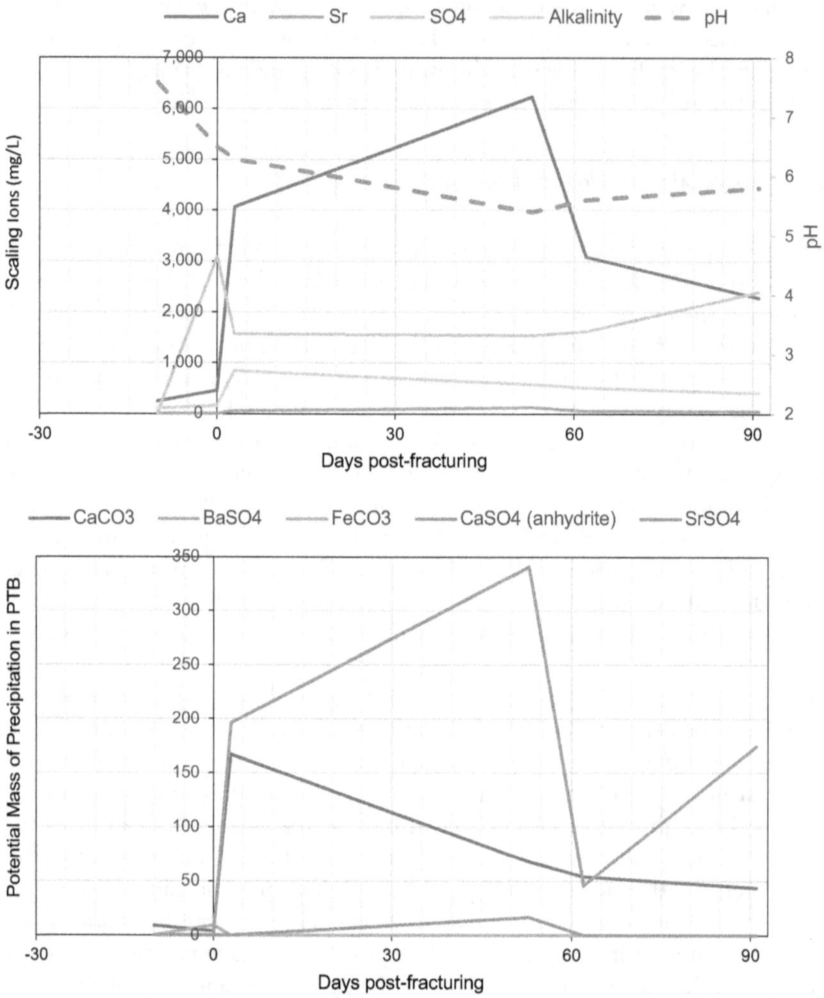

Figure 10.6. Variation in water chemistry and the associated scaling tendency in fracturing flowback water.

10.2.4.4 Diverse Water Chemistries

The minerals dissolved in flowback or produced waters play a critical role in determining scaling in oil and gas production. However, the composition shows significant variations among geographic regions due to diverse geological conditions and well operating designs. Figure 10.7 illustrates the differing physical appearances of representative flowback water samples collected in various plays. Notably, the hazy dark green color in samples from Marcellus and Bakken indicates the presence of significant iron and high TDS, while clearer water from Barnett and Niobrara generally correlates with lower TDS levels.

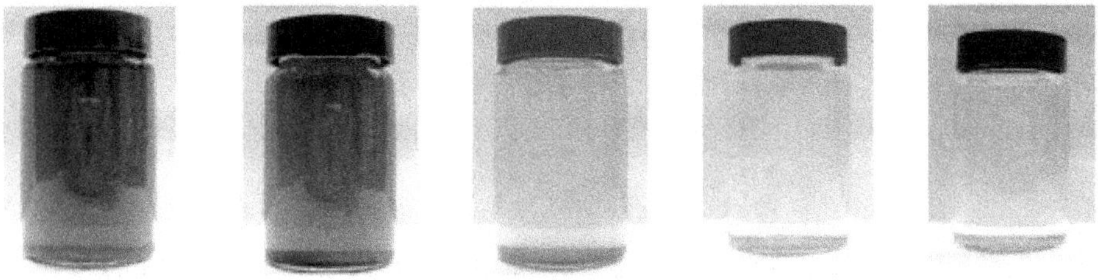

Figure 10.7. Physical appearance for flowback waters from shale plays.

Shaffer *et al.* summarized the high-frequency TDS ranges for produced water from four shale plays as follows: Bakken-Williston Basin (200,000–300,000 ppm), Marcellus-Appalachian Basin (150,000–300,000 ppm), Permian Basin (30,000–150,000 ppm), and Niobrara-Powder River Basin (5,000–15,000 ppm).[28]

While acknowledging that TDS provide only a limited perspective for scaling potentials, the more critical aspect lies in the speciation composition of these scaling ions. The U.S. Geological Survey establishes a national geochemical database for produced water, comprising around 110,000 water analyses gathered from wells across the country spanning from 1905 to 2014. Each entry not only includes the chemical composition but also crucial information such as collection date, location, formation, basin, and well types (shale gas, tight oil, conventional hydrocarbon, *etc.*). By the conclusion of 2023, the total count of usable entries for unconventional wells stood at 11,400, represented by dots in orange, maroon, and purple in Figure 10.8. In comparison to the majority of 98,200 conventional wells (depicted as white dots), their numbers are still relatively modest. The current version (V2.3) of the database surprisingly includes only tens of unconventional wells in plays known for high fracturing activities, such as Permian, Eagle Ford, and Haynesville. One potential explanation could be related to data entry labeling, and it is anticipated that this issue will be addressed in the next update. By analyzing all the accreditable entries, detailed water compositions can be categorized across four basins, as presented in Table 10.1. Furthermore, it facilitates a direct comparison of water chemistries between conventional and unconventional wells in the respective regions.

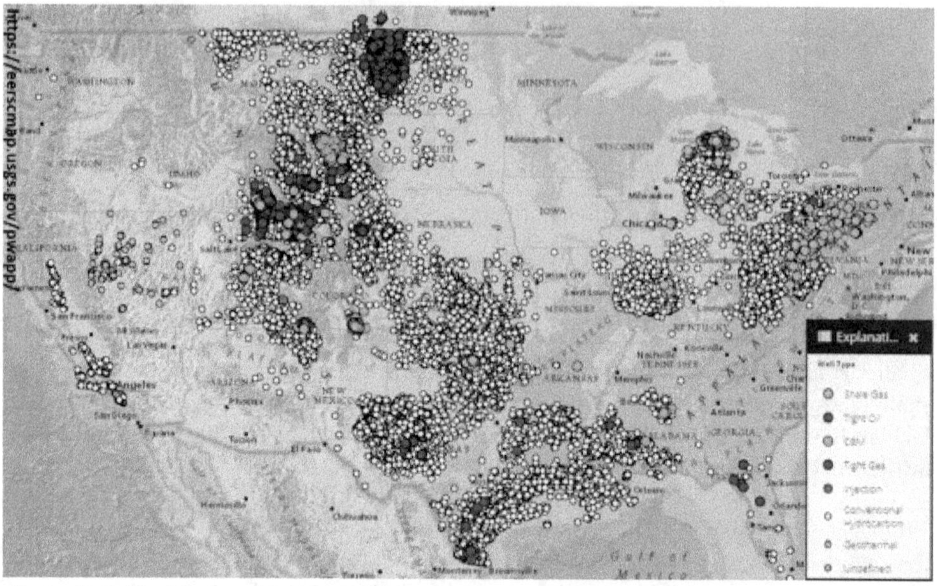

Figure 10.8. The interface of USGS National Produced Water Geochemical Database. Reproduced from database v2.3, https://eerscmap.usgs.gov/pwapp/.

Table 10.1: A comparison of water chemistry from conventional and unconventional wells from four major plays.

	Well Types	Well Counts	TDS	Ca	Sr	Ba	Fe (Total)	SO_4^{2+}	HCO_3^-	pH
Appala-chian-Marcellus	Conv.	321	175,600	17,700	1,380	571	102	267	167	6.2
	Shale G.	429	106,300	9,600	1,870	2,260	63	80	169	6.6
Williston-Bakken	Conv.	10,109	162,800	12,300	595	44	39	1,500	257	6.7
	Tight oil	466	263,300	13,000	832	67	90	793	291	6.2
Permian	Conv.	15,771	101,900	5,360	206	31	74	1,960	494	7.0
	Tight oil	23	99,100	2,270	540	-	73	549	704	7.4
Haynes-ville	Conv.	1,887	74,700	4,970	235	41	94	805	-	6.9
	Tight oil	33	86,400	1,990	-	19	12	22	-	7.1

Note: "Conv." refers to conventional well, whereas "Shale G." means shale gas wells.

In the Marcellus Basin, produced water from the conventional wells displays higher levels of TDS and total soluble iron, whereas unconventional wells exhibit relatively lower levels, possibly due to the widespread use of freshwater in hydraulic fracturing operations. Conversely, barium levels the transition from moderate in the conventional wells to exceptionally high in the unconventional wells. Effective scale management in the region may need to concentrate on addressing incompatible water and managing scale types such as barite, calcite, and iron carbonate.

In the Bakken formation, produced water from the unconventional wells shows elevated levels of all scaling ions except sulfate, compared to conventional wells. Moreover, the higher reservoir temperature in the Bakken poses additional challenges for scaling risk in the shale development. Similar considerations should be taken into account as in the Marcellus Basin. Extra attention is needed to ensure the compatibility of scale inhibitors with local high-TDS/iron brine and the widespread use of cationic friction reducers in the region.

These synopses offer a broad perspective, and it is necessary to acknowledge the inherent heterogeneity within individual shale plays and, indeed, at each well site. Furthermore, ongoing progress in water management and treatment methodologies constitutes a pivotal factor in mitigating challenges associated with produced water in shale operational contexts.

10.3 Challenges for Scale Inhibitor Application in Fracturing Fluids

Scale deposition can be mitigated through appropriate water management and/or well intervention. However, scale inhibitors are strongly recommended as a preventive measure when mineral oversaturation is expected based on the assessment or field experience. In unconventional operations, a scale inhibitor is commonly included in fracturing fluids at fixed rates typically ranging from 100 ppm to 500 ppm. There are two major types of fracturing fluids: gels and slickwater. Gels have a dynamic viscosity greater than water, allowing them to carry the proppant further into the fractures. Slickwater uses friction reducers to lower fluid viscosity, improving overall efficiency in the transportation of proppant. Besides these two categories of chemicals, hydraulic fracturing fluid can consist of any number of other additives, including surfactants, clay stabilizers, breakers (oxidizers, acids, enzymes, viscosity stabilizers), biocides, acids, salts, fluid loss control additives, corrosion inhibitors, iron control agents, foamers and antifoaming agents, $H_2S/O_2/CO_2$ scavengers, viscosifiers, weighting agents, emulsifiers, pH control additives, chelating agents, mutual solvents, and so on. Scale inhibitors may interact with many of the above substances, due to the negatively charged nature. Understanding these chemical interactions is crucial for developing effective scaling mitigation strategies in fracturing.[20–22]

Moreover, as indicated in Table 10.1, soluble iron is frequently present in the produced water from numerous U.S. shale basins. Consequently, there is a need for scale inhibitors that demonstrate sound tolerance to iron and effective inhibition performance against the common scales besides siderite, particularly in regions like Marcellus and Bakken.[11]

10.3.1 Scale Inhibitor Tolerance with Iron-Rich Water

The origin of iron involves the chemical or geochemical leaching from iron-rich formation rocks such as siderite, chlorite, ankerite, and pyrite. Additionally, iron can emerge as a by-product from the corrosion of carbon steel pipelines. The presence of soluble iron poses significant challenges in scale control, leading to the formation of iron-related scales and disrupting the performance of various scale inhibitor chemistries. However, the mechanism behind the interaction of iron-scale inhibitors remains a topic of debate, especially in scenarios involving mixed scaling, such as in calcium/iron carbonate scale systems and iron carbonate/barite matrices.

In earlier publications, it was observed that both Fe(II) and Fe(III) could slow down the growth of calcite crystals and delay the onset of calcite formation in dynamic scaling tests.[29] Under defined conditions, they themselves can act as "scale inhibitors" for calcite. In the presence of classic scale inhibitors, such as diethylenetriamine penta(methylene phosphonic acid), their inhibition performance on barite could be improved. Graham *et al.* attributed this phenomenon to an in situ pH drop caused by the formation of secondary siderite scale over the active growth sites on the majority calcite matrix.[30] Another theory suggests that the incorporation of iron into the calcite crystal distorts the lattice, thereby reducing the overall scale growth.[30, 31] It is worth noting that the observed marginal benefits of soluble iron in controlling calcite and barite were obtained through testing brines with relatively low iron concentrations, typically at sub-ppm levels.

Contemporary iron-containing brine from U.S. oilfields commonly exhibits iron concentrations exceeding 10 ppm (approximately a 50% chance), sometimes falling within the range of 10–100 ppm (approximately 30%).[32] In the Bakken Basin, water composition typically includes iron concentrations ranging from 50 ppm to 300 ppm, alongside calcium concentrations ranging from 10,000 ppm to 25,000 ppm. A series of studies have explored the impacts of iron on common scale inhibitors under more realistic oilfield conditions. For instance, Graham et al. reported that the addition of Fe(II), even in concentrations as low as 10 ppm, can significantly impair the performance of all scale inhibitors, including the previously considered iron-tolerant poly(vinylsulphonate). This led to up to a 100-fold increase in the minimum effective dose for some phosphonate inhibitors.[30] Zhang conducted a comprehensive study examining the factors affecting the Fe-scale-inhibitor system, including the trace levels of dissolved oxygen, scale inhibitor chemistries (phosphonates vs. polymers), scaling temperature, pH, and the inclusion of synergistic iron chelators (citric vs. EDTA).[32] Shen *et al.* investigated a scenario using Marcellus brine, which contains elevated levels of calcium and barium, coupled with 200-ppm iron. Inhibition performance dropped sharply for common inhibitors, including amino tri(methylene phosphonic) acid, polycarboxylic acid, and carboxymethyl inulin. However, specially formulated inhibitors were nearly immune to the high-iron stress.[33] Similarly, researchers from Halliburton[11] focused on the scale challenges in Bakken formation water with both high calcium and iron (up to 235 ppm) and relatively higher temperatures. Through a range of inhibitor chemistries, one novel phosphonate was screened out and demonstrated the super scale inhibitor-brine tolerance and excellent inhibition performance. Based on these investigations, there is a consensus that the presence of soluble iron in typical oilfield brine has an adverse effect on the effectiveness of scale inhibitors in controlling calcium carbonate and barite.

The detrimental impacts of high levels of iron on scale inhibitors can be explained from a few aspects. First, iron is considered to have a "poisoning effect" on the common scale inhibitors, due to its strong affinity for the inhibitor molecules. Nitrilotriacetic acid (NTA), a well-known metal chelator sharing similar functional groups with many phosphonate-based scale inhibitors, has chelation stability constants (in Log_{10}) of 4.83, 6.41, 5.46, and 4.98 for Ba^{2+}, Ca^{2+}, Mg^{2+}, and Sr^{2+} (the four most abundant scaling cations), respectively. However, they are 8.84 and 15.87 for Fe(II) and Fe(III), respectively. The higher values for Fe indicate that the functional groups of scale inhibitors prefer to complex with iron, compared to other cations in the brine.[34] Therefore, a higher inhibitor dosage would be required to overcome the presence of iron, especially in the form of Fe(III).[31] Secondly, iron atoms insert into the growing crystal structures of other scales, leading to coprecipitations—essentially new forms of scales, which imposes new performance requirements for common scale inhibitors.[35] Furthermore, the complexes of iron with common scale inhibitors appear not only unfavorably stable but also poorly soluble in brine. The scale inhibitor can form pseudo precipitations with many divalent cations, when the levels of cations and inhibitor are both high. This phenomenon is typically referred to as scale inhibitor brine compatibility or tolerance. As shown in Figure 10.9, a common scale inhibitor is only compatible with typical Bakken water up to 1,000 ppm and can cause precipitate at 2,000 ppm. In contrast, specially crafted inhibitor chemistry demonstrates wider brine compatibility up to 10,000 ppm. Inhibitor precipitating at high rates not only reduces the inhibitor performance but, more importantly, may cause damage to the formation or blockage on the injection ports. It is essential to consider both the scale inhibitors' tolerance in high-iron water and scale inhibition performance in the presence of iron.[36]

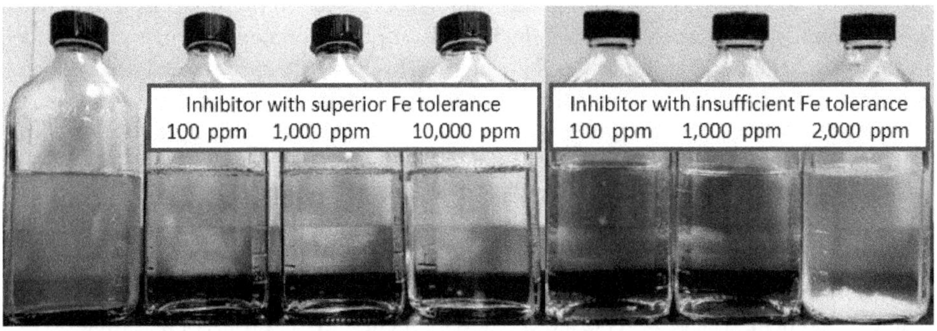

Figure 10.9. Inhibitor compatibility testing with iron-containing brine. The fluffy precipitate in the last photo indicates an incompatible scenario under the given conditions. From left to right: Blank, iron-tolerant scale inhibitors at 100/1,000/10,000 ppm, regular scale inhibitor at 100/1,000/2,000 ppm. Conditions: typical Bakken brine with 200 ppm iron, 250°F for 4 hours.

10.3.2 Scale Inhibitor Compatibility with Cross-Linkers

Cross-linkers play a crucial role in polymer-based fracturing fluids by interlinking linear gel molecules, creating a robust 3D structure. This process significantly increases viscosity and enhances the proppant transport capability of the fluid. Hydraulic fracturing gelling fluids utilize two categories of cross-linking additives: boron-based and metallic cross-linkers.

Boron-based cross-linkers, such as borate, are the most prevalent type. In this process, borate and the guar or its derivatives form the cross-linking network at high pH levels (typically between 8.5 and 11.5) and can be delinked when the fluid pH is lowered, or breakers (such as oxidizers) are activated. Boron cross-linking relies on hydrogen bonds, making it relatively weak, reversible, and re-healable. Boron and its common additives do not have an apparent affinity for scale inhibitors and are not expected to contribute to the formation of common scale types in the oilfield. Therefore, the presence of boron has only limited impact on the scale inhibitor performance. However, the cross-linking process requires elevated pH levels, introducing additional risks for carbonate scale in calcium-rich water and brucite, when seawater is used as part of the fracturing fluids.[19] On the opposite side, the acidic nature of many scale inhibitors can affect the viscosity and break times of frac-pack fluids. Overall, by optimizing fluid formulations, these challenges can be addressed to meet the designed rheological and inhibition performance.[37]

Several studies have been published incorporating various scale inhibitors into borate cross-linked gels for different applications. For instance, Marquez *et al.* discussed the development process to incorporate a "pentaphosphonate" scale inhibitor into a seawater-based frac-pack operation. Adjustments had to be made to the cross-linking package, polymer loading, and fluid pH to solve the incompatibility between the inhibitor, cross-linking agents, and buffering conditions.[38] Patterson *et al.* reported a deepwater West African field with water-sensitive formations experiencing a high-barite tendency due to seawater breakthrough. To increase the volumes of scale inhibitor deployed in preproduction treatments while minimizing the water injection, high levels of phosphonate inhibitor were included in the frac-packing gel and injected following relatively small conventional scale squeeze treatments. The authors reported that this innovative approach extended well treatment life from 190,000–550,000 bbl of produced water to 860,000–1,000,000 bbl of produced water.[39] Ping *et al.* deployed a traceable polymeric scale inhibitor as part of a combined fracture package to squeeze a high-temperature reservoir (200°C) in an Australian gas field. The scale inhibitor features high efficacy in the prevention of carbonate and sulfate scales, great thermal stability, and low formation damages to the reservoir with low permeability, low reservoir pressure, extremely high temperature, and water sensitivity.[40] In a recent development, Yue *et al.* reported the development and field trials of a seawater-based fracturing fluid package integrated with a polymeric scale inhibitor to treat barium-rich formations in the Gulf of Mexico offshore field. The study emphasized solving technical challenges for long-term (2–6 months) barite inhibition due to the extended field shut-in after well completion and maintaining fluid rheology while maximizing inhibitor loading with minimum gel debris to reduce formation damage risk.[20–22] In terms of scale inhibitor placement, borate cross-linking gel systems can be considered relatively benign matrices.

Another type of cross-linking agent falls into the category of metallic cross-linkers, including salts of aluminum, chromium, iron, titanium, and especially zirconium. Metallic cross-linkers create much stronger covalent bonds, boosting gel viscosity significantly. Fluids cross-linked with metallic agents have advantages like maintaining high viscosity at elevated temperatures (up to 350°F) and tolerating a broader pH range (3.5–10.5). In formations using CO_2 foaming or energized fluids, metallic cross-linkers are preferred, because boron-based cross-linking typically occurs at $pH > 9$ and breaks the gel at lower pH levels.

However, metallic cross-linker will directly interfere with scale control. Unlike borate, almost all cations from metallic cross-links can independently form mineral precipitation or coprecipitate with other scale forms. Additionally, they can cause severe compatibility issues with scale inhibitors at levels much higher than borate. In terms of scale inhibition chemistries, most inhibitors disrupt the nucleation of scaling cations (Ca^{2+}, Mg^{2+}, or Ba^{2+}) using chelation functional groups like aminocarboxylic acids or aminophosphonic acids in their molecular structures. These functional groups also have high affinities for Zr and other transition metals. For example, an amino-bis-methylphosphonate compound can form a strong insoluble Zr complex with the ability to adsorb more transition metals. The feasibility of such metal-ligand reactions can be measured using chelation stability constants. Common aminocarboxylic-based chelation agents, such as ethylenediaminetetraacetic acid (EDTA) and nitrilotriacetic acid (NTA), show higher stability constants for Zr^{4+} compared to other common scaling cations (Ba^{2+}, Ca^{2+}, Mg^{2+}, Sr^{2+}). For instance, EDTA has stability constants of 7.76, 10.96, 8.69, and 8.63 for Ba^{2+}, Ca^{2+}, Mg^{2+}, and Sr^{2+}, respectively, while the value for Zr^{4+} is 29.9. Similar trends are seen for NTA, with stability constants of 4.83, 6.41, 5.46, 4.98, and 20.80 for Ba^{2+}, Ca^{2+}, Mg^{2+}, Sr^{2+} and Zr^{4+}, respectively.[34] The higher values for Zr suggest that the functional groups of scale inhibitors tend to form complexes more readily with Zr compared to other cations in brine. For instance, Alsaiari *et al.* reported that the presence of a 10-ppm Zr-cross-linker reduced the calcium sulfate inhibition performance for all six scale inhibitors involved in the study.[41]

On the other side, the competition for metallic cross-linkers is raised by the presence of scale inhibitors, reducing their availability for the cross-linking process. In a fracturing gelling system, this interaction can weaken the covalent bonds between the cross-linker metals and guar or guar derivatives, ultimately compromising the rheological performance of the cross-linked fluid. Because of this, liquid scale inhibitors are often introduced in the uncross-linked early stages or are even excluded entirely from the fracturing treatment. During these treatments, fracturing fluids from other stages and reservoirs that contact these fluids are exposed to scale deposition risks. An alternative approach involves the use of solid scale inhibitors or inhibitors impregnated in proppant. Solid types of scale inhibitors can gradually release active ingredients that are preadsorbed on porous matrix materials. The time-release mechanism prevents a sudden delivery of inhibition chemicals at a high dosage, thereby significantly reducing compatibility issues. However, the types of chemistry and the total amount of ingredients that can be reversibly adsorbed to the solid matrix are limited. For instance, a drilling company reported that solid scale inhibitors were effective but could only protect a small portion of the total volume of water expected in their deepwater projects.[37] These concerns, coupled with the higher cost, limit the widespread use of solid scale inhibitors in the industry.

Few studies in the literature have attempted to address the challenge posed by the interaction between liquid scale inhibitors and metallic cross-linked gels. Yue *et al.* conducted a study, where two phosphonate and seven polymeric scale inhibitors were screened for their application in a Zr-based gelling system. They identified two polymer chemistries that exhibited manageable impacts on gel rheology and tolerated the detrimental effects of the optimized dosage of Zr-cross-linker on calcite inhibition.[34] Similarly, Wylde *et al.* explored these effects in the development of a Zr-gelling seawater system and selected two polymeric scale inhibitors with minimal mutual impacts to and from the cross-linking fluids.[42] Despite these efforts, challenges persist in reconciling the compatibility between liquid scale inhibitors and metallic cross-linked fluids. A thorough qualification process is necessary, and balanced requirements should be targeted, when an inhibitor metallic-cross-linked fluid is needed.

10.3.3 Scale Inhibitor Interaction with Friction Reducers

Friction reducers are one of the key chemical additives in the slickwater fracturing fluid. Upon introduction into water, it demonstrates the capability to significantly reduce water friction pressure by up to 70%, thereby alleviating stress on pumps and augmenting the overall efficiency of proppant transportation and distribution during operations. Friction reducers are typically incorporated into fracturing fluid at concentrations ranging from 0.2 to 2 gallons per 1,000 gallons.

The fundamental component in the formulation of friction reducers is long-chain polyacrylamide-based copolymers modified with additional monomers to optimize their performance under diverse water salinities. Classification of friction reducers can be based on their charges, leading to distinctions as anionic, cationic, or amphoteric. Anionic friction reducers, which include acrylate and/or acrylamidomethylpropane sulfonate in the polymer backbone, are predominantly employed in most U.S. shale plays. Cationic friction reducers are typically synthesized with acryloyloxyethyltrimethylammonium chloride or acrylamidopropyltrimethylammonium chloride. Amphoteric friction reducers involve the polymerization of both anionic and cationic monomers.[43]

Given that the majority of scale inhibitors carry negative charges, it is anticipated that they will remain inert, when combined with anionic friction reducers but can interact with cationic or amphoteric friction reducers. This interaction gains particular significance in scenarios where polymeric scale inhibitors and cationic friction reducers are used simultaneously. In a typical slickwater design, the friction reducer is selected first, followed by the determination of other components, such as scale inhibitors. Regarding friction reduction performance, anionic friction reducers demonstrate compatibility with nearly all types of scale inhibitors. However, cationic friction reducers will have limited options for compatible scale inhibitor chemistries.[43] Inappropriate selection of these two chemicals can compromise friction reduction performance and lead to the precipitation of "gummy-bear" structures, posing a risk of formation damage. Surprisingly, the impact of friction reducers on scale inhibitor performance has been a subject of debate. In a study examining such impacts from both anionic and cationic friction reducers, the minimum inhibition concentrations were found to increase—two to five times for all five scale inhibitors in the presence of friction reducers, irrespective of charge.[44] The mechanism behind this observation remains unclear, possibly linked to specific testing conditions. As an industry standard practice, it is recommended to validate the friction reduction and scale inhibition performance for the complete suite of the chemical additive package selected for the slickwater fracturing.

10.4 Scale Control Innovations for Unconventional Operation

10.4.1 Solid Scale Inhibitors

Application of scale inhibitors in liquid form is a common practice to mitigate scale deposition during fracturing operations. Typically, these inhibitors are introduced either through premixing or via real-time mixing, also known as the "on-the-fly" process, with the fracturing fluids during the injection stages. Alternatively, solid scale inhibitors can also be applied in real time alongside the proppant pumping process. These solid products are usually available as inhibitor-impregnated proppants, capsules, solidified substances, or nanoparticle carriers.[45] Solid scale inhibitors provide a slow release of scale control ingredients, reducing fluid incompatibility issues during well completion and offering long-term scale protection throughout the well's production period. The fundamentals of solid scale inhibitors will be addressed in another chapter and are not discussed in detail here.

10.4.2 Environment-Friendly Inhibitor Package

10.4.2.1 Chemical Disclosure

Shale gas operators face ongoing demands to provide comprehensive disclosure of the composition of fracturing fluids, including information on additive types, CAS registry numbers, and concentrations. For instance, FracFocus serves as a public database for "the national hydraulic fracturing chemical disclosure registry" in the United States. Managed by a partnership between the U.S. Ground Water Protection Council and the Interstate Oil and Gas Compact Commission, FracFocus has included reports from more than 1,600 companies and detailed chemicals used over 189,000 hydraulic fracturing jobs in the United States as of March 1, 2020. FracFocus disclosures provide information on well IDs, names, locations, and details about the injected chemicals, including trade names, suppliers, product purposes, and associated hazardous ingredients. Detailed disclosures can be found on the FracFocus website (http://fracfocus.org). While requirements for chemical disclosure vary by region or country, there is a clear trend toward stricter regulations against hazardous chemicals in treatment packages.

10.4.2.2 Green Chemistries for Scale Inhibitors

For scale management, there has been a noticeable emergence of environment-friendly chemistries in hydraulic fracturing, addressing public apprehensions about the use of hazardous chemicals. Scale inhibitors can be considered "environmentally friendly," if they possess some or all of the following characteristics: (1) renewability—utilizing materials derived from natural organic molecules, such as plant extracts or biomass raw materials; (2) nontoxicity; (3) biodegradability and nonbioaccumulation in ecosystems. In addition to these green criteria, new chemistries must meet fundamental requirements for scale inhibitors, such as inhibition performance, thermal stability, brine compatibility, and cost-effectiveness. Ideally, green scale products should combine all these features, but practically they may need to be ranked under a weighted matrix, balancing performance with toxicity and biodegradability thresholds. Green chemistries for general oilfield scale management will be addressed in another chapter and are not discussed in detail here.

10.4.2.3 Ranking of Entire Inhibitor Package

Scale products used in fracturing are typically formulated packages in which active scale control ingredients account for a relatively small portion. A significant part of these products includes water or other chemical additives like solvents (*e.g.*, alcohols, glycols, mutual solvents), acids, neutralizers, biocides, surfactants, and so on. These additives serve multiple functions, including antifreeze, formulation stabilization, wettability alteration, and antifouling. However, many of these additives are classified as hazardous and may have a greater environmental impact than the scale inhibitor itself. Effective ranking models consider the overall effect of these chemical packages.

For instance, Halliburton and its partners introduced the Chemistry Scoring Index (CSI) system to rank potential Health, Safety, and Environmental hazards for products used in the oil and gas industry. The CSI tool assigns hazards to all product components based on their intrinsic hazard properties and generates numerical scores based on the individual component hazard assignments and their percent composition. The scoring matrix summarizes assessments from three major categories (Environmental, Physical, and Health) and includes over 50 subcategories such as acute aquatic toxicity, hazardous air pollutants, biodegradation, endocrine disruption, flammability, organic peroxide presence, corrosivity, carcinogenicity, acute toxicity, reproductive toxicity, mutagenicity, and so on. For example, CSI scores are 0 for water and greater than 360 for formulations containing nonhazardous active ingredients but 10%–29.9% methanol. In peer bidding processes, a product with a lower CSI score is considered a safer alternative within its use group.[46, 47] Similarly, Southwestern Energy Company's "The Right Products Program" assesses all chemical products intended for hydraulic fracturing operations. This program provides a standardized workflow to review all hazardous components before their use.[48] Therefore, developing "green scale products" requires an overall consideration of the environmental impact of the entire chemical treatment package, rather than focusing solely on the scale inhibitor ingredient itself.

10.4.3 Scale Squeeze while Fracturing

Scale inhibitors have been integrated into fracturing fluids to afford initial scale squeeze protection, during the early stages of water breakthrough. Therefore, the engineering team has a better chance to respond with a better designed conventional squeeze. The incorporation of scale inhibitors at percentage levels into a diverse hydraulic fluid poses a nontrivial challenge. A significant hurdle involves addressing the interactions between scale inhibitors and other components of fracturing fluids, including friction reducers, cross-linkers, gel polymers, breakers, pH modifiers, surfactants, and biocides, among others. Special attention must be given to achieving synergistic effects rather than antagonistic interactions within an integrated scale squeeze fracturing package.

Furthermore, traditional squeeze design models, primarily reliant on parameters from conventional wells, need adjustments or the development of new theories to accommodate the unique characteristics of squeezing fractured wells with extended horizontal laterals.

10.4.4 Digitalization

The future of shale development may see the emergence of the digital oilfield trend. Scale control can be seamlessly incorporated into various aspects, including a shared database for unconventional water composition, advanced modeling to predict reservoir scaling based on the formation mineralogy and injection water, and integrated platforms for remote monitoring and control of scale-related processes.

A recent development by BP details the integration of a suite of Azure cloud-based "digital twin" tools, providing online, real-time calculation of scale risk and monitoring the health of barriers deployed to manage these risks on a well-by-well basis. Published case studies describe how this digital dashboard optimizes wash water injection rates to manage halite in assets with produced water handling restrictions and employs a squeeze life tracker to enhance treatment scheduling.[49] This pioneering work demonstrates how the adoption of digital tools facilitates rapid decision-making, minimizes overtreatment of wash water (or scale inhibitors as a forward-looking direction), and mitigates impacts on production.

10.5 Summary

This chapter addresses the challenges of scale prevention in shale gas and oil production, with a focus on unconventional reservoirs and hydraulic fracturing operations. It examines the characteristics of various shale formations, including the Permian, Marcellus, Utica, Bakken, Eagle Ford, Haynesville, Barnett, and Niobrara shales. Key aspects discussed include their mineralogy, reservoir conditions, and water usage, along with the intensified rock-fluid interactions that occur during shale development, particularly in horizontal drilling and hydraulic fracturing. The chapter highlights the impact of subsurface fluid interactions on brine composition and the potential for mineral scale depositions, emphasizing the sensitivity to scaling in low-permeability shale formations and the associated risk of formation damage.

The complexities of water management in unconventional oil and gas operations are thoroughly explored, including challenges related to scale formation and water chemistry. Data are provided on the increasing water usage in hydraulic fracturing over the years, with details on the average water used per well in different regions. The document differentiates between flowback water and produced water, emphasizing the risks of commingling different water sources and discussing the interactions between scaling ions in frac flowbacks and formation rocks and brines. Brine compositions from various shale basins are analyzed, categorizing scale challenges by geological regions and underscoring the importance of understanding specific scale control challenges in shale production.

Innovations and challenges in the application of scale inhibitors in fracturing fluids are also covered, including the types of fracturing fluids commonly used and potential adverse interactions between scale inhibitors and frac fluid additives. The importance of effective scale control strategies is emphasized, along with the compatibility and performance of scale inhibitors in the presence of other chemicals. The chapter concludes by stressing the need for ongoing advancements in water management and treatment methodologies to address the challenges associated with produced water in unconventional operations. It advocates for integrated approaches and digital tools to optimize scale management strategies in the shale production industry.

Nomenclature

AMPS	Acrylamidomethylpropane sulfonate
bbl	Barrels
CAS	Chemical Abstracts Service
CSI	Chemistry Scoring Index
CMPW	Cubic meters per well
EDTA	Ethylenediaminetetraacetic acid
EIA	Energy Information Administration
EOR	Enhanced oil recovery
FP water	Flowback and produced water
FW	Flowback water
HF	hydraulic fracturing
HSE	Health, Safety, and Environmental
mD	Micro-Darcy
MIC	Minimum inhibition concentration (for chemical inhibitors)
MRC/ERC	(Wells with) maximum reservoir contact and extreme reservoir contact
NTA	Nitrilotriacetic acid
O&G	Oil and gas
PAM	Polyacrylamide
POP	Put-on-production
PTB	Pounds per 1,000 barrels
ppm	Parts per million
PW	Produced water
SI	Saturation index
TDS	Total dissolved solids
μm	Micrometer
USGS	United States Geological Survey

References

1. S. Zendehboudi, A. Bahadori, *Shale Oil and Gas Handbook: Theory, Technologies, and Challenges* (Elsevier, 2017).
2. E. Reinsalu, I. Aarna, "About Technical Terms of Oil Shale and Shale Oil," *Oil Shale* 32 (2015): pp. 291–292.
3. *World Energy Resources: 2013 Survey* (London: World Energy Council, 2013).
4. J.E. Oddo, M.B. Tomson, "Why Scale Forms in the Oil Field and Methods to Predict It," *SPE Production & Facilities* 9 (1994): pp. 47–54.
5. W. Wang, W. Wei, Y. Chao, "Scale Prediction and Management for Unconventional Production," SPE-200668-MS, SPE International Oilfield Scale Conference and Exhibition, held June 24–25, 2020 (Virtual).
6. M.S. Blondes, J.L. Shelton, M.A. Engle, et al., "Utica Shale Play Oil and Gas Brines: Geochemistry and Factors Influencing Wastewater Management," *Environmental Science & Technology* 54 (2020): pp. 13917–13925.
7. P.E. Dresel, A.W. Rose, *Chemistry and Origin of Oil and Gas Well Brines in Western Pennsylvania* (University Park: The Pennsylvania State University, 2010).
8. R.M. Horner, C.B. Harto, R.B. Jackson, et al., "Water Use and Management in the Bakken Shale Oil Play in North Dakota," *Environmental Science & Technology* 50 (2016): pp. 3275–3282.
9. X. Wang, E.J. Mackay, "Simulation Study for Scale Management During Shale Gas Production," SPE-190717-MS, SPE International Oilfield Scale Conference and Exhibition, held June 20–21, 2018 (Aberdeen, UK).
10. M. Freyman, *Hydraulic Fracturing & Water Stress: Water Demand by the Numbers* (Boston: Ceres, 2016).
11. Y. Peng, Z. Yue, C. Ozuruigbo, et al., "Carbonate Scale Control Under High Level of Dissolved Iron AND Calcium in the Bakken Formation," SPE- 173714-MS, SPE International Symposium on Oilfield Chemistry, held April 13–15, 2015 (The Woodlands, Texas, USA).
12. K. He, Z. Yue, X. Liang Xu, "Using Sacrificial Agents to Enhance Surfactant Performance in the Eagle Ford Shale," SPE-180156-MS, 78th EAGE Conference and Exhibition, held May 30–June 2, 2016 (Vienna, Austria).
13. Y. Peng, L. Shi, Z. Yue, et al., "Combination Package Development of Scale Inhibitors and Hydrogen Sulfide Scavengers for Sour Gas Production in Barnett shale," SPE-174018-MS, SPE Western Regional Meeting, held April 27–30, 2015 (Garden Grove, California, USA).
14. W. Wang, W. Wei, D. Leach, et al., "Rock-Fluid Interaction and Its Application in Unconventional Production," URTeC-1050, SPE/AAPG/SEG Unconventional Resources Technology Conference, held July 20–22, 2020 (Virtual).
15. H.J. Khan, E. Spielman-Sun, A.D. Jew, et al., "A Critical Review of the Physicochemical Impacts of Water Chemistry on Shale in Hydraulic Fracturing Systems," *Environmental Science & Technology* 55 (2021): pp. 1377–1394.
16. J.E. Thompson, C.S. DeVine, "Fracturing Fluid Interactions With Formation Minerals and Their Subsequent Effect on Formation Permeability," SPE-29500-MS, SPE Production Operations Symposium, held April 2–4, 1995 (Oklahoma City, Oklahoma, USA).
17. A.D. Jew, Q. Li, D. Cercone, et al., "Barium Sources in Hydraulic Fracturing Systems and Chemical Controls on Its Release Into Solution," URTEC-2899671-MS, SPE/AAPG/SEG Unconventional Resources Technology Conference, held July 23–25, 2018 (Houston, Texas, USA).
18. K.N. Baruah, N.M. Al-Otaibi, F.M. Al-Subaie, "Challenges in Scale Inhibitor Treatments in Multilateral Complex Wells—A Production Engineering Perspective With Case Study," SPE 164264-MS, SPE Middle East Oil and Gas Show and Conference, held March 10–13, 2013 (Manama, Bahrain).

19. W. Wang, L. DeVoe, W. Wei, et al., "Tackling a Critical Challenge in Shale Development in the Delaware Basin: An Interdisciplinary Field Case Study on Subsurface Scale Diagnosis and Control and Impact on Production Performance," URTEC-3708626, SPE/AAPG/SEG Unconventional Resources Technology Conference, held June 20–22, 2022 (Houston, Texas, USA).
20. Z. Yue, A. Slocum, X.L. Tian, et al., "An Integrated Scale Protection Package for Offshore Fractured Wells Under Designed Shut-In Extension," SPE-204363-MS, SPE International Conference on Oilfield Chemistry, held December 6–7, 2021 (The Woodlands, Texas, USA).
21. L. Ke, Z. Yue, A. Slocum, et al., "A Novel Seawater-Based Fracturing Fluid for Streamlined Logistics, Long-Term Scale Protection, and Enhanced Oil Recovery," IPTC-23106-MS, International Petroleum Technology Conference, held March 1–3, 2023 (Bangkok, Thailand).
22. A. Slocum, Z. Yue, L.P. Ke, et al., "Case Studies of Novel Seawater-Based Fracpac Fluid Used in Deepwater Gulf of Mexico With Barium-Rich Formation Fluids Present," SPE 218713-MS, SPE Oilfield Scale Symposium, held June 5–6, 2024 (Aberdeen, UK).
23. P.N. Nelson, "Pore-Throat Sizes in Sandstones, Tight Sandstones, and Shales," *American Association of Petroleum Geologist Bulletin* 93 (2009): pp. 329–340.
24. A. Abouie, M. Tagavifar, A. Sanaei, et al., "Production Loss by Scale Formation in Unconventional Reservoirs," SPE-190148-MS, SPE Western Regional Meeting, held April 22–26, 2018 (Garden Grove, California, USA).
25. A. Abouie, A. Sanaei, K. Sepehrnoori, "Simulation Study of Scale Management During Hydraulic Fracturing in Unconventional Reservoirs," SPE-193570-MS, SPE International Conference on Oilfield Chemistry, held April 8–9, 2019 (Galveston, Texas, USA).
26. J.F. Valder, R.R. McShane, J.N. Thamke, et al., "Estimates of Water Use Associated With Continuous Oil and Gas Development in the Permian Basin, Texas and New Mexico, 2010–19," USGS Scientific Investigations Report (Washington, D.C., 2021).
27. A.J. Kondash, N.E. Lauer, A. Vengosh, "The Intensification of the Water Footprint of Hydraulic Fracturing," *Sci. Adv.* 4 (2018): p. aar598.
28. D.L. Shaffer, L.H. Chavez, M. Ben-Sasson, et al., "Desalination and Reuse of High-Salinity Shale Gas Produced Water: Drivers, Technologies, and Future Directions," *Environ. Sci. Technol.* 47 (2013): pp. 9569–9583.
29. J.L. Katz, M.R. Reick, R.E. Herzog, et al., "Calcite Growth Inhibition by Iron," *Langmuir* 9 (1993): pp. 1423–1430.
30. G.M. Graham, R. Stalker, R. McIntosh, "The Impact of Dissolved Iron on the Performance of Scale Inhibitors Under Carbonate Scaling Conditions," SPE-80254-MS, SPE International Symposium on Oilfield Chemistry, held February 5–7, 2003 (Houston, Texas, USA).
31. Z. Zhang, F. Zhang, Q. Wang, et al., "Ferrous Iron Impact on Phosphonate and Polymeric Scale Inhibitors at Temperature Ranging From 25 to 70°C," SPE-173770-MS, SPE International Symposium on Oilfield Chemistry, held April 13–15, 2015 (The Woodlands, Texas, USA).
32. P. Zhang, Z. Zhang, Y. Liu, et al., "Investigation of the Impact of Ferrous Species on the Performance of Common Oilfield Scale Inhibitors for Mineral Scale Control," *Journal of Petroleum Science and Engineering* 172 (2019): pp. 288–296.
33. D. Shen, D. Shcolnik, R. Perkins, et al., "Evaluation of Scale Inhibitors in Marcellus High-Iron Waters. SPE-141145-PA," *Oil & Gas Fac* 1 (2012): pp. 34–42.
34. Z. Yue, Q. Fu, N. Lang, et al., "Liquid Scale Inhibitors for Metallic-Crosslinked Gel Fracturing Systems," SPE 169806-MS, SPE International Oilfield Scale Conference and Exhibition, held May 14–15, 2014 (Aberdeen, UK).
35. H. Davies, M. Toole, R. Higgins, "Development and Field Application of an Iron Tolerant Scale Inhibitor for Use in Fracturing Completions and Production," SPE-190728-MS, SPE International Oilfield Scale Conference and Exhibition, held June 20–21, 2018 (Aberdeen, UK).

36. H. Lu, F. Alattal, J. Ruggeri, et al., "Iron Effects on Scale Inhibitor Performances for Laboratory and Field Operations," NACE-2016-7771, CORROSION 2016, held March 6–10, 2016 (Vancouver, British Columbia, Canada).
37. A.M. Fitzgerald, L.G. Cowie, "A History of Frac-Pack Scale-Inhibitor Deployment," SPE-112474-MS, SPE International Symposium and Exhibition on Formation Damage Control, held February 13–15, 2008 (Lafayette, Louisiana, USA).
38. M. Marquez, L. Schafer, W.D. Norman, "Preemptive Scale Management: Treating With Scale Inhibitor While Frac Packing a Well," SPE-147006-MS, SPE Annual Technical Conference and Exhibition, held October 30–November 2, 2011 (Denver, Colorado, USA).
39. D. Patterson, W. Williams, M. Jordan, et al., "Preproduction-Deployed Scale-Inhibition Treatments in Deepwater West Africa," *SPE Prod & Oper* 32 (2017): pp. 333–342.
40. P. Chen, S. Rawlins, T. Hagen, et al., "Developing a Polymer Scale Inhibitor for a Combined Fracture and Inhibitor Squeeze Treatment for High-Temperature Reservoirs," SPE-194929-MS, SPE Middle East Oil and Gas Conference, held March 18–21, 2019 (Manama, Bahrain).
41. H.A. Alsaiari, M. Al-Khaldi, I. Al-Taie, et al., "Effect of Crosslinkers on the Performance of Calcium Sulfate Scale Inhibitors at High Temperature: Impact of Zirconium," SPE-179878-MS, SPE International Oilfield Scale Conference, held May 11–12, 2016 (Aberdeen, UK).
42. J.J. Wylde, A. Mahmoudkhani, "Development of a Scale Inhibitor for Zr-Crosslinked Seawater Systems: A Case History of Successful Testing to Failure and Field Applications," SPE-169806-MS, SPE International Oilfield Scale Conference and Exhibition, held May 11–12, 2016 (Aberdeen, UK).
43. L. Ke, J. Chano, M. Weston, et al., "Dry cationic Friction Reducers: New Alternative for High TDS Slickwater," SPE-204286-MS, SPE International Conference on Oilfield Chemistry, held December 6–7, 2021 (The Woodlands, Texas, USA).
44. C.Q. Ozuruigbo, C. Fan, "Influence of Friction Reducer on Scale Inhibitor Performance and Scaling Tendency in Slickwater Fracturing," SPE-171661-MS, SPE/CSUR Unconventional Resources Conference, held September 30–October 2, 2014 (Calgary, Alberta, Canada).
45. A.T. Kan, Z. Dai, M.B. Tomson, "The State of the Art in Scale Inhibitor Squeeze Treatment," *Petroleum Science* 17 (2020): pp. 1579–1601.
46. T. Verslycke, K. Reid, T. Bowers, et al., "The Chemistry Scoring Index (CSI): A Hazard-Based Scoring and Ranking Tool for Chemicals and Products Used in the Oil and Gas Industry," *Sustainability* 6 (2014): pp. 3993–4009.
47. Z. Yue, Y. Peng, K. He, et al., "Multifunctional Fracturing Additives as Flowback Aids," SPE-181383-MS, SPE Annual Technical Conference and Exhibition, held September 26–28, 2016 (Dubai, UAE).
48. M. Boothe, S. Videlock, D. Lincicome, et al., "Choosing the Right Products," SPE 189891-MS, SPE Hydraulic Fracturing Technology Conference and Exhibition, held January 23–25, 2018 (The Woodlands, Texas, USA).
49. H.M. Bourne, S. Hosein, G. Keilty, et al., "Integrating Digital Tools To Optimise Scale Management—Case Studies," SPE-218700-MS, SPE Oilfield Scale Symposium, held June 5–6, 2024 (Aberdeen, UK).

Multifunctional Production Chemicals

Sunder Ramachandran[1] and Johnathon Brooks[1]
[1]Baker Hughes, Sugar Land, TX, USA

11.1 Introduction

Production chemicals are used to provide flow assurance, integrity management, and increased production in oil and gas production. Use of these chemicals reduces capital and operating costs of oil and gas production. Flow assurance is the prevention of interruption of flow due to the precipitation of solids. The solids could be inorganic salts, organic hydrates, waxes, or asphaltenes. Chemicals used to provide flow assurance in oil and gas production are listed in Table 11.1.

Table 11.1: Production chemicals for flow assurance and solubility.

Flow Assurance Need	Production Chemical	Solubility
Prevent scale precipitation	Scale inhibitor	Water-soluble
Prevent asphaltene precipitation	Asphaltene inhibitor	Oil-soluble
Disperse asphaltenes	Asphaltene dispersant	Oil-soluble
Dissolve asphaltenes	Asphaltene solvent	Oil-soluble
Prevent paraffin formation	Paraffin inhibitor	Oil-soluble
Dissolve paraffin	Paraffin solvent	Oil-soluble
Prevent hydrate formation	Hydrate inhibitor	Water-soluble
Disperse solids	Dispersant	Oil- and water-soluble
Dissolve solids	Dissolver	Oil- and water-soluble

Integrity management is the preservation of the existing facility. The facility can be corroded by acid gases, oxygen, and microbes. Production chemicals used to provide integrity management are listed in Table 11.2.

Table 11.2: Production chemical for integrity management and solubility.

Production Need	Production Chemical	Solubility
Prevent corrosion	Corrosion inhibitor	Oil- and water-soluble
Remove hydrogen sulfide from production fluid	H_2S scavenger	Oil- and water-soluble
Remove oxygen from production fluid	Oxygen scavenger	Water-soluble
Kill harmful microorganisms	Biocides	Water-soluble

Chemicals are used to remove water from oil, remove oil from water, reduce drag, or change flow regimes. Production chemicals used to increase and optimize production are listed in Table 11.3.

Table 11.3: Production chemicals to increase and optimize production and solubility.

Production Need	Production Chemical	Solubility
Decrease drag in wells and pipelines	Drag reducer	Oil- and water-soluble
Unload gas wells or deliquefy pipelines	Foamer	Oil- and water-soluble
Allow the separation of oil with low water content	Demulsifier	Oil- and water-soluble
Allow separation of water with low oil content	Water clarifier	Oil- and water-soluble

Production chemicals need to be compatible with system fluids and facilities, and work within the system operating condition envelope.[1] Production chemicals require separate chemical injection facilities and tanks. These facilities can be expensive in some cases, especially for deepwater fields.[2] Subsea manifold systems, as illustrated in Figure 11.1, are difficult and expensive to deliver chemicals to. In this figure, one can see the number of umbilical lines to the subsea manifold system. Conventional umbilical can cost between \$500 k and \$800 k per kilometer to install.[3] Production chemicals with many different functionalities can decrease the facility cost. Decreasing the number of umbilical lines reduces the capital cost of a subsea tie back well.

Combining different chemistries into a single product can be difficult. Multifunctional production chemicals must be compatible with each other and not separate. The product needs to be thermally stable. The different functionalities should not affect the performance of each other. The product must have very low amounts of suspended solids. Despite the difficult product conditions, multifunctional production chemicals of different types are used in many fields. In this chapter, we will look at different types of multifunctional production chemicals, the difficulties of combining different products, experimental procedures for thermal stability and compatibility, and field experience with different multifunctional production chemicals.

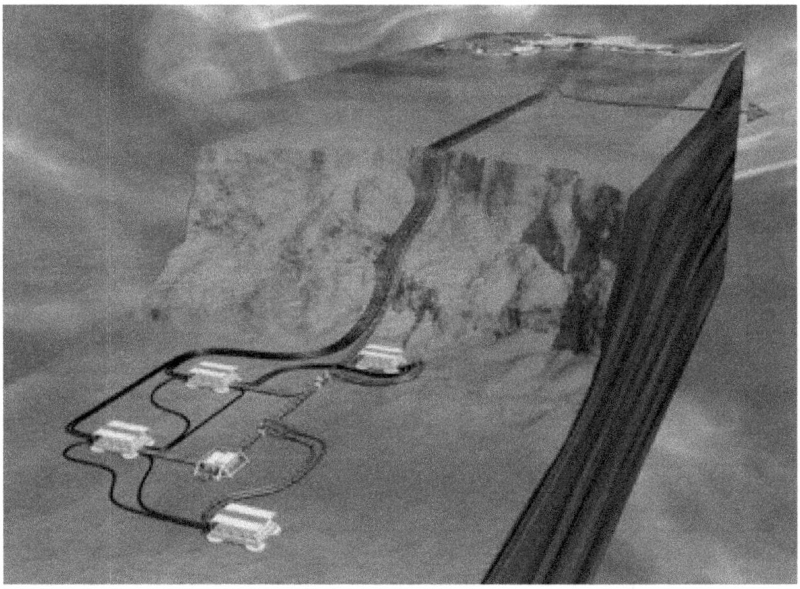

Figure 11.1. Schematic drawing of a subsea manifold system.

11.2 Types of Multifunctional Chemicals

There are many different types of multifunctional production chemicals. Some are binary combinations of different functionalities, while others are ternary combinations of different functionalities. Some combinations are exclusively for flow assurance, integrity management, or production improvement and optimizations. Other multifunctional production chemicals are combinations of these functions.

11.2.1 Binary Combinations for Flow Assurance

Flow assurance chemicals are used to keep oil and gas production flowing. The chemicals range from scale inhibitors, asphaltene inhibitors, asphaltene dissolvers, paraffin inhibitors, hydrate inhibitors, dispersants, and dissolvers. Some binary combinations exclusively for flow assurance are listed in Table 11.4.

Table 11.4: Exclusive binary combination products for flow assurance.

Combination Product	Component 1	Component 2	Comments
Scale inhibitor and asphaltene dissolver	Scale inhibitor	Asphaltene dissolver	Scale inhibitor is water-soluble. Asphaltene dissolver is oil-soluble
Scale inhibitor and paraffin inhibitor	Scale inhibitor	Paraffin inhibitor	Scale inhibitor is water-soluble. Paraffin is oil-soluble and water-dispersible
Scale inhibitor and hydrate inhibitor	Scale inhibitor	Hydrate inhibitor	Both components are normally water-soluble
Scale inhibitors and dispersant	Scale inhibitor	Dispersant	Both components are normally water-soluble
Scale inhibitor and dissolver	Scale inhibitor	Dissolver	Both components are normally water-soluble
Asphaltene inhibitor and paraffin inhibitor	Asphaltene inhibitor	Paraffin inhibitor	Both components are oil-soluble
Asphaltene dispersant and asphaltene solvent	Asphaltene dispersant	Asphaltene solvent	Both components are oil-soluble
Paraffin inhibitor and paraffin solvent	Paraffin inhibitor	Paraffin solvent	Both components are oil-soluble

11.2.1.1 Scale Inhibitors and Asphaltene Dissolvers

Squeeze treatments have been developed that combine scale inhibitors with asphaltene dissolvers.[4] For combination products, it is difficult to combine scale inhibitors and asphaltene dissolvers, as most scale inhibitors are water-soluble, while asphaltene dissolvers are oil-soluble.

11.2.1.2 Scale Inhibitors and Paraffin Inhibitors

Stable formulations are possible with 10%–20% polyacrylic acid scale inhibitor and a new aqueous paraffin inhibitor dispersion.[5] This feature allows the product to be compatible with fluids used during fracturing. Normally, scale inhibitor and paraffin inhibitor combination products are difficult to develop, due to the different solubility properties of the different inhibitors.

11.2.1.3 Scale Inhibitors and Hydrate Inhibitors

Methanol is used frequently to prevent hydrate formation. Introduction of methanol in a brine can increase barite scale precipitation if no additional scale inhibitors are added.[6] Ethylene glycol has a less adverse effect than methanol on scale formation.[7] Tri-ethylene glycol (TEG) has a less adverse impact on barite scale formation than monoethylene glycol (MEG).[8] Kinetic hydrate inhibitors have been formulated with scale inhibitors and used on offshore fields.[9]

11.2.1.4 Scale Inhibitors and Dispersant

Scale inhibition normally occurs due to nucleation prevention. Dispersants allow the transport of small solids. Polymeric dispersants of carboxymethyl cellulose (CMC) can prevent the deposits of sulfide, sulfate, and carbonate scales.[10] CMC combined with phosphonate scale inhibitors can enhance the inhibition of phosphonate scale inhibitors.[10]

11.2.1.5 Scale Inhibitor with Dissolvers

There are several types of scale dissolvers. Acids dissolve calcium carbonate. Tetrakis (hydroxymethyl) phosphonium sulfate (THPS) dissolves iron sulfide (FeS). A blend of THPS with D-amino acids completely inhibited FeS formation for 2 h and dissolved iron sulfide scales up to 40%.[11] Iron sulfide solids were quickly dissolved by 30% in 10 min, and then slowly dissolved in the next 100 min to reach a maximum.[11]

11.2.1.6 Asphaltene and Paraffin Inhibitors

Paraffin deposits consist of linear and branched chain hydrocarbons in the range of $C_{18}H_{38}$ to $C_{60}H_{122}$.[12] Paraffin inhibitors are chemicals that incorporate into and alter wax crystal structures.[13] Asphaltenes are the fraction of crude oil that is soluble in toluene but insoluble in n-heptane.[14] In some instances, there exists an affinity between paraffin waxes and asphaltenes.[15] Combined use of asphaltene and paraffin inhibitors is effective, when asphaltene and wax co-precipitation occurs.[16]

11.2.1.7 Asphaltene Dispersant and Solvent

There are several asphaltene solvents, such as xylene and toluene. Di-hexyl and di-octyl ethers in combination with long-chain branched surfactants prevent asphaltene deposition by adsorbing on the asphaltene crystals and preventing agglomeration.[17] In some cases, the performance of the combination can be improved using an anionic surfactant.

11.2.1.8 Paraffin Inhibitor and Solvent

Typically, paraffin inhibitors are polymers that are diluted in paraffin solvents.[13] The polymers alter and disrupt crystallization.[13] Temperature and pressure can significantly affect the phase behavior of the polymer solution. Specific solvent combinations are chosen to optimize the performance of the fluid.[18]

11.2.1.9 Paraffin dispersant and solvent

Solvents such as benzene, toluene, ethyl benzene and xylene, heavy aromatic naphtha, or diesel are effective in dealing with wellbore/formation damage caused by asphaltenes and paraffins. These solvents are undesirable, due to their unfavorable di-hexyl and di-octyl ethers in combination with long-chain branched surfactants also prevent wax deposition.[17] Appropriate combinations of dispersants with solvents mitigate paraffin precipitation.

11.2.2 Binary Combinations for Integrity Management

Integrity management chemicals are used to protect system facilities. The chemicals inhibit corrosion, react with corrosive gases, or kill microbes that cause corrosion. The chemicals range from corrosion inhibitors, H_2S scavengers, oxygen scavengers, and biocides. Some binary combinations exclusively for integrity management are listed in Table 11.5.

Table 11.5: Exclusive binary combination products for integrity management.

Combination Product	Component 1	Component 2	Comments
Corrosion inhibitor and H_2S scavenger	Corrosion inhibitor	H_2S scavenger	Both components can be water- or oil-soluble
Corrosion inhibitor and oxygen scavenger	Corrosion inhibitor	Oxygen scavenger	Both components are normally water-soluble
Corrosion inhibitor and biocides	Corrosion inhibitor	Biocide	Both components are normally water-soluble

11.2.2.1 Corrosion Inhibitors and H_2S Scavengers

Film-forming corrosion inhibitor actives have been combined with the H_2S scavenger (3,3′-methylenebis(5-methyloxazolidine)), to form a combination corrosion inhibitor and H_2S scavenger.[2] Corrosion inhibitors were also combined with the H_2S scavenger, ethylenedioxy (dimethanol), (EDDM), to form a combination corrosion inhibitor and H_2S scavenger.[19]

11.2.2.2 Corrosion Inhibitors and Oxygen Scavengers

The addition of corrosion inhibitors with oxygen scavengers substantially reduces corrosion in air foam drilling.[20] Combination corrosion inhibitors with oxygen scavengers are recommended for use, when fluids typical of hydrostatic fluids are kept in the pipeline for a long period of time.[21] Polyamine-based corrosion inhibitors provide corrosion inhibition and some oxygen scavenging for boiler water treatments.[22]

11.2.2.3 Corrosion Inhibitors and Biocides

Amine-based corrosion inhibitors and phosphonate biocides are compatible with each other.[23] This combination of corrosion inhibitor and biocide has been used in a Mexican field. Combination corrosion inhibitors and biocides are used in pipeline pigging operations.[24]

11.2.3 Binary Combinations for Flow Assurance and Integrity Management

Flow assurance chemicals are used to keep oil and gas production flowing. The products prevent precipitation of organic and inorganic scales. Integrity management chemicals are used to protect system facilities. Some combinations for flow assurance and integrity management are listed in Table 11.6.

Table 11.6: Binary combination products for flow assurance and integrity management.

Combination Product	Component 1	Component 2	Comments
Scale inhibitor and corrosion inhibitor	Scale inhibitor	Corrosion inhibitor	Both components can be water- or oil-soluble
Hydrate inhibitor and corrosion inhibitors	Hydrate inhibitor	Corrosion inhibitors	Both components can be water- or oil-soluble
Dispersants and corrosion inhibitors	Dispersants	Corrosion inhibitors	Both components can be water- or oil-soluble

11.2.3.1 Scale and Corrosion Inhibitors

Combination scale and corrosion inhibitors have formed some of the earliest multifunctional combination products. Ideal corrosion inhibition occurs on clean metal surfaces of parent pipeline material.[25] Simultaneous evaluation of corrosion and scale has been achieved using a modified bubble cell apparatus in which pH, turbidity, and linear polarization are measured in the same apparatus.[26] Synergistic additives can be used to prevent localized corrosion of umbilical and capillary materials during chemical injection.[27] Combination corrosion and scale inhibitors are among the most popular combination products in the industry.

11.2.3.2 Scale Inhibitors and H_2S Scavenger

Scale inhibitors have been used to mitigate the scaling effects of H_2S scavengers. Direct injection of H_2S scavengers in mixed production systems with high calcium content can result in calcite deposition. A combination of a monoethanol amine (MEA) triazine and a proprietary phosphonate scale inhibitor minimized calcium carbonate formation.[28]

11.2.3.3 Hydrate and Corrosion Inhibitors

Corrosion and hydrate inhibitor combination products have been frequently made. Combinations of low-dose hydrate inhibitors and corrosion inhibitors have been successfully used in the Gulf of Mexico, since the beginning of this century.[29] A large amount of effort has been made to develop combination sour gas corrosion inhibitors with hydrate inhibitors.[30–32]

11.2.3.4 Dispersants and Corrosion Inhibitors

Sulfur deposition in low-velocity gas systems can cause severe corrosion problems, due to the deposition of sulfur.[33] In some cases, existing liquid hydrocarbon can dissolve sulfur and lower corrosion rates by decreasing iron ion diffusion through scale.[34] Appropriate hydrocarbon solvents as a carrier fluid for inhibitors have been used to reduce corrosion of carbon steel alloys in a highly sour system.[35, 36] Solvent packages were developed using tests that simulated operating conditions of sulfur deposition.[35–37] Surfactants such as nonyl phenol ethoxylates can work synergistically with sour corrosion inhibitors.[38] A modified fatty acid product was found to be environmentally friendly and act as both a cleaner and a corrosion inhibitor.[39]

11.2.4 Binary Combinations for Flow Assurance and Production Optimization

Flow assurance chemicals are used to keep oil and gas production flowing. Production optimization chemicals are used to increase and optimize production. Some combinations for flow assurance and production optimization are listed in Table 11.7.

Table 11.7: Binary combination products for flow assurance and production optimization.

Combination Product	Component 1	Component 2	Comments
Scale inhibitor and foamer	Scale inhibitor	Foamer	Both components can be water- or oil-soluble

11.2.4.1 Scale Inhibitors and Foamers

Foamers and scale inhibitors have been used in several applications. A foamer is used to prevent gas well liquid loading. The scale inhibitor is used to prevent inorganic scales. The combination products are comparatively easier to keep compatible. They are often used in gas wells with brines that have a high total dissolved solids (TDS) content.

11.2.5 Binary Combinations for Integrity Management and Production Optimization

Integrity management chemicals are used to protect system facilities and allow these facilities to be built with less expensive materials. Production optimization chemicals are used to increase and optimize production. Some combinations for flow assurance and production optimization are listed in Table 11.8.

Table 11.8: Binary combination products for integrity management and production optimization.

Combination Product	Component 1	Component 2	Comments
Corrosion inhibitor	Corrosion inhibitor	Drag reducer	Both components can be water- or oil-soluble
Corrosion inhibitor and foamer	Corrosion inhibitor	Foamer	Both components can be water- or oil-soluble
Corrosion inhibitor and demulsifier	Corrosion inhibitor	Demulsifier	Both components can be water- or oil-soluble

11.2.5.1 Corrosion Inhibitors and Drag Reducers

The corrosion inhibitor cetyl trimethylammonium bromide was chemically modified to cetyl trimethylammonium salicylate (CTAS) by reacting with sodium salicylate.[40] CTAS showed some drag reduction and some corrosion inhibitor performance.[40, 41] A formulation of fatty acid imidazoline, organic acids, amino-carboxylates, a cosolvent, and an emulsion of a water-soluble polymer in oil was found to be effective in providing corrosion inhibition and drag reduction.[42]

11.2.5.2 Corrosion Inhibitors and Foamers

Combination foamer and corrosion inhibitor surfactant products are used to prevent liquid loading and corrosion.[43] Liquid loading occurs frequently in shale gas reservoirs. Often these reservoirs contain carbon dioxide and hydrogen sulfide. A combination corrosion inhibitor foamer product unloads gas wells and prevents corrosion.[43]

11.2.5.3 Corrosion Inhibitors and Demulsifiers

Corrosion inhibitors frequently cause water separation problems. An oil-soluble water-dispersible corrosion inhibitor was blended with a demulsifier for a deepwater application.[2] Synergistic interactions between specific corrosion inhibitors and demulsifiers were found in very early work.[44]

11.2.6 Ternary Combinations

Ternary combinations are more difficult to create, but there are several combination products that are used. Both products are popular in shale oil and gas reservoirs. Some of the products are listed in Table 11.9.

Table 11.9: Ternary combination production chemical.

Combination Product	Component 1	Component 2	Component 3
Scale inhibitor, scale dissolver, and corrosion inhibitor	Scale inhibitor	Scale dissolver	Corrosion inhibitor
Scale inhibitor, corrosion inhibitor, and foamer	Scale inhibitor	Corrosion inhibitor	Foamer

11.2.6.1 Scale Inhibitor, Scale Dissolver, and Corrosion Inhibitor

A multifunctional chemical was developed to dissolve iron sulfide, inhibit mild steel corrosion, and prevent calcium carbonate and barite scale formation.[45] The product utilized THPS to dissolve iron sulfide and had corrosion inhibitors to prevent the corrosion of mild steel by carbon dioxide and hydrogen sulfide and scale inhibitors to inhibit calcium carbonate and barium sulfate precipitation.[45] A combination product with a different iron sulfide dissolver was recently made.[46]

11.2.6.2 Scale Inhibitor, Corrosion Inhibitor, and Foamer

Combination foamer, corrosion, and scale inhibitors have been used to improve productivity in tight gas fields.[47] In this case, the combination product unloaded gas wells, prevented corrosion of mild steel by carbon dioxide, and inhibited calcium carbonate and barite scale formation.[47] Some combination products claim the ability to reduce iron sulfide deposition.[48]

11.3 Challenges in Chemical Formulation

There are many difficulties in combining different production chemicals. These range from reactivity of the different chemicals, decrease in performance, difficulty of creating a homogeneous solution of the chemicals, thermal stability, compatibility with metals, compatibility with elastomers, gas hydrate formation in a chemical delivery line, and overdosage of production chemicals.

11.3.1 Reactivity of Different Chemicals

Some chemicals react with each other. Oxygen scavengers react with biocides. For this reason, these products are never kept together. When one has an acidic compound, they may react with basic compounds. The reaction product may have different properties than the two reactants. This is the reason that performance testing is done on the final formulation of the multifunctional product.

11.3.2 Performance Decreases due to Component Interference

There are a few times where the different types of chemicals can interfere with performance. The operating pressure can decrease kinetic hydrate inhibitor efficiency.[49] Corrosion inhibitors and hydrate inhibitors can interfere with each other and can cause a decrease in performance.[50, 51] It has been proposed by some that corrosion inhibitor adsorption to a polymeric hydrate inhibitor can change the structure of the polymeric hydrate inhibitor causing a loss of hydrate inhibition.[51] Thermodynamic hydrate inhibitors such as MEG and methanol decrease calcite dissolver performance.[52] Increasing concentrations of hydrate inhibitors decrease halite inhibition efficiency at 10 wt% hydrate inhibitor concentration.[53]

Oily deposits interfere with the adsorption of corrosion inhibitors on the metal surface.[34] There is the potential of harmful incompatibility between anionic foaming agents and cationic corrosion inhibitors.[54]

Injection of H_2S scavengers of high pH can have an adverse effect on scale inhibitor performance.[55] Increasing scale inhibitor dosage may not solve the scaling problem. It is important to understand the effect of H_2S scavenger on scale inhibitor performance to take the appropriate remedial step.

11.3.3 Difficulty in Creating a Homogeneous Solution

Many production chemicals are polymers with large molecular weights. Some of the production chemicals that can be polymers are listed in Table 11.10.

Table 11.10: Production chemicals that can be polymers.

Production Chemical	Comment
Scale inhibitor	Many scale inhibitors are polymeric
Asphaltene inhibitor	Many asphaltene inhibitors are polymeric
Paraffin inhibitor	Many paraffin inhibitors are polymeric
Hydrate inhibitor	Some hydrate inhibitors are polymeric
Drag reducer	Most drag reducers are polymeric

The amount of polymer that can be placed in solution depends on the thermodynamic phase behavior of the polymer.[56] Combining two different polymers in solution or having different molecules that interfere with the solvent keeping the polymer in solution will force the formulator to have a low concentration of actives.

The varied hydrophobic and hydrophilic nature of different active molecules also makes the formulation difficult for multifunctional chemistries. Computer modeling is often used to optimize blend formulations.[57] In some instances, the new formulations can be microemulsions.[58]

11.3.4 Thermal Stability

Some solutions have a lower critical solution temperature (LCST). This temperature is the critical temperature, below which all components are miscible. Above this temperature, the formulation becomes miscible. As an example, some formulations with hydrate inhibitors and corrosion inhibitors have a LCST.[59]

The threshold temperature for the thermal breakdown is 120°C for many oilfield chemicals.[60] The mechanism for high temperature thermal breakdown is radical initiated pyrolysis of the molecular structure and high temperature hydrolysis catalyzed by acids or bases. A review came to the significant findings listed in Table 11.11 for these important intermediates.

Table 11.11: Significant findings for thermal stability on important oilfield chemical intermediates.

Intermediate	Comment
Alkylthiols	Reactive in water solutions, Sensitive to hydrolysis. Some formulations are stable to 204°C
Amidoimidazolines	Sensitive to hydrolysis at high temperatures. Some formulations are stable to 232°C
Imidazolines	Among the most stable high-temperature intermediate. Some formulations are stable to 232°C
Phosphate esters	Sensitive to hydrolysis. Some formulations are stable to 177°C
Polymers	Polymers were stable in many high-temperature formulations in the temperature range of 120°C-177°C
Quaternary amines	Sensitive to water often fail at 120°C. There are some formulations that have survived up to 177°C

11.3.5 Metal Compatibility

Several production chemicals are acidic. Some chemicals can be corrosive to the metals used in the capillary string. Low-dosage hydrate inhibitors (LDHI) have been observed to cause pitting corrosion of super duplex stainless steel umbilical tubes.[61] Ideally appropriate testing procedures are needed to ensure that the production chemical does not corrode metals used in the capillary string.

11.3.6 Elastomeric Compatibility

Chemical degradation of elastomer material can occur with exposure to production chemicals. Inhibitors and biocides have damaged some types of elastomeric seals.[62] High temperature can shorten elastomers structural rigidity and pressure retaining performance. Testing procedures for multifunctional chemicals should have an evaluation of the elastomer with chemical exposure.

11.3.7 Gas Hydrate Formation in Chemical Delivery Line

Cases have been reported in the North Sea, where water-based chemical formulations have resulted in hydrate caused blockages of the chemical delivery line. Water-based chemical formulations are typically modified by including thermodynamic inhibitors such as methanol, MEG, or TEG to avoid this problem.[63] The composition of the solvent package can be frequently determined using commercial software packages that determine when hydrate formation occurs.

11.3.8 Overdosing of Production Chemicals

A multifunctional production chemical will have a fixed ratio of the different types of production chemicals. The fixed ration may be appropriate and optimal for a given set of production conditions. As the production conditions change, one will have to dose at appropriate conditions for a set of operating conditions, this will necessarily lead to an overdosage of one type of chemical. The overdosage is compensated by the decrease in capital costs in the production facility.

11.4 Laboratory Assessment

There are different experimental procedures used to assess thermal stability and compatibility. One procedure is thermogravimetric analysis (TGA).[60] Another procedure is capillary screening.[64] It is important to test performance at the required temperature. One important procedure to facilitate rapid screening of intermediates is to use aged samples.[65] High-pressure viscosity measurements are important for umbilical and capillary injection systems.[13] Experiments are needed to ensure the chemicals are compatible with the metals during chemical injection and storage. Elastomeric compatibility tests ensure that the chemicals do not damage elastomeric sealing systems.

11.4.1 TGA

TGA measures the weight of sample as it is heated in a small container. It is recommended to use nitrogen as a blanketing atmosphere, and in one study, samples were heated at rates of 5°C/min to 500°C. An example of the readings obtained for an unstable intermediate is shown in Figure 11.2.

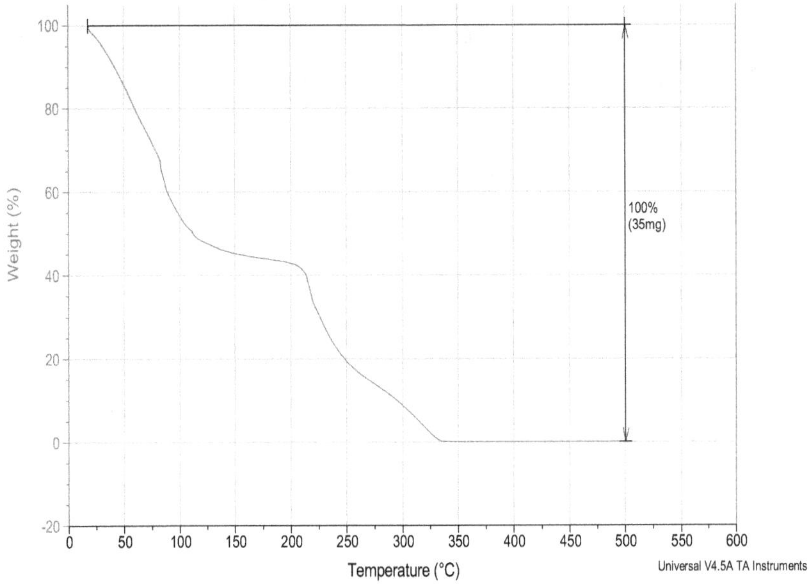

Figure 11.2. Results of thermogravimetric analysis of a temperature unstable intermediate.

The initial weight loss is due to the loss of solvent. At 210°C, the active ingredient starts to decompose. TGA results were used to identify the degradation temperatures of eight oilfield intermediates.[60]

11.4.2 Capillary Screening

Capillary screening is an important requirement for combination products to be used in umbilical and capillary injection systems. One example of a capillary stability test apparatus is shown in Figure 11.3.

The differential pressure across the loop is measured with time. Ideally, there should not be an increase in this differential pressure with prolonged heating of the formulation.

Figure 11.3. Capillary stability test apparatus with capillary loop placed in an oven.

11.4.3 Screening of Aged Intermediates

One rapid means of screening intermediates for thermal stability is to run simple kettle tests at low temperatures and run the same tests of the same chemical that have been aged at different temperatures.[65] One can compare the performance of the room temperature and aged products. Normally, with the increasing temperature, a deterioration of performance is observed. There are examples of some aged corrosion inhibitors seeing better performance than the product kept at room temperature.[65]

11.4.4 High-Pressure Viscosity

Deepwater systems operate at high pressures. It is not uncommon for the viscosity of a production chemical formulation to increase by a factor of 3–4 going from ambient pressure to 15,000 psi at 40°F.[13] High-pressure viscosity measurements are useful in determining appropriate formulations for deepwater applications.[13]

11.4.5 Chemical Compatibility with Metals

The NACE International standard TM0169-2000 can be used as a screening procedure to assess compatibility of the formulation for different metals.[66] Cyclic potentiodynamic polarization techniques can be used to assess the potential of the formulation to create pitting corrosion on the metal used for chemical injection or storage.[67, 68]

11.4.6 Chemical Compatibility with Elastomers

One recommendation for basic compatibility screening of elastomeric materials in liquid environments is to use two stages of immersion testing considering system temperature, pressure, and time in five cycles.[62] Some assessments of chemical compatibility of elastomers are performed using procedures involving a modification of TM0169-2000.

11.5 Field Applications

11.5.1 Scale and Asphaltene Control

One Latin American well exhibited a production decline of 80% over a 7-month period.[4] Deposition of both asphaltenes and calcium carbonate was determined from deposit analysis. Squeeze treatments for both asphaltene and scale were initiated resulting in a 240% increase in oil production. Squeeze treatments varied from 2 months to 2 years. Finally, the well was treated continuously through a gas lift system.

11.5.2 Scale and Corrosion Inhibition

A combination scale and corrosion inhibitor facilitated a reduction in the number of umbilical lines to provide flow assurance in the Galley field of the U.K. sector.[69] In this instance, the product needed to inhibit vaterite and barium sulphate formation and prevent corrosion. In the North Sea, combination products need to be created with no ingredients in the North Sea production chemical substitution list.[70] Environmentally friendly combined scale and corrosion inhibitors have been developed that mitigate galvanic corrosion of weldments under high-velocity conditions.[71] Combination corrosion and scale corrosion inhibitors have been developed for high-temperature systems (150°C) and produced water systems with high-calcium concentrations.[64]

11.5.3 Scale and H_2S Control

A proprietary phosphonate scale inhibitor was formulated with MEA triazine and used on the Coevorden-26 well in the Netherlands.[28] When proper two-phase atomization nozzles were used, sufficient reduction of H_2S was obtained at a high product dosage of 20 L/kg H_2S. There was the complete absence of scale deposits or sludge with the application of the product with the two-phase atomization nozzles. Chemical costs of U.S.$11 per kg H_2S were seen with the use of this product.

A water-soluble amine-aldehyde condensate with pH of 10 was formulated with a scale inhibitor that was either phosphonate-based or polymeric was developed for a Gulf of Mexico field.[72] The product reduced H_2S concentrations to below 4 ppm, and there was no evidence of calcium carbonate scale. The product was used field wide with chemical cost savings over the previous chemical treatment program of over $1 million dollars per year.

11.5.4 Scale Control and Corrosion Inhibition

A multifunctional chemical that dissolved iron sulfide, inhibited mild steel corrosion, and prevented mild steel corrosion was applied in three different production scenarios with the results shown in Table 11.12. It can be seen from the table that the product was successful in all three applications.[45]

Table 11.12: Application of combination scale inhibitor, scale dissolver, and corrosion inhibitor in three different production scenarios.

Scenario	Injection into Production Tubing Through a Capillary Tube Above an Electrical Submersible Pump (ESP)	Injection into the Wellbore Fluids at the End of the Production Tubing Through a Capillary Tube, in a Well Producing with the Support of Gas Lift	Injection into the Wellbore Fluids at the End of the Production Tubing Below an ESP
Bottom hole temperature (°F)	190	160	160
Bottom hole pressure (psi)	1,000	2,000	2,000
Wellhead temperature (°F)	90	110	110
Bottom hole pressure (psi)	100	250	250
Problems before application	Deposits in the interior of production tubing, corrosion, and under deposit corrosion	Iron sulfide in lateral sections of the well. Under deposit corrosion	Iron sulfide in lateral sections of the well. Under deposit corrosion
Status after application of multifunctional chemical	Chemical drastically reduced corrosion and deposition of solids	Formulation prevented iron-based solids from depositing. Low corrosion rates were seen on wellhead corrosion coupons.	Chemical drastically reduced corrosion and deposition of solids

11.5.5 Scale Inhibition for Low-Pressure Gas Wells

Scale inhibitors and foamers have been used in both batch treatment and squeeze treatments for low-pressure German gas wells.[73] Foam and scale inhibitor squeeze treatments have allowed operators in Germany to maximize their gas production from water flooded fields. Many foamer applications incorporate other inhibitors, such as scale inhibitors in their formulation.

11.5.6 Scale and Corrosion Inhibition in Low-Pressure Gas Wells

Combination foamer, corrosion, and scale inhibitors have been used to improve productivity in tight gas formations.[47] The product was used in the Barnett Shale region. Before the use of the product, wells had high corrosion due to carbon steel, as is shown in Figure 11.4. In addition, barite scale was observed in these wells (Figure 11.5). Calcium carbonate scaling was also seen on some wells. An example of the type of calcium carbonate scaling is shown in Figure 11.6.

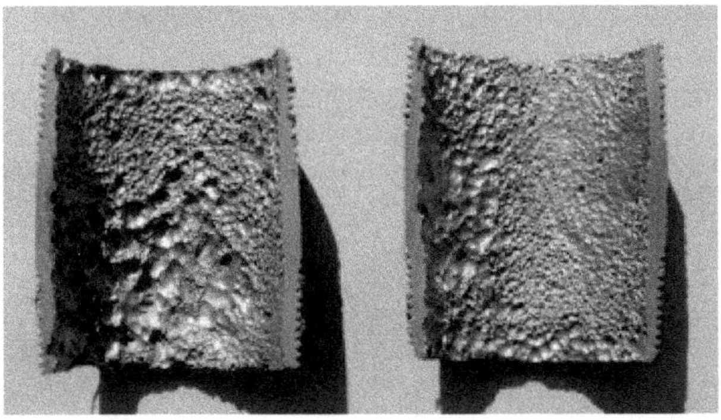

Figure 11.4. Typical corrosion of mild steel coupons seen in the Barnett Shale before the application of combination scale inhibitor, corrosion inhibitor, and foamer product.

Figure 11.5. Typical barite scaling seen in the Barnett Shale before the application of combination scale inhibitor, corrosion inhibitor, and foamer product.

Figure 11.6. Typical calcium carbonate scaling seen in the Barnett Shale before the application of combination scale inhibitor, corrosion inhibitor, and foamer product.

The product was applied on 43 wells at a rate of two gallons per day and on five wells at an average rate of one gallon per day.[47] The production increase in one well using the product is shown in Figure 11.7. In this instance, the production in the well improved from 20 MCFPD to 100 MCFPD. Use of the product went field wide with very successful results. The use of this product allowed a gain in productivity, ultimate recovery, and asset integrity.

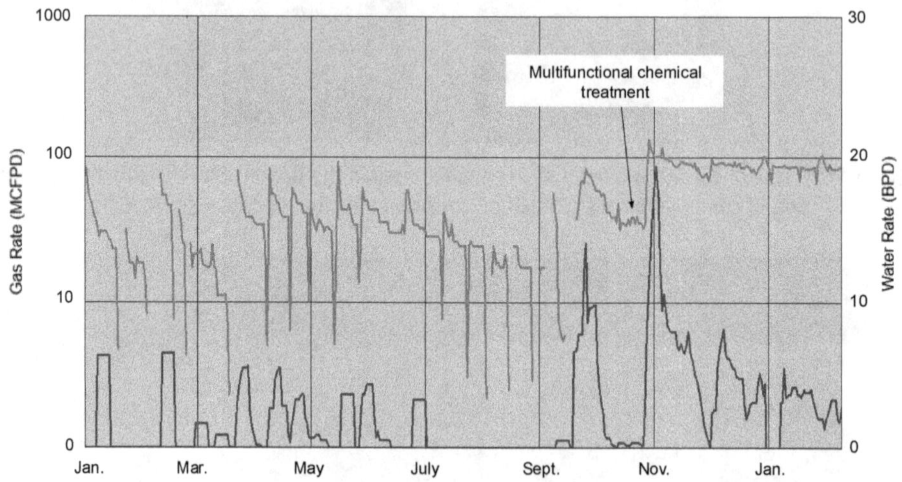

Figure 11.7. Production increase seen in the Barnett Shale well with the application of combination scale inhibitor, corrosion inhibitor, and foamer (red line shows gas production, while the blue line shows water production).

11.5.7 Corrosion and H_2S Control

A formulation of an amine-based corrosion inhibitor and EDDM has been used in a Norwegian North Sea field.[19] Higher production rates from this field using the formulation as the product allowed higher production rates from high H_2S wells without any modification of the chemical injection system or topside process system.

11.5.8 Corrosion Inhibitor and Biocide Treatment

A formulation of 100-ppm phosphonate-based biocide and 25-ppm amine corrosion inhibitor is being used in Mexican pipelines, where there are corrosion problems induced by microbial activity.[23] Combination corrosion inhibitor and biocide formulations are frequently used in pipeline pigging operations.[24]

11.5.9 Hydrate and Corrosion Prevention

The use of combination hydrate and corrosion inhibitors has enabled a successful repurposing of a non-insulated carbon steel service line into a production line without any plant shutdown under several operating conditions.[74] Kinetic hydrate inhibitors (KHI) and corrosion inhibitors (CI) have been used in the Southern North Sea successfully.[59] Kinetic hydrate inhibitors with corrosion inhibitors are used in the Qatar North Field that produces 1.6 billion cubic feet of gas per day to prevent hydrate blockages and sour gas corrosion of mild steel.[32] A combined KHI/CI application was used in an offshore U.K. sector application eliminating the previous use of MEG and TEG.[75] The application afforded the operator over $3 million in cost savings after the first year of deployment. A similar application was applied on an unmanned platform in the offshore Dutch sector. The product was deployed for 3 years until the end of life.[75]

11.5.10 Dispersion and Corrosion Inhibition

Oil-soluble water-dispersible combination cleaners and corrosion inhibitors have been applied to water injection lines.[76] The product was used to improve water injectivity, prevent sticky organic deposition, and improve corrosion protection.

11.5.11 Corrosion Inhibition and Demulsification

A multifunctional corrosion inhibitor/demulsifier was needed for a deepwater field located offshore Brazil, due to a limitation on the number of chemical umbilicals in the system.[2] A blend product of alkoxylated resins and polyalcohols with corrosion inhibitor actives was developed and used for this application.[2]

11.5.12 Corrosion Inhibitor Treatment of Low-Pressure Wells

Combination foamer and corrosion inhibitor surfactant products have been used to prevent liquid loading and corrosion, due to carbon dioxide in wells with declining gas production.[43] The products have been used to increase production in a gas well and simultaneously reduce corrosion. The products can be used with coiled tubing or velocity strings.[43] A gas well with 41% H_2S was treated with simultaneous injection of foamer and corrosion inhibitor down the annulus to prevent corrosion and liquid unloading.[77]

11.6. Summary

Multifunctional production chemicals are used to decrease the number of umbilical lines and capillary tubes used in a facility. This decreases the capital cost of the facility. Many multifunctional production chemicals illustrate the wide variation of production chemical needs throughout the world.

Many multifunctional production chemicals are difficult to formulate. This is due to the different solubility properties of the different production chemical intermediates and the polymeric nature of some of the intermediates. The polymer solutions may have a LCST. Developing high-temperature multifunctional production chemicals is difficult, due to the temperature instability of the intermediates and the ease of hydrolysis at high temperatures. It is important to have multifunction production chemicals that do not have high viscosities at high pressure and are compatible with system metals and elastomers.

In this chapter, a variety of multifunctional production chemicals used throughout the world in different applications are reviewed. The review should aid the reader in investigating, if a multifunctional production chemical provides advantages for use in their facilities.

Nomenclature

BPD	Barrels per day
BTEX	Benzene, toluene, ethyl benzene, and xylene
CI	Corrosion inhibitor
CMC	Carboxymethyl cellulose
CTAB	Cetyl trimethylammonium bromide
CTAS	Cetyl trimethylammonium salicylate
EDDM	Ethylenedioxy(dimethanol)
FeS	Iron sulfide
H_2S	Hydrogen sulfide
HAN	Heavy aromatic naphtha
KHI	Kinetic hydrate inhibitor
LCST	Lower critical solution temperature
LDHI	Low-dosage hydrate inhibitor
MBO	3,3′-methylenebis(5-methyloxazolidine)
MCFPD	Thousand cubic feet per day
MEA	Monoethanol amine
MEG	Monoethylene glycol
TDS	Total dissolved solids
TEG	Triethylene glycol
TGA	Thermogravimetric analysis
THPS	Tetrakis (hydroxymethyl) phosphonium sulfate

References

1. S. Ramachandran, M.B. Ward, K.A. Bartrip, et al., "Corrosion Inhibitor Considerations for Deepwater Systems," SPE-50758-ms, SPE International Symposium on Oilfield Chemistry, held February 16–19, 1999 (Houston, Texas, USA).

2. A. Jenkins, A. McRae, W. Cardwell, et al., "Formulation of Novel Combination Production Chemicals for Deepwater Oil and Gas Fields," C2022-7608, AMPP Annual Conference & Expo, held March 6–10, 2022 (San Antonio, Texas, USA).
3. A.J. Schroeder, T.A. Gay, J.E. Chitwood, et al., "Qualification and Case Studies Subsea Chemical Storage and Injection Unit," OTC-29307-MS, Offshore Technology Conference, held May 6–9, 2019 (Houston, Texas, USA).
4. B.K. Lambert, "Novel Chemical Applications for Enhancing Production From Mature Assets," OTC-18421-MS, Offshore Technology Conference, held May 1–4, 2006 (Houston, Texas, USA).
5. L. Xu, S. Sirak, H. Schenderlein, et al., "Newly Developed Aqueous Paraffin Inhibitor Provides Flow Assurance During and After Simulation," SPE-210317-MS, SPE Annual Technical Conference and Exhibition, held October 3–5, 2022 (Houston, Texas, USA).
6. A.T. Kan, G. Fu, M.A. Watson, et al., "Effect of Hydrate Inhibitors on Oilfield Scale Formation and Inhibition," SPE-74657-MS, SPE Oilfield Scale Symposium, held January 30–31, 2002 (Aberdeen, UK).
7. M.B. Tomson, A.T. Kan, G. Fu, et al., "Scale Formation and Prevention in the Presence of Hydrate Inhibitors," *SPE Journal* 11 (2006): pp. 248–258.
8. M.B. Tomson, A.T. Kan, G. Fu, "Inhibition of Barite Scale in the Presence of Hydrate Inhibitors," *SPE Journal* 10 (2005): pp. 256–266.
9. E. Hills, A. Hase, K. Turner, et al., "Hydrate—Inhibited Scale Inhibitor for Use on Subsea Templates," SPE-100472-MS, SPE International Oilfield Scale Symposium, held May 30–June 1, 2006 (Aberdeen, UK).
10. S. Ko, Y. Zhao, X. Wang, et al., "Deposit Prevention of Mineral Scales Using a Universal Dispersant of Carboxymethyl Cellulose," *SPE Journal* 28 (2023): pp. 173–183.
11. S. Ko, X. Wang, C. Dai, et al., "Deposition Prevention and Removal of Sulfide Scales Using Novel Chemicals," C2020-14934, NACE Annual Conference, held June 14–18, 2020 (Virtual).
12. M.E. Newberry, K.M. Barker, "Formation Damage Prevention Through Control of Paraffin and Asphaltene Deposition," SPE-13796-MS, SPE Production Operation Symposium, held March 10–12, 1985 (Oklahoma City, Oklahoma, USA).
13. D.W. Jennings, M.E. Newberry, "Paraffin Applications in Deepwater Offshore Developments," IPTC-12127-MS, International Petroleum Technology Conference, held December 3–5, 2008 (Kuala Lumpur, Malaysia).
14. O.C. Mullins, "Molecular Structure and Aggregation of Asphaltenes and Petroleomics," SPE-95801-MS, SPE Annual Technical Conference, held October 9–12, 2005 (Dallas, Texas, USA).
15. A. Punase, J. Aquiar, R. Smith, et al., "Importance of Comprehending Waxphaltenes and the Need to Develop New Inhibitors to Prevent Its Deposition," SPE-196368-MS, SPE/IATMI Asia Pacific Oil and Gas Conference and Exhibition, held October 29–31, 2019 (Bali, Indonesia).
16. T. Yoshimoto, M. Kurusowa, J. Fukuda, et al., "Impact of Wax on Jet Pump on Asphaltene Deposition Under Jet Pump Operation As Revealed By Asphaltene Inhibitor Field Test," SPE-203483-MS, International Petroleum Exhibition and Conference, held November 9–12, 2020 (Abu Dhabi, UAE).
17. C. Stanciu, J.M. Fernandez, "Paraffin/Asphaltene Cleaning Formulations—Lab Design and Case Studies," SPE-199259-MS, SPE International Conference on Formation Damage Control, held February 19–21, 2020 (Lafayette, Louisiana, USA).
18. D.W. Jennings, "Paraffin Inhibitor Compositions and Their Use in Oil and Gas Production," U.S. Patent 7514315B2, 2009.
19. M. Stipancev, O. Birkeitveit, V.H. Kvalheim, et al., "Multifunctional H_2S Scavenger and Corrosion Inhibitor: Addressing Integrity Challenges and Production Output of the Mature Field," SPE-190911-MS, SPE International Oilfield Corrosion Conference, held June 18–19, 2018 (Aberdeen, UK).

20. Y. Meng, L. Wan, X. Chen, et al., "Discussion of Foam Corrosion in Air Foam Drilling," SPE-94469-MS, SPE International Symposium on Oilfield Corrosion, held May 13, 2005 (Aberdeen, UK).
21. D.E. Powell, B. Melacon, R.H. Winters, "Guidelines When Conducting Hydrostatic Pressure Tests and Management of Extended Dwell Times Through Dry/Wet Lay Ups," C2011-11082, NACE Annual Conference, held March 13–17, 2011 (Houston, Texas, USA).
22. R. Crovetto, A.M. Rossi, E. Murtagh, "Research Evaluation of Polyamine Chemistry for Boiler Treatment: Corrosion Protection," C2011-11391, NACE Annual Conference, held March 13–17, 2011 (Houston, Texas, USA).
23. A. Alamina, I. Salgado, J. Ghenno, et al., "Effect of Phosphonates in the Microbiological Activity of SRB and APB in Oilfields," NACE-5689, NACE Annual Conference, held March 15–19, 2015 (Dallas, Texas, USA).
24. D. Armstrong, H. Sun, Y. Lu, et al., "Development of an Efficient Microbial Influenced Corrosion (MIC) Mitigation and Control Strategy in Pipeline Pigging Operations," C2022-17632, AMPP Annual Conference, held March 6–10, 2022 (San Antonio, Texas, USA).
25. S. Turgoose, G.E. Dicken, J.W. Palmer, "Review of Issues Associated With Inhibition of Scale Deposit-Covered Pipelines," SPE-100728, SPE International Oilfield Corrosion Conference, held May 30, 2006 (Aberdeen, UK).
26. L. Sanders, X. Hu, E. Mavredaki, et al., "Combined Inhibitors—A New Experimental Method for a Simultaneous Assessment of Scale and Corrosion Phenomena," SPE-155462-MS, SPE International Conference and Exhibition on Oilfield Scale, held May 30–31, 2012 (Aberdeen, UK).
27. C.M. Menendez, Z. Liu, S. Ramachandran, "New Synergistic Combination of Additives for the Prevention of Localized Corrosion of Capillary Materials During Chemical Injection," NACE-13058, NACE Annual Conference, held March 24–28, 2019 (Nashville, Tennessee, USA).
28. J.G.R. Eylander, H.A. Holtman, T. Salma, et al., "The Development of Low Sour Gas Reserves Utilizing Direct Injection Liquid Hydrogen Sulphide Scavengers," SPE-71514-MS, SPE Annual Technical Conference, held September 30–October 3, 2001 (New Orleans, Louisiana, USA).
29. S.E. Campbell, L.M. Frostman, W. Li, et al., "Corrosion and Hydrate Inhibition in High Shear, Low Temperature Conditions," NACE-02290, NACE Annual Conference, held April 7–11, 2002 (Denver, Colorado, USA).
30. J.J. Moloney, W.Y. Mok, C.G. Gamble, "Corrosion and Hydrate Control in Wet and Sour Gas Transmission Systems," SPE-115074-MS, SPE Asia Pacific Oil and Gas Conference and Exhibition, held October 20–22, 2008 (Perth, Australia).
31. J.J. Moloney, W.Y. Mok, C.G. Gamble, "Compatible Corrosion and Kinetic Hydrate Inhibitors for Wet Sour Gas Transmission Lines," NACE-09350, NACE Annual Conference, held March 22–26, 2009 (Atlanta, Georgia, USA).
32. H. Pit, R. Unsal, S. Mc Farland, "Development of Corrosion Inhibitor and Kinetic Hydrate Inhibitor for Pearl GTL Project," IPTC-12405-MS, International Petroleum Conference, held December 3–5, 2008 (Kuala Lumpur, Malaysia).
33. M.R. Gregg, J. Slofstra, D. Thill, et al, "Corrosion Experiences and Inhibition Practices Managing Wet Sour Salty Gas Pipeline," NACE-03174, NACE Annual Conference, held March 16–20, 2003 (San Diego, California, USA).
34. S.N. Smith, R.S. Pakalapati, "Thirty Years of Downhole Experience at Big Escambia Creek: Corrosion Mechanisms and Inhibition," NACE-04744, NACE Annual Conference, held March 28–April 1, 2004 (New Orleans, Louisiana, USA).
35. M.C. Place, "Corrosion Inhibition for Severely Corrosive Gas Wells," *Corrosion*, 48 (1992): pp. 341–352.
36. J.A. Dougherty, B.A. Alink, D.F. Ho, et al., "Corrosion Inhibition of Super Sour (90+% H2S) Gas Production," NACE-940222, NACE Annual Conference, held February 28–March 4, 1994 (Baltimore, Maryland, USA).

37. H. Sun, H. Fang, J. Davis, et al., "Elemental Sulfur Corrosion and Inhibition in the Presence of Sulfur Solvent," NACE-11125, NACE Annual Conference, held March 13–17, 2011 (Houston, Texas, USA).
38. L. Chen, N. Obeyesekere, J. Wylde, "The Effect of Surfactant Additions on the Mechanism of Sulfur Corrosion Inhibition in Sour Systems," NACE-3828, NACE Annual Conference, held March 9–13, 2014 (San Antonio, Texas, USA).
39. D.I. Horsup, R. Meyer, L. Tiwari, et al., "Integrity in Heavily Fouled Systems—Phase II Environmentally Sustainable Solutions," NACE-08623, NACE Annual Conference, held March 16–20, 2008 (New Orleans, Louisiana, USA).
40. V. Jovancicevic, S. Ramachandran, Y.S. Ahn, et al., "Corrosion Inhibition and Drag Reduction in Multiphase Flow," NACE-01052, NACE Annual Conference, held March 11–16, 2001 (Houston, Texas, USA).
41. S.E. Campbell, V. Jovancicevic, "Performance Improvement From Drag Reducers," SPE-65021-MS, SPE International Symposium on Oilfield Chemistry, held February 13–16, 2001 (Houston, Texas, USA).
42. B.M. Miksic, A.Y. Furman, M.A. Kharshan, "Effectiveness of the Corrosion Inhibitors for the Petroleum Industry Under Various Flow Conditions," NACE-09573, NACE Annual Conference, held March 22–26, 2009 (Atlanta, Georgia, USA).
43. S. Campbell, S. Ramachandran, K. Bartrip, "Corrosion Inhibition/Foamer Combination Treatment to Enhance Gas Production," SPE-67325-MS, SPE Production and Operations Symposium, held March 24–27, 2001 (Oklahoma City, Oklahoma, USA).
44. M. Yang, A.C. Stewart, G.A. Davies, "Interactions Between Chemical Additives and Their Effects on Emulsion Tendency," SPE-33617-MS, SPE Annual Technical Conference and Exhibition, held October 6–9, 1996 (Denver, Colorado, USA).
45. P. Rodgers, B. Lundy, S. Ramachandran, et al., "Multifunctional Chemical for Simultaneous Dissolution of Iron Sulfide, Corrosion Inhibition and Scale Inhibition," SPE-193619-MS, SPE International Conference on Oilfield Chemistry, held April 8–9, 2019 (Galveston, Texas, USA).
46. A. Ferrer, B. Bennett, "Remediating Oilfield Deposition From the Laboratory Bench to the Field-Development of a Multifunctional Product," IPTC-22249-MS, International Petroleum Conference, held February 21–22, 2022 (Riyadh, Saudi Arabia).
47. S. Ramachandran, M. Duckworth, K. Johnson, et al., "Use of a Combination Foamer, Corrosion and Scale Inhibitor to Improve Productivity in Tight Gas Formations," Gas Well De-Liquification Workshop, held March 1, 2005 (Denver, Colorado, USA).
48. J.J. Wylde, N. Turner, M. Austill, et al., "Development, Testing, and Field Application of a Novel Combination Foamer-Iron Sulfide Scale Inhibitor and Corrosion Inhibitor in East Texas," SPE-184584-MS, SPE International Conference on Oilfield Chemistry, held April 3–5, 2017 (Montgomery, Texas, USA).
49. J.L. Peytavy, P. Glenat, P. Bourg, "Kinetic Hydrate Inhibitors-Sensitivity Towards Pressure and Corrosion Inhibitors," IPTC-11233-MS, International Petroleum Technology Conference, held December 4–6, 2007 (Dubai, UAE).
50. B. Fu, "Development of Non-Interfering Corrosion Inhibitors for Sour Gas Pipelines With Co-Injection of Kinetic Hydrate Inhibitors," NACE-07666, NACE Annual Conference, held March 11–15, 2007 (Nashville, Tennessee, USA).
51. J.A. Moore, L. Ver Vers, "Understanding Hydrate Inhibitor and Corrosion Inhibitor Interactions," OTC-19869-MS, Offshore Technology Conference, held May 4–7, 2009 (Houston, Texas, USA).
52. H. Williams, T. Herrmann, M. Jordan, et al., "The Impact of Thermodynamic Hydrate Inhibitors MEG and Methanol on Scale Dissolver Performance," SPE-179864-MS, SPE International Oilfield Scale Conference, held May 11–12, 2016 (Aberdeen, UK).

53. S.A. Hosseini, P. Torino, E. Joonaki, et al., "Impact of Gas Hydrate Inhibitors on Halite Scale Precipitation—An Experimental and Morphological Investigation," SPE-190707-MS, SPE International Oilfield Scale Conference and Exhibition, held June 20–21, 2018 (Aberdeen, UK).
54. M. Pakulski, R. Martin, "Synergism and Antagonism of Foaming Agents and Corrosion Inhibitors," SPE-65016-MS, SPE International Symposium on Oilfield Chemistry, held February 13–16, 2001 (Houston, Texas, USA).
55. M. Sumestry, H. Tedjawidja, "Case study—Calcium Carbonate-Cale Inhibitor Performance Degradation Because of H_2S Scavenger Injection in Semoga Field," *Oil and Gas Facilities* 2 (2013): pp. 40–45.
56. D.W. Jennings, M.E. Newberry, "Application of Paraffin Inhibitor Treatment Programs in Offshore Developments," OTC-19154-MS, Offshore Technology Conference, held May 5–8, 2008 (Houston, Texas, USA).
57. R.J. Ferguson, "Optimizing Inhibitor Blends Using Corrosion Modeling," NACE-07061, NACE Annual Conference, held March 11–15, 2007 (Nashville, Tennessee, USA).
58. L. Quintero, R. Ponnapati, M.J. Felipe, "Cleanup of Organic and Inorganic Wellbore Deposits Using Microemulsion Formulations: Laboratory Development and Field Applications," OTC-27653-MS, Offshore Technology Conference, held May 1–4, 2017 (Houston, Texas, USA).
59. P. Weber, N. Morales, P. Conrad, et al., "Development of a Dual Functional Kinetic Hydrate Inhibitor," SPE-164107-MS, SPE International Symposium on Oilfield Chemistry, held April 8–10, 2013 (The Woodlands, Texas, USA).
60. T. Jackson, J. Brooks, Z. Liu, et al., "Guidance for Thermal Stability of Common Oilfield Chemicals," C2022-17745, AMPP Annual Conference, held March 6–10, 2022 (San Antonio, Texas, USA)
61. H.A. Suarez, R.J. Franco, W.E. Bond, et al., "Chemical Induced Pitting Corrosion of Super Duplex Stainless Steel Umbilical Tubes," MACE-11349, NACE Annual Conference, held March 13–17, 2011 (Houston, Texas, USA).
62. J.C. Price, "Options in Material, Coating and Chemical Selection to Mitigate Failure in Hydrocarbon Production," ISOPE-I-12-600, Twenty-second International Offshore and Polar Engineering Conference, held June 17–22, 2012 (Rhodes, Greece).
63. M.M Jordan, N.D. Feasey, "Meeting the Flow Assurance Challenges of Deepwater Developments: From Capex Development to Field Start Up," SPE-112472-MS, SPE North Africa Technical Conference, held March 12–14, 2008 (Marrakech, Morocco).
64. C.M. Menendez, J. Stewart-Ayala, Z. Liu, et al., "Development of New High Temperature Scale and Corrosion Inhibitor Combination Product for Capillary Injection," NACE-13190, NACE Annual Conference, held March 24–28, 2019 (Nashville, Tennessee, USA).
65. Z. Liu, T. Jackson, "Understanding Thermal Stability and Inhibition Effectiveness of Corrosion Inhibitors At High Temperatures," NACE-7535, NACE Annual Conference, held March 6–10, 2016 (Vancouver, British Canada).
66. *NACE Standard TM0169-2000: Laboratory Corrosion Testing of Metals* (Houston: NACE International, 2020).
67. Y.H. Tsang, T. Garza, C.A. Burger, et al., "Localized Corrosion Testings of Stainless Steel in Low-Dose Hydrate Inhibitor By Cyclic Potentiodynamic Polarization," OTC-22539-MS, Offshore Technology Conference, held October 4–6, 2011 (Rio de Janeiro, Brazil).
68. Y. De-Abreu, A. Frerrer, G. Rana, et al., "Cyclic Polarization and Electrochemical Impedance Techniques to Rapidly Screen and Assess Pitting Tendencies of Corrosion Inhibitor Actives in Carbon Steel," C2023-19389, AMPP Annual Conference, held March 19–23, 2023 (Denver, Colorado, USA).
69. M.M. Jordan, I. Archibald, L. Donaldson, et al., "Deployment, Monitoring and Optimisation of a Combined Scale/Corrosion Inhibitor Within a Subsea Facility in the North Sea Basin," SPE-80214-MS, SPE International Symposium on Oilfield Chemistry, held February 5–7, 2003 (Houston, Texas, USA).

70. N.D. Feasey, M.M. Jordan, M. Budge, et al., "Development and Deployment of Improved Performance "Green" Combined Scale/Corrosion Inhibitor for Subsea and Topside Application, North Sea Basin," SPE-100355-MS, SPE Oilfield Corrosion Symposium, held May 30, 2006 (Aberdeen, UK).
71. A. Jenkins, A. Fraser, L. Nolasco, et al., "Development of an Environmentally Acceptable Combined Scale/Corrosion Inhibitor for a North Sea Oilfield With High Velocity Flow," C2012-0001257, NACE Annual Conference, held March 11–15, 2012 (Salt Lake City, Utah, USA).
72. C.D. Sitz, D.K. Barbin, B.J. Hampton, "Scale Control in a Hydrogen Sulfide Treatment Program," SPE-80235-MS, SPE International Symposium on Oilfield Chemistry, held February 5–7, 2003 (Houston, Texas, USA).
73. O. Rauf, S. Gerdes, N.R. Lummer, "Foam and Scale Inhibitor Squeeze With Natural Gas: A Lifeline for Water Flooded and Damaged Gas Wells," SPE-165098-MS, SPE European Formation Damage and Exhibition, held June 5–7, 2013 (Noordwijk, The Netherlands).
74. M. Montini, S. Brioschi, A. Bianco, et al., "Debottlenecking of Deepwater Production Network By Converting an Uninsulated Into a Production Line: Engineering Analysis and Successful Application," OMC-203-401, OMC Med Energy Conference and Exhibition, held May 23–25, 2023 (Ravenna, Italy).
75. A. MacDonald, M. Petrie, J.J. Wylde, et al., "Field Application of Combined Kinetic Hydrate and Corrosion Inhibitors in the Southern North Sea: Case studies," SPE-99388-MS, SPE Gas Technology Symposium, held May 15–17, 2006 (Calgary, Canada).
76. M.G. Hilfinger, S.A. Stewart, A.M. Ansari, "Water Injection Enhancement Through a Novel Oil Soluble/Water Dispersible Cleaner and Corrosion Inhibitor," SPE-142876-MS, Brasil Offshore Conference, held June 14–17, 2011 (Macae, Brazil).
77. S. Ramachandran, K. Bartrip, S. Mancuso, et al., "Important Considerations for Corrosion Inhibitors in the Laboratory and the Prediction of Field Performance Based on Laboratory Data," NACE-08627, NACE Annual Conference, held March 16–20, 2008 (New Orleans, Louisiana, USA).

12 Scale Inhibitor Analysis

Stephen Heath[1]
[1]Creative Chemical Solutions, Chester, UK

12.1 Introduction

The use of chemical scale inhibitors (SIs) to prevent scale forming in oilfield production facilities, through either continuous chemical injection and/or periodic batch squeeze treatments, is the most preferred method of protecting against scale deposition and maintaining well productivity. It is essential to have reliable data on SI residuals in produced fluids to ensure optimum efficiency of continuous injection treatments and enable maximum lifetime for scale squeeze treatments, especially when operating in complex operating environments (like subsea deepwater scenarios). The accurate and precise analysis of both topside injection and downhole squeeze inhibitors, therefore, has a very important role in making key scale management decisions, in conjunction with other field data, including ion analysis, total suspended solids, and well productivity index.

SI squeeze treatments provide one of the most common and efficient methods for preventing the formation of carbonate and sulfate scales in the near wellbore, tubing, and topside facilities of production wells. For any scale management strategy where scale squeeze treatments are deployed, the objective is to achieve the longest possible lifetime for maximizing oil production and minimizing the number of well interventions required, especially in subsea and deepwater production scenarios, where well intervention treatments are very costly. To achieve the required longevity for squeeze treatments, it is essential to utilize a SI that is fit-for-purpose for the proposed scale squeeze application, that is, with good scale inhibition performance, nondamaging, good retention/release properties, and good treatment design, and readily detected to low levels in produced fluids. The accurate and precise detection of the SI in the produced fluids to low levels at or below the minimum inhibitor concentration (MIC) is one of the key factors to enable long treatment lifetimes and prevent wells from scaling.

The inhibitor return concentration in scale squeeze treatments gives an indication of the scale squeeze performance in conjunction with other field data, such as ion analysis, total suspended solids, and productivity index. However, if there are interferences in measuring the residual inhibitor concentrations, the efficiency and lifetime of the squeeze treatment can sometimes be difficult to assess.

A typical example of this is where similar products are deployed subsea or topside and are found to interfere with the scale squeeze residuals. For example, in conventional vertical platform wells, this occurs, if the fluid sampling point is downstream of the topside SI injection point, and the topside inhibitor is not switched off before sampling. In the case of similar topside and downhole products for both vertical and subsea wells, this problem can be solved by changing the chemistry of the subsea/topside product so as not to interfere, that is, phosphonate downhole and non-phosphorus-containing polymer subsea or topside.

Another way of solving these issues is to use SI tags. By incorporating a tag into the polymeric SI backbone, this allows for selective detection of the scale squeeze inhibitor.[1–4]

In another example, a subsea well is squeezed with a phosphonate inhibitor and a phosphorus (P)-containing corrosion inhibitor (CI) is injected subsea. It was not possible to change the CI which negated the use of P analysis via inductively coupled plasma optical emission spectroscopy (ICP-OES) to measure the downhole phosphonate residuals. In this case, it was necessary to develop specialist phosphonate detection techniques based on the time-resolved fluorescence (TRF).[5–7]

The industry desire for long squeeze treatment lifetimes based on accurate and precise detection of the scale squeeze residuals to MIC below 1 ppm (especially for polymeric scale squeeze inhibitors) has led to significant research on SI detection over the last 30 years.[4, 8–22]

A range of SI detection techniques now exist for the detection of phosphonates, phosphate esters, and polymeric SIs which are employed in the oil and gas industry and could also be used for detection of SI residuals in geothermal fluids, where SI injection is used for scale control. These techniques consist of simple methods based on the wet chemical analysis techniques for phosphonates and polymers (including P-tagged products), such as Hach and Hyamine, respectively[8, 9, 11, 13, 16] and P detection via ICP-OES[23, 24] and ICP-OES mass spectroscopy (ICPMS) for phosphonates and P-tagged polymers.[25]

In addition, fluorescence methods for tagged polymers[3, 4, 26] and TRF for phosphonates and sulphonated polymers[4, 6, 7, 22] are now commercially available. Ion chromatography (IC), for phosphonate assay and P speciation[27, 28] has been developed and IC coupled with mass spectroscopy (IC-MS) has been used for multiple phosphonate inhibitors in the same sample.[20] For complex polymer assay methods, high-performance liquid chromatography (HPLC)[29–31] and HPLC-MS techniques have been developed.[25, 32, 33]

The development of subsea and deepwater fields has placed even more focus on improving the detection of SI residuals in produced water, especially for comingled well streams, where there is a need to detect more than one type of scale squeeze inhibitor in the same water sample. More recent work has focused on the development of readily detectable chemicals,[3–5, 26] using more advanced methods of separation and detection[11, 25, 31] and using natural functional groups in SIs, for example, quaternary amines, vinyl, and allyl sulphonates, as tracers for multiple SI detection.[32, 33]

All these techniques mentioned above, including combinations thereof, are currently in use except HPLC-MS for polymeric SIs (method discontinued at development company).

One major aspect of this chapter on SI analysis is to list the various common techniques and their suitability for assaying SI residuals by highlighting the benefits and illustrating the limitations and sources of interference for each method. The impact of P speciation in P-tagged polymers and for thermal degradation of phosphonates and phosphate esters will also be considered as part of this chapter.

Examples will show how the analysis results can be misinterpreted if the wrong sample stabilization and analysis techniques are used. In addition, specific scenarios in North Sea fields for treating conventional and comingled sub-sea wells will be discussed to highlight the use of SI analysis techniques to aid chemical selection based on chemical retention, MIC value, detection limits, and well production conditions.

12.2 Overview

This overview of SI analysis methods is based on proven existing methods such as Hach, Hyamine, and ICP-OES. It will also include alternative approaches and recent developments in both HPLC techniques, such as the second-, third-, fourth-, and fifth-generation methods for polymer speciation.[25, 32, 33] In addition, C18 ion-pair approaches[31] and improved benchtop polymer analysis for sulphonated polymers[11] and more recent methods such as TRF and other fluorescence approaches[4–7, 26] will also be included. A detailed review of SI analysis techniques is presented with a description of how these techniques can be used to help with scale management decisions in simple to complex production scenarios through the selection and use of the best SI analysis techniques available.

The main premise of this chapter is that there is still not one SI detection method that is fit for purpose for the analysis of all types of SIs, or indeed, one SI under all production environments. Although various methods are available for SI residual analysis, most suffer interference of some degree or another and where a preferred method may be very sensitive and accurate under a particular set of conditions, it is unlikely to be effective under all production environments. The SI analyst therefore requires a large toolbox of alternative approaches which can supplement the original methodologies, and the selection and application of the right analysis techniques are crucial to obtain reliable residual data.

In many cases, inappropriate analysis techniques can render expensive core flood testing and field trials, thereby leading to (i) the risk of qualification and selection of inappropriate products in the laboratory[12] and (ii) poor performance when applied in the field. For field residuals, the use of an inappropriate method has obvious implications which lead to increased risk of well failures due to uninhibited scaling as well as rendering subsequent field optimization via near-wellbore modeling ineffective, when utilizing scale squeeze design codes.[34]

Sample preservation is also critical to the results obtained from the chosen analytical method.[14, 35–38] Using the wrong stabilization agent prior to shipping the sample to the analysis laboratory can render the results invalid. A discussion on sample preservation is given in Section 12.4.

Residual analysis is, unfortunately, often labor-intensive and based on the lowest cost approach available, to avoid the use of time-consuming and laborious techniques. For example, direct assay of phosphonates based on the ICP-OES for total P can be a factor of 10 times lower cost to perform than labor-intensive bench top wet chemical approaches for polymeric and phosphonate inhibitors. Due to this and the desire to improve accuracy, large arrays of tagged polymers containing a more readily assayable fluorescent tag or various P- or amine-containing polymers have been developed. The idea being that the assay of the polymer can be readily achieved by assaying the tag, thereby, rendering the analysis fast, routine, and low cost, although this is often not the case as some methods require detailed sample preparation and complex equipment, like HPLC-MS is needed to measure some molecular tags.

Automated approaches are often preferred in the industry, though often there is a mistaken belief (or desire) that an automated approach is inherently more sensitive, reproducible, and accurate than the manual method methods. In some cases, new automated methods are promoted by the supply industry, primarily to demonstrate new technology and therefore provide a technical edge over competitors. This may be true in some cases, but what is often overlooked is that the reliability of most automated approaches still relies on effective sample clean up and pretreatment which, if not conducted properly, has obvious consequences.

In a simplistic view, such aspects may include:

- Analysis techniques applicable to fresh waters or sea waters may suffer considerable interference in higher salinity formation waters.
- Methods that may be available and routine for clean laboratory samples (e.g., core flood effluent samples) may not be applicable to dirty contaminated field samples.
- Methods which analyze specific moieties or analyzable tags may not be applicable to squeeze return samples, where the reservoir core material acts as a chromatography column, unless it can be demonstrated that the tag or additive is fully incorporated within the active polymer and that its level is not so low as to suffer interference from other naturally occurring species. In short, it is critical to assay the active scale-inhibiting component present in any produced water and not simply detect a readily analyzable moiety.

- For phosphonate species, many impurities are often present in commercial phosphonates. In extreme cases, the impurities can contribute both to the inhibitory effect and interfere with the retention and release properties of the main inhibitor. This may mean that for some cases, the impact of such other P species needs to be accountable. However, for individual phosphonate squeezed products, this is not a significant issue, since the active SI generally dominates. IC can assist to differentiate one P-containing species from another, although this technique has its limitations which will be described later in this chapter.
- For certain applications, thermal stability can be a significant issue, particularly for phosphate ester chemicals at moderate temperatures (80–100°C) or conventional phosphonates at elevated temperatures (120–180°C), depending on the phosphonate chemistry and the production environment (pH, brine composition, etc.).
- The presence of interfering impurities, or other production chemicals, often means that the simplest approach (*e.g.*, total P or total polymer) may again not be relevant.

Several examples will be presented ,which deal with interferences in production brines, sample preservation, detectable tags, and P speciation. In addition, the use of the more complex techniques (like HPLC-MS) will be discussed regarding both individual and multiple SI analysis in comingled well environments, respectively.

A summary of the range of techniques commonly adopted or proposed for residuals analysis for phosphonate, polyacrylate, and sulphonated polymers in different production environments will be listed, along with advantages and disadvantages of these methods in conjunction with the types, that is, wet chemical (benchtop), automated or semi-automated, which also infers the degree of operator labor. In addition, tagged polymers with detection methods will be listed comparing some of the attributes of the different methods along with detection limits and the potential to separate between different polymer species by using different tags and sequentially using more than one technique.

12.3 Interferences

There are multiple interferences present in the assay of oilfield SIs, some of which are often overlooked that simply relate to the produced water chemistry. Many are well understood, such as the impact of high chlorides on benchtop wet chemical polymer analysis using Hyamine, see Table 12.1, and the impact of formation waters containing barium and strontium cations on certain wet chemical phosphonate analysis protocols (due to interferences with standard reagents). In addition, interferences also arise for phosphonate detection, due to natural background P as well as other production chemicals present in the system, which can often respond in a similar manner to the SI of interest.

Table 12.1: Impact of chloride on Hyamine analysis of polymeric scale inhibitors.

Brine Type	Inhibitor Concentration (mg/L)	UV Absorbance (AU)
Distilled water	0	0
	5	0.3
	10	0.6
Formation water (5,000 mg/L of chloride ions)	0	>1
	5	>1
	10	>1

Similarly, when considering ICP-OES analysis for total P, which is the main method for phosphonates and P-tagged polymers, potential interference from certain brine ionic species is often overlooked, as is the critical importance of matrix matching. Typically, two reference wavelengths are used for the analysis of P, namely 177.495 nm and 178.287 nm, in produced brines, although there can also be some interference from sulfur (S), platinum (Pt), and iodine (I) ionic species present in saline oilfield brines.[17, 18]

Calibration standards for ICP-OES must be matrix matched to the produced waters, as inaccurate assays can result due to interference. More importantly, the presence of salt has a large effect on the intensity of the plasma torch, which impacts sensitivity and causes instrumental drift. Therefore, matrix matching is essential, when assaying phosphonates and polymers via P analysis using ICP-OES at residual level (mg/L).

12.4 Sample Preservation

One of the most critical aspects of the accurate assay of residual SI concentrations in produced water relates to the preservation or stabilization of oilfield samples prior to shipment to the analytical laboratory, whether that laboratory be located locally (at the field/platform) or remotely in an offshore/onshore analytical laboratory. Poor sample preservation can render analysis inaccurate.

In many cases, onsite filtration and acidification provide a simple and accurate manner to preserve samples and allow accurate residual analysis. The presence of acid prevents postsampling carbonate precipitation, minimizes bacteria activity, drives off other dissolved gases such as H_2S, and prevents some iron sulfides and other iron oxide/hydroxide salts from precipitating.

However, in some situations, this can also render the analysis useless. For example, if using a polymaleic acid inhibitor (or some of the greener, more readily degraded polymers), storage in acid solutions can lead to hydrolysis of the polymers and effective unzipping of the polymer, which can result in assay recording near zero levels.[17, 18]

Moreover, simple acid addition does not prevent sulfate precipitation. Therefore, in cases where barium sulfate is expected, additional additives (such as a noninterfering SI) may be required. Ethylenediaminetetraacetic acid (EDTA) is often used to stabilize cations (such as barium, calcium, *etc.*) and again can be used, when subsequent assay of phosphonates or P-tagged polymers is required by ICP-OES; but it is not recommended for P-tagged polymers at low levels, if preconcentration is required. Also, the use of sodium NaEDTA can often result in the precipitation of iron salts at brine pH ≥ 10. It is better to use the ammonium salt (NH_4) to prevent this and minimize interference with sodium (Na) or potassium (K) analysis by ICP-OES or IC. In addition, the use of NaEDTA or KEDTA can be inappropriate for polymeric inhibitors assayed via wet chemical approaches, since the initial stage of the sample preparation for analysis is to reduce the pH which results in precipitation of EDTA. This also induces consumption of the SI. Even without acid addition, the use of EDTA can impact the adsorption of certain polymeric SIs on the C18 solid phase extraction cartridges and, in addition, interfere with UV detection (depending on the wavelength).

A group of authors have carried out investigations to determine the best sample preservation methods.[14, 34–37] This work was mainly performed on fields in the Norwegian sector of the North Sea, and, therefore, the preservation was primarily in relation to the water produced from wells that were squeezed with environmentally acceptable polymeric SIs, although phosphonates were also considered.[14, 37]

Additional works based on U.K. and Middle East fields investigated the use of EDTA, HCl, and vinyl sulphonate copolymers (Vs-Co) or other non-P-containing polymeric SIs for produced brines with both calcium carbonate and barium sulfate scaling tendency and containing high iron (Fe).[38]

It was also summarized that no preservation of produced waters was required, when the returned concentrations of polymer squeeze SI were high; but when residual concentrations are close to MIC, a phosphonate SI could be added to stabilize the sample, provided the polymer was not P-tagged and measured by ICP-OES.[35, 36] It should be noted that the fields investigated in this case had a relatively mild barium sulfate scaling environment. This work also noted that where phosphonate SI residual analysis was concerned, different preservation additives, such as HNO_3, HCl, EDTA, poly vinyl sulphonate (PVS) SI (50% dilution), gave a range of results. EDTA was identified as giving the highest phosphonate residual concentration.

Other works on sample preservation explored the relationship of the laboratory analysis of samples to operational observations in the field.[14] In this work, it was found that preserving samples for ion and polymer SI residual analysis with a multidentate phosphonate (typically from four to five phosphonate groups per molecule) resulted in the formation of a calcium phosphonate complex in some field waters with low-calcium concentrations of around 700 mg/L. This complex formed a precipitate, leading to decreased concentrations of calcium and magnesium in the ion analysis and artificially increased polymeric SI residuals by Hyamine method.

Reducing the concentration of phosphonate inhibitor preservative also led to the precipitation of calcite scale, which also impacted the calcium ion results and polymer SI residual levels. The addition of this preservative was found to increase the concentration of total suspended solids (TSS) in the samples, when compared to unpreserved samples. Since most formation waters typically contain more than 700 ppm Ca, it was recommended to use a more calcium-tolerant phosphonate, should this method still be used.

An additional study investigated the use of phosphonates as preservatives.[35, 36] However, in this case, a more calcium-tolerant, lower molecular weight phosphonate molecule (two phosphonate groups per molecule) was used. It was confirmed that the addition of the multidentate phosphonate resulted in an increase of total suspended solids, due to Ca phosphonate complex formation in produced waters containing over 700-mg/L Ca. This solid formation resulted in overly high results for polymer SI analysis carried out by the turbidimetric Hyamine 1622 method. On the other hand, use of the low-molecular-weight phosphonate did not increase the TSS compared to unpreserved samples and contained less than the multidentate phosphonate preserved samples. In some of the cases, it was found that the use of the low-molecular-weight phosphonate appeared to have a better effect on stabilizing Fe in produced waters than the addition of HCl, although the effect of the HCl on iron concentration should be the same as for the low-molecular-weight phosphonate.[35] Also, acid should not be added to a sample already preserved with phosphonate, as the low pH caused protonation of the phosphonate and rendered it ineffective as a preservation agent. The use of low-molecular-weight phosphonate for all samples containing Fe, those for polymer inhibitor residual analysis, and where TSS analysis was required along with ions to detect seawater breakthrough.[35]

12.4.1 Case Studies

12.4.1.1 Wells with High Fe in Produced Water

A seawater flooded field in the Middle East with a high-iron produced water was being treated for a downhole barium sulfate scale problem by SI squeezing. A phosphonate inhibitor was employed, and the residual analysis was carried out by P analysis via ICP-OES. The initial SI return profile initially looked normal, and then the residual SI concentrations quickly dropped below anticipated.

Examination of both unpreserved samples and samples stabilized with NaEDTA showed a precipitate which was obviously removing some of the phosphonate SI from solution. The precipitates observed were soluble in HCl, and remeasured SI was then significantly higher. When future samples were stabilized with HCl or KEDTA, no precipitation was observed, and the highest SI residuals were observed with HCl. However, since barium ion analysis was also a requirement, KEDTA preservation proved to be the best option for both ion and SI analysis.[38]

12.4.1.2 Well with Moderate-Barium Sulfate Scale

A field in the central North Sea produces oil from a Forties type reservoir uses seawater for pressure support. The formation water contains up to 270-ppm barium and up to 800-ppm strontium, so sulfate scales were expected at the production wells. Wells with seawater breakthrough were treated with a phosphonate SI squeeze to control barium sulfate and calcium carbonate scale formation. According to the data initially observed, the unpreserved sample analysis suggested that the SI residual had fallen below the MIC of 2.5 ppm after 30,000 bbl of water produced, equating to a production time of only 4 days. The squeeze treatment has been designed to last for 180 days to MIC of 2.5 ppm, so this was unexpected and required further investigation.

New samples were taken and preserved 1:1 with NaEDTA. The phosphonate residuals were increased significantly. Based on these results, the old samples were reanalyzed after the addition of 1:1 NaEDTA. The phosphonate residuals were again much higher, although not as high as the new samples. In this case, the stabilization of the water sample with NaEDTA was preventing barium sulfate formation and parasitic consumption of the phosphonate SI, as presented in Table 12.2.

Table 12.2: Comparison of EDTA- and non-EDTA-stabilized samples for phosphonate scale squeeze.

Water Produced After Squeeze (bbl)	Old Samples with EDTA (mg/L)	Old Samples with No EDTA (mg/L)
1,000	70,000	1,000
10,000	90	70
30,000	50	2
50,000	30	1.8
75,000	9.5	1
90,000	7	<1
110,000[(A)]	~50	NA
130,000[(A)]	50	NA

[(A)]EDTA added on all fresh samples taken after 110,000 bbl of produced water.

For wells squeezed with phosphonate SIs, the samples can also be stabilized with the addition of ~5,000 ppm of a non-P-containing SI such as Vs-Co. The sulphonated SI has the same effect as the NaEDTA and provides a considerable improvement in the SI residual return profile compared to unstabilized samples and samples stabilized with HCl, see Table 12.3.

Table 12.3: Comparison of EDTA- and non-EDTA-stabilized samples for phosphonate scale squeeze.

Sample Date	Inhibitor Concentration (mg/L)		
	Unpreserved	HCl	Vs-Co
April 4, 2020	0	0	-
May 4, 2020	2	5	-
July 4, 2020	2	10	23
December 4, 2020	0	2.5	22
May 4, 2021	5	10	18.5
July 4, 2021	2.5	3	16
September 4, 2021	2	3	24
December 4, 2021	-	-	17
February 4, 2022	-	-	14
April 4, 2022	-	-	12.5

12.4.1.3 Wells with Harsh Barium Sulfate Scale

In the case where wells have harsh barium sulfate scaling and are squeezed with a non-P-containing polymer, the samples can be stabilized with 5,000 ppm of a phosphate ester, as this will not interfere in either HPLC, fluorescence, or Hyamine analysis.[38] For P-tagged polymers analyzed directly by P via ICP-OES, the samples can be stabilized by 2,000- to 5,000-ppm Vs-Co, provided the dose of Vs-Co is fully compatible.

12.4.2 Guidelines

Unfortunately, due to the great variety of brine types, different scale squeeze SI chemistry, produced water compositions, and challenges brought by each field scenario, it is not possible to design a "one for all" sample preservation regime. Therefore, sample preservation must reflect the specific nature of the analyte being analyzed. Otherwise, sample preservation can, and in many cases does, negate the residual analysis of SIs.

Some basic guidelines for sample preservation for phosphonates and polymeric SIs are given below. It should be noted that these guidelines are by no means definitive. All sample preservation methods should first be checked for their impact on the measurement of accurate and precise SI residuals.

12.4.2.1 Preservation of Samples for Phosphonate

If carbonate scales or high concentrations of iron are present in the produced water, HCl or nitric acid (HNO_3) should be added to prevent entraining of SI in calcite or oxidized iron solids. Where chloride ion (Cl^-) analysis is required, the use of nitric acid is recommended, as it does not interfere unlike HCl. In the case where mild to moderate masses of sulfate scales are anticipated, the addition of up to 5,000-ppm non-P-containing polymeric SIs (such as Vs-Co) should be considered. In the case of severe sulfate scale supersaturation or for sulfate scale in the presence of iron, the addition of K_4EDTA or NH_4EDTA is advised.

12.4.2.2 Preservation of Samples for Polymers

If carbonate scale or high concentrations of iron are present in the produced waters, either HCl/HNO_3 or 2,000–5,000 ppm of a low-molecular-weight and calcium- and iron-tolerant phosphonate should be added to prevent entraining of SI in calcite or oxidized iron solids. If a P-tagged polymer is used, then samples should be stabilized with HCl/HNO_3.

Where mild to moderate masses of sulfate scales are anticipated with a non-P-containing polymer, the addition of 2,000–5,000 ppm of a calcium-tolerant phosphonate or phosphate ester could be considered. In the case of severe sulfate scale supersaturation or for sulfate scale in the presence of iron, the addition of 2,000–5,000 ppm phosphate ester is advised, as this does not interfere with HPLC or Hyamine.

For a P-tagged polymer, the use of a non-P-tagged polymer (like Vs-Co) could also be considered at 2,000–5,000 ppm for direct P analysis via ICP-OES or even K_4EDTA. If pretreatment through C18 cartridges is required to upconcentrate P to detectable levels at low-SI residual levels, the use of EDTA is not recommended, and it is advised to check on any interference from Vs-Co prior to use.

12.5 Detectable Tags

The issue of analyzable moieties or tags within polymeric SIs has been covered in several publications.[17, 18, 25] This is highlighted in Figure 12.1, where a P-containing copolymer, typically containing 0.5% P in the stock as supplied polymer (this compares with typically 2%–8% P for most phosphonates), is analyzed. Residual inhibitor concentrations in post-flush samples were determined by both P content via ICP-OES and the active polymer content using dialysis/hyamine.

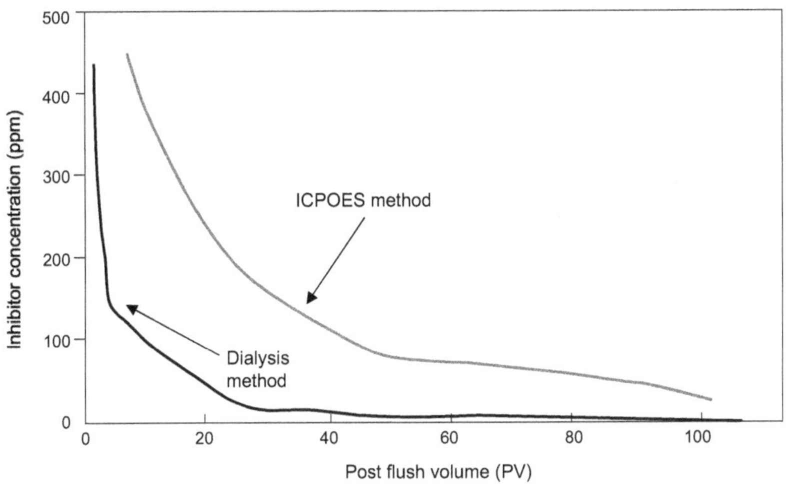

Figure 12.1. Comparison of core flood return profiles for a phosphorus-tagged copolymer using wet chemical Hyamine vs. total phosphorus analysis via inductively coupled plasma optical emission spectroscopy.

In this case, the dialysis/Hyamine results represent the active polymeric SI component, whereas the P appears to be a combination of product impurities (unreacted P-containing impurities) as well as residual background P produced from the formation. Since ICP-OES gives a total P result and cannot differentiate between the different ionic P species, the net effect is that the assay based on the total P by ICP-OES was not suitable for this chemical. This gives much higher results, which are misleading for the perceived squeeze treatment lifetime.

In these cases, the P component is behaving as if separated from the main polymer and retaining/releasing from the reservoir at different rates. It is noted that this effect can be overcome by cleanup/preseparation benchtop approaches, for example, C18 cartridge for acrylate polymers and strong anion exchange (SAX) cartridge for P-tagged sulphonated polymers, prior to the assay by ICP-OES. These approaches add considerable time (and therefore cost) to the analytical approach and result in a lower response, if the P is not well bound to the polymer. They should be more comparable to Hyamine results. However, without cleanup/preseparation, this can result in a significantly higher response and give a "false" impression to the level of scale protection and potential squeeze treatment lifetime, as shown in Figure 12.1.

This example also serves to highlight the well-known issues, when squeezing blended products, because the chromatographic nature of the reservoir will cause the two components to separate in the reservoir, and they will also be returned at different rates. Therefore, the squeeze returns, especially in the critical lower concentration tail, will typically contain only one of the blended components. Blended products are therefore not generally recommended for squeeze application for this reason, although they can be very effective for continuous injection applications.

In a separate example (shown in Figure 12.2) for a different P-containing terpolymer, the P analysis results via ICP-OES give a lower response than for the polymer assay by Hyamine. In this case, the P analysis more closely represents the active SI compound and could be used for the residual assay of this product, providing the P background interference from the reservoir is not sufficiently high to interfere.

Fluorescent (F)-tagged polymer products have also been developed, and a comparison of the analysis of the initial field returns by fluorometer and HPLC (active polymer) is presented in Figure 12.3. There was quite a good agreement between the fluorescence and HPLC techniques, although some deviation between the two techniques was observed. Further work on sample cleanup remedied that deviation, and a good correlation was obtained between the two techniques,[3] indicating that the analysis of the F tag is representative of the active polymeric SI.

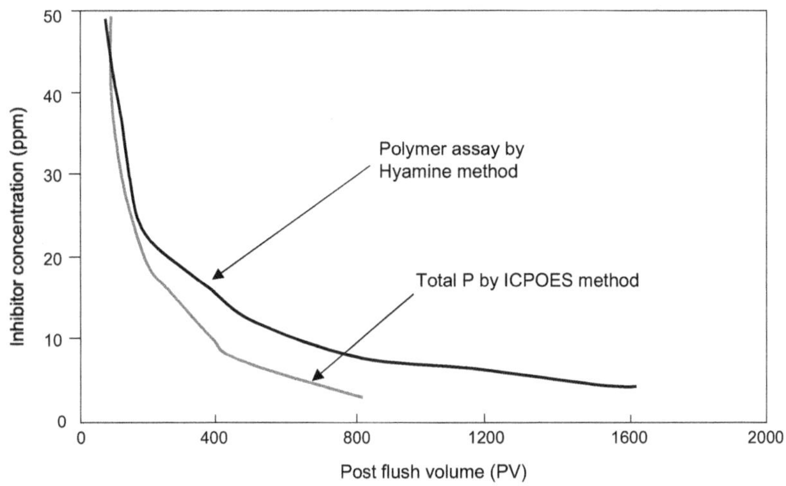

Figure 12.2. Comparison of scale inhibitor return profiles for a phosphorus-tagged terpolymer using wet chemical Hyamine vs. total phosphorus analysis via inductively coupled plasma optical emission spectroscopy.

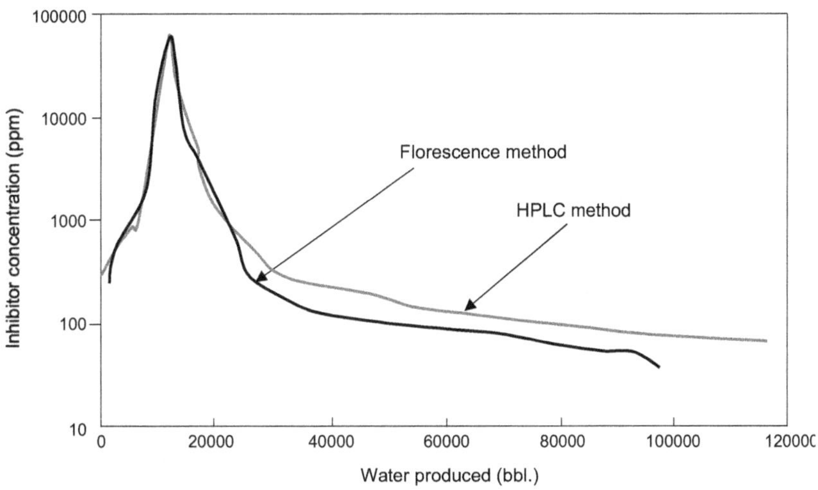

Figure 12.3. Comparison of initial scale squeeze inhibitor residuals for fluorescent-tagged polymeric scale inhibitor detected by both fluorescence and high-performance liquid chromatography methods.

A separate field trial on another North Sea field also gave good SI return data for an F-tagged polymer which followed the typical trend of a scale squeeze treatment with a polymeric SI, as shown in Figure 12.4.

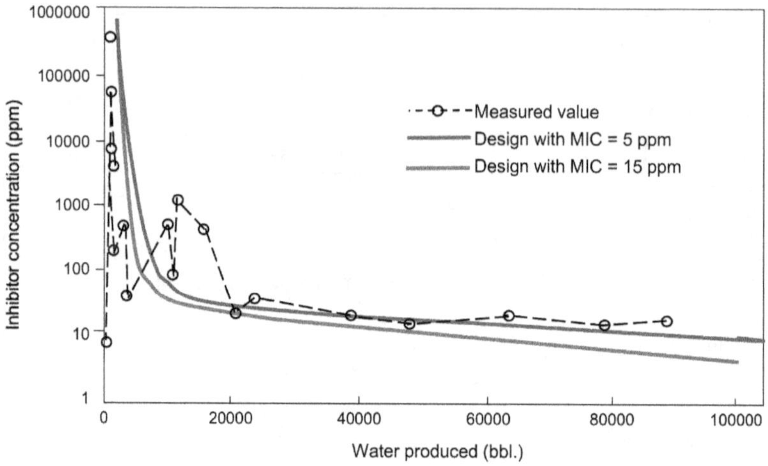

Figure 12.4. Scale inhibitor return profile from a North Sea field for a fluorescent-tagged polymer detected by fluorescence.

The main conclusion for tagged polymers at this stage is that it is critical that the chosen analytical approach assays the active SI component present in any produced water sample and does not simply detect some readily analyzable moiety within the original supplied SI.[17, 18]

12.6 Inhibitor Speciation

12.6.1 Importance of P Speciation

As illustrated earlier, many commercial phosphonate chemicals contain a range of components in addition to the main SI component. Furthermore, in many production environments, several P-containing chemicals may be applied into the system. Therefore, it is necessary to distinguish between the total P (obtained by ICP-OES) and the actual SI component of interest.

Similarly, in high-temperature environments (or moderate-temperature environments, when using phosphate esters), thermal degradation of the phosphonate may mean that the returns are dominated by degradation products such as orthophosphate (PO_4^{3-}) and lower molecular weight phosphonate components with minimal active inhibitor present.

Phosphonates and phosphate esters are not often considered for treating comingled wells in the same flowline, because of the difficulty to differentiate between the different P-based products. SIs such as penta-phosphonates (diethylenetriaminepenta phosphonic acid [DETPMP]), phosphate esters, and P-tagged polymer SI can be analyzed by ICP-OES with quantification routinely to sub-ppm concentrations,[23, 24] especially if combined with MS detection.[25] This automated method boasts rapid turnaround times with minimum sample preparation. However, the major limitation is that no speciation information can be obtained. This method can only give the total P concentration of all components present in the sample.

12.6.1.1 Phosphorus Speciation in Deta-Phosphonate, Diethylenetriaminepenta Phosphonic Acid, by Ion Chromatography

In this example, scaling was observed in a field production environment, despite inhibitor returns indicating high levels of phosphonate inhibitor being returned in the produced brine. IC was utilized to check the stability of a phosphonate SI. The main aspect here was to assess whether the analysis of the effluent samples by ICP-OES for total P was detecting the active sale inhibitor component.

IC is successful in separation of cationic or ionic species by using specific ion-exchange columns and mobile phases. For phosphonate analysis, the ion-exchange column is packed with a high-capacity exchange resin which has a high affinity for phosphonates. The detection method incorporates a flow injection analysis method based on the phosphomolybdenum postcolumn reaction, see Figure 12.5. A gradient of increasing nitric acid concentration was used to elute the phosphonate species, in order of increasing charge and molecular weight. Since DETPMP has five phosphonate groups, it will have a high affinity to the column and will take longer to elute from the column; whereas smaller molecules with lower amounts of P ionic groups (like orthophosphate, PO_4^{3-}) will be retained less and elute much quicker. This methodology gives speciation information allowing quantification of multiple P-containing species in a single sample.

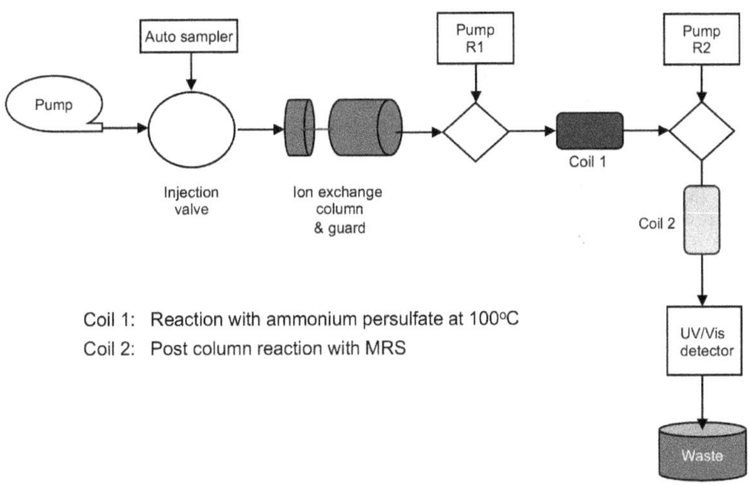

Figure 12.5. Set up for ion chromatography for phosphonate speciation.

In this example, selected field samples from a DETPMP squeeze in a North Sea field were initially analyzed by ICP-OES for total P content, and then normalized by appropriate dilution so that each sample contained the same amount of total P prior to any further analysis. The normalized samples were then analyzed by IC to identify which P species were present.

For the purposes of this study, the detection of the different P species was qualitative. However, IC can also be used for quantitative analysis, although there are limitations of this technique for phosphonates, as they are not very UV-responsive even when complexed. To achieve a good quantitative response, a solid phase extraction (SPE) and/or ion-pairing technique or a good selective detection technique (like MS) is generally required.[20]

The most recent methods for phosphonate analysis use aminopropyl cartridges for upconcentration coupled with MS-MS detection.[19, 20] Ion pairing with Fe(III) has also been used in conjunction with UV detection and methylation with MS detection.

The ion chromatogram of the field DETPMP samples showed two different peaks indicating some speciation[6, 7, 17, 18] with peaks being observed for both orthophosphate (degradation product that elutes first) and the DETPMP. As return samples progressed chronologically, the response from DETPMP decreased, and the degradation products increased. For this example, each sample examined by IC was adjusted to contain the same equivalent amount of total P.

To summarize, the levels of the active inhibiting species decreased as samples progressed chronologically (Table 12.4), indicating thermal degradation of DETPMP as time progressed. This example highlights the limitation of ICP for total P. Interferences are due to (i) the presence of other P-containing chemicals, (ii) the background residual inorganic P produced from the reservoir, and (iii) the thermal degradation of the SI. Similar results have been obtained when using IC as a qualitative technique to evaluate the impact of product pH on thermal stability of phosphonates, where the impact of degradation was also confirmed by performance testing.[27]

Table 12.4: Ratio of degradation products determined by qualitative IC.

Sample No.	Ratio of Peak Area for DETPMP vs. Orthophosphate
1	1:0
2	1:0
3	1:5
4	1:8
5	0:8

12.6.2 Polymer Speciation

12.6.2.1 Isolate Polyacrylate from Polyacrylate/Sulphonated Polymers Mix (Simple Approach)

Polyacrylate-based polymers can be readily separated from sulphonated species by utilizing C18 solid phase extraction. The single use disposable cartridge containing an octadecylsilane stationary phase retains neutral species. For polyacrylate inhibitors, this is achieved by lowering the pH and subsequently suppressing the ionization. As the solution of polyacrylate/sulphonated species is passed through the cartridge, only the neutralized polyacrylate polymer retains. The charged species, that is, sulphonated polymers (and brine salts), pass through the cartridge, although a tiny amount of sulphonated polymer can be retained. After washing the cartridge to ensure all interferences have passed through into the waste, the pH is then increased, causing ionization of the polymer and subsequent release from the cartridge. Now free from the sulphonated species (and brine salts), the polyacrylate polymer is isolated and easily quantifiable. In the standard wet chemical approach, this would be achieved by complexing with Hyamine (a quaternary ammonium salt) forming a turbid solution and analyzing the absorbance at 500 nm.

12.6.2.2 Isolate and Quantify P-Tagged and Non-P-Containing Polyacrylate Polymers

As described in the approach above, C18 SPE is utilized to retain polyacrylate species and offers a simple approach for discrimination, when sulphonated polymers are present in the mix. However, in the presence of two polyacrylate species (one P-tagged and one P-free, *e.g.*, PPCA vs. PAA), subsequent steps are required following retention on the C18 cartridge. Since both species in this example retain to C18, the total polymer concentration is achieved by eluting the chemicals from the C18 cartridge, once free from interference and complexing with Hyamine to quantify the total polymer concentration, that is, there is no discrimination between the P-tagged polyacrylate and non-P-containing polyacrylate at this stage. Following this, ICP-OES can be utilized to determine the P content of the P-tagged chemical samples pretreated using C18 SPE cartridges. The resulting P content then determines the concentration of the P-tagged polymer present, and this can be deducted from the total polymer concentration to determine the concentration of the P-free polyacrylate polymer.

12.6.2.3 Isolate and Quantify P-Tagged and Non-P-Containing Sulphonated Copolymers

In a similar approach for polyacrylate species, a solid phase extraction method can be utilized, as both a method of determining the total polymer content and as a precursor method to ICP-OES to determine the P content and subsequently deduct and quantify both polymers. SAX solid phase extraction offers a simple benchtop wet chemical approach for sulphonated species analogous to that of C18 for polyacrylate polymers. The charged sulphonic acid groups (unable to retain on C18 SPE cartridges) retain to the SAX stationary phase, brine salts are not retained and pass through into the waste. Free from interferences, the polymer is released from the column and can subsequently be analyzed by Hyamine. Following the determination of P in the sample by ICP-OES, the concentration of P-tagged sulphonated copolymer (P-VS-Co) can be deducted from the total polymer to calculate the non-P-containing sulphonated copolymer (Vs-Co) concentration.

12.6.2.4 High-Performance Liquid Chromatography and Mass Spectroscopy

HPLC is a similar technique to ion exchange. The separation of the species to be analyzed is achieved by either molecular weight (gel permeation chromatography) or adsorption/desorption from a specific column combined with detection by ultraviolet/visible (UV/VIS) absorbance, fluorescence, charged aerosol detection (CAD), or MS. Eluents, such as water/methanol/acetonitrile with pH buffer control, are used to control the elution and separation of the different polymer molecules. HPLC techniques with UV/VIS detection are often used for the analysis of polymeric SIs (like Vs-Co) in produced brines. But they were restricted to practical detection limits of between 5 ppm and 10 ppm and, in addition, suffered interference from high-saline brines.

Major advances have been recently made with improvements in sample processing, salt separation columns, ion-pairing methods with C18 reverse phase columns, and detection using CAD[30] and MS.[25, 32] These new methods have significantly lowered the detection limits of polymeric SIs to <1 ppm and improved accuracy and precision. Indeed, the use of HPLC provided an effective technique for the detection of maleic acid copolymer SI using reverse phase liquid chromatography (RPLC) and UV/VIS detection with a limit of detection (LOD) of 0.2 ppm. There was also no interference from the presence of phosphonate inhibitors with this method.

A method using salt separation columns and UV/VIS detection had been able to detect sulphonated co-polymers in field brines with LOD of 1 ppm. This method has been able to demonstrate that good squeeze lifetimes can indeed be obtained in the field,[34] for this type of chemistry which was not previously deployed in the field due to poor detection by dialysis/Hyamine method. Vs-Co had always been perceived to demonstrate very poor retention properties. Phosphonate SIs also do not interfere in this method, if the right elution solvent composition is used. The elution solvent usually constitutes a water-based mobile phase with the addition of solvents, for example, methanol or acetonitrile. However, if other polymeric inhibitors are present in the produced brine sample, they will interfere and will need to be removed using the preseparation approaches listed under polymer speciation.

The analysis of Vs-Co and P-tagged versions has since been further improved using MS detection, and methods have been developed with LOD of 100 ppb for both sulphonated polymers and P-containing sulphonated polymers.[33] The new methods can also differentiate between both types of polymer SI based on the type of sulphonate functional group.

MS as a detection method has several advantages to other detection techniques. It is selective for analysis with functional groups in polymers used as specific tracers, that is, cationic amine and sulphonated groups can be used as selective tracers to differentiate between different polymer types.[32, 33] Amine-based polymers can be selectively detected with a LOD of 50 ppb in North Sea field brines in the presence of other polymers such as polyaspartate- and P-containing polyamines. This has also been successfully reported for sulphonated polymers.

IC with MS-MS detection has been shown to be successful in quantifying squeezed phosphonate inhibitors in the presence of P-containing SI from continuous injection.[20] IC-MS technique has been used for analysis of a SI and CI in a multicomponent sample.[7] Both chemicals were successfully quantified to ppb concentrations, highlighting the sensitivity. This technology allows individual inhibitor species to be successfully quantified in the presence of multiple production chemicals with subsequent quantification of all species.

Like most methods, there are some limitations with MS. The instrument costs are quite high (even for portable benchtop versions), and interpretation of results requires significant experience and expertise which can increase the overall analysis cost.

Matrix interferences and the presence of many chemical species in produced brines can make quantification of residual chemical species more difficult.[30] The assay can be limited to two or three SIs in the presence of each other. However, MS still offers an advantage over conventional methods (like HPLC, Hyamine, or ICP-OES), which traditionally detect one SI, and often the detection limits are an order of magnitude lower.

MS detection methods can also have issues with polymeric analytes, as no single ion is generated, making "fingerprint" assay and interpretation difficult. This can be overcome by using MS-MS or secondary ion mass spectroscopy) for specific molecular fragments of cationic amines[32] and sulphonated molecular groups.[33] The ionization process can also be influenced by factors including spectrometer voltages, flow rates, matrix and mobile and phase effects, and method development can take some time.

Once suitable methods have been developed, MS is quite a powerful tool for the low-level SI detection. The high analysis costs can be outweighed by the benefits of extended squeeze lifetimes at low MIC. Indeed, in the following example, the use of HPLC-MS for the detection of a P-tagged sulphonated copolymer has demonstrated significant benefits for extending the scale squeeze lifetime, compared to using HPLC or P detection via ICP-OES.

In this case, a field in the North Sea producing from both conventional vertical and horizontal wells required SI squeeze treatments in several of the wells to control barium and strontium sulfate scale. The wells were originally squeezed with phosphonate-based SIs before being replaced with a more environmental-friendly polymeric SI. A commercially available P-containing sulphonated copolymer was selected for squeeze treatments and a sulphonated carboxylate copolymer was also applied topside for continuous injection. The sampling point for the squeeze inhibitor was downstream of the topside product application and often the topside SI was not switched off, when taking the squeeze samples. Normally, this would not matter, since the topside product does not contain any P, and ICP-OES could be used to measure the returns of the P-tagged sulphonated copolymer. However, the residual data from the measurement of the squeeze polymer seemed to give very high returns. It was realized from testing blank samples before squeezing with the P-tagged polymer that there was still a lot of background P interference from the reservoir from previous squeezes with phosphonate inhibitors.

The scale squeezes were lasting much shorter than expected, because the P analysis data indicated that 100% of the applied SI had been returned quite quickly (within 40–50,000 m^3 of produced water). Since the operator did not have absolute confidence in the data, the wells were resqueezed after much shorter periods than should have been required.

Clearly, a new analysis technique was required for distinguishing between two types of sulphonated SIs and not being impacted by background P from the reservoir. HPLC and colorimetric analysis were investigated, but interference was found from the topside polymeric SI. Thus, these methods were not really any use, except for when the residual concentrations of P-tagged polymer were >500 ppm.

A new HPLC-MS technique was applied. It was found to be suitable for accurate detection of the P-tagged polymer to 1 ppm. This offered a great improvement in the confidence the operator had in the SI return profiles. By using this technique for the active polymer, the scale squeeze lifetimes were extended to more than double the previous squeeze lifetimes, without changing the applied squeeze design, as shown in Table 12.5.

Table 12.5: Scale squeeze returns comparison for ICP-OES and HPLC-MS with a P-tagged sulphonated copolymer.

Water Produced (m^3)	Inhibitor Concentration (mg/L)		% Returned Chemical	
	ICP-OES	HPLC-MS	ICP-OES	HPLC-MS
2,000	70,000	40,000	50	30-50
4,000	8-10,000	8-10,000	85	54
20,000	200	100		57
40,000	90	30		58
60,000	20	25	100	60
90,000	40	60	100	65
115,000	-	40	-	68
130,000	-	25	-	70

12.7 Fluorescence

Fluorescence is a form of luminescence and is described as the emission of light by a substance that has absorbed light or other electromagnetic radiation. When a substance absorbs light, the electrons are excited to a higher energy state. When the excited electrons return to the original ground state, that energy is emitted. The excited state in direct fluorescence exists for a finite time, typically 1–10 ns. When the excitation energy discharges, the sample emits photons that can be measured with a specific detector. The amount of energy emitted is dependent on the structure and concentration of the fluorescent molecule (fluorophore), the solvent type, pH, and temperature. Therefore, fluorescence can be used to accurately quantify the concentration of fluorophores to ppm or even sometimes ppb level, depending on the sample matrix and the molecular composition.

When fluorescence occurs, some energy is lost. The emitted light is therefore at a higher wavelength than the excitation wavelength. Fluorophores absorb and emit light at a specific wavelength, enabling different F tags to be distinguished from each other. Fluorescence spectra can be measured with a spectrofluorometer that has an excitation light source and wavelength filters to isolate emission photons from excitation photons, combined with a detector that can record emission photons and produce a reading. The fluorescence emission signal is proportional to the fluorescent molecule or F-tagged SI concentration, over a dynamic range of the measurement.

Fluorescence measurement is very sensitive. In case of high concentration of a tagged SI chemistry, a sample may need dilution for the dynamic range of the measurement. The fluorometric detection may use simply excitation and emission wavelength peak maxima, or a range of excitation and emission wavelengths may be used in the absence of interfering fluorescence signals in the detection wavelength ranges. The response from F tags can also be dependent on solution pH, as this can help with the detection of different tags, since some tags do not give a response at pH < 2–3 and only fluoresce at $>$pH 5.

There are also many interferences with fluorescent measurements, including high brine salinity (which quenches the fluorescent signal) and crude oil which contains many fluorescent species. These interferences have been overcome by the development of sample pretreatment methods,[39] such as filtration with Whatman filter paper (0.45 μm) to remove oil contamination, sample dilution and using a NAP, C18 or weak anion exchange cartridge as a final treatment stage to remove any extra interfering species like other production chemicals and organic acids.

Good calibration curves and low LOD of 1–1.5 ppm for different tagged products are readily achievable in field brines, and good results have been obtained when using F-tagged polymers in the field, as shown in Figure 12.4. The concentration of the tag in the polymer can also be increased to increase the LOD, if required. Care must be taken with molecular design, because if the tag concentration in the polymer molecule exceeds 2.5%, then it is classed as a new product and would require a full suite of environmental tests. In addition, by increasing the concentration of the fluorophore in the polymer, this might also change its performance as a SI and its thermal stability. This would also then need a series of laboratory tests to confirm the new product is still a suitable SI.

Some applications of F-tagged polymers are presented below.

- Squeezing a subsea well with F-tagged polymer with continuous injection of untagged polymer

In the following example, the use of fluorescence spectroscopy for the detection of an F-tagged sulphonated copolymer has demonstrated significant benefits for improving the confidence in residual data and extending the scale squeeze lifetime compared to using

HPLC. In this case, a subsea horizontal well in a low-temperature (78°C) North Sea field required scale squeeze treatments to treat against a moderate barium sulfate scale problem.

A sulphonated copolymer was qualified for scale squeeze treatments across the field, and the same SI chemistry is also applied at the subsea wellhead in subsea wells. The polymer is normally detected by HPLC, but in this case, the injection of additional polymer at the wellhead was found to interfere in the scale squeeze residuals, giving little confidence in the performance of the squeeze treatment at low-SI residual levels and therefore leaving no room for optimization of treatment design. This presented an issue for downhole scale management in this subsea well, which was solved by changing the chemistry of the downhole product by using an F-tagged polymer which could be detected by fluorescence spectroscopy without any interference from the untagged polymer.

The F-tagged product could be reliably detected to ~ 1 ppm in this specific field brine, and the ability to accurately detect the scale squeeze residuals at low levels in the presence of the untagged polymer gave the operator much improved confidence in the performance of the squeeze treatments compared to squeezing the untagged polymer using HPLC detection. Indeed, the squeeze treatments were considered successful and lasted between 1.5 and 2 years.

- Squeezing a subsea well with F-tagged polymer with continuous injection of combined scale/CI and wax inhibitor

In this case, a subsea well comingled with two to three other wells required scale squeeze treatments to treat against a mild- to moderate-barium sulfate scale problem. A sulphonated polymer was initially selected for scale squeeze, since it was not possible to select a phosphonate or P-tagged polymer due to the continuous wellhead injection of a combined scale/corrosion (SI/CI) inhibitor which contained a P-containing SI.

The initial proof of concept for detection of the sulphonated copolymer in synthetic brine indicated that HPLC would be suitable for low-level detection of the scale squeeze inhibitor with no interference from the combined SI/CI. However, when the analysis was repeated in samples of field brine, an unexplained interference was found which significantly increased the detection limits to unacceptable limits. Further investigation revealed that the interference came from certain components of the wax inhibitor which had partitioned into the brine phase. This issue was overcome by using suitable pretreatment with a C18 cartridge and selecting an F-tagged sulphonated polymer which was readily detectable to low levels without any interference from the soluble components of the wax inhibitor or the combined SI/CI.

- Two F-tagged polymers by fluorescence and total polymer by TRF

Three SI polymers, one untagged and two fluorescent-tagged versions, were dosed to produced water collected from a North Sea field. TRF was tested for the determination of total SI concentration. All tested polymers had the same polymer backbone, only the fluorophores were different. The total SI concentration varied between 100 ppm and 1,000 ppm. The fractions of individual products were varied to verify that the untagged and tagged products give the same TRF response in total polymer measurement. The blank sample contained only produced water.

Before the measurement, the samples were pretreated using a developed protocol. The produced water samples were diluted with synthetic brine using a dilution of 1:150. The samples were transferred to the measuring cuvettes containing the reagents needed and mixed. After mixing the samples were measured with a TRF reader. The measurement is based on the formation of a loose chelate between the SI polymer and europium ion (Eu^{3+}). The TRF signal of Eu^{3+} is amplified, when the chelate is formed, and the signal amplification is proportional to the polymer concentration.

The response was found to be linear with the tested 100–1,000 ppm concentration range indicating that the total polymer measurement is accurate and samples containing different ratios of three polymers gave the same TRF signal. The tested protocol could therefore be used for the measurement of total SI polymer concentration in oilfield produced water samples.

Before the direct fluorescence measurement, the produced water samples were pretreated using SPE cartridges. SPE cartridges were used to remove all unwanted constituents that might cause interference during the measurement. Produced water samples typically contain, for example, hydrocarbon residues and production chemicals. Some of these compounds have natural fluorescence properties potentially influencing the measurement. The interfering compounds need to be removed by sample pretreatment, before measuring the tagged SIs with a fluorometer. Product-specific excitation and emission wavelengths were used for tagged Polymer-1 and tagged Polymer-2. Different wavelength areas enable the accurate detection of both chemicals in the same sample, and the calibration curves were measured in the same North Sea produced water used for TRF tests. The emission intensities were recorded as a sum of intensities from the emission spectra. The calibration curves are linear over the range 0–75 ppm with a good correlation coefficient.

The concentrations of tagged Polymer-1 and tagged Polymer-2 were measured from produced water samples that also containing untagged polymers. The product concentrations measured vs. actual concentrations are presented in Table 12.6.

Table 12.6: Measured vs. actual concentrations for F-tagged polymers in produced water.

Sample No.	Untagged Polymer (ppm)	F-Tagged Polymer-1 (ppm)		F-Tagged Polymer-2 (ppm)	
		Measured	Actual	Measured	Actual
1	25	27.4	25	3.4	3
2	25	50.5	50	3.4	3
3	25	73.9	75	3.6	3
4	25	3.6	3	27	25
5	25	3.1	3	51.9	50
6	25	3.5	3	81.8	75

The analyzed SI concentrations are close to the real values. The sample pretreatment method is effectively removing interfering compounds from the produced water sample. The results are in line with earlier results of F-tagged polymers[3, 39] and confirm the direct fluorescence detection method is reliable with high accuracy and precision. The untagged polymer concentration can be calculated by subtraction of the F-tagged polymers measured by direct fluorescence and the total polymer concentration measured by TRF.

The advantage of fluorescence detection is that it can be used on-site, with a portable fluorescence spectrometer[4, 26] coupled with easy sample pretreatment methods.[39] It can also be combined with other methods (like TRF) to determine multiple polymers that are both F-tagged and untagged.[4, 26]

12.8 Time-Resolved Fluorescence

TRF techniques have been actively used in the biomedical industry for ~35 years. This technology has been transferred across to the oil and gas industry in the last 10 years to investigate the analysis of the active polymeric SIs and phosphonates in oilfield produced brines.

TRF measurement is very similar to the standard fluorescence measurements. The main difference is the timing of the excitation/measurement process. Fluorescence is a very quick process, and emission is measured simultaneously or very quickly after the excitation process. In TRF, there is delay in the excitation pulse, resulting in lower measurement backgrounds than in the standard fluorometric process.[40]

TRF relies on the use of very specific fluorescent species, such as lanthanide ions like Europium (Eu) and Terbium (Tb), which exhibit peculiarly long Stokes's shifts and fluorescence lifetimes. Alone, lanthanide ions have low TRF signal, due to the low yield of the excitation energy. However, the fluorescence yield is enhanced, when the lanthanide ions are complexed with other species. The TRF signal can be measured even milliseconds after excitation, compared to most standard fluorescent molecules that emit within a few nanoseconds of being excited.

In TRF measurement, the liquid solution containing the lanthanide complex is excited, and the fluorescence signal is measured after a short delay. The time delay of TRF measurement eliminates common short-lived fluorescence interfering signals, resulting in a very low background signal. The short-lived fluorescence exists only for a few nanoseconds, whereas the lifetime of TRF signals is much longer and typically recorded with >100 μs delay times, see Figure 12.6. This feature of TRF results in detection of analytes with superior sensitivity.[40]

TRF has been reported to be suitable for on-site detection of polymeric SIs[4, 40, 41] and is reported to be less susceptible to interferences than the conventional standard fluorescence measurements. Typical detection limits are reported to be 1–2 ppm for polymers and <0.5 ppm for phosphonates. TRF also offers the potential for analyzing both polymeric and phosphonate SIs in the same produced water sample.[4, 5, 22]

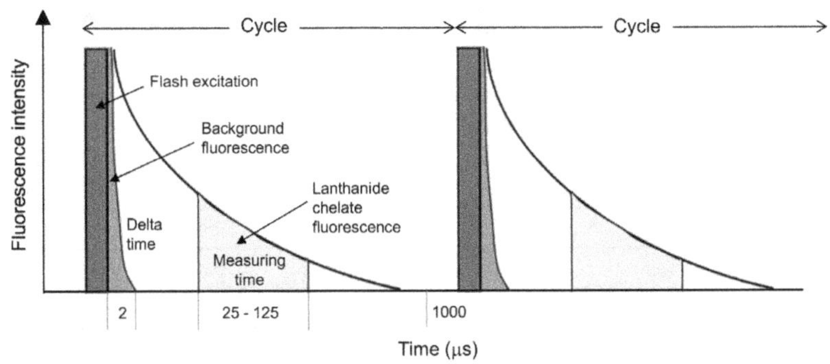

Figure 12.6. Illustration of the principles behind time-resolved fluorescence.

TRF can also be used in conjunction with other techniques (like fluorescence) to detect multiple polymers present in one sample. Fluorescence can be used to detect up to three F-tagged polymers. If any untagged polymer is present (which will not respond to conventional fluorescence technique), TRF can be used to quantify the total polymer concentration present; the untagged polymer can then be calculated by mathematical subtraction.[26]

TRF is currently at an advanced stage of development at the time of writing this chapter. The equipment has now become commercially available and has been used for the on-site detection of both polymeric and phosphonate SIs.[4]

A comparison of TRF vs. fluorescence techniques detailing the applications and limitations of each technique for both polymers, tagged polymers and phosphonates is given in Table 12.7.

Two case studies are summarized below to highlight the use of TRF technique.

- Phosphonate measured with TRF in a sample matrix-containing sulphonated copolymer SI and P-containing CI

In this example from a North Sea subsea field, it was necessary to be able to measure the concentration of a phosphonate scale squeeze inhibitor to 1–2 ppm in the presence of 200 ppm of P-containing CI and 60 ppm of sulphonated copolymer, both of which are injected continuously at the subsea wellhead.

ICP-OES analysis of the P content of the phosphonate and the CI showed P contents of 11.48% and 0.55%, respectively. The amount of P in the CI was enough to give substantial interference in the measurement of the phosphonate by ICP-OES at the low levels required, and detection limits for the SI were >20 ppm. IC was also investigated, and although the phosphonate could be separated from the polymer and the CI, the LOD was 10 ppm. This would mean much shorter squeeze lifetimes than would be possible at LOD of 1–2 ppm.

Table 12.7: Comparison of fluorescence vs. time resolved fluorescence.

Comparison Factor	TRF	F
Interferences	Salinity, crude oils, and other production chemicals. Removed by dilution and/or suitable pretreatment with NAP, C8, and C18 cartridges Chelants may interfere	Salinity, crude oils, and other production chemicals. Removed by dilution and/or suitable pretreatment with NAP, C8, and C18 cartridges. Chelants may interfere
On-site analysis	Yes	Yes
Proven applications	Yes	Yes
Phosphonates	Yes	No
Sulphonated polymers	Yes. The presence of phosphonates requires pretreatment	No
F-tagged sulphonated polymers	Yes. The presence of phosphonates requires pretreatment	Yes
P-tagged sulphonated polymers	Yes. The presence of phosphonates requires pretreatment	No
Polyphosphino carboxylate	Yes. The presence of phosphonates requires pretreatment	No
Total polymer analysis	Yes	No
Multiple polymer analysis in the same sample	Possible but requires further work	No
Multiple F-tagged polymer analysis in the same sample	No	Yes
Multiple P-tagged polymer analysis in the same sample	Possible but requires further work especially in the presence of untagged polymers	No
Multiple phosphonate analysis in the same sample	Possible but requires further work	No
Phosphonate and polymer in the same sample	Yes. Suitable pretreatment required	No
Phosphonate in the presence of phosphate ester	Yes	No

An alternative method was therefore required, and TRF was used for the measurement of low concentrations of downhole phosphonate SI in produced water samples containing up to 200-ppm P CI and ~60-ppm sulphonated copolymer SI.

A good calibration graph for the phosphonate was obtained in produced water with no interference from the polymeric SI or the P-containing CI. The concentration of phosphonate SI was as expected (Table 12.8), and the phosphonate squeeze SI can be quantified to 0.3 ppm with repeatability of ±0.1 ppm.

Table 12.8: Experimental results for 5 mg/L VS-Co and DETPMP by TRF after pretreatment.

Sample No.	Vs-Co (mg/L)	DETMP (mg/L)
1	5.8	4.9
2	3.5	5
3	4.3	5
4	4.2	4.7

The method is currently in use and provides the operator with improved confidence in the squeeze residuals at low levels. This allows the squeeze treatments to run to their yearly designs at 1 ppm MIC and negates the need for extra treatments and high costs associated with extra subsea well interventions.

- On-site analysis of a polymeric SI squeeze inhibitor in a remote desert location

In this example, several conventional wells in a steam flood field in a remote desert location, with harsh calcium sulfate scale issues and bottomhole temperature of 200°C, were squeezed with a polymeric SI. The polymeric SI was selected as the best product for both downhole injection and scale squeeze under the high-temperature steam flood conditions.

In view of the harsh calcium sulfate scaling environment, the operator requested a quick turnaround of scale squeeze residuals. However, this proved difficult initially, since the samples needed to be analyzed by HPLC, and this technique was not available on site and would involve several weeks of delay, while samples were sent to the nearest local laboratory with HPLC capability.

TRF was therefore selected as a possible alternative method for on-site analysis using a robust, portable spectrometer. Calibration curves were prepared in the produced water and a good linear plot over the range of 0–5,000 ppm was observed with a detection limit of ~1–2 ppm. In this case, the MIC for the polymeric SI was 5 ppm, so the detection limits were considered acceptable.

Five wells were squeezed with the polymeric SI, and the scale squeeze residuals were monitored with TRF. The data from all five wells followed the generic trends expected from scale squeeze treatments with a high initial return concentration followed by a long tail at lower concentrations.

To validate the TRF analysis data for modeling purposes, the samples collected were also sent to the laboratory for HPLC analysis. The results showed a good correlation in the data trends between TRF and HPLC methods.

For the initial squeeze returns, higher values were recorded for the TRF samples, but good agreement was obtained for the concentrations below 250 ppm, which were considered more important for key decisions on whether the well is protected. TRF is therefore proven to be suitable for on-site analysis of polymeric scale squeeze inhibitors and provides reliable data for squeeze modeling purposes.

12.9 Squeezing Comingled Wells in the Same Flowline

Several factors need to be considered when squeezing comingled wells in the same flowline. The main challenge is to squeeze up to three or four comingled wells with different SIs, respectively, and to achieve selective detection of these SIs in the same sample in the presence of the subsea and topside continuously injected SI, which is often a polymer.

The approach to solving this challenge has been to use separate and selective analysis techniques for each product, coupled with specialist sample preparation and chromatographic separation techniques to remove saline interferences and provide upconcentration and some separation of the selected SIs. This has involved various approaches for the detection of multiple products in the same sample, including the use of P- and F-tagged inhibitors,[1–3] HPLC and ICP-OES combined with complex multiple solid phase extraction steps[42] and more recently HPLC-MS techniques.[25, 32, 33]

In addition to the challenge of detecting the various families of SIs in the presence of one another, the effect of uneven produced water rate from the wells must also be considered. In a comingled flow line, this influences the detection possibilities of the different SIs. If the extent of produced water dilution is such that the SI concentration is likely to be reduced to below its detectable limits, this is accounted for in the overall scale squeeze design and the product selection for these wells.

For example, a SI may have an MIC of 2 ppm and a detection limit of 2 ppm. This means that for the single well, it would be achievable to detect the SI to its MIC level in the produced water, and a treatment could be designed for this accordingly. However, if the brine was diluted by a factor of two in a comingled stream, this would mean that for a detectable level of 2 ppm in the comingled brine stream, the actual amount detectable in the original well fluid would then be equivalent to 4 ppm; and a new squeeze design would therefore be required at this resqueeze detection limit to get the same squeeze lifetime as was designed for the individual well at 2 ppm MIC. If the detection limit of the product was lower than the MIC, say 1 ppm, then in this case, the treatment could run down to its original design of 2 ppm MIC, as this would also be the resqueeze detection limit in the comingled brine.

It is therefore desirable to have the lowest possible detection limits for all the selected scale squeeze inhibitors, preferably <1 ppm and lower than the MIC, to compensate for any dilution effects and to provide increased confidence in the accuracy and precision of the SI residual concentrations at low levels, especially when at or below the MIC level.

In comingled brine environments, the dilution factors are often two to three, although it may be possible to get 8–10 dilution for some wells in certain circumstances, especially when wells from different fields are comingled. The extent of dilution and the detection limits will therefore have a big impact on the product that is selected for scale squeeze in a particular well. In this case, a high-retaining product like a phosphonate could be selected for the high water cut wells, where less dilution is encountered; and for low water cut and high dilution wells, a polymer with very low detection limits could be considered. However, this does not preclude phosphonates for use in low water cut wells, because their good adsorption/retention characteristics may compensate for dilution impacts to a certain extent.

The choice of products for comingled well environments will depend very much on the water composition, the scaling scenarios and the detection methods available for the products that may be fit for scale squeeze application. In some cases, only polymeric SIs may be suitable, depending on the scaling severity and environmental legislation in the region concerned.

The selection of product and its detection limits could impact the subsequent squeeze lifetimes for the individual products, although the treatment lifetimes can be compensated for to a certain extent in the treatment design based on known dilution factors and detection limits. However, it is also desirable to have efficient SI retention and release characteristics to achieve the longest possible treatment lifetimes and to reduce the number of well interventions.

The selection of products for squeezing comingled wells may be aided by the information detailed in this chapter. The review of analytical options available for specific SIs is summarized in the section below.

12.10 Summary of Options for Scale Inhibitor Analysis: Benefits, Limitations, and Common Interferences

A summary of the range of techniques commonly adopted or proposed for SI residual analysis for different SI types in different production environments is presented in Tables 12.9–12.11.

Table 12.9: Scale inhibitor analysis options for phosphonates and phosphate esters.

Method	Type	Advantages	Disadvantages	Comments
Wet chemical phosphomolybdenum blue	Wet chemical	Records all P, rapid and accurate approach Minor modification provides effective assay for orthophosphate content (e.g., degradation of phosphonates) and other inorganic P species Separate from total P phosphonate content	Standard method suffers interference in high-salinity brines, especially in the presence of dissolved Fe	Calibration typically 1-5 ppm active phosphonate (1-3 ppm orthophosphate) Standard Hach approach cannot be used in the presence of sulfate, alternative reagents are therefore used routinely

Table 12.9 Cont'd

Method	Type	Advantages	Disadvantages	Comments
Inductively coupled plasma optical emission spectroscopy	Automated	Quick turnaround time. Records total phosphorus content in sample. Less labor-intensive than benchtop methods	Measures total P. Cannot give information on P speciation. Also subject to interferences from different metal ions found in oilfield brines	Matrix matching of produced water samples important for residual levels—often not conducted Method should not be used for samples containing other P-based inhibitors (unless conc. of other P species can be independently assessed) Method should not be used for direct assay of residual levels of scale inhibitors which contain very low P content (*e.g.* P-tagged polymers—impact of other low-level P contaminants can result in significant errors unless inhibitor is preseparated)
Ion chromatography	Automated	Can discriminate between different P-containing species, that is, different phosphonate scale inhibitors/orthophosphate, and so on Applicable to HT samples that have undergone thermal degradation and for mixtures of different P-containing production chemicals, since P-containing species are discriminated	Less routine than ICP-OES or wet chemical approach Can suffer interference in high-salinity brines, so high dilution may be required unless SPE pretreatment used. Sensitivity is reduced with dilution	Very useful for P speciation but less routine and more expensive than ICP or wet chemical approaches Unsuitable for measuring below 5 ppm has become susceptible to interference from ionic species unless using MS detection
TRF	Wet chemical	Good method for single phosphonate species with LOD < 0.5 ppm	No interference from polymers and phosphate esters, but Fe and chelants will interfere unless prtreatment methods applied	Very useful for single phosphonate species, but the method requires more work for multiple phosphonate species
SERS	Wet chemical	Will detect phosphonates to 1 ppm	Interference from polymers, phosphate esters, and Fe to be established	Cannot distinguish between multiple phosphonate species without further method development. Handheld instrument for field measurement tends to be more costly than TRF and other methods. Not used as a full commercial method

Table 12.10: Scale inhibitor analysis options for polyacrylates.

Method	Type	Advantages	Disadvantages	Comments
Standard Hyamine without polymer separation	Wet chemical	Simple technique	Does not distinguish between polymer species. Subject to interferences especially salinity. So, without separation can only measure high concentrations which require dilution	Dilute to 1–30 ppm into fresh water to give <10,000 mg/L TDS
C18/Hyamine (using one disposable C18 cartridge per sample)	Wet chemical	Simple technique suitable for all PAA/PMA/PPCA polymers. Opportunity to concentrate low concentrations of polymer. Ability to discriminate between polymers	Labor-intensive, if not automated with greater operator error than automated methods PMA method has been automated with LOD 1 ppm	If sulphonated polymers are present C18 will separate them from the mix. Retention of the sulphonated components will need to be investigated as some species are retained on C18 cartridge and minimal recovery will be needed to remove interference Interference from organic acids can be removed by pretreatment with C18 cartridge at neutral pH
SAX/Hyamine	Wet chemical	Standard method records all types of polymers including sulphonated, PAA, PMA, PPCA, and PASP		Nor generally used as first choice for PAA polymers is C18 cartridge
WAX/Hyamine	Wet chemical	Not generally recommended, unless the brine is low salinity	Very poor recovery of polymer in moderate salinity brines	Not recommended
Dialysis/Hyamine	Wet chemical	Simple technique. Often very effective and accurate with clean samples	Time consuming and laborious. Poor sensitivity of 2 ppm. Method does not allow up concentration	Can be used routinely in the laboratory but C18 method preferred for PAA/PMA/PPCA
HPLC: C18 postcolumn derivatization/Hyamine	Semiautomated	Automates standard C18 approach above. Therefore, less time consuming	Not fully automated Still requires significant operator time	Often used to automate where direct detectors are unavailable
Benchtop C18/flow injection/CAD/ELSD	Semiautomated	Provides direct accurate assay	C18 HPLC columns lower cost than GPC CAD less robust than ELSD but more sensitive. Therefore, ELSD preferred if sufficiently sensitive	Useful if HPLC not available, but detectors and flow injection system are available

Table 12.10 Cont'd

Method	Type	Advantages	Disadvantages	Comments
HPLC C18 with ELSD	Automat-ed	Provides direct accurate assay	Cannot use with CAD detector since meth-od involves change in gradient	Recommended
HPLC GPC column with aerosol de-tection (CAD, ELSD First-gen-eration HPLC/GPC approach	Automated	Provides Direct Accu-rate Assay First-generation GPC incorporating more sen-sitive aerosol detectors. Originally developed for sulphonated polymers to improve on dialysis approaches	Does not discriminate between species, because SI polymers are of similar low molecular weight Sample preparation is time consuming and often critical CAD less robust than ELSD but more sensi-tive. Therefore, ELSD preferred if sufficiently sensitive	Second- and third-gen-eration methods allow speciation Due to time consuming sam-ple preparation C18 HPLC or C18 benchtop approaches often preferred for PAA/ PMA/PPCA
Advanced HPLC/GPC approach-es (often referred to as third-, fourth-, and fifth-gen-eration approaches)	Automated	Provides direct assay and enables discrimi-nation Allows speciation of polymers by MS or other advanced detection systems	Sample preparation still critical Extensive analysis experience and exper-tise required	C18 HPLC or wet chemical C18/Hyamine preferred unless polymer speciation required
ICP-OES for P measurement in P-tagged or P-containing polymers	Automated	Provides direct analy-sis, but only if sufficient-ly sensitive and polymer preseparated from impurities	See earlier notes on phosphonate analysis Should not be used for P-tagged polymers unless extensive preseparation performed to remove impurities	Recommended for polymers with sufficient P but pre separation still required For P-tagged polymers with low P content it is not recom-mended unless extensive pre separation is performed
Fluorescence for F-tagged or fluorescent polymers	Wet chemical	Direct assay, if F tag is part of the polymer LOD <1 ppm depending on the F tag content Allows polymer spe-ciation using different F tags in the same polymer and in different polymers No interference from phosphonates	Does not work for non- fluorescent species. Requires sample clean up prior to analysis with NAP, C8 or C18 cartridges	Field proven method for F-tagged polymers Can be used for multiple comingled wells in the same flowline with up to three tags being detected in the same sample

Table 12.10 Cont'd

Method	Type	Advantages	Disadvantages	Comments
TRF	Wet chemical	Not fully evaluated for PAA but works for sulphonated acrylate copolymers	No polymer speciation but offers on-site detection Maybe some interference from chelants and Fe	Method needs more development work
SERS	Wet chemical	Not effective	Not evaluated properly, but poor signal response obtained for polymers at low concentrations	Not recommended

Table 12.11: Scale inhibitor analysis options for sulphonated polymers.

Method	Type	Advantages	Disadvantages	Comments
SAX/Hyamine	Wet chemical	Can deliver rapid, reliable, and accurate data Ability to upconcentrate samples	Interference from other polymers Greater potential for operator error in comparison to automated techniques	Preseparate PAA by C18 allows speciation between PAA and sulphonated polymers Need to check retention of sulphonated species on C18 for effective polymer separation SAX does offer good retention of sulphonated polymers and is similar to C18 for PAA Preferred over dialysis
Dialysis/Hyamine	Wet chemical	Simple technique that it is effective with clean samples	Time consuming and laborious. Poor sensitivity of 2 ppm. Method does not allow upconcentration.	Can be used routinely in the laboratory for sulphonated polymers, but C18 method preferred for PAA/PMA/PPCA
HPLC–C18 approaches	Automated	Not effective	Not effective	Not effective
WAX Hyamine	Wet chemical	Not recommended Only applicable to low-salinity brines	Not recommended Ver poor response in moderate-salinity brines	Not recommended
C18 ion pair chromatography with ELSD	Automated	Provides direct analysis	Sample preparation essential to avoid drifts from start to end of run	More details in SPE 179902

Table 12.11 Cont'd

Method	Type	Advantages	Disadvantages	Comments
HPLC GPC column with aerosol detection (CAD, ELSD) First-generation HPLC/GPC approach	Automated	Provides direct accurate assay First-generation GPC incorporating more sensitive aerosol detectors. Originally developed for sulphonated polymers to improve on dialysis approaches	Does not discriminate between species, because SI polymers are of similar low molecular weight Sample preparation is time-consuming and often critical CAD, less robust than ELSD but more sensitive. Therefore, ELSD preferred, if sufficiently sensitive	Second- and third-generation methods allow speciation Benchtop SAX/Hyamine or C18-IP maybe preferred for single polymers
Advanced HPLC/GPC approaches (often referred to as third-, fourth-, and fifth-generation approaches)	Automated	Provides direct assay and enables discrimination Allows speciation of polymers by MS or other advanced detection systems	Sample preparation still critical Extensive analysis experience and expertise required	Methods allow speciation SAX/Hyamine and C18-IP preferred unless polymer speciation required
ICP-OES for P measurement in P-tagged or P-containing sulphonated polymers	Automated	Provides direct analysis but only, if sufficiently sensitive and polymer preseparated from impurities	See earlier notes on phosphonate analysis Should not be used for P-tagged polymers unless extensive preseparation performed to remove impurities	Recommended for polymers with sufficient P, but preseparation still required For P-tagged polymers with low P content, it is not recommended unless extensive preseparation is performed
Fluorescence for F-tagged or fluorescent polymers	Wet chemical	Direct assay, if F tag is part of the polymer LOD < 1 ppm depending on the F tag content Allows polymer speciation using different F tags in the same polymer and in different polymers No interference from phosphonates	Does not work for non-fluorescent species Requires sample cleanup prior to analysis with NAP, C8, or C18 cartridges	Proven field technique Can be used for multiple comingled wells in the same flowline with up to three tags being detected in the same sample
TRF	Wet chemical	Direct analysis method. LOD ~1 ppm	On-site detection technique Needs sample pretreatment in the presence of phosphonates	Proven field technique
SERS	Wet chemical	Poor response	Poor response	Not recommended

12.11 Conclusions

This chapter on SI analysis has demonstrated that using the right SI analysis technique can deliver several scale management benefits for wells in both conventional, subsea and deepwater fields.

- A variety of SI analysis techniques exist for phosphonates, phosphate esters, and polymeric SIs including UV/VIS, Hyamine, ICP-OES, fluorescence, HPLC, MS, and TRF.
- Most analytical methods suffer interference of some degree or another and where a preferred method may be very sensitive and accurate under one set of conditions, it is unlikely to be effective under all production environments.
- Using inappropriate analysis techniques can render expensive core flood testing and field trials ineffective, thereby leading to the qualification and selection of inappropriate products in the laboratory and poor performance when applied in the field.
- It is essential to monitor the active SI component returned from SI squeeze treatments and not simply detect some readily analyzable moiety within the original supplied SI unless good agreement has been obtained with an alternative technique which is known to measure the active SI.
- The ability to monitor with improved confidence to low MIC allows squeeze treatments to run to longer lifetimes in conjunction with other field data.
- P speciation in phosphonates and phosphate esters can impact the analysis results, and for some cases, especially where thermal degradation is an issue, the impact of other P species needs to be accounted for.
- The analysis technique for chemicals in comingled well streams must also account for the impact of dilution on detection limits and the subsequent impact on treatment lifetimes and chemical selection.
- Discrete and accurate analysis of multiple SIs in one sample offers improved scale management in both conventional, subsea and deepwater fields.

In summary, there is still not a single method that is fit for purpose for any single or multiple inhibitors across all potential production environments. Many interferences are possible and the impact of these interferences on the accuracy and reliability of a particular technique changes from case to case. Fluorescence and TRF and combinations thereof now allow speciation and measurements in the field, but the analyst still requires an extensive toolbox at their disposal.

Nomenclature

AU	Absorbance unit
bbl	Barrels
Ca	Calcium
CAD	Charged aerosol detection
CI	Corrosion inhibitor
DETPMP	Diethylenetriaminepenta phosphonic acid
EDTA	Ethylenediaminetetraacetic acid
F	Fluorescence
Fe	Iron
HPLC	High-performance liquid chromatography
IC	Ion chromatography
ICP-OES	Inductively coupled plasma optical emission spectroscopy
K	Potassium
KEDTA	Potassium EDTA
LOD	Limit of detection
MIC	Minimum inhibitor concentration
MRS	Molybdenum reagent solution
MS	Mass spectroscopy
Na	Sodium
NaEDTA	Ethylenediaminetetraacetic acid disodium salt dihydrate
nm	Nanometers
P	Phosphorus
PAA	Polyacrylate acid
ppb	Parts per billion
PPCA	Polyphosphino carboxylic acid
ppm	Parts per million
PV	Pore volume
PVS	Poly vinyl sulphonate
RPLC	Reverse phase liquid chromatography
SAX	Strong anion exchange
SI	Scale inhibitor
SPE	Solid phase extraction
TRF	Time-resolved fluorescence
TSS	Total suspended solids
UV/VIS	Ultraviolet/visible
Vs-Co	Vinyl sulphonate copolymers
WAX	Weak anion exchange
μm	Micrometers
μs	Microseconds

References

1. A. Leontieff, N. Poynton, D. Dugue, et al., "Development of a Phosphorus Tagged Polymeric Scale Inhibitor Chemistry to Enable Squeeze and Topside Scale Control via the Same Inhibitor Chemistry," Chemistry in the Oil Industry IX, held October 31–November 2, 2005 (Manchester, UK).
2. H. Montgomerie, P. Chen, T. Hagen, et al., "Development of a New Polymer Inhibitor for Downhole Squeeze Application," *SPE Prod & Oper* 24 (2009): pp. 459–564.
3. N. Poynton, A. Molliet, A. Leontief, et al., "Development of a New Tagged Polymeric Scale Inhibitor With Accurate Low-Level Residual Inhibitor Detection for Squeeze Applications," SPE-155187-MS, SPE International Conference on Oilfield Scale, held May 30–31, 2012 (Aberdeen, UK).
4. S.M. Heath, S. Green, A. Badawy, et al., "On-site Analysis of Scale Squeeze Inhibitors in Remote Environments Using Fluorescence and Time Resolved Fluorescence," SPE-200682-MS, SPE International Conference on Oilfield Scale, held June 24–25, 2020 (Aberdeen, UK).
5. S.M. Heath, S. McClenaghan, G. Graham, et al., "Development of a Novel Residual Analysis Technique for Phosphonate Scale Squeeze Inhibitors to Enhance Scale Squeeze Management in Sub-Sea Wells," Tekna Oilfield Chemistry Symposium, held March 26–29, 2017 (Geilo, Norway).
6. S.M Heath, S. Ewing, G. Graham, et al., "Phosphonate Scale Inhibitor Residual Assay in Multi Component Blends—The Importance of Speciation and the Potential Benefit of TRF Techniques," Tekna Oilfield Chemistry Symposium, held March 4–6, 2018 (Geilo, Norway).
7. S. Ewing, G. Graham, K. Whittingham, et al., "Phosphonate Speciation Challenges and Solutions," SPE-190698-MS, International Oilfield Scale Conference, held June 20–21, 2018 (Aberdeen, UK).
8. G.M. Graham, K.S. Sorbie, L.S. Boak, "Development and Application of Accurate Detection and Assay Techniques for Oilfield Scale Inhibitors in Produced Water Samples," SPE-28997-MS, SPE International Symposium on Oilfield Chemistry, held February 14–17, 1995 (San Antonio, Texas, USA).
9. G.M. Graham, K.S. Sorbie, L.S. Boak, "Development and Accurate Assay Techniques for Poly Vinyl Sulphonate (PVS) and Sulphonated Co-Polymer (VS-Co) Oilfield Scale Inhibitors," Tekna Oilfield Chemistry Symposium, held March 22–25, 1995 (Geilo, Norway).
10. C.E. Anderson, G. Ross, G.M. Graham, "Automated Instrumental Analysis of Residual Polymeric Scale Inhibitor Species," Tekna Oilfield Chemistry Symposium, held April 1–4, 2001 (Geilo, Norway).
11. G.M. Graham, I. Munro, N. Harvison, et al., "Improved Scale Inhibitor Assay for Sulphonated Polymers in Oilfield Brine," SPE-131100-MS, SPE International Conference on Oilfield Scale, held May 26–27, 2010 (Aberdeen, UK).
12. N.P. Chilcott, D.A. Phillips, M.G. Sanders, et al., "The Development and Application of an Accurate Assay Technique for Sulphonated Polyacrylate Co-Polymer Oilfield Scale Inhibitors," SPE-60194-MS, SPE International Conference on Oilfield Scale, held January 26–27, 2000 (Aberdeen, UK).
13. L.S. Boak, K. Sorbie, "New Developments in the Analysis of Scale Inhibitors," *SPE Prod & Oper* 25 (2010): pp. 533–544.
14. K. Ramstad, H.C. Rohde, T. Tydal, et al., "Scale Squeeze Evaluation Through Improved Sample Preservation, Inhibitor Detection and Minimum Inhibitor Concentration Monitoring," *SPE Prod & Oper* 24 (2009): pp. 530–542.
15. K. Marshall, G. Graham, I. Munro, et al., "Challenges Facing Residual Chemical (Polymer) Scale Inhibitor Assay in Oilfield Waters," Tekna Oilfield Chemistry Symposium, held March 22–25, 2009 (Geilo, Norway).

16. N. Harvison, K. Marshall, "Improved Scale Inhibitor Assay for Sulphonated Polymers in Oilfield Brines," Tekna Oilfield Chemistry Symposium, held March 15–17, 2010 (Geilo, Norway).
17. S. Heath, G. Ross, G. Graham, et al., "The Importance of Scale Inhibitor Analysis in Scale Management—State of the Art Overview to Provide Cost Effective Scale Control from Simple to Complex Production Scenarios," SPE 179900-MS, SPE International Conference on Oilfield Scale, held May 11–12, 2016 (Aberdeen, UK).
18. S. Heath, G. Ross, G. Graham, et al., "State of the Art Overview of Scale Inhibitor Chemistry and Analysis Techniques to Aid Chemical Selection and Scale Management for Simple to Complex Production Scenarios," Tekna Oilfield Chemistry Symposium, held March 14–16, 2016 (Geilo, Norway).
19. Y. Zhang, J.K. Daniels, J. Hardy-Fidoe, et al., "Scale Inhibitor Residuals Analysis: Twenty-First Century Approach," SPE-169773-MS, SPE International Oilfield Scale Conference, held May 14–15, 2014 (Aberdeen, UK).
20. L. Cheng, C. Durnell, S. Linares-Samaniego, et al., "Development and Improvement of a Novel IC-MS Technique for Phosphonate Scale Inhibitors," SPE-190719-MS, SPE International Oilfield Scale Conference, held June 20–21, 2018 (Aberdeen, UK).
21. S. Murugesan, D. Ahoor, L. Souza, et al., "Next Generation On-Site Scale Inhibitor Analysis—Surface Enhanced Nanotechnology Detection," SPE-200702-MS, SPE International Oilfield Scale Conference, held June 24–25, 2020 (Aberdeen, UK).
22. S. Peat, D. Nichols, "Time Resolved Fluorescence Analysis for Oilfield Production Chemicals," Chemistry in the Oil Industry XVI, held November 4–6, 2019 (Manchester, UK).
23. G.M. Graham, K.S. Sorbie, A. Johnston, et al., "Complete Chemical Analysis of Produced Water by Modern Inductively Coupled Plasma Spectroscopy (ICP)," Tekna Oilfield Chemistry Symposium, held March 17–20, 1996 (Geilo, Norway).
24. C.R. Jones, M.J. Hails, B.L. Downward, "New Convenient Analytical Method for Biodegradable Squeeze Scale Inhibitors," NACE-07062, Corrosion Annual Conference, held March 11–15, 2007 (Nashville, Tennessee, USA).
25. S.M. Heath, O. Selle, E. Storås, et al., "Squeezing Sub-Sea Wells Co-Mingled in the Same Flowline on the Norne Field," SPE-169793-MS, SPE International Conference on Oilfield Scale, held May 14–15, 2014 (Aberdeen, UK).
26. S.M. Heath, S. Toivonen, V. Vuori, et al., "Application of Advanced Fluorescence Detection Technology to Improve Scale Management in Both Conventional and Sub-Sea Fields," SPE 193632-MS, SPE International Conference on Oilfield Chemistry, held April 8–9, 2019 (Aberdeen, UK).
27. S.J. Dyer, C.E. Anderson, G.M. Graham, "Thermal Stability of Amine Methyl Phosphonate Scale Inhibitor," *Journal of Petroleum Science and Engineering* 43 (2004): pp. 259–270.
28. K.G. Cook, S.J. Deas, H. Taylor, et al., "Fully Automated Low-Level Detection of Phosphonate-Based Scale Inhibitors Using Ion Chromatography," SPE Oilfield Scale Symposium, held January 27–28, 1999 (Aberdeen, UK).
29. A. Thompson, K. Burnett, "Development and Validation of a Novel Method for Quantification of Polymeric Scale Inhibitors and a Comparison of Obtained Data With Current Commercial Techniques," Chemistry in the Oil Industry X, held November 5–7, 2007 (Manchester, UK).
30. A. Thompson, A. Gangstad, H.K. Kotlar, "Oil Field Data/Return Analysis: A Comparison of Scale Inhibitor Return Concentrations Obtained With a Novel Analytical Method and Current Commercial Techniques," SPE-114049-MS, SPE International Oilfield Scale Conference, held May 28–29, 2008 (Aberdeen, UK).
31. K. MacKinnon, S.J. Ewing, G.M. Graham, "New Robust Analytical Development for Sulphonated Polymers in Oilfield Brines," SPE-179902-MS, SPE International Oilfield Scale Conference, held May 11–12, 2016 (Aberdeen, UK).

32. A Thompson, T.E. Gundersen, A.T. Andresen, et al., "Emerging 4th Generation Analytical Technology for the Individual Analysis of Multiple Families of Co-Mingled Scale Inhibitors in Deepwater Wells," SPE-155171-MS, SPE International Conference on Oilfield Scale, held May 30–31, 2012 (Aberdeen, UK).
33. S.M. Heath, C. Johnstons, T. Gundersen. et al., "Application of Advanced Mass Spectroscopy Techniques for Improved Scale Management in Conventional and Sub-Sea Fields," SPE-173744-MS, SPE International Conference on Oilfield Chemistry, held April 13–15, 2015 (The Woodlands, Texas, USA).
34. G. Ross, A. Strachan, "Dramatically Reducing OPEX Cost by Careful Monitoring and Optimizing a Squeeze Treatment in a Subsea Well," Tekna Oilfield Chemistry Symposium, held March 17–20, 2013 (Geilo, Norway).
35. E. Myhr, K. Ramstad, T. Tydal, et al., "Preservation of Produced Water Samples for Better Scale Management and Extended Squeeze Lives," Tekna Oilfield Chemistry Symposium, held April 1–4, 2001 (Geilo, Norway).
36. E. Myhr, K. Ramstad, T. Tydal, et al., "Preservation of Produced Water Samples for Better Scale Management—An Update Including Non-Squeezed Wells," Tekna Oilfield Chemistry Symposium, held March 18–20, 2002 (Geilo, Norway).
37. K. Alnes, K. Ramstad, H.R. Rohde, "Improved Scale Management With Novel Sample Preservative," Tekna Oilfield Chemistry Symposium, held March 15–17, 2010 (Geilo, Norway).
38. C.J. Johnstone, M.M. Jordan, "The Procedures That Time Forgot: Sample Preservation, Back to Basics," SPE 190734-MS, SPE International Oilfield Scale Conference, held June 20–21, 2018 (Aberdeen, UK).
39. V. Vuori, V. Nuutinen, T. Honkanen, et al., "Accurate Detection of Tagged Polymeric Scale Inhibitors in Oilfield Produced Water Samples," SPE-179908-MS, SPE International Conference on Oilfield Scale, held May 11–12, 2016 (Aberdeen, UK).
40. J. Johnstone, S. Toivonen, R. Griffin, et al., "Novel Method for Real-Time Monitoring of Scale Control Products at the Site of Use," SPE-167797-MS, SPE International Oilfield Scale Conference, held May 14–15, 2014 (Aberdeen, UK).
41. M. Martini, T. Brichart, A. Marais, et al., "How to Monitor Scale Inhibitor Squeeze Using Simple TRF Tracers," SPE-173768-MS, SPE International Conference on Oilfield Chemistry, held April 13–15, 2015 (The Woodland, Texas, USA).
42. J. Wylde, A. Savin, J. Kerr, et al., "Determination of Multiple Scale Inhibitor Residuals in Single Produced Waters: A Case History From the North Sea Differentiating Four Chemistries," SPE-164079-MS, SPE International Symposium on Oilfield Chemistry, held April 8–10, 2013 (The Woodlands, Texas, USA).

13 Scale Monitoring Techniques

Abdur Rehman[1] and Ole Magnar Drønen[1]
[1]Scale Protection AS, Harstad, Norway

13.1 Introduction

One of the paramount concerns within the oil industry pertains to scale formation, with deposits varying in severity from minor to significant. The expenses associated with oilfield scale are elevated due to a significant decline in oil and gas production; frequent removal of downhole equipment for replacement; reperforation of intervals affected by scaling, reaming, and redrilling of obstructed oil wells; stimulation of plugged oil-bearing formations; and various other remedial workover procedures. Scale prevention is important to ensure continuous production from existing reserves that produce brine. Several saturation indices and computer algorithms have been devised to ascertain the likelihood, timing, and locations, where scale may develop. Nevertheless, predicting the scale formation remains challenging as forecasts from various models frequently diverge significantly, particularly under extreme conditions. Wells could be abandoned prematurely, because of poor management of scale. As scale accumulates around the wellbore, the porous formation becomes obstructed, potentially rendering it impermeable to any fluids. This phenomenon results in formation damage and a subsequent reduction in hydrocarbon production.[1–6]

Numerous techniques have been devised for inhibiting and remediating scale formation. A more effective management strategy is needed to address the scale issue. Instead of allowing scale to accumulate before implementing corrective measures, a proactive scale prevention approach should be adopted from the outset. It's worth noting that descaling can be financially costly. Therefore, employing methods that can restrict or eliminate scale deposition in the early stages is advisable. Understanding the concept of scale formation is crucial to avoid making suboptimal corrective decisions that could jeopardize production.[4, 6] Therefore, the monitoring of scale is deemed crucial and should be anticipated in advance to avert potential substantial losses.[7, 8]

13.2 Scale Monitoring

There are two types of approaches for scale monitoring: proactive and reactive,[9] as listed in Table 13.1. The proactive approach is based on the observation of future scaling potential or at least noticing scaling at an early stage. The reactive approach aims to limit damage by responding to parameters after the scale is precipitated.

Table 13.1: Scale monitoring methods.

Proactive Approach	Reactive Approach
Scale potential modeling	Well productivity
Water ion monitoring	System pressure
Inhibitor residual measurements	Pipeline scale growth
Suspended solid analysis	Scale coupon

The proactive indications are often more uncertain, compared to the reactive. However, a proactive approach could contribute to avoiding more extensive scale damage. Although a reactive approach has the advantage of lower uncertainty, a key disadvantage is that scale indications will not appear before damage has already occurred. A strategy should be made in advance and should specify whether the objective is scale prevention, repair, or a balance of these two. This strategy will guide monitoring processes. If the strategy is to repair, there is no need to analyze solids.

Adopting a scale preventive strategy will always be the most costeffective, because monitoring costs are significantly lower than repair costs. This is definitely the case in an offshore environment; it is also better to adopt a preventive strategy onshore to avoid any production loss. The type of scale is also of importance, since repair costs can be much greater, if the scale cannot be removed chemically, which is the case for sulfate scale.

13.2.1 Proactive Scale Monitoring

There are often uncertainties when using the proactive approach. These uncertainties are whether the scale precipitation takes place and if the precipitation can cause a problem. The scale precipitation could take place inside the reservoir without any danger, or it may be such a slow process that the water is produced out of the well before any scale is seen. Combined use of these parameters is therefore vital to get the most reliable and comprehensive understanding.

13.2.1.1 Scale Modeling

Simulation tools can accurately show the potential for scale precipitation, as presented in Chapters 2–4.

Anticipating the scaling tendency is a complex undertaking due to numerous sensitive parameters, including hydrodynamic factors, variations in temperature and pressure, ionic strength, pH levels, impurities, and the mingling of incompatible waters. An effective approach to preventing scale formation involves the utilization of thermodynamic methods for predicting the scale potential in oil fields. Various techniques have been employed to forecast scaling phenomena in oilfield operations,[10] as delineated in Chapter 2. However, these tools are primarily based on the thermodynamics and are not capable of predicting kinetics or the share of scale that is growing on the surfaces.[11]

13.2.1.2 Water Ion Monitoring

Water ion monitoring is widely used to analyze water composition to get indications of scale precipitation. Numerous ion monitoring methods are being used. The assessment of cations and anions, encompassing sodium, potassium, magnesium, calcium, strontium, barium, iron, chloride, bromide, nitrate, carbonate, and sulfate, is commonly carried out through gravimetric, volumetric, photometric, atomic absorption spectrophotometry, atomic emission spectrophotometry, and ion chromatographic methods.[12]

Water ion monitoring method is especially effective for sulfate scale, as a sudden drop in dissolved barium ions (Ba^{2+}) is a reliable indicator of scale precipitation. A confounding variable for this method is that injected water often does not contain Ba^{2+} ions, so an increase of injected water will also cause the concentration of Ba^{2+} ions to drop.

This can be accounted for by calculating a reference Ba^{2+} level based on Ba^{2+} in formation water and the share of injected water from other ions than Ba^{2+}.[13] It is important to note that a decrease in Ba^{2+} concentration cannot discriminate where the scale has precipitated, that is, whether the scale has precipitated within the formation or inside the wellbore.

13.2.1.3 Inhibitor Residual Measurements

For wells where inhibitor is squeezed, monitoring the remaining inhibitor concentration can be valuable. The concentration of back-produced inhibitor will slowly decline after an inhibitor squeeze. This information is then compared to a minimum inhibitor concentration (MIC) acquired by lab measurement. A concentration in the well below the MIC could result in scaling. It is important to understand that a MIC from a lab may differ from a field MIC, since the laboratory experimental conditions do not perfectly replicate the real scaling process.

This analysis does not consider the distribution of inhibitors across well zones. MIC will always be represented as the well's average. Uneven inhibitor distribution could lead to zones being left unprotected, despite the produced inhibitor concentration appearing sufficient to prevent scale.

13.2.1.4 Suspended Solid Analysis

In this method, environmental scanning electron microscopy coupled with energy dispersive X-ray analysis (ESEM-EDX) needs to be implemented to analyze samples, due to the very small particle size of these solids. The principle of ESEM-EDX is illustrated in Figure 13.1.

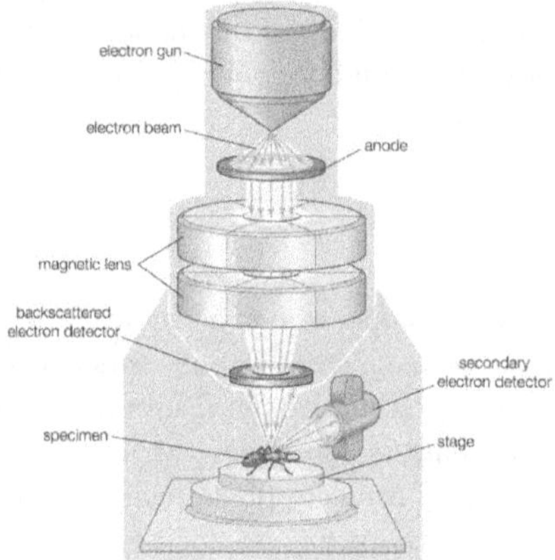

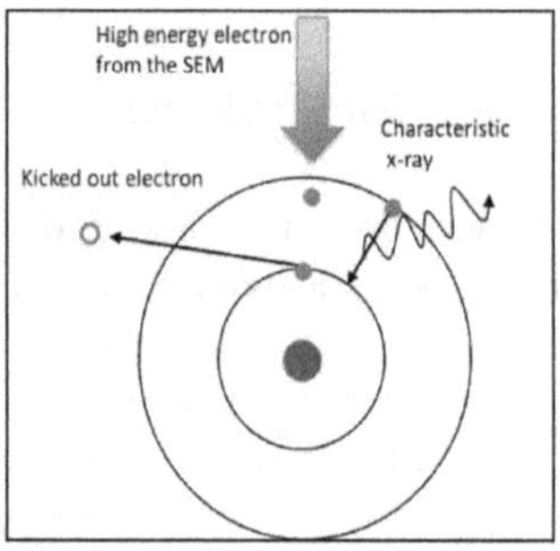

Figure 13.1. Principle of environmentally scanning electron microscopy coupled with energy dispersive X-ray analysis method.

It is a more direct analysis to look for solids in the produced water. The greatest advantage of this method is that the result is a direct observation about scaling; it shows real scale particles precipitated in the well itself. Scale in the produced water indicates the occurrence of precipitation and thus the greater risk of scale growth. However, this measurement does not provide information about the proportion of actively growing scale compared to how much of it is produced through the bulk.[14] This will be discussed in greater detail in the forthcoming sessions.

13.2.2 Reactive Scale Monitoring

13.2.2.1 Thickness Shear-Mode Resonator

Scale formation can be determined by using a thickness shear-mode resonator (TSMR). It consists of a piezoelectric wafer sandwiched between two thin metal electrodes.[15] The application of an electric potential across the wafer results in its shear deformation at a resonant frequency dependent on the properties of the water. TSMRs are useful as mass sensors, because the adherence of mass on the electrode surface can alter its resonant frequency. In the case of scale monitoring, a decrease in the resonant frequency is an indication of scale deposition.

This method demonstrates promising potential as an online monitoring tool, applicable in both surface equipment and downhole scenarios. The clear advantages include automated chemical feed and prompt identification of scaling attributed to process variations, such as the onset of seawater breakthrough.[16, 17]

The sensor can, however, be rendered ineffective for continuous monitoring, due to fast saturation and loading limits. Adhesion of organics and any other types of inorganic materials, including oil droplets, biomass, corrosion inhibitors, corrosion byproducts, sand and clay particles from reservoirs, can affect the results, if they are present, which limits its applicability. Moreover, TSMR cannot measure dispersed scale, as it is a point measurement and can only measure the scale adherent to the surface.

13.2.2.2 Gamma Ray Transmission

Gamma ray transmission technique is another technique for scale detection inside the pipes by providing a density profile of the studied pipeline section. Cobalt-60 and cesium-137 stand out as the two frequently employed radioactive sources. Given its higher energy, cobalt-60 is typically employed for pipelines with thick walls or larger diameters. Conversely, cesium-137 is utilized for pipelines with thin or standard thickness walls.[18] The procedure initiates by employing radioactive sources on one side of the examined material, such as barium. As the source emits gamma rays at various energy levels, the attenuation of gamma rays is then detected through a detector strategically positioned on the opposite side of the material.

The fundamental principle of the application of this technique is based on the fact that the atoms composing the scale are heavier than the carbon, hydrogen, and oxygen atoms found in water and hydrocarbons. Consequently, the presence of scale leads to a higher attenuation level. The utilization of triple energy is another technique that demonstrates increased robustness and faster capabilities compared to the use of dual energy.

These measurements do not disrupt the flow and offer valuable information about scale deposits, encompassing details about their nature, quantity, thickness, and rate of accumulation. However, these methods come with certain limitations. The portability of the device is hindered by the weight of the shielding layer in the instruments. Additionally, the presence of solid particles and dust can impact gamma attenuation. Identifying the exact location of scale is challenging, and this method is unsuitable, when pipelines are buried or challenging to access from both sides. This method is also not applicable, when the pipe diameter is large, primarily due to the high attenuation effect. Typically, it necessitates multiple measurements at different points to accurately reflect the true condition of the pipe, posing another potential challenge.[19]

13.2.2.3 Heat-Based Sensor

Decrease in heat flow in the scaled pipe can be observed, because the scale is less thermoconductive than the steel. Based on this principle, heat-based sensors have been invented to determine scale deposition by measuring the temperature change or the decline in the heat transfer coefficient. Thermal imaging is an example of a heat-based sensor, as exhibited in Figure 13.2, along with operating steps.

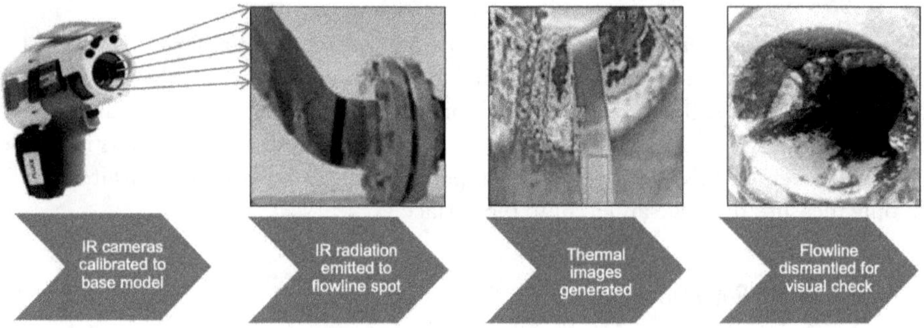

Figure 13.2. Flowline scale inspection using heat base sensor.

Thermal imaging harnesses infrared (IR) radiation to provide insights into the composition of a material by highlighting variations compared to a baseline model. This information encompasses changes in composition, temperature, moisture content, and the degree of consolidation. The imaging process is carried out using specially designed thermal imaging cameras and equipment capable of quantifying temperature variations through the measurement of transmitted, detected, and emitted IR energy.

Typically, this equipment is integrated with computers featuring preprogrammed software for conducting analysis and generating thermal images of the object. The operation is straightforward and requires only a few personnel to capture temperature readings. Strategic spots, such as those near surface valves, joints, elbows, and potential locations of flow restriction, can be selected. Subsequently, all temperature profiles are compared to a base model, which contains a thermal image of the same spot depicting a healthy flowline.

Deviations in temperature readings signal the abnormalities within the flowline and are often associated with scale and solid deposition. The thickness and uniformity of the scale are primarily correlated with the profile shape. Typically, these scale particles function as insulators, resulting in a cooler temperature than anticipated. As this method depends on heat change, if there is no heat transfer, this method cannot be used, and it is only limited to the downstream facilities.[20]

13.2.2.4 Scale Coupon

The coupon is rectangular in shape and is attached to a strip coupon holder. It can be retrieved from a production stream and investigated for the presence of scale. Notably, the coupon features a series of holes of varying sizes machined into it. Figure 13.3 lists typical coupons that are used in scale monitoring. Scale is most likely to form in cavities, and therefore forms most readily in the smallest size hole. If a pipeline medium has a tendency to form scale, the strength of this tendency can be determined by the largest hole that has accumulated scale. The greater the concentration of scaling compounds, the larger the hole size that is likely to accumulate scale.

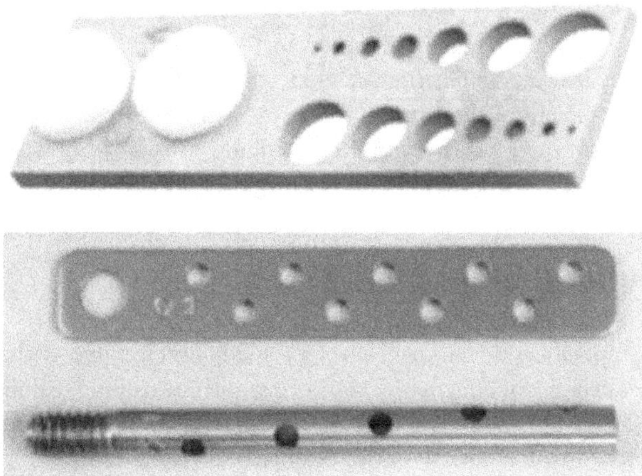

Figure 13.3. Typical scale coupons.

Figure 13.4 illustrates a standard retrievable installation of coupons in a flowline. Manual insertion of coupons is feasible, if the pressure is below 50 psi. However, at higher pressures, mechanical methods of insertion must be employed.

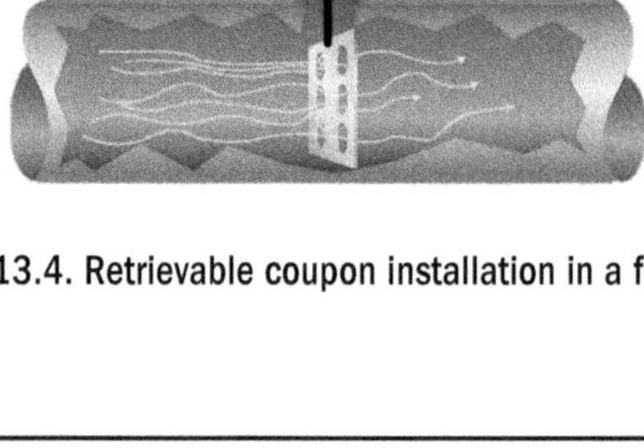

Figure 13.4. Retrievable coupon installation in a flowline.

The weighed coupons are mounted in the flowing stream for a specific duration. Upon retrieval, the coupons undergo a drying process and are subsequently reweighed. In this state, the scale is subjected to visual examination, X-ray diffraction analysis, and chemical analysis.

The optimal approach for cleaning the scaled equipment, particularly for certain types of scale, involves descaling the coupons using acids or cleaning agents. Mechanical removal is deemed most effective for eliminating deposits of strontium sulfate and barium sulfate. It includes carefully scraping and wire brushing the coupon. The quantity of deposited scale is assessed by calculating the difference in dry weights between the coupons before and after exposure.

While the coupon method is widely accepted in the industry, it does have several drawbacks. For instance, it must be positioned where the scale is actively forming; scaling can occur during data collection; the sampling process can be time-consuming; there's a risk of corrosion to the coupon; and it can be challenging to remove or dissolve the scale for analysis. Additionally, the coupon may not be fully representative of the entire system.[21]

13.3. Suspended Solid Analysis

Analyzing the composition and morphology of solids precipitated within formation and produced waters is crucial for effectively monitoring and controlling scaling in the near-wellbore zone, wells, subsea installations, and topside facilities. Ideally, any chosen monitoring method should provide information on the extent of particulate formation, that is, percentage coverage, morphology including shape and size, composition, and, if feasible, the origin of the scale whether it is actively forming, modified, or transported.

Analysis of filtered particles from produced water using ESEM-EDX) is an efficient method for scale monitoring and the visualization of scale particles formed. There are a variety of analysis methods on the market.[14, 21–24] The one referenced here is Bulk Scale Analysis (BSA™), provided by Scale Protection AS.

13.3.1 Bulk Scale Analysis Principle

Figure 13.5 shows the BSA workflow from sampling to a risk-based result. A predetermined volume of produced water is retrieved, and suspended solids are captured on a specialized filter. Solid particles are then prepared for instrumental analysis. The analysis result summarizes the proportion of each mineral and its morphology.

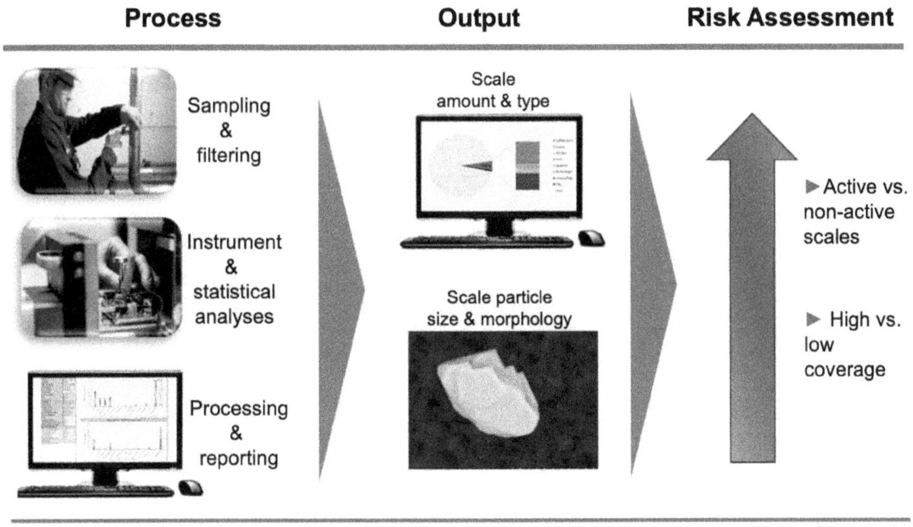

Figure 13.5. A schematic showing of Bulk Scale Analysis workflow.

The statistical analysis is acquired through an advanced workflow. The SEM is programmed to analyze particles at randomly selected areas on the filter, and then the data are reported for all the analyzed particles with each particle's elemental composition and size. The data are used to classify each particle's mineral species and to calculate coverage per mineral. The "coverage" is the percentage of the investigated area covered by a specific mineral. This is the major parameter used in this analysis method to estimate scale potential (Figure 13.6). It is not straightforward to relate coverage to concentration in the water, weight of scale per meter, or similar measurements. Coverage is therefore best utilized as a longitudinal measurement.

A definitive weakness in the measurement is that the coverage is proportional to the scale in the bulk, not the scale that is growing on valves and pipe walls. The ratio between these is unknown, and parameters that affect this ratio are not very well understood. It is, however, an indication that the proportion of growing scale on surface is larger at low-scale potential conditions. This might be related to the fact that the process where a particle is formed is slower, which allows for more growth on pipe walls. The scale crystal morphology can provide additional information related to scale risk. This is described in later sections.

13.3.2 Differentiation Between Scale in Reservoir and Wellbore

Figure 13.7 illustrates the difference between water and solids analysis. For scale types like barite and other sulfate scales, a drop in scaling ion concentration like Ba^{2+} and Sr^{2+} is a clear indication of scale precipitation. For other scales, such as calcium carbonate, the high concentration of Ca^{2+} ions in the water complicates this method. Even within the uncertainty range of measuring Ca^{2+} concentration, a lot of scale can be precipitated.

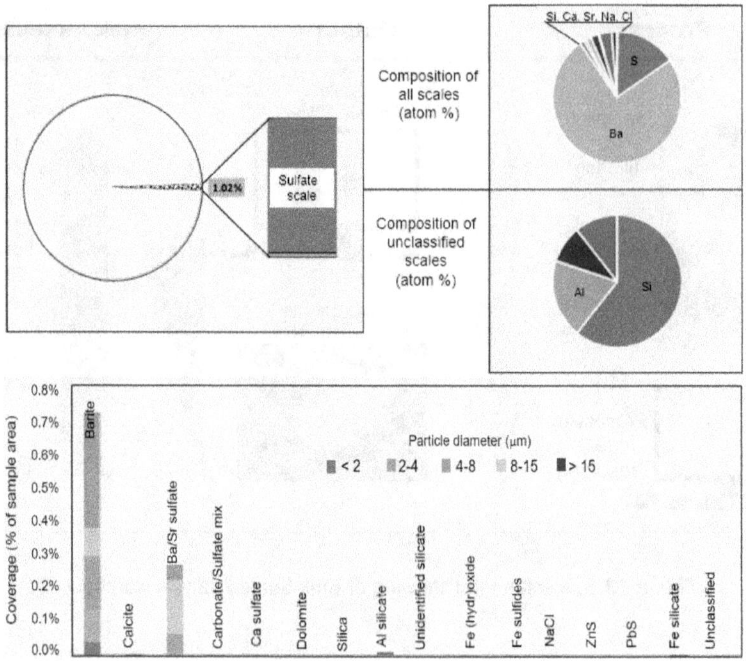

Figure 13.6. Bulk Scale Analysis statistical result presentation.

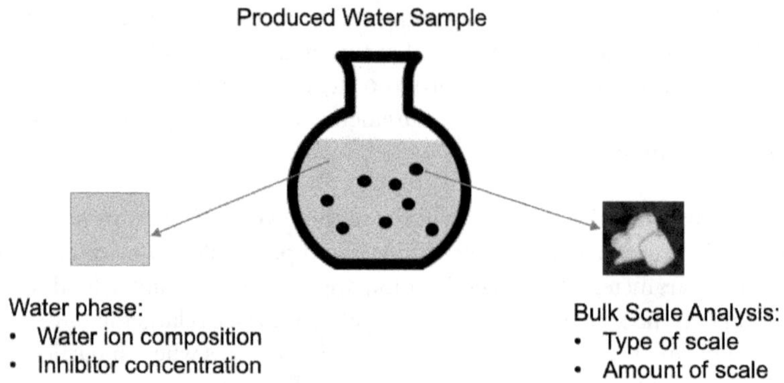

Figure 13.7. Illustration of water and solids analysis.

Water analysis is also very useful to monitor the mixing of injected water. It will also be a vital input for modelling purposes. A rapid drop in Ba^{2+}, as illustrated in Figure 13.8, is likely due to the replacement of formation water with injected water or scale precipitation. The proportion of injected water can be estimated from other ions like Mg^{2+} and also SO_4^{2-}. Given that the drop in these ion concentrations is caused by scale precipitation, it can be due to a mixture of sulfate-rich water and formation water in between injector and producer, in the reservoir, or it can be in the well. The resulting Ba^{2+} curve will differentiate between these two cases, reservoir or well scale precipitation.

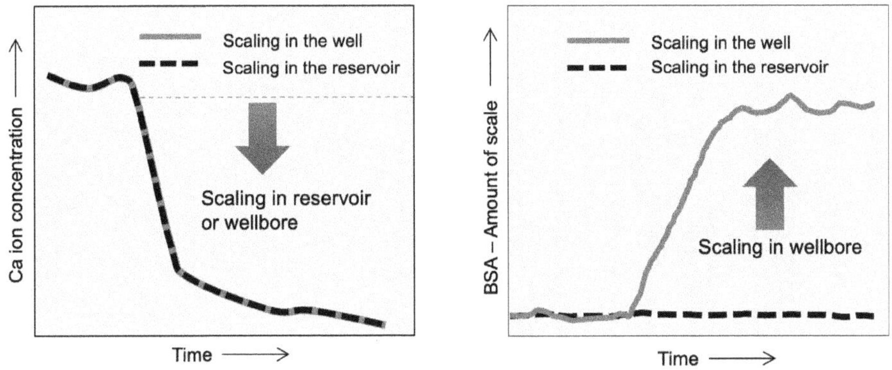

Figure 13.8. Illustration on how solids analysis will differentiate between reservoir and well scaling.

This is different when looking at solids. Scale particles will not flow into the pore structure within the formation; they will be trapped. A reservoir scale situation will therefore not show scale particles in the produced water. Scale on the filter will always have formed from within the well itself. However, the results cannot differentiate where in the well they have formed; it could have occurred from the perforation tunnels all the way to wellhead. This illustrates how the combined use of water and BSA can provide a more complete understanding of scale risk.

13.3.3 Zonal Resolution

As stated in the overview section, the data from water analysis are always an average of the contribution from all the well's zones. A drop in Ba^{2+} in one zone may easily be masked in such an average, and even more likely, lack of inhibitor in one zone may not be detected due to an increased amount from the others.

Using BSA could in some cases resolve this lack of zonal resolution. In a case where all zones are prevented by inhibitor except one, it will show up as a coverage on the filters. An example is shown in Figure 13.9. The coverage of sulfate scale analyzed by BSA for a specific well is plotted as a function of time. The blue line shows the inhibitor concentration during the same periods. Despite an inhibitor concentration far above MIC, and a repeated squeeze, marked as Squeeze #1, the sulfate coverage remains high. Another squeeze, marked as Squeeze #2, was performed using higher pumping pressure. The BSA shows that there was no longer scale in the produced water after high-pressure pumping. This case illustrates that an uneven squeeze placement cannot be detected from the water analysis but shows up in the BSA.

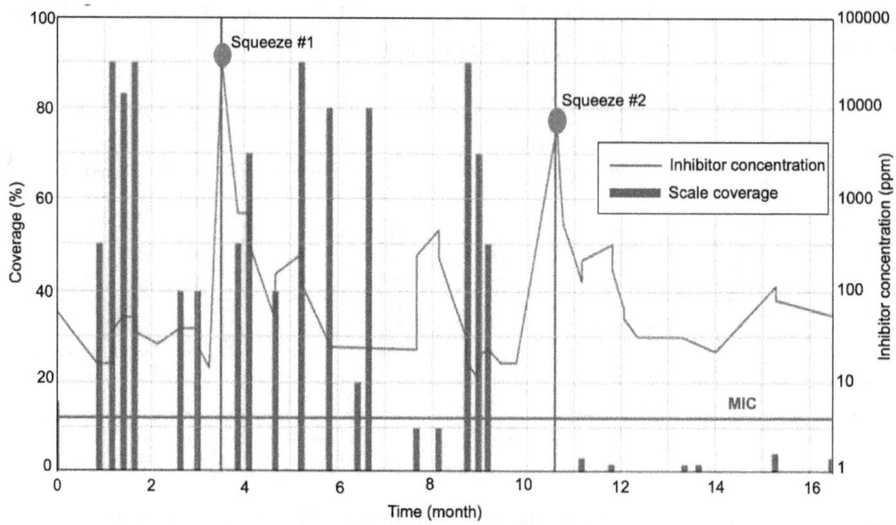

Figure 13.9. Example of zonal resolution.

13.3.4 Squeeze Timing Optimization

Figure 13.10 shows an example where inhibitor concentration, MIC, and BSA coverage data are combined. Usually, the resqueeze timing is selected to be whenever the inhibitor concentration declines close to or below the MIC. Using BSA, it is possible to secure a safe production beyond this, and then be able to extend the lifetime of each squeeze significantly. In many cases, it is used to ensure no scale shows up before resqueezing. In other cases, like the one displayed in Figure 14.15, the resqueeze is done as soon as a peak in scale coverage is observed. A significant gain in squeeze lifetime is achieved, without adding risk.

13.3.5 Solid Morphology

The inhibitors are designed to both prevent scale from forming and halt the growth of precipitated particles. Despite the inhibitor reducing the formation of scale, in most cases, scale will still be present. However, their morphology is most often altered. These morphological differences are widely understood to be important, when monitoring scale risk.[15]

Based on morphology, scale can be divided into three main categories, that is, active scale, transported scale, and modified scale. Actively growing scale particles can be identified by their unique morphological features. These solids often have a simple orthorhombic form, planar symmetry, or straight and uninterrupted edges, as shown in Figure 13.11.

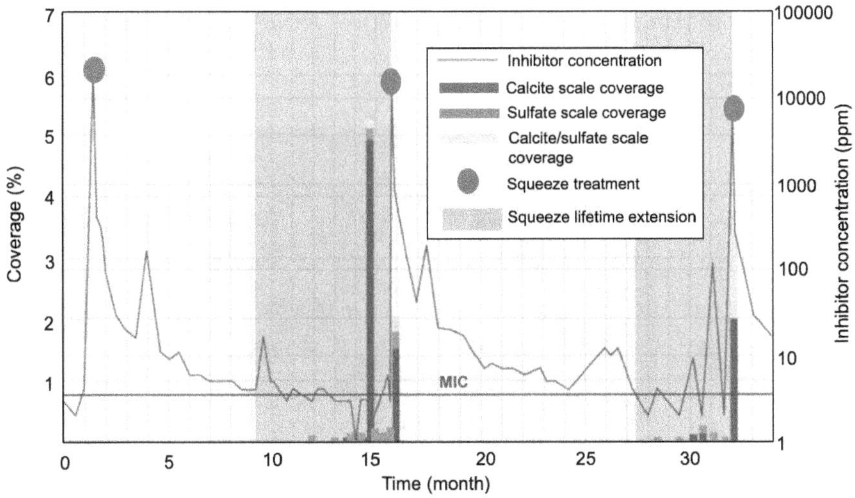

Figure 13.10. Example of using bulk Scale Analysis to optimize timing for resqueezing treatment.

Figure 13.11. Morphology of typical active scale crystal.

Scale particles that have been grown on a surface and then ripped off will have another shape with less sharpened edges. This group is called transported or old scale, as exhibited in Figure 13.12.

Figure 13.12. Example of a transported scale crystal.

In the presence of scale inhibitors, the morphology of active scale particles is altered. This modified scale has a more irregular, rounded, and sinuate form compared to the sharp appearance of active scale, as shown in Figure 13.13.

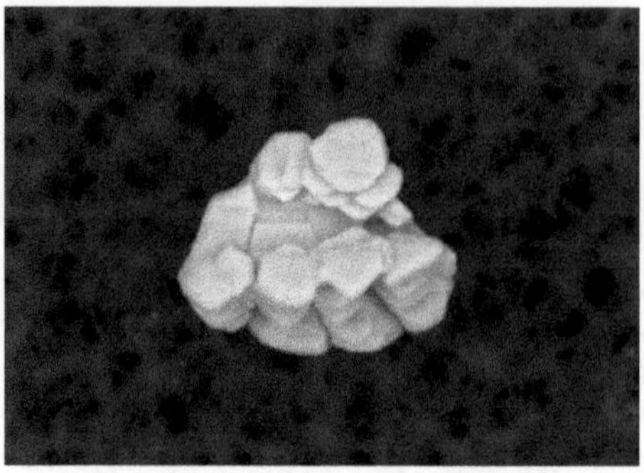

Figure 13.13. Morphology of typical modified scale crystal.

13.3.5.1 Overlapping

Like many other techniques, the BSA method also has some limitations. Although there are reasonable differences between active, transported, and modified crystal morphologies, it can occasionally be challenging to discriminate between them. This is particularly true for transported and modified particles, and their morphological changes in the presence and absence of scale inhibitors. However, it's certain that active particles will always have a sharp-edged shape, which gives a clear indication of scaling in the wellbore using BSA method.

13.3.5.1.1 Overlap between the Presence and Absence of Inhibitor

For nonactive scale, crystals can have the same morphology, regardless of inhibitor presence. For example, the absence of scale inhibitor does not necessarily mean crystals will grow in a sharp-edge shape like active scale; they can grow in many different shapes depending on many other factors as well. All other crystal shapes other than typical active crystal shapes are considered less dangerous, as shown in Figure 13.14.

13.3.5.1.2 *Overlap between the Modified and Transported Particles*

Another limitation of BSA method is the difficulty in consistently distinguishing between modified and transported scale particles. Because modified and transported scale particles do not always follow the same sharp-edge pattern as active scale particles, they can be found in many different shapes, as illustrated in Figure 13.15. In this case, it can be challenging to differentiate, if the crystals have been modified due to inhibitors or have been transported due to some other unknown factors.

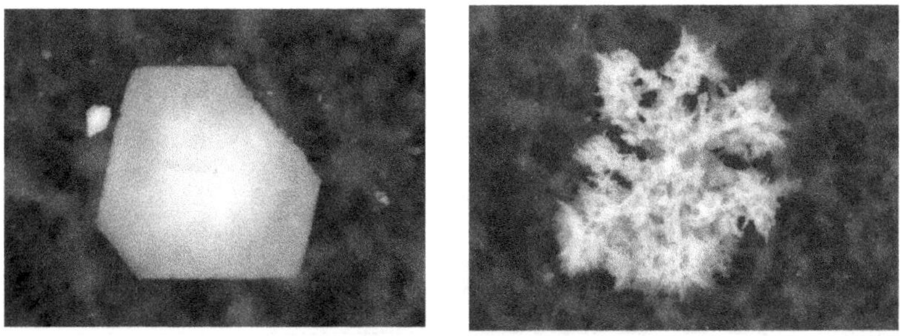

No inhibitor

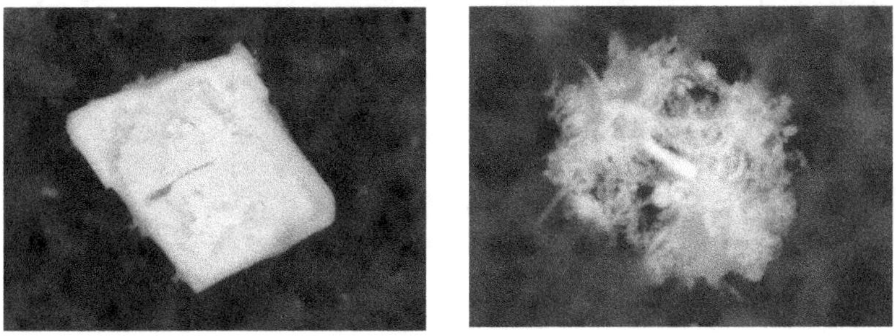

With inhibitor

Figure 13.14. Morphological change uncertainty in the presence and absence of scale inhibitor.

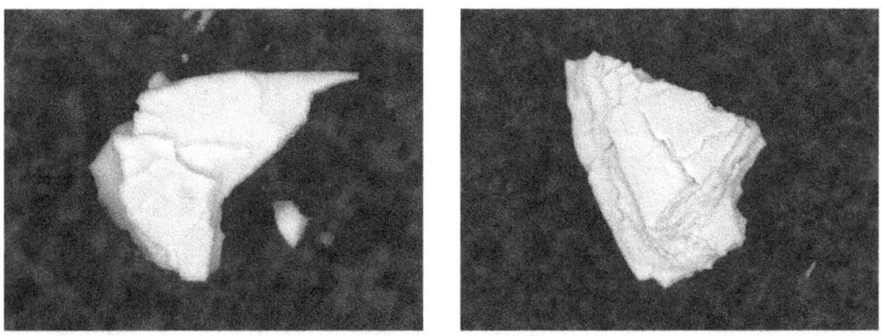

Modified

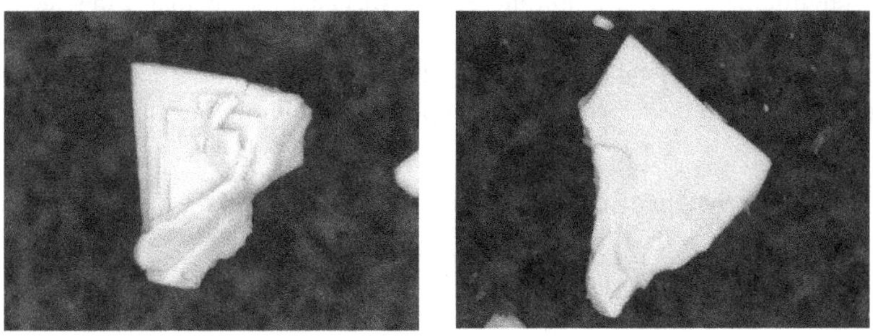

Transported

Figure 13.15. Overlap between modified and transported scale.

13.3.5.2 Coverage vs. Morphology

For the BSA method, a risk indicator chart is plotted for coverage and morphology (Figure 13.16). Coverage is defined by the number of scale particles measured on the filter, and morphology signifies the different shape of scale crystals.

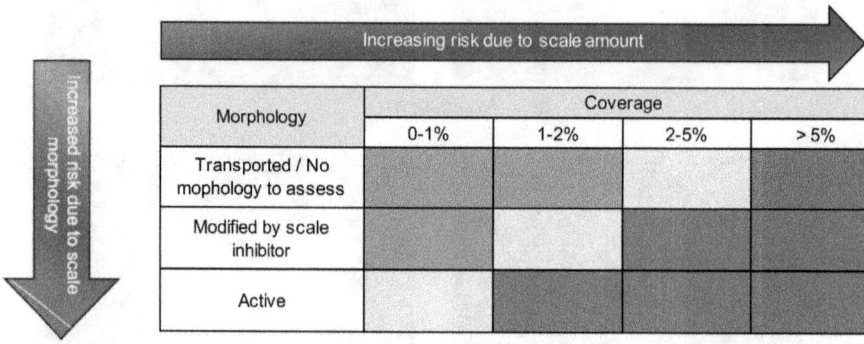

Figure 13.16. Risk indicator chart between morphology vs. coverage.

There is a difference in opinion as to whether the coverage should be included as a risk factor, if the analyzed particles are modified or transported. However, regarding active crystal morphology, it is generally agreed that it must be considered as an alarm. If coverage exceeds a certain level, it will also be regarded as a potential threat. High coverage means the scale is being formed regardless of active, modified, or transported morphology. Hence,

- High coverage indicates risk, regardless of morphology.
- Active morphology is a warning signal, despite low coverage.

13.3.5.3 Inhibition after Scale Formation

A question that is often raised concerning the interpretation of morphology is whether an active crystal that has been formed downhole can become modified, when it enters an inhibited environment such as a flowline. Figure 13.17 shows the images of scale formation without inhibitor and scale inhibitor addition after scale crystal has formed. The morphology examples show that crystal modification/inhibition can only occur during crystal formation. Once the crystals have formed, it is not possible to inhibit them using scale inhibitors.

13.3.5.4 Static vs. Dynamic Scale Conditions

It has also been a question, whether the morphology of scale particles changes based on the static or dynamic conditions. Figure 13.18 compares morphology from some of the static jar tests with a rotating cylinder at the same conditions. It seems like for the blank experiments, there is no difference in crystal morphology, which means that static or dynamic conditions do not play a large role in crystal shape.

Figure 13.17. Scale crystals without inhibitor (left) and effect of scale inhibitor after scale crystals have formed (right).

Figure 13.18. Crystal formation under static (left) and dynamic (right) conditions.

13.4. Summary

BSA is a direct analysis to look for solids in the produced water. The advantage of this method is that the result is a direct observation about scaling; it shows real scale particles precipitated in the well itself. It is being used across the globe along with other scale monitoring tools and has been proven to be a valuable addition to the conventional scale monitoring techniques.

The resqueeze timing is selected to be whenever the inhibitor concentration declines close to or below the MIC. Using BSA, it is possible to secure a safe production beyond this, and then be able to extend the lifetime of each squeeze significantly.

Scale particles will not flow into the pore structure within the formation; they will be trapped. In this way, BSA is more robust for wellbore scale monitoring compared to monitoring water, which will not be able to distinguish between wellbore and reservoir scaling.

Using BSA could, in some cases, resolve the lack of zonal resolution. In a case where all zones are prevented by inhibitor except one, it will show up as coverage on the filters.

Like many other techniques, the BSA method also has some limitations. Scale in the produced water indicates the occurrence of precipitation and thus the greater risk of scale growth. However, this measurement does not provide information about the proportion of actively growing scale compared to how much of it is produced through the bulk.

Although there are reasonable differences between active, transported, and modified crystal morphologies, it can occasionally be challenging to discriminate between them. This is particularly true for transported and modified particles, and their morphological changes in the presence and absence of scale inhibitors.

Nomenclature

BSA	Bulk Scale Analysis
EDX	Energy dispersive X-ray analysis
ESEM	Environmental scanning electron microscopy
MIC	Minimum inhibitor concentration
TSMR	Thickness shear-mode resonator
IR	Infrared

References

1. A.T. Kan, M.B. Tomson, "Scale prediction for oil and gas production," *SPE Journal* 17 (2012): pp. 362–378.
2. M.I. Refaei, I. Al-Kandari, "Oil fields scale deposition prediction methodology," SPE-126745-MS, SPE Kuwait International Petroleum Conference and Exhibition, held December 14-16, 2009 (Kuwait City, Kuwait).
3. A.T. Kan, J.Z. Dai, G. Deng, et al., "Recent advances in scale prediction: approach and limitations," *SPE Journal* 24 (2019): pp. 2209–2220.
4. J. Patel, A. Nagar, "Oil field scale in petroleum industry," *International Journal of Innovative Research in Engineering & Management* 9 (2022): pp. 288–293.
5. M. Kelland, *Production Chemicals for The Oil and Gas Industry*, 2nd ed. (Boca Raton: CRC Press, 2014).
6. M.F. Mady, A. Rehman, M.A. Kelland, "Synthesis and study of modified polyaspartic acid coupled phosphonate and sulfonate moieties as green oilfield scale inhibitors," *Industrial & Engineering Chemistry Research* 60 (2021): pp. 8331–8339.

7. I. Meric, G.A. Johansen, S.D.B. Melo, et al., "Monitoring of scale deposition in petroleum pipelines by means of photon scattering: a preliminary study," INIS-BR-14335, International Nuclear Atlantic Conference, held November 24–29, 2013 (Recife, PE, Brazil).
8. M.F. Mady, A. Rehman, M.A. Kelland, "Synthesis and antiscaling evaluation of novel hydroxybisphosphonates for oilfield applications," *ACS Omega* 6 (2021): pp. 6488–6497.
9. W. Wang, W. Wei, C. Yan, "Scale prediction and management for unconventional production," SPE-200668-MS, SPE International Oilfield Scale Conference and Exhibition, held June 24–25, 2020 (Virtual).
10. M. Salman, H. Qabazard, M. Moshfeghian, "Water scaling case studies in a Kuwaiti oil field," *Journal of Petroleum Science and Engineering* 55 (2007): pp. 48–55.
11. A.A Olajire, "A review of oilfield scale management technology for oil and gas production," *Journal of Petroleum Science and Engineering* 135 (2015): pp. 723–737.
12. N.A. Silva, I.M Raimundo, "Determination of scaling ions in oilfield produced water by laser-induced breakdown spectroscopy," *Journal of Analytical Atomic Spectrometry* 38 (2023): pp. 1450–1460.
13. M. El-Said, M. Ramzi, T. Abdel-Moghny, "Analysis of oilfield waters by ion chromatography to determine the composition of scale deposition," *Desalination* 249 (2009): pp. 748–756.
14. S. Heath, M. Stamnes, O.M. Drønen, "Improving scale management using ESEM Bulk Scale Analysis," AMPP-2022-18215, AMPP CORROSION, held March 6–10, 2022 (San Antonio, TX, USA).
15. N.D. Feasey, E. Freiter, M.M. Jordan, R. Wintle, "Field experiences with a novel near real time monitor for scale deposition in oilfield systems," NACE-00117, NACE Annual Corrosion Conference, held March 26–31, 2000 (Orlando, FL, USA).
16. D.H. Emmons, G.C. Graham, S. Holt, et al., "On-site, near-real-time monitoring of scale deposition," SPE-56776-MS, SPE Annual Technical Conference and Exhibition, held October 3–6, 1999 (Houston, TX, USA).
17. D.H. Emmons, S.A. Smith, T. Weathers, et al., "Deployment of real-time scale deposition monitoring equipment to optimize chemical treatment for scale control during stimulation flowback," SPE-113804-MS, SPE International Oilfield Scale Conference and Exhibition, held May 28–29, 2008 (Aberdeen, UK).
18. P. Rostron, "Critical review of pipeline scale measurement technologies," *Indian Journal of Science and Technology* 11 (2018): pp. 1–18.
19. P. Jean-Pierre, S. Gérard, T. Eric, "Real-time method for the detection and characterization of scale," SPE-74659-MS, International Symposium on Oilfield Scale, held January 30–31, 2002 (Aberdeen, UK).
20. A.A. Al-Ghamdi, I. Mohamad El-Zefzafy, M.H. Al-Malki, et al., "First flowline scale inspection in Ghawar field using a robust infrared technology offers a reliable and cost effective approach," SPE-181356-MS, SPE Annual Technical Conference and Exhibition, held September 26–28, 2016 (Dubai, UAE).
21. A. Hussein, *Essentials of Flow Assurance Solids in Oil and Gas Operations: Understanding Fundamentals, Characterization, Prediction, Environmental Safety, and Management*, 1st ed. (Cambridge: Gulf Professional Publishing, 2022).
22. M.M. Jordan, J. Buckman, C.J. Johnston, "Direct monitoring of the performance of scale control programs across the produced water life cycle via suspended solids analysis," SPE-93892-MS, SPE European Formation Damage Conference and Exhibition, held May 25–27, 2005 (Sheveningen, The Netherlands).
23. M.F. Mady, M.A. Kelland, "Review of nanotechnology impacts on oilfield scale management," *ACS Applied Nano Materials* 3 (2020): pp. 7343–7364.
24. B. Kaasa, B. Somme, R. MacDonald, et al., "The Oda Field: From scale prediction to reality-why squeezing was not necessary despite a high calcium sulphate risk," SPE-209507-MS, SPE International Oilfield Scale Conference and Exhibition, held May 25–26, 2022 (Aberdeen, UK).

14 Sulfide Scales

Jonathan J. Wylde[1,2]
[1]SPL Inc., Katy, Texas, USA
[2]Heriot-Watt University, Edinburgh, Scotland

14.1 Introduction

Some years ago, oilfield sulfide mineral scales were considered exotic, and one can still claim this as true for sulfides of lead (galena, PbS) and zinc (sphalerite, ZnS). Iron sulfide scales though, nowadays, seem to be far more common, especially in, for example, mature fields, and those with large waterfloods. While not perhaps exotic in their prevalence, their treatment still is because they have remarkably low solubilities, multiple orders of magnitude less than that of, for example, $BaSO_4$ (barite),[1–3] see Table 14.1. This translates to a formation that is thermodynamically exceedingly favorable (essentially instantaneous) and tough to prevent.

Table 14.1: Solubility of key sulfide minerals and other common oilfield scales at 25°C.

Mineral	Formula	Solubility Product Constant (K_{sp})
Halite	NaCl	3.6
Calcite	$CaCO_3$	3.4×10^{-9}
Siderite	$FeCO_3$	3.2×10^{-11}
Barite	$BaSO_4$	8.9×10^{-11}
Pyrite	FeS_2	8.5×10^{-26}
Mackinawite	FeS_{1-x}	2.9×10^{-18}
Pyrrhotite	$Fe_{1-x}S$	2.7×10^{-19}
Sphalerite	ZnS	1.6×10^{-25} to 2.5×10^{-22}
Galena	PbS	8×10^{-28}

The reader of this chapter is already very likely to understand that conventional mineral scales deposit in tubulars, flow lines, pipelines, just about anywhere, and cause flow assurance challenges. Sulfides, just like these scales, can also be found almost universally in the oilfield from the reservoir (near-wellbore scaling), perforations, downhole tubing and equipment (*e.g.*, safety valves), wellhead, choke, separator, flow lines, and generally anywhere in the produced water handling plant.

Up until the last 15–20 years, only sporadic publication and research can be found on sulfide scales from the oil and gas industry. This is no longer the case, and it is interesting to speculate on the reasoning for this (in no particular order):

- Management of iron sulfide has been focused on removal methods versus preventative techniques, and these have been successful. Efforts spent on remediation has been perceived as a worthwhile investment with a tangible return, whereas inhibitors have been viewed commonly as ineffective.[4]
- Zinc and lead sulfide solids have been even less studied than iron sulfides due to their rarity being associated with high-pressure, high-temperature, and high-salinity (HP/HT/HS) reservoir conditions. These types of reservoir conditions and the sulfide scales associated with them have become more common as the industry proceeds to develop these more challenging resources.
- Slight amounts of sulfide scale do not form a flow assurance challenge and has been viewed by many engineers to form a protective layer.[5]
- Sulfide scales, and in particular iron sulfide, are very challenging to work with in the laboratory. The necessity to maintain the oxidation potential (Eh) in experiments to ensure $[Fe^{2+}]$ is aligned to field conditions and that $[HS^-]$ remains speciated as such.

This chapter reviews the different types of sulfide scale, their formation mechanisms, and occurrence to give the reader a thorough understanding of sulfide scale nucleation, growth, and surface adhesion. This chapter will also discuss sulfide scale control, prevention, and remediation methodologies—gleaned not only from the literature but from the decades of industry-leading experience of the author.

14.2 Mineral Chemistry

14.2.1 Lead Sulfide

The mineral scale galena is seldom observed in oil and gas production and has most often been aligned with high-temperature, high-salinity production environments. Galena has several crystalline structures with the most common being cubic, where the lead atoms coordinate octahedrally with six sulfur atoms (and vice versa). Impurities most often appear as mineral phases and cations such as zinc, cadmium, arsenic, copper, manganese, iron, and bismuth are all known. At high temperatures, an elemental solid solution exists between galena and clausthalite (PbSe). An extreme density is characteristic of galena (7.4–7.6 g/cm^3), as is the softness (2-½ on Mohs scale).

14.2.2 Zinc Sulfide

In nature, sphalerite is an abundant species, but in oil and gas operations, it is more uncommon. Like galena, it has a cubic habit but shows tetrahedral and dodecahedral forms along with crystal twinning on {1 1 1}. Sphalerite is the low-temperature polymorph of ZnS with wurtzite the high-temperature polymorph which tends to have a hexagonal close packed structure—to the authors knowledge, the presence of wurtzite has not been reported in oil and gas operations. In the structure of sphalerite, each zinc atom is tetrahedrally coordinated to each sulfur atom and vice versa. Predominantly covalent in structure, the bonds do however display partial ionic tendencies. Coprecipitation with iron is almost universal and can give a further indication as to the formation environment.

14.2.3 Iron Sulfide

Iron sulfide is more complex than lead and zinc sulfide, as there are many different polymorphs—summarized in Table 14.2. The different polymorphs have different iron to sulfur stoichiometry and crystalline habit resulting in drastic differences in behavior, especially when it comes to control and remediation methods as we will see later. There are four polymorphs that appear more regularly in oil and gas deposits: pyrite, pyrrhotite, mackinawite, and marcasite. The exact species precipitated in each system depends on a host of factors (pressure, temperature, coprecipitation to name a few), and knowledge of this is critical to developing appropriate management strategies. To complicate matters, oxidation and partial oxidation of iron sulfides can occur leading to formation of (for example) various oxides of iron.

Table 14.2: Description of the known polymorphs of iron sulfide.

Polymorph	Formula	Properties
Ferric sulfide	FeS	Amorphous or noncrystalline, 4.3 g·cm^{-3}, sulfur equivalent of hematite, intermediate compound, and the first solid product from the reaction of sulfide and Fe^{2+}
Greigite	Fe_3S_4	Cubic, 4.0 g·cm^{-3}, sulfur equivalent of magnetite (magnetic), bacterial origins, and a well-known precursor to pyrite
Kansite	Fe_9S_8	Cubic, 3.1 g·cm^{-3}, magnetic, and identified from corrosion deposits in oil and gas
Mackinawite	FeS_{1-x}	Tetragonal, 4.2 g·cm^{-3}, corrosion product–formed under low-H_2S and low-temperature environments so very common in the oilfield
Marcasite	FeS_2	Orthorhombic, 4.9 g·cm^{-3}, hydrothermal and sedimentary deposits, and metastable iron disulfide phase
Pyrite	FeS_2	Cubic, 5.0 g·cm^{-3}, most abundant FeS on Earth's surface, and often called as "fool's gold"
Pyrrhotite	$Fe_{1-x}S$	Monoclinic, hexagonal, 4.6 g·cm^{-3}, most abundant FeS on earth, weakly magnetic, and usually formed at high temperature
Smythite	$Fe_{3+x}S_4$	Rhombohedral, 4.1 g·cm^{-3}, strongly magnetic, and usually associated with nickel
Troilite	$Fe_{1-x}S$	Monoclinic, 4.6 g·cm^{-3}, iron-rich pyrrhotite analogue, and very rare on earth (common on the Moon and meteors)

14.2.3.1 (Amorphous) Ferric Sulfide and Mackinawite

The term *amorphous FeS* is usually applied to the first solid precipitates observed in laboratory simulations, as the scale is challenging to characterize due to its noncrystalline or extremely fine-grained (micro-) crystalline structure. It is typically characterized in this way, as when classic X-ray diffraction (XRD) is performed, the results come back as amorphous due to an inability to measure any crystalline structures. More detailed studies, however, using more sensitive techniques, such as extended X-ray absorption fine structure elucidates a very fine, tetragonal mackinawite structure. This is highly unstable and can transform into fully crystalline mackinawite via ripening (aging).[6–8] The earliest example of mackinawite identification was as a corrosion deposit of steel in solutions of H_2S and since then understanding has evolved, and the upper limit of (thermal stability) deposition is believed to be 130°C.[9, 10]

14.2.3.2 Pyrite and Marcasite

Pyrite is the most abundant iron sulfide mineral found in nature and is often called *fool's gold.* It has a distinctive structure analogous to that of sodium chloride. Pyrite (and pyrrhotite) are generally the scales formed from corrosion processes and after a long-term exposure in the system. The exact reaction pathway (mechanism) for pyrite formation is contentious with different research yielding varied results. For instance, there is no direct evidence that pyrite can form directly from an aqueous solution, and an intermediate phase is necessary and must be dissolved and reprecipitated; although other studies suggest that mackinawite converts to pyrite via a direct oxidation process.[11, 12]

Marcasite is grouped with pyrite, as it is the orthorhombic variant and is less common in the oilfield, presumably as experimentally, it has been shown to favor an acidic formation environment.[11] Other studies have demonstrated the influence of supersaturation and marcasite has been shown to form preferentially in low-pH or sulfide-deficient solutions (saturation index <1,000), and pyrite preferentially forms at highly supersaturated solutions (saturation index >1,000).[13]

14.2.3.3 Greigite

The iron sulfide equivalent of magnetite (Fe_3O_4), having a similar structure, and it is accepted that greigite forms indirectly from solution via mackinawite.[12] This transformation can occur in aqueous solutions, where dissolution and reprecipitation processes are likely involved, or it can happen in anhydrous conditions, where the processes are infinitely more complex and varied. Oxidation has been proposed to promote the formation of a ferric iron layer on existing mackinawite, which then undergoes a solid state transformation to greigite.[14] Alternatively, it has been suggested that greigite forms from a reaction of mackinawite with anoxic water and occurs rapidly at temperatures >70°C.[6]

14.2.3.4 Pyrrhotite and Troilite

High-formation temperature environments are characteristic of pyrrhotite, where in nature, it is more often associated with hydrothermal deposits.[6] Furthermore, experimentally, it has been demonstrated that it is challenging to prepare in aqueous solutions <100°C.[15] The two forms of pyrrhotite also yield temperature formation information, as monoclinic pyrrhotite is more stable than hexagonal pyrrhotite, when the temperature is <200°C.[16]

Troilite is a member of the pyrrhotite group and is the name given to the 100% monoclinic endmember and has no vacancies in its structure; thus, $x = 0$ in the $Fe_{1-x}S$ empirical formula. Troilite has been reported as a common low-temperature corrosion deposit.[15] This observation was corroborated in much later work, where it was detected as a corrosion deposit of carbon steel in medium-salinity brines, and it can coprecipitate with pyrrhotite at higher temperatures in the same corrosion test.[17]

14.2.3.5 Kansite

Kansite (Fe_9S_8) was initially described as a new iron sulfide mineral by the Phillips Petroleum Corp. in 1953–1955, after being removed from an oil well in Kansas and was identified using XRD.[18] This data was subsequently challenged, and in fact refuted as XRD data identified a face-centered cubic crystal, when in fact, it was later proved to be tetragonal mackinawite.[19, 20] It was therefore proposed in the work by Milton to drop the name kansite, and instead, mackinawite is more appropriately deployed. Noteworthy is that mackinawite has been confused over the years (especially before 1963) as the rarer valleriite ($CuFeS_2$). It is tough to find any other literature resource on this mineral, and it is therefore included here for completeness.

14.2.3.6 Smythite

The mineral smythite has been described as a rhombohedral iron sulfide (Fe_3S_4) occurring both naturally and has been synthesized via a reaction between aqueous $FeCO_3$ and Na_2S at standard conditions.[21] Initially found as inclusions in calcite crystals in Bloomington, Indiana, smythite has similar properties to that of pyrrhotite and appeared as thin hexagonal flakes. Smythite is strongly magnetic, and its similarity is so strong to that of pyrrhotite that they can only be distinguished from one another via XRD. It appears this is not a mineral that appears in oil and gas operations commonly, if ever, as it has seemingly not been reported in the literature.

14.3 Origin and Formation

Sulfide scales form in the oilfield, because of the mixing of incompatible ions of lead, zinc, or iron (all collectively designated as M^{2+}) with a source of bisulfide or sulfide (bisulfide at usual oilfield pH) and can be expressed as

$$M^{2+} + HS^- \rightleftharpoons MS(s) + H^+, \quad (14.1)$$

$$M^{2+} + S^{2-} \rightleftharpoons MS(s). \quad (14.2)$$

In the case of iron, equation *x* – 1 is a shortened version used in place of articulating the hexa-aqua species $[Fe(H_2O)_6]^{2+}$, as formation of iron sulfide is a ligand substitution reaction, per:

$$\left[Fe(H_2O)_6\right]^{2+} + HS^- \rightleftharpoons FeS(s) + H^+ + 6H_2O. \quad (14.3)$$

When studying the origin of sulfide scale deposition, one should focus on three aspects:

- The source of Pb^{2+}, Zn^{2+}, and/or Fe^{2+}
- The source of H_2S or HS^-
- Process conditions that create a favorable environment for precipitation (and speciation of sulfur)

Lead and zinc are found naturally in reservoir formation waters, and while iron too can originate authigenically, it typically originates from corrosion of the carbon steel found so very often in oil and gas piping and tubulars. Concentrations of these metals are often very low in formation water, but extreme cases are present in the literature, for example, lead of 172 mg/L and zinc of 452 mg/L were measured from produced water from the North Sea's Culzean field, while in the U.S. Gulf of Mexico, iron has been measured as high as 400 mg/L.[22] Typically, the concentrations of these ions are 1, if not two orders of magnitude lower than these extreme examples.

The rate of iron scale deposition can be a strong indicator for the source of iron. When deposition is very fast, it is likely that iron originates from the reservoir, where it is typically found at a higher concentration than when borne from corrosion processes, where deposition is ordinarily slow and steady. It should be understood, however, that even a few mg/L of iron (and/or lead, and/or zinc) is more than adequate to yield a supersaturation with respect to the sulfide mineral.

Turning our attention to the source of sulfur, there is also more than one possible origin. The presence of H_2S can occur naturally within a reservoir, where conditions allow the dissolution of sulfide-rich minerals—which enriches the formation water in H_2S or HS^-—a process termed thermochemical sulfate reduction. Often however, the presence of H_2S originates from biological (bacterial) activity in the reservoir (or other part(s) of the system). The presence of H_2S is induced by the oil and gas production process itself, for example, seawater injection in the reservoir for pressure support provides sulfate-reducing bacteria (SRB) with a source of copious sulfate. The literature has multiple examples of field- and laboratory-based evidence supporting SRB production as the source of sulfur in oilfield sulfide scales with various techniques available to categorize and enumerate the bacterial presence to the point, where specific SRB species can be identified and tracked.[23]

As well as supersaturation of the scaling ions, it is necessary for the correct conditions to be present so that precipitation occurs. Fluid pH is the critical parameter to focus on as it is this that, along with Eh, determines the speciation state of the sulfur atom in an aqueous system. Furthermore, when a sulfide mineral precipitates, it releases a proton (Equation [14.1]), and therefore, the system pH decreases. The basic equilibrium reactions for sulfur that occur at typical oilfield Eh conditions have been written in Equation (14.4). Eh is the redox potential of a system and is a measure of the propensity to lose or gain electrons. When a system has a low (more negative) Eh, then it is more reducing and likely to gain electrons and thus move to a more oxidized state. It is usual for oilfield system Eh to be low due to anoxic conditions being the norm. There are multiple equilibria which can be expressed as follows:

$$H_2S(g) \rightleftharpoons H_2S(aq) \rightleftharpoons HS^- + H^+ \rightleftharpoons S^{2-} + 2H^+. \tag{14.4}$$

The critical pH range is between 6.5 and 7.5, which is where the equilibrium between H_2S and HS^- disproportionately changes.[24] At approximately pH 7, the fractions of H_2S and HS^- are equal at 50% each. Increasing the pH to 8 sees almost 100% speciation to HS^- away from H_2S and decreasing the pH to 6 results in almost 100% speciation to H_2S, away from HS^-. As we know, if HS^- is absent then metal sulfide scales will not form. The criticality of pH should be the focus of system reviews/field surveys, where the passage of fluids through a given system can be subject to changes in pH around the critical values described. Degassing of CO_2 from water as it depressurizes can increase pH sharply and specifically across chokes or restrictions. Often plants recycle fluids, and this can cause inconsistencies with system pH, especially if production chemicals of extreme pH are used, for example, H_2S scavengers and corrosion inhibitors.

14.4 Occurrence in the Field

14.4.1 Lead and Zinc Sulfides

Reported field occurrences of both lead and zinc scales are few and far between with only a handful of case histories available. One of the best covered examples in the literature concerns the North Sea Elgin-Franklin fields, where challenges associated with the co-precipitation of lead and zinc sulfide have been reported.[25] These fields are in the Central Graben Aera of the North Sea, some 250 km east of Aberdeen. These are deep wells with an average reservoir temperature of 200°C (in some instances 215°C) and pressures of up to 16,000 psi. $CaCO_3$ scale occurred in wells at depths between 2,000 m and 5,400 m—and as it formed slowly, it was manageable via acid pickling of the tubing (organic and not mineral acids). In the upper reaches of the tubing (2,000 m or shallower), the temperature was cooler (125°C), and both lead and zinc sulfide scales deposited and posed a significant risk to well operability, due to small amounts of scale-causing inoperable downhole safety valves. The challenges for this application are compounded by several factors:

- Ultra-high temperature: results in challenging depositional environments (enhances kinetics of scale precipitation) and causes issues for removal and prevention using chemistry due to compatibility and corrosivity concerns.

- High salinity: formation water more than 250 g/L total dissolved solids and >20 g/L Ca^{2+} results in a severe chemical deployability challenge, due to incompatibility. Furthermore, the brine is enriched in 450 mg/L of iron ions which complexes the scale coprecipitation environment further and can create other compatibility challenges with prevention chemistry.
- Scale coprecipitation: the presence of conventional scales along with sulfide scales makes for an entirely exotic (and unique) situation, where conventional chemistry and solutions are simply unsuitable.

Other examples of the more exotic lead and zinc sulfide scales exist in the literature, although none quite as extensively as the previous example:

- In the Gulf of Mexico's Mobile Bay, an unspecified operator and field experienced a combination of barium sulfate with both zinc and lead sulfides.[26] Very little detail is given on the field scenario itself as the article focuses on the selection of appropriate control chemistry. It is revealed though that prior to application of preventative measures, the platform's overboard water line required cleanout every two months due to scale buildup. Furthermore, whenever the surface coolers and separators were inspected, zinc sulfide scale was reported in large quantities.
- A second example from the Gulf of Mexico Mobile Bay, presumably operated by Apache, and presumably different from the previous example, describes a downhole case of lead and zinc sulfide deposition.[27] The Incoloy 825 well completion required frequent well repairs and intervention caused by the deposition of these exotic scales combined with the high-downhole temperature. The well in question experienced a rapid production decline that was being treated with a combination of acid washes and mechanical milling operations.

A common theme that emerges from the literature for zinc and lead sulfide deposition is that these scales are rarely seen in isolation and coprecipitate with themselves and other scale types. They also seem to be associated with high-temperature reservoirs that are isolated to certain geographies (geologies). This is in quite stark contrast to iron sulfide scales, which are both far more common and have a much different origin and occurrence.

14.4.2 Iron Sulfide

Iron sulfide scale formation seems almost indiscriminate and has been reported as far upstream as in the formation and throughout the downhole equipment and surface facilities.[28–30] Iron sulfide scales are oleophilic (preferentially, oil wetting) which means, unlike lead and zinc sulfides, deposits can occlude copious organic material into the matrix. Furthermore, free-floating iron sulfide particles can accumulate at the oil-water interface, thereby causing stable emulsions. Finally, it is noteworthy that iron sulfide scale can be pyrophoric, and caution is required when handling and preparing samples that should be transported wet (ideally in produced water), as spontaneous combustion can occur, if samples are allowed to dry out.

While there are many field studies to choose from, there are none quite as detailed as the examples that come from the Middle East, and more specifically, from Saudi Arabia and the prolific Ghawar field that produces from the Khuff formation.[31] The reservoir is highly sour (% levels) and is believed to originate via thermochemical sulfate reduction, where anhydrite and dissolved sulfate is reduced at the expense of methane and (likely) other hydrocarbons[32]:

$$CaSO_4 + CH_4 \rightarrow CaCO_3 + H_2S + H_2O. \tag{14.5}$$

While corrosion has not been a major concern, presumably due to a passivating layer of scale depositing in the tubing, the formation of scale has been one of the largest factors affecting production potential. Scales formed downhole in the tubulars and restricted flow and downhole operations. Scales-deposited downhole were also observed to dislodge and transport downstream and accumulated near to the wellhead, thus blocking the flow line.

One of the most valuable contributions from this work is the incredible data set presented on the compositions of 90 individual scale samples gathered from well workovers and interventions, and the insights this provides to the complexity of the formation process for iron sulfide scales. The following observations were made:

- There were a wide range of mineral phases present, and their distribution was just as varied with nearly all field scales containing several different mineral phases.
- Iron sulfide was nearly always the dominant phase and included polymorphs of (in order of instances of occurrence): pyrrhotite > pyrite > marcasite > mackinawite > troilite > greigite.
- Ferric iron scales were also commonly present and included hematite (Fe_2O_3), magnetite (Fe_3O_4), goethite (α-FeOOH), akaganeite (β-FeO(OH,Cl)), and lepidocrocite (γ-FeOOH).
- Ferrous compounds, such as hibbingite ($Fe_2(OH)_3Cl$), ferrous chloride tetrahydrate ($FeCl_2 \times 4H_2O$), and wüstite (FeO), were identified.
- Additionally, to further complicate composition, calcite ($CaCO_3$), barite ($BaSO_4$), gypsum ($CaSO_4 \times 2H_2O$), anhydrite ($CaSO_4$), and dolomite ($Ca,Mg(CO_3)_2$). In addition, elemental sulfur, silica, and clays were also detected.

The mineralogy of scales exhibited a wide-ranging variation from well to well, sample to sample. There were predictable variations with well depth, and the length of time the scale had been present in the well. Structured layering was also present in many scale samples with distinct composition (and formation environment).

The examples described above are largely confined to high-temperature gas wells, and a further example of iron sulfide scale has been drawn from low-temperature, mature, water-flooded fields of California in the San Joaquin Valley in the Bakersfield/Taft area.[33] The water flood that operated in this field was super complex and comprised over 2,000 wells (the field had some 5,000 producing wells). Iron sulfide scale deposition caused a serious operational issue through the produced water handling system and ultimately caused injectivity issues, which in turn yielded poorer production.

Like the previous example in Saudi Arabia, there were many analyses of scale samples performed over the years (largely using XRD), and the compositions consisted primarily of iron sulfide (pyrite, pyrrhotite, and mackinawite) with coprecipitated calcium carbonate and iron carbonate and occasionally akageneite. The deposits were nearly always associated with organic (hydrocarbon) material and presented as flow line blockages, filter media plugging, equipment (meters and valves) blockages, and injectivity decline.

The deposition mechanism for the iron sulfide in this water injection system was classic for this type of deposit and can be explained via three causal factors:

- Iron came from corrosion occurring mainly from the producing wells and piping upstream of the water injection system.
- Sulfide originated from biological activity and the reducing of sulfate to sulfide via SRB.
- Organic content was from crude oil and poor separation that yielded excursions entering the water plant and adsorbing to the iron sulfide scale surfaces.

Scale was dispersed throughout the water injection system, and it was highly desirable to restore injectivity as this was easily relatable to oil production. The dispersed nature of the scale and the vastness of the pipework of the injection system focused the remediation to first optimize the process and operability of the plan. For instance, an aggressive integrity management process was put in place on the producing wells with both continuous and batch corrosion inhibitor application on new wells and retrofitted to many older wells. Application of biocide to the whole of the water injection system was meticulously managed to ensure H_2S production was kept to a minimum. A sidestream device was used to determine remedial chemical strategies and was constructed from transparent plastic pipe rated to the relatively low pressures of the water injection handling system. As different parameters and chemicals were trialed, the success or failure could be discerned easily and quickly, as it was visual. Furthermore, different chemical types and amounts could be injected sequentially along the length of the sidestream to inform operations about the best approach to be implemented. There were many different chemical approaches trialed this way, and the most successful saw wider field implantation to determine improvement to overall injectivity of water and lowering of (injection) well failure.

Iron sulfides in general then are a far more ubiquitously encountered scale in the oil and gas industry than zinc or lead sulfides. There are many different polymorphs that can form, all of which tell a different story of deposition and require a customized approach to remediation and management. Iron sulfide can form just as commonly in higher pressure and higher temperature downhole environments than it can in low-pressure, low-temperature water process systems. The root causes across many cases yield the same causal mechanisms of corrosion sourcing the iron and microbial activity the sulfides with the oilfield creating the perfect set of conditions for deposition and fouling to occur.

14.4.3 Black Powder

Black powder is perhaps one of the lesser appreciated, or even understood, forms of scale that occurs specifically in gas or oil pipeline systems. It is itself, a generic term, coined in the late 1990s, to describe regenerative (typically, iron-based) deposits observed in oil and/or gas transportation pipelines.[34] It is also a "mid-streamer" issue and has arguably not been something commonly discussed in typical "upstream" circles, and it is proposed that critical links and, therefore, understanding can be obtained through studying the multitude of information available on black powder. When one considers the key observations in the Khuff and California case histories previously discussed, one can instantly see alignments between the observations for black powder deposition.

Black powder takes many compositional forms and is borne from the reaction of the pipeline alloy with the transported media. In other words, black powder, is generated from low-level corrosion that occurs in the pipeline—this, of course, is the same general mechanism for a lot of iron sulfide seen in upstream oil and gas operations. The literature describes black powder as comprising three major groups, and these have been summarized in Table 14.3. In some instances, a dominance of iron sulfide minerals,[34] in others, a dominance of oxides,[35] and then in further instances equimolar amounts of sulfides and oxides.[36] What is striking from this table is that it is nearly identical to that described in the upstream Saudi case history. Naturally occurring radioactive materials such as Pb^{210} have been described as associated with black powder, more often in gas vs. liquid transportation pipelines.[37] The presence of mercury, while not radioactive, is equally challenging to handle, and has been described at levels >1,000 ppm.[38]

Table 14.3: Common scale species found in black powder.

Sulfide Species	Oxide and Hydroxide Species	Other Species
"Generic iron (ferric) sulfide" (Fe_2S_3)	Hematite (Fe_2O_3)	Siderite ($FeCO_3$)
Mackinawite (Fe_9S_8)	Magnetite (Fe_3O_4)	Cementite (Fe_3C)
Pyrite (cubic FeS_2)	Goethite (α-FeOOH)	Elemental sulfur (S_8), Hg, Fe
Pyrrhotite (Fe_7S_8)	Akaganeite (β-FeOOH)	Silicates, sand (SiO_2), clay, dirt (soil minerals)
Marcasite (orthorhombic FeS_2)	Lepidocrocite (γ-FeOOH)	Other common scale minerals and salts including NORM
Smythite (rhombohedral Fe_3S_4)		
Greigite (cubic Fe_3S_4)		

The exact composition of black powder varies system to system and in general terms is related to the composition of the transported media. For instance, a pipeline that has a known and significant H_2S content in the gas it transports is likely to observe sulfide species in the black powders, rather than oxides (or hydroxides) of iron. These anticipated field observations are supported by detailed experimental studies. In one study, a mixture of iron sulfides, iron oxides, siderite, and elemental sulfur form in the presence of CO_2 (1 bar), H_2S (0.1 mbar), and O_2 (12 mbar). In the same study, siderite and iron oxides formed in the same gas mixture, when H_2S was absent, and only iron sulfides formed in CO_2 and H_2S.[39] These results are corroborated by a more recent, and similar, study, where the CO_2, H_2S and O_2 partial pressures were systematically varied along with different water condensation rates.[40]

It is worth, at this point, to digress into the complexity of the occurrence of both iron sulfide and iron oxide species together in black powder deposits as it can, at times, seem incongruous. Furthermore, and as previously stated, this is an analogous situation to many upstream deposition scenarios. Ferric iron oxides form consequently to aqueous oxidation of the iron in the pipeline steel wall. Oxygen ingress into systems occurs via leaks, as oil and gas production and transmission systems are usually designed to be anoxic. It is well-known that even ingress of single-digit parts per million of oxygen can cause severe corrosion.[41, 42] One study performed a survey of 44 gas pipeline companies in North America on the oxygen concentrations in pipeline systems. The typical allowable oxygen present ranged from 0.01 mol% (noncorrosive) to 0.1 mol% (highly corrosive), and many systems had typical measured values of 0.02 mol%.[43] Oxygen corrosion can be expressed via:

$$2Fe^0 + H_2O + 1\tfrac{1}{2}O_2 \rightarrow 2\alpha,\beta,\gamma\text{-FeO(OH)}. \qquad (14.6)$$

If this occurs in aqueous, low-oxygen conditions (which most oilfields are), then γ-FeO(OH) typically does not result, as it is unstable and converts to magnetite (Fe_3O_4) via the following reaction:

$$8\gamma\text{-FeO(OH)} + Fe \rightarrow 3Fe_3O_4 + 4H_2O. \qquad (14.7)$$

All polymorphs of FeO(OH) have been documented in black powder deposits removed from pipelines and hematite forms preferentially to magnetite in aqueous systems that have higher (*i.e.*, saturated) oxygen contents.[44] Bacterial action has been reported to generate various iron oxides and is particularly associated with acid-producing bacteria and iron-oxidizing bacteria.[45]

Preexisting deposits of iron carbonate or iron sulfide can be oxidized *in situ* to generate magnetite. Despite being a very slow reaction, and it can also be constrained to the first few atomic layers of a mineral surface, magnetite can be generated from oxidation of preexisting deposits of iron carbonate or iron sulfide.[46] The reaction for conversion of siderite proceeds according to:

$$FeCO_3 + \tfrac{1}{6}O_2 \rightarrow \tfrac{1}{3}Fe_3O_4 + CO_2. \qquad (14.8)$$

The conversion of iron sulfide on the other hand is more rapid and progresses during the wet cycles in a pipeline according to

$$2Fe_9S_8 + 9H_2O + 13\tfrac{1}{2}O_2 \rightarrow 18\gamma\text{-FeO(OH)} + 2S_8 \qquad (14.9)$$

followed by the transformation of γ-FeO(OH) into magnetite, as described previously in Equation (14.7).

A final note is that if oxygen ingress occurs and the pipeline is in sour service, elemental sulfur can occur. Elemental sulfur is known to cause severe localized corrosion and requires a specially tailored integrity management program.[47] It forms according to the following reaction:

$$2H_2S + O_2 \rightarrow 2H_2O + 2S, \tag{14.10}$$

$$3FeS + 2O_2 \rightarrow Fe_3O_4 + 3S. \tag{14.11}$$

Returning to black powder, and in general given that most transportation pipelines are constructed of steel that is susceptible, at least in part, to corrosion, it would seem the deposition of black powder is inevitable. Certainly, the publications of disclosed geographies would support this statement. Case histories are available from every continent from Brazil to the United States, from Japan to Greece, and from China to Oman.[48–50] Bountiful examples are also prevalent outside the oil and gas industry entirely, for example, in transportation of domestic water.[51, 52]

Black powder forms *in situ* and invariably becomes self-limiting and may not present an operational challenge; that is, until the black powder becomes entrained in the produced media of the pipeline and are transported downstream. Movement of black powder therefore is a key topic, not only to understand the mobility during steady state operation, but also so that pigging and cleaning operations can be optimized. Parameters that are influential to transportability of particles in a pipeline include gravity, buoyancy, lift, and drag. Other parameters to consider include particle shape, size, composition, and roughness, for example, a well-rounded particle will roll, especially, if the surface is also smooth. The chemical composition and makeup also can be influential with "stickiness" becoming a factor in some systems. Much literature is devoted to this topic with excellent examples of both experimental studies and real field observations.[53]

The prevalence of SRB in black powder deposits has been investigated and alludes to a general biocorrosion mechanism for formation. Some studies suggest that it is not just SRB that are responsible for influencing the amount and type of black powder. For instance, the influence of methanogens and metal-reducing bacteria has been revealed as significant, as well as iron oxides yielded from the action of acid-producing bacterial or iron-oxidizing bacteria.[34] When one amalgamates the literature, it appears that bacteria are found associated with black powder deposits for a minority of cases (<25%), and an absence of detectable cultures is predictably reported.

It is finally noteworthy that the volume of black powder deposited in each pipeline system can collectively be substantial, especially, when one considers it is usual of it to be collected (pigged) into a confined space (pig trap). Black powder occurrence is usually nonuniform, and the assumption that finding black powder in two spots and assuming it is present all places in between is imprudent. Quantified examples have been provided in the literature, and a relatively clean pipeline can exhibit an average of 10–30 g/ft^2 of black powder; taking a conservative value of 15 g/ft^2, for 100 km of 18″ pipe, this is a little over 22 metric tons of black powder.[54]

14.5 Operational Challenges Caused by Sulfide Scale

Sulfide scale, like nearly all mineral scales, deposition can occur anywhere in a production process—it can occur in the reservoir and near wellbore, through the separation system and produced water handling system and even encroach into the gas plant and certainly into the oil export system. Many examples exist in the literature, several of which have already been elaborated in this chapter. In a worst case entire wells or parts of process systems can be lost, sometimes irretrievably to sulfide scale deposition. When this happens, costs can become astronomical especially if one now must account for remedial (workover) costs. Finally, it is the location and composition of the sulfide scale that impacts most the cost of remediation and deferred oil losses.

One unique aspect to sulfide scale, in fact, to iron sulfide scale, is the oleophilic nature it possesses—it is like a proverbial sponge for oil. This gives this scale a unique set of additional challenges caused by its occurrence, especially when this occurs, as it invariably does, in a multiphase environment. There are two considerations that need to be made:

- *Wettability effects* concern how the surfaces of the scales preferentially react to the presence of oil and water. Oil-wet iron sulfide particles can better agglomerate. Oil-wet black powder is highly immobile compared to a dry black powder—this has led to several descriptors emerging in the literature, such as the notorious *schmoo* or (even worse) *black sticky stuff*.[55]
- *Emulsion effects* concern a fine dispersion of any scale can influence oil and water separation, whether that be the resolution of water-in-oil or oil-in-water emulsions. Iron sulfide can have an affinity for the oil-water interface and create a pseudo interfacial crowding effect and thus stabilize emulsions, or at least, reduce the efficacy of demulsifiers.

The type of operational challenge faced by the precipitation of sulfide scales does depend upon the location of formation:

Near-wellbore sulfide deposition includes scales occurring within several feet of the perforations (into the reservoir) and across the perforations (but before fully entering the production tubing). It is less common to see sulfide scale in these locations vs. both other scale types that typically result from mixing of incompatible waters and sulfide scale in other locations further downstream. It is not impossible, however, for a near wellbore to have been made sour via poor drilling and completion practices and provide a source of H_2S. It is also not impossible to have a source of authigenic iron, zinc, or lead in the correct concentrations and under the right conditions for a skin of scale to form in the near wellbore with an obvious and dilapidating effect on production—in many cases, requiring a (re)stimulation.

- *Well production and flow line* scaling of sulfide minerals is perhaps the most widely reported (along with the water handling plant), and it really only comes down to where the source of lead, zinc, or iron is coming from to determine whether the scale occurs downhole (where the cations typically come from the reservoir, or in the case of iron very aggressive corrosion) or further downstream in the process (typically where light, general corrosion sources the iron ions). As one moves through the production tubing, wellhead and into the flow line, there is often a lot of areas of pressure drop and high turbulence that are hot spots for scale formation, for example, across downhole safety valves, chokes, gas lift valves, and the geometry of the wellhead itself.

- *Production facility* scale formation is the downstream extension of flow line scaling with areas of pressure drop and turbulence being a higher risk, such as across valves, pipe bends, filters, and pumps. Recirculation of fluids is something that needs to be considered to prevent unwanted mixing of incompatible waters.
- *Water injection system* scaling for sulfide scales is associated more with the recirculation of produced water vs. the injection of other types of water, for example, seawater, lake water, or aquifer water. Low-level corrosion can occur within the water injection handling system and typically presents as injector well plugging and near-wellbore skin formation.
- *Oil and gas export pipeline* scaling concerns the formation of black powder, although not exclusively. It is the mobility of black powder in oil pipelines that can create additional challenges such as erosion, loss of valve functionality along with a steady increase in differential pressure caused by deposition. Gas pipelines are not immune to black powder, and even prior to the pipeline, there have been cases of black powder in the gas processing equipment such as in compressors, scrubbers, filters, and driers.[56] The presence of black powder can affect in-line inspection tool efficiency, for example, magnetic flux leakage surveying had a poor signal strength, speed, girth weld readings, rotation, and inclination readings.[57]

14.6 Prevention and Remediation

In this section, the methods of managing the impact of sulfide scales will be discussed with a particular focus on chemical methods. The presence of several root causes to the deposition of sulfide scale results in multiple methods available for control and prevention. It is the experience of the author that the solution to a successful field strategy is to truly understand the source of metal ions and sulfide and why the conditions are good for scale formation. Once this is understood, all the preventative and control methods should become apparent. The chemical methods for control that can be employed include:

- Corrosion inhibitor (to lower total iron in system)
- Surfactants (for water wetting)
- Biocide (to target SRB and source of H_2S)
- Sulfide scale inhibitor/dispersant chemistry (for crystal growth inhibition)
- Chelating agents (for iron sequestration)
- Dissolver chemistry (for post deposition removal)

We shall discuss, in this chapter, the last three items on this list, as they concern the most direct use of chemistry in the control and removal of sulfide scales.

14.6.1 Prevention by Chemical Inhibition

The development of new and effective sulfide scale inhibitor chemistry, driven by an industry need for iron sulfide control products, is one of the last remaining innovation areas for scale research and development efforts. Much of the research efforts that are older than 10–15 years focused on the use of classic, or moderately adapted, inhibitor chemistry (typically used at abnormally high concentrations). Furthermore, much research has focused on the application of continuously injected chemicals. Squeeze deployable sulfide scale inhibitors are few and far between and seldom arise in the literature, although cases do exist. Given the formation of lead and zinc sulfide scales often occurs deep downhole, the development of these types of chemistries seems highly desirable.

Early studies around the turn of the century were restricted to evaluate classic scale inhibitor chemistries of phosphonate, phosphate esters, and polymeric species composed largely of acid monomers of acrylic acid or 2-methyl-2-(prop-2-enamido)propane-1-sulfonic acid (otherwise known as AMPS) or on amine and quaternary amine or phosphonic acid structures.[58–61] Only some of these cases really worked well (only some of the time), and it was clear that a gap in the production chemistry arsenal was present. It was the advent of severe lead and zinc sulfide scales in offshore assets (in the mid- to late-00s) that drove the innovation effort; it was not driven by the occurrence of iron sulfide, as these scales have several indirect chemical solutions available. Focusing on the more modern research and novel chemistry to inhibit sulfide scale at very low ppm levels, there are several key developments that require review.

- The use of hydroxethylacrylate/acrylic acid copolymer that has an average molecular weight of 1,000–20,000 AMU, and the treatment is claimed to work between 2-ppm and 100-ppm active polymer.[62]
- The use of alkyldiphenylether sulfonate (alkyl is between 4 and 30 carbon atoms and can be branched or linear) for sulfide scale combined with a phosphonate or phosphate ester for conventional calcite control has been reported[63]
- The use of a free-radical copolymer of α-olefin (C_{10} to C_{36}) with maleic anhydride has been revealed.[64] The final polymer has anhydride moieties on an unhydrolyzed copolymer backbone.
- A cationic copolymer was described containing acrylamide and diallyldimethylammonium salt moieties, and the examples disclose that these products also treat for conventional scales.[65] The copolymers have an average molecular weight ranging from 500,000 AMU to 5,000,000 AMU, and their applied dose rate seems be to very high with wt% disclosed.
- Polymers composed from monomers of acrylate, acrylamide, and methacrylamide and terpolymers of dimethyldiallylammonium salt, 2-hydroxypropyl acrylate and acrylic acid have been reported.[66] The polymers within this invention have average molecular weights from 5,000 AMU to 5,000,000 AMU.
- The use of a copolymer based upon AMPS with additional comonomers of either vinylphosphonic acid, acrylate, and N-vinylamide has been disclosed as a highly efficient polymer.[67] The polymers have an average molecular weight of 500,000–5,000,000 AMU. The invention also describes blending of the copolymer species with

conventional scale inhibitors, and the work by the same group discloses several such formulations.[68, 69]

- Finally, a polymeric scale inhibitor composition, comprising 80–82 mol% of AMPS, 2–18 mol% of a monomer selected from N-vinyl formamide, N-vinyl pyrrolidone, and diallyl dimethyl ammonium chloride; and 2–18 mol% of monomer selected from acrylic acid, methacrylic acid, esters of acrylic acid, or methacrylic acid with an alcohol having one to four carbon atoms has been revealed in the patent literature.[70] The average molecule weight of this terpolymer is between 300,000 AMU and 1,200,000 AMU. Further information about this chemistry is available in other publications.[71, 72]

One can see from this review of the patent literature that chemical sulfide inhibition has been focused around similar, large, complex polymer chemistry which suggests a mechanism that is founded on the principles of dispersion rather than classic threshold nucleation or growth inhibition.

It is worth mentioning briefly that laboratory test methods for sulfide scale inhibitors are not standardized across the industry, and a few research groups have set the bar with certain methodologies. Tests are challenging because both Eh and pH are required to be strictly controlled so that the correct supersaturation conditions are realized. This can be discerned in the literature, where tests cannot be corroborated—for example, in some early work, researchers report that regular scale inhibitor displays a degree of efficacy to lead and zinc sulfide scales, whereas latter work claims absolutely zero efficacy.[3, 26] Like conventional scale though, test methods for sulfide scales are combinations of dynamic scale loop tests and static jar tests.[73, 74]

When delving into the literature resource concerning black powder, it yields that prevention is not a topic that is classically treated using scale control chemistry. As this section began talking about the root causes of sulfide scale deposition, we return to this again. As we know, the root cause of black powder formation is from corrosion of the pipeline itself yielding in situ, regenerative deposits. The chemical prevention of these deposits is usually part of a larger integrity management process and concerns the use of corrosion inhibitors to arrest the release of iron ions into the produced fluids.[75] Even an acceptable corrosion rate of 4 mpy (0.1 mm/y) can yield many tons of black powder per year, and lowering the corrosion rate even slightly to 2 mpy (0.05 mm/y) can have a profoundly positive effect on the volume of black powder and therefore pipeline operability. Application of chemistry to achieve such a goal has been reported in the literature with the application of batch corrosion inhibitors as "black powder inhibitors" for gas pipelines.[76] Similarly, the use of volatile (top of the line) corrosion inhibitors has been applied specifically to lower the black powder production.[77] It is generally observed that corrosion inhibitors are deployed to pipelines that transport wet gas or liquids with appreciable aqueous content; it is not typical to deploy corrosion inhibitors to dry (gas) pipelines.

14.6.2 Prevention by Chelation Chemistry

The action of chelants is to sequester the metal scaling ions from produced water, therefore, rendering them unavailable for scale formation. The amounts of chemical required are stoichiometric, and usually this strategy is employed, where there is a low level of iron sulfide risk caused by mild corrosion. The most common ligands are ethylenediaminetetraacetic acid (EDTA), nitrilotriacetic acid (NTA), and diethylenetriaminepentaacetic acid (DTPA); other amino polycarboxylates are known but are not as economical to produce as the others mentioned.[78] As the levels of chemical required are stoichiometric, this type of preventative treatment is deemed too expensive for systems that comprise large volumes of water. More recently, the use of glutamic acid diacetate (GLDA) has been reported as well as other synergistic blends of complexing agents and ligands.[79] The challenge with all these species is that they are not specific to the scaling ions that are desirable to be sequestered, and therefore. even more chemical has to be added to overcome of the other (monovalent) ions species very often present. When compared to older scale inhibitor technologies, which must be deployed at almost the same concentration though, and in the right conditions of salinity, so on, chelation chemistry does have its place. These specificity shortcomings are overcome by the last chelant to mention which is tetrakis (hydroxymethyl) phosphonium sulfate (THPS). Much literature is available on this species as an iron sulfide dissolver, and this has been elaborated in detail in subsequent sections of this chapter. The mode of action of THPS is one of chelation, as it is specific to iron, and efficacy can be enhanced with the addition of ammonium ions. One potential advantage of THPS is that it can be injected continuously to not only prevent the deposition of iron sulfide, but it will actually slowly remove existing deposits, even when injected at low-level (ppm) concentrations[30].

14.6.3 Remediation with Mechanical Methods

To mention physical methods briefly, they have been divided into two parts: *downhole* and *pipeline* remedial methods. In downhole methods, there are generally one of two methods:

- Drilling of scale can be achieved using a drill string or coil tubing to rotate the bit thus cutting the scale out of the tubulars. Fluids and/or air is used to circulate and lift cuttings, and many examples are available in the literature describing these operations.[80, 81]
- Cutting of scale using simpler wireline deployable brushes, cutting heads, scrapers, and so on, is performed as a matter of routine, and programs are tailored to field needs.[82]

Removal of sulfide scale from large pipelines concerns black powder, and the techniques used differ greatly from those used for downhole scenarios and involves the use of separation, filtration, pigging, or a combination thereof[83]:

- Separation systems for black powder are used exclusively for gas pipelines and are usually cyclonic in design working on the principles of centrifugal forces. Particles need to be quite large for systems to work effectively (up to 10 μm).[48]
- Filtering at the end of a pipeline prior to it, feeding onward is the most common method of physically preventing black powder. There are a variety of parameters to consider for filter specification such as particle size distribution, physical filter media, filter media fill, filter elements, differential collapse pressure, and so on. It can be the case, however, that the black powder particles shear to create smaller particles, which can then pass through filtration systems.[35]

- Pipeline pigging for black powder removal requires a knowledge of the type of solids, their volume, and location. The topography and geometries of the pipeline are important to understand to select the correct train of pigs and fluids. Pigging is usually more successful, when low quantities of scale are present, based on the ability to move the scale down the pipeline. More typically, heavy bidirectional pigs are used composed of polyvinyl disc pigs and/or brushes.[49]

Examples exist in the literature of nonchemical methods to deal with sulfide scales within surface operations and processes, where various techniques including jetting, blasting, reaming, and so on.[59, 84]

14.6.4 Remediation Using Scale Dissolvers

The application of chemical sulfide scale dissolvers has received much attention over the years, particularly for iron sulfide.[41, 84] The dissolution of iron sulfide is a complex business and is intensified by the fact that iron sulfide is oleophilic, and often there is a requirement to deal with an oil/hydrocarbon coating, in addition to the mineral scale dissolution. Furthermore, the specific type of iron sulfide polymorph(s) will dictate how successful any given dissolver is. There are five classes of dissolver for iron sulfide:

- *Acids*, such as mineral and organic acids, are the most used fluid, with hydrochloric acid being the most common within these. Some polymorphs of iron sulfide (pyrite and marcasite) have been shown as sparingly soluble in HCl, whereas others (those with lower stoichiometric amount of sulfur) are more soluble.[85] Common to all acid dissolvers is that their action on sulfide scales yields the release of copious amounts of H_2S gas, and that they are all highly corrosive to the metallurgy of the systems into which they are deployed.
- *Oxidizers* such as hydrogen peroxide and chlorine dioxide avoid some of the drawbacks of acid application (*e.g.*, release of H_2S) but bring their own challenges including deployment issues, corrosivity, and production of elemental sulfur.[86]
- Chelants have been discussed in the context of inhibition, when used at lower concentrations than the neat forms used for dissolution of scales. Their use has been well documents and includes EDTA, DTPA, hydroxyethylethylenediaminetriacetic acid, NTA, and/or GLDA.[87] Compared to mineral and organic acids, they perform poorly and require a high pH for full efficacy realization.
- *Acrolein* (CH_2CHCHO) is a colorless liquid with a piercing, acrid smell, and while is a challenging material to handle, it can be an effective iron sulfide dissolver.[88] Acrolein is not just a dissolver of sulfide scale, but it will also act as an H_2S scavenger and biocide, thus imparting a useful multifunctionality. One advantage of acrolein is that it can be continuously injected at low (ppm) concentrations to slowly remove the deposit of sulfide scale, but it is less effective against troilite and better against sulfur-rich forms of iron sulfide.[85]
- THPS has previously been mentioned in the control section and has found application as a dissolver chemistry for iron sulfide scales long ago, as this is a very effective chemistry. The mechanism of function is via chelation and efficacy is majorly enhanced when ammonium ions are present due to the formation of an intermediate complex.[86]

Addition of other synergistic additives such as phosphonate scale inhibitors and other scale inhibitor polymers has also been reported in the literature.[30, 87] THPS products are very often of a low pH (as the THP^+ molecule is only stable at a pH < 3), and therefore, liberation of H_2S can still be a challenge with this product class.

Zinc and lead sulfides are a separate narrative to iron sulfides, as they are the most insoluble and challenging to remove scales in the oilfield. Many of the solutions that work effectively enough for iron sulfide only work sparingly. The use of HCl or acetic acid form the basis of the commonly used products and have been reported multiple times in the literature ranging from a case history in the Gulf of Mexico to lead sulfide in the North Sea.[88]

14.7 Summary

In conclusion, it is understood that sulfide mineral scales in the oil and gas industry concern insoluble inorganic salts of iron, zinc, and lead. These sulfide scales can form almost anywhere in the oil and gas system, ranging from the reservoir, downhole perforations, tubulars, through the process, and into the pipeline system. Sulfide scales are dependent on the correct concentration of metal ions and sulfur and the correct conditions of pH and Eh, so that the sulfur atom is in the correct speciation state.

The different types of sulfide minerals that form have been reviewed, including an in-depth review of the many different polymorphs of iron sulfide. A comparison of mechanical and chemical methods for control (both prevention and removal) yields the use of chemistry as the most common primary methods employed. There is still a need for innovation in this area, in particular for high-efficacy products that can be delivered and perform downhole.

Nomenclature

AMPS	2-Acrylamido-2-methyl-1-propanesulfonic acid
AMU	Atomic mass unit
DTPA	Diethylenetriaminepentaacetic acid
EDTA	Ethylenediaminetetraacetic acid
Eh	Oxidation-reduction potential
GLDA	Glutamic acid diacetate
HEDTA	Hydroxyethylethylenediaminetriacetic acid
mpy	Mils per year (one mil = one thousandth of an inch)
NTA	Nitrilotriacetic acid
SRB	Sulfate-reducing bacteria
THPS	Tetrakis (hydroxymethyl) phosphonium sulfate
XRD	X-ray diffraction

References

1. C.W. Blount, "Barite Solubilities and Thermo-Dynamic Quantities up to 300°C and 1400 Bars," *American Mineralogist* 62 (1977): pp. 942–957.
2. C. Monnin, "A Thermodynamic Model for the Solubility of Barite and Celestine in Electrolyte Solutions of Seawater to 200°C and to 1 kbar," *Chemical Geology* 153 (1999): pp. 187–209.
3. C. Okocha, K.S. Sorbie, "Scale Prediction for Iron, Zinc and Lead Sulphides and Its Relation to Scale Test Design," SPE-164111-MS, SPE International Symposium on Oilfield Chemistry, held April 8–10, 2013 (The Woodlands, TX, USA).
4. J.L. Przybylinski, "Ferrous Sulfide Solid Formation and Inhibition at Oxidation-Reduction Potentials and Scaling Indices Like Those That Occur in the Oil Field," SPE-80260-MS, SPE International Symposium on Oilfield Chemistry, held February 5–7, 2003 (Houston, TX, USA).
5. E.J. Classen, "Iron Sulfide as a Water-Deposited Scale in Sour Gas Wells," NACE-191, Corrosion Annual Conference, held March 21–25, 1988 (St. Louis, MI, USA).
6. D. Rickard, G.W. Luther, "Chemistry of Iron Sulfides," *Chemistry Reviews* 107 (2007): pp. 514–562.
7. A.R. Lennie, D.J. Vaughan, "Spectroscopic Studies of Iron Sulfide Formation and Phase Relations at Low Temperatures," *Mineral Spectroscopy* 5 (1996): pp. 117–131.
8. M. Wolthers, S. Van der Gaast, D. Rickard, "The Structure of Disordered Mackinawite," *American Mineralogist* 88 (2003): pp. 2007–2015.
9. R.A. Berner, "Tetragonal Iron Sulfide," *Science*, 137 (1962): pp. 669–670.
10. P. Taylor, "The Stereochemistry of Iron Sulfides; A Structural Rationale for the Crystallization of Some Metastable Phases From Aqueous Solution," *American Mineralogist* 65 (1980): pp. 1026–1030.
11. M.A.A. Schoonen, H.L. Barnes, "Reactions Forming Pyrite and Marcasite From Solution: I. Nucleation of FeS_2 Below 100°C," *Geochimica Cosmochimica Acta* 55 (1991): pp. 1495–1504.
12. S. Hunger, L.G. Benning, "Greigite: A True Intermediate on the Polysulfide Pathway to Pyrite," *Geochemical Transactions* 8 (2007): pp. 1–20.
13. G. Qian, F. Xia, J. Brugger, et al., "Replacement of Pyrrhotite by Pyrite and Marcasite Under Hydrothermal Conditions Up to 220°C: An Experimental Study of Reaction Textures and Mechanisms," *American Mineralogist* 96 (2011): pp. 1878–1893.
14. A.R. Lennie, S. Redfern, P.E. Champness, et al., "Transformation of Mackinawite to Greigite: An In Situ X-ray Powder Diffraction and Transmission Electron Microscope Study," *American Mineralogist* 82 (1997): pp. 302–309.
15. A.G. Wickjord, T.E. Rummery, F.E. Doern, et al., "Corrosion and Deposition During the Exposure of Carbon Steel to H_2O-H_2S Solutions at Elevated Pressures," *Corrosion Science* 20 (1980): pp. 651–671.
16. D.J. Vaughan, J.R. Craig, *Mineral Chemistry of Metal Sulfides* (Cambridge: Cambridge University Press, 1978).
17. S. Zheng, L. Liu, C. Zhou, et al., "Effects of H_2S-Containing Corrosive Media on the Crystal Structures of Corrosion Product Films Formed on L360NCS," *Industrial Journal of Electrochemistry Science* 8 (2013): pp. 1434–1442.
18. F.H. Meyer, O.L. Riggs, R.L. McGlasson, et al., "Corrosion Products of Mild Steel in Hydrogen Sulfide Environments," *Corrosion* 14 (1958): pp. 109–115.
19. H.T. Evans, C. Milton, E.C.T. Chao, et al., "Valleriite and the New Iron Sulfide, Mackinawite," US Geological Survey Prof. Paper 475-D, Article 133, pp. 64–69, 1964.
20. C. Milton, "Kansite" = mackinawite, FeS," *Corrosion* 22 (1966): pp. 191–193.
21. R.C. Erd, H.T. Evans, D.H. Richter, "Smythite, A New Iron Sulfide, and Associated Pyrrhotite From Indiana," *American Mineralogist* 42 (1957): pp. 309–333.

22. R.A. McCartney, S. Brice, A. Chalmers, "High Pb and Zn Concentrations in Formation Water From the Culzean Field: Real or Artefact?," SPE-179906-MS, SPE International Oilfield Scale Conference and Exhibition, held May 11–12, 2016 (Aberdeen, UK).
23. J. Kilbane, J.J Wylde, A. Williamson, "Investigation of Microorganisms in a West Texas Oilfield Using Growth and Genetic Testing," SPE-173709-MS, SPE Oilfield Chemical Symposium, held April 12–15, 2015 (The Woodlands, TX, USA).
24. E. Lien, G. Valsvik, J.V. Nordstand, et al., "The SeaRAS AquaSense™ System: Real-Time Monitoring of H_2S at Sub µg/L levels in Recirculating Aquaculture Systems (RAS)," *Frontiers in Marine Science* 9 (2022): p. 894414.
25. S. Baraka-Lokmane, C. Hurtevent, O. Tillemont, et al., "Development and Qualification of New Zinc and Lead Sulphide Scale Inhibitors for Application Under Harsh Conditions," SPE-173761-MS, SPE International Symposium on Oilfield Chemistry, held April 13–15, 2015 (The Woodlands, TX, USA).
26. T.H. Lopez, M. Yuan, D.A. Williamson, et al., "Comparing Efficacy of Scale Inhibitors for Inhibition of Zinc Sulfide and Lead Sulfide Scales," SPE-95097-MS, SPE International Symposium on Oilfield Scale, held May 11–12, 2005 (Aberdeen, UK).
27. G. Darby, D. Uhlig, J. Steckler, et al., "Scale Prevention Application of a Chemical Technology THAT Has Been Effectively and Economically Applied to a 420°F, 13,500-ft-Deep, Gulf of Mexico (GOM) Well to Prevent Lead and Zinc Sulfide Deposition," SPE-124155-MS, SPE Annual Technical Conference and Exhibition, held October 4–7, 2009 (New Orleans, LA, USA).
28. M.S. Kamal, I. Hussein, M. Mahmoud, et al., "Oilfield Scale Formation and Chemical Removal: A Review," *Journal of Petroleum Science and Engineering* 171 (2018): pp. 127–139.
29. H.A. Nasr-El-Din, B.A. Fadhel, A.Y. Al-Humaidan, et al., "An Experimental Study of Removing Iron Sulfide Scale From Well Tubulars," SPE-60205-MS, SPE International Symposium on Oilfield Scale, held January 26–27, 2000 (Aberdeen, UK).
30. J.J. Wylde, I.G. Winning, "Challenges and Solutions Associated with the Development of an Iron Scale Dissolver Chemistry," NACE-04730, Corrosion Annual Conference, held March 28–April 1, 2004 (New Orleans, LA, USA).
31. Q. Wang, J. Leal, I. Syafii, et al., "Iron Sulfide and Removal in Scale Formation Sour Gas Wells," SPE-179869-MS, SPE International Oilfield Scale Conference and Exhibition, held May 11–12, 2016 (Aberdeen, UK).
32. K.O. Temeng, M.J. Sadeg, W.A. Mulhim, "Compositional Grading in the Ghawar Khuff Reservoirs," SPE-49270-MS, SPE Annual Technical Conference and Exhibition, held September 27–30, 1998 (New Orleans, LA, USA).
33. J.J. Wylde, "Sulfide Scale Control in Produced Water Handling and Injection Systems: Best Practices and Global Experience Overview," SPE-169776-MS, SPE Oilfield Scale Conference and Exhibition, held May 14–15, 2014 (Aberdeen, UK).
34. R.M. Baldwin, "Black Powder Control Starts Locally, Works Back To Source," *Pipeline and Gas Industry* 82 (1998): pp. 81–87.
35. S. Balhasan, M. Omar, H. Alhamoudi, "Black Powder Removal From Oil Pipelines," SPE-198105-MS, SPE Kuwait Oil & Gas Show and Conference, held October 13, 2019 (Mishref, Kuwait).
36. T.S. Khan, M.S. Al-Shehhi, S. Stephen, et al., "Characterization and Preliminary Root Cause Identification of Black Powder Content in a Gas Transmission Network—A Case Study," *Journal of Natural Gas Science and Engineering* 27 (2015): pp. 769–775.
37. A.P. Schmidt, "Lead Precipitates From Natural Gas Production Installations," *Journal of Geochemical Exploration* 62 (1998): pp. 193–200.
38. H. Sirnes, T. Gundlach, "Black Powder in Pipeline, Cleaning Program," NACE-2012-1528, Corrosion Annual Conference, held March 11–15, 2012 (Salt Lake City, UT, USA).
39. G. Svenningsen, M. Foss, R. Nyborg, et al., "Laboratory Simulation of Black Powder Formation in Flowing Gas With Water Condensation," NACE-11086, Corrosion Annual Conference, held March 13–17, 2011 (Houston, TX, USA).

40. M. Colahan, D. Young, R.P. Nogueira, "Black Powder Formation by Dewing and Hygroscopic Corrosion Processes," *Journal of Natural Gas Science and Engineering* 56 (2018): pp. 358–367.
41. Q. Wang, T. Chen, "Oilfield Iron Sulfide Scale Formation and Mitigation," in *Water-Formed Mineral Scales and Deposits: Fundamentals and Mitigation Strategies*, eds. Z. Amjad, K. Demadis (New York: Elsevier, 2015), pp. 307–323.
42. O. Yepez, N. Obeyesekere, J.J. Wylde, "Development of novel phosphate based inhibitors effective for oxygen corrosion. Part I: Rotating cage autoclave study," C2015-6163, Corrosion Annual Conference, held March 15–19, 2015 (Dallas, TX, USA).
43. N. Sridhar, D.S. Dunn, A.M. Anderko, et al., "Effects of Water and Gas Compositions on the Internal Corrosion of Gas Pipelines, Modeling and Experimental Studies," *Corrosion* 57 (2001): pp. 221–235.
44. A. Bhardwaj, B. Kumar, S. Prasad, et al., "Characterization of Black Powder in Gas Pipelines," *Materials Performance* 55 (2016): pp. 54–59.
45. R.M. Baldwin, "The Characteristics of Black Powder in Gas Pipelines and How to Combat the Problem," *Corrosion Prevention and Control* 47 (2000): pp. 95–102.
46. A.M. Sherik, B.R. Davis, "Thermodynamics Help Determine Underlying Black Powder Processes," *Oil and Gas Journal* 108 (2010): pp. 35–42.
47. L. Chen, J.J. Wylde, N. Obeyesekere, "The Effect of Surfactant Additions on the Mechanism of Sulfur Corrosion in Aqueous Sour Systems," C2014-3828, Corrosion Annual Conference, held March 9–12, 2014 (San Antonio, TX, USA).
48. A.M. Sherik, S.R. Zaidi, E.V. Tuzan, et al., "Black Powder in Gas Transmission Systems," NACE-11094, Corrosion Annual Conference, held March 13–17, 2008 (New Orleans, LA, USA).
49. N.A. Tsochatzidis, K.E. Maroulis, "Methods Help Remove Black Powder From Gas Pipelines," *Oil and Gas Journal* 105 (2007): pp. 52–58.
50. D.E. Powell, R.H. Winters, M.A. Mercer, *Field Guide for Managing Iron Sulfide (Black Powder) Within Pipelines or Processing Equipment: For Corrosion Control and Operations Personnel* (Houston: NACE International, 2018).
51. H. Tong, P. Zhao, H. Zhang, et al., "Identification and Characterization of Steady and Occluded Water in Drinking Water Distribution Systems," *Chemosphere* 119 (2015): pp. 1141–1147.
52. S. Jia, Y. Tian, W. Zhao, "Field Study on the Characteristics of Scales in Damaged Multi-Material Water Supply Pipelines: Insights Into Heavy Metal and Biological Stability," *Journal of Hazardous Materials* 424 (2022): p. 127324.
53. M. Wicks, "Transport of Solids at Low Concentration in Horizontal Pipes," in *Advances in Solid-Liquid Flow in Pipelines and its Applications*, ed. I. Zandi (New York: Pergamon Press, 1971), pp. 101–124.
54. S. Arrington, "Pipeline Debris Removal Requires Extensive Planning," *Pipeline and Gas Journal* 233 (2006): pp. 61–62.
55. V. Eroini, M.C. Oehler, B.K. Graver, et al., "Investigation of Amorphous Deposits and Potential Corrosion Mechanisms in Offshore Water Injection Systems," NACE-2017-9433, Corrosion Annual Conference, held March 26–30, 2017 (New Orleans, LA, USA).
56. K.A. Al-Jumayiah, "The Impact of Black Powder on Gas Plant Safety Relief Valve Isolation Valve," SPE-164198-MS, SPE Middle East Oil and Gas Show and Conference, held March 10–13, 2013 (Manama, Bahrain).
57. J.P. Perez, "Affect of Black Powder on Inline Inspection Accuracy," NACE-11091, Corrosion Annual Conference, held March 13–17, 2011 (Houston, TX, USA).
58. M. Guzmann, Y. Liu, R. Konrad, et al., "Copolymers as Scale Inhibitors," Patent WO2007/125073A1, 2007.
59. M. Lehmann, F. Firouzkouhi, "A New Chemical Treatment to Inhibit Iron Sulfide Deposition," SPE-114065-MS, SPE International Oilfield Scale Conference, held May 28–29, 2008 (Aberdeen, UK).

60. H.L. Becker, "Polymers for Oilfield Applications," U.S. Patent 2010/0105580, 2009.
61. E. Greyson, J. Manna, S.C. Mehta, "Scale and Corrosion Inhibitors for High Temperature and Pressure Conditions," U.S. Patent 2011/0046023, 2010.
62. D. Emmons, G.R. Chesnut, "Hydroxyethylacrylate/Acrylate Copolymers as Zinc Sulfide Scale Inhibitors," U.S. Patent 4762626, 1988.
63. R.I. Kaplan, "Scale Deposition Inhibitor," U.S. Patent 5171459, 1992.
64. S.L. Fisher, B.D. Mercer, "Iron Sulfide Dispersing Agents," U.S. Patent 5332491, 1994.
65. M Ke, Q. Qu, "Method for Inhibiting or Controlling Inorganic Scale Formations," U.S. Patent 7159655, 2003.
66. M. Lehmann, F. Firouzkouhi, M.P. Squicciarini, et al., "Dispersing Sulfide Scales in Oil and Gas Production Systems," U.S. Patent 8703655, 2014.
67. M. Todd, C. Strachan, G. Moir, et al., "Process for Inhibition of Sulfide Scales," Patent WO2013/152832, 2013.
68. A. Savin, B. Adamson, J.J. Wylde, et al., "Sulfide Scale Control: A High Efficacy Breakthrough Using an Innovative Class of Polymeric Inhibitors," SPE-169777-MS, SPE Oilfield Scale Conference and Exhibition, held May 14–15, 2014 (Aberdeen, UK).
69. J.J. Wylde, C. Okocha, "Composition and Method for Inhibition of Sulfide Scales," U.S. Patent 10633573, 2020.
70. L. Petrozziello, C.W. Kayser, C. Okocha, et al., "Iron Sulfide Inhibitor Suitable for Squeeze Application," U.S. Patent 11466196, 2022.
71. L. Petrozziello, C.W. Kayser, C. Okocha, et al., "Iron Sulfide Scale Inhibition: Squeeze Life Extension Through Improved Interaction Between Scale Inhibitor and Rock," SPE-200709-MS, SPE Oilfield Scale Symposium, held June 24–25, 2020 (Aberdeen, UK).
72. C. Okocha, A. Thornton, T. Chen, et al., "Development of a Squeezable Iron Sulfide Scale Inhibitor for Ultra-Tight Sandstone," AMPP-2022-17896, AMPP Annual Conference, held March 6–10, 2022 (San Antonio Texas, USA).
73. S. Dyer, K. Orski, C. Menezes, et al., "Development of Appropriate Test Methodologies for the Selection and Application of Lead and Zinc Sulphide Inhibitors for the Elgin/Franklin Field," SPE-100627-MS, SPE International Oilfield Scale Symposium, held May 31–June 1, 2006 (Aberdeen, UK).
74. H. Williams, E. Temple, M.M. Jordan, et al., "Selection of a Sulphide Scale Inhibitor for Application in North Sea Fields," TEKNA Oil Field Chemistry Symposium, held March 26–29, 2017 (Geilo, Norway).
75. K. Cattanach, V. Jovancicevic, S. Ramachandran, et al., "Development of New Corrosion Inhibitor to Prevent Black Powder Formation Using Quartz Crystal Microbalance Technique," NACE-11093, Corrosion Annual Conference, held March 13–17, 2011 (Houston, TX, USA).
76. A.M. Sherik, A.L. Lewis, A.H. Rasheed, et al., "Black Powder Inhibitors—Performance Study," NACE-11094, Corrosion Annual Conference, held March 13–17, 2011 (Houston, TX, USA).
77. Z. Belarbi, B. George, N. Moradighadi, et al., "Volatile Corrosion Inhibitor for Prevention of Black Powder in Sales Gas Pipeline," C2018-10962, Corrosion Annual Conference, held June 5–8, 2018 (New Orleans, LA, USA).
78. J.C. Cowan, D.J. Weintritt, *Water-Formed Scale Deposits* (Houston: Gulf Publishing, 1976).
79. H.B. Bakar, I.C. XianLung, M.Z. Nadzri, et al., "Offshore Field Application of a Low Corrosive Fluid Designed for De-Scaling of Well With ESP Completion," SPE-166335-MS, SPE Annual Technical Conference and Exhibition, held September 30–October 2, 2013 (New Orleans, LA, USA).
80. W. Jelinek, K. El-Mabrouk, E. Biezen, et al., "Well Interventions in Depleted Gas Wells Made Economical With Use of Small Size (3/4 in.) Coiled Tubing," SPE-143295-MS, SPE/ICoTA Coiled Tubing & Well Intervention Conference and Exhibition, held April 5–6, 2011 (The Woodlands, TX, USA).

81. S. Kumar, T.K. Naiya, T. Kumar, "Developments in Oilfield Scale Handling Towards Green Technology: A Review," *Journal of Petroleum Science and Engineering* 169 (2018): pp. 428–444.
82. R. Rahman, P.S. Kurniawati, M.N. Jamal, "Cost Effective Long Interval Wellbore Cleanout Using Wireline Technology: Field Application in Peciko Field, East Kalimantan," SPE-196454-MS, SPE/IATMI Asia Pacific Oil & Gas Conference and Exhibition, held October 29–31, 2019 (Bali, Indonesia).
83. A.T. Nikolaos, E.M. Konstantinos, "Methods Help Remove Black Powder From Gas Pipelines," *Oil and Gas Journal* 105 (2007): pp. 52–58.
84. J.J. Wylde, C. Okocha, R. Smith, et al., "Dissolution of Sulfide Scale: A Step Change With a Novel, High Performance, Non-Mineral Acid Chemical," SPE-179880-MS, SPE International Oilfield Scale Conference and Exhibition, held May 11–12, 2016 (Aberdeen, UK).
85. T. Chen, Q. Wang, F. Chang, et al., "Recent Development and Remaining Challenges of Iron Sulfide Scale Mitigation in Sour Gas Wells," IPTC-19315-MS, International Petroleum Technology Conference, held March 26–28, 2019 (Beijing, China).
86. M.A. Mahmoud, M. Kamal, B.S. Bageri, et al., "Removal of Pyrite and Different Types of Iron Sulfide Scales in Oil and Gas Wells Without H_2S Generation," IPTC-18279-MS, International Petroleum Technology Conference, held December 6–9, 2015 (Doha, Qatar).
87. R. Ramanathan, H.A. Nasr-El-Din, "A Comparative Experimental Study of Alternative Iron Sulfide Scale Dissolvers in the Presence of Oilfield Conditions and Evaluation of New Synergists to Aminopolycarboxylic Acids," *SPE Journal* 26 (2021): pp. 693–715.
88. P. Rodgers, B. Lundy, S. Ramachandran, et al., "Multifunctional Chemical for Simultaneous Dissolution of Iron Sulfide, Corrosion Inhibition, and Scale Inhibition," SPE-193619-MS, SPE International Conference on Oilfield Chemistry, held April 8–9, 2019 (Galveston, TX, USA).

15

Halite Scale

Jonathan J. Wylde[1,2]
[1]SPL Inc., Katy, Texas, USA
[2]Heriot-Watt University, Edinburgh, Scotland

15.1 Introduction

Halite (NaCl) scale has been recognized for some time as a flow assurance challenge in the oil and gas industry. It has often been considered an exotic mineral scale, because of its lack of prevalence compared to scales of carbonate and sulfate. In the past 20 years, there has been a steady and marked increase in the occurrence of this scale type, linked to increased production from deeper and hotter gas and gas condensate fields with low water cut, and high-total-dissolved-solids-produced brines.

Halite scale presents much like any other mineral scale and can catastrophically block the flow of hydrocarbon production by plugging downhole tubulars, topside equipment, pumps, valves, and so on. It is (unsurprisingly) associated with high-salinity produced water and the link to high-temperature high-pressure reservoirs stems from the two major driving forces for halite scale precipitation: cooling and depressurization.[1, 2] Halite is different to conventional scale types (carbonates and sulfates of calcium, strontium, and barium), as the driving force for precipitation is often very small, yet the concentration of ions is huge, and therefore, even very small degrees of supersaturation can lead to a resultant scale buildup many times larger in volume.[3]

Halite deposition is challenging to monitor operationally and therefore to predict. Consequently, the literature describes removal techniques as being the norm. The common method of removal is to inject fresh (or low salinity) water either continuously or via batch treatments to the targeted zones. While being effective and easy to apply, this approach is not always the cheapest and most practicable, as freshwater scarcity can give challenges. It is in these occurrences where application of chemical halite inhibitors could be a more valid operational option.

15.2 Halite Scale Form and Occurrence

15.2.1 Halite Mineralogy

The word *halite* derives from the Greek word *halos* meaning *salt*. Halite has several unique identifying features. It has a distinctive cubic crystal habit with other forms exceedingly rare. The occurrence of *rock salt* can also appear massive, granular, or compact. It has a perfect {0 0 1} cleavage, presents a specific gravity of 2.16, and is very soft, just 2-½ on Mohs scale of hardness (same hardness as a human fingernail). Pure NaCl is composed of 39.3% Na and 60.7% Cl but commonly contains impurities of calcium and magnesium chloride and/or sulfates, which can change the color from a clear/white to shades of red, blue, or yellow.[4]

The geological deposition of halite is highly relevant to the oil and gas industry, due to its impermeable nature and its ability to flow under only moderate pressure. As well as being able to flow downward as glaciers (e.g., continental slope of the Gulf of Mexico), salt can rise upward and intrude into overlying formations as a salt dome or diapir. Diapirs can trap hydrocarbon accumulations and are often associated with oil and gas exploration. Salt dome traps are responsible for many large fields being produced today in the Southern United States (and Gulf of Mexico), Brazil, Canada, Angola, Iran, Southern North Sea, Iraq, and Germany. Salt is also able to "self-heal," if fractures develop, which make it ideal for man-made underground storage for hydrocarbons or waste (salt caverns).[5]

15.2.2 Field Occurrence

For many years, halite scale was considered an exotic (or perhaps nonconventional) scale, but with the trend toward field developments in hotter, deeper, higher pressure, gas condensate fields, there has been more industry (and literature) exposure to halite. The earliest literature occurrences for the oil and gas industry available include a patent application[6] and a review article.[7] The study of NaCl is not, however, constrained to the oil and gas industry, and many literature works on the occurrence (and inhibition) of NaCl are available from decades prior,[8–10] and even earlier examples of crystallization experiments using NaCl exist.[11, 12] There is little doubt in this author's mind that halite scale occurred in the oil and gas industry long before these highlighted publications, it just did not make it into the prevalent scientific or patent literature.

Theoretically, halite scale has the potential to form anywhere in the process system that handles substantial amounts of produced water. This statement is made, because temperature and pressure are key drivers for supersaturation, and the changes that oilfield fluids endure from reservoir to the end of the process drive supersaturation. When halite forms downhole, it can rapidly plug the tubulars and even the near wellbore, and while this may be true of any scale, the aggressive speed of growth of halite compared to "conventional" carbonate and sulfate scales can see total well loss in a matter of days.[13, 14] In the case, where gas stripping is the driving force for supersaturation, salt deposition is normally experienced in gas wells with low formation water to gas ratios (<5 standard barrels [stb]/million standard cubic feet [MMscf]). Salt deposition is likely to occur in the near-wellbore and bottom parts of the downhole tubing and a characteristic sawtooth pattern of gas production results. Surveillance can be challenging, as samples taken at the wellhead can be deceivingly low in salt content, due to not only salt deposition downhole but also dilution from condensed water. Where decreases in temperature are the driving force for supersaturation and the formation water to gas ratio is typically >20 stb/MMscf, salt deposition would be expected in the upper reaches of the tubing and perhaps even into the topsides itself, and water samples taken at the wellhead are likely to be saturated with NaCl. An individual well can experience both mechanisms of deposition, during their lifetime as liquid production (and therefore, formation water to gas ratio) increases with well age.[2] When halite forms in the topside (surface) facilities, the most common areas of deposition include level control valves, gauges, pumps, meters, separator bridles, bends, tees, and so on.[15, 16] Halite scale has been also encountered in carbon capture and storage operations caused by the injection of supercritical CO_2, causing a variety of physical and chemical processes causing supersaturation of halite in the saline aquifers.[17] Halite scale rarely forms in isolation as the highly saline brines that they form from often also contain high levels of other monovalent and divalent cations. Coprecipitation of halite with other scales has been described in the literature by numerous authors, and there are even examples, where injection of "fresh" water downhole on the one hand solves the halite issue but at the same time creates a calcite scaling issue due to mixing of otherwise incompatible waters.[1, 16, 18]

15.2.3 Formation Mechanisms

Unlike sulfate and carbonate scales, the driving force for halite scale precipitation is small but is more than compensated by the vast ionic concentration.[3] There are several literature studies that are focused specifically on halite scale precipitation mechanisms and crystal growth,[13] modeling studies,[19] and thermodynamics.[20]

In this review, we shall consider halite scale using classic crystal nucleation and growth theory. The chart in Figure 15.1 is a representation of the supersaturation concept.[21] The chart displays two scenarios, an unseeded and seeded route to crystallization. The *x*- and *y*-axes define increasing concentration of precipitants, so that at the axes-origin solution conditions are undersaturated, defined as a zone, where no crystallization will occur and, in fact, crystals placed in the solution will dissolve. Increasing concentration of the crystallizing components will eventually reach a critical point called saturation. At this exact point, homogeneous nucleation does not occur, and additional crystals added will neither grow nor dissolve. After saturation has been achieved, there are three different zones:

1. Metastable zone: homogeneous nucleation will not occur in conventional time periods, but if seed crystals are added, they will grow.

2. Labile zone: homogeneous nucleation can occur.
3. Precipitation zone: supersaturation is so high that amorphous precipitation can occur.[22]

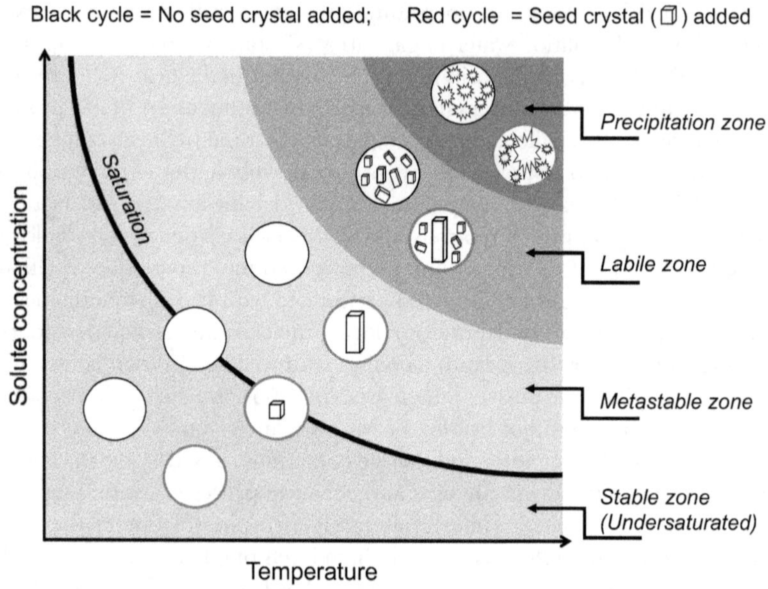

Figure 15.1. Schematic representation of the concept of supersaturation with respect to nucleation and crystal growth.

These fundamental observations show that ideal conditions for crystal growth may not be the same as those for crystal nucleation evidenced by the observation that the latter requires a higher level of super saturation. This explains how increasing energy in the system via agitation (*i.e.*, a flowing system vs. a static system) or increasing temperature can induce nucleation—the likelihood of viable nuclei forming is increased, as aggregation of molecules increases due to more interactions.

Supersaturation of halite in oilfield waters is brought about, because of changing conditions caused by production operations. Sodium chloride is by far the most common salt to be encountered in the oilfield with high-salinity brines of >200,000 mg/L NaCl not uncommon. High-salinity brines have a strong potential for halite to form either by cooling the hot, saturated brines, or evaporation of water caused by depressurization.[23] More specifically, it is the dependence of the solubility product of NaCl on pressure and temperature, and then also the phase behavior of water which can reside in either the gaseous or aqueous phases through evaporation (depressurization) and/or condensation (falling temperatures). The saturation of NaCl in pure water as a function of temperature is plotted in Figure 15.2. Cooling from 100°C to 20°C results in a 4-wt% loss in solubility. Obviously, it is only pure H_2O that undergoes the phase change leaving behind the salt to either concentrate up or become diluted down—the changes to supersaturation can be significant and upward of 30%, depending on the exact operating conditions, see the example in Figure 15.3. A good example of how these supersaturation mechanisms can manifest in the field is provided by Kirk and Dobbs.[24] In mature gas reservoirs where pressures have declined, gas rates were no longer effective at lifting the produced water out of the well. Consequently, gas will bubble through the water in the bottom of the well and carry water vapor to the surface, which leaves a concentration of salt behind. This supersaturates the brine in the bottom of the well causing salt bridging and lost production. Gas composition has a significant impact on the water content in that given gas—this is especially true of the common acid gases CO_2 and H_2S.[25]

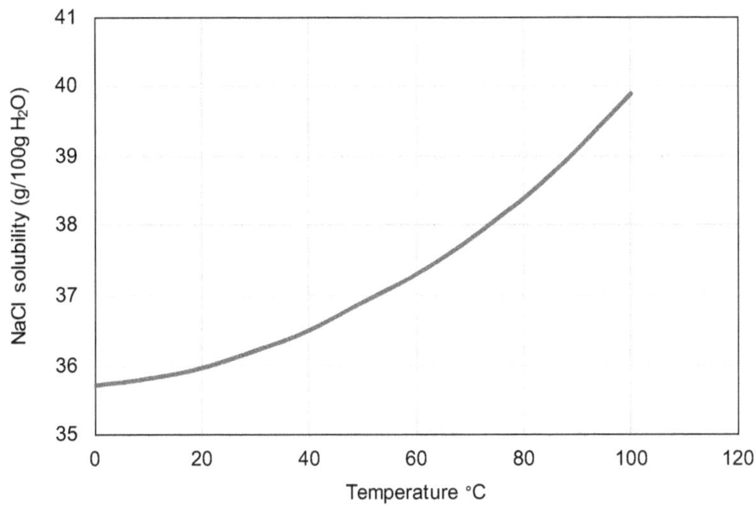

Figure 15.2. Solubility behavior of NaCl in pure water as a function of temperature.

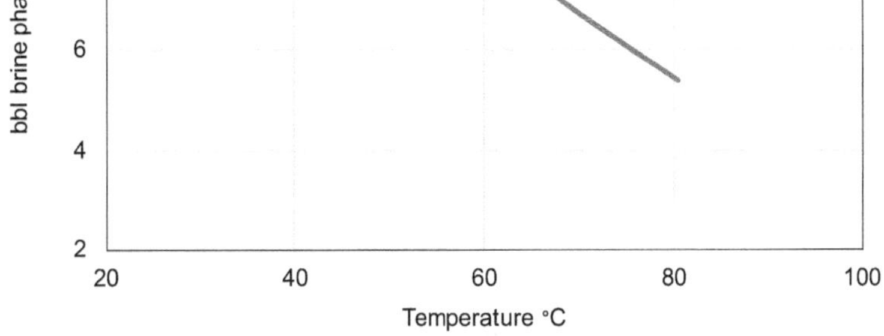

Figure 15.3. Example of volume of water remaining after evaporation into dry gas, assumes 10 bbl of formation water leaving the reservoir at 12,800 psi and 179.5°C. The bubble point is 4000 psi, and 1 MMscf of methane is generated at this point, and the chart defines the remaining water at this pressure.

The importance of seeding in the real oilfield system cannot be understated and is critical to appreciate to garner robust laboratory to field correlations. There are many potential seeding sources in the oilfield, and the major source is the presence of other mineral species. It is, for example, important to consider other salt species in the brines as well as the conditions and way supersaturation. For example, the driving force for supersaturation with respect to halite scale can be a decrease in temperature, yet in the same system, the driver for calcite ($CaCO_3$) scale is depressurization and ultimately an increase in pH. These two drivers of supersaturation are unlikely to occur at the same point in time, and precipitation of one scale can seed the heterogeneous nucleation of the other. Surprisingly, there seems to be little work performed in this area. One study performed a mixed modeling and experimental review of mixed scale formation and showed clearly that coprecipitation is an important effect to account for when modeling and predicting halite formation.[26] Experimentally, it is not clear cut as the driving forces for these different types of scale can be vastly different (*e.g.*, evaporation rate of water vs. cooling effect and which is more prevalent at a given time). Another influence on halite scale formation that appears in the literature often is that of thermodynamic hydrate inhibitors (THI). This is a repeated topic in the literature, because of the shared challenge of hydrates and halite scale that occurs in hotter, gas-condensate field. It has been found that the presence of THIs exacerbates halite scale formation and that generally, ethylene glycol has much less impact on mineral solubility than methanol.[20, 27]

15.3 Halite Scale Prediction

15.3.1 Modeling of Halite Scale

The value that accurate halite prediction provides to project engineers is obvious, however what is not so obvious is how to resolve the challenge of the many competing forces in a complex and "real" oilfield water system such as

1. Evaporation
2. Condensation
3. Phase changes
4. Dynamic flow conditions
5. Temperature changes
6. Pressure changes
7. Coprecipitation of scale
8. Presence of THI

In a field system, many of these parameters are constantly and dynamically changing, while being complexly intertwined with one another. There are a multitude of models (commercial and otherwise) available that have attempted to resolve aspects of this complexity and excellent reviews of their evolution and development are given by various authors.[15, 19] The efficacy of these models depends not only on how they predict the solubility of the salt species in the given brine as a function of temperature, pressure, and all the other species present in the mixture, but also on how well partitioning of species is modeled between the various phases.[28] In particular, the importance of water vaporization rates and the series of complex interactions that make this up are covered as well as entrapment of water that can occur in (for example) depleted gas wells—the volume of studies on this topic alone in the literature is almost overwhelming.[14, 29–31] Indeed, it has been pointed out that one of the major shortcomings is to assume a given model correctly predicts water vapor transfer between phases—previous work has noted deviations between codes.[19] Nuances in predictions began to emerge gaining further accuracy, for example, where the expansion of water due to gas production and consequent decrease in reservoir pressure which in turn leads to an increase in water volume and saturation thus mobilization of water.[32] Models have also been developed to replicate the precipitation of salts in the near wellbore of a water injector due to mixing of different waters, therefore incorporating interactions of different scale species in the same system.[33] As with all scale modeling, equilibrium conditions of homogeneous nucleation are assumed in most cases, and the effect of kinetics is poorly understood. This is especially true of halite scale, where metastable supersaturation is not quantitatively understood.[28]

The fact models rely on laboratory-generated data from a host of experimental methods also gives pause for thought. There are some cases where matching to field conditions is reported but often this is not the case and concerns exist about the true ability of models to predict accurately all field scenarios.[19] Given the ambiguity in the literature when discussing the exact and correct method to reproduce the supersaturation in experimental procedures, one can envisage why there could often be a mismatch with predicted (modeled) halite scale and what ends up occurring in the field. A further shortcoming with halite prediction models is the accuracy of data being input. The degree of supersaturation for halite scale is critical and on a knife-edge; very small changes can make the difference between scale predicted to form or not. Consider for a moment the methods employed to gather preproduction formation water samples and the quantification of the (very high concentrations of) sodium and chloride that are present and all the possible error that can arise. Furthermore, the kinetic aspects of halite scale formation are poorly understood and challenging to therefore quantify—this is especially true, if supersaturation is in the metastable zone.[34]

15.3.2 Field Monitoring: Leading and Lagging Indicators

As is the case for many scales, leading indicators for deposition are often challenging to integrate into a successful field surveillance program, and it is often lagging indicators that prevail. A realistic challenge for an engineer is to get the lagging indicators as close to the commensuration of deposition as possible.

The analysis of produced water for ionic composition is the most common method for halite scale monitoring. The original concept of water analysis trending for ionic composition was first summarized in the 1960s.[35, 36] Brine scaling ions are measured and trended over time with a reduction of a given scaling ion in reference to the formation water composition could indicate a potential scaling event. In the specific case of halite scale, the susceptibility for error is high, because of the very high ionic concentrations of scaling ions in question. Even specific gravity can be used in some instances as an indicator for scaling, such as the density of highly saturated brine. A comprehensive field study[32] from Northern Germany gas reservoirs gave great examples of ion tracking as a useful tool for early detection of halite precipitation. Their work found for the gas wells under study that Na/Li, Na/K, and Na/(Ca + Cl) were demonstrably affected after salt precipitation occurred and may provide indicators for the recognition of salt precipitation. These ideas were derived from the knowledge of natural evaporitic deposits in the Zechstein saliniferous successions.[37]

Other methods that can be used, and shall only be mentioned here, include in-line scale coupons,[38] analysis of suspended solids,[39, 40] infrared thermography,[41] and monitoring of process trends, for example, differential pressures across runs of pipe, level control valve functionality, rate of recycling of pumps, and so on.

15.4 Halite Scale Prevention

There is a rich body of literature resource concerning the practical management of halite scale in the field and an evolution in understanding with practical application methods can be picked out. Given the mechanisms of formation drive toward increased supersaturation and that NaCl is readily soluble in (fresh) water, it was identified a long, long time ago that dilution is absolutely a solution! In this section, we shall review the continuous injection of so called "wash water" for prevention as well as batch application of fresh water for removal operations. Use of chemical inhibitors will be reviewed in Section 15.5.

15.4.1 Nonchemical Methods

Nonchemical (or mechanical) methods are used to limit halite scale and typically are confined to process parameters that can limit supersaturation. For example, limiting drawdown in gas wells is a common method used to limit halite scale deposition or cause it to occur further downstream in a location that may be more convenient to clean. Being able to avoid deposition downhole vs. the ability to have a controlled deposition in surface process equipment would be an attractive prospect to any oil and gas operation. One example of how this was achieved using specialized downhole completions was in Oman, where velocity string installation (tubing size reduction to increase gas flow rate) enabled the liquid loaded wells to continually produce their water and not endure excessive dehydration and supersaturation downhole.[42]

15.4.2 Continuous Injection of Wash Water

The most common way to remediate halite deposition is to circulate fresh water to dissolve the salt. As well as having a sufficiently low salinity, injected *wash water* may need to be further treated to avoid secondary issues such as corrosion and inducing precipitation of other scale species. Addition of corrosion inhibitors, anti-clay swelling agents, scale inhibitors for $CaCO_3$, and oxygen scavengers have all been reported.[16, 26, 43] While this method sounds quick, easy, and cheap—it can in fact get expensive, for example, fresh water might not be readily available and transportation of the large volumes of wash water required along with well downtime and lifting costs due to fluid circulation.[24] For example, in the U.S. Bakken, a case is described, where fresh water (also in this case called *maintenance water*) injection was performed through the annulus and was successful in decreasing workover and intervention frequency, yet significantly contributed to associated water use and disposal costs along with deferred oil production.[44] The regional challenge was such that the number of saltwater disposal wells increased from 292 in 2006 to 487 in 2014, and projections estimated that water demand in the Bakken area will reach 380 million bbl by 2035 (almost the same as the water demand for fracturing up to 426 million bbl).

Often, a conservative approach is taken when adding wash water, and the usage is likely to be significantly higher than required to maintain undersaturated conditions. Optimization of fresh water is challenging for a variety of reasons including lack of reliable reservoir fluid composition data (and rates), inconsistency of produced water composition, and so on.[44] One example of optimization of freshwater treatments is provided from Eastern Siberia, where the field surveillance program specifically acted on the volume of freshwater injection and optimized the freshwater treatment design frequency and production regimes.[45] An excellent and detailed case history of downhole wash water injection being preplanned into a well completion design is available from the Southern North Sea.[46] In this study, three capillary lines were run through a production packer to chemical injection values deployed on a preperforated tailpipe inside the lower completion. Injection of up to 8 bbl/d of freshwater occurred to successfully control halite and was modeled to be 20% more than required to arrest the worst case predicted halite scale.

15.5 Use of Prevention Chemicals

15.5.1 Halite Inhibitors

The discovery that additives can affect the crystal habit of halite crystals was first demonstrated and recorded in the 1700s, where the application of urea (and formamide) enabled the precipitation of octahedral halite crystals vs. the usual cubic form.[11] These initial observations were enriched over the many years that have followed with experimental quantification, and molecular modeling technique applied to determine that urea preferentially interacts with the 1 1 1 face (capped with chloride ions) compared to the 0 0 1 and 1 1 1 surfaces (capped with sodium ions).[5, 47, 48]

Application of scale inhibitors usually applied for carbonate and sulfate species is nearly always not relevant for halite scale and strong hints of alternate chemistry came initially from the utility of salt anticaking agents. This alternate industry to oil and gas largely forms the basis for early understanding of salt inhibition mechanisms and identification of suitable chemistries. By the middle of the twentieth century, only a handful of inhibitor agents had been identified in the literature and focused on chemistries such as salts of cadmium, antimony, zinc, and manganese and the famous ferrocyanides.[49–52] One study screened over 500 different organic and inorganic compounds and identified sodium hexametaphosphate in combination with aluminum salts which, when in the correct ratio, altered the crystal habit of halite during evaporation experiments.[8] Other notable studies (outside of the oil and gas industry) identified chemistries such as polysaccharides, carrageenan, sodium alginate, sodium dodecyl sulfate, polyethylene glycol, and nitrilotriacetic acid.[53–55] None of these chemicals, however, have been reported as used in the oil and gas industry for this purpose. There are two classes of chemical that come-up time and time again in the oil and gas literature: ferrocyanide-based products and nitrilotrialkanamides. Discovering the origins of the two classes of chemical has proven somewhat elusive but has been attempted below.

15.5.1.1 Ferrocyanide (Hexacyanoferrate) Salts

This is the name given to the anion $[Fe\{CN\}_6]^{4-}$ (see Figure 15.4)—these are not to be confused with the ferricyanide anion $[Fe\{CN\}_6]^{3-}$ which is formed from Fe(III) and not Fe(II). There are a multitude of salts of the ferrocyanide ion, but two are of particular interest in oil and gas: sodium ferrocyanide and potassium ferrocyanide. Noteworthy is these species are considered much less toxic than other cyanide salts, as they tend not to release free cyanide and for instance, ferrocyanides are among the multiple United Nations approved anticaking agents for table salt and animal feeds—sodium ferrocyanide (anticaking agent 535), potassium ferrocyanide (anticaking agent 536), and calcium ferrocyanide (anticaking agent 538) are all in use today.[56, 57]

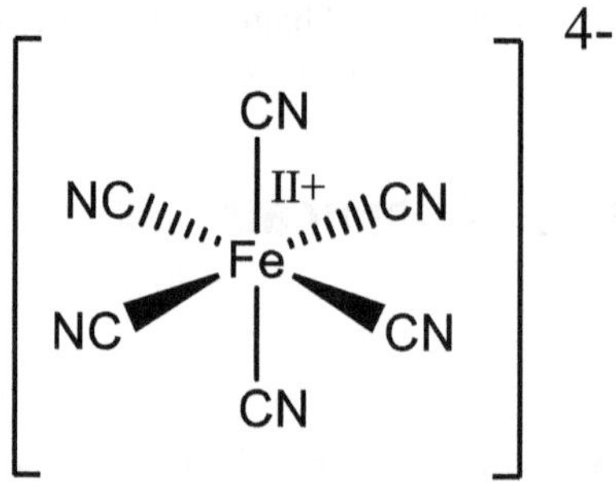

Figure 15.4. Ferrocyanide ion structure.

The use of ferrocyanide chemistry was imported into the oil industry, and early usage can be traced as far back as the 1930s, with the preparation of ammonium ferrocyanide from Prussian blue (or royal blue) treated with ammonium hydroxide[58]:

$$Fe_4\left(Fe(CN)_6\right)_3 + 12NH_4OH \rightarrow 4Fe(OH)_3 + 3(NH_4)_4\,Fe(CN)_6 \qquad (15.1)$$

Application of ferrocyanides as a general anticaking agent was patented in the 1950s and 1960s,[50, 52] with a flurry of turnkey and detailed studies following.[9, 59, 60] Application in the oil and gas industry appears to have first been reported in the 1980s (although one assumes it had been applied in oil and gas long before) in drilling mud application[7] and in production as late as the 2000s, where excellent detail is provided on the performance and field monitoring.[61] In the same field study, it was noted that performance in the field (25 ppm) was lower than in the laboratory (10 ppm) and was caused by chemical incompatibility with high levels of iron present in the produced water (or due to excessive liquid phase evaporation). During the three-month-long field trial, halite precipitation was not completely suppressed but equipment performance increased that would have otherwise quickly failed. Potassium ferrocyanide has been shown by other workers to be poisoned by ferrous ions leading to secondary deposition of insoluble reaction products; formulations have been reported that incorporate suitable chelating agents such as citric acid or nitrilotriacetic acid to prevent this.[10, 23, 62] Since these early and historical applications, the use of ferrocyanide has been reported in the literature and is often included as a standard benchmark in halite scale studies.[13, 63, 64]

15.5.1.2 Nitrilotrialkanamides and Quaternary Salts

There are two species from this class of chemistry that have utility in halite control (see Figure 15.5); both nitrilotriacetamide and nitrilotripropionamide which can be expressed as $N[(CH_2)_nCONH_2]_3$, with $n = 1$ and 2, respectively,[60] as well as their salts which have been claimed in the oil and gas patent literature since the 1960s.[6, 65, 66] In general, the application of NTAA seems to be similar in extent to that of ferrocyanide-based products—although this is a general assumption in the experience of this author. NTAA seems to be a more modern approach in the oil and gas industry, when compared to ferrocyanide, despite being patented originally in the 1960s, where 1–300 parts of product were claimed to be effective,[6] which has been corroborated by a field-based study which suggested 40–400 ppm of a (presumed) NTAA-based product was effective.[24]

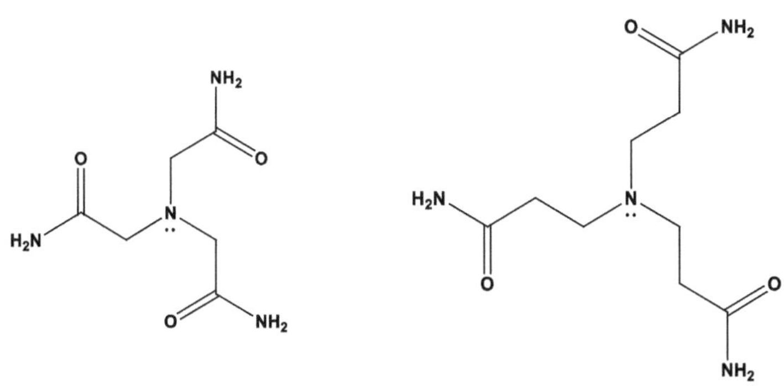

Figure 15.5. Structures of nitrilotriacetamide (left) and nitrilotripropionamide (right).

NTAA has been widely reported to have successfully inhibited halite formation, and while considered by some to be expensive to manufacture, is high temperature-stable with a relatively favorable biodegradation profile.[6, 13, 65] Despite being reported extensively as a salt control agent, the use of NTAA was only patented for application in oil and gas wells decades later.[66] NTAA has also been reported as a component in multifunctional formulations in combination with gas well deliquification foamer chemistries[24] and (presumably) used by the same institution for downhole reservoir application.[67]

15.5.1.3 Other Chemistries for Halite Control

There are a multitude of other chemistries that have been used to control halite beyond the two benchmark chemistries. Ferrocyanide- and NTAA-based products were the mainstay until midway through the 2000s, when alternative chemistries began to emerge in the (industry) literature. For instance, the application of carboxylic acid-based polymers, quaternary amine-based polymers, and polymers and phosphonate blends were described in a series of field case studies.[40] Unfortunately, for this example, and many like it, the exact identity of the chemistries used is (understandably) redacted, which only allows for hints of functional group structure-activity determination. One study from outside the oil and gas industry disclosed a great deal of chemical information on a variety of monomers and polymers ranging from polyvinyl alcohol, polyacrylic acid, polyacrylamide, and so on.[68] This work is somehow corroborated by oil and gas researchers, and quaternary amine-based, polyamide-based, and carboxylic acid-based products are described.[69] Patent literature is also an excellent resource to garner more insights into the actual chemistry used in some of these studies. For instance, small amides such as urea, methyl urea, methyl formamide, methyl acetamide, formamide in combination with metal salts selected from cadmium chloride, cadmium fluoride, cadmium bromide, cadmium iodide, and combinations thereof are claimed and likely pertain to the work published by the same authors around the same time.[13, 70] The application of sodium polyaspartate chemistry is well known to those skilled in the art of halite scale control. Polymers that have a molecular weight of between 6,000 AMU and 8,000 AMU have particular application to halite scale. The use of these chemistries is not directly mentioned in the literature, but the environmental profile described in one case history from the U.S. Rockies is likely to be describing this class of chemistry.[16] The application of sodium alkyl naphthalene sulfonate has been disclosed in another case history along with other chemicals tested such as acrylic copolymers, sulfonated polycarboxylates, and polyacrylic acid homopolymer.[64] The use of high-molecular-weight polymers was corroborated by separate workers, and a carboxylate-sulfonate copolymer has been described as a field deployed product after screening various polyacrylamide, polycarboxylate, polysulfonate, and their copolymers.[71] The use of sodium dodecyl sulfate and polyethylene glycol has been investigated and suggested as salt inhibitors and that 400 ppm of the former was required for halite control.[55] Finally, useful hints and perhaps the widest screening of different products available in the literature are present in the work of industry sponsored consortia, where over 40 different chemistries, presumably containing the best commercial products from all the usual large suppliers of chemistry and members of the consortia.[34, 72] The chemistries tested in these works are unfortunately only generically described and include classes of polymer such as aspartate, carboxylate (second largest family represented), sulfonate, phosphonate, carboxylate/sulfonate copolymer (largest family represented), carboxylate/phosphonate copolymer, carboxylate/sulfonate/phosphonate copolymer, and various polymeric species with unknown functional groups. The best products identified were all carboxylate/sulfonate copolymers, and all extended scale induction time significantly and slowed down the formation of scale. A ferrocyanide benchmark was tough to beat performing very well, and several of the commercial products tested struggled with calcium compatibility, especially at high temperature. This led to mixing of products to identify synergies that existed between various combinations of chemistry.

15.5.2 Mechanisms of Halite Inhibition

The mechanistic explanation of halite scale inhibition is similar to any crystal nucleation and/or growth inhibition mechanism, and several literature works specifically deal with this topic. Halite scale inhibition has been studied for quite some time as the manipulation of NaCl stretches far beyond the oil and gas industry and the action of common chemicals such as ferrocyanate and NTAA has long been known.[73, 74] In general terms, these studies explained the inhibitive properties of halite control additives as growth inhibitors, because of the obvious crystal morphology changes brought about in their presence. The inhibitors were postulated to absorb to the crystals, perhaps altering the surface electrical charge. This monolayer absorption theory cannot account for the crystal habit changes observed, or the supersaturation limits that can be exceeded which is more akin to a nucleation type mechanism. For this, a more elaborate heterogeneous nucleation inhibition mechanism was proposed.[17]

Later workers described a more complex, multifaceted mechanism for halite inhibition which involved three aspects[63]:

- *Inhibition of nucleation and growth*: experimentation found that the size of halite crystals formed in the presence of inhibitors was smaller than those in the absence of inhibitor showing growth inhibition causes more and smaller crystals.
- *Crystal distortion*: this involves an increase in nucleation stability as the inhibitor adsorbs into the growing nuclei and changes the crystal structure during growth. The modified halite crystal can allow different adhesion to a surface as well as a different agglomeration ability.
- *Dispersion*: explained as an indirect inhibition mechanism and is the ability of the inhibitor to reduce agglomeration and settling. It is argued that by including relatively large, polymeric inhibitor species in the crystalline lattice, scale can be prevented from developing further into larger crystals or adhering to surfaces. Important to understand is that in this mechanism, inhibitors do nothing to prevent nucleation and growth of halite crystals, and in static (jar) testing, products can appear to be poor performers when compared to dynamic test methods.[69]

One additional study yielded useful insights into the application of polysaccharides (carrageenan and sodium alginate) to the inhibition of halite.[53] In this study, the morphology of growing halite crystals was carefully studied under controlled conditions of evaporation; the authors speculated that adsorption of the polysaccharides occurred on the growing crystal edges to present a different (dislocation) growth pattern to uninhibited cases. Consequently, they reported heterogeneous and secondary nucleation processes were strongly affected. In another study, it was supposed the most likely mechanism for effective prevention of halite was nucleation inhibition, although the same work also observed crystal morphology modification proving that halite inhibitors work via multiple mechanisms.[13] Three new compounds were studied (two inorganic and one organic nitrogen-based product) and benchmarked against NTAA and potassium ferrocyanide. All products saw a change in halite crystal morphology from cubic crystals to much smaller (dispersed) crystals with a variety of shapes; the new chemistries showed improved static inhibition efficiency against

NTAA and potassium ferrocyanide (although this was reported as potassium ferricyanide throughout this work but has been assumed incorrectly labeled).

The influence that divalent cations (Ca^{2+}, Mg^{2+}, Zn^{2+}) have on the efficacy of various halite inhibitors has been studied and reported.[3] Mixed results were seen; Ca^{2+} enhanced the activity of an inorganic inhibitor yet decreased the efficacy of an organic inhibitor. The effect reported for Mg^{2+} was negligible on all inhibitors tested, while Zn^{2+} had its own inhibitory influence. Unfortunately, no detail was provided on the exact chemical nature of the (formulated and commercial) inhibitors tested.

15.5.3 Chemical Application Methods

There are, as with most scale inhibitor chemicals, a limited number of options to deliver the product to where it is required; the exact treatment delivery method depends on the configuration of the plant or well. Injection into surface operations is a straightforward concept and when compared to injection into a well is obviously much preferred. Unlike with conventional sulfate and carbonate scales, with halite scale it is not always (theoretically) necessary to inject the chemical upstream of the point of supersaturation. Practically speaking, it is advisable to do so as the kinetics of precipitation are challenging to accurately predict. Treatment of halite deposition downhole is a little more challenging and, frankly, is one of the most common places for halite scale to form, given this is the location where the most cooling and depressurization typically occur. One of the most effective methods is to deliver the chemical continuously downhole via a capillary. These are not always available, and chemical can be delivered down the annulus of a well with a flush of fresh or produced water to slipstream the chemical into the produced fluids, or sometimes, even applied with continuously injected wash water to reduce the requirement.[16, 64] Sometimes, even this is not possible, and chemical can be injected into the reservoir via a squeeze treatment or even deployed as part of a hydraulic fracture job prior to completing the well.[24, 40, 75] For hydraulic fracturing, even more engineered approaches are available; one such example is halite inhibitor infused into a porous ceramic proppant and used as part of the stimulation treatment.[76]

15.6 Halite Scale Testing

There are many methods that have been historically reported in the literature for testing halite scale growth and inhibition—almost to the point where there seems to be no standard method available. The ambiguity, and frankly, sometimes confusion, exists in understanding the true driving force for supersaturation in each field application. Previously in this chapter, we dealt with the various causes of supersaturation for halite scale. Temperature decreases and pressure changes induce phase changes (evaporation or condensation) and gas to water ratio and degree of water saturation of the gas are all highly relevant to consider and replicate in laboratory testing. One final point is that halite inhibitor application in the field can work best when products are injected into fluids that have already achieved supersaturation with respect to NaCl. It stands to reason that laboratory testing needs to exactly replicate the way supersaturation occurs in the field, because this is highly likely to affect different inhibitor products in specific ways. Some workers have argued that inducing supersaturation outside of what is occurring in the field and not embracing the multiple driving forces in testing is likely to result in unrepresentative conclusions.[28] For example, when only NaCl is used and other species are omitted or when they are introduced in unrepresentative amounts (*e.g.*, $LiCl_2$ addition), or omission of Fe^{2+} which allows for (easier) aerobic testing but could miss out important compatibility concerns with inhibitor chemistries. Further, it is suggested that excessive cooling to achieve supersaturation often occurs in the laboratory, further deviating from field reality. Finally, testing is often performed at ambient pressures (especially in static methods). Setting aside the manner that supersaturation is replicated for a moment, there are three broad types of tests reported in the literature: static (jar) tests, dynamic (loop) tests, and kinetic turbidity tests (KTTs).

15.6.1 Static Testing

Static jar testing is a widespread methodology used to evaluate halite crystal growth and inhibition with a plethora of references available.[7, 13] The method involves supersaturating a given brine in a closed environment of a static vessel in the absence and presence of halite inhibitors, and then comparing the relative abundance and form of halite scale that develops. Many researchers have used cooling to induce supersaturation.[61, 64, 71] The simplest method is to create a high-temperature (>75°C) solution of >35 g NaCl to 100 g of deionized water and cool to room temperature, after which the resultant crystals can be isolated, dried, weighed, and examined. This method however may not be representative of the supersaturation conditions occurring in the field; for instance, the coupled effects of temperature and time have every bearing upon the kinetics of precipitation, and if not replicated, exactly can lead to false conclusions.[61] This was addressed very recently in work that utilized very strict thermal conditions and control of temperature to investigate the effects this had.[77] Other researchers have achieved supersaturation by evaporating water from the saturated solutions; this however has limitations with regard to the practical levels of supersaturation achievable.[10] To reach higher levels of supersaturation, an alternative method uses mixing of incompatible brines, for example, a saturated NaCl solution mixed with a solution containing $CaCl_2$ plus NaCl.[3, 61] As well as gravimetrically determining the efficacy of halite inhibition in these tests, it is possible to observe halite crystal morphology modification thus enabling a determination of how inhibitors affect the size, shape, growth patterns, and number of crystals forming in an inhibited and noninhibited environment.[13] Another methodology reported the use of methanol addition to drive supersaturation combined with the use of ultrasonic baths to introduce energy into the system to induce precipitation, borne from the observation that shaking an NaCl brine saturated to the metastable zone can induce nucleation.[69] Addition of methanol induces supersaturation as mineral scales are well-known to be less soluble in brine and cosolvent mixtures.[78] No real explanation is provided as to why this method is better than others, although excellent reproducibility was reported.

15.6.2 Dynamic Testing

Testing for halite in flowing scenarios requires the use of a dynamic scale (flow) rig and is perhaps even more confusing than static techniques, in that no standard methodology exists. Principally, the equipment used for dynamic halite scale testing is the same as that used for sulfate and calcite scales.[63] The key to dynamic methods, as is the case with the corresponding static testing, is the way supersaturation is brought about and how closely this replicates what is occurring in the field.

In its simplest form, dynamic testing involves pumping a heated, saturated NaCl brine through a lower temperature capillary, while measuring the differential pressure in the presence and absence of halite inhibitor chemistries. As we have learned previously, there are several different methods to induce supersaturation of NaCl in a brine, and the same methods can be employed in dynamic testing. For example, the addition of other chloride salts to a nearly saturated NaCl brine has been reported.[34, 64, 69] Typically, this comprises pumping two or more fluids in a dynamic flow loop containing a first brine of nearly saturated NaCl solution and the second a $CaCl_2$, $LiCl_2$, or $MgCl_2$ brine. An optional third brine can contain nearly saturated NaCl solution (same ionic strength as the first brine) and halite scale inhibitor. The ratio of the brines can be varied, so that the scale inhibitor is stepwise reduced to determine the minimum effective dose required to inhibit scale formation in the loop.[71] The use of methanol to induce supersaturation in dynamic tests is reported (alongside the equivalent static tests using the same supersaturation method).[69]

Some workers have attempted to better replicate conditions in the field and have used the co-injection of dry gas and NaCl-laden brine in a dynamic scale rig to induce supersaturation.[15, 26, 28] The principal function of this technique is not that much different to that of a conventional scaling rig except that gas and brine are co-injected through a microbore scaling coil, and the differential pressure is monitored. The ratio of gas to brine was increased, as the test continued (which increases evaporation and therefore supersaturation in the brine) and at some point reached a critical point, where precipitation occurred and blocked the microbore coil. Overall, this approach found limited chemical performance, even at a low supersaturation; many chemicals did not perform <1,000 mg/L, which seems out of synchronization with other testing methods and field observations of performance. The authors themselves critiqued their methods and suggested improvements to bridge the lab to field correlation gap further, such as the use of smaller step changes in brine injection to examine saturation ratios closer to unity and to highlight the critical ratio for halite deposition related to the field.[28] Another approach to using evaporation caused by the flow of dry gas aimed to mimic deposition in the reservoir by performing experiments in a sand pack apparatus, where a combination of CO_2 and air was used to condition the pack and dehydrate the saturated brine within it.[15]

15.6.3 Kinetic Turbidity Testing

The KTT was introduced to the oil and gas industry, approximately a decade ago.[79] This recent addition can be thought of as an advanced version of the static jar test, where small (*ca.* 5 mL) water samples are heated and stirred, and a ultraviolet beam is continuously used to measure small changes in turbidity. To conduct a test, a cuvette (custom vial to fit in the heating block) with deionized water is used to establish a turbidity baseline. The anionic brine components are added to fresh cuvettes along with various concentrations of inhibited anionic brine. The test begins on addition of the cationic brine to each cuvette. Readings are taken by the spectrophotometer every 2 min, with a 5-s measurement time per cuvette over a 2-h period.[18] Two functions were described for the application of KTT to halite scale: a kinetic function and a thermal function. The kinetic application method evaluates absorbance over time to determine the rate of reaction, that is, as scale forms, the solution becomes more turbid. In the presence of scale inhibitors, the rate of turbidity development and scale growth are affected and give an indication to performance. The thermal function compares absorbance with temperature change to determine the scale formation temperature. The thermal function can help simulate field conditions more accurately and provide more realistic information on scale inhibition performance. Other researchers using this method have claimed that it is possible to detect scaling events related to both nucleation and growth, therefore bridging the historic static and dynamic techniques. The use of KTT has been specifically reported for evaluating halite scale.[75, 80]

15.6.4 Test Method Summary

So which test method is the correct one and most representative of the field? Like many aspects of production chemistry, the answer is not clear cut or straightforward. In one study by Chen *et al.*,[63] a dynamic halite loop test was developed and compared to traditional static cool down methodology. They reported inconsistencies between the two methods and concluded that the dynamic loop test was more representative of field conditions. When identical brines were tested, the dynamic loop resulted in scale, whereas the jar tests did not, until they were vigorously shaken, and crystallization spontaneously occurred. It was speculated that in the dynamic test, due to the increased energy provided by the flowing system, supersaturation moved from the metastable zone through the labile zone and ultimately into precipitation. The answer to the initial question is that to truly replicate the field conditions, one must first understand exactly how deposition is occurring in the field, then to design, and implement the correct experiment that mimics the correct combination of supersaturation driving forces. The most successful lab to field correlations will use this approach, and thus lower the risk of failure, when it comes to ultimate field trial and application.

15.7 Halite Removal

The major removal method of halite has already been established as the relatively "easy" deployment of fresh water to dissolve the salt blockages. In the same way as continuous injection of wash water is effective, the batch application of fresh, or low salinity, water is the most common method of removing halite scale. This is a very efficient approach but, at the same time, might not be entirely practical, as large volumes of fresh water may be required and unavailable, or it may not be possible to introduce the water in the required volume to the affected zones. An excellent review of removal methods and the operations that are required for downhole deployment is provided in the literature.[14] In this study, the relative dissolution rates of pure water, 2% KCl solution, and 2% KCl with 3% butyl glycol solution are provided. Inclusion of 2% KCl was to prevent formation damage (e.g., clay mineral swelling), and 3% butyl glycol was used to decrease interfacial tension between oil and water to facilitate treatment backflow. A further example details the deployment of remediation operations in an offshore scenario, where infield flow lines were becoming blocked by halite scale deposition.[81] In this case history, regular flushing and pigging operations were used to keep the flow lines free and clear of salt. Large amounts of freshwater were pumped into the flow line in the first treatments, and up to 1,500 bbl was used. This was later optimized based on the location of the salt plug by using pressure buildup testing. The optimized operation still took 15 h in total (7 h pumping, 5 h soaking, and 3 h to pig), which while optimized, still represented a significant production downtime and loss.

15.8 Case Histories

15.8.1 Green Multifunctional Scale Inhibitor in the U.S. Rockies

This case history is drawn from a published work that details the development and application of a continuously injected halite inhibitor applied to wash water to control a multitude of scale types, including halite.[16]

A high-salinity brine was being produced by a field in the U.S. Rockies with significant cooling occurring from the reservoir (bottom-hole temperature of 280°F) to surface and further through the process. Halite scale deposited if left unchecked, and the operator had been injecting "transported fresh water" as a standard practice to dilute the produced water. However, as can be seen in Table 15.1, the formation water was very concentrated in calcium ions, and the injection water in bicarbonate ions—therefore, calcite scaling was a challenge created by the mitigation of halite. Calcite occurred downhole in the lift systems (both in electronic submersible pumps and jet pumps) which led to a loss in production and downtime, while the wells were periodically remediated.

Table 15.1: Ionic compositions of formation water and injection water in U.S. Rockies case history.

Ion (mg/L)	Formation Water	"Fresh" Water Injection
Sodium (Na^+)	109,843	800
Potassium (K^+)	8,265	2.5
Magnesium (Mg^{2+})	1,740	0.9
Calcium (Ca^{2+})	24,653	3.8
Strontium (Sr^{2+})	986	0.2
Barium (Ba^{2+})	55	0.2
Iron (Fe^{2+})	3	0.1
Chloride (Cl^-)	202,950	220
Bicarbonate (HCO_3^-)	153	1,342
pH	Not measured	8.7

The operator attempted to reduce the freshwater injection rate which successfully reduced the calcite propensity but saw the return of halite scale deposition downhole with the formation of bridging plugs. There was no injection rate that gave an optimal balance between the two scale types. A different approach was deemed necessary, and the application of chemicals was considered.

Laboratory testing was undertaken to identify a single product that when injected would control both halite and calcite scales, and the work focused almost exclusively on this product. Unfortunately, no detail on the chemistry used is given—it could be a single molecule or a blend of molecules formulated into a single product. The testing performed utilized static jar tests to evaluate the halite scale control, with cooling used as the method for supersaturation; for calcite conventional dynamic scale, tube blocking tests were performed. Halite inhibition efficacy was (only) reported for the multifunctional product at 75 ppm (99.9% halite control, 50 ppm gave only 15%) and for tube blocking tests, a minimum inhibitor concentration (MIC) was 4 ppm—very comparable with other classic chemistries tested such as diethylenetriamine phosphonate, phosphate ester, Vs-Co, and polyphosphinocarboxylic acid.

Excellent details on the field trial are provided, which give good guidance for operators and engineers to successfully control the halite scale challenge. The well in question was producing 150 barrels of water per day (BWPD), 25 barrels of oil per day (BOPD), and 30-thousand standard cubic feet per day (Mscf/d) gas and was rod pumped with no packer installed. Freshwater was introduced downhole via annular (or backside) injection, passed over the perforations, and mixed with produced water. Typically, 130 bbl/d of freshwater was injected, as it was necessary to keep the ratio this high to prevent halite deposition (0.8:1 freshwater to produced water); the presence of calcite scale was confirmed using an environmental scanning electron microscope (ESEM) technique.

Injection of the multifunctional scale inhibitor commenced at a conservative 100 ppm into an unchanged 130-BWPD freshwater injection. After confirming this approach controlled halite and calcite scales, there was a stepwise risk-managed approach implemented. After stabilization of the initial reduction, freshwater injection was reduced to 75% of the original amount, and after six weeks reduced further to 50% of the original amount. No halite scale was discerned downhole (it would appear that no leading indicators were available for surveillance), and calcite scale was kept in check, as determined by regular ESEM analyses.

A second well trial is detailed, where production was 900 BWPD, 163 BOPD, and 219-Mscf/d gas and was completed with a jet pump. Freshwater was injected at 125 bbl/d, and previous attempts to reduce this had led to halite deposition and lost production. Similarly, a stepwise approach was taken to introduce the multifunctional chemical to this well; the same 100 ppm was initially applied and after stabilization, the injected water rate was reduced first to 75%, and then 50% of the original. On this well, due to the reduced water injection rates, and therefore favorable hydrostatics on the well, there was a modest increase in production which more than offset the cost of the chemical program.

15.8.2 Squeezing Halite Inhibitor into High-Temperature Reservoirs

This case history comes from North Texas, in the U.S., and details squeezing of a halite inhibitor into a reservoir that had a temperature of 420°F (215.6°C).[75] The Bald Prairie

Field produced from a variety of tight sand reservoirs. Production was very prolific from the Cotton Valley Sand and Bossier formations, which themselves are characterized by major salt features including salt anticlines, piercement domes, and pillows. This makes the Bossier sands, in particular, overpressured, high temperature with very saline formation water (>25 wt% NaCl).

KTT combined with ultra-high temperature stability testing was used to evaluate three candidate halite scale inhibitors described as:

- Product A: sulfonated polycarboxylate copolymer (Vs-Co) 40% active
- Product B: polyacrylic acid homopolymer 40% active
- Product C: acrylic copolymer 40% active

The blank tests showed significant halite formation within 10 min, and Vs-Co-based Product A showed the best performance in the KTT evaluation but with an exceptionally high MIC of 3,000 ppm, which was attributed to overly harsh conditions being purposely used in the laboratory to differentiate products. Thermal stability testing utilized Fourier-transform infrared spectroscopy to determine any breakdown of the polymer, and results showed an acceptable match of aged and unaged chemicals.

The field application of Product A (40% active Vs-Co) occurred into the ultra-hot well (420°F, 17,460 ft) which produced 20 MMscf/d and 150 BWPD. It was a cased-hole completion, which made this well an excellent candidate for squeeze application. Water washing occurred every 48–72 h to counter the erratic gas production caused by severe halite deposition. The water washes were up to 500 bbl, and an aminopolycarboxylic acid halite inhibitor was added into the wash water. The cost and frequency of these treatments along with the lost production required a more efficient treatment option to be implemented. A multitude of options were considered, including varying the shut-in periods, large washes with larger "halite inhibitor" pill size, reduced gas production rate, and several acidizing and coil tubing treatments. None of these approaches helped, and ultimately halite inhibitor squeeze application was selected. Unfortunately, not many details are provided on the size or application of the treatments, but the results are elaborated in detail. The first treatment extended the retreatment life to over 21 days and was unchanged from the water wash and used 350 bbl of water with just 100 ppm of the new Product A inhibitor added. Approximately, 10 or so subsequent treatments were presented and ultimately led to continuous production for 40 days on average between squeezes, as halite inhibitor concentration was increased to 200 ppm.

15.8.3 Deepwater Halite Scale Deposition in the Gulf of Mexico

This case history concerns a deepwater field flow line in the Mississippi Canyon area of the Gulf of Mexico, in the U.S., operated in 5,600 ft of water, 150 miles southeast of New Orleans. There were a mixture of platform wells (dry tree) and subsea tiebacks (SSTB), which were connected via 6-mile-long 6″ and 10″ flow lines.

One of the SSTBs flowed a commingled production of Well D and Well G fluids, and it is this part of the system that makes up this case history. The Well D wellhead and downhole process trends are displayed in Figure 15.6 and incorporate several important background aspects. Well D was completed with the ability to produce from multiple zones, and from January to April, production was from the Red Sand, and during March, production rates declined sharply as indicated by the dramatic fall in wellhead temperature from 120°F to 85°F and associated decline in downhole pressure. At the beginning of April, production was switched from the Red Sand to the Purple Sand, and rates were restored, even enhanced, as wellhead temperatures edged over 130°F. As can be seen though, after this initial flush of production, rates dropped dramatically, and by May, the well had to be shut-in, pending an intervention. Note, the downhole pressure trend flattens, then rises, and upstream production pressure at the tree (grey) flattens, then eventually falls off trend in line with bottom-hole pressure. The well intervention attempted to remove and replace the choke, but it was unsuccessful, and it was concluded that a blockage had occurred through the choke which restricted flow and prevented its removal. This was followed by multiple attempts to cycle the well and choke during May until the well was shut-in during June.

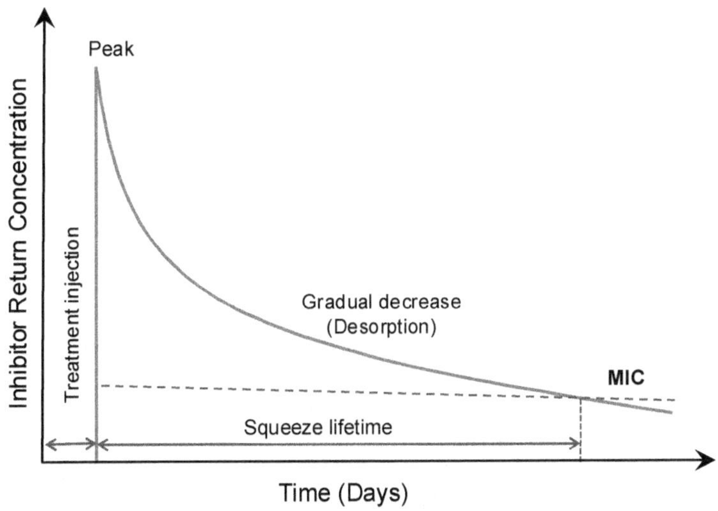

Figure 15.6. Wellhead temperature trends (yellow) and downhole pressure (purple) for a six-month period (January to June) showing declining pressure and temperature indicative of flow restrictions.

What followed was an extensive flow assurance evaluation into determining what the root cause of the blockage was and then how to remediate it and was run in parallel with mechanical assessments for failure. A review of the pressure-volume-temperature (PVT) analysis of Well D fluids indicated that organic deposits would not be an issue, and water analyses cross-referenced to the actual operating conditions did not suggest severe scaling or hydrate formation.

No Well D reservoir water samples had been taken since the original PVT analyses, but there had been many samples on the flowline produced water sample which was a comingle of Well D and Well G. The salinity of the comingled water had been giving a consistent 140,000–190,000 mg/L chloride concentration. The consistency of analyses had led to the assumption that the salinities of the two comingled wells were similar. A produced water sample taken in January has been included in Table 15.2. Scale modeling using the production rates, field conditions, and the comingled water analysis showed only a minor potential for barite scale (<0.1 lb/d).

Table 15.2: Ionic composition of flowline (comingled Well D and Well G) produced water and the calculated most probably Well D produced water in U.S. Gulf of Mexico case history.

Ion	Flowline Produced Water (mg/L)	Well D Produced Water (mg/L)
Sodium (Na^+)	115,000	134,500
Potassium (K^+)	450	525
Magnesium (Mg^{2+})	560	660
Calcium (Ca^{2+})	3,150	3,675
Strontium (Sr^{2+})	160	185
Barium (Ba^{2+})	4.4	5.1
Iron (Fe^{2+})	43	50
Chloride (Cl^-)	185,000	216,000
Sulfate (SO_4^{2-})	300	350
Bicarbonate (HCO_3^-)	120	145

After Well D was shut in due to the inoperable choke, an intensive sampling campaign began on the flowline produced water which now contained only Well G fluids. The findings were startling and suggested much lower chloride concentrations from Well G ranging from 36,000 mg/L to 40,000 mg/L, significantly lower than originally anticipated. This meant that the salinity of Well D must have been much higher and necessitated a revaluation of the scaling scenarios. Based on the estimated production ratios between Well D and Well G, a most probable composition was derived for Well D and has been also placed in Table 15.2. Considering the salinity of the Well G, produced water was about 25% of the combined mixture, the calculations appeared reasonable and appropriate. The chlorides from Well D were now increased to 218,000 mg/L (an approximate 20% increase). When remodeled, halite scale formation was predicted using Well D fluids at an equilibrium deposition rate of 710 lb/d, obviously, ample scale could account for the malfunctioning subsea choke and flowline restrictions (as there was evidence, the restriction occurred between the downhole gauge and the wellhead as the differential pressure increased prior to shutting in Well D, as the flowrates were decreasing).

The best remediation option identified was to deploy fresh water to the scale, but this was made more challenging in the deepwater environment because adding water to a low-temperature, high-pressure system could induce the formation of gas hydrates. While methanol is a better thermodynamic inhibitor for hydrates, it can have a negative impact on scaling tendencies and could restrict the ability of the water to solubilize the deposited halite. For this reason, monoethylene glycol (MEG) was selected as a cosolvent for the remedial halite wash water. Hydrate modeling showed that a 40-wt% MEG mixture with water was sufficient to prevent a hydrate and gives the best chance for successful halite dissolution.

The remediation injected the MEG/water mixture through an umbilical tube and into the wellhead in between the wing and master valves. The upper 1,000 ft of the wellbore was displaced with this mixture, and then the production jumper was displaced followed by a 10- to 12-h soak time. After this, the well was rocked to see, if dissolution had occurred. This operation was repeated five times and upon rocking the well, the choke functionality suddenly was 100% restored, and the steady state flow suggested no restriction remained. A continuous injection of water/glycol was then established to keep the well constantly clear from halite. This approach was very successful and remained in force for many years.

15.9 Final Remarks

Halite scale is an unconventional mineral deposit that requires a specific approach when compared to other common oil and gas scales. The specific approach is required because of the way supersaturation occurs in the field determines not only the depositional features of the scale but also how and what halite control methods and chemicals might work or not. Comparatively speaking, calcite scale can be more easily controlled using low levels of a plethora of suitable chemistries and is somewhat technically commoditized.

The development of chemistries for halite scale seems to not be the issue, there are a wide variety of products that all seemingly provide adequate protection. The challenge, again, is where and when to deploy them in the most cost-effective and -efficient manner—related once again to the kinetics of precipitation and the metastable supersaturation state that can exist for this scale.

The key missing piece in the science of halite precipitation is the understanding of when crystallization will occur after supersaturation, that is, the kinetics of precipitation and perhaps the influence that seeding has upon this. Prediction models largely assume equilibrium and cannot account for the kinetic aspects, whereas experimental techniques can introduce turbulence and therefore energy to induce nucleation, but mimicking the exact way this occurs in the field is challenging.

Acknowledgments

I would like to thank Grahame Taylor for the drawings, Ahmad Afif Abdul-Majid for references and proofreading services, and my wife Sue for all the love and support to whom I dedicate this chapter.

Nomenclature

bbl	Barrels
BHT	Bottom-hole temperature
BOPD	Barrels of oil per day
BPD	Barrels per day
BWPD	Barrels of water per day
CCS	Carbon capture and storage
EDTA	Ethylenediaminetetraacetic acid
ESEM	Environmental scanning electron microscope
FTIR	Fourier-transform infrared spectroscopy
KTT	Kinetic turbidity test
MEG	Monoethylene glycol
MIC	Minimum inhibitor concentration
MMscf	Million standard cubic feet
Mscf	Thousand standard cubic feet
NATT	Nitrilotriacetamide
PPCA	Polyphosphinocarboxylic acid
PVT	Pressure-volume-temperature
stb	Standard barrels
SWD	Saltwater disposal
TDS	Total dissolved solids
THI	Thermodynamic hydrate inhibitor
Vs-Co	Sulfonated polycarboxylate copolymer

References

1. R. Wat, A.M.B. Iversen, Y.H. Belsvik, "The Diagnosis and Decline and Successful Recovery of PI in a 'Dry' HPHT Gas Well That Has Been Affected By In-Situ CaCO3 Scale and Salt Deposition," SPE-130521-MS, SPE International Conference on Oilfield Scale, held May 26–27, 2010 (Aberdeen, UK).
2. Y. Zhang, E. Isaj, "Halite Envelope for Downhole Salt Deposition Prediction and Management," SPE-174206-MS, SPE European Formation Damage Conference and Exhibition, held June 3–5, 2015 (Budapest, Hungary).
3. J.K. Smith, J.L. Przybylinski, "The Effect of Common Brine Constituents on the Efficacy of Halite Precipitation Inhibitors," NACE-06387, Corrosion Annual Conference, held March 12–16, 2006 (San Diego, CA, USA).
4. C. Klein, C.S. Hurlbut Jr., *Manual of Mineralogy*, 21st ed. (New York: John Wiley and Sons Inc., 1993), pp. 681.
5. D. Aquilano, F. Otálora, L. Pastero, et al., "Three Study Cases of Growth Morphology in Minerals: Halite, Calcite and Gypsum," *Progress in Crystal Growth and Characterization of Materials* 62 (2016): pp. 227–251.
6. P.H. Ralston, L.J. Persinski, "Inhibiting Salt Deposition," U.S. Patent 3367416, 1968.

7. S.L. Earl, J.J. Nahm, "Use of Chemical Salt Precipitation Inhibitors to Maintain Supersaturated Salt Muds for Drilling Salt Formations," SPE-10097-MS, Annual Technical Conference and Exhibition, held October 4–7, 1981 (San Antonio, TX, USA).
8. R.S. Ploss, "Sodium Chloride: Modification of Crystal Habit By Chemical Agents," *Science* 144 (1964): pp. 169–170.
9. E.G. Cooke, "The influence of added impurities on the growth and nucleation of sodium chloride," *Kristall und Technik* 1 (1966): pp. 119–126.
10. A. Glasner, M. Zidon, "The Crystallization of NaCl in the Presence of [Fe(CN)6]4– Ions," *Journal of Crystal Growth* 21 (1974): pp. 294-304.
11. J.-B.-L.R. De L'Isle, *Cristallographie ou déscription des formes propres à tous les corps du regne minéral, dans l'état de combinaison saline, pierreuse ou métallique*, 2nd ed. (France: de l'Imprimerie de Monsieur, 1783), pp. 670.
12. O. Lehmann, *Molekularphysik mit Besonderer*, 1st ed. (Leipzig: Verlag von Wilhelm Engelmann, 1888), pp. 697.
13. H. Guan, R. Keatch, C. Benson, et al., "Mechanistic Study of Chemicals Providing Improved Halite Inhibition," SPE-114058-MS, SPE International Oilfield Scale Conference, held May 28–29, 2008 (Aberdeen, Scotland, UK).
14. W. Kleinitz, M. Koehler, G. Dietzsch, "The Precipitation of Salt in Gas Producing Wells," SPE-68953-MS, SPE European Formation Damage Conference, held May 21–22, 2001 (The Hague, Netherlands).
15. Q.T. van Dorp, M. Slijkhuis, P.I.J. Zitha, "Salt Precipitation in Gas Reservoirs," SPE-122140-MS, European Formation Damage Conference, held May 27–29, 2009 (Scheveningen, The Netherlands).
16. K.J. Spicka, M.C. Hirst, G.E. Hochhalter, et al., "Successful Deployment of a Green Multifunctional Scale Inhibitor, A Case History From the Rockies," SPE-153952-MS, International Oilfield Scale Conference, held May 30–31, 2012 (Aberdeen, UK).
17. G. Bacci, A. Korre, S. Durucan, "Experimental Investigation Into Salt Precipitation During CO_2 Injection in Saline Aquifers," *Energy Procedia* 4 (2011): pp. 4450–4456.
18. H. Lu, J. Penkala, J. Brooks, et al., "Novel Laboratory and Field Approaches to Control Halite and Other Problematic Scales in High Salinity Brine From the Bakken," SPE-174788-MS, Annual Technical Conference and Exhibition, held September 28–30, 2015 (Houston, Texas, USA).
19. R. Jasinski, D. Frigo, "The Modelling and Prediction of Halite Scale," IBC International Conference, Advances in Solving Oilfield Scaling, held January 22–23, 1996 (Aberdeen, Scotland, UK).
20. H.J. Shipley, A.T. Kan, G. Fu, et al., "Effect of Hydrate Inhibitors on Calcite, Sulfates, and Halite Scale Formation," SPE-100522-MS, SPE International Oilfield Scale Conference, held May 30–June 1, 2006 (Aberdeen, Scotland, UK).
21. J.R. Luft., G.T. De Titta, "A Method to Produce Microseed Stock for Use in the Crystallization of Biological Macromolecules," *Acta Crystallographica* D55 (1999), pp. 988–993.
22. J. Dalmolen, "Synthesis and Application of New Chiral Amines in Dutch Resolution: Family Behavior in Nucleation Inhibition" (Ph.D. diss., University of Groningen, 2005).
23. M.C. Place Jr., J.T. Smith, "An Unusual Case of Salt Plugging in a High-Pressure Sour Gas Well," SPE-13246-MS, Annual Technical Conference and Exhibition, held September 16–19, 1984 (Houston, Texas, USA).
24. J.W. Kirk, J.B. Dobbs, "A Protocol to Inhibit the Formation of Natrium Chloride Salt Blocks," SPE-74662-MS, SPE Oilfield Scale Symposium, held January 30–31, 2002 (Aberdeen, UK).
25. J.J. Carroll, "The Water Content of Acid Gas and Sour Gas From 100° to 220°F and Pressures to 10,000 PSIA," 81st Annual GPA Convention, held March 11–13, 2002 (Dallas, Texas, USA).
26. N. Goodwin, G.M. Graham, D.M. Frigo, "Halite Deposition—Prediction and Laboratory Evaluation," SPE-179861-MS, International Oilfield Scale Conference and Exhibition, held May 11–12, 2016 (Aberdeen, UK).

27. E. Djamali, H. Lu, A.T. Kan, et al., "Effects of Hydrate Inhibitors on the Solubility of Barite and Halite in Produced Water," SPE-155042-MS, International Oilfield Scale Conference, held May 30–31, 2012 (Aberdeen, UK).
28. T. Tjomsland, O.M. Selle, A.F. Mitchell, et al., "Halite Deposition—Thermodynamic Oversaturation and Chemical Qualification," Oil Field Chemistry Symposium, held March 13–16, 2016 (Geilo, Norway).
29. E. Morin, F. Montel, "Accurate Predictions for the Production of Vaporized Water," SPE-30719-MS, SPE Annual Technical Conference & Exhibition, held October 22–25, 1995 (Dallas, Texas, USA).
30. E. Zuluaga, N.I. Muñoz, G.A. Obando, "An Experimental Study to Evaluate Water Vaporisation and Formation Damage Caused By Dry Gas Flow Through Porous Media," SPE-68335-MS, SPE International Symposium on Oilfield Scale, held January 30–31, 2001 (Aberdeen, UK).
31. R. Tang, Y. Etzion, "Comparative Studies on the Water Evaporation Rate From a Wetted Surface and That From a Free Water Surface," *Building and Environment* 39 (2004): pp. 77–86.
32. W. Kleinitz, G. Dietzsch, M. Koehler, "Halite Scale Formation in Gas-Producing Wells," *Chemical Engineering Research and Design* 81 (2003): pp. 352–358.
33. A.A. Rocha, M. Frydman, S.A.B. da Fontoura, et al., "Numerical Modeling of Salt Precipitation During Produced Water Reinjection," SPE-68336-MS, SPE International Symposium on Oilfield Scale, held January 30–31, 2001 (Aberdeen, UK).
34. S. Paudyal, G. Ruan, J-Y. Lee, et al., "Development of Dynamic Tube Blocking Test Method to Study Halite Scale Deposition and Inhibition," SPE-204389-MS, SPE International Conference on Oilfield Chemistry, held December 6–7, 2021 (The Woodlands, TX, USA).
35. J.M. Tinsley, R.M. Lasater, J.A. Knox, "Design Techniques for Chemical Fracture-Squeeze Treatments," *SPE Journal of Petroleum Technology* 19 (1967): pp. 1493–1499.
36. B.D. Park, J.G. Elkins, S.R. Finklea, "A Single Treatment Controls Scale, Emulsion, and Corrosion From the Producing Formation to the Disposal Sand," SPE-2602-MS, SPE of AIME Fall Meeting, held September 28–October 1, 1969 (Denver, CO, USA).
37. A.B. Carpenter, "Origin and Chemical Evolution of Brines in Sedimentary Basins," SPE-7504-MS, SPE Annual Fall Technical Conference and Exhibition, held October 1–3, 1978 (Houston, TX, USA).
38. O.J. Vetter, "An Evaluation of Scale Inhibitors," *Journal of Petroleum Technology* 24 (1972): pp. 997–1006.
39. E. Delbos, A.R. Thornton, K. Kolnes, et al., "Post-Squeeze Particle Mobilisation in a High Temperature North Sea Chalk Well: A Study Performed via Advanced Particle Classification Scanning Electron Microscopy," Paper 9, 23rd TEKNA International Oil Field Chemistry Symposium, held March 18–21, 2012 (Geilo, Norway).
40. C. Johnston, M. Bhandari, J. Fidoe, "Halite Removal and Control, Extending Production Uptime and Reducing Treatment Cost: Case Studies From Onshore Europe and USA," Oil Field Chemistry Symposium, held March 26–29, 2017 (Geilo, Norway).
41. V.S. Agarwala, P.L. Reed, S. Ahmad, "Corrosion Detection and Monitoring—A Review," NACE-00271, Corrosion Annual Conference, held March 26–31, 2000 (Orlando, FL, USA).
42. A. Al Rashidi, B. Ghaempanah, A. Al Hijri, et al., "Reviving Production From Over 100 Deep Gas Wells Suffering From Halite Deposition in Oman Thru Velocity String," SPE-213501-MS, SPE Middle East Oil, Gas and Geosciences Show, February 19–21, 2023 (Manama, Bahrain).
43. M.A. Kelland, *Production Chemicals for the Oil and Gas Industry*, 2nd ed. (New York: CRC Press, 2014), pp. 454.
44. N. Zhang, D. Schmidt, W. Choi, et al., "Halite Challenges and Mitigation in the Bakken—Experience of Managing High Saline Produced Water From Hydraulically Fractured Wells," SPE-190739-MS, SPE International Oilfield Scale Conference and Exhibition, held June 20–21, 2018 (Aberdeen, UK).

45. A.D. Andryushchenko, "Halite Precipitation in Brine Reservoirs: Prediction and Control By Numerical Model, Optimization of the Fresh Water Treatments and Well Production Regimes," SPE-206645-MS, SPE Russian Petroleum Technology Conference, held October 12–15, 2021 (Virtual).
46. D. Maxwell, K. Young, "Concept, Design and Implementation of a Halite Mitigation Completion in a Gas Well," SPE-169764-MS, SPE International Oilfield Scale Conference and Exhibition, held May 14–15, 2014 (Aberdeen, UK).
47. A.J. Retzel, "The Force of Crystallization of Sodium Chloride and Its Dependence on the Solvent," *Zetschrift fuer Kristollographic und Mineralogie* 49 (1912): pp. 152–192.
48. P.E. Smith, "The Effect of Urea on the Morphology of NaCl Crystals: A Combined Theoretical and Simulation Study," *Fluid Phase Equilibria* 290 (2010): pp. 36–42.
49. H.E. Buckley, *Crystal Growth*, 2nd ed. (New York: Wiley, 1952), pp. 571.
50. W.E. May, T.R. Scott, "Production of Three-Dimensional Dendritic Sodium Chloride Crystals," U.S. Patent 2642335, 1953.
51. D.W. Kaufmann, *Sodium Chloride, Reinhold Publishing Company* (New York, 1960), pp. 743.
52. D.W. Kaufmann, "Method and Means for Inhibiting Freezing of Wetted Granular Materials," U.S. Patent 3090756, 1963.
53. J.D. Birchall, R.J. Davey, "The Crystallization of Sodium Chloride From Aqueous Solution in the Presence of Polysaccharides," *Journal of Crystal Growth* 54 (1981): pp. 323–329.
54. M.A.S. Khan, A. Singh, S. Haldar, et al., "Can Nitrilotriacetic Acid (NTA) Act as a Habit Modifier for Rock Salt Crystals? An Answer From Com putational and Experimental Studies," *Crystal Growth & Design*, 11 (2011): pp. 1675–1682.
55. S.A. Hosseini, M. Asemani, F. Esmaeilzadeh, et al., "Experimental Investigation of Inhibitors Injection to Control Salt Precipitation Using Wetted Wall Column," *Journal of Petroleum Science and Engineering* 122 (2014): pp. 667–674.
56. European Commission Health, Consumer Protection Directorate-General, *Opinion of The Scientific Committee for Animal Nutrition on The Safety of Potassium and Sodium Ferrocyanide Used as Anticaking Agents* (December 3, 2001).
57. P. Kuhnert, "Foods, 3. Food additives," in *Ullman's Encyclopedia of Industrial Chemistry*, ed. C. Ley (New York: John Wiley and Sons, 2016).
58. E.C.H. Davies, "Growth of Salt Flowers on Coal and Other Solids," *Journal of Chemical Education* 11 (1934): p. 409.
59. M.A. van Damme-van Weele, "Influence of Additives on the Growth and Dissolution of Sodium Chloride Crystals" (Ph.D. thesis, Technical University Twente, Enschede, 1965).
60. L. Phoenix, "How Trace Additives Inhibit Caking of Inorganic Salts," *British Chemical Engineering*, 11 (1966): p. 34.
61. D.M. Frigo, L.A. Jackson, S.M. Doran, et al., "Chemical Inhibition of Halite Scaling in Topsides Equipment," SPE-60191-MS, International Oilfield Scale Conference, held January 26–27, 2000 (Aberdeen, UK).
62. N. Zaid, "Well Treating Composition and Method of Use," U.S. Patent 5396958, 1995.
63. T. Chen, H. Montgomerie, P. Chen, et al., "Understanding the Mechanisms of Halite Inhibition and Evaluation of Halite Scale Inhibitor By Static and Dynamic Tests," SPE-121458-MS, SPE International Symposium on Oilfield Chemistry, held April 20–22, 2009 (The Woodlands, TX, USA).
64. J.J. Wylde, J.L. Slayer, "Halite Scale Formation Mechanisms, Removal and Control: A Global Overview of Mechanical, Process and Chemical Strategies," SPE-164081-MS, International Symposium on Oilfield Chemistry, held April 8–10, 2013 (The Woodlands, TX, USA).
65. D.E. Boone, P.H. Ralston, "Stabilized Drilling Mud," U.S. Patent 3544461, 1970.
66. J.W. Kirk, "Methods for the Inhibition of Salt Blockages in Oil and Gas Wells," U.S. Patent 7028776, 2006.

67. S. Szymczak, R. Perkins, M. McBryde, et al., "Salt Free: A Case History of a Chemical Application to Inhibit Salt Formation in a North African Field," SPE-102627-MS, SPE International Symposium on Oilfield Chemistry, held February 28–March 2, 2007 (Houston, TX, USA).
68. E.R. Townsend, W.J.P. Enckevort, J.A.M. Meijer, et al., "Polymer Versus Monomer Action on the Growth and Habit Modification of Sodium Chloride Crystals," *Crystal Growth & Design* 15 (2015): pp. 5375–5381.
69. K. Ho, T. Chen, P. Chen, et al., "Development of Novel Test Methodology to Understand the Mechanisms of Halite Inhibition and Environmentally Acceptable Halite Scale Inhibitors for High Temperature Application," SPE-169803-MS, International Oilfield Scale Conference and Exhibition, held May 14–15, 2014 (Aberdeen, UK).
70. R. Keatch, H. Guan, "Method of Inhibiting Salt Precipitation From Aqueous Streams," U.S. Patent 2011/0024366, 2011.
71. S. Wang, A. Miller, C. Haugen, et al., "Multifunctional Polymeric Scale Inhibitor for High Brine Systems: Development, Implementation and Evaluation," SPE-189246-MS, SPE Symposium: Production Enhancement and Cost Optimisation, held November 7–8, 2017 (Kuala Lumpur, Malaysia).
72. G. Ruan, A.T. Kan, Z. Dai, et al., "Fast and Accurate Screening and Evaluation of Halite Scale Inhibitors By Two Novel Static Bottle Testing Methods," SPE-184593-MS, International Conference on Oilfield Chemistry, held April 3–5, 2017 (Montgomery, TX, USA).
73. S. Sarig, A. Glasner, J.A. Epstein, "Crystal Habit Modifiers: I. The Relationship Between the Structure of the Additive and the Crystal Lattice," *Journal of Crystal Growth* 28 (1975): pp. 295–299.
74. C. Selwitz, E. Doehne, "The Evaluation of Crystallization Modifiers for Controlling Salt Damage to Limestone," *Journal of Cultural Heritage* 3 (2002): pp. 205–216.
75. C. Okocha, S. Wang, A. Kaiser, et al., "Establishing Stable Gas Production By Squeezing a Novel Halite Inhibitor into High Temperature Texas Gas Wells," SPE-190736-MS, SPE International Oilfield Scale Conference and Exhibition, held June 20–21, 2018 (Aberdeen, UK).
76. J. Hebert, J. Leasure, P. Saldungaray, "Prevention of Halite Formation and Deposition in Horizontal Wellbores: A Multi Basin Developmental Study," SPE-181735-MS, SPE Annual Technical Conference and Exhibition, held September 26–28, 2016 (Dubai, UAE).
77. C. Navarathna, C. Leschied, X. Wang, et al., "A Novel Experiment Setup to Model the Effects of Temperature on Halite Scaling and Inhibition," SPE-213849-MS, SPE International Conference on Oilfield Chemistry, held June 28–29, 2023 (The Woodlands, TX, USA).
78. M.B. Tomson, A.T. Kan, G. Fu, et al., "Scale Formation and Prevention in the Presence of Hydrate Inhibitors," SPE-80255-MS, SPE International Symposium on Oilfield Chemistry, held February 5–7, 2003 (The Woodlands, Texas, USA).
79. T.D. Baugh, J. Lee, K. Winters, et al., "A Fast and Information-Rich Test Method for Scale Inhibitor Performance," OTC-23150-MS, Offshore Technology Conference, held April 30–May 3, 2012 (Houston, TX, USA).
80. J. Brooks, H. Lu, M. Barber, "Kinetic Turbidity Test Method for Scale Inhibitor Evaluation on Multifunctional Scales," NACE-2021-16959, Corrosion Annual Conference, held April 19–30, 2021 (Virtual).
81. A.M. Darwish, A.K. Khalil, M. El-Hussein El-Dessouky, et al., "Preventing Halite Scaling in Offshore Pipelines Using Integrated Scale Management System and Modeling—Case Study from Gulf of Suez, Egypt," OTC-31455-MS, Offshore Technology Conference Asia, held March 22–25, 2022 (Kuala Lumpur, Malaysia).

16

Silica and Silicate Deposits

Samridhdi Paudyal[1] and Chunfang Fan[1]
[1]SLB, Houston, TX, USA

16.1 Introduction

Silicon is one of the most abundant elements in the lithosphere and comprises about 25%.[1] Silica, also known as silicon dioxide (SiO_2), is available in nature in the form of mineral quartz and makes up the majority of the earth's crust and sand. Silicon is also one of the major constituents of water which is detected as dissolved silica (SiO_2) and colloidal or particulate silicates (SiO_4^- and SiO_3^{2-}) form.[2, 3] It links up with oxygen to form silica, which is the most common type of Si-bearing minerals (Figure 16.1).

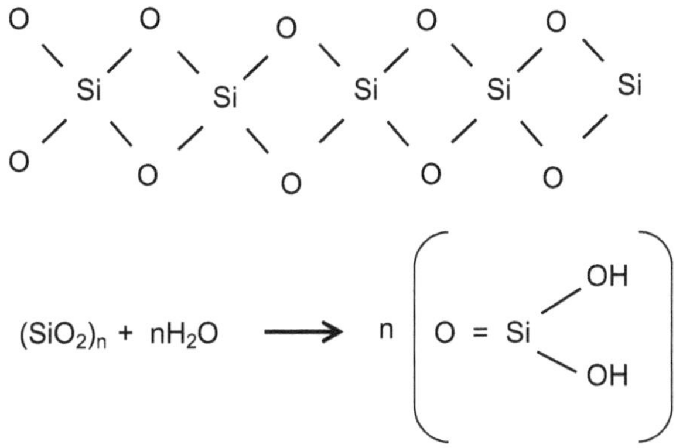

Figure 16.1. Amorphous silica (top) and silicic acid (H_2SiO_3) (bottom) models.

Silica is found as dissolved silica or suspended silicate particles in most natural water, and in crystalline forms as in quartz. Silica concentration in freshwater is commonly below 30 mg/L, but sometimes, it is also recorded to be 120 mg/L.[4] The concentration of the silica can exceed 1,000 mg/L in brines and brackish water.[5] It combines with other oxides and metals in a variety of metal silicates in amorphous form.[6] Soluble silicon dioxide is derived from the dehydration polymerization reaction of silicic acid $Si(OH)_4$.[2] Reversely, in aqueous conditions, silicon dioxide prevalently depolymerizes in water, and then hydrolyzes to form silicic acid either in monomeric form as mono silicic acid or polymerized state as ortho silicic acid. Ortho silicic acid is commonly referred to in the literature as the model (H_4SiO_4) that follows the tetrahedral orientation and predicts very weak acidity, which reasonably explains the formation of soluble silica at higher pH.[3] When the pH of water is greater than 10, silica remains in water-soluble monomeric form, which is referred to as monosilicic acid, and it contains only one silicon atom.[7] When the pH drops, these monomeric silica ions undergo condensation and polymerization processes to form complex silicates by sharing hydroxide ions.[3] Reversely, the dissolution of silicic acid is explained by depolymerization and rehydration processes and is driven by the temperature and pH of the aqueous system. There is 100-fold less dissolution, when the pH of a system is 2–3 than the one with a higher pH of 11. At higher pH, 50% of the mono silicic acid is found in ionized form, compared to 10% at a lower pH of 8.5.[7]

Silica can be classified as a crystalline form as quartz and an amorphous form including dissolved silica, polymerized silica, colloidal silica, and suspended silica. Amorphous silica, a general term for any silica lacking crystal structure or having a lattice-like structure is found in aqueous systems in a reactive state as dissolved or polymerized silica and unreactive state as colloidal or suspended particles.[4] Dissolved silica is referred to as the reactive form of silica, whereas the unreactive form as colloidal silica particles are in a suspended phase of an aqueous system. The solubility of silica is dependent on several factors including pH, temperature, water chemistry, and system-operating conditions. During the polymerization process, the monosilicic acids are transformed into low polymers and further into colloidal particles.[7] Amorphous silica occurs in the metastable form of crystalline quartz. When the silica solubility limit is exceeded, due to the changes in temperature and pH, it polymerizes through the dehydration process. Polymerization by dehydration often accompanies cross-linking reactions and aggregations, which lead to the form of soft gels that gradually turn into hard gels.[4]

Silica and silicates are one type of inorganic scale deposits often found in oil and gas production systems, industrial water systems, and geothermal systems. Inorganic scale deposits are formed, when a brine/water is supersaturated with scaling ions such as Ba^{2+}, Ca^{2+}, Sr^{2+}, SiO_4^-/SiO_3^{2-}, and SO_4^{2-}, the system tends to achieve solubility equilibrium by precipitating these forms of inorganic scale. Several thermodynamic parameters like pH, temperature, pressure, and dissolved ions impact the saturation index of the system.[8] These parameters can impact the solubility limit of the aqueous system of producing wells at a considered condition, and the solubility can be changed due to the natural mixing of the brines of two different zones or artificial mixing of formation water with injected water, implementing well treatment plans such as acidification, alkaline polymer flooding, water-alternating gas-enhanced oil recovery (EOR), and so on.[9]

Common inorganic mineral scales in the aqueous systems are calcium carbonate, species of calcium sulfate, barium sulfate, calcium phosphates, silica, and metal silicates.[4] Out of these mineral scales, silica and metal silicates are one of the major challenges, because of their amorphous nature and the difficulty in preventing their formation. Silica and metal silicates are common in other industries including evaporative cooling systems, semiconductor manufacture, and boilers, besides aforementioned industries as well.[10] The problem is much worse in the U.S. regions mainly in Texas, Arizona, New Mexico, some parts of California, southern parts of Europe, the Pacific Rim, and Latin America.[10, 11] The complexity of the scale increases with the composition of metal ions and their concentration along with other coexisting salt concentrations in the brine and feedwater used from these sources. In this chapter, we will discuss factors that affect the solubility of silica and metal silicates, types of silica, and metal silicates usually encountered in oil and gas production, industrial water, and geothermal systems with some case studies, and common methods that can be applied to evaluate and control these types of scales.

16.2 Silica Scale

Silica precipitation is problematic, even when the silica concentration of water is on the lower side, due to its low solubility in water. The dissolution and degradation of natural silica-containing minerals is the main way to introduce silica in fresh water, and the pH of the water is a primary factor followed by temperature to control silica solubility in water. The solubility of silica in water is minimally altered in the pH range of 6–8. Beyond pH 8.5, the solubility of silica is drastically increased. When the pH exceeds 10.7, all the solid phase of amorphous silica dissolves to form soluble silicate, and no amorphous solid phase remains in equilibrium.[4] Temperature also affects the solubility of silica. Its solubility increases with the increase in temperature, until reaching a plateau around 350°C at vapor saturation pressures.[12] The temperature effect on solubility is responsible for silica deposits being more severe in relatively colder areas of a system.[13] For example, when a brine saturated with silica at 90°C cools to 20°C, the mass of silica precipitated to reach its solubility limit of 120–150 ppm at 20°C could be much higher than a brine saturated with silica at 60°C that cools down to 20°C.[4, 14] This can be a huge problem for any industrial system with high-operating temperatures, where the feedwater contains higher silica than the solubility limit. The solubility of amorphous silica is increased with the temperature, until it reaches 200°C. The silica solubility gradually decreases with the increased temperature beyond 200°C.[12] The solubility of the silica is also impacted by other ions' composition and concentration in the water. Amorphous silica solubility is decreased with increased concentration of the inorganic electrolytes such as $NaHCO_3$, Na_2SO_4, NaCl, KCL, $CaCl_2$, $MgCl_2$, $LiNO_3$, and LiCl, due to the "salting out" effect at lower temperature. However, the process can be reversed to "salting in," when the temperature is increased above 150°C.[15]

Based on the solubility and reactivity, amorphous silica and silicates are categorized into four major states which are dissolved silica or reactive form, polymerized form, colloidal form, and suspended form.[4] Monosilicic acid is the reactive form of silica and is often referred to as orthosilicic acid or silicic acid monomer, and it is a weak tetravalent acid with pK_a of 9.9, 11.8, and 12.[2] Monosilicic acid is in a deionized state at a neutral pH range and remains in a monomeric state for a long period of time at 25°C, when the concentration in aqueous condition remains below the solubility limit. At higher concentrations, monosilicic acid polymerizes to form polysilicic acid (Figure 16.2).

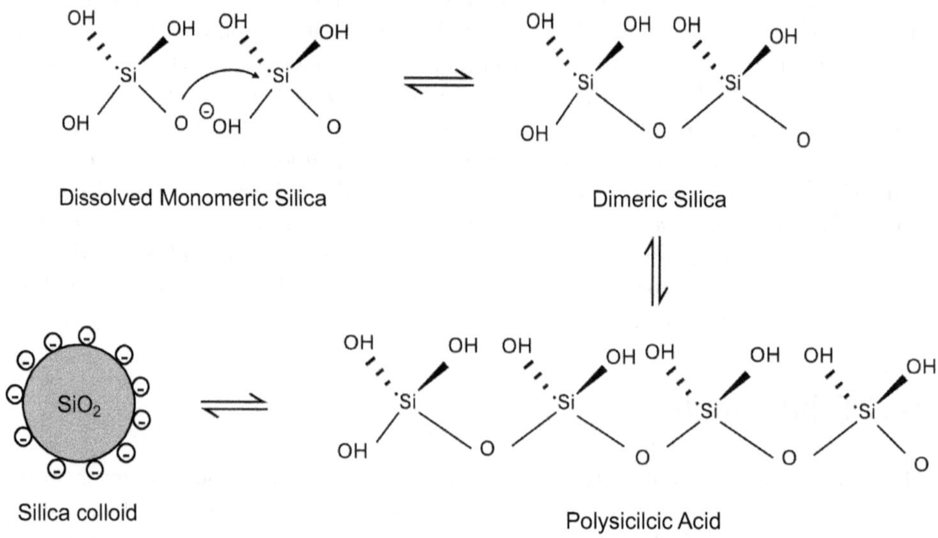

Figure 16.2. Polymerization of monomeric silica to polysilicic acid.

pH is the primary factor affecting the polymerization process and ionization level of silicic acid. At pH of 8.5, silicic acid is ionized by 10%, and at pH 10 by 50%.[7] In alkaline conditions at pH 10 or above, the silicic acid is ionized, fully ionized giving off soluble monomeric silica (H_3SiO_4).[16] The higher the ionization percentage, the lower will be the polymerization rate. The reactive form of silica readily reacts with ammonium molybdate to form a yellow-colored silicomolybdate complex.[4] Due to this readily complex-forming behaviors with silicic acid or silicate in acidic conditions, the ammonium molybdate reaction method is used as an analytical measuring technique to evaluate active silica in the aqueous phase. The silica monomers link to form oligomeric form and eventually grow to form the polymeric colloidal particles with sizes ranging from 10 Å to 50 Å.[16] The silica polymerization process usually occurs at higher pH ranging from 6 to 9 and continues until the aqueous condition reaches an equilibrium level for amorphous silica.

There are two common nucleation mechanisms, homogeneous nucleation and heterogeneous nucleation, generally described for inorganic scale formation which can be applied to silicate scale formation and growth.[17] During the heterogeneous nucleation process, monomeric silica forms bonds with elements on solid surfaces, such as pipe or equipment surfaces. Such polymerization process happens via diffusion of monomeric silica to form a polymeric chain. On the other hand, homogeneous nucleation happens, when silicate polymers are developed in the aqueous phase in a solution as a spontaneous process, when the brine is supersaturated.

16.3 Metal Silicates

Silica and metal silicate scale deposits are distinguished based on their compositions. Pure silica as SiO_2 is the form when metal ions are not incorporated in the particles. In the presence of divalent metal ions in the connate water, amorphous metal silicate scales can form as the divalent ions create bridges between colloidal silicate particles. Metal silicates are found in various forms, such as silica with magnesium (Mg/SiO_2), silica with aluminum (Al/SiO_2), silica with iron (Fe/SiO_2), and silica with zinc (Zn/SiO_2).[6] The polymerization process is strongly affected by the pH and temperature of the brine. For instance, when the pH of the system is higher than 8.5, and the water has a significant amount of magnesium in the form of magnesium hydroxide and silicate ions, magnesium silicate is very likely to occur.[16]

Metal silicates are formed, when the brine contains a high level of metal cations relative to silicon. Silicon-to-metal ratios are commonly found between 4:1 and 7:1. However, 1:1 or 1:2 is also observed in some conditions.[1] The silicic acid preferably interacts with OH groups as offered by many of the metal hydroxides.[18] Silicate scales are commonly observed with mixtures of different metals like Fe, Al, Ca, and Mg, when they are present in the same matrix in their reactive forms as metal hydroxides such as $Fe(OH)_3$, $Al(OH)_3$, $Ca(OH)_2$, and $Mg(OH)_2$.[2] At suitable high pH conditions, these hydroxides interact with SiO_2 and readily participate in the copolymerization resulting in the incorporation of Fe, Al, Ca, and Mg into complex anhydrous silicate structures.[4] When there is metal hydroxide (MOH) on the surface, where M is metal, silicic acid reacts with MOH to form silicate at favorable pH (Figure 16.3) and builds up as a film.[7] The insoluble metal silicate precipitation process follows the order of preferred perceptibility with copper, zinc, manganese, cadmium, lead, nickel, silver, magnesium, and calcium.[2] However, at one given concentration for these metal ions, coprecipitation of metal silicate prefers metal ions such as Mg^{2+}, Ca^{2+}, and Fe^{2+}over others.[4]

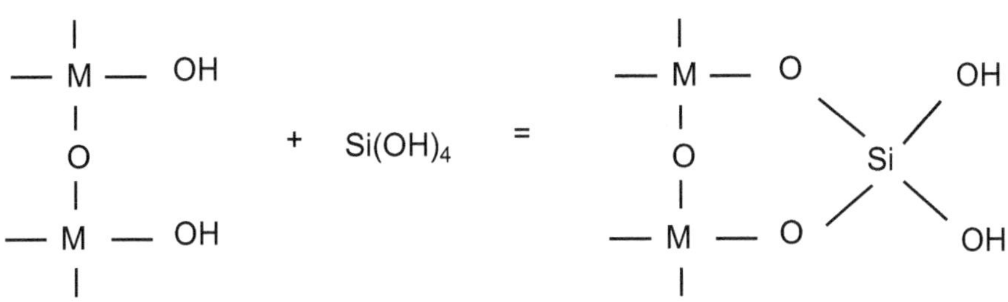

Figure 16.3. Reaction of metal hydroxide with silicic acid to form metal silicates.

16.3.1 Magnesium Silicate

Among the metal silicates, magnesium silicate is the most common one to be found in oil and gas production systems and water purification systems. When the pH is sufficiently reached at a higher level above 8, due to flashing off of noncondensable gases such as H_2S

and CO_2,[19] aqueous magnesium, and silica concentrations in the water are above the saturation limit. The saturated brine with magnesium and silica tends to maintain its equilibrium state leading to the formation of magnesium silicate scale. When the solution contains similar oversaturation with respect to magnesium silicate and brucite ($MgOH_2$), magnesium silicate precipitation occurs more easily than brucite.[16] Concentrations of soluble silica and ionized form of magnesium as magnesium hydroxide are higher in higher pH at 10 or above contributing to the precipitation as magnesium silicate. However, magnesium hydroxide precipitation occurs at a pH higher than 11, eliminating the chances of coprecipitation with magnesium silicate.[16, 20] Though silica solubility is increased with temperature, the magnesium silicate precipitation process is favored by high temperatures above 60°.[21] Once magnesium silicate scales are formed, it is difficult to remove them, due to low solubility in such conditions. The rate of magnesium silicate formation is affected by the presence of calcium carbonate ($CaCO_3$), calcium phosphate, and other metal ions in the matrix such as (Fe^{2+}, Fe^{3+}) or aluminum (Al^{3+}) and their hydroxides.[22] Small amounts of magnesium (~2.5% by weight) ions are also involved in magnesium oxide formation that can coprecipitate with other amorphous forms of metal silicates, such as iron silicate or aluminum silicate. The process is affected by other divalent and monovalent cations such as Mg^{2+}, Ca^{2+}, Na^+, and K^+ in the scale matrix. In most cases, the ratio of magnesium to silicon observed is 1:1 to 1:2.[1]

In industrial systems, magnesium silicate is often found in heat exchangers and condensers of the power plant.[20] Another observation of magnesium silicate formation was made in alkaline-surfactant-polymer (ASP) flooding with a pH of 11 or higher, when monomeric silica ($Si(OH_3)O^-Na^+$) is dissolved, as it passes through the reservoir. When the flood solution is mixed with connate water with neutral pH, the dissolved silica is transformed into colloidal silica along with magnesium via a cross-linking process to form amorphous magnesium silicate scales.[23] Magnesium silicate is also observed in the boilers used in once-through-steam-generation (OTSG). When the feeder water high in silica content is evaporated, there is a high chance of silica/silicate scale precipitation in the OTSG boiler. Silica chemistry is very complicated and can produce three-dimensional polymeric structures.[24] The target pH of the feedwater to the steam generator is set to be from 7 to 11 to mitigate the corrosion issues and keep silica in the soluble form to avoid inorganic scale formation.[25] However, when the silica concentration is higher than 150 ppm,[26] the chance of forming metal silicate is very high. Supersaturation of magnesium silicate formation event has been observed during the mixing of geothermal water, which is high in pH, and silica mixed with other water sources, which has additional magnesium amount resulting in magnesium silicate precipitation.[1]

16.3.2 Iron Silicate

Iron silicate is another metal silicate commonly found in iron-rich water such as geothermal power plants. When the brine rich in calcium, magnesium, and silica encounters iron oxide due to corrosion, precipitation of magnesium iron silicate and calcium magnesium iron silicate is commonly observed.[22] When ionic iron Fe^{3+}/Fe^{2+}, iron oxides, and iron sulfides interact with Si ions in the matrix at a pH higher than 8.5, silica colloids can form through coagulation, flocculation, and/or deposition.[27] The formation of iron silicates is controlled by the concentration of $Fe(OH)_2$ or $Fe(OH)_3$ in the solution. These forms of iron hydroxides react with amorphous silica with the incorporation of magnesium ions to form various forms of iron and iron/magnesium silicates.[28]

For example, iron silicate problems have been observed in steam generators and producer wellhead chokes in the Cymric Field, San Joaquin Valley, California. The scaling problems, especially, the formation of metal silicates, were noticed during steam flooding in the opal-rich siliceous shale reservoirs in this field. The scale consists of iron silicates and complex forms of metal silicates such as $NaFe(SiO_3)_2$ and vermiculite $(Mg, Fe^{2+}, Fe^{3+})_3[(Al, Si)_4O_{10}](OH)_2 \cdot 4H_2O$.[29] In some brines such as those from the Salton Sea which consist of unusually high iron (>500 ppm) that interacts with silicon to form iron silicates, due to the mixing of brines, changes in pH, and/or temperature.[28] Other evidence of iron silicate formation observed in steam generators has been reported. Even though the target pH of 7–11 for the feedwater is maintained to control the corrosion problem,[25] there is still an additional amount of Fe and Mn sourced from the generator tube and potential corrosion by-products in the system. When the SiO_2 concentration in the feedwater from a softener is higher than the recommended level (<150 ppm) to interact with a trace amount of ionic form of iron, the pH favors the iron silicate precipitation.[25, 26] Corrosion and scale formation are common phenomena in geothermal systems. Geothermal fluids contain hot water, steam, and corrosive gases such as ammonia, carbon dioxide, hydrogen sulfide, and methane. Not only do these dissolved gases exacerbate corrosion problems in the geothermal plants but also provide precursors for iron silicate scale deposition in various equipment and parts of the plants, when a geothermal brine is higher in silica amount.[30]

16.3.3 Aluminum Silicate

Clay minerals are a group of hydrous aluminum silicates with a sheet-like structure (phyllosilicates), which adsorb water on their surfaces[31] and combine with silica by occupying unique positions and evenly distributed throughout the metal silicate deposits. The presence of a trace amount of aluminum in a silica gel could make the gel reactive for some hydrocarbon reactions, which otherwise is inactive for such hydrocarbon reactions.[32] Produced water often contains a high amount of aluminum ions and can interact with silica ions in the soluble phase to form aluminum silicates. Zeolites or aluminosilicate gels ($5SiO_2Al_2Na_2O$) are generated by ion exchange properties, where sodium ions are exchangeable.[28] Ion concentrations especially NaCl concentration, temperature, and pH affect aluminum silicate formation. Compared to other metal silicates, the amorphous aluminum silicate scale is preferably formed at high temperatures and is less soluble than amorphous silica. It has been reported in one study that aluminum silicate polymerization and precipitation rates are reduced, when the pH is lower than 6.[33]

16.3.4 Other Metal Silicates

Besides the abovementioned common metal silicates, manganese and zinc silicates are also observed in industrial and geothermal systems, though these scales are rare. Manganese is not found at high concentrations in most geothermal brines. Manganese silicates are generally noticed in trace levels incorporated with silica scales.[1] Zinc silicate issues are found in systems like cooling towers and heat exchangers, where the brass and galvanized steel parts are used.[6]

16.4 Occurrences

The silica level and its solubility as amorphous silica in the feedwater are important factors that need to be considered in various industrial processes that are dependent on groundwater resources. The groundwater source used in industrial applications sometimes contains high silica concentrations (>150 ppm) from the dissolution of quartz of formation rock resulting in an amorphous form of silica and silicate scale precipitation and deposition.[22] The silica and silicate issues in the water resources are common in areas including Texas, New Mexico, Arizona, California, the southern part of Europe, the Pacific Rim and Mexico, and some parts of Latin America.[22, 34] When feedwater with higher silica content is used in an industrial process of water treatment, silica and metal silicate deposits are often observed. The problems are mostly observed in water purification and clarification systems that include cooling water systems (heat exchangers, reverse osmosis [RO] units, and boilers), oil and gas production systems, and geothermal systems.[7, 26]

16.4.1 Water Purification and Clarification Systems

In many of the desalination units of RO systems, silica fouling has been commonly observed as a challenge. Amorphous silica precipitates occur on the surface of RO membranes, when the dissolved level of silica is higher than the equilibrium level at the condition. The silicic acid or polysilicic acid will readily react with metal hydroxide of aluminum and iron to form insoluble metal silicates such as aluminum and iron silicates.[6] By nature, polymerized silicate solutions are viscous and form a gel-like layer on the surface of RO membranes and heat exchangers. Over time, it hardens to form a hard solid deposit that can interfere with heat transfer and fluid flow.[35] Such fouling can occur in the form of bulk precipitation and colloidal or particulate forms which can be related to the onset of any other inorganic scales either as bulk precipitation or slow-growing particles on the surface as deposition. To avoid these problems in RO systems, the industry standard has been set to 120 mg/L for feedwater at 25°C.[7] Analyses of RO membrane foulants indicated the presence of silica on 66% of the membranes analyzed.[7] In one of the cases, the major elements observed were aluminum and iron in the silica deposits, indicating the existence of metal silicates rich in Al and Fe. These elements may be naturally occurring in the feedwater source and/or introduced into the system as coagulants.[27]

The sources of water such as recirculation water or makeup water in cooling water systems may contain silicic acid around 100 mg/L as silicon dioxide, which is below the number (150 mg/L at ambient temperature), as provided by the traditional industrial guidelines to control silica/silicate deposits silica and metal silicate deposits. However, when the water reaches the surface of the high-temperature outlet of heat exchangers, inlet pipes of the heat exchangers, and the surface of the cooling tower fillers, amorphous silicates are found to be deposited due to the change in the saturation level with increased temperature of certain parts. These deposits generally consist of metal elements including aluminum, iron, zinc, and magnesium, with variable chemical compositions.[6] In one case, gritty scale deposition was noticed on the immersed or wetted galvanized or polyvinyl chloride (PVC) and stainless steel surfaces of cooling towers that had morphology similar to magnesium silicate scales. This silicate was found coprecipitated along with calcium carbonate on the surface. Microscopic analyses reveal that these gritty scales are often transparent and glass-like crystals. The elemental composition revealed the scale containing predominantly silicon and calcium followed by iron, magnesium, aluminum, and manganese. However, magnesium is a kinetically favored element among all to form magnesium silicate.[27, 32] These metal silicates (magnesium, calcium, and iron) are also found in low-pressure boilers, when silica is not completely removed with pretreatment processes such as ion exchange and RO.

16.4.2 Oil and Gas Production Systems

In oil and gas production, the major problems regarding silica/silicate scaling are noticed in the Pacific Rim with the volcanic area, which is referred to as the "ring of fire" region. Often of the times, the silica levels in these regions are recorded higher than the silica solubility limit of 50–100 ppm as SiO_2 at 25°C and pH 8.5.[36] There is a high tendency to form silica and silicate scales in these areas, when pH-based EOR technologies such as ASP floods are implemented during oil and gas production. Such scaling problems have also frequently occurred in sandstone reservoirs of the Daqing field of China, Canada, and the Malay Basin of Malaysia.[37, 38] The precipitation or deposition of silicate scales can create near-wellbore formation damage, constriction of flow lines, pump failure, and safety valve and choke failure.[26, 38] In these reservoirs, the flood pH typically attains higher than 11 resulting in the dissolution of monomeric silica. When this pH is reduced after mixing with connate water with neutral pH, the dissolved silica starts forming polymerized colloidal silica. If metal hydroxide ions are present in the matrix, amorphous metal silicates often form. For instance, magnesium ions in water can bridge the colloidal silicate particles forming hard-to-dissolve amorphous magnesium silicate scales. Also, the problem is worsened, when the water is supersaturated with calcite, which acts as a seed to promote the formation and growth of silicates.[37]

The coexistence of silica scale deposition and corrosion events has been reported in the fields, where EOR with steam flooding is performed. When the water used in the steam generation contains silica and high amounts of metal ions like Ca, Mg, and Fe, metal silicates were often observed along with other inorganic scales. Various equipment and flow lines such as boilers and tubes were found fouled by inorganic scale and corrosion issues. Among the inorganic scales deposited, amorphous silica and metal silicates were noticed significantly higher in steam flood boiler systems, when feeding water with high silica concentration was used during the dissolution cycles of EOR. Lab analysis on the precipitates with X-ray diffraction (XRD) and X-ray fluorescence indicated the presence of a significant amount of unidentified amorphous or microcrystalline species. Elemental analyses on these species suggested the presence of iron magnesium silicates that are usually formed at higher pH, when iron hydroxides $Fe(OH)_2$ or $Fe(OH)_3$ reacted with dissolved amorphous silica in the water.[28]

16.4.3 Geothermal Systems

In geothermal systems at 200–350°C, dissolved silica concentration in aquifer waters is generally high around 300–700 ppm. When the water travels from very hot regions to the surface, there will be a huge drop in temperature up to the range of 100–200°C.[18] The resulting supersaturated brine is pushed toward equilibrium by forming amorphous silica or metal silicates. These deposits are often found in power plants and installations, when there is a high concentration of metal ions such as magnesium, calcium, iron, and aluminum in the geothermal brine.[39] Measuring amorphous silica in a metastable phase at geothermal conditions is very difficult. The concentrations of soluble silica vary even, if there is a slight change in parameters such as temperature and pH, which affect the solubility of silica. Most silica solubility studies reported in the literature were performed at temperatures up to 100°C in laboratory conditions, even though the geothermal system represents temperatures higher than 200°C. Due to the lack of data at such high temperatures, the solubility data of silica and metal silicates at higher temperatures are calculated by using multiple linear regression models based on the data obtained at lower temperatures. Some of the work in the geothermal conditions has discussed the prediction of solubility of silica up to 350°C using the mathematical models that utilize aqueous silica as neutral ortho silicic acid H_4SiO_4.[21]

Magnesium silicate is one of the common metal silicates that is found in inorganic geothermal deposits. The metal silicate solubility is dependent on the pH of the brine. Precipitation can happen in the pH ranges of 8.5–10. This range of pH favors optimized protonation of monomeric silicic acid to form silicate ions $[Si(OH)_3O^-]$. The silicate ions interact with Mg^{2+} cations in the system by virtue of electrostatic attraction and form the magnesium silicate.[39] Evidence of the formation of amorphous to poorly crystalline metal silicates was reported in geothermal installations in Iceland.[40] The concentration of magnesium ions in the groundwater from springs and drillholes used in geothermal energy generation is fairly low. When there was a mixing of geothermal brine of pH approximately 9 or above with groundwater containing a fair amount of magnesium, the resulting mixed brine attains high temperature and favorable pH to form metal silicate.[12] In some other fields such as the Fushime geothermal area of the Yamagawa field in the Satsuma peninsula of Japan, SiO_2 concentration measured at atmospheric temperature is reported in the range of 540–1,250 mg/L, and the temperature is recorded higher than 300°C. Formation of magnesium silicate was observed in some of the wells in this field even at neutral pH with a moderate amount of magnesium, when SiO_2 is extremely high.[41] Another common metal silicate that occurs in geothermal systems is iron silicate. Iron silicate was a serious problem for the geothermal field of the Salton Sea that hindered the smooth operation of brine processing. The amorphous form of scales is observed in various parts of surface equipment, when the operating temperature is around 200°C. The scale consisted of a mixture of silica, iron oxide, and alkaline earth oxide.[42]

16.5 Mitigation

Once silica is deposited, primary removal methods that have been widely practiced are mechanical and chemical cleaning and sometimes a combination of both. Common silica scale inhibition and control methods applied are hot brine reinjection, change in pressure, adjustment of brine pH, and use of inhibitors and dispersants.[18, 43]

16.5.1 pH Control

The solubility of the amorphous form of silica and metal silicate is directly affected by the pH of the brine. Silica scales usually form, when the pH is from 8 to 8.5. Metal silicates such as magnesium silicates are formed, when the pH exceeds 8.5 and remains below 10. Above pH 10, silica solubility is increased, and silicic acid remains soluble in the fully ionized form as monomeric silica $H_3SiO_4^-$ aqueous phase. The silica polymerization process reaches optimum levels, when the pH is from 6–9.[16, 35] However, the solubility of the silica scale does not significantly change in a lower pH range of 6–7.[10] Silicic acid is predominantly found in lower pH below 6–7, and the change in thermodynamic parameters such as temperature and pressure would not affect much in the polymerization process. However, the silica solubility and equilibrium concentration shift, when hydroxide-forming metal ions such as iron, magnesium, or aluminum are present in the brine due to shift in pH.[35]

Adjusting the pH of the brine is widely used and one of the most successful methods in controlling these silicate scales. The pH of industrial aqueous systems is often controlled by an acidification process using acids such as sulfuric acid (H_2SO_4), ammonia bifluoride (NH_4HF_2), or hydrofluoric acid.[15, 44] In laboratory testing, two different pH buffers including ammonia with NH_4Cl and NH_3 and boric acid with H_3BO_3 were used to control the pH of the brine.[21, 35] The other study about the dissolution of silica scales reported the process by maintaining an acidic brine environment with solvents containing hydrofluoric acid or alkaline solution using NaOH. The dissolution rate is observed faster in alkaline media and responds in a dose-dependent manner with NaOH.[45] Acidification of the brine slows down the kinetics of silica polymerization. It has been reported that when the pH is as low as 3, the amount of silica-rich scale formed is significantly lower, and this was found to be very effective in controlling iron-rich silicate scales. For instance, ferric silicate scaling in commercial operations at the Salton Sea was controlled by injecting a small amount of hydrochloric acid into a near-wellhead separator, and it was found effective.[42]

16.5.2 Silica Removal by Precipitation

The colloidal form of silica is formed by polycondensation of silicate monomers to form dimers, oligomers, and colloidal silica subsequently via protonation. The process is affected by the concentrations of certain divalent ions including calcium, magnesium, aluminum, and iron. They can function like catalytic agents at neutral to alkaline pH (7–10) to enhance the polymerization rate of silicate ions and promote the formation of metal silicate precipitates. This alkaline condition also promotes the formation of metal hydroxide, which stimulates the silicate polymerization process.[33] Thus, silica precipitation usually involves complex forms with divalent ions rather than SiO_2.

The most common method of soluble and colloidal silica removal through precipitation and filtration is by applying polyvalent metals that include $Fe(OH)_3$, $Al(OH)_3$, and $Mg(OH)_2$.[7] Colloidal forms of silica are highly polymerized species or particles that are larger than 50 Å. They are removed using lime softeners, hot process softener, anion resins, upflow filters with chemical feed, crossflow microfiltration with chemical feed, and ultrafiltration.[7] Iron, aluminum, and calcium ions are poor choices of precipitants for the precipitation method, due to temperature requirements and water quality issues. Coprecipitation with magnesium ions is thermodynamically favored and a conventional method to precipitate silica out of the concerning brine.[3] Either using already precipitated solid magnesium oxide for ex situ or using the ionic form of Mg to precipitate $MgOH_2$ *in situ* are the two options in the precipitation method.

Reactive or dissolved silica can be removed using lime softeners with increased pH and with magnesium hydroxide, hot process softeners, RO, and desilicizers.[7] In this lime-softening method, dissolved silica will be adsorbed on insoluble magnesium hydroxide at a pH of 9–10, when the solubility of the $MgOH_2$ is lowest. Similarly, it can be removed by applying magnesium bicarbonate and sodium aluminate. Sodium aluminate contains sodium oxide and aluminum oxide, which forms Al_2O_3 precipitates, and dissolved silica can adsorb onto it. In addition, it forms OH^- ions in water that promote the formation of $MgOH_2$, an adsorbent for silica as explained in the lime softening method.[46] Desilicizers are demineralizers using a strong cation exchanger in sodium form, followed by a strong base anion exchanger in hydroxide form.[3]

Silica scales in geothermal water are formed, when amorphous silica is supersaturated. Since the amorphous silica formation is kinetically favored by higher temperatures, rapid cooling of brine can be done to achieve less deposition. The process reduces amorphous silica deposition rate by allowing dissolved silica to polymerize. Geothermal fields such as Svartsengi, Namafjall, and Nesjavellir in Iceland and Olakaria in Kenya have implemented this method to control the silica deposition. When the temperature of the brine is instantly dropped down from 160°C to 70°C, events of deposition were not observed in this field. This rapid cooling method was enough to quench the amorphous silica deposition reaction on the equipment surface for most practical purposes. However, mechanistic study on inhibition of deposition rate and pathways has not been extensively done.[40]

16.5.3 Ion Exchange Technologies

RO technology adopts ion exchange technology to remove the nonreactive form of silica. Cellulose acetate and thin film composite materials are used to filter out nonreactive silica at higher pH and are a very effective technology to remove weakly ionized forms of silica.[3]

16.5.4 Chemical Inhibition

One of the common ways to prevent colloidal and amorphous silica is to use inhibitors and chemical additives. Nonpolymeric inhibitors including phosphonates and phosphate esters and polymeric inhibitors such as polyphosphonate, polycarboxylates, acrylic acid- or maleic acid-based homo- and copolymers are commonly used to treat inorganic scales in aqueous conditions.[47] These types of anionic inhibitors can effectively control carbonate, fluorite, fluorspar, and sulfate scales. Anionic additives that include functional groups with carboxylates, phosphonates, or both are also used to control inorganic scales including silica and silicates.[48] However, it is difficult to control the silica scales with such anionic chemistries, because silica and silicates are formed as polymerized amorphous solids. Additionally, since silicates or metal silicates are predominantly formed at higher pH in the range of 8–10, which is often not the favorable pH range for these types of inhibitors, most of the inhibitors' efficiency studies performed within this pH range show limited success in controlling silicate scales. Therefore, controlling silica and silicate precipitation and deposition on equipment surfaces in industrial water systems remains a challenging problem.

Lately, biodegradable products that contribute to a low-carbon footprint are environmentally favored. The dissolution work on silica suggested that some of the natural products including carboxymethyl inulin, phenylalanine, histidine, and maleic acid have mild effects in the dissolution of silica.[48] However, the rate of dissolution is prolonged, and the dosage rate reported to achieve 250 ppm of silica is above 2,500 ppm of the products in 72 h.[48] Comparatively, additives with catecholate moieties (*e.g.*, 3,4-dihydroxybenzoic acid [catechuic acid, DHBA]) showed a faster rate of dissolution under the same testing conditions, and the dissolution rates are dose-dependent.[48] In some geothermal systems, organic inhibitors are used as alternative solutions, because strong acids are corrosive.[49] Another study reported that combining several chemistries—such as phosphonic acid—and nonionic polymers—such as polyvinylpyrrolidone, polyethylene oxide, and polyacrylamide—can show increased inhibition of the polymerization of silicic acid.[50]

Several cationic-based polymers, copolymers, and surfactants are very effective in inhibiting silica polymerization process or dispersing amorphous silica in various temperature ranges including geothermal system conditions.[44, 51, 52] It has been reported that polymers that contain hydroxyl phosphonic acetic acid and copolymers with acrylic acid and allyl hydroxyl propyl sulfonate can control silica deposition on equipment surfaces in industrial water systems such as cooling, desalination, and geothermal.[35] Polyaminoamide-based dendrimers are also effective silica inhibitors for cooling water applications, and the branching of the dendrimers affects their performance. The dendrimers accommodate and stabilize silica inside the structure and prevent silicic acid polymerization. However, they are not very effective in dispersing the silicate particles.[48] The silica deposition was controlled in the heat exchangers in the presence of the formulated product, whereas untreated ones had a large amount of silica and silicate depositions.[48] These dendrimers show a synergistic effect, when they are applied together with biomolecules like carboxymethyl inulin, and can have a dispersant effect on colloidal silica.[48] The formulation contains hydroxyl phosphonoacetic acid and a copolymer of acrylic acid: allyl-hydroxy sulfonate.[53]

For silica and metal silicate control, it is reported high-molecular weight (MW, <20 kDa) polymeric dispersants have been used to keep amorphous silica particles dispersed and reduce the risk of deposition on equipment in oil and gas production and geothermal energy generation systems.[35] Good chemical reagent candidates for silica control should have the efficiency to disperse silica and metal silicates and other foulants such as carbonate and sulfate scales, which may provide the foundation for silicate scale nucleation. These impurities such as inorganic scales, metal ions—such as aluminum, iron, manganese, and zinc—and suspended particles impact the efficiency of the inhibitors or dispersants and act antagonistically. Reduced efficiency of chemicals is reported, when the water is being cotreated with coagulants and flocculants such as ferric chloride or diallyl-dimethyl ammonium chloride providing the source for inorganic flocculants as Al^{3+} and Fe^{3+}.[44]

Another possible way is to control other inorganic scales such as $CaCO_3$ which is one of the inducing factors for silicate nucleation. One study shows that applying inhibitors, such as l-hydroxy ethylidene I-I-diphosphonic acid and polyamino polyether methylene phosphonic acid, that are suitable to control calcite scale increased the recovery rate of both soluble silica and calcium in the system.[13] Commonly available phosphonates or polymers such as polyacrylic acid or polymaleic acid are used to cotreat calcite and silica or silicates in cooling water towers and have MW in the range of 1,000–1,300 Da.[54] Several acrylic acid/hydroxylated lower alkyl-acrylate-based copolymers were patented and claimed to have inhibitory or dispersive effects on magnesium and calcium silicates.[54] The same report has mentioned the copolymers using (meth)acrylic, (meth)acrylamide, and alkoxylated primary alcohol ester of (meth)acrylic acid to control silica and silicates.[27] Some copolymers and terpolymers containing (meth)acrylic or maleic acid that have been reported have a MW from 1,000–25,000 Da indicating that these products have dispersive action.[54]

Alkylsilanes (silanes) are saturated compounds that contain silicon atoms and are used to modify a wide variety of surfaces including silicates by working as cross-linking or dispersing agents to modify polymerization process. These compounds follow the same pathway of silica polymerization by forming Si-O-Si bonds and modify the adhesive properties of the silica and silicate polymers. Similarly, polyethylene glycol (PEG) polymers can interact with silica or silicates by forming a bond between hydroxyl groups (–OH) and their backbone. The PEG polymer chains inhibit the growth of silicate by forming a layer around the silicate nuclei. A study showed that applying PEG and PEG-functionalized silane (silane-PEG) can inhibit the silicate formation.[55] The better performance was observed, when the polymer is longer and MW is above 5 kDa. Combination of PEG 20000 and PEG-silane-5200 showed higher efficiency and acted in a dose-dependent manner.[55]

Most common scale inhibitors are not very effective in controlling silica or silicates due to their amorphous nature. In iron silicates, the ferric silicate scale, which consists of ferric ions (Fe^{3+}), is more insoluble in water than the iron silicate scale containing ferrous (Fe^{2+}) ions. In such cases, applying a reducing agent such as formic acid, sodium formate, potassium iodide, sodium dithionite, thioglycolic acid, and dextrose can reduce trivalent iron to divalent iron in anoxic conditions and has demonstrated good efficiency in controlling ferric silicate.[56] Metal chelators such as ethylenediaminetetraacetic acid (EDTA) are widely known and used in various industrial processes including water purification. This can be one of the options to control metal silicates including iron silicate, calcium silicate, and magnesium silicate problems by limiting the available divalent ions in the water. However, the intactness of the metal-EDTA complex is uncertain at high temperatures and may cause a reversed effect, if the complex decomposes by supplying ferric or ferrous ions to the geothermal system at very high temperatures.[26] Some plant-based biopolymers including lignosulfonates can disperse iron oxide due to the presence of the sulfonate ($-SO_3H$) group that may contribute to controlling iron silicate by limiting available iron in the system for iron silicate formation.[27]

One study showed that molecular modeling can be adopted to predict interfacial interaction between the silica and magnesium silicate polymers and high-MW additives such as polyethylene glycols (PEGs, 1,550–20,000 Da) and polyvinyl alcohols (11,000–31,000 Da).[57] This molecular simulation predicts the interfacial interaction between the polymers. It is predicted that PEGs can stabilize silicic acid by forming hydrogen bond with silanol groups (–Si–OH) of monosilicic acid molecules and ether oxygens of backbone of PEG polymer. Similarly, PVA can interact the same silanol groups with its hydroxyl (–OH) groups on its backbone but cannot contribute to stabilize the silicic acid. The molecular modeling results were further verified by laboratory tests and concluded that PEGs showed good inhibition efficiency. It was also demonstrated that the performance is improved with higher MW polymers, when compared between 1,550 and 10,000 Da.[57]

16.6 Detection and Evaluation Methods

The aqueous amorphous silica remains stable in pH below 7, and thus, neutral pH is the suitable pH condition to measure the solubility at a wide temperature range from 8°C to 310°C. The aqueous silica in this condition is detected as monomeric neutral species H_4SiO_4. Almost 42% of silica has been observed in dimeric form or colloidal silica as $H_6Si_2O_7$ at 25°C. Both the monomeric and dimeric forms are generally considered during solubility studies. Silica gel is being used as one of the options to understand the equilibrium of silica at these temperatures via diffusion. To understand and determine the amorphous or crystalline form of silica, XRD and scanning electron microscope techniques are used. Silica gel analyses with these techniques before and after the solubility tests can confirm the properties of aqueous silica.[12] A case study on a failed steam generator showed that the quantitative XRD pattern was used to analyze various types of minerals in scale deposits inside the tube, when the sample was scanned at an angular range of 5–65°C.[43] Several forms of silica and silicate scales were detected at different layers. The top layer showed a negligible amount of quartz (SiO_2), while the bottom layer had several other forms of silicate and complex metal silicates including quartz (SiO_2) 1.6%, zirconium silicate ($ZrSiO_4$) 1.7%, natrosilite ($Na_2Si_2O_5$) 16%, tremolite ($Ca_2Mg_5Si_8O_{22}(OH)F$) 10.6%, natrolite ($NaAl_4Si_6O_{20}(H_2O)_4$) 5.6%, pectolite ($HNaCa_2Si_3O_9$) 20.6%, and wollastonite ($CaSiO_3$).[26, 43]

Silicate stock solutions and synthetic brines containing silica are generally prepared from sodium metasilicate in polyethylene bottles for static bottle tests to avoid silica contamination from glass-made bottles. Since the pH has a direct impact on the solubility of amorphous silica, polymerization experiments are performed with pH-controlled supersaturated brine.[35] The polymerization and precipitation kinetics of metal silicates are slow processes.

Therefore, a relatively longer reaction time is needed to observe a significant amount of precipitation than the time for other inorganic minerals.

The titration method by adding NaOH solution has been reported to be used to measure the solubility of metal silicate such as magnesium silicate at different pH conditions.[58] When the pH of a brine is gradually increased, the concentrations of magnesium and silica in aqueous form decrease accordingly. The depletion of magnesium is faster than the depletion of silica. Elemental analyses can be achieved using filtered samples through membrane filters with 0.2 μm or smaller pores and running inductively coupled plasma atomic emission spectroscopy (ICP-AES) programs to determine changes in magnesium or silica.[21] Similarly, the EDTA titration method can also be used to determine the concentration of calcium and magnesium in a brine.[35]

To measure silica concentrations, the absorption of spectra for reactive silica in aqueous form is measured using colorimetric techniques with ultraviolet-visible spectrophotometry at a wavelength of 400 nm. In this method, yellow silicomolybdic acid assay is used, which has a shorter reaction time with reactive silica to form a yellow molybdenum silicate complex.[59] The acidification of sample solutions will form a yellow crystalline precipitate that contains molybdate and silicate, which is referred to as molybdosilicate.[55] When molybdosilicate is reduced using reducing agents like hydrazinium sulphate or 1-amino-2-napthol-4-sulphonic acid, a blue color is produced.[59, 60] This blue silicomolybdic assay is preferably run at the wavelength of 810 nm, and it has a relatively longer reaction time (~2 h) to develop color. This method is widely used to determine amorphous silica. The major limitation of this molybdate reactive silica test method is exposed number of silicon atoms in the polymer surface.[36] Only a few silicon atoms are available to react with ammonium molybdate providing partial quantification of the silica concentration in ambient conditions.

Soluble silicic acid that can react with molybdate to form a compound that can be detected using the Hach method with spectrophotometry analyses.[33] Though these are available methods to analyze the silica and silicate scales, they cannot provide information for quantitative analyses on the exact amount of aqueous form of silica, because all the silicon atoms are not exposed from the chain of polymerized silica. This polymerization process gives off "giant silica" or colloidal form of silica, which are typically larger sized particles.[60] These larger particles can be removed through an ion exchange method or using microfilters of 0.1 μm or less. ICP-AES analyses of the samples after filtration (0.1 μm) can provide some quantitative numbers on amorphous silica content to some extent.[12]

16.7 Final Remarks

Silica and metal silicate scales have been observed in various industrial systems such as water purification, geothermal, and oil and gas production systems. They can deposit on the surfaces of equipment, membranes, tubing, valves, and vessels and cause malfunction of equipment. Preventing silica and silicate scales is more complicated than controlling carbonate or sulfate scales.

As discussed in this chapter, silica could exist in reactive form as dissolved or polymerized silica and in unreactive form as colloidal or particulate silica suspension. Characteristically, these deposits are amorphous in nature and polymerize into colloidal particles and larger and heavier flocculates by adding several silicic acid monomers in the chain. The particles initially form as pure silica without metal or silicates with metal ions incorporated in the chain. Since the monomeric form of silica has an affinity toward the –OH moiety of solid surfaces and condenses on surfaces, it can form a complex polymer with metal(s) by interacting with metal hydroxides present. In the presence of the divalent metal ions and their hydroxide forms in connate water or feedwater, amorphous metal silicate scales can be polymerized into colloidal silicate particles by creating a bridge with M-OH. The polymerization process is strongly affected by brine pH and temperature conditions. Divalent metal ions, such as Ca^{2+}, Mg^{2+}, $Fe^{2+/3+}$, Zn^{2+}, and Al^{3+}, monovalent ions such as Na^{+} and K^{+}, and inorganic scales such as $CaCO_3$ and $Ca_3(PO_4)_2$ can affect the complexation of silica and silicates significantly. Although there are many types of silicates, the most common silicates are magnesium silicate, aluminum silicate, iron silicate, calcium silicate, and zinc silicate.

In this chapter, some common practices in controlling silica or silicate scales have also been reviewed. Maintaining the silica or metal ions level and adjusting the pH of the fluid system are often used in water purification and geothermal systems. Another common practice is applying chemical reagents in the form of additives, inhibitors, dispersants, and dissolvers depending upon the feasibility of application processes. Some cationic-based chemicals act as inhibitors preventing silica oligomerization and polymerization and keeping silica soluble in water, while some of them work as dispersants such as chemicals that contain hydroxyl phosphonic acetic acid and copolymers with acrylic acid and allyl hydroxyl propyl sulfonate may prevent agglomeration and flocculation of the particles and control deposition rate on the surface. However, very limited chemistries have shown excellent inhibition efficiency against silica and silicates. Especially, at high temperatures (>200°C), those common inhibitors such as phosphonates, phosphate esters, polyphosphonate, polycarboxylates, acrylic acid- or maleic acid-based homo- and copolymers could not achieve the same level of inhibition efficiency as they do for controlling carbonate and sulfate scales.

Limited studies were performed on preventative and remedial chemical treatments that are cost-effective and feasible to apply in field conditions, in which temperatures are often very high. Even though some chemicals are claimed to perform well at lower temperature ranges, little data have shown that those chemistries can effectively treat silica and silicate issues at high temperatures. Besides chemistries, robust methods are still needed for analytical and chemical evaluation processes in static and dynamic conditions that can simulate field silica/silicate scaling scenarios including geothermal conditions.

Nomenclature

ASP	Alkaline-surfactant-polymer
CMI	Carboxymethyl inulin
Da	Dalton
DHBA	3,4-Dihydroxybenzoic acid
EDTA	Ethylenediaminetetraacetic acid
EOR	Enhanced oil recovery
HEDP	Hydroxyethylidene diphosphonic acid
HIST	Histidine
ICP-AES	Inductively coupled plasma atomic emission spectroscopy
kDa	KiloDalton
MAL	Maleic acid
MOH	Metal hydroxide
MW	Molecular weight
nm	Nanometers
OSTG	Once-through-steam-generation
PAPEMP	Polyamino polyether methylene phosphonic acid
PEG	Polyethylene glycol
PHAL	Phenylalanine
ppm	Parts per million
PVA	Polyvinyl alcohol
QXRD	Quantitative X-ray diffraction
RO	Reverse osmosis
SEM	Scanning electron microscope
WAG	Water-alternating gas
XRD	X-ray diffraction
XRF	X-ray fluorescence
	Angstrom (=0.1 nm)
µm	Micrometers

References

1. M.A. Todd, M.J. Bluemle, "Formation and Mitigation of Mineral Scaling in Geothermal Power Plants," in *Water-Formed Deposits: Fundamentals and Mitigation Strategies,* eds. Z. Amjad, K.D. Demadis (Elsevier, 2022), pp. 269–282.
2. R.Y. Ning, "Discussion of Silica Speciation, Fouling, Control and Maximum Reduction," *Desalination* 151 (2003): pp. 67–73.
3. P. Meyers, "Behavior of Silica: Technologies Available and How They Rate," *Water Conditioning & Purification International Magazine* (March 2004), pp. 22–24.

4. P. Zhul, Z. Zahid, "Solution Chemistry Impact on Silica Polymerization by Inhibitors," in *Mineral Scales in Biological and Industrial Systems*, ed. Z. Amjad (New York: CRC Press, 2014), pp. 173–200.
5. F.N. Kemmer, *The NALCO Water Handbook* (New York: McGraw-Hill Book Company, 1988).
6. I. Nishida, "Silica and Metal Silicate Deposits," in *Water-Formed Deposits: Fundamentals and Mitigation Strategies*, eds. Z. Amjad, K.D. Demadis (Elsevier, 2022), pp. 195–211.
7. R. Sheikholeslami, S. Tan, "Effects of Water Quality on Silica Fouling of Desalination Plants," *Desalination* 126 (1999): pp. 267–280.
8. A.T. Kan, M.B. Tomson, "Scale Prediction for Oil and Gas Production," *SPE Journal* 17 (2012): pp. 362–378.
9. S. Paudyal, Z. Dai, C. Dai, et al., "Introduction of a New BCC Continuous Stirred Tank Reactor Inhibitor BCIn Screening Method for Chemicals in Controlling Inorganic Scales in Oil/Gas Production," SPE-209477-MS, SPE International Oilfield Scale Conference, held May 25–26, 2022 (Aberdeen, Scotland).
10. K.D. Demadis, G. Skordalou, I. Aristodemou, "Distinct Mechanisms of Silica Scale Formation: Silicic acid Polycondensation and Silica Particle Growth," NACE-2020-14290, NACE Annual Conference, held June 14–18, 2020.
11. R.W. Henley, "pH and Silica Scaling Control in Geothermal Field Development," *Geothermics* 12 (1983): pp. 307–321.
12. I. Gunnarsson, S. Arnorsson, "Amorphous Silica Solubility and the Thermodynamic Properties of $H_4SiO°_4$ in the Range of 0° To 350°C At P_{sat}," *Geochimica et Cosmochimica Acta* 64 (2000): pp. 2295–2307.
13. J. Gill, "Silica Scale Control," NACE-98226, NACE Annual Conference, held March 22–27, 1998 (San Diego, California, USA).
14. J.D. Willey, "Effects of Aging on Silica Solubility: A Laboratory Study," *Geochimica et Cosmochimica Acta* 44 (1980): pp. 573–578.
15. M. Azaroual, C. Fouillac, J.M. Matray, "Solubility of Silica Polymorphs in Electrolyte Solutions, II. Activity of Aqueous Silica and Solid Silica Polymorphs in Deep Solutions From the Sedimentary Paris Basin," *Chemical Geology* 140 (1997): pp. 167–179.
16. R.A. Sazali, K.S. Sorbie, L.S. Boak, "The Effect of pH on Silicate Scaling," SPE-174193-MS, SPE European Formation Damage Conference and Exhibition, held June 3–5, 2015 (Budapest, Hundary).
17. G.M. Graham, C.P. McMahon, "The Effect of Scale Inhibitor Performance Against Bulk (Homogeneous) and Surface (Heterogeneous) Scale Nucleation and Growth By the Addition of Film Forming Corrosion Inhibitors," NACE-02315, NACE Annual Conference, held April 7–11, 2002 (Denver, Colorado, USA).
18. M. Al Saadi, N. Al Haddabi, "The Root of Silica Scale Formation and Its Remedy," *Global Journal of Engineering Sciences* 3 (2019): pp. 1–3.
19. K. Lichti, "Materials Selection Challenges for Geothermal Energy Projects," NACE-2017-9258, NACE Annual Conference, held March 26–30, 2017 (New Orleans, Louisiana, USA).
20. P.R. Puckorius, "Cooling Water Deposits. Silica and Magnesium Silicate Prediction Tools: Comparison and Guidelines," *Proceedings International Water Conference* 62 (2001): pp. 288–293.
21. I. Gunnarsson, S. Arnorsson, S. Jacobsson, "Precipitation of Poorly Crystalline Antigorite Under Hydrothermal Conditions," *Geochimica et Cosmochimica Acta* 69 (2005): pp. 2813–2828.
22. K.D. Demadis, "Water Treatment's 'Gordian Knot,'" *Chemical Processing* 66 (2003):29–34.
23. J. Arensdorf, S. Kerr, K. Miner, et al., "Mitigating Silicate Scale in Production Wells in an Oilfield in Alberta," SPE-141422-MS, SPE International Symposium on Oilfield Chemistry, held April 11–13, 2011 (The Woodlands, Texas, USA).
24. S. Dong, K. Chao, S. Margan, "Laboratory Evaluation of Scale Inhibitors Used in OTSG Boilers for Control of Silica Related Scale," AMPP-2023-18761, AMPP Annual Conference (Denver, Colorado, USA).

25. J. Xie, X. Liu, W. Xie, "Technology and Application of Recycling Produced Water From Heavy Oil Production to Steam Generators," IPTC-13572-MS, International Petroleum Technology Conference, held December 7–9, 2009 (Doha, Qatar).
26. W. Wang, W. Wei, N. Ferrier, et al., "Scale Formation and Inhibition Study for Water Injection Wells," SPE-190732-MS, SPE International Oilfield Scale Conference and Exhibition, held June 20–21, 2018 (Aberdeen, UK).
27. P.G. Koutsoukos, P.D. Natsi, Z. Amjad, "Review of the State of the Art of Siliceous Scale Management in Industrial Systems," AMPP-2022-17731, AMPP Annual Conference, held March 6–10, 2022 (San Antonio, Texas, USA).
28. D.L Gallup, C.J. Henrichsen, "Control of Silicate Scales in Steam Flood Operations," SPE-114042-MS, SPE International Oilfield Scale Conference, held May 28–29, 2008 (Aberdeen, UK).
29. E.J. Daniels, W.E. Frank, G.S. Orton, "Formation and Prevention of Iron-Silicate Scales in Steam Generators, Cymric field, California," SPE-37272-MS, International Symposium on Oilfield Chemistry, held February 18–21, 1997 (Houston, Texas, USA).
30. K. Jarrahian, E. Mackay, M. Singleton, et al., "Scale Control in Geothermal Wells—What Are the Options for Effective and Economic Scale Management?," SPE-218737-MS, SPE Oilfield Scale Symposium, held June 5–6, 2024 (Aberdeen, Scotland, UK).
31. P. Spponer, "Lifting the Fog of Confusion Surrounding Clay and Shale in Petrophysics," SPWLA-2014-VV, SPWLA 55th Annual Logging Symposium, held May 18–22, 2014 (Abu Dhabi, UAE).
32. M.W. Tamele, L.B. Ryland, L.D. Rampino, et al., "Alumina-Silica Cracking Catalyst–Influence of Aluminum Content on Activity," WPC-4310, 3rd World Petroleum Congress, held May 28–June 6, 1951 (The Hague, The Netherlands).
33. M. Kamaratou, K.D. Demadis, "Investigating Alternative Mechanisms for Metal-Silicate Scale Formation: Do Metal Hydroxides Play A Role?," AMPP-2022-17702, AMPP Annual Conference, held March 6–10, 2022 (San Antonio, Texas, USA).
34. J.H. Bardsley, S.T. Robertson, W.M. Hann, et al., "Silica Scale Inhibition," Patent EP0654449B1, 1998.
35. Z. Amjad, R.W. Zuhl, "Laboratory Evaluation of Process Variables Impacting the Performance of Silica Control Agents in Industrial Water Treatment Programs," AWT Annual Convention, held November 5–8, 2008 (Austin, Texas, USA).
36. K.D. Demadis, "Bioinspired "Green" Scale Inhibitors for Mitigation of Silica Scales," NACE-2016-7332, NACE Annual Conference, held March 6–10, 2016 (Vancouver, British Columbia, Canada).
37. J. Arensdorf, D. Hoster, D. McDougall, et al., "Static and Dynamic Testing of Silicate Scale Inhibitors," SPE-132212-MS, International Oil and Gas Conference and Exhibition in China, held June 8–10, 2010 (Beijing, China).
38. H.V. Yee, N.H. Bt Halim, I.K. Bt Salleh, et al., "Managing Chemical EOR (ASP) Effects on Formation Damage and Flow Assurance in Malay Basin, Malaysia," IPTC-16777-MS, International Petroleum Technology Conference, held March 26–28, 2013 (Beijing, China).
39. K.D. Demadis, A. Spinthaki, M. Kamaratou, et al., "Control of Metal Silicate Scaling by Chemical Additives," NACE-2019-12863, NACE Annual Conference, held March 24–28, 2019 (Nashville, Tennessee, USA).
40. S. Arnorsson, "Injection of Waste Geothermal Fluids: Chemical Aspects," World Geothermal Congress, held May 28–June 10, 2000 (Kyushu-Tohoku, Japan).
41. M. Yagi, Y. Yamada, H. Okada, et al., "Occurrence and Variation of Clay Minerals as Hydrothermal Scale in the Yamagawa Geothermal Wells, Kyusu, Japan," World Geothermal Congress, held May 28–June 10, 2000 (Kyushu-Tohoku, Japan).

42. D.L. Gallup, "Silicate scale formation and inhibition at the Salton Sea geothermal field," *Geothermics* 18 (1989): pp. 97–103.
43. W. Wang, W. Wei, "Silica and silicate scale formation and control: Scale modeling, lab testing, scale characterization, and field observation," SPE-179897-MS, SPE International Oilfield Scale Conference, held May 11–12, 2016 (Aberdeen, UK).
44. Z. Amjad, R.W. Zhul, "The role of water chemistry on preventing silica fouling in industrial water systems," NACE-10048, NACE Annual Conference, held March 14–18, 2010 (San Antonio, Texas, USA).
45. S.A. Gharaibeh, T.R. McCartney, R.A. Shank, "A novel non-toxic method for the decontamination of silicate scales—A case study," NACE-2017-9087, NACE Annual Conference, held March 26–30, 2017 (New Orleans, Louisiana, USA).
46. R. Sheikholeslami, I.S. Al-Mutaz, T. Koo, et al., "Pretreatment and the Effect of Cations and Anions on Prevention," *Desalination* 139 (2001): pp. 83–95.
47. M.A. Kelland, *Production Chemicals for the Oil and Gas Industry*, 2nd ed. (Boca Raton: CRC Press, 2014).
48. K.D. Demadis, E. Mavredaki, M. Somara, "Additive-Driven Dissolution Enhancement of Colloidal Silica. 2. Environmentally Friendly Additives and Natural Products," *Industrial & Engineering Chemistry Research* 50 (2011): pp. 13866–13876.
49. D.L. Gallup, "Investigations of Organic Inhibitors for Silica Scale Control in Geothermal Brines," *Geothermics* 31 (2002): pp. 415–430.
50. I. Nishida, "Inhibitor and Method for Preventing Silica-Based Scale," U.S. Patent 8927654, 2015.
51. Z. Amzad, B. Zhul, "Silica Scale Control Additives for Industrial Water Systems," *Materials Performance* 48 (2009): pp. 48–52.
52. J.E. Harrar, F.E. Locke, C.H. Otto, Jr., et al., "Field Tests of Organic Additives for Scale Control at the Salton Sea Geothermal Field," *SPPE Journal* 22 (1982): pp. 17–27.
53. Z. Amjad, C.M. Fellows, "Polymers for Industrial Water Systems: Synthesis, Characterization, and Applications," in *Water-Formed Deposits: Fundamentals and Mitigation Strategies*, Z. Amjad, K.D. Demadis, eds. (Elsevier, 2022), pp. 369–396.
54. J.H. Bardsley, W.M. Hann, S.T. Robertson, et al., "Silica Scale Inhibition," Patent Application EU95100758, 1991.
55. F. Meng, C. Aften, "Polyethylene Glycol PEG Functionalized Silane for Highly Efficient Silicate Scale Control," SPE-184570-MS, SPE International Conference on Oilfield Chemistry, held April 3–5, 2017 (Montomery, Texas, USA).
56. D.L. Gallup, "The Use of Reducing Agents for Control of Ferric Silicate Scale Deposition," *Geothermics* 22 (1993): pp. 39–48.
57. E. Korhatzis, K. Xanthopoulos, K.D. Demadis, "Polymeric Silica Scale Inhibitors: Computational Predictions and Experimental Studies on Inhibitor Efficiencies," AMPP-2022-17706, AMPP Annual Conference, held March 6–10, 2022 (San Antonio, Texas, USA).
58. J.G. Southwick, "Solubility of Silica in Alkaline Solutions: Implications for alkaline flooding," *SPE Journal* 15 (1985): pp. 857–864.
59. T. Coradin, D. Eglin, J. Livage, "The Silicomolybdic Acid Spectrometric Method and Its Application to Silicate/Biopolymer Interactions Studies," *Journal of Spectroscopy* 18 (2004): pp. 567–576.
60. P. Meyers, "Behavior of Silica in Ion Exchange and Other Systems," IWC-99-64, IWC Conference, held October 18–20, 1999 (Pittsburgh, Pennsylvania, USA).

17

Green Scale Inhibitors

Tao Chen[1] and Ping Chen[2]
[1]EXPEC Advanced Research Center, Saudi Aramco, Dhahran, Saudi Arabia
[2]CNPC R&D (DIFC) Company Ltd., Dubai, United Arab Emirates

17.1 Introduction

Chemical scale inhibition is one of the most popular scale management strategies. A process of chemical selection is performed to determine the most suitable and effective scale inhibitors for the particular conditions encountered in the oilfields. The conventional scale inhibitors commonly used in the oil industry include phosphonates, phosphonate esters, sulfonates, and polymers such as homo- and copolymers comprising phosphonate, sulfonate, and/or carboxylate groups.

Phosphorus-based scale inhibitors are widely used in the oil and gas industry. The typical organophosphates (also known as phosphate esters) are a class of organophosphorus compounds with the general structure $O{=}P(OR)_3$. The typical phosphonates or phosphonic acids are organic chemistry characterized by the PO_3H group on an aliphatic carbon or nitro chain. The typical phosphonate scale inhibitors applied in oilfields are 1-hydroxyethylidenediphosphonic acid (HEDP),[1] tris(phosphonomethyl)amine, nitrilotrimethylphosphonic acid, aminotris(methylphosphonic acid) (ATMP), bis(hexamethylenetriaminepenta(methylenephosphonic acid)), and diethylenetriaminepenta(methylenephosphonic acid) (DETPMP). The phosphonate species are less effective at preventing initial nucleation. However, once nucleation has started, they are effective at stopping further crystal growth by adsorbing at active growth sites on the scale crystal lattice.[2]

There are many polycarboxylic acids that are commonly used to prevent scale formation. These are characterized by the COOH group on an aliphatic carbon chain, and their molecular weights range from a few hundred to millions of mg/mol.[3] Polyacrylates (PAA) and polymaleic acid are two common polycarboxylic acids. The small polyelectrolytes work mainly as nucleation inhibitors. After the initial nucleation, they continue to retard the crystal growth; they do not block it entirely and become less effective with time, as they are consumed in the crystal lattice.[1]

Polyelectrolytes, including homo-, co-, and tripolymers, have been developed to control scale deposition. The polymers are not only the carboxyl or phosphate group; there are other groups such as hydroxyl, sulfonate, and phosphonate. Polyphosphinocarboxylic acid, polyvinyl sulfonate and polyacrylic acid copolymer, phosphonocarboxylic acid, and 2-phosphono-butane-1,2,4-tricarboxylic acid (PBTC) are some typical polyelectrolyte scale inhibitors.[4] Due to their multiple functionalities, the polyelectrolytes show properties of all the groups, PO_3H_2, COOH, and SO_3H, on scale inhibition.

The key characteristics of the conventional scale inhibitors applied in the oil and gas industry are summarized in Table 17.1.

Table 17.1: Summary of general characteristics of conventional scale inhibitors.

Inhibitor	Thermal Stability	Nucleation Inhibition	Inhibition of Crystal Growth	Squeeze Treatment
Phosphate ester	Up to 90°C	Less effective	Effective	Less effective
Phosphonates	Up to 160°C	Less effective	Effective	Effective
Polycarboxylic acid or polyelectrolytes	Up to 200°C or higher	Effective	Less effective	Less effective

Since the 1970s, environmental regulations have noticeably shifted and established emphasis in response to ecosystem pollution by man-made chemicals. The General Environmental Regulation sets out wide-ranging prohibitions of pollution and contamination of air, land and water, with particular reference to all parties involved in services, industry, or other economic activities. In the past two decades, regulatory bodies and scientific communities have confronted new aspects of environmental issues, such as the global impact of pollution, prevention of pollution through applications of "greener" chemicals, and the development of new computer modeling tools for assessing environmental impact.[5]

The growing concerns about the utilization of environmentally safe chemicals and new environmental legislation control have led to the development of greener or environmentally friendly scale inhibitors to be applied in the oil and gas industry. Three parameters, such as toxicity, bioaccumulation, and biodegradability, are usually used to determine whether a scale inhibitor is "green" or environmentally friendly.[6–8]

The most commonly applied scale inhibitors, phosphorus-containing inhibitors, are highly effective against most of the types of scales. However, they are nutrients after discharging into the sea which leads to eutrophication.[9] The conventional polyacrylate-, polyphosphonate- and polysulphonate-based scale inhibitors have low biodegradability and fail to meet the minimum biodegradation requirements of the increasingly strict environmental legislation in the regions with stricter environmental regulations.[10] The poor ecotoxicity has severely impeded the use of conventional phosphorus- and polymer-based scale inhibitors, especially in regions with stricter environmental regulations.

This chapter is mainly focused on the development and application of environmentally acceptable scale inhibitors to meet the newly applied environmental regulations in the oil and gas industry.

17.2 Environmental Regulations

Oilfield chemicals are developed to solve specific technical problems encountered in oilfields, from reservoirs to surface facilities, such as stimulation, water, sand, integrity, and scale mitigation. These chemicals may contain ingredients that may be harmful, if discharged into the environment.

The oil and gas industry has been facing severe restrictions concerning the discharge of oilfield chemicals into the environment. More and more severe environmental regulations have been legislated to constrain the usage and disposal of these kinds of harmful chemicals. Improving the characteristics of these chemicals to reduce risk or damage to marine life requires changing some of the conventional acceptable oilfield chemicals, such as eliminating restricted materials and incorporating components with improved ecotoxicity.[5]

Scale inhibitors, as one of the typical oilfield chemicals, need to pass a series of environmental tests to meet the environmental regulations prior to application in the regions with specific environmental regulations, as well as to show acceptable scale inhibition performance under application conditions. The scale inhibitors' toxicity to mammals, freshwater or marine species, and the tendency of bioaccumulation and biodegradation of the scale inhibitors must be assessed, which are critical to their acceptance in specific regions.

A number of international regulations have been legislated and implemented for the protection of the environment, especially for marine ecosystems. The most known are the four major international environmental jurisdictions of the European Union (EU) for the North Sea (United Kingdom, Norway, Denmark, and the Netherlands), the United States for the U.S. Gulf Coast oil and gas fields, Canada, and Australia.

17.2.1 Oslo and Paris Convention Regulations

While the requirements are variable depending on the legislative body that oversees the use and approval of oilfield chemicals, one regulation that stands out as being at the forefront of environmental legislation is the Convention for the Protection of the Marine Environment of the North-East Atlantic (the "Oslo and Paris Convention (OSPAR) Convention"),[11] opened for signature at the Ministerial Meeting of the Oslo and Paris Commissions in Paris on September 22, 1992, and entered into force on March 25, 1998. In this decade, the Convention will be implemented through OSPAR's North-East Atlantic Environment Strategy 2030.[11] The Convention has been signed and ratified by all of the contracting parties, including Belgium, Denmark, Finland, France, Germany, Iceland, Ireland, Luxembourg, The Netherlands, Norway, Portugal, Spain, Sweden, Switzerland, and the United Kingdom, together with the EU. The OSPAR Convention was introduced to better prevent and eliminate pollution from the marine environment of the north-east Atlantic from sea, land, and air and to protect the sea area against the adverse effects of human activities. It also sought to safeguard human health, conserve marine ecosystems, and, where practicable, restore marine areas that had already been impacted by human activities such as the oil and gas industry.[12] It covers a series of annexes on the prevention and elimination of pollution from land-based sources, by dumping or incineration, and from offshore sources; the assessment of the quality of the marine environment; and the protection and conservation of the ecosystems and biological diversity of the maritime area.

OSPAR regulates the offshore use of chemicals in the covered regions. Although the regulations are often referred to as North Sea regulations, OSPAR is actually the Convention for the Protection of the Marine Environment of the Northeast Atlantic, which is much wider than just the North Sea. The OSPAR Convention maritime area covers the internal waters and territorial seas of its contracting parties, including Arctic waters, the Greater North Sea, the Celtic Seas, the Bay of Biscay, the Iberian Coast, and the Wider Atlantic.

OSPAR implemented the Harmonized Mandatory Control Scheme (HMCS) to protect the marine environment by identifying those chemicals used in offshore oil and gas operations with the potential for causing an adverse environmental impact and restricting their use and discharge to the sea. Under these programs, measurement methods were developed to identify, prioritize, monitor, and control the emissions, discharges, and losses of hazardous substances that reach or could reach the marine environment.

The Convention requires a standard ecotoxicity testing data set for each chemical substance with test results from approved laboratories accredited to the Good Laboratory Practice standard. The data required for each substance include results for biodegradation, bioaccumulation, and toxicity.[13, 14]

The standard biodegradation test accepted in all OSPAR territories is the OECD306: biodegradation in seawater using either the shake flask dissolved organic carbon (DOC) method or the closed bottle biological oxygen demand (BOD) method. The test specifically shows how a material biodegrades in seawater. Biodegradation is calculated as the ratio of the amount of oxygen consumed during the degradation period to the calculated theoretical oxygen demand. The test is normally carried out over a 28-day period. The current minimum requirement is >20% biodegradation. Ideally, chemicals should be readily biodegradable, with >60% biodegradability.

Bioaccumulation is linked to the relative solubility of a chemical in lipids (found in fatty tissues) and water and can be predicted based on the relative solubility of octanol (representing fatty tissue) and water. The partition coefficient is defined as the ratio of the equilibrium concentrations of a dissolved substance in a two-phase system consisting of two largely immiscible solvents. A standard bioaccumulation test accepted in OSPAR territories is the OECD107 Partition Coefficient (n-octanol/water) using the shake flask method. The partition coefficient is expressed as the logarithm of the octanol-water partition coefficient: log Pow. The range of log Pow is normally between –2 and 4 (occasionally up to 5). If the coefficient is greater than or equal to 3, the chemical is assumed to be a bioaccumulation risk unless experimental testing of a bioaccumulation factor (BCF) indicates otherwise.[14]

To provide harmonized data sets from the testing of offshore substances and preparations, OSPAR agrees on four ring tests to evaluate the toxicity of chemicals to be applied in the offshore oil industry, namely, growth inhibition test using the marine alga *Skeletonema costatum*; acute toxicity test using the marine copepod *Acartia tonsa*; a sediment bioassay using an amphipod *Corophium* sp.; and protocol for a fish acute toxicity test. Acute toxicity is measured by exposure to *Acartia tonsa* copepods according to OECD TG 203 as measured by the International Organization for Standardization (ISO) Protocol 14669. The objective is to determine the initial concentration that kills 50% of *Acartia tonsa* copepods with an acceptable maximum 48-h LC50 of 10 ppm.[13, 14]

The HMCS has developed two major tools for evaluating the environmental properties of chemicals: the Prescreening Scheme and the Chemical Hazard and Risk Management (CHARM) model.

The prescreening schemes allow authorities to identify the hazardous substances to be substituted and ultimately phase them out and to regulate and control the other substances where necessary. It should be noted that prescreening applies only to chemicals used or discharged as a result of offshore activities. The criteria used in the prescreening are acute toxicity to marine species, biodegradation and bioaccumulation potential, or a combination of these properties, where biodegradation and bioaccumulation play dominant roles. The replacement criteria for the chemicals based on bioaccumulation and biodegradation are shown in Figure 17.1. Based on the outcome of prescreening, the regulatory agencies may decide whether to give the chemicals permission to use, request substitution, give temporary permission, or refuse permission.

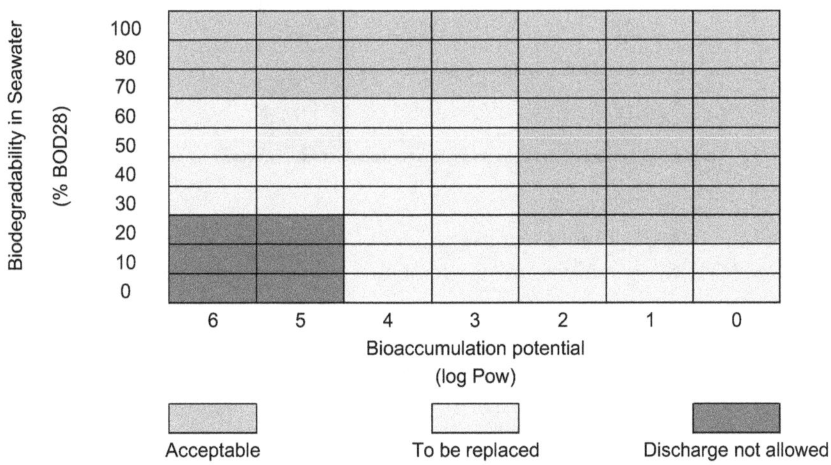

Figure 17.1. Replacement criteria for chemicals based on bioaccumulation and biodegradation.

Products that pass the prescreening scheme must be ranked using the CHARM mathematical model, which calculates the hazard quotient (HQ), a ratio between the predicted environmental concentration and the predicted no-effect concentration, using toxicity, biodegradation, and bioaccumulation data provided by suppliers on the Harmonized Offshore Chemical Notification Format (HOCNF) form.[15]

$$HQ = PEC / PENC \quad (1)$$

The HQ is calculated for an "average platform" under "realistic worst-case conditions." Many default values are used, and safety factors and worst-case scenarios have been built in. If the HQ is calculated to be less than 1, then no adverse health effects are expected as a result of exposure. If the HQ is greater than 1, then adverse health effects are possible. The HQ is converted to a color banding under the Offshore Chemical Notification Scheme (OCNS), as illustrated in Table 17.2.

Table 17.2: OCNS HQ and color bands.

Minimum HQ Value	Maximum HQ Value	Color Banding	
>0	<1	Gold	Lowest hazard ↓ Highest hazard
≥1	<30	Silver	
≥30	<100	White	
≥100	<300	Blue	
≥300	<1,000	Orange	
≥1,000		Purple	

The use of the CHARM model for assessment of chemical products was phased in March 2002. The HMCS regulations are implemented differently for the countries within the North Sea sector.[16]

The United Kingdom classifies the products into Group A through E. During the hazard assessment process, each individual substance is ranked by applying the OCNS ranking scheme. The overall ranking for a product is determined by the product substance which has the worst-case OCNS ranking. The ranking system contains an initial OCNS group and an adjustment for the final OCNS group.

The initial group is determined using Table 17.3, where the aquatic toxicity refers to the Algae EC50, Crustacean LC50, and Fish LC50 toxicity tests (units = ppm or mg/kg), and the sediment toxicity refers to the Sediment reworker LC50 test (units = ppm or mg/kg).

Table 17.3: Initial OCNS grouping.

Initial Grouping	A	B	C	D	E
Result for aquatic toxicity data (ppm)	<1	>1-10	>10-100	>100-1,000	>1,000
Result for sediment toxicity data (ppm)	<10	>10-100	>100-1,000	>1,000-10,000	>10,000

All submitted toxicity data for each substance are compared against the table. The most toxic response is used as the initial group for the substance.[5]

The final grouping is determined using Table 17.4 as a guide. Select the column that applies to the candidate substance and adjust the initial group accordingly. If the classification should theoretically move beyond Group A or E, the product will be assigned to that group.

Table 17.4: Adjustment criteria for OCNS grouping.

Increase by Two Groups (*e.g.*, from C to E)	Increase by One Group (*e.g.*, from C to D)	Do Not Adjust Initial Grouping	Decrease by One Group (*e.g.*, from C to B)	Decrease by Two Groups (*e.g.*, from C to A)
Substance is readily biodegradable and is nonbioaccumulative	Substance is inherently biodegradable and is nonbioaccumulative	Substance is not biodegradable and is nonbioaccumulative or	Substance is inherently biodegradable and bioaccumulates	Substance does not biodegrade and bioaccumulates
		Substance is readily biodegradable and bioaccumulates		

The definitions of the terms are:

- Readily biodegradable: results of >60% biodegradation in 28 days (OECD 306, 301B-F method), >70% in 28 days (OECD 301 A, 301E) to an OSPAR HOCNF accepted ready biodegradation protocols. Freshwater testing is allowable in the U.K. sector.
- Inherently biodegradable: results of >20% and <60% with the OECD 306, 301B-F method (or <70% in the OECD 301 A/301E test) according to an OSPAR HOCNF accepted ready biodegradation protocol.
- Not biodegradable: results from the OSPAR HOCNF accepted ready biodegradation protocol or inherent biodegradation protocol are <20%, or half-life values derived from aquatic simulation tests indicate persistence.
- Nonbioaccumulative: log Pow <3, or BCF ≤100; the molecular weight is ≥700.
- Bioaccumulative: log Pow ≥3, or BCF >100, the molecular weight is <700, or if the conclusion of a weight-of-evidence expert judgment under OSPAR Agreement 2008-5 is negative.

Group A represents the most environmentally hazardous products, whereas Group E represents the products with the least environmental hazards based on marine toxicity, biodegradation, and bioaccumulation. The groups are often reviewed and used by other countries. They give a ready indication of the environmental hazard; however, they do not indicate the potential risk of the product.

In the above section, it has been explained how hazard assessment of an offshore chemical is conducted in the United Kingdom, based on the test data provided by the chemical supplier on the HOCNF. However, the reproducibility of the tests involved is imperfect.

To provide the most accurate prescreening of the chemicals submitted for assessment, the Centre for Environment, Fisheries, and Aquaculture Science (Cefas), an Executive Agency of Defra (the U.K. Government's Department of Environment, Food, and Rural Affairs), therefore takes account of the data it holds for similar substances. This is known as The Global Approach, which is based on the weight of evidence principles described by ECHA Practical Guide 2: How to Report Weight of Evidence and in Sections R.4 and R.7a-R.7c of the Guidance on Information Requirements and Chemical Safety Assessment provided under Registration, Evaluation, Authorization and Restriction of Chemicals (REACH) legislation.

17.2.1.1 Biodegradation

Where multiple biodegradation screening test results of equivalent quality exist, a positive result is accepted over negative results. This application of the global approach may result in the removal of a substitution warning.

17.2.1.2 Toxicity

Where there are less than three values for the same end point and they are within one order of magnitude, the lowest value will take precedence. Where there are greater than three values for the same end point and they are within one order of magnitude, the geometric mean will be calculated and used in the assessment.

17.2.1.3 Bioaccumulation Factor Tests

A BCF value will take precedence over any other data and will be compared to HMCS criteria. Where more than one BCF value occurs, the highest value will take precedence.

Consideration of global BCF data could remove or apply a substitution warning based partly on log Pow data, since BCF data always take precedence over log Pow data.

17.2.1.4 Bioaccumulation (Log Pow)

Where only log Pow data are available and there are less than three values, the highest value will take precedence. Where there are more than three values, the geometric mean will be calculated and used in the assessment. The relevance of weight of evidence approaches conducted under REACH or HMCS will be evaluated on a case-by-case basis.

It is stressed that the applicability of the global approach is only valid, where the data being compared relate to the same substance. Expert judgment is used to ensure that this is the case, assisted by database searches and discussions with suppliers, where required.

Chemicals that are hazardous to the marine environment are subject to substitution warnings under the Harmonized Mandatory Control Scheme (HMCS).

An offshore chemical has a substitution warning, if it contains a substance that is on the OSPAR list of chemicals for priority action; or contains a substance that is on the authorization list established pursuant to REACH (Annex XIV) and the Sunset Date in that Annex has passed; or its offshore use is covered by restrictions under Annex XVII to REACH; or it is considered by the authority to be of equivalent concern for the marine environment as the substances covered by the previous subparagraphs; or it is inorganic has a LC50 or EC50 less than 1 mg/L; or it has an ultimate biodegradation of less than 20% in OECD 306, Marine BODIS or any other accepted marine protocols or less than 20% in 28 days in freshwater (OECD 301 and 310); or half-life values derived from simulation tests submitted under REACH (EC 1907/2006) are greater than 60 days and 180 days in marine water and sediment, respectively (*e.g.*, OECD 308, 309 conducted with marine water and sediment as appropriate); or it meets two of the following three criteria:

1. Biodegradation: $<60\%$ in 28 days (OECD 306 or any other OSPAR-accepted marine protocol); or in the absence of valid results for such tests: less than 60% (OECD 301B, 301C, 301D, 301 F, freshwater BODIS) or less than 70% (OECD 301A, 301E).
2. Bioaccumulation: BCF > 100 or log Pow ≥ 3 and molecular weight < 700, or if the conclusion of a weight of evidence judgment under Appendix 3 of the OSPAR Agreement 2008-5 is negative.
3. Toxicity: LC50 < 10 mg/L or EC50 < 10 mg/L; if toxicity values <10 mg/L are derived from limit tests to fish, actual fish LC50 data should be submitted, and a less hazardous (or preferably nonhazardous) substitute is available.

The Norwegian Sector interprets the regulations similarly to those in the United Kingdom based on the criteria for chemicals' biodegradation, bioaccumulation (log Pow), and toxicity.[17] However, the Norwegian sector only accepts the OECD 306 Marine biodegradation test instead of both the OECD 306 Marine and OECD 301A and E freshwater biodegradation tests accepted in the U.K. sector. The seawater test OECD 306 has a similar test setup to OECD 301A and E (shake flask DOC die away and closed bottle analysis). However, no microorganisms are added to the test water besides the indigenous organisms in the seawater, which makes the OECD 306 test more stringent than the OECD 301 test.

In the Norwegian sector, chemicals used in the oil and gas industry are categorized by color: black, red, yellow, and green environmental categories, as shown in Table 17.5.

Table 17.5: Categorization of chemicals used in oilfields in the Norwegian sector.

Colour	Criteria	Priority list from Stortingsmelding nr. 21
	BOD28 < 20% + Kow > 4.5 BOD28 < 20% + (LC50/EC50 < 10 mg/L) Substances that are genotoxic/reprotoxic	OSPAR list of chemicals for priority action
	BOD28 < 20%Or two of the following criteria BOD28 < 60% Log Pow > 3 and Mw < 700 LC50 or EC50 < 10 mg/L	Priority list for substitution with less harmful alternatives
	BOD28 > 20% Log Pow < 3 EC50 or LC50 > 10 mg/L	Does not fit in black and red list Not listed on the OSPARs PLONOR list
	Pose little or no risk to the environment	listed on the OSPARs PLONOR list

Reproduced from EPA.[18]

Chemicals in the black category are not readily biodegradable (BOD28 < 20%) in the marine environment, with a high potential for bioaccumulation and acute toxicity, which are listed in the OSPAR List of Chemicals for Priority Action.

Chemicals in the red category are slowly biodegraded, have a potential for bioaccumulation, and/or are acutely toxic. Organic chemicals are classified in red when the biodegradation measured as BOD28 is ≤20%, or if two of the following three criteria are fulfilled: biodegradation measured as BOD < 60%, log Pow ≥ 3, acute toxicity ≤ 10 mg/L. Chemicals in the red category can be harmful to the environment and should be prioritized for substitution with less harmful alternatives.

Chemicals in the yellow category are the ones that do not qualify for the red or black category and are not listed on the OSPAR PLONOR list (pose little or no risk to the environment). For the yellow category, the chemicals must meet the requirements of BOD28 > 20%, log Pow < 3, and EC50 or LC50 > 10 mg/L. The yellow designation is further subcategorized into Y1, Y2, and Y3 under the SKIM guidelines 7, which state that Y1 is for the chemical expected to biodegrade completely; Y2 is for the chemical expected to biodegrade to products that are not environmentally hazardous; and Y3 is for the chemical expected to biodegrade to products that may be environmentally hazardous.

Chemicals in the green category are the ones listed on the OSPAR PLONOR list with little or no risk to the environment.

In general, North Sea legislation demands that all chemicals be determined for their ecotoxicity according to a number of standardized tests that determine biodegradation, bioaccumulation, and toxicity to a number of marine organisms. All chemicals that are used and discharged into the sea must be tested in accordance with the requirements that are specified in the HMCS and OECD.[17] The results of the tests are reported using the HOCNF. Failure to meet the required criteria will eliminate from consideration the use of a particular chemical.[10]

17.2.2 USA Regulations

Authorized by the Clean Water Act (CWA), the National Pollutant Discharge Elimination System (NPDES) permit program controls water pollution by regulating point sources that discharge pollutants into the waters of the United States.

Offshore waste discharges, regulated by NPDES, are pursuant to Section 402 of the CWA, administered by the Environmental Protection Agency (EPA). Regulations focus on controlling the pollutant properties in effluent discharges resulting from well chemical treatments used in offshore oil and gas operations.[18–23]

The EPA's approach with regard to oil and gas exploration and production is to regulate flowback offshore discharges that originate from these processes. Assessing environmental impact is based on discharge volumes; a discharge permit is issued with prescribed monitoring requirements and effluent limitations. The well treatment, completion, and workover fluids should follow limitations: (1) free oil: no free oils shall be discharged, monitoring shall be performed using the static sheen test; (2) oil and grease: a daily maximum of 42 mg/L, and a monthly maximum of 29 mg/L must not be exceeded; (3) priority pollutants: prohibition of discharge of toxic pollutants, pursuant to statute 40, Code of Federal Regulations, Part 401.15 list of 65 chemicals.

The EPA publishes laboratory analytical methods (test procedures) that are used by industries and municipalities to analyze the chemical, physical, and biological components of wastewater and other environmental samples required by the CWA.

17.2.3 Canada Regulations

In Canadian offshore oil and gas activity regions, the National Energy Board (NEB), the Canada-Newfoundland Offshore Petroleum Board (C-NOPB), and the Canada-Nova Scotia Offshore Petroleum Board (C-NSOPB) promulgated and executed the regulations of offshore drilling and production activities under the jurisdiction of Newfoundland and Nova Scotia in the most prominent Canadian. These regulations combine elements of EU and USA offshore effluent discharge regulations, including the selection of chemicals and waste ocean disposal guidelines. Two fundamental guidelines cover recommended practices and standards: the Offshore Chemical Selection Guidelines and the Offshore Waste Treatment Guidelines in Canada.[24, 25] The Offshore Chemical Selections Guidelines, legislated in 1999, are more relevant to the application of oilfield scale inhibitors, which will be introduced in this chapter.

The Offshore Chemical Selection Guidelines provide a consistent framework for chemical selection as part of the environmentally responsible management of chemicals used in offshore drilling and production. The regulatory framework specified by the Guidelines is based on the legislation authorized by NEB, C-NOPB, and C-NSOPB, which are collectively referred to as the Energy Legislation.

The process of obtaining authorization requires operators to demonstrate results from the selection of chemicals equivalent to those achieved when applying the guidelines. Selection criteria include a number of steps, as listed in Table 17.6.

Table 17.6: Canadian offshore chemical selection criteria.

Step	Concern/Action	Description
1	Propose chemical for application	Determines the overall chemical acceptability
2	Chemical used as a biocide?	
3	Listed on domestic substance list (DSL)?	
4	Hazardous to humans?	Criterion of carcinogenicity or mutagenicity
5	Discharge to environment?	
6	Canadian Environmental Protection Act (CEPA) listed?	Prohibited, toxic, and restricted list?
7	OSPAR HOCNF	Product on tint list?
8	OSPAR List A or B	Product on List A or B?
9	PARCOM (Paris Commission)	OCNS hazard rating
10	Microtox	EC50 (15) >75%?
11	Discharge quantity?	
12	Hazard analysis?	The chemical will be approved or rejected.
13	Hazard analysis process	

Detailed chemical and toxicological information are submitted to the appropriate regulatory representative, and a monitoring schedule of pollutant levels is set in place.

17.2.4 Australia Regulations

The Australian offshore oil and gas activity is regulated by the National Offshore Petroleum Safety and Environmental Management Authority (NOPSEMA) and Western Australia Department of Mines and Petroleum, covering the Commonwealth offshore water and the Western Australia water state, respectively. Each authority selects different environmental regulatory models to suit that purpose and regulates offshore oil and gas activities differently.[26]

NOPSEMA regulates operators to submit an environmental plan detailing all the efforts put into place to protect the marine environment without clarifying the specific obligation regarding registration and toxicity acceptance for chemicals to be used. Some operators adapt the North Sea environmental regulation model and use Cefas-rated OCNS D or

E and CHARM gold or silver products. One of the major issues of this approach is the mismatch of the import or manufacture product list authorized by the Australian Government Department of Health and Cefas-rated OCNS D or E and CHARM gold or silver products.[27] Some of the products authorized by the Australian Government Department of Health but not classified as Cefas-rated OCNS D or E and CHARM gold or silver products. On the contrary, some of the Cefas-rated OCNS D or E and CHARM gold or silver products are not authorized to import or manufacture locally by the Australian Government Department of Health. This restricted the application of the Cefas-rated OCNS D or E and CHARM gold or silver products in the authorized regions.

The Western Australia Department of Mines and Petroleum authorizes the Western Australia water state, detailed in Environmental Risk Assessment of Chemicals used in WA Petroleum Activities Guidelines. The regulation is applicable for oil and gas activity onshore and offshore, which requests to conduct an environmental risk assessment including subsurface applications of chemicals in oilfields. The chemicals should be evaluated for the potential health hazard and environmental hazard, acute and chronic ecotoxicity, biodegradation, and bioaccumulation potential should be evaluated. It is acceptable to apply the data on a substance level, if data are not available on a product level.[28] The PLONOR substances are exempt from risk assessment, when used offshore. The concern of this requirement is the acute ecotoxicity data could be available for products or substances registered for use offshore in the North Sea, but chronic ecotoxicity, in addition to acute/chronic toxicity, is not always available.

17.2.5 Summary of the Regulations

Scale inhibitors play a major role in flow assurance to mitigate oilfield scale deposition from reservoirs to surface facilities. They have been used primarily through squeeze treatment and continuous injection and could be discharged into the sea after field applications. In recent years, there has been increasing concern over potential damage to the environment due to the discharged chemicals and their impact on marine ecosystems.[5, 29]

International guidelines and regulations have been put in place in various parts of the world to constrain the usage and disposal of harmful chemicals and minimize their negative effects on marine ecosystems. However, there are differences of opinion, even between countries using the same regulations.[29]

A globally harmonized system needs to be put in place. If a global agreement cannot be reached, then it should be incumbent on operators, service providers, and chemical manufacturers to voluntarily ensure that the best chemicals and practices are used to protect the environment.

17.3 Chemistries and Applications

The majority of conventional scale inhibitors are potentially harmful to the environment. Phosphonate-based scale inhibitors have been widely used as effective scale inhibitors for oil and gas fields, which are characterized as effective scale inhibitors with low concentration, easy residual detection, and a good adsorption and desorption profile on rock with an expected long squeeze life.[30–33] However, phosphonate-based scale inhibitors can serve as nutrients and cause eutrophication.[34, 35] Moreover, the limited biodegradation made many of these scale inhibitors environmentally unacceptable when assessed for the HMCS. These traditional phosphonate inhibitors are considered unsuitable for use in both the United Kingdom and Scandinavia due to environmental constraints, for example, DETPMP.[36] They must be substituted for applications in OSPAR countries due to their ecotoxicity issues. Although highly effective, a number of the polymer-based scale inhibitors, polyacrylates, polyphosphonates, and polysulphonates, have less than the required minimum 20% biodegradation according to OECD 306 test and hence can no longer be considered for application in OSPAR countries. Poor ecotoxicity has severely impeded the use of these phosphonate-based and many conventional polymer-based scale inhibitors, which fail to meet minimum biodegradation requirements.[10]

The need for novel green scale inhibitors is highly influenced by the newly issued environmental regulations, which varies across different business regions. The need for environmentally friendly scale inhibitors is increasing, and "greener" scale inhibitors are sought to meet the requests of environmental regulations.[37, 38]

In the last couple of decades, researchers have been working on developing green scale inhibitors that may replace conventional scale inhibitors. There have been quite considerable advances in the development of green scale inhibitors for controlling deposits in the oil and gas industry. A series of improved polymeric scale inhibitors have been developed in accordance with the ecotoxicity, bioaccumulation, and biodegradability standards set by the OSPAR. The conventional phosphorus inhibitors have first been replaced by less toxic organic phosphorus compounds such as carboxy hydroxyl methyl phosphonic acid and dibutyl phosphor dithoic acid. Then, new classes of non-phosphorus-based polymers have been introduced into the industry, such as polyacrylates and derivatives and various maleic and sulfonic acids, homo- and copolymers, which have less ecotoxicity. Some new chemicals were also proposed, including natural compounds and carboxylated plant polysaccharides such as carboxylated inulins. A few popular, environmentally acceptable scale inhibitors, polyaspartate, inulin derivatives, a hybrid polymer, and biodegradable phosphonate-based scale inhibitors will be introduced in detail in this chapter. These inhibitors are highly effective for control of a variety of scales and perform comparably to nonbiodegradable equivalents, while exceeding the standards required for yellow-banding status in oil production for the Norwegian sector of the North Sea, one of the strictest environmentally regulated regions globally.

17.3.1 Polyaspartate

Polyaspartate (PASP), also known as polyaspartic acid (PASA), was developed in the early 1990s as a green multifunctional inhibitor as both a scale and a corrosion inhibitor in the oil and gas industry.[39–41] It is a biodegradable, water-soluble condensation polymer based on the amino acid aspartic acid.[42, 43] It is non-nitrogenous, nonphosphorus, and has excellent biodegradable properties, making it one of the most promising environmentally acceptable alternative inhibitors to the conventional scale inhibitors. PASP has excellent scale inhibition performance against calcium carbonate, calcium sulfate, barium sulfate, and so on.[44]

Polyaspartate is derived from petrochemicals. Polyaspartate is made from L-aspartic acid through three different synthesis paths that are based on intermolecular dehydration: (1) thermal condensation of aspartic acid; (2) catalyzed polymerization of aspartic acid; and (3) thermal polymerization of maleic acid and ammonium hydroxide. The structural formula of PASP is shown in Figure 17.2.[40, 45, 46]

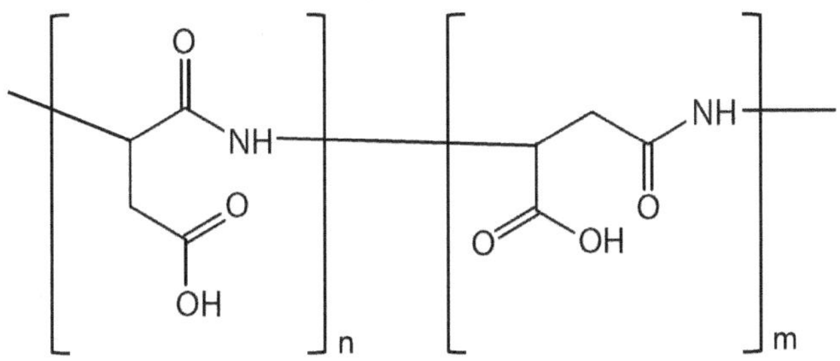

Figure 17.2. Molecular structure of polyaspartate.

Estievenart *et al.* synthesized PASP using different ways of thermal polymerization, bulk polycondensation, and bulk polycondensation with different catalysts.[46] All polymers contain 75% beta form (the alpha and beta forms are shown in Figure 17.2). Their inhibition efficiency on both the nucleation and growth of calcium carbonate ($CaCO_3$) deposits has been studied. The molecular weight and biodegradability of PASP and $CaCO_3$ nucleation and scaling times are summarized in Table 17.7.[47–49]

Table 17.7: Synthesis and biodegradability of polyaspartates and performance of $CaCO_3$ inhibition.

	PASP	Molecular Weight (g/mol)	Biodegradability (%)	t_n (s)	t_s (s)
Thermal polymerization in mass	NC2	7,300	37	377	2,640
Thermal polymerization in mass	NC4	10,800	33	590	3,310
Bulk polycondensation	NM7	7,300	83	472	2,870
Bulk polycondensation	MS10	23,000	90	708	4,760
Bulk polycondensation + Catalyst A	MS8	30,000	91	n/a	n/a
Bulk polycondensation + acid catalyst	MSter	3,600	n/a	5,390	7,140

The most biodegradable polyaspartates are those synthesized by bulk polycondensations, which have 80–92% biodegradability. Those from thermal polymerizations are only partially biodegradable, with ~37% biodegradability. The different biodegradability of PASP synthesized by bulk polycondensation and thermal polymerization could be due to the terminal groups induced by each synthetic route. The molecular weight influences only the biodegradability of polyaspartates obtained by thermal polymerizations in mass; the longer the polymer chain, the lower its biodegradability.

The inhibition performance was evaluated using a quartz crystal microbalance. The t_n and t_s are the nucleation time and scaling time, respectively. For all the tested polyaspartates, nucleation and scaling times are lower than for the blank solution, where t_n and t_s are 0 s and 2,640 s, respectively. Increases in both t_n and t_s suggested that polyaspartates inhibited both the nucleation and growth of calcium carbonate formation processes.

In Estievenart's study,[47] the molecular weight influences the inhibition performance of thermal polymerizations in mass or bulk polycondensation: the longer the polymer chain, the higher the inhibition performance with longer t_n and t_s. However, this is in contradiction with the conclusion drawn by other studies[50–54] that there is no obvious effect of the molecular weight on the scale inhibiting efficiency of PASP and that there is a drop in inhibition performance with the increase in molecular weight of PASP as the molecular weight increases and the number of molecules in a given weight of polymer decreases. The polyaspartates synthesized by bulk polycondensation in the presence of an acid catalyst are the most performing, with the longest t_n and t_s at 5,390 s and 7,140 s, respectively.

In addition, the tested polyaspartates affect the morphologies of the $CaCO_3$ scale. PASP inhibits calcite nucleation and favors the nucleation and growth of the vaterite form.

Liu *et al.* synthesized PASP by thermal polycondensation with maleic anhydride and ammonium carbonate as raw materials, which is a low-cost and environmentally friendly method. The optimum molar ratio of maleic anhydride to ammonium carbonate is 1.0:1.2, and the reaction temperature and duration are 180°C and 2.0 h. The synthesized PASP showed an excellent inhibition efficiency of 95% against calcium carbonate and 90% against calcium sulfate in the presence of 5 mg/L PASP. In addition, PASP showed an excellent dispersion effect and prevented the agglomeration of scale particles.[44]

Salima evaluated the performance of rock interaction with polyaspartate to determine the suitability of polyaspartate as a scale inhibitor for squeeze application in sandstone. Core flood tests were performed with polyaspartate in sandstone core plugs. The retention of PASP is lower than that of DETPMP. The concentration of PASP drops much more rapidly during the flowback than the conventional phosphonate-based scale inhibitor DETPMP. Although the coreflooding inhibitor return profile of PASP is not as good as DETPMP, the coreflooding results obtained from the PASP scale inhibitor are promising as a scale inhibitor for squeeze applications. Another criterion for a scale inhibitor for squeeze application is that the inhibitor should be ready for analysis accurately at the application dose rate. As polyaspartate is a non-phosphor-containing inhibition and cannot be accurately detected by inductive coupled plasma (ICP) as applied for DETPMP, the hyamine analysis method could be used for polyaspartate analysis. However, the hyamine method should be replaced by a more accurate and less time-consuming method of inhibitor detection.[55]

In general, polyaspartate is a biodegradable green scale inhibitor with excellent scale inhibition performance against various types of oilfield scales. The synthetic routes of polyaspartate could influence molecular weight, biodegradability, and scale inhibition efficiency.

17.3.2 Carboxymethyl Inulin

Polysaccharides, or polycarbohydrates, are long-chain polymeric carbohydrate structures that are made of monosaccharide units bound together by glycosidic linkages. They are the most abundant carbohydrates found in food. Polysaccharides are nontoxic, biodegradable, and renewable, which meet the requirements of environmentally friendly materials.[56, 57] Carboxymethyl inulin (CMI) is one of the polysaccharides that can be isolated from the roots of Inula helenium. It has been developed and used as an oilfield scale inhibitor to replace poorly biodegradable scale inhibitors such as phosphonate and acrylate-based scale inhibitors, which are being phased out by stringent environmental regulations executed in the North Sea.[58]

The molecular structure of carboxymethyl inulin is shown in Figure 17.3. The carboxymethyl inulins are derivatives of inulin, which is a natural β (2-1) poly-fructoside with a glucose unit at the reducing end. Carboxylate groups are introduced into the polysaccharide by carboxymethylation with sodium monochloro acetate as a reagent in alkaline medium.[59]

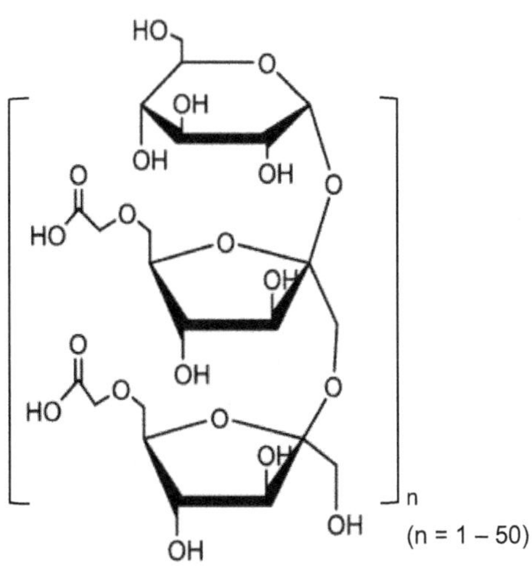

Figure 17.3. Molecular structure of carboxymethyl inulin.

CMI is inherently biodegradable, with more than 20% biodegradation according to the OECD 306 test method and an excellent ecotoxicity profile. It contains no phosphorus or nitrogen and has very low aquatic toxicity.[57, 60, 61] Its potential applications as a scale inhibitor have attracted much research attention.

CMI is highly effective against calcium carbonate crystallization, as a carboxylic group is present in its structure. Verraest[59] observed the effect of CMI on calcium carbonate crystallization at a low concentration of 0.1–200 mg/L. Boels[62] studied the inhibition effect of CMI on calcite crystallization in comparison with that of trilotris (methylenephosphonic acid) (NTMP) and HEDP. CMI exhibited an equally strong inhibition effect on the seeded growth of calcite as NTMP and HEDP. Their study also demonstrated that CMI might be a good inhibitor of calcite crystallization. Moreover, CMI has been studied as a crystal growth inhibitor for calcium carbonate inhibition. It has been shown that CMI is a very effective inhibitor of calcite crystal growth.

Zhang studied the degree of polymerization and substitution of CMI on the inhibition performance of calcium carbonate. It shows that the higher the polymerization and substitution, the more effective the calcium carbonate inhibition performance.[63] The degree of polymerization and substitution, or the average number of carboxylate moieties per fructose unit, has a significant role in the inhibition performance of calcium carbonate.[45] The interaction between CMI (n = 1, 3, and 5) with different degrees of polymerization and the growth of calcite crystals was systematically investigated by molecular dynamics simulation. The study showed that CMI (n = 1, 3, and 5) molecules can all interact closely with the calcite crystal surfaces, affecting calcite crystal growth and possessing superior inhibition performance. For the same amount of mass of CMI (n = 1, 3, and 5), CMI exhibits different abilities to inhibit the growth of calcite crystals, in the order CMI ($n = 5$) > CMI ($n = 3$) > CMI ($n = 1$) for the calcite. Kirboga and Öner[61] also studied the growth rate of calcium carbonate in the presence of CMI. The average number of carboxylic moieties per fructose unit was in the range of 1.5 (CMI-15) to 2.5 (CMI-25). A similar conclusion was drawn that the higher number of negative charge functional groups in CMI has greater effectiveness toward calcium carbonate precipitation.

Temperature affects the interactions between CMI and calcite surfaces. Interactions between CMI and the calcite surface increase with an increase in temperature. The structures of CMI are deformed during their interaction with the calcite surfaces. The interaction between CMI and calcite surfaces is obviously larger than that between CMI and vaterite, which indicates that vaterite could be more easily formed in the presence of CMI. The same conclusion was drawn from Kirboga's study[64] that CMI could induce the formation of different polymorphs during the calcite crystal growth process.

CMI is also a threshold scale inhibitor for sulfate scales such as calcium sulfate and barium sulfate.[65]

Bazin studied the performance of CMI on $BaSO_4$ scale inhibition, along with phosphonate (DETPMP), polyaspartate (PASP), and polyacrylate (PAA) scale inhibitors. The results generated in static bottle tests and dynamic scale loop tests showed that CMI could effectively inhibit $BaSO_4$ scale with comparable inhibition performance as DETPMP and can be an effective alternative green $BaSO_4$ inhibitor. The CMIs with a higher degree of substitution or molecular weight provided better $BaSO_4$ inhibition.

CMI also showed inhibition performance on calcium sulfate but was not as effective as the conventional scale inhibitors. Kan evaluated the calcium sulfate inhibition performance in the presence of different types of scale inhibitors. The tested pentaphosphonates, phosphinopolycarboxylates and maleic acid terpolymers showed better inhibition performance than carboxymethyl inulin.

CMI is highly soluble in water and has excellent calcium tolerance, while having low viscosity in an aqueous solution, which allows its application in various oilfield conditions as an anti-scalant.[59, 66]

The performance of rock interaction with CMI was evaluated for the suitability of CMI as a scale inhibitor for squeeze applications. Baraka-Lokmane[55] evaluated the CMI's interaction with sandstone, along with PASP and DETPMP. The results of coreflood performed with a polyaspartate (PASP) scale inhibitor are much better than those performed with a CMI scale inhibitor. The retention of PASP, although lower than that of DETPMP, is higher than that of CMI scale inhibitors. The PASP levels drop much more rapidly than the DETPMP levels, but the squeeze lifetime is much longer than the CMI corefloods. It indicates that CMI may not be a suitable scale inhibitor for scale squeeze application in sandstone. Bazin[60] evaluated the CMI's interaction with limestone, along with polyacrylate (PAA). The results showed that polyacrylate and CMI have similar adsorption and desorption profiles. However, PAA has a relatively poorer performance for squeeze application than conventional phosphonate (DETPMP), whose squeeze treatment life was more than five times longer than PAA in similar test conditions. These indicate that CMI may not be a suitable scale inhibitor for scale squeeze application in both sandstone and limestone reservoirs.

In summary, CMIs are biodegradable green scale inhibitors with excellent scale inhibition performance against different types of oilfield scales. The synthetic routes of CMI could influence molecular weight and scale inhibition efficiency. CMIs are high-potential alternative green-scale inhibitors for conventional phosphonates. However, CMI may not be a suitable scale inhibitor for scale squeeze applications due to its weak interaction with mineral rocks.

17.3.3 Hybrid Polymer

Polyaspartates and carboxymethyl inulins are "green" polymers from naturally formed monomers. Their anti-scaling functional group chemistry, carboxylate, is fixed, with only molecular weight or degree of substitution as variables to optimize the performance of anti-scaling.[10]

Different from polymers synthesized from naturally formed monomers, such as polyaspartates and carboxymethyl inulins, hybrid polymer scale inhibitors are a new type of inhibitor based on a process of synthetic or natural polymer hybridization. Different-natured polysaccharides with various functional characteristics, or polydiallyldimethylammonium chloride (polyDADMAC), can be introduced into the synthesis of hybrid polymers. The synthetic anti-scaling functional groups, such as acrylic, carboxylic, sulfonate, and phosphonate, can be synthesized and chemically bound to the polysaccharide or polyDADMAC components. The synthesis process of hybrid polymers is extremely flexible. Free radical polymerization of water-soluble reactive monomers in aqueous solution is the standard method of production of polyacrylic acids and other acrylate, sulfonate, or phosphonate polymers. The hybrid polymers can be co- or terpolymerized, and the functional group density can be manipulated to target different types of oilfield scale under various field conditions.

A series of environmentally acceptable hybrid scale polymers synthesized based on polyDADMAC is introduced in this chapter to demonstrate the concept, synthesis, and anti-scaling performance of hybrid copolymers.

A typical structure of the hybrid polymeric scale inhibitor is shown in Figure 17.4. Like other polymer inhibitors widely used in the oil industry, the monomer contains anti-scaling functional groups (R1, R2, and R3), such as acrylic, carboxylic, sulfonate, and phosphonate groups. Different from the conventional polymer scale inhibitors, a quaternary amine monomer, polyDADMAC, with a positive charge was introduced to the hybrid polymer.[67]

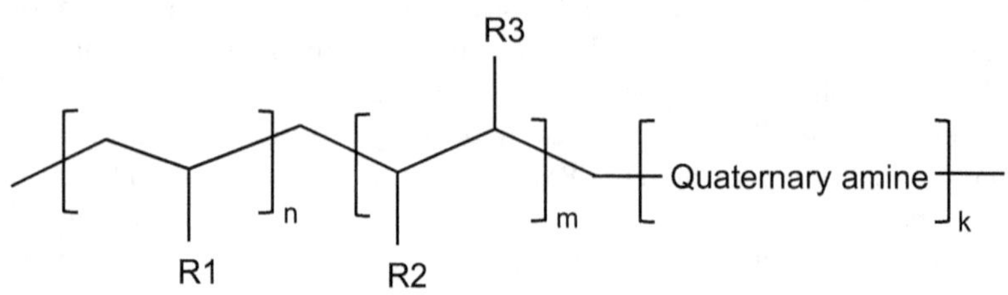

Figure 17.4. Typical structure of polydiallyldimethylammonium chloride-based hybrid polymer. R1, R2, and R3 are anti-scaling functional groups, sulfonate, carboxylic, acrylic, and phosphonate groups.

This type of polyDADMAC-based hybrid polymer demonstrates favorable environmental properties. The toxicity and biodegradation of the active component of polyDADMAC-based hybrid polymers were assessed independently using PARCOM-approved methods to determine the likely environmental impact of their deployment in the field. The active components of these polyDADMAC-based hybrid polymers were nontoxic and highly biodegradable. The LC50s for algae (*Skeletonema costatum*), crustaceans (*Acartia tonsa*), and fish (juvenile turbot, *Scophthalmus maximus*) were >10,000 mg/L. The polyDADMAC-based hybrid polymers were tested using the OECD 306 test and showed more than 20% biodegradation, and a certain modified polyDADMAC showed >60% biodegradation. This type of polyDADMAC-based hybrid polymer is categorized as "gold" with no substitution warning in the United Kingdom and "yellow" in Norway.[68]

While displaying excellent environmental properties, the polyDADMAC-based hybrid polymers also showed good performance results through conventional scale laboratory screening tests. It has comparable scale inhibition performance against calcium sulfate scale to a conventional sulphonate-based polymer tested at 155°C. With modified anti-scaling functional groups, the polyDADMAC-based hybrid polymers demonstrate excellent inhibition performance on barium sulfate and calcium carbonate scales.[68]

The suitability of the scale inhibitors for squeeze treatment was considered at the initial concept stage of the inhibitor development. The improved interaction of these polyDADMAC hybrid polymers is achieved through two approaches: the introduction of positively charged quaternary amine monomers and P tags. The introduction of the positively charged quaternary amine monomer would offer a favorite property for the inhibitor to be adsorbed to the surfaces of the negatively charged sandstone formations. This is particularly useful for the inhibitor to be applied as a squeeze scale inhibitor in downhole applications. It is thought that by "fixing" the polymers to the surface of the formation (*i.e.,* by electrostatic attraction), the inhibitor adsorption can be enhanced. The amount of the positively charged polymer amine monomer had to be optimized during the polymerization in order to allow the adsorbed inhibitors to be released into the formation brine. Otherwise, the retention becomes irreversible. The instruction of P tags embodied in the molecular structure of polyDADMAC hybrid polymers is another approach to improving the interaction between the rock and scale inhibitors, which was found particularly useful to enhance the treatment life when applied as a squeeze treatment.[69–72] The squeeze life of a P-tagged polyDADMAC hybrid polymer and a conventional sulphonate-based scale inhibitor was evaluated using coreflood test in clean sandstone cores. The conventional sulphonate-based scale inhibitor gave 800 kbbl of water protection, compared to 1,400 kbbl of water protection by the P-tagged polyDADMAC hybrid polymer at a minimum inhibitor concentration (MIC) of 2.5 ppm. This indicates that P-tagged polyDADMAC hybrid polymer is a suitable scale inhibitor for scale squeeze application in clean sandstone reservoirs.[52] In addition, the P-tagged polymers are readily detected by the ICP technique, offering an additional advantage over conventional polymeric scale inhibitors as squeezable scale inhibitors. The P-tagged polyDADMAC hybrid polymer has been successfully deployed through squeeze treatment in a long horizontal well in the North Sea with satisfactory results. The treatment life was achieved for up to 6 months in a clean sandstone reservoir.

Moreover, polymer-based scale inhibitors have enhanced thermal stability and calcium tolerance.[73] Holt's study showed that the tested hybrid polymer scale inhibitor has similar thermal stability as a tested conventional polymer-based scale inhibitor. Moreover, the hybrid polymer scale inhibitor has significantly improved calcium tolerance up to two orders of magnitude versus a comparable conventional polymer.[10] These agree with Chen's finding that the P-tagged polyDADMAC hybrid polymer has excellent compatibility with high total dissolved salts and high Ca-containing brines and is thermally stable at high temperatures.[68]

In summary, hybrid polymer scale inhibitors are types of inhibitors based on a process of synthetic or natural polymer hybridization. The hybrid polymer scale inhibitors are biodegradable and environmentally accepted green scale inhibitors to be applied in the North Sea. The introduction of different anti-scaling functional groups and hybrid polymers can provide excellent scale inhibition performance against different types of oilfield scales. With modified functional groups, the hybrid polymers have significantly improved interaction properties with rock and can be easily detected by ICP at low concentrations, which makes them suitable for squeeze treatment in clean sandstones. With additional excellent thermal stability and calcium tolerance, hybrid polymer scale inhibitors are promising alternatives to the conventional scale inhibitors currently used in the oil and gas industry.

17.3.4 Biodegradable Phosphonate-Based Scale Inhibitors

Organophosphorus scale inhibitors have been widely applied to mitigate oilfield scale deposition, especially through squeeze treatments. The phosphonate groups ($-PO_3H_2$) play key roles in oilfield scale inhibition performance, in achieving long squeeze lifetimes through excellent adsorption and desorption characteristics on formation rocks, and easy and accurate scale inhibitor detection at a low concentration. However, most of the conventional polyphosphonate scale inhibitors failed to meet the minimum biodegradation requirements of the increasingly strict environmental legislation. The poor ecotoxicity has severely impeded the use of conventional phosphorus based scale inhibitors, especially in regions with stricter environmental regulations.

In the last decade, researchers have reported on the development of biodegradable phosphonate-based scale inhibitors with better ecological behavior with respect to conventional phosphonate scale inhibitors.

Fisher[74] evaluated a number of selected mono- and diphosphonates, which are not traditionally used in oilfield applications. Some of the tested phosphonates exhibit improved or acceptable biodegradation using the BOD biodegradation test method. The tested phosphonate, characterized chemical bonding and biodegradation test results are summarized in Table 17.8.

Table 17.8: Characterized chemical bonding and biodegradation of mono- and diphosphonates.

Type of Phosphonate	Characterized Chemical Bonding	Typical Inhibitor	Biodegradation (BOD Method)
Monophosphonates	Only one C-P bond per molecule	Phosphonoacetic acid (PA)	>80% in 28 days
		Analogue of PA, with a longer alkyl sequence between the phosphonic acid and COOH groups	25% in 63 days
		Phenylphosphonic acid (PhPA)	23% in 63 days
		(Dimethylamino)methylphosphonic acid	<20% in 63 days
		Bis(2-hydroxyethyl)amino)methylphosphonic acid	<20% in 63 day
Diphosphonates	P-C-P	HEDP	24% in 28 days; 34% in 63 days
	P-C-N-C-P	MEA and ATMP phosphonates	-
		2-HPBMP, 3-HPBMP and a number of other variants, where R contained one or more carboxylic acid group(s)	up to 44% in 28 days; 89% in 63 days

Monophosphonates contain only one phosphonic acid group per molecule, usually only one C-P bond per molecule. They are the only type of phosphonate found in nature. Monophosphonates have a relatively low molecular weight and relatively few charged (ionizable) groups per molecule, which could potentially contribute to better biodegradation. Phosphonoacetic acid (PA), an example of a naturally occurring monophosphonate, has a biodegradation value >80% in 28 days using the marine BOD method, as shown in Table 17.8. PA shows excellent biodegradation performance. However, the single phosphonic acid unit per molecule could reduce the potential of their application in the oilfield due to their limited scale inhibition performance, especially for sulfate scales.

Phenylphosphonic acid, a low-molecular-weight monophosphonate with a phosphonic acid group, exhibits a maximum value of 23% biodegradation in a 63-day marine BOD test, significantly lower biodegradation than PA, indicating that the presence of the phosphonic acid group typically reduces marine biodegradation.

Fisher[74] evaluated the effect of chemical groups of phosphonic acid, COOH, an analogue of PA with a longer alkyl sequence between the phosphonic acid and COOH groups and terminal hydroxyl groups and cyclization on the biodegradation performance. The longer alkyl chain analogue has a maximum of 25% biodegradation in 63 days using marine BOD method. (dimethylamino) methylphosphonic acid is another simple monophosphonate that shows biodegradation <20% in 63 days. The monophosphonate (bis(2-hydroxyethyl)amino) methylphosphonic acid, a terminal hydroxyl group with the phosphonic acid to yield mainly the cyclized structure, exhibits <20% biodegradation over the 63 days. It is hypothesized that the cyclization of the phosphonate further ties up the C-P bond from being effectively cleaved by microorganisms in the marine biodegradation test, even though the total charge of the molecule may be reduced.

This indicates the selected monophosphonates can exhibit biodegradation >60% in 28 days tested by an approved marine BOD test method. However, most of the selected low molecular weight monophosphonates do not exhibit acceptable biodegradation. In addition, generally, the low-molecular-weight monophosphonates do not have very good performance versus sulfate scales, for example, BaSO4 or CaSO4, even though they show some efficacy versus carbonate scales. In general, higher molecular weight, more substituted and highly charged phosphonates typically show low levels of biodegradation.

Diphosphonates contain two phosphonate groups, P-C-P. HEDP, a typicla diphosphonate containing a free amine group, shows a maximum of 24% biodegradation in 28 days and a maximum of 34% biodegradation in 63 days tested by the marine BOD method.[74]

Diphosphonates containing the P-C-N-C-P structural fragment are widely used as oilfield scale inhibitors, such as MEA phosphonate (containing two methyl phosphonate groups per molecule) and ATMP (containing three methyl phosphonate groups per molecule). Fisher[74] synthesized a number of diphosphonates containing the P-C-N-C-P moiety from commercial sources with modified groups containing alcohol (–OH) group(s), carboxylic acid (–COOH), and other substituents. Two P-C-N-C-P diphosphonates, 3-hydroxypropyl bismethylenephosphonic acid (3-HPBMP) and 2-hydroxypropyl bismethylenephosphonic acid (2-HPBMP), and two P-C-N-C-P diphosphonates containing R-OH groups were synthesized and evaluated, where 2-HPBMP is mainly the cyclized phosphonate product while 3-HPBPB is mainly the linear form.

The 3-HPBMP is mainly a linear analogue of MEA diphosphonate; the key difference between the molecules is the length of the intermediate alkyl group and its tendency for cyclization. A number of other variants, where R contained one or more carboxylic acid group(s) were also investigated, and the candidate does exhibit biodegradation up to 89% evaluated by the BOD marine test over 63 days. With fewer alkyl phosphonate groups, for example, diphosphonates with P-C-P, or diphosphonates with P-C-N-C-P type structures, some scale inhibitors exhibit promising biodegradation performance. These inhibitors also show good inhibition performance against different types of scales.

To improve the biodegradation performance of phosphonate-based scale inhibitors, Fisher[74] developed phosphonate scale inhibitors based on the structure of conventional phosphonate inhibitors with modified alkyl, alcohol (–OH) group(s), carboxylic acid (–COOH), and other substituents. Mady *et al.*[75] synthesized and developed a series of phosphonated polyetheramines based on polyetheramine via the Moedritzer-Irani reaction. The general synthesis reaction of modified polyetheramine scale inhibitors is shown in Figure 17.5. The phosphoric functional groups are introduced, bind to the amine groups, and form phosphonated polyetheramines with various molecular weights.

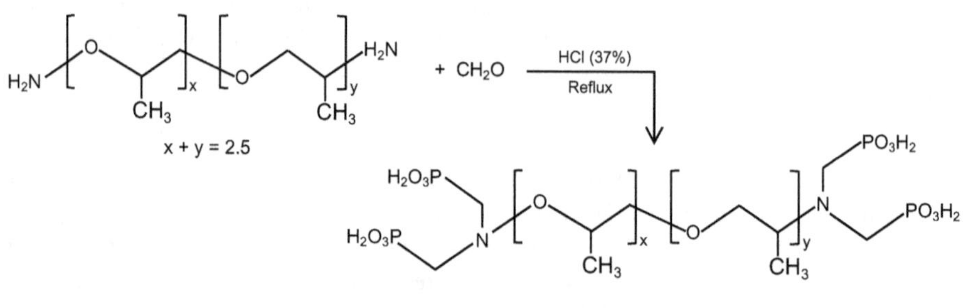

Figure 17.5. Synthesis of phosphonated polyetheramine derivatives.

Five phosphonated polyetheramine scale inhibitors were synthesized and evaluated for biodegradation using the approved OECD 306 method. The results are summarized in Table 17.9, along with the conventional phosphonate inhibitor, DETPMP, as a benchmark.[75, 76]

Table 17.9: Chemical structure and biodegradation of phosphonated polyetheramines.

Scale Inhibitor	Chemical Structure	Biodegradation (OECD 306) 28 Days
SI-1	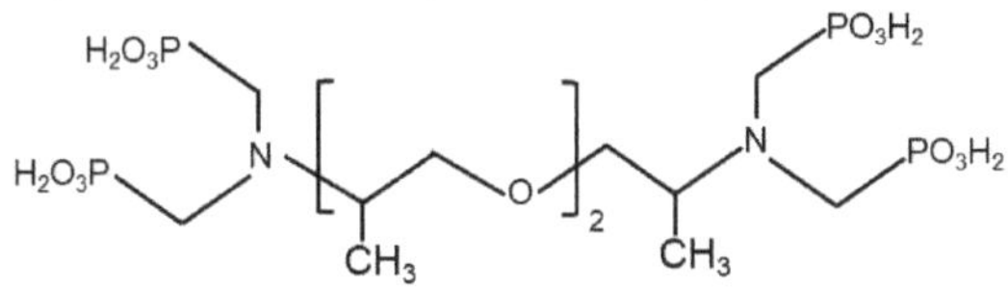	13%
SI-2	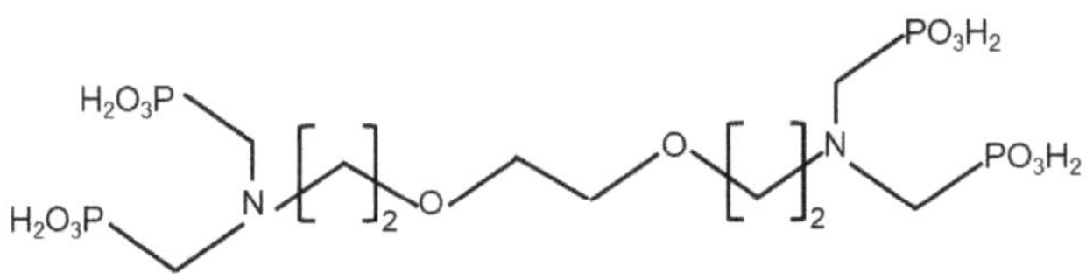	47%
SI-3	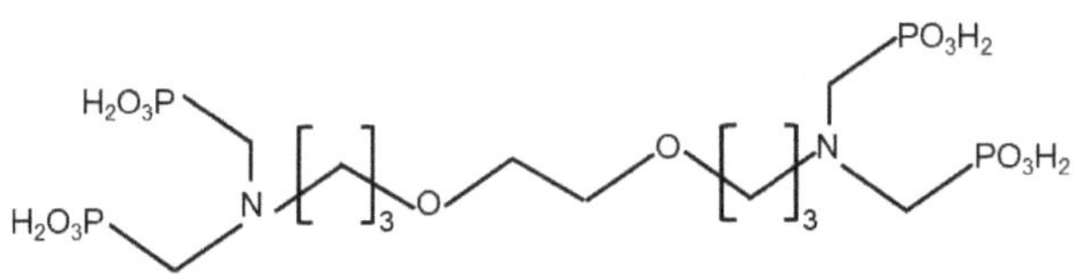	24%
SI-4	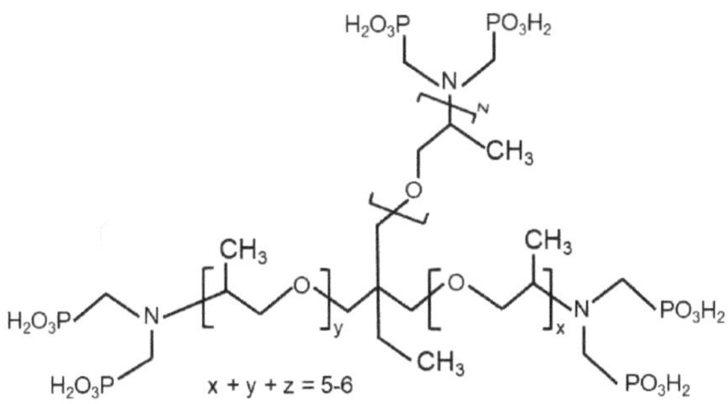	21%
SI-5	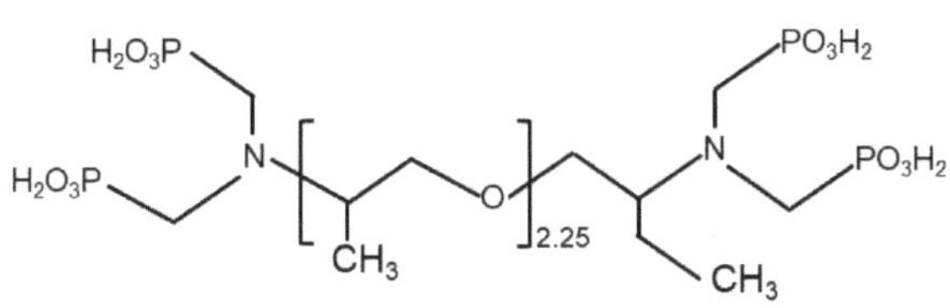	18%
DETPMP	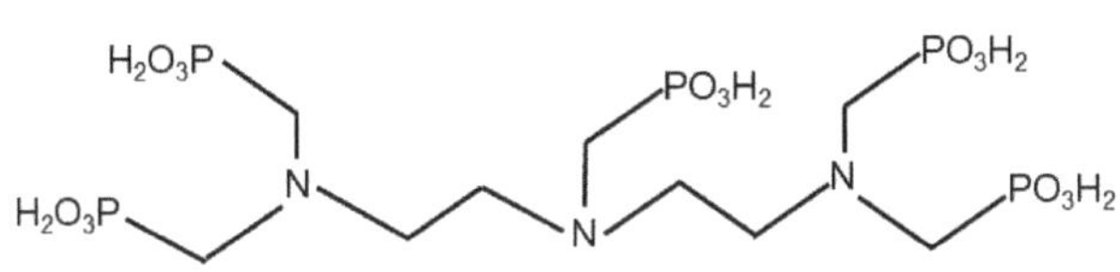	15%

The conventional phosphonate, DETPMP, showed a poor degradation of 15% in 28 days. Most of the phosphonated polyetheramine inhibitors show improved biodegradation performance, except for SI-1. The linear phosphonated polyetheramines SI-2 and SI-3, both having ethoxylate polyether linkages rather than propoxylate groups, give BOD28 values of 47% and 24%, respectively. Both inhibitors can be classified as environmentally acceptable inhibitors in many offshore regions. SI-4 and SI-5, containing branching and a tertiary carbon atom, show a lower BOD28 biodegradation of 21% and 18%, respectively. As the branching in the chain structure has a role in determining the biodegradation rates, it could be assumed that the linear/unbranched backbone of polyetheramines provides better seawater biodegradation performance, and inhibitors with monobranched alkyl groups exhibit lower biodegradation performance than the linear ones. Along with the improved biodegradation performance, these phosphonated polyetheramine inhibitors show very good inhibition capability against both calcite and barite scales, excellent compatibility with Ca^{2+} up to 10,000 ppm, and good thermal stability tested at 130°C. With further tests of rock-inhibitor interaction evaluation, these scale inhibitors could be promising environmentally acceptable alternatives to be applied in the oil and gas industry.[75]

In summary, biodegradable phosphonate-based scale inhibitors are types of inhibitors based on modified alkyl, alcohol (–OH) group(s), carboxylic acid (–COOH), and other substituents based on conventional phosphonate-based scale inhibitors or the introduction of a phosphoric group on the biodegradable polymer backbone. With modified chemical structures, the scale inhibitors could have improved biodegradation performance and meet environmental regulations. With high concentrations of phosphoric groups in the molecules, these inhibitors have excellent inhibition performance against different types of scales, improved interaction properties with rock, and are easily detected by ICP at low concentrations, which makes them as promising environmentally acceptable inhibitors for squeeze treatment in the oil and gas industry.

17.3.5 Other Green Scale Inhibitors

In the last couple of decades, many researchers have reported on the development of new scale inhibitors with better ecological behavior with respect to conventional scale inhibitors.

Natural molecules obtained from plant extracts may be a potential source of green scale inhibitors. Several researchers have worked with various plant materials for scale inhibitor preparation.

Abdel-Gaber *et al.* reported a scale inhibitor extracted from the fig leaves (*Ficus carica* L.). The inhibitor's anti-scaling performance was evaluated on a calcite scale. The results showed this fig leaf-extracted scale inhibitor can increase the nucleation time of calcite and impede calcite deposition in the supersaturated brine.[77]

Boris *et al.* identified seaweed polysaccharides and soybean oil methyl ester as potential scale inhibitors. These inhibitors can be a good choice due to their excellent environmental profiles. Along with scale protection, the inhibitors provide a certain level of additional corrosion protection to the metallic substrates. An additional benefit of the soybean oil methyl ester is that it has better solubility than petroleum-derived scale inhibitors.[78]

Abdel-Gaber *et al.* extracted and prepared green scale inhibitors from the hull and leaf of pomegranates. The inhibitors contain ellagic acid, which could form a calcium complex in the brine to prevent calcium carbonate precipitation as nucleation inhibitors.[79]

Suharso *et al.* investigated gambier extracts as calcium carbonate scale inhibitors. Gambier extracts exhibit anti-scaling function due to the presence of certain chemicals such as tannic acid, catechin, and quercetin, which have proved an anti-scaling effect for calcium carbonate scale. The Gambier extracts showed around 40%–100% inhibition of the growth rate of $CaCO_3$ scale tested at 80°C.[80]

Plant extracts are relatively less expensive, environmentally friendly, degradable, and sustainable. However, plant extract is susceptible to decomposition by microorganisms, and it can be prevented, if biocides are used.

Naturally available organic matter could be another potential source of green-scale inhibitors. McCartney and Alexander[81] found proteinaceous materials and long-chain polymers with carboxy side chains were effective in reducing the rate of formation of calcium carbonate and calcium sulfate scales. The effect of natural organic molecules such as gelatin in low concentrations can reduce the growth of calcium sulfate dihydrate crystals. Tang *et al.*[82] developed the physicochemical and functional properties of hemp (*Cannabis sativa* L.) protein isolate as a green scale inhibitor and compared its anti-scaling performance with that of soy protein isolate. Although hemp protein isolate did not show anti-scaling properties, the dried hemp may be used to extract organic polymers, which might have the potential to be used as an efficient scale inhibitor due to its presence of cellulose, lignin, hemicellulose, and pectin.

Kumar *et al.*[83] worked on a new natural material, pteroyl-L-glutamic acid (PGLU), as a green scale inhibitor. PGLU is a water-soluble vitamin M, which is highly biodegradable and nontoxic and thus has an excellent environmental profile. PGLU showed 100% calcium carbonate inhibition at 120 ppm dose rate using the NACE TM 0374-2016 static jar testing method at 70°C and effective inhibition at 160 ppm dose rate tested using a dynamic tube blocking test at 110°C, which concluded its promising scale inhibition properties as a green scale inhibitor.

Marques *et al.*[84] studied a calcium carbonate scale inhibitor developed from a water-soluble carboxymethyl chitosan (CMC). CMC is a chitosan derivative with enhanced properties like good biocompatibility due to its biodegradability attributed to the carboxymethyl groups on the copolymer backbone.[85] CMC was evaluated under the conditions of oil wells in the northeast of Brazil and presented a MIC of 170 ppm at 70°C. CMC can deform the calcium carbonate crystals, attributed to the effective interactions of CMC through its carboxylate ions and lone pair of electrons on OH and NH_2 groups with calcium ions.[86]

Dickinson and Wilson[14, 87] developed biodegradable equivalent scale inhibitors, based on the conventional basic polycarboxylic, polysulphonated, and polycarboxylic scale inhibitors. The detailed chemistries and synthesized processes were not released. The developed scale inhibitors are environmentally acceptable to apply in the North Sea and meet the standards for "yellow banding" in the Norwegian sector. The inhibitors display equivalent performance in controlling barium sulfate scale to their nonbiodegradable equivalents with both nucleation and crystal growth inhibition mechanisms. The inhibitors also show good inhibition performance against various oilfield scales, such as strontium sulfates and calcium carbonate, along with excellent thermal stability up to 180°C.

Moreover, chemical synthesis is widely applied to develop environmentally acceptable scale inhibitors. Chemical substances containing heteroatom linkages in the main chain, such as ether and/or ester groups, displayed improved biodegradation properties.[88] Lei *et al.*[89] developed polyepoxysuccinic acid as a green scale and corrosion inhibitor for the oil industry. Mady *et al.* developed a new class of modified aliphatic polycarbonates (PCA–COOH and PCA–PO_3H_2–COOH) as environmentally friendly oilfield scale inhibitors for both carbonate and sulfate scales. Wang *et al.*[90] synthesized a series of PBTCA-modified hyperbranched polyethers as biodegradable corrosion and scale inhibitors, and Liu *et al.*[91] developed a polyether copolymer as an environmentally friendly scale and corrosion inhibitor.

In summary, much effort has been put into developing environmentally acceptable scale inhibitors. Various sources, such as natural molecules extracted from plants, natural organic materials, and biodegradable equivalent materials based on conventional polymer inhibitors, have been used to develop newly developed green scale inhibitors. In this direction, more work is still required to be done to meet these specific challenges in the oil and gas industry.

17.4 Summary

The general public has become increasingly aware of the concept of sustainability. Environmental restrictions legislated by regulatory authorities play an important role in shifting uses from conventional scale inhibitors to green scale inhibitors to handle oilfield scaling problems.

The development of an appropriate green scale inhibitor to prevent mineral scales in the oil and gas industry is still an industrial challenge due to the diversity of parameters that should be considered. The environmental regulation applied in the North Sea, one of the strictest environmental regulations applied globally, is shown in Table 17.10 to demonstrate the challenges of developing green scale inhibitors.

Table 17.10: Diversity of parameters of scale inhibitor selection for oilfield applications.

Field Deployment Method	Key Performance Parameters of Scale Inhibitor	Key Environmental Regulation Applied in North Sea
Continuous injection	• Brine compatibility at, especially in high-TDS water with high concentration of calcium ions • Inhibition performance against various oilfield scales, such as carbonate, sulfate, sulfide, and so on • Thermals stability at application temperature up to 170°C	• Biodegradable >20% (OECD 360 28-day test) • Bioaccumulation log Pow < 3 • Toxicity EC50 or LC50 >10 mg/L
Squeeze treatment	• No formation damages • Good inhibitor interaction with rock • Easy and accurate inhibitor analysis • Long squeeze life	

The scale inhibitor must be compatible with the field brines and relatively stable to thermal degradation at reservoir conditions. Brine and scale inhibitor compatibility is a major issue of concern, since premature precipitation of inhibitor complexes during injection may lead to the formation of a pseudoscale with associated fines plugging. Thus, for squeeze treatment inhibitors, it is sufficient not only that a particular species inhibits scale but also that it has the correct interaction with the formation to give sufficiently extended return curves without formation damage.

The development of green scale inhibitors for oilfields should be considered very seriously and balance the anti-scaling performance with the stricter environmental regulations. With the increased biodegradability and excellent scale inhibition performance of a range of green scale inhibitors, oil and gas producers could control oilfield scales, while meeting environmental regulations applied in the North Sea and other regions around the world at a certain level. However, the development of better anti-scaling performance and greener and more cost-effective green scale inhibitors is still one of the industrial challenges in developing and applying green scale inhibitors globally.

Moreover, a number of international environmental regulations with different requests have been legislated and implemented for protection of regional environment. A global harmonized environmental regulation system should be put in place. If global agreement cannot be reached, oil and gas operators and chemical manufacturers should voluntarily ensure that the best chemicals and practices are used to protect the environment. In countries where no regulations are in place, self-regulation through the deployment of chemicals that meet the most stringent regulations that have been introduced in other parts of the world should be adhered to.

Nomenclature

2-HPBMP	2-Hydroxypropyl bismethylenephosphonic acid
3-HPBMP	3-Hydroxypropyl bismethylenephosphonic acid
ATMP	Aminotris(methylphosphonic acid
BCF	Bioconcentration Factor Database
BHEAMP	Bis(2-hydroxyethyl)amino)methylphosphonic acid
BOD	Biological oxygen demand
CEFAS	Centre for Environment, Fisheries, and Aquaculture Science
CHARM	Chemical Hazard and Risk Management
CMC	Carboxymethyl chitosan
CMI	Carboxymethyl Inulin
C-NOPB	Canada-Newfoundland Offshore Petroleum Board
C-NSOPB	Canada-Nova Scotia Offshore Petroleum Board
CWA	Clean Water Act
DADMAC	Diallyldimethylammonium chloride
DETPMP	Diethylenetriaminepenta(methylenephosphonic acid)
DOC	Dissolver organic carbon
ECHA	European Chemical Agency
EPA	Environmental Protection Agency
HEDP	1-Hydroxyethylidenediphosphonic acid
HQ	Hazard Quotient
HMCS	Harmonized Mandatory Control Scheme
ICP	Inductive coupled plasma
ISO	International Organization for Standardization
MEA	Methylene phosphonic acid
MIC	Minimum inhibitor concentration
NEB	National Energy Board
NPDES	National Pollutant Discharge Elimination System
NTMP	Tris(phosphonomethyl)amine, Nitrilotrimethylphosphonic acid, Aminotris (methylphosphonic acid)
OCNS	Offshore Chemcial Notification Scheme
OECD	Organisation for Economic Co-operation and Development
OSPAR	Oslo and Paris Convention
PA	Phosphonoacetic acid
PAA	Polyacrylates
PASA	Polyaspartic acid
PASP	Polyaspartate
PBTC	Phosphono-butane-1, 2, 4-tricarboxylic acid
PCA	Polycarbonates

PGLU	Pteroyl-L-glutamic acid
PhPA	Phenylphosphonic acid
PSAP	Polyaspartate
REACH	Registration, Evaluation, Authorization and Restriction of Chemicals
WOE	Weight of evidence

References

1. J.D. Sallis, W. Juckes, M.E. Anderson, "Phosphocitrate—Potential to Influence Deposition of Scaling Salts and Corrosion," in *Mineral Scale Formation and Deposition*, eds. Z. Amjad (New York: Plenum Press, 1995), pp. 87–98.
2. N. Bonnett, E.R. Fieler, J. Hen, "Application of a novel squeeze scale inhibitor in the Beryl Field," SPE-23107-MS, SPE Offshore Europe, held September 3–6, 1991 (Aberdeen, UK).
3. P.A. Thomas, M.A. Mullins, "A Current Review of Polymeric Structures and Their Practical Significance In Cooling Water Treatment," NACE-130, Corrosion Annual Conference, held June 10–12, 1985 (Gaithersburg, Maryland, USA).
4. S. Patel, A.J. Nicol, "Development a Cooling Water Inhibitor With Multifunctional Deposit Control Properties," *Materials Performance* 35 (1996): pp. 41–47.
5. D.G. Hill, K. Dismuke, W. Shepherd, et al., "Reducing Risk of Oilfield Chemicals to Marine Environments—Development Practices, Achievements and Benefits," SPE-74085-MS, SPE International Conference on Health, Safety and Environment in Oil and Gas Exploration and Production, held March 20–22, 2002 (Kuala Lumpur, Malaysia).
6. S. Kumar, T.K. Naiya, T. Kumar, "Developments in Oilfield Scale Handling Towards Green Technology—A Review," *Journal of Petroleum Science & Engineering* 169 (2018): pp. 428–444.
7. G. Liu, M. Xue, H. Yang, "Polyether Copolymer as An Environmentally Friendly Scale and Corrosion Inhibitor in Seawater," *Desalination* 419 (2017): pp. 133–140.
8. M.F. Mady, P. Charoensumran, H. Ajiro, et al., "Synthesis and Characterization of Modified Aliphatic Polycarbonates as Environmentally Friendly Oilfield Scale Inhibitors," *Energy & Fuels: An American Chemical Society Journal* 32 (2018): pp. 6746–6755.
9. B. Zhang, L. Zhang, F. Li, et al., "Testing the Formation of Ca–Phosphonate Precipitates and Evaluating the Anionic Polymers as Ca–Phosphonate Precipitates and CaCO3 Scale Inhibitor in Simulated Cooling Water," *Corrosion Science* 52 (2010): pp. 3883–3890.
10. S. Holt, J. Sanders, K. Rodrigues, et al., "Biodegradable Alternatives for Scale Control in Oilfield Applications," SPE-121723-MS, SPE International Symposium on Oilfield Chemistry, held April 20–22, 2009 (The Woodlands, Texas, USA).
11. A. Carpenter, "OSPAR Review of the State of the North Sea: Oil Inputs and Their Impacts on The Marine Environment of the North Sea," in *Oil Pollution in the North Sea. The Handbook of Environmental Chemistry*, ed. A. Carpenter (Springer, 2015), pp. 255–282.
12. OSPAR Commission, *OSPAR Commission for the Protection of the Marine Environment of the North-East Atlantic* (London, UK: OSPAR, 2021).
13. C.F. Henry, F.M. Andrew, H.B. Scot, et al., "Progress Towards Biodegradable Phosphonate Scale Inhibitors," 20th International Oil Field Chemistry Symposium, held March 22–25, 2009 (Geilo, Norway).
14. W. Dickinson, L. Sanders, L. Chris, "Development and Performance of Biodegradable Antiscalants for Oilfield Applications," OTC-21788-MS, Offshore Technology Conference, held May 2–5, 2011 (Houston, Texas, USA).

15. M. Thatcher, M. Robson, L.R. Henriquez, et al., *A User Guide for the Evaluation of Chemicals Used and Discharged Offshore* (Aberdeen, UK: European Oilfield Speciality Chemicals Association (EOSCA), 1999).
16. CEFAS, *Guidelines for the UK Revised Offshore Chemical Notification Scheme* (Essex, UK: Burnham-on-Crouch, 2001).
17. OECD, *OECD Guidelines for the Testing of Chemicals* (Paris, France, 1993).
18. EPA, *Office of Pollution Prevention and Toxics: Pollution Prevention (P2) Framework.* EPA-748-B-00-001 (Washington, D.C., USA, 2000).
19. EPA, *Integrated risk information system (IRIS), National Center for Environmental Assessment* (Washington, D.C., USA, 2001).
20. EPA, *CORMIX: Cornell Mixing Zone Expert System (Version 3120).* EPA-823-C-97-002 (Washington, D.C., USA: Office of Water, 1995).
21. EPA, *AQUATOX: A Simulation Model for Aquatic Ecosystems (Release 1). EPA-823-C-00-001* (Washington, D.C., USA: Office of Water, 2000).
22. EPA, *Better Assessment Science Integrating Point and Nonpoint Sources (BASINS) (Version 3).* EPA-823-C-01-01-001 (Washington, D.C., USA: Office of Water, 2001).
23. EPA, *Final NPDES General Permit for New and Existing Sources and New Discharges in the Offshore Subcategory of Oil and Gas Extraction Category for the Western Portion of the Outer Continental Shelf of the Gulf of Mexico (GMG290000) and Notice of a Proposed Modification to that Permit* (Washington, D.C., USA: Environmental Protection Agency, 2023).
24. National Energy Board, Canada-Newfoundland Offshore Petroleum Board, Canada-Nova Scotia Offshore Petroleum Board, *Offshore Chemical Selection Guidelines for Drilling & Production Activities on Frontier Lands* (Canada, 2009).
25. National Energy Board, Canada-Newfoundland Offshore Petroleum Board, Canada-Nova Scotia Offshore Petroleum Board, *Offshore Waste Treatment Guidelines* (2010).
26. C.R. Johnson, N.D. Shindgikar, M.M. Beurel, "Overcoming Environmental and Technical Challenges for Well Integrity—A Global Perspective," SPE-172792-MS, SPE Middle East Oil & Gas Show and Conference, held March 8–11, 2015 (Manama, Bahrain).
27. Department of Health, *Australia Inventory of Chemical Substances* (Canberra, Australia, 2014).
28. Department of Mines and Petroleum, *Environmental Risk Assessment of Chemicals used in WA Petroleum Activities Guideline (version 1a)* (Government of Western Australia, Perth, Australia, 2013).
29. D.G. Hill, M.A. Malwitz, "Green Corrosion Inhibitor Achievements—Assessment of Performance vs. Reality of Regulation," NACE-06488, Corrosion Annual Conference, held March 12–16, 2006 (San Diego, California, USA).
30. K.P. Davis, S.D. Fidoe, G.P. Otter, et al., "Novel Scale Inhibitor Polymers With Enhanced Adsorption Properties," PE-80381-MS, International Symposium on Oilfield Scale, held January 29–30, 2003 (Aberdeen, UK).
31. N. Fleming, G. Ross, C. Hobden, et al., "End Capped Polymer Technology: A Mean of Custom Synthesizing Enhanced Environmental & Scale Inhibitor Performance?," 15th International Oil Field Chemistry Symposium, held March 28–31, 2004 (Geilo, Norway).
32. M.A. Singleton, J.A. Collins, N. Poynton, "Developments in Phosphonomethylated Polyamine (PMPA) Scale Inhibitor Chemistry for Severe BaSO4 Scaling Conditions," SPE-60216-MS, International Symposium on Oilfield Scale, held January 26–27, 2000 (Aberdeen, UK).
33. P. Chen, B. Juliussen, O. Vikane, et al., "Development of Non-Aqueous/Low Density Scale Inhibitor Package for Downhole Squeeze Treatments," SPE-130686-MS, SPE International Conference on Oilfield Scale, held May 26–27, 2010 (Aberdeen, UK).
34. B. Zhang, L. Zhang, F. Li, et al., "Testing the Formation of Ca–Phosphonate Precipitates and Evaluating the Anionic Polymers as Ca–Phosphonate Precipitates and CaCO3 Scale Inhibitor in Simulated Cooling Water," *Corrosion Science* 52 (2010): pp. 3883–3890.
35. S. Lattemann, T. Hopner, "Environmental Impact and Impact Assessment of Seawater Desalination," *Desalination* 220 (2008): pp. 1–15.

36. S. Baraka-Lokmane, K. Sorbie, N. Poisson, et al., "Can Green Scale Inhibitors Replace Phosphonate Scale Inhibitors?: Carbonate Coreflooding Experiments," *Petroleum Science and Technology* 27 (2009): pp. 427–441.
37. P.T. Anastas, J.C. Warner, *Green Chemistry: Theory and Practice* (Oxford: Oxford University Press, 1998).
38. D. Darling, R. Rakshpal, "Green Chemistry Applied to Corrosion and Scale Inhibitors," NACE-98207, Corrosion Annual Conference, held March 22–27, 1998 (San Diego, California, USA).
39. A. Zeino, I. Abdulazeez, M. Khaled, et al., "Mechanistic Study of Polyaspartic Acid (PASP) as Eco-Friendly Corrosion Inhibitor on Mild Steel in 3% NaCl Aerated Solution," *Journal of Molecular Liquids* 250 (2018): pp. 50–62.
40. I.R. Collins, "Surface Electrical Properties of Barium Sulphate Modified By Adsorption of Poly, Aspartic Acid," *Journal of Colloid and Interface Science* 212 (1999): pp. 535–544.
41. C.E. Inches, K.S. Sorbie, C. Christophe, et al., "Thermal Stability of Selected Green Scale Inhibitors," 18th International Oil Field Chemical Symposium, held March 25–28, 2007 (Geilo, Norway).
42. T. Klein, R. Moritz, R. Graupner, *Ullmann's Encyclopedia of Industrial Chemistry* (Weinheim: Wiley-VCH, 2008).
43. H. Adelnia, H. Tran, P. Little, et al., "Poly(Aspartic Acid) in Biomedical Applications: From Polymerization, Modification, Properties, Degradation, and Biocompatibility to Applications," *ACS Biomaterials Science & Engineering* 7 (2021): pp. 2083–2105.
44. Z. Liu, Y. Sun, X. Zhou, et al., "Synthesis and Scale Inhibitor Performance of Polyaspartic Acid," *Journal of Environmental Sciences* 23 (2011): pp. S153–S155.
45. D. Hasson, H. Shemer, A. Sher, "State of the Art of Friendly "Green" Scale Control Inhibitors: A Review Article," *Industrial & Engineering Chemistry Research* 50 (2011): pp. 7601–7607.
46. N. Kohler, F. Ropital, C. Estievenart, et al., "Polyaspartates: Influence of Synthetic Routes on Scale Inhibition and Biodegradability," NACE-02412, Corrosion Annual Conference, held April 7–11, 2002 (Denver, Colorado, USA).
47. C. Estievenart, N. Kohler, F. Ropital, et al., "Mechanisms of Scale and Corrosion Inhibition by Polyaspartates," NACE-04536, Corrosion Annual Conference, held March 28–April 1, 2004 (New Orleans, Louisiana, USA).
48. T. Nakato, M. Yoshitake, K. Matsubara, et al., "Relationships Between Structure and Properties of Poly(aspartic acid)s," *Macromolecules* 31 (1998): pp. 2107–2113.
49. M. Freeman, Y. Paik, G. Swift, et al., "Biodegradability of Polycarboxylates: Structure-Activity Studies," in *Hydrogels and Biodegradable Polymers for Bioapplication, American Chemical Society Symposium Series 627*, eds. R.M. Ottenbrite, S.J. Huang, K. Park (Washington, D.C., 1996), pp. 118–136.
50. L. Addadi, J. Moradian-Oldak, S. Weiner, "Macromolecule-Crystal Recognition in Biomineralization; Studies Using Synthetic Polycarboxylate Analogs," in *Surface Reactive Peptides and Polymers: Discovery and Commercialization, American Chemical Society Symposium Series 444*, eds. C.S. Sikes, A.P. Wheeler (Washington, D.C., 1991), pp. 13–27.
51. C. Sikes, A. Wierzbicki, "Atomic Force Microscopy and Molecular Modeling of Branched and Unbranched Polyaspartate Bound to Calcite and Mica," NACE-97166, Corrosion Annual Conference, held March 9–14, 1997 (New Orleans, Louisiana, USA).
52. I. Collins, "Scale inhibition with poly(amino-acids)," Oilfield Scale Symposium, held January 30–31, 1999 (Aberdeen, UK).
53. R. Ross, K. Low, J. Shannon, "Polyaspartate Scale Inhibitors, Biodegradable Alternative to Polyacrylates," NACE-96162, Corrosion Annual Conference, held March 24–29, 1996 (Denver, Colorado, USA).
54. R.J. Ross, K.C. Low, J.E. Shannon, "Polyaspartated Scale Inhibitors-Biodegradable Alternatives to Polyacrylates," *Materials Performance* 36 (1997): pp. 53–57.

55. S. Baraka-Lokmane, K.S. Sorbie, N. Poisson, et al., "Application of Environmentally Friendly Scale Inhibitors in Carbonate Coreflooding Experiments," SCA2008-05, International Symposium of the Society of Core Analysts, held October 29–November 2, 2008 (Abu Dhabi, UAE).
56. J. Liu, S. Willför, C.L. Xu, A Review of Bioactive Plant Polysaccharides: Biological Activities, Functionalization, and Biomedical Applications," *Bioactive Carbohydrates and Dietary Fibre* 5 (2015): pp. 31–61.
57. D. Pro, S. Huguet, M. Arkoun, et al., "From Algal Polysaccharides to Cyclodextrins to Stabilize a Urease Inhibitor," *Carbohydrate Polymers* 112 (2014): pp. 145–151.
58. B.L. Knudsen, M. Hjelsvold, T.K. Frost, et al., "Meeting the Zero Discharge Challenge for Produced Water," SPE-86671-MS, SPE International Conference on Health, Safety, and Environment in Oil and Gas Exploration and Production, held March 29–31, 2004 (Calgary, Alberta, Canada).
59. D.L. Verraest, J.A. Peters, H. Bekkum, et al., "Carboxymethyl Inulin: A New Inhibitor for Calcium Carbonate Precipitation," *Journal of the American Oil Chemists' Society* 73 (1996): pp. 55–62.
60. B. Bazin, N. Kohler, A. Zaitoun, et al., "A New Class of Green Mineral Scale Inhibitors for Squeeze Treatments," SPE-87453-MS, SPE International Symposium on Oilfield Scale, held May 26–27, 2004 (Aberdeen, UK).
61. J. Liu, S. Willför, C. Xu, "A Review of Bioactive Plant Polysaccharides: Biological Activities, Functionalization, and Biomedical Applications," *Bioactive Carbohydrates and Dietary Fibre* 5 (1): pp. 31–61.
62. L. Boels, G. Witkamp, "Carboxymethyl Inulin Biopolymers: A Green Alternative for Phosphonate Calcium Carbonate Growth Inhibitors," *Crystal Growth Design* 11 (2011): pp. 4155–4165.
63. H. Zhang, X. Luo, X. Lin, et al., "Biodegradable Carboxymethyl Inulin as a Scale Inhibitor For Calcite Crystal Growth: Molecular Level Understanding," *Desalination* 381 (2016): pp. 1–7.
64. S. Kirbogaa, M. Öner, "Investigation of Calcium Carbonate Precipitation in the Presence of Carboxymethylinulin," *CrystEngComm* 15 (2013): pp. 3678–3686.
65. A.T. Kan, G. Fu, C. Fan, et al., "Quantitative Evaluation of Calcium Sulfate Precipitation Kinetics in the Presence and Absence of Scale Inhibitors," SPE-121563-MS, SPE International Symposium on Oilfield Chemistry, held April 20–22, 2009 (The Woodlands, Texas, USA).
66. F.R. Johannsen, "Toxicological Profile of Carboxymethyl Inulin," *Food and Chemical Toxicology* 4 (2003): pp. 49–59.
67. P. Chen, T. Hagen, H. Montgomerie, et al., "Scale Inhibiting Well Treatment," Patent AU2006274668A1, 2011.
68. T. Chen, P. Chen, T. Hagen, et al., "Development of an Environmentally Friendly Polymer Squeeze Inhibitor for Short Perforation and High Water Production ESP Wells," NACE-2013-2661, Corrosion Annual Conference, held March 17–21, 2013 (Orlando, Florida, USA).
69. P. Chen, T. Hagen, H. Montgomerie, et al., "A Scale Inhibitor Chemistry Developed for Downhole Squeeze Treatments in a Water Sensitive and HTHP Reservoir," Oil Field Chemistry Symposium, held March 9–12, 2008 (Geilo, Norway).
70. H. Montgomerie, P. Chen, T. Hagen, et al., "Development of a New Polymer Inhibitor Chemistry for Downhole Squeeze Applications," SPE-113926-MS, SPE International Oilfield Scale Conference, held May 28–29, 2008 (Aberdeen, UK).
71. P. Chen, H. Montgomerie, T.H. Hagen, et al., "Meeting the Challenges in Downhole Scale Control for High-Water-Production Horizontal Wells in an Environmentally Sensitive Part of the Norwegian North Sea," *SPE Production & Operation* 27 (2012): pp. 382–389.
72. H. Montgomerie, P. Chen, T. Hagen, et al., "Development of a New Polymer Inhibitor Chemistry for Downhole Squeeze Applications," *SPE Production & Operation* 24 (2009): pp. 459–564.
73. A.A. Younes, H.H. El-Maghrabi, H.R. Ali, "Novel Polyacrylamide-Based Solid Scale Inhibitor," *Journal of Hazardous Materials* 334 (2017): pp. 1–9.

74. H.C. Fisher, "Progress Towards Biodegradable Phosphonate Scale Inhibitors," 20th International Oil Field Chemistry Symposium, held March 22–25, 2009 (Geilo, Norway).
75. M.F. Mady, P. Bayat, M.A. Kelland, "Environmentally Friendly Phosphonated Polyetheramine Scale Inhibitors-Excellent Calcium Compatibility for Oilfield Applications," *Industrial & Engineering Chemistry Research* 59 (2020): pp. 9808–9818.
76. M.F. Mady, A. Bagi, M.A. Kelland, "Synthesis and Evaluation of New Bisphosphonates as Inhibitors for Oilfield Carbonate and Sulfate Scale Control," *Energy Fuels* 30 (2016): pp. 9329–9338.
77. A.M. Abdel-Gaber, B.A. Abd-El-Nabey, E. Khamis, et al., "Investigation of Fig Leaf Extract as a Novel Environmentally Friendly Antiscalent for $CaCO_3$ Calcareous Deposits," *Desalination* 230 (2008): pp. 314–328.
78. Boris, A.M., Margarita, A.K., Alla, Y.F., "Vapor Corrosion and Scale Inhibitors Formulated From Biodegradable and Renewable Raw Materials," Paper 83, European Symposium on Corrosion Inhibitors, held September 4–8, 2005 (Ferrara, Italy).
79. A.M. Abdel-Gaber, B.A. Abd-El-Nabey, E. Khamis, et al., "A natural Extract as Scale and Corrosion Inhibitor for Steel Surface in Brine Solution," *Desalination* 278 (2011): pp. 337–342.
80. B. Suharso, B. Syaiful, E. Teguh, "Gambier Extracts as an Inhibitor of Calcium Carbonate ($CaCO_3$) Scale Formation," *Desalination* 265 (2011): pp. 102–106.
81. E.R. McCartney, A.E. Alexander, "The Effect of Additives Upon the Process of Crystallization: I. Crystallization of Calcium Sulfate," *Journal of Colloid Science* 13 (1958): pp. 383–396.
82. C.H. Tang, Z. Ten, X.S. Wang, et al., "Physicochemical and Functional Properties of Hemp (Cannabis sativa L.) Protein Isolate," *Journal of Agricultural and Food Chemistry* 54 (2006): pp. 8945–8950.
83. T. Kumar, S. Vishwanatham, S.S. Kundu, "A Laboratory Study on Pteroyl-L-Glutamic Acid as a Scale Prevention Inhibitor of Calcium Carbonate in Aqueous Solution of Synthetic Produced Water," *Journal of Petroleum Science and Engineering* 71 (2010): pp. 1–7.
84. R.G. Macedo, N.N. Marques, J. Tonholo, et al., "Water-Soluble Carboxymethylchitosan Used as Corrosion Inhibitor for Carbon Steel in Saline Medium," *Carbohydrate Polymers* 205 (2019): pp. 371–376.
85. S. Islam, S. Rahman, T. Ahmed, et al., "Chapter 23-Chitosan and Chitosan-Based Biomaterials for Wound Management," in *Handbook of Chitin and Chitosan*, eds. S. Gopi, S. Thomas, A. Pius (Elsevier, 2020), pp.721–759.
86. R. Gabriela, N. Marques, L. Paulucci, et al., "Water-Soluble Carboxymethylchitosan as Green Scale Inhibitor in Oil Wells," *Carbohydrate Polymers* 215 (2019): pp. 137–142.
87. D. Wilson, H. Kelly, "Development of a 'Green' Hydrothermally Stable Scale Inhibitor for Topside and Squeeze Treatment," NACE-10135, Corrosion Annual Conference, held March 14–18, 2010 (San Antonio, Texas, USA).
88. D. Coleman, N. Gathergood, "Biodegradation Studies of Ionic Liquids," *Chemical Society Reviews* 39 (2010): pp. 600–637.
89. W. Lei, F.Y. Wang, M.Z. Xia, et al., "Synthesis and its Scale Inhibition Effect of Green Scale Inhibitor Polyepoxysuccinic Acid," *Journal of Industrial and Engineering Chemistry* 57 (2006): pp. 2207–2213.
90. Y.T. Wang, H.L. Chen, Z.J. Zhang, et al., "Synthesis and Characterization of PBTCA-Modified Hyperbranched Polyether Corrosion and Scale Inhibitors," *Journal of Applied Polymer Science* 136 (2019): p. 48041.
91. G. Liu, M. Xue, H. Yang, "Polyether Copolymer as an Environmentally Friendly Scale and Corrosion Inhibitor in Seawater," *Desalination* 419 (2017): pp. 133–140.

18 Nanomaterials for Scale Inhibition

Yuanyuan Wang[1] and Ping Zhang[1]
[1]Department of Civil and Environmental Engineering, Faculty of Science and Technology, University of Macau, Taipa, Macau

18.1 Introduction

Squeeze treatment using SIs is the most common downhole technique used in oilfields. It can be very money saving and provide long-period inhibition.[1–4] While SIs can effectively delay the scale formation, they often face challenges in long-distance transport within reservoir formations. This limitation can result in a restricted area of scale protection within the formation. Additionally, chemical inhibitors employed in squeeze treatments often exhibit a relatively brief lifespan, and a substantial portion of these inhibitors may be prematurely depleted from the formation. Indeed, many squeeze SIs exhibit one or more disadvantages, including but not limited to low performance, poor biodegradability, short-squeeze lifetime, high cost, and limited thermal stability.[1] These limitations can pose challenges in effectively mitigating the scale formation in oilfield operations. Hence, some researchers have proposed the concept of developing scale inhibitor nanomaterials (SINMs) as a potential alternative to conventional chemical inhibitors for controlling scale in oilfield operations.[5–7] Following suitable surface chemical modifications, SINMs have the potential to significantly prolong the squeeze lifetime by facilitating their migration over long distances within the formation.[7, 8] It is important to note that there is no universally accepted definition of nanomaterials in the literature. Generally, nanomaterials are described as materials with at least one dimension measuring within 100 nm in size.[1]

18.2 Potential for Scale Inhibition

In recent years, there has been a growing body of research focused on exploring the applications of nanotechnology in the oil and gas industry.[1] The aim of these studies is to provide remedies for the typical challenges faced in the oil and gas industry, thereby contributing to technological advancements in various areas such as enhanced oil recovery (EOR), drilling and completion, fracturing, flow assurance, refining, and processing, among others.[9–14] Olayiwola and Dejam have conducted comprehensive studies on the interaction of nanomaterials with low-salinity water and surfactants as a means to enhance oil production from sandstone and carbonate fields.[15] Metal oxide nanoparticles (NPs) are utilized in laboratory studies to inhibit asphaltene deposition for the purpose of ensuring the smooth flow of fluids in the oil and gas industry.[16] Furthermore, NPs are also employed in the process of tubing dewaxing in the oil and gas industry.[17] Compared to conventional chemicals, nanomaterials possess a significant advantage: their superior ability to transport through porous formation media without causing substantial risks of formation damage.[6] There are a lot of literature and numerous publications related to the subject of the migration of various NPs in reservoir formations.[18–20] Nanomaterials exhibit excellent properties in various aspects, including long-distance migration, good diffusion, high retention, and chemical stability in reservoir wells.[21] These properties are beneficial for the application of SINMs in squeeze treatments.

A significant challenge in oilfield scale control is the efficient delivery of SIs to the intended zone. In certain locations, such as wellheads or upstream, SIs can be effectively delivered through continuous injection to achieve the objective of scale control.[22–24] Compared to continuous injection, delivering SIs to the bottom of a well or downhole formation is a significantly more complex process. Therefore, a squeeze treatment is typically employed, which involves injecting a specific volume of inhibitor through the production tubing into the wellbore formation to achieve scale control.[6]

Squeeze treatment is a significant engineering solution for mitigating downhole and wellbore scaling threats. Conventional scale squeeze treatments often fail to achieve the desired effect. This is because the adsorption and/or precipitation reactions limit the migration capacity of the chemical inhibitors, making it challenging to protect a relatively broad area from scale formation near the wellbore.[6, 25, 26] It is evident that the greater the migratory range of chemical inhibitors, the broader the zone surrounding the wellbore that can be safeguarded. Furthermore, it is commonly observed that shortly after the resumption of production, roughly one-third of the introduced inhibitor will swiftly be purged from the reservoir formation.[6] This is primarily because a significant portion of the injected inhibitors moves freely and disperses within the pore space of the formation, instead of being held by the formation material. The squeeze lifetime is significantly shortened due to the aforementioned phenomenon.[6] Therefore, an effective approach to placing SIs further away from the wellbore is required to extend the squeeze lifetime. During traditional squeeze treatments, the SIs injected into the well inevitably engage in interactions with divalent metal species. This interaction leads to the formation of metal-inhibitor complexes or precipitates. Hence, it is conceivable that in a laboratory setting, it is possible to synthesize a solid material consisting of metal-inhibitor compounds with precisely controlled particle sizes and tailored surface chemistry. To successfully introduce the prepared solid material into the deeper formation, a crucial criterion is that the material must exhibit stability, when exposed to brine solutions under representative oilfield conditions. This implies that the material should maintain its structural integrity, chemical composition, and desired functionality without undergoing significant degradation or dissolution in the presence of brine solutions.[6] By carefully controlling the particle size and modifying the surface chemistry of the solid material, it becomes possible to achieve the desired stability in brine solutions under representative oilfield conditions. Precise particle size control ensures that the material maintains its structural integrity, while surface chemistry modification enables the material to resist dissolution or degradation when exposed to brine. These strategies collectively contribute to enhancing the stability of the solid material and making it suitable for successful delivery into deeper formations. Furthermore, it is essential that the prepared material possesses the capability to systematically release scale inhibitor chemicals. In simpler terms, the material should have the ability to release inhibitors steadily and gradually into the produced water under conditions representative of an oilfield, ensuring that the concentration upon return remains above the minimum inhibitor concentration (MIC) for a long time. This controlled release mechanism ensures effective inhibition of scale formation over a prolonged duration, providing reliable protection against scale deposition in the oilfield system.[6] As elaborated above, SINMs exhibit exceptional properties such as fine particle size, strong surface potential, adjustable surface chemical characteristics, morphologies, and so on. These attributes render the prepared SINMs highly suitable for fulfilling the requirements of squeeze treatment in oilfields for scale control.

18.3 Common Scale Inhibitor Nanomaterials

18.3.1 Amorphous Ca-Diethylenetriamine Pentakis (Methylene Phosphonic Acid) Scale Inhibitor Nanomaterials

An early example of a metal-inhibitor scale inhibitor was likely introduced by Shen and colleagues in 2008.[7] In this research, the study involved the combination of a commonly used oilfield phosphonate inhibitor known as diethylenetriamine pentakis (methylene phosphonic acid) (DETPMP) with calcium ions to create a solid precipitate called Ca-DETPMP. This Ca-DETPMP solid was then transformed into a SINM through a centrifugation process while utilizing a polymer phosphino-polycarboxylic acid (PPCA), as illustrated in Figure 18.1. PPCA is an effective polymeric scale inhibitor on its own. When PPCA is introduced into the system, it has two notable effects on the Ca-DETPMP SINM. Firstly, it reduces the particle size of the SINM, bringing it within the range of 100–200 nm. Secondly, the introduction of PPCA leads to an increase in the zeta potential of the particles, which becomes approximately −30 mV. This change in zeta potential indicates improved aqueous stability, making the SINM more desirable for practical applications. According to X-ray diffraction (XRD) characterization, this material appears to be in an amorphous state. Just like in this crucial study, metal-phosphonate SINMs were also synthesized using a surfactant-assisted synthesis approach, as detailed by Zhang *et al.* in their 2010 work.[25] Additionally, these SINMs were created using a silica-templated method as well.[8]

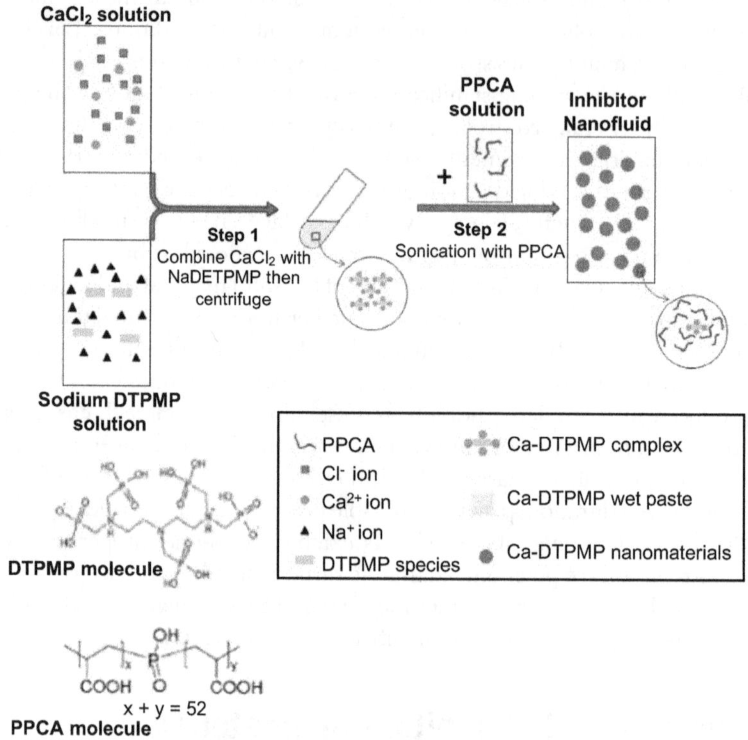

Figure 18.1. Illustration depicting the synthesis process of Ca-diethylenetriamine pentakis (methylene phosphonic acid) nanofluid modified with phosphino-polycarboxylic acid. Reproduced with permission from Ref. [5]

After the successful synthesis of SINMs, a critical aspect that needs investigation is the ability of these NPs to transport within the formation media. To address this, a laboratory column setup was created to evaluate the migration capacity of SINMs within a porous medium that mimics common oilfield conditions. The result from Figure 18.2 reveals a significant enhancement in the transport of these materials within a calcite (calcium carbonate) porous medium, when PPCA is present, as opposed to virtually no transport observed in the absence of PPCA. Calcite was chosen as the column-packing material due to its significance as the primary reactive component with phosphonate inhibitors. This selection was made, because calcite is known to interact prominently with phosphonate-based inhibitors, making it a suitable medium to simulate and study the behavior of phosphonate inhibitors in practical applications.[25, 27] Moreover, following the conclusion of the transport study, the distribution of DETPMP solid phase in the calcite medium was thoroughly examined. The findings indicate that the Ca-DETPMP SINM, modified with PPCA on its surface, exhibits a relatively uniform distribution throughout the full extent of the packed column. This indicates that surface modification of Ca-DETPMP SINM with PPCA effectively enhances its uniform dispersion within calcite media. The mechanism of PPCA action is reflected in three aspects: firstly, it alters the surface morphology of Ca-DETPMP NPs, potentially leading to changes in their surface physical and chemical properties. Secondly, it increases the Zeta potential of Ca-DETPMP SINM, resulting in changes in particle surface charge properties. Thirdly, PPCA enhances the electrostatic repulsion between Ca-DETPMP SINM and calcite particles, promoting the transport of Ca-DETPMP SINM within calcite media, while preventing NP aggregation and increasing their dispersion.

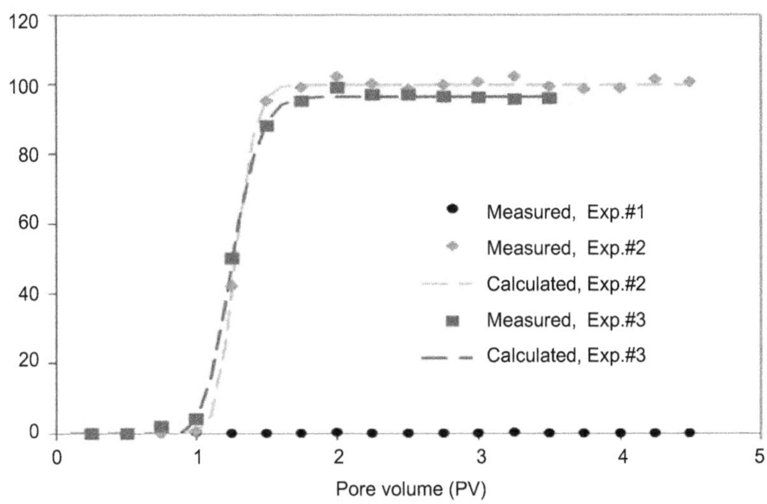

Figure 18.2. Breakthrough curves of scale inhibitor nanomaterials in a calcite medium. The dots, diamonds, and squares correspond to the breakthrough levels measured experimentally in transport Experiments #1 to #3, respectively. The dashed lines indicate the breakthrough levels calculated using a one-dimensional advection-dispersive equation (1-D ADE). Reproduced with permission from Ref. [5]

In summary, PPCA appears to have a dual role in modifying the surface properties of Ca-DETPMP SINM, increasing their zeta potential and facilitating electrostatic repulsion between the SINM and calcite particles, which collectively enhance the transportability of the SINM in the porous medium. These effects are crucial in understanding and optimizing the performance of SINMs in oilfield applications.

Certainly, another important characteristic of SINMs is their capacity to gradually release SIs into the produced water, ensuring that the inhibitor concentration remains consistently higher than the MIC for an extended duration, while maintaining control over the release process. The laboratory squeeze simulation experiment is a testing method conducted in a controlled laboratory environment to assess the effectiveness of a scale inhibitor product in terms of its ability to inhibit scale formation over an extended period and its long-term performance.[28] In this research, squeeze simulation testing replicates the same key steps as field operations, including preflush, pill injection, overflush, shut-in, and resumed production. The synthetic brine employed in this study was carefully balanced to maintain equilibrium with calcite, ensuring that there would be no dissolution or precipitation of calcite solids during the test, which is consistent with conditions expected during reservoir flowback. To ensure the preservation of formation integrity and effectively control clay expansion, two Ca-DETPMP nanofluids were selected and prepared using high-concentration potassium chloride.[28] Prior to testing, the two samples were subjected to ultrasonic treatment using a probe sonicator device to enhance their effective flow and dispersion capabilities within the formation. In the squeeze simulation experiment, a 1% KCl solution was first used to flush the sandstone columns, simulating the common preflushing treatment used in field squeeze operations. Whereafter, they introduced 0.5 pore volume (PV) of the nanofluid into the formation, followed by the injection of the remaining half PV of a 1-M NaCl solution, a standard field overflush fluid. The purpose of this overflush treatment was to drive the nanofluid toward the opposite end of the column. In the wake of completing the overflush treatment, they extracted the sandstone material from the column to measure the distribution of DETPMP solid phase within it. The results indicate that in the case of both tested nanofluids, a significant portion of the injected SINM was effectively transported to the latter half of the column. This is evident from the notably higher DETPMP solid phase content in the second half of the column compared to the first half. These findings strongly imply that the SINM in question exhibits excellent transportability within the sandstone column, enabling it to be pushed deeper into the formation for effective scale control purposes.

The authors conducted comprehensive squeeze simulation tests in Frio sandstone columns.[28] Following the overflush treatment, the columns were closed off for an 18-h period, replicating a common field practice aimed at facilitating the adherence of the inhibitor to the surfaces of the formation. Subsequently, the sandstone columns were flushed with a synthetic brine equilibrated with calcite, but in a reverse direction, mimicking the resumption of production in a field setting. For comparative purposes, the authors also executed two additional squeeze simulation tests, one employed an acidic DETPMP pill with a pH of 1.5 and the other with a pH of 6.7. As a result, the neutralized pill exhibits less robust attachment to the formation surfaces when contrasted with the acidic pill. This difference arises from the fact that a portion of the acidic pill undergoes a chemical reaction with divalent metals, primarily calcium ions (Ca^{2+}), forming divalent metal-phosphonate complexes that bind to the surfaces of the formation. The return of phosphonates from the nanomaterial is primarily dominated by the dissolution of these calcium-DETPMP complexes, rather than the release of DETPMP from the surfaces of the formation. The extended duration of the squeeze using SINM compared to the acidic pill is due to only a partial interaction of DETPMP in the acidic pill, creating complexes with divalent metals. The remaining inhibitor persists as unbound phosphonate. These findings also indicate that it might be possible to adjust the composition of nanofluids to adjust the duration of the squeeze and modify the recovery characteristics of these inhibitor nanomaterials.[28] As outlined in the preceding discussions, a specific volume of inhibitor nanofluid can be introduced through the production well into the deeper layers of the formation. Following a designated shut-in period, these delivered nanomaterials have the capacity to migrate back along with the produced water, effectively mitigating scale deposition.

18.3.2 Crystalline Ca-Diethylenetriamine Pentakis (Methylene Phosphonic Acid) Scale Inhibitor Nanomaterials

In 1994, Kan and his team conducted a groundbreaking study, which marked the initial exploration of a crystalline Ca-DETPMP material.[29] They found that this crystalline form exhibited an aqueous solubility significantly lower by several orders of magnitude when compared to its amorphous counterpart. This observation was conducted under conditions that effectively simulate typical oilfield scenarios, encompassing a diverse range of concentrations for both calcium and DETPMP. In a different research, both static and dynamic dissolution assessments indicated that the transition of Ca-DETPMP from a solid amorphous state to a crystalline state is a gradual process.[27] The observed alteration was identified to involve an interim stage marked by a shift from an amorphous to a microcrystalline structure, as evident from XRD and scanning electron microscopy examinations.

Following the laboratory investigations of Ca-DETPMP synthesis mentioned earlier, practical experiments were conducted to synthesize low-solubility crystalline Ca-DETPMP solid-state inorganic NPs.[30] The benefit of this material lies in its low solubility, enabling it to release the DETPMP inhibitor at significantly lower concentrations upon return. This characteristic implies that, for a specific quantity of inhibitor injected into the formation, the lifespan of the squeeze operation can be substantially prolonged. The initial documented synthesis of crystalline Ca-DETPMP involved the use of silica nanomaterials as a template. The synthesis method involves combining a solution containing DETPMP with another solution of Ca-silica slurry, resulting in the formation of a white precipitate of Ca-DETPMP. After the creation of the Ca-DETPMP slurry, efforts were made to transform this material into a crystalline phase by utilizing a diafiltration cell configuration. The diafiltration cell, which combines dialysis and filtration, permits a consecutive passage of brine containing calcium via the formed solid Ca-DETPMP, facilitating the achievement of a phase transition. Experiments on transport indicate that the crystalline material can effectively traverse a porous medium within the formation, and the point of breakthrough was significantly influenced by the velocity of the pores.[30] Laboratory squeeze simulation tests were conducted to investigate the long-period inhibitory performance of the synthetic crystalline Ca-DETPMP solid inhibitor.[31]

To assess how well the crystalline Si-Ca-DETPMP nanomaterials perform during the flowback process and to replicate the interaction between these nanomaterials and the reservoir formation under realistic conditions, researchers conducted a sequence of laboratory simulations mimicking the squeeze process.[31–33] In these tests, authors introduced the Si-Ca-DETPMP nanofluid into both Louise sandstone and calcite columns using a completely contained squeeze protocol. Over the course of a 24-h shut-in period, the nanomaterials were permitted to bind to the surfaces of the formation materials through mechanisms such as sorption or surface complexation. Following this period, it was reversed the flow direction and flushed the column with a synthetic brine. This reverse elution process aimed to mimic the production process in an oilfield, with connate fluids flowing back through the formation. Figure 18.3 illustrates the return curves obtained from the three-squeeze simulation experiments conducted under similar conditions. In the experiment involving an acidic pill solution, it was observed that the concentrations of DETPMP returned were initially quite high, reaching levels of a thousand mg/L or even higher within the first several pore volumes. Subsequently, the concentration of DETPMP drops gradually to levels below 1 mg/L within a few hundred pore volumes. The proposed mechanism for the retention of phosphonate inhibitors within formation materials rich in calcite involves the formation and retention of calcium-phosphonate salt through precipitation. This occurs when a significant quantity of phosphonate pill solution is introduced into the formation materials, resulting in the creation and retention of these calcium-phosphonate compounds.[31–33]

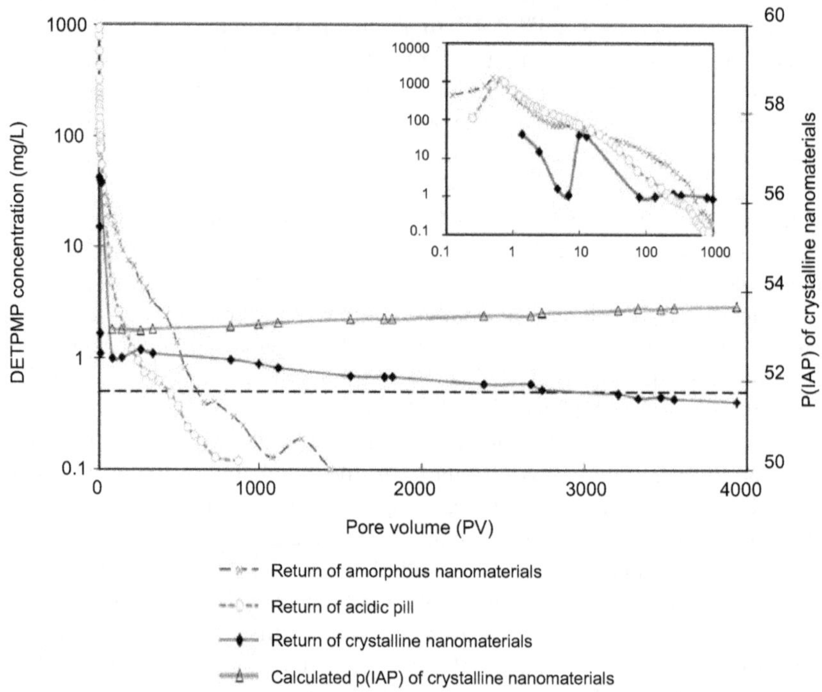

Figure 18.3. The extended flowback behavior of the crystalline Ca-diethylenetriamine pentakis (methylene phosphonic acid) (DETPMP) scale inhibitor nanomaterial (SINM) in the squeeze simulation test conducted in a sandstone medium. Three distinct return curves were generated: one from a crystalline SINM, another from an acidic pill solution, and the third from an amorphous nanomaterial fluid. The inset graphically depicts the concentrations of DETPMP returns within the initial 1,000 PVs. The dotted line indicates a concentration of 0.5 mg/L for DETPMP. Reproduced with permission from Ref. [31]

The findings depicted in Figure 18.3 indicate that the crystalline inhibitor consistently sustains a DETPMP return concentration within the range of 0.5–1 mg/L for an impressive span of almost 4,000 PVs. In contrast, the acidic DETPMP pill and the amorphous Ca-DETPMP solid inhibitor can only maintain their inhibitory effectiveness for a shorter period. The acidic pill remains effective for several hundred PVs, while the amorphous Ca-DETPMP inhibitor shows effectiveness for approximately 1,200 PVs. The initial decrease in inhibitor return concentration during the early stages of the return of crystalline SINM can be attributed to its limited ability to dissolve in water. Crystalline SINM has significantly lower solubility when compared to conventional pills and amorphous SINM. Consequently, only a small quantity of crystalline SINM dissolves into the brine water. As a result, the release of DETPMP inhibitor from crystalline SINM into the brine occurs slowly and continuously throughout the squeeze simulation test. This gradual release extends the duration of effective inhibition.

Building upon the earlier research on the synthesis of crystalline SINM, more recent research has introduced a simplified one-pot synthesis approach for producing low-solubility Ca-DETPMP material.[34] This method involves the preparation of low-solubility metal-phosphonate colloidal inhibitors (MPCIs) using a citrate-assisted approach. During the preparation process, citrate initially forms complexes with Ca^{2+}, which are then mixed with a DETPMP solution to create Ca-DETPMP precipitates. Because of the substantial decrease in the concentration of free Ca^{2+} ions in the liquid phase, the obtained Ca-DETPMP compound exhibits minimal solubility within the liquid phase. Laboratory transport experiments have revealed a significant drop in effluent efficiency, declining from 100% to 10% after just 0.6 PVs of brine flushing, and further decreasing to less than 4% after an additional 0.6 PVs of flushing.[34] This finding strongly suggests that the attachment of the MPCI material onto the sand surface during transport is a reversible process, primarily governed by physical adsorption phenomena. It is reasonable to speculate that during hydraulic fracturing procedures, the resulting Ca-DETPMP can be transfused into shale formations alongside proppant materials. The resulting Ca-DETPMP can be transported to deeper layers of the formation and subsequently adsorbed by the formation materials. However, it's important to note that if the Ca-DETPMP material's adsorption to the formation is overly strong, it could severely limit its migration within the formation medium. The reversibility of Ca-DETPMP material transport in sand formations indicates a great capacity for migration. This characteristic contributes to an extended duration of effective squeeze in the squeeze simulation test.

In a field-scale squeeze treatment, an inhibitor solution is introduced into the formation and allowed to remain within it. Subsequently, the inhibitor retained on the formation surface is released. It is crucial to achieve two key objectives: (1) maximize the retention of the inhibitor on the surface of formation to minimize the release of a substantial amount during the initial return phase; (2) ensure that the concentration of returned inhibitor remains above the MIC level for an extended squeeze duration. In this research, after completing the transport experiment, the authors conducted a laboratory-scale squeeze simulation test in a sand medium to assess how the prepared Ca-DETPMP behaves under oilfield conditions.[34] The findings indicate that following a 24-h shut-in treatment, the initial return concentration of DETPMP, after 0.4 PVs of return, was approximately 1,700 mg/L. Within the first 1.2 PVs, the DETPMP return accounted for less than 3% of the total amount of the inhibitor injected. For comparison, in the control study where a 2% aqueous DETPMP pill was used, the DETPMP return concentration was significantly higher, reaching 15,700 mg/L after 0.4 PVs of return. This return concentration closely matched the DETPMP concentration in the pill and was more than nine times higher than the initial return concentration observed in the prepared Ca-DETPMP study. Consequently, in the

DETPMP test, more than 80% of the injected DETPMP was recovered after only 1.2 PVs. The concentration of inhibitor returned in the early stages of the squeeze treatment was primarily determined by the mass of unretained "free" inhibitor present in the pores of the formation. Irrespective of the retention mechanism at play, the quantity of unretained inhibitor typically gets expelled from the formation's pore spaces soon after the start of the inhibitor return, leading to the elevated concentration observed during the initial flowback phase. In the control study, the sand material exhibited limited capability in retaining the DETPMP inhibitor, allowing only a restricted amount to be held within its structure. This clarifies the result of an exceptionally high-DETPMP return concentration in the control study. Conversely, the MPCI material exhibited a significant reduction in the release of DETPMP during the initial return, indicating that a substantial portion of the injected Ca-DETPMP materials was effectively retained in the sand pore space. This suggests that the prepared Ca-DETPMP possesses favorable transport characteristics within the sand medium, implying that the adsorption of Ca-DETPMP onto the surface of sand appears to be relatively weak during the transport process. This underscores the effectiveness of Ca-DETPMP in adhering to the sand medium and mitigating the early release of the inhibitor. A shut-in period is intentionally included in the field squeeze treatment process, occurring from the end of the inhibitor injection until the commencement of the inhibitor return phase. The primary purpose of this shut-in period is to facilitate the bonding of the injected inhibitor to the surface of the formation materials. This, in turn, helps reduce the quantity of inhibitor that remains unretained within the pore spaces of the formation. It is plausible that the aqueous stability of Ca-DETPMP was adversely affected during the 24-h shut-in period, resulting in the retention of Ca-DETPMP complexes on the surface of the sand material. This study proposes a simpler and cost-effective approach. It significantly broadens our knowledge regarding colloidal inhibitor materials and paves the way for their practical application in oilfield scale squeeze treatments.

18.3.3 Carbon-Based Scale Inhibitor Nanomaterials

In the past few decades, carbon-based nanoparticles (CBNs) have emerged as an emerging area of nanotechnology, unveiling a plethora of promising applications.[35] This burgeoning field has presented numerous opportunities, setting the stage for innovative applications and advancements across various industries. CBNs can be classified into three main types: graphite-based, carbon-based dots, and hybrids.[36] Among the three CBNs, carbon nanotubes (CNTs) are unique CBNs that exhibit a one-dimensional structure. CNTs can be classified based on the number of layers they possess, leading to single-walled CNTs, double-walled CNTs, and multiwalled carbon nanotubes (MWCNTs).[37]

CNTs hold significant potential in the oilfield industry due to their outstanding properties, including high strength, flexibility, excellent conductivity, and chemical stability.[14, 38–41] Researchers have introduced an innovative technique called nano-assisted squeeze treatment (NAST) aimed at enhancing the efficiency and durability of squeeze treatments in the oilfield.[42, 43] In the NAST process, the procedure involves the adsorption of CNTs onto the area near the wellbore, leading to a permanent modification of this area. Following this step, the injection of a SI solution leads to the adsorption of PPCA onto the previously adsorbed CNTs. The CNTs employed in this study had specific characteristics, including a diameter of less than 8 nm, a length ranging from 0.5 μm to 2 μm, and a surface area of 500 m^2/g. In simpler terms, NAST entails the initial binding of CNTs to the near-wellbore area, creating a lasting modification. Subsequently, a PPCA solution is introduced, which adheres to the CNTs. The CNTs utilized in this study were of a specific size and surface area. Tests showed that a higher amount of PPCA was adsorbed onto the inner surface of the core when CNTs were used, as opposed to the conventional coreflood test. Furthermore, in the conventional coreflood, the core plug reached saturation with PPCA after 3 PV; in contrast, the core plug treated with CNTs did not achieve full saturation even after 10 PV. The research extensively investigated how this CNT-capped PPCA interacts with the rock's surface chemistry. This approach holds potential benefits for reservoir management and EOR processes. The characteristics of calcium carbonate crystals formed through biomimetic mineralization with untreated and modified CNTs were analyzed using advanced techniques, including transmission electron microscopy (TEM), XRD, and Fourier transform infrared spectroscopy (FT-IR).[44, 45] These methods provided detailed insights into the morphology and structure of the mineralized crystals, shedding light on the influence of original and modified CNTs on the mineralization process. The findings from the study demonstrated that carboxyl-functionalized CNTs had a positive impact on the formation and stability of calcium carbonate ($CaCO_3$) crystals.

Takizawa *et al.* found that multiwalled carbon nanotube-co-polyamide (MWCNT-PA) nanocomposite reverse osmosis desalination membranes, exhibit significant anti-scaling performance.[46] The introduction of a small amount of MWCNTs can effectively be incorporated into the membrane, preventing scaling efficiently without altering the overall structural integrity of the membrane. Ishtiaq *et al.* improved the efficiency of squeeze treatment by enhancing the adsorption performance of SIs using graphene oxide (GO) and/or CNTs.[47] In the nano-carbon-enhanced squeeze treatment (NCEST) experiments, GO and/or CNTs were employed to improve the adsorption of ethylenediaminetetraacetic acid (EDTA) on rock formations. The results indicated a higher degree of adsorption of the scale inhibitor in rock formations treated with nanomaterials. This is attributed to the significant increase in the number of active adsorption sites for the scale inhibitor on the rock matrix due to the nanomaterials. The lower final concentration of EDTA in the effluent suggests effective retention of EDTA in the formation, thereby prolonging the lifespan of the squeeze treatment. Furthermore, conducting coreflood experiments using GO as a nanomaterial, it was observed that during static adsorption, the adsorption capacity of EDTA on GO was around 590 mg/g. However, in dynamic adsorption tests, the capacity decreased to approximately 180 mg/g. This reduction can be attributed to the fact that not all of the surface area of GO is directly exposed to EDTA due to the attachment of the nanomaterial to the rock. Therefore, the observed adsorption value is lower in the dynamic test compared to the static test, where nearly the entire surface of GO is available for direct contact with EDTA. It is important to recognize that the coreflood test is the most realistic representation of the squeeze treatment process under reservoir conditions. Consequently, the adsorption value observed in this test is a more accurate reflection of how this process would perform in real-world scenarios.

Carbon-based quantum dots (CQDs) have gained more and more attention owing to their unique properties. In one study, Hao *et al.* produced and utilized carboxyl-modified CQDs (CCQDs) as an ecofriendly scale inhibitor to it control the formation of gypsum and barite scales.[48] The findings revealed that in the absence of any additives, calcium sulfate crystals exhibited a rod-like structure. However, with the introduction of 20 mg/L of CCQDs, a noticeable alteration in the morphology of calcium sulfate crystals occurred. When SIs were not present, barite scales formed as brick-like crystals. Conversely, with the introduction of CCQDs as SIs, the morphology of the barite scales changed into intricate flower-shaped structures resembling "fleshy plants." Static scale inhibition experiments demonstrated that at a low dosage, CCQDs exhibited exceptional scale inhibition performance for both gypsum and barite scales. The inhibition mechanism suggested by the aforementioned findings is proposed. When CCQDs are dissolved in water, carboxyl anions with a negative charge form chelates with Ca^{2+}, thereby increasing the solubility of inorganic salt. This process disrupts the normal growth of inorganic salt crystals, leading to a reduction in the formation of salt scale. Additionally, CCQDs adsorb onto the microcrystals of inorganic salt, enhancing repulsion between particles. This effect impedes their coalescence, maintaining a well-dispersed state that prevents or minimizes scale formation. Furthermore, there is a significant increase in Brownian motion when using smaller sized CCQDs. The rapid motion of these carbon quantum dots enhances the disruption of calcium sulfate crystals, leading to irregular growth and distortion of the lattice structure. Consequently, the calcium sulfate scale undergoes fragmentation and disorder.

18.3.4 Metal/Metal Oxide Scale Inhibitor Nanomaterials

The use of metals and metal oxide nanoparticles (M/MO-NPs) has garnered significant attention in recent years across multitudinous industries, environmental applications, catalysis, and biomedical fields.[49] Particularly, the functionalization of M/MO-NPs with various linkers, like polymers and surfactants, has become a focal point in the oil and gas industry. These modified NPs show substantial promise for applications in EOR and flow assurance management.[41, 50] In the context of EOR, functionalized NPs can be integrated as additives in injection fluids to enhance the recovery of oil. They possess the capability to influence the wetting properties of reservoir rocks, alter the interfacial tension between oil and water, and facilitate the displacement of trapped oil. The surface chemistry of these functionalized NPs can be customized to interact selectively with specific components within the reservoir, enabling targeted and efficient oil recovery strategies. In the realm of flow assurance management, which is essential for preventing blockages and disruptions in production pipelines, functionalized NPs come into play. They can effectively address issues such as wax deposition, asphaltene precipitation, and hydrate formation by either inhibiting or modifying the formation of these undesirable substances. The functionalized NPs act as dispersants, inhibitors, or modifiers, preventing the aggregation or adhesion of these materials and thereby ensuring a smooth and uninterrupted flow within the pipelines. The tailored functionalization of M/MO-NPs enables them to engage specifically with components present in the oil and gas industry, offering effective solutions to the challenges encountered in EOR and flow assurance management.

Magnetic NPs have garnered growing interest in their application within the oil and gas industry. They have been found utility in a range of areas, including heavy oil recovery, targeted adsorption, reservoir sensing, and imaging.[50] In the field of scale control, Do *et al.* achieved successful synthesis of magnetic iron oxide NPs encapsulated within maleic-coacrylamido-2-methyl-1-propanesulfonate copolymers. These NPs exhibited an average size distribution ranging from 20 nm to 30 nm.[51] The authors first prepared oleic acid-coated iron oxide nanoparticles (OMNPs) through the coprecipitation of Fe^{2+} and Fe^{3+} ions with the assistance of ultrasound irradiation. Maleic acid and sodium 2-acrylamido-2-methyl-1-propanesulfonate (MA-co-AMPS), as a copolymer, were subsequently employed to encapsulate the aforementioned OMNPs using inverse mini-emulsion polymerization. The encapsulation process resulted in the formation of OMNPs encapsulated with MA-co-AMPS, leading to the production of MA-co-AMPS-OMNPs. They point out the core-shell structure can be formed during the synthesis process of MA-co-AMPS-OMNPs. The TEM micrographs, as well as calculations using the Debye-Scherrer equation, revealed that the average size of MA-co-AMPS-OMNPs was approximately 20–30 nm. Additionally, the MA-co-AMPS-OMNPs demonstrate excellent thermal stability under inert conditions. The results of the scale inhibition experiment indicated that when MA-co-AMPS-OMNPs were used as a scale inhibitor, the resulting calcium carbonate precipitates were in the aragonite form. Furthermore, the calcium inhibition efficiencies of magnetic NPs and pure copolymer were determined to be 63.6% and 65.0%, respectively, under the conditions of 90°C with a 5 ppm concentration of SI. In addition to exhibiting scale inhibition activity comparable to the parent copolymer, magnetic nanomaterials used as oilfield SIs possess a significant advantage: after their application in oilfields, these magnetic NPs can be readily detected in produced water through magnetic monitoring devices.[51]

To advance the research on oilfield scaling control, a novel type of scale inhibitor nanoparticles (SINPs) called Al-SPCA (sulfonated polycarboxylic acid) was synthesized using an environmentally friendly method. In an effort to advance research in the control of scaling in oilfields, a novel type of SINPs called Al-SPCA was synthesized using an environmentally friendly method.[52] The study investigated the effect of various factors, such as precursor concentration, reaction time, and ionic strength, on the particle size of Al-SPCA NPs. After optimizing the preparation conditions, Al-SPCA NPs with a particle size of approximately 80 nm were successfully produced. The flowback performance column test of the Al-SPCA nano inhibitor demonstrated a remarkable long-term return efficiency of over 3,800 PVs. This indicates the effectiveness of Al-SPCA NPs in inhibiting scale formation and facilitating the recovery of oilfield production. Furthermore, when Al-SPCA NPs were used in combination with PPCA, they achieved nearly complete breakthrough in Iceland spar calcite formation. This suggests that Al-SPCA NPs, when combined with appropriate compounds, can be highly effective in controlling and mitigating scaling issues in the petroleum industry.

Following the aforementioned study, boehmite NPs (-AlO(OH)) with a particle size ranging from 3 nm to 10 nm were subjected to modification.[53] This involved cross-linking with SPCA to obtain cross-linked polymeric SIs, denoted as AlOOH-SPCA SINPs. The developed AlOOH-SPCA SINPs can be further optimized to achieve a reduced particle size ranging from 20 nm to 60 nm while maintaining a narrow size distribution. The stability of these particles is affected by salinity, pH, and temperature. The results of the AlOOH-SPCA SINPs squeeze test demonstrated a return performance of more than 2,000 PVs. During the test, it was observed that 73% of the SPCA content could be retained in the Iceland spar calcite during the initial 3 PVs. Even after flowing a synthetic brine equivalent to 2,000 PVs, approximately 51% of the total injected SPCA was still retained in the column. These findings highlight the effective retention and inhibition capabilities of the AlOOH-SPCA SINPs, indicating their potential for long-term scale control in the presence of challenging brine conditions. Additionally, it was observed that the introduction of Ca^{2+} ions led to the cross-linking of free SPCA molecules. Consequently, the presence of Ca^{2+} ions improved the sorption of SPCA onto the AlOOH NPs. This interaction between Ca^{2+} ions and SPCA molecules likely contributed to enhanced binding and stability of the scale inhibitor on the NPs, which can have significant implications for their performance in inhibiting scale formation in practical applications.

Aluminum oxide (Al_2O_3) nanofluids and commercial SIs like 2-phosphonobutane-1,2,4-tricarboxylic acid (PBTCA) were also used for calcium carbonate ($CaCO_3$) scale control.[54, 55] Wang *et al.* found that the introduction of Al_2O_3 NPs has ability to shift $CaCO_3$ from calcite to aragonite. Al_2O_3 NPs exhibited a more pronounced influence on the alteration of the morphology of calcium carbonate crystals. While Al_2O_3 NPs may potentially affect the ability of PBTCA to retain calcium ions, the surface anti-scaling efficiency of Al_2O_3 NPs exceeds 80%, making it a promising candidate material for controlling scaling applications.

Jiang *et al.* explored the efficiency and mechanism of scaling inhibition, particularly the formation of $CaCO_3$ (calcite) scale, using a superhydrophobic CuO (copper oxide) nanowire layer.[56] The nanomaterials were prepared by first anodizing copper in a sodium hydroxide (NaOH) solution, to grow oxide layers on Cu surfaces. Subsequently, the surface was modified using 1H,1H,2H,2H-perfluorooctyltriethoxysilane (FAS-17). The obtained copper surface has high hydrophobicity. The result of inhibiting $CaCO_3$ scale test indicated the superhydrophobic CuO-incorporated FAS-17 surface significantly inhibited the formation of calcium $CaCO_3$ crystals. It was proposed that the excellent scale inhibition performance of the superhydrophobic CuO-incorporated FAS-17 surface can be attributed to the low surface energy of the modified surface and the ability to retain an air film, both of which reduce the adhesion and nucleation of $CaCO_3$ crystals.

18.3.5 Silicon-Based Scale Inhibitor Nanomaterials

Silica NPs have been extensively investigated for diverse applications, including EOR, scaling control, and flow assurance in recent years.[14, 50, 57, 58] For instance, the influence of silica NPs on the scale inhibition performance of DETPMP against calcium sulfate scaling was assessed.[59] Various scenarios were compared involving silica NPs alone, DETPMP inhibitor alone, and blends of silica NPs and DETPMP. The experimental results demonstrate that the exclusive application of silica NPs effectively retards the decline in solution conductivity. Silica NPs serve as inhibitors for the crystal growth of calcium sulfate scale. Meanwhile, when combined with the DETPMP inhibitor, silica NPs play a substantial role in reducing the rate at which solution solubility decreases. An optimal concentration of silica NPs combined with the scale inhibitor DETPMP can markedly slow down the decline in solution conductivity. Consequently, this dual treatment effectively mitigates the deposition of scale, addressing a critical challenge in the oil industry. It is quite clear that the strategic use of silica NPs and DETPMP can exert significant control over the precipitation of calcium sulfate scale.

Furthermore, Tavakoli *et al.* recently conducted experimental studies investigating the impact of silica NPs on the kinetics of barite scale deposition in the processes of water injection.[60] The findings suggest that silica NPs are important in impeding and delaying the development of barite scale in water injection operations. Silica NPs are effective in preventing scale formation, but their effectiveness is limited. The ideal concentration of silica NPs as a scale inhibitor was determined to be 0.1 wt%. Reducing the size of SiO_2 NPs results in a notable rise in the density of surface charges or the number of surface charges per unit of surface area. Consequently, this leads to the creation of a robust network of surface charges across the NP surface. This network is crucial in solution dynamics, as it effectively impedes the formation of scale. Additionally, another study explored the application of silica NPs in reducing calcite sulfate scale formation at elevated temperatures, within the context of water injection scenarios involving both high- and low-salinity levels was also explored by Moghadasi *et al.*[61] It indicates that silica NPs exhibit greater effectiveness in mitigating calcium sulfate scale in conditions of higher salinity. However, it should be noted that the efficacy of silica NPs tends to diminish as the temperature of the system rises. Overall, these results highlight the potential of silica NPs in controlling scale formation during water injection processes, with their effectiveness being influenced by factors such as salinity levels and system temperature.

Zhang *et al.* conducted a study, where they employed a silica-templated process to create silica-zinc diethylenetriaminepenta (methylene phosphonic acid) surface-modified NPs (Si-Zn-DETPMP SINPs).[62] In this research, they used commercially available SiO_2 NPs with a diameter of 22 nm and a surface area of 135 m^2/g. The SiO_2 NPs were used at a concentration of 30 wt%. The resulting Si-Zn-DETPMP SINPs had a diameter of approximately 100 nm and exhibited a zeta potential of −55 mV. TEM images showed that these nanomaterials had a spherical shape and were evenly dispersed in the presence of anionic sodium dodecyl benzene sulfonate (SDBS) surfactants. The results obtained from FT-IR spectra revealed that the inclusion of silica NPs did not exert a substantial influence on the configuration of Zn-DETPMP precipitates. This suggests that the incorporation of silica NPs did not alter the chemical composition or structural properties of the Zn-DETPMP particles. This outcome indicates that the combination of silica NPs and Zn-DETPMP was successful, and the properties of the resulting Si-Zn-DETPMP SINPs were in line with the desired characteristics. To evaluate the transport performance of Si-Zn-DETPMP SINPs in carbonate and sandstone formation media, a column breakthrough experiment was also conducted. The results indicated that in the absence of SDBS surfactants, the Si-Zn-DETPMP SINPs were unable to achieve a complete breakthrough. This suggests that the NPs faced challenges in effectively transporting through the media and reaching their desired destination. However, the study also found that the transport performance of the Si-Zn-DETPMP SINPs could be improved by preflushing the system with an SDBS surfactant solution. This preflushing process likely facilitated the dispersion and mobilization of the NPs, enabling them to transport through the porous structure of the formation media more effectively. The enhanced transport performance observed after preflushing with the SDBS surfactant solution indicates that the presence of SDBS surfactant helped overcome the transport limitations of the Si-Zn-DETPMP SINPs. It likely improved the dispersion and stability of the NPs, enabling them to achieve better breakthrough and enhance their transport efficiency within the system. The inhibitor-return curves for Si-Zn-DETPMP nanofluid were obtained by the laboratory squeeze simulation test. During the inhibitor return phase, the Si-Zn-DETPMP nanomaterials, in their stable state, gradually released phosphonate inhibitors into the brine solution throughout the entirety of the squeeze simulation process. The results involving the newly developed Si-Zn-DETPMP nanomaterials have shown enhancements in the long-term flowback performance of the inhibitors and an extended squeeze lifetime.

18.3.6 Polymer Scale Inhibitor Nanomaterials

Polymer nanocomposites are a recent innovation with potential applications across various industries, notably in the oil and gas sector. Sheikhi *et al.* introduced a groundbreaking development by producing cellulose fibrils at the nanoscale, enabling precise control over scale formation.[63, 64] These novel nanocellulose materials, environmentally friendly by nature, were assessed for their ability to inhibit $CaCO_3$ scale through a global potential-controlled electrochemical technique known as chronoamperometry. To prepare dicarboxylated cellulose (DCC), cellulose underwent a process involving periodate and chlorite oxidation. The process resulted in DCC with a substantial number of carboxyl units attached to the cellulose backbone, reaching as high as 12 mmol/g. Furthermore, three distinct types of electrostatically stabilized nanocrystalline cellulose (ENCC): ENCC1, ENCC2, and ENCC3 were produced and each characterized by different lengths of "hairs"—long hairs, medium-sized hairs, and short hairs, respectively. Calcite scale inhibition experiment results demonstrated that these hairy nanocelluloses exhibited enhanced performance in comparison to traditional cellulose nanocrystals (CNC) as well as commercially available phosphonated and carboxylated SIs. In contrast, CNC displayed poor scale inhibition performance, leading to the deposition of calcite, even at high additive concentrations. ENCC2 and ENCC3 showed relatively modest performance in inhibiting calcite scaling. However, both DCC and ENCC1 demonstrated excellent efficacy in preventing calcite scale formation, when compared to commercial phosphonated SIs.

Polyvinyl sulfonate (PVS) is a highly efficient scale inhibitor used in the oilfields for managing barite scale. Veisi *et al.* prepared polyelectrolyte complex (PEC) NPs (PECNPs) to extend the effective duration of PVS treatment in reservoirs during squeeze operations.[65] In this research, PECNPs with a positive charge were synthesized by combining a polycation, polyethylenimine (PEI), and the polyanionic PVS. This approach was employed to encapsulate the PVS within the NP structure. Multiple PECNPs were synthesized by varying the mass ratios of PEI and PVS solutions, with the aim of identifying the most suitable NPs for use in squeeze treatment applications. The results obtained from static and dynamic adsorption experiments demonstrate that PVS, when encapsulated within PECNPs, exhibited a rapid and robust adsorption capacity on the surface of Berea sandstone. The effectiveness of scale inhibition and the durability of squeeze treatments involving PVS-loaded PECNPs were evaluated through a combination of core flooding experiments and a dynamic tube blocking assessment conducted on Berea core samples. The findings indicated that encapsulating PVS within PECNPs extended the duration of squeeze treatments by a notable 22% compared to treatments with unencapsulated PVS. Furthermore, subjecting the system to ionic strength shocks resulted in an additional enhancement, increasing the release of PVS and extending the lifetime of squeeze treatments involving PECNP-entrapped PVS by 40%, when contrasted with the use of unentrapped PVS.

Cross-linked SIs can be employed to enhance the longevity of scale inhibitor squeeze treatments in downhole applications. Liang *et al.* obtained a patent for a nanoscale cross-linked polymeric SIs, aimed at extending the duration of oilfield scale inhibition treatments.[66] The NPs were created through a two-step synthesis process. In the initial step, a solution polymerization reaction of commercially available monomers was conducted to produce the cross-linked polymeric scale inhibitor in a gel form. Subsequently, in the second step, the obtained cross-linked polymeric SIs were mixed with an anionic solution to achieve the targeted NP size for the cross-linked SINPs. To assess the binding of cross-linked SINPs to the geological formation rock, both static and dynamic adsorption experiments in sandstone and carbonate reservoir rocks were conducted. These experiments utilized a quartz crystal microbalance with dissipation monitoring to measure the adsorption behavior. The findings indicated that cross-linked SINPs exhibited superior performance in scale inhibition, leading to an extended squeeze lifetime in the formation rock when compared to conventional cross-linked scale inhibitor gel.

Zhou *et al.* investigated the viability of using PEC NPs to augment the adsorption of anionic surfactants on the quartz sand grains with negatively charged surface.[67] Both the speed and degree of adsorption of an anionic sulfate surfactant were quantified on untreated quartz sand grains in comparison to sand grains that underwent pretreatment with PECs. The results indicate a substantial increase in the adsorption level of the anionic sulfate surfactant on quartz sand grains treated with PEC, compared to minimal adsorption on untreated sand grains. It further proposed a hypothesized four-stage adsorption mechanism, elucidating the intensified binding of anionic surfactants to quartz sand particles modified with PEC NPs and exhibiting a negatively charged surface.

Polymeric SIs are considered environmentally friendly and fall under the category of "green inhibitors," because they typically have low- or no-phosphorus content, which distinguishes them from aminophosphonates, a group of SIs with potentially higher environmental impact.[68–70] A research study has been conducted to explore the synthesis method and laboratory testing of a novel colloidal material based on polystyrene sulfonate (PSS) as an inhibitor.[71] Researchers used both precipitation and surfactant surface modification techniques to develop a colloidal inhibitor material known as barium-PSS (BaPCI). This material has an average particle size in the range of several hundred nanometers. The results from the scale inhibition tests clearly demonstrate that the inclusion of BaPCI effectively manages the formation of barite scale nuclei. Furthermore, higher concentrations of BaPCI were found to extend the duration of scale inhibition. An essential measure of the effectiveness of the prepared BaPCI material is its ability to migrate within subsurface formation media. The transport findings reveal that the BaPCI material prepared in the study exhibits a satisfactory capability to move within the formation medium. In laboratory squeeze simulation experiments involving BaPCI, it was observed that a significantly greater quantity of PSS inhibitor was flushed out from the column during the initial stage of inhibitor return when the PSS inhibitor was used, as compared to when BaPCI was employed. In contrast to the behavior observed with a PSS solution, the BaPCI material exhibits a more favorable tendency to adhere to the surface of calcite, resulting in a decreased amount of inhibitor that being washed out from the column. Despite the BaPCI material showing good mobility within a calcite medium during transport experiments, it was effectively retained by the calcite medium during a 24-h shut-in period. The objective of this process was to facilitate the binding of the injected inhibitor material to the surface of the formation medium. The prolonged squeeze lifetime observed for BaPCI can be attributed to two main factors: (1) strong adherence to the surface of the calcite medium; and (2) limited solubility in the synthetic brine solution, keeping it in the medium for an extended period. Extensive discussions have contended that the behavior of inhibitor return is significantly influenced by the aqueous solubility of the colloidal inhibitor materials. It is proposed that the inhibitor materials retained on the surfaces of the formation medium will gradually release, with the rate of release being controlled by their solubility in a brine solution.[6, 34] A reduced solubility in aqueous solutions results in a slower and more gradual release of the inhibitor materials, ultimately leading to a longer lasting squeeze effect.

18.4 Summary

Mineral scale buildup can pose significant risks to the safety and integrity of oil and gas operations, especially in shale and deepwater fields. To combat this scale threat, some chemical inhibitors are commonly employed in both onshore and offshore operations. While traditional inhibitors are effective in mitigating scale issues, they suffer from limitations related to their short-range dispersion and the limited duration of their effectiveness due to their inherent chemical properties. This is particularly evident during squeeze treatments aimed at controlling downhole and near-wellbore scale issues. Considering that SIs tend to form complex compounds or precipitates with the metal ions present in the formations, one can consider the concept of creating preprecipitated metal-inhibitor SINMs within a laboratory environment. Over the last decade, extensive experimental research has been conducted to create and assess various types of SINMs for controlling scale in oilfields. SINMs can be broadly classified based on their chemical compositions, including Ca-DETPMP SINMs, metal/metal oxide SINMs, carbon-based SINMs, silica-based SINMs, polymer SINMs, and so on. The evaluation of these nanomaterials centers around some key aspects, including their synthesis, testing, effectiveness in inhibiting scale formation, ability to move within the formation medium, and performance in terms of inhibitor return. Concerning inhibition effectiveness, SINMs can either directly prevent scale formation or impact the performance of chemical inhibitors on the basis of their chemical makeup. In terms of SINM migration, research findings underscore that several factors play a pivotal role in influencing the movement of SINMs within porous formation media. These factors encompass properties of the medium itself, the specific composition and surface chemistry of SINMs, as well as the use of preflushing and overflush fluids. When it comes to the performance of these inhibitors upon return, laboratory simulation tests indicate that amorphous metal-inhibitor SINMs surpass conventional inhibitors in terms of effectiveness. Furthermore, the development of crystalline metal-inhibitor SINMs from amorphous materials has shown a remarkable extension of the squeeze lifetime. It is evident that SINMs hold substantial promise for real-world applications in oilfields, where they can effectively mitigate mineral scale issues. These materials offer an enticing alternative to traditional inhibitor chemicals, characterized by their superior mobility within formation media and prolonged effectiveness. Nevertheless, it is imperative to conduct supplementary experimental studies to verify the full compatibility of these materials under the conditions experienced in oilfield operations. In an ideal scenario, it is crucial to conduct transport and squeeze simulation tests under conditions that closely resemble those encountered in oilfields, meaning high-temperature and high-pressure conditions, using actual core materials. This necessitates further laboratory endeavors to explore innovative compositions and synthesis methods for SINMs. These SINMs should exhibit the ability to endure the harsh reservoir environment while demonstrating effective transport characteristics and a prolonged squeeze lifetime. As expounded in this review, it becomes evident that SINMs have the capacity to enhance our capabilities in addressing the issue of oilfield scale formation. They hold promise as an alternative engineering solution for large-scale control campaigns. The SINMs showcased in this chapter serve as prime examples of our ongoing progress in harnessing advanced nanotechnology to tackle real-world engineering challenges within the petroleum industry.

Nomenclature

ADE	Advection dispersive equation
AMPS	2-Acrylamido-2-methyl-1-propanesulfonate
CBN	Carbon-based nanoparticle
CCQD	Carbon quantum dot
CNT	Carbon nanotube
CQD	Carbon-based quantum dot
Ca	Calcium
DETPMP	Diethylenetriamine pentakis (methylene phosphonic acid)
EDTA	Ethylenediaminetetraacetic acid
EOR	Enhanced oil recovery
FT-IR	Fourier transform infrared spectroscopy
GO	Graphene oxide
M/MO-NPS	Metal and metal oxide nanoparticle
MA	Maleic acid
MIC	Minimum inhibitor concentration
MWCNT	Multiwalled carbon nanotube
NAST	Nano-assisted squeeze treatment
NCEST	Nano-carbon enhanced squeeze treatment
OMNP	Oleic acid-coated iron oxide nanoparticle
PBTCA	2-Phosphonobutane-1,2,4-tricarboxylic acid

PECNP	Polyelectrolyte complex nanoparticle
PPCA	Phosphino-polycarboxylic acid
PV	Pore volume
PVS	Polyvinyl sulfonate
RO	Reverse osmosis
SEM	Scanning electron microscopy
Si	Silicon
SINM	Scale inhibitor nanomaterial
SINP	Scale inhibitor nanoparticle
SPCA	Sulfonated polycarboxylic acid
SWCNT	Single-walled carbon nanotube
TEM	Transmission electron microscopy
XRD	X-ray diffraction

References

1. M. Mady, M. Kelland, "Review of Nanotechnology Impacts on Oilfield Scale Management," *ACS Applied Nano Materials* 3 (2020): pp. 7343–7364.
2. Y. Bai, Q. Bai, *Subsea Engineering Handbook* (New York: Elsevier, 2010): pp. 915-919.
3. Z. Amjad, *The Science and Technology of Industrial Water Treatment* (Boca Raton: CRC Press, 2010): pp. 337-339.
4. M. Kelland, *Production Chemicals for the Oil and Gas Industry* (Boca Raton: CRC Press, 2014): pp. 51-81.
5. P. Zhang, D. Shen, G. Ruan, et al., "Phosphino-Polycarboxylic Acid Modified Inhibitor Nanomaterial for Oilfield Scale Control: Synthesis, Characterization and Migration," *Journal of Industrial and Engineering Chemistry* 45 (2017): pp. 366–374.
6. P. Zhang, "Review of Synthesis and Evaluation of Inhibitor Nanomaterials for Oilfield Mineral Scale Control," *Frontiers in Chemistry* 8 (2020): p. 576055.
7. D. Shen, P. Zhang, A. Kan, et al., "Control Placement of Scale Inhibitors in the Formation With stable Ca-DTPMP Nanoparticle Suspension and Its Transport Porous Media," SPE-114063-MS, SPE international oilfield scale conference, held May 28–29, 2008 (Aberdeen, UK).
8. P. Zhang, A. Kan, C. Fan, et al., "Silica-Templated Synthesis of Novel Zinc-DTPMP Nanomaterials: Their Transport in Carbonate and Sandstone Media During Scale Inhibition," *SPE Journal* 16 (2011): pp. 662–671.
9. K. Hoelscher, G. De Stefano, M. Riley, et al., "Application of Nanotechnology in Drilling Fluids," SPE International Oilfield Nanotechnology Conference and Exhibition, held June 12–14, 2012 (Noordwijk, The Netherlands).
10. A. Alhuraishawy, R. Hamied, H. Hammood, et al., "Enhanced Oil Recovery for Carbonate Oil Reservoir by Using Nano-Surfactant: Part II," SPE-198666-MS, SPE Gas & Oil Technology Showcase and Conference, held October 21–23, 2019 (Dubai, UAE).
11. B. Ju, T. Fan, M. Ma, "Enhanced Oil Recovery By Flooding With Hydrophilic Nanoparticles," *China Particuology* 4 (2006): pp. 41–46.
12. D. Wasan, A. Nikolov, "Spreading of Nanofluids on Solids," *Nature* 423 (2003): pp. 156–159.
13. M. Chaudhury, "Spread the Word About Nanofluids," *Nature* 423 (2003): pp. 131–132.
14. J. Yang, S. Ji, R. Li, et al., "Advances of Nanotechnologies in Oil and Gas Industries," *Energy Exploration & Exploitation* 33 (2015): pp. 639–657.

15. S. Olayiwola, M. Dejam, "A Comprehensive Review on Interaction of Nanoparticles With Low Salinity Water and Surfactant for Enhanced Oil Recovery in Sandstone and Carbonate Reservoirs," *Fuel* 241 (2019): pp. 1045–1057.
16. M. Mohammadi, M. Akbari, Z. Fakhroueian, et al., "Inhibition of Asphaltene Precipitation by TiO_2, SiO_2, and ZrO_2 Nanofluids," *Energy & Fuels* 25 (2011): pp. 3150–3156.
17. Z. Haindade, A. D. Bihani, S. Javeri, et al., "Enhancing Flow Assurance Using Co-Ni Nanoparticles for Dewaxing of Production Tubing," SPE-157119-MS, SPE International Oilfield Nanotechnology Conference and Exhibition, held June 12–14, 2012 (Noordwijk, The Netherlands).
18. T. Zhang, M. Murphy, H. Yu, et al., "Investigation of Nanoparticle Adsorption During Transport in Porous Media," SPE-166346-MS, SPE Annual Technical Conference and Exhibition, held September 30–October 2, 2013 (New Orleans, Louisiana, USA).
19. G. Kini, J. Yu, L. Wang, et al., "Salt-and Temperature-Stable Quantum Dot Nanoparticles for Porous Media Flow," *Colloids and Surfaces A: Physicochemical and Engineering Aspects* 443 (2014): pp. 492–500.
20. J. Wu, J. He, O. Torsæter, et al., "Effect of Nanoparticles on Oil-Water Flow in a Confined Nanochannel: A Molecular Dynamics Study," SPE-156995-MS, SPE International Oilfield Nanotechnology Conference and Exhibition, held June 12–14, 2012 (Noordwijk, The Netherlands).
21. A. Mehmani, S. Kelly, C. Torres-Verdín, "Review of Micro/Nanofluidic Insights on Fluid Transport Controls in Tight Rocks," *Petrophysics* 60 (2019): pp. 872–890.
22. Z. Zhang, P. Zhang, Z. Li, et al., "Laboratory Evaluation and Mechanistic Understanding of the Impact of Ferric Species on Oilfield Scale Inhibitor Performance," *Energy & Fuels* 32 (2018): pp. 8348–8357.
23. P. Zhang, Y. Liu, N. Zhang, et al., "A Novel Attach-and-Release Mineral Scale Control Strategy: Laboratory Investigation of Retention and Release of Scale Inhibitor on Pipe Surface," *Journal of Industrial and Engineering Chemistry* 70 (2019): pp. 462–471.
24. P. Zhang, N. Zhang, Y. Liu, et al., "Application of a Novel Tube Reactor for Investigation of Calcium Carbonate Mineral Scale Deposition Kinetics," *Chemical Engineering Research and Design* 137 (2018): pp. 113–124.
25. P. Zhang, D. Shen, C. Fan, et al., "Surfactant-Assisted Synthesis of Metal-Phosphonate Inhibitor Nanoparticles and Transport in Porous Media," *SPE Journal* 15 (2010): pp. 610–617.
26. A. Stamatiou, K. Sorbie, "Analytical Solutions for a 1D Scale Inhibitor Transport Model With Coupled Adsorption and Precipitation," *Transport in Porous Media* 132 (2020): pp. 591–625.
27. P. Zhang, D. Shen, G. Ruan, et al., "Mechanistic Understanding of Calcium–Phosphonate Solid Dissolution and Scale Inhibitor Return Behavior in Oilfield Reservoir: Formation of Middle Phase," *Physical Chemistry Chemical Physics* 18 (2016): pp. 21458–21468.
28. P. Zhang, D. Shen, A. Kan, et al., "Phosphino-Polycarboxylic Acid Modified Inhibitor Nanomaterial for Oilfield Scale Control: Transport and Inhibitor Return in Formation Media," *RSC Advances* 6 (2016): pp. 59195–59205.
29. A. Kan, J. Oddo, M. Tomson, "Formation of Two Calcium Diethylenetriaminepentakis (Methylene Phosphonic Acid) Precipitates and Their Physical Chemical Properties," *Langmuir* 10 (1994): pp. 1450–1455.
30. P. Zhang, C. Fan, H. Lu, et al., "Synthesis of Crystalline-Phase Silica-Based Calcium Phosphonate Nanomaterials and Their Transport in Carbonate and Sandstone Porous Media," *Industrial & Engineering Chemistry Research* 50 (2011): pp. 1819–1830.
31. P. Zhang, A. Kan, M. Tomson, "Enhanced Transport of Novel Crystalline Calcium-Phosphonate Scale Inhibitor Nanomaterials and Their Long Term Flow Back Performance in Laboratory Squeeze Simulation Tests," *RSC Advances* 6 (2016): pp. 5259–5269.
32. M. Tomson, A. Kan, G. Fu, "Control of Inhibitor Squeeze via Mechanistic Understanding of Inhibitor Chemistry," SPE-87450-MS, 6th International Symposium on Oilfield Scale, held May 26–27, 2004 (Aberdeen, UK).

33. A. Kan, G. Fu, M. Tomson, "Prediction of Scale Inhibitor Squeeze and Return in Calcite-Bearing Formation," SPE-93265-MS, SPE International Conference on Oilfield Chemistry, held February 2–4, 2005 (The Woodlands, TX, USA).
34. G. Ruan, A. Kan, M. Tomson, et al., "Facile One-Pot Synthesis of Metal-Phosphonate Colloidal Scale Inhibitor: Synthesis and Laboratory Evaluation," *Fuel* 282 (2020): p. 118855.
35. D. Maiti, X. Tong, X. Mou, et al., "Carbon-Based Nanomaterials for Biomedical Applications: A Recent Study," *Frontiers in Pharmacology* 9 (2019): p. 1401.
36. L. Hoa, M. Vestergaard, E. Tamiya, "Carbon-Based Nanomaterials in Biomass-Based Fuel-Fed Fuel Cells," *Sensors* 17 (2017): p. 2587.
37. I. Jeon, D. Chang, N. Kumar, et al., "Functionalization of Carbon Nanotubes," *Carbon Nanotubes - Polymer Nanocomposites* 6 (2011): pp. 91–110.
38. S. Hajiabadi, H. Aghaei, M. Kalateh, et al., "An Overview on the Significance of Carbon-Based Nanomaterials in Upstream Oil and Gas Industry," *Journal of Petroleum Science and Engineering* 186 (2020): p. 106783.
39. M. Alsaba, M. Al Dushaishi, A. Abbas, "A Comprehensive Review of Nanoparticles Applications in the Oil and Gas Industry," *Journal of Petroleum Exploration and Production Technology* 10 (2020): pp. 1389–1399.
40. A. Lu, E. Salabas, F. Schüth, "Magnetic Nanoparticles: Synthesis, Protection, Functionalization, and Application," *Angewandte Chemie International Edition* 46 (2007): pp. 1222–1244.
41. C. Negin, S. Ali, Q. Xie, "Application of Nanotechnology for Enhancing Oil Recovery–A Review," *Petroleum* 2 (2016): pp. 324–333.
42. N. Ghorbani, M. Wilson, N. Kapur, et al., "Adsorption of Polyphosphinocarboxylic Acid (PPCA) Scale Inhibitor on Carbon Nanotubes (CNTs): A Prospective Method for Enhanced Oilfield Scale Prevention," *Journal of Petroleum Science and Engineering* 150 (2017): pp. 305–311.
43. N. Ghorbani, M. Wilson, N. Kapur, et al., "Carbon Nanotubes: A New Methodology for Enhanced Squeeze Lifetime CNTs," SPE-169763-MS, SPE International Oilfield Scale Conference and Exhibition, held May 14–15, 2014 (Aberdeen, UK).
44. W. Li, C. Gao, "Efficiently Stabilized Spherical Vaterite $CaCO_3$ Crystals by Carbon Nanotubes in Biomimetic Mineralization," *Langmuir* 23 (2007): pp. 4575–4582.
45. C. Wan, L. Wang, J. Sha, et al., "Effect of Carbon Nanoparticles on the Crystallization of Calcium Carbonate in Aqueous Solution," *Nanomaterials* 9 (2019): p. 179.
46. Y. Takizawa, S. Inukai, T. Araki, et al., "Effective Antiscaling Performance of Reverse-Osmosis Membranes Made of Carbon Nanotubes and Polyamide Nanocomposites," *ACS Omega* 3 (2018): pp. 6047–6055.
47. U. Ishtiaq, A. Muhsan, A. Rozali, et al., "Graphene Oxide/Carbon Nanotubes Nanocoating for Improved Scale Inhibitor Adsorption Ability Onto Rock Formation," *Journal of Petroleum Exploration and Production Technology* 10 (2020): pp. 149–157.
48. J. Hao, L. Li, W. Zhao, et al., "Synthesis and Application of CCQDs As a Novel Type of Environmentally Friendly Scale Inhibitor," *ACS Applied Materials & Interfaces* 11 (2019): pp. 9277–9282.
49. J. Pal, T. Pal, "Faceted Metal and Metal Oxide Nanoparticles: Design, Fabrication and Catalysis," *Nanoscale* 7 (2015): pp. 14159–14190.
50. K. Zhou, X. Zhou, J. Liu, et al., "Application of Magnetic Nanoparticles in Petroleum Industry: A Review," *Journal of Petroleum Science and Engineering* 188 (2020): p. 106943.
51. B. Do, B. Nguyen, H. Nguyen, et al., "Synthesis of Magnetic Composite Nanoparticles Enveloped in Copolymers Specified for Scale Inhibition Application," *Advances in Natural Sciences: Nanoscience and Nanotechnology* 4 (2013): p. 045016.
52. C. Yan, A. Kan, W. Wang, et al., "Synthesis and Size Control of Monodispersed Al-Sulphonated Polycarboxylic Acid (Al-SPCA) Nanoparticles With Improved Squeeze Performance and Their Transport in Porous Media," SPE-155627-MS, SPE International Oilfield Nanotechnology Conference and Exhibition, held June 12–14, 2012 (Noordwijk, The Netherlands).

53. C. Yan, A. Kan, W. Wang, et al., "Sorption Study of AlO(OH) Nanoparticle-Crosslinked Polymeric Scale Inhibitors and Their Improved Squeeze Performance in Porous Media," *SPE Journal* 19 (2014): pp. 687–694.
54. L. Wang, H. Ge, Y. Han, et al., "Effects of Al_2O_3 Nanoparticles on the Formation of Inorganic Scale on Heat Exchange Surface With and Without Scale Inhibitor," *Applied Thermal Engineering* 151 (2019): pp. 1–10.
55. F. Wang, H. Ge, K. Wu, et al., "Effects of Al_2O_3 Nanoparticles on the Crystallization of Calcium Carbonate in Aqueous Solution," *Journal of Nanoscience and Nanotechnology* 19 (2019): pp. 3471–3478.
56. W. Jiang, J. He, F. Xiao, et al., "Preparation and Antiscaling Application of Superhydrophobic Anodized CuO Nanowire Surfaces," *Industrial & Engineering Chemistry Research* 54 (2015): pp. 6874–6883.
57. H. Lau, M. Yu, Q. Nguyen, "Nanotechnology for Oilfield Applications: Challenges and Impact," *Journal of Petroleum Science and Engineering* 157 (2017): pp. 1160–1169.
58. M. Safari, A. Golsefatan, M. Jamialahmadi, "Inhibition of Scale Formation Using Silica Nanoparticle," *Journal of Dispersion Science and Technology* 35 (2014): pp. 1502–1510.
59. A. Golsefatan, M. Safari, M. Jamialahmadi, "Using Silica Nanoparticles to Improve DETPMP Scale Inhibitor Performance As a Novel Calcium Sulfate Inhibitor," *Desalination and Water Treatment* 57 (2016): pp. 20800–20808.
60. H. Tavakoli, M. Jamialahmadi, S. Kord, et al., "Experimental Investigation of the Effect of Silica Nanoparticles on the Kinetics of Barium Sulfate Scaling During Water Injection Process," *Journal of Petroleum Science and Engineering* 169 (2018): pp. 344–352.
61. R. Moghadasi, A. Rostami, A. Tatar, et al., "An Experimental Study of Nanosilica Application in Reducing Calcium Sulfate Scale at High Temperatures During High and Low Salinity Water Injection," *Journal of Petroleum Science and Engineering* 179 (2019): pp. 7–18.
62. P. Zhang, A. Kan, C. Fan, et al., "Silica-Templated Synthesis of Novel Zinc-DTPMP Nanomaterials: Their Transport in Carbonate and Sandstone Media During Scale Inhibition," SPE-130639-MS, SPE International Conference on Oilfield Scale, held May 26–27, 2010 (Aberdeen, UK).
63. A. Sheikhi, A. Kakkar, T. Van De Ven, "Nanoengineering Colloidal and Polymeric Celluloses for Threshold Scale Inhibition: Towards Universal Biomass-Based Crystal Modification," *Materials Horizons* 5 (2018): pp. 248–255.
64. A. Sheikhi, A. Kakkar, T. Van De Ven, "Biomimetic Scale-Resistant Polymer Nanocomposites: Towards Universal Additive-Free Scale Inhibition," *Journal of Materials Chemistry A* 6 (2018): pp. 10189–10195.
65. M. Veisi, S. Johnson, K. Peltier, et al., "Application of Polyelectrolyte Complex Nanoparticles to Increase the Lifetime of Poly Vinyl Sulfonate Scale Inhibitor," SPE-189564-MS, SPE International Conference and Exhibition on Formation Damage Control, held February 7–9, 2018 (Lafayette, LA, USA).
66. J. Liang, H. Guan, "Nanosized Scale Inhibitors for Increasing Oilfield Scale Inhibition Treatment Lifetime," U.S. Patent 20200385629A1, 2020.
67. X. Zhou, J. Liang, C. Andersen, et al., "Enhanced Adsorption of Anionic Surfactants on Negatively Charged Quartz sand Grains Treated With Cationic Polyelectrolyte Complex Nanoparticles," *Colloids and Surfaces A: Physicochemical and Engineering Aspects* 553 (2018): pp. 397–405.
68. S. Li, C. Guo, X. Wang, et al., "Corrosion Inhibition Coating Based on the Self-Assembled Polydopamine Films and Its Anti-Corrosion Properties," *Polymers* 14 (2022): p. 794.
69. Y. Liu, P. Zhang, "Review of Phosphorus-Based Polymers for Mineral Scale and Corrosion Control in Oilfield," *Polymers* 14 (2022): p. 2673.
70. B. Martínez-Sánchez, A. Quintero-Jaime, F. Huerta, et al., "Synthesis of Phosphorus-Containing Polyanilines by Electrochemical Copolymerization," *Polymers* 12 (2020): p. 1029.
71. H. Wang, H. Dong, X. Liu, et al., "Preparation and Laboratory Testing of Polymeric Scale Inhibitor Colloidal Materials for Oilfield Mineral Scale Control," *Polymers* 14 (2022): p. 4240.

19 Engineering Coatings and Surfaces

Manuel Marya[1]
[1]SLB, Houston, Texas, USA

19.1 Introduction

The fouling of surfaces, that is the deposition and accumulation of materials such as inorganic scales (*e.g.*, carbonates, sulfates, sulfides, silicates, halites), organic scales (*e.g.*, asphaltenes, paraffins, waxes), hydrates (*e.g.*, methane, salts), and combinations thereof, have been a long-lasting and costly flow assurance challenge for oil and gas infrastructures. For inorganic (mineral) deposits, the latter is often recognized as a "water problem," indicating that the waters and their compositions cause scaling.[1, 2] The most severe consequence of unwanted thick and adherent deposits, whether solid or semisolid (viscous), is a flow restriction resulting in blockage or plugging. When such settlements develop in less accessible wellbore spaces, such as the inner cavities of critical well completion equipment, the normal functioning, reliability, integrity, and lifespan of the downhole hardware are all impaired.[3–7] The consequences for subsurface safety valves (SSSVs), flow control valves (FCVs), inflow control devices (ICDs) and valves, gas lift valves (GLVs), sand screens, and others can be dire, also resulting in hydrocarbon production losses, when extended to electric submersible pumps (ESPs).[8, 9] Thick deposits also restrict well access to servicing equipment for survey (*e.g.*, multifinger calipers, electromagnetics, sonics, or gamma rays) and well intervention. These deposits vary as broadly as the compositions of the waters (brines) and hydrocarbons. The inorganic deposits usually form by crystallization and precipitation from saturated hard waters. Calcium carbonate ($CaCO_3$), barium sulfate (barite, $BaSO_4$), and strontium sulfate (celestine, $SrSO_4$) are prevalent scales of hydrocarbon production infrastructures. Carbonate scaling in its diverse allotropic forms (*e.g.*, calcite, aragonite, and valerite) has been investigated and reported the most. It occurs when a change in temperature and pressure results in the release of carbon dioxide (CO_2) from a supersaturated water.[10–12] Carbonate scaling is thus a "water problem" not surprisingly shared by oil and gas production, desalination, water treatment, water transportation, geothermal exchangers, power generation, hydrometallurgy, among others.[6] In wells, sulfate deposition arises when partially immiscible waters, such as formation water, commingle with an injected water.[1, 13] Inorganic scaling increases in severity with water production (water cut) and declines in the presence of oil wetting of the metallics along the flow path. Enhanced oil recovery and steam-assisted gravity drainage (SAGD) tend to aggravate solid deposition.[14] In corrosive reservoirs, particularly highly sweet and sour, scaling can also originate from corrosion products. For example, the loss of wall thickness in a lower completion can result in the accumulation of carbonates or sulfides over the upper completion inner walls. In sour wells, metal sulfides, such as iron sulfide (Fe_2S), lead sulfide (PbS), and zinc sulfide (ZnS), have become a major concern in high-pressure high-temperature formation waters.[15] Organic scales, such as asphaltenes, paraffins, and waxes, are derived from high-molecular-weight and nonvolatile crude oils. Similar to carbonate, asphaltenes segregate and aggregate on the metallic walls, where the oils undergo major pressure drops. Waxes usually deposit when the tubing temperatures dip below the so-called wax appearance temperature. At last, in high-pressure wells, hydrate plugging should not be overlooked, whenever formation water is believed to dissolve heavily in oil, and then coagulate as the tubing temperature falls up the production tubing.[16]

No single solution effectively and permanently addresses multiple deposits and therefore solves the multitude of flow assurance challenges. The traditional methods for mitigating the fouling of solids include: (a) chemical treatments by acids for acid-soluble deposits (*e.g.*, hydrochloric acid; sulfuric acid), solvents, and chelating agents; (b) removal treatments by mechanical or electromechanical methods, where the deposits are scrapped, blasted off, or drilled through by bottom-hole assembly runs; and (c) the batch or continuous chemical injection of inhibitors, such as inorganic phosphates, organic polymers, organophosphorus compounds, phosphate ethers, and others in use for scales.[17, 18] Many of these approaches are associated with lost production time and involve workovers tailored to each deposit, without a strong guarantee of blockage recurrence. Chemical treatments are also increasingly constrained by regulations that demand green inhibitors and innovative antifouling solutions. By altering the physicochemical nature of the surfaces, advanced engineering coatings and surface treatments offer a simple strategy for mitigating inorganic and organic deposits.[12, 16, 19–22] These can be applied to the inner surfaces of tubulars and well completion equipment through well-established and broadly available manufacturing processes. To develop such engineering solutions, the key parameters of the surfaces must be recognized first. As a minimum, these include: (a) the surface energy, which is often sufficiently and simplistically represented by the water contact angle in air; (b) the surface topology, which is captured by the average roughness (Ra); and (c) the environmental (corrosion and wear) resistance of the exposed surfaces.[10–13, 15, 19, 20] Although occasionally overlooked by the literature on fouling, the latter is critical for assessing any complementary coating or surface treatment to deliver long-term benefits worth any additional upfront expenditures.

Antifouling coatings, even more so than surface treatments, have been offered since the 1980s for oilfield tubulars, and recently, there has been renewed interest in completion equipment.[23–25] Oilfield tubulars, namely in oil country tubular goods (OCTGs), come with internal diameters between 140 mm (5-½ in) and 508 mm (20 in) and can normally accommodate thick "high-performance" polymer coatings or liners. For the most advanced thermoplastics, their thicknesses are between 5 mils (0.13 mm) and 25 mils (0.64 mm). All polymer coatings comprise fluid-compatible materials applied either as liquid or powder. These polymers are described and selected for their environmental resistance and their ability to reduce fluid drag by as much as 50% vs. uncoated surfaces, resulting in improved hydraulic pumping efficiency.[24, 26, 27] Epoxies, phenolics, and Novalacs are the major thermoset resins found in thick OCTG tubulars with a thickness of up to approximately 0.38 mm (15 mils). Among these, the phenolics are generally stable up to approximately 200°C (392°F) in brines for CO_2 service; in contrast, the Novalac resins are restricted between 120°C (248°F) and 149°C (300°F), depending upon modifications. Urethanes or polyurethanes can be both thermoset and thermoplastic; they are also used as thick coatings, with good resistance to water, oil, and hydrocarbons up to 200°C (392°F). Nylons are low-end thermoplastics found up to approximately 107°C (225°F), for general oil and gas applications; they are exclusively found on casing and tubing strings with a thickness of up to 0.63 mm (25 mils). The so-called high-performance coatings also fall under the two distinct groups of thermoset and thermoplastic polymers, and extend to nonmetallics, including ceramic coatings. These coatings are found on completion equipment with thicknesses of up to 0.07 mm (3 mil). High-performance thermoplastic coatings, such as polyphenylene sulfide (PPS), polytetrafluoroethylene (PTFE), perfluoro alkoxy (PFA), and polyether ether ketone (PEEK), possess high-temperature resistance, culminating near 260°C (500°F) for PEEK and 200°C (392°F) for fluoropolymer coatings. When fully fluorinated or when a fluorinated component is added to another resin, these thermoplastic coatings exhibit low frictional and anti-fouling characteristics. The sol gels, and recent carbon-thin films, including diamond-like carbon (DLC) and polymer-like carbon (PLC) coatings, are indistinctly a special class of specialty ceramic coatings.[28] Compared with metals (alloys), both polymers and ceramic materials do not noticeably suffer from weight loss or localized corrosion and may be considered inert. Both types of materials have nonstick characteristics. They can be hydrophobic, when repelling water; oleophobic, when repelling oil; or omniphobic, when acting against both liquid types.[29–31] Solid surfaces that deter liquids and solids have low surface energies and may also be perceived as self-cleaning and fouling resistant.[26]

In completion, particularly for SSSVs, similar nonmetallic coatings using PPS and PEEK have been employed for approximately four decades. These coatings are true surface films less than 0.05-mm (2-mil) thick and are formulated with friction reducers such as PTFE, PFA, and/or disulfides such as molybdenum disulfide (MoS_2). The main requirements of such coatings are to control friction under contact load (stress), lower the risk of galling, and ultimately establish a range of actuation forces. Several antifouling coatings from the 1990s for SSSVs and GLVs include PTFE along with other solid lubricants. With the advent of deeper and hotter wells associated with greater hydrocarbon and solid production, newer coatings and surface engineering solutions have been developed. These include carbon-thin films, among which are DLC and PLC coatings (P in PLC in reference to polymer when carbon percentage is reduced over that of silicon).[28] With the increasing demand for electrified equipment, a newer driver has thus emerged for the further development of thin and long-lasting antifouling coatings. The reduction of available actuation force and power compared with hydraulics is such that electrically operated equipment such as FCVs,

SSSVs, or GLVs will demand advanced coating and surface engineering solutions. Figure 19.1 shows examples of mechanical components from various subsurface valves that are complemented with a black proprietary antifouling coating for scale mitigation.

Figure 19.1. Examples of mechanical components from downhole valves supplemented by a thin and black antifouling coating, typically requested by end customers (courtesy of SLB).

In this chapter, the deposition mechanisms of typical inorganic solids are described as a foundation for understanding and engineering antifouling surfaces. Next, coatings and surface engineering among nonmetallics and metallics are detailed and examined for their documented antifouling performance, particularly against inorganic scales. When carefully selected, these surface engineering solutions can minimize unwanted sticky deposits within tubulars, and subsurface equipment that functionally relies on smooth and reliable actuation cycles.[7, 20] A proper surface engineering solution can lower field development costs through reductions in umbilical tubing and tubing cleanouts, chemical injection systems, hydrates, and thermal insulation coatings.[27] Coatings and surface treatments should be considered as supplementary means to the existing equipment. They are part of a general multifaceted strategy geared toward reducing conventional remedial measures to minimize operational losses.[3, 4, 7] Jitka *et al.* reported that surface engineering solutions can reduce fouling with a staggering 90% effectiveness, highlighting that coatings and surface treatments should not be overlooked.[17] As remark to the reader, many comparisons between coatings and surface treatments are attempted in this chapter. They are accomplished utilizing data from published articles as well as the author's experience in the field. Because the data were not produced under identical, comparable, or standardized conditions, the reader is invited to pay more attention to trends and relative performances rather than the absolute data values associated with different surfaces. This chapter intentionally ignores the laboratory and field test conditions, being irrelevant to the intents of this chapter. If the reader is interested in further understanding the fouling conditions and parameters, the cited references should be consulted. Finally, the reader should recognize that coatings, even more so than surface treatments, are often poorly described in the literature, because of their proprietary nature to oilfield manufacturers.

19.2 Solid Deposition Mechanisms

The formation of deposits along the production flow involves complex physicochemical events fundamentally originating from single or multiple solute supersaturations in one or more fluid solvents, such as water-based (aqueous) or hydrocarbon. Supersaturation occurs when the solute concentration exceeds its maximum solubility at equilibrium for some extended time. When a fluid is in a supersaturated state and characterized as having a saturation ratio greater than unity, solid precipitation becomes a thermodynamic possibility. Before any observable fouling occurs, kinetics comes into play by effectively introducing an induction time before any precipitation or deposition is detected. With inorganic scales such as barium sulfate, supersaturation can exist over extended periods; however, the greater the supersaturation or driving force for deposition, the shorter the induction time.[6, 32] Mechanistically, the deposition of common solids on surfaces may be divided into at least two consecutive mechanisms. First, there is a deposition process in which heterogeneous nucleation and growth occur over a surface, particularly at micro-asperities, lattice discontinuities (*e.g.*, grain boundaries and dislocations), and chemical heterogeneities (*e.g.*, impurities and distinct phases). Second, there is an adhesion process in which preexisting deposits acquire new crystals of various sizes dispersed in the flow. For inorganic scaling, Figure 19.2 illustrates, with more details, these processes, starting with crystal nuclei spontaneously forming by ion pairing in a supersaturated brine. These nuclei then cluster and aggregate to form larger microcrystals and eventually agglomerate into even larger crystals that deposit on the surface. Through a second and complementary process, other crystals nucleate heterogeneously on preferential sites of the surface, and then grow over time from the accumulation of new microcrystals. The two processes coexist and follow different kinetics that are worth recognizing, when reviewing inorganic deposition data.[10, 13, 20] In Figure 19.2, key factors that define a surface, such as macro- and micro-asperities, crystallographic imperfections at the nanoscopic scale, and chemical heterogeneities down to the scale of inclusions, are also depicted. All these key parameters may be represented by more simplistic measures such as surface roughness and surface energy, as extensively discussed in the remainder of this chapter.

To comprehend the contributions of coatings and surface treatments, some more fundamentals of solid deposition are worth exploring before moving into the subject of engineering surfaces. For inorganic scales that rely on water as a precursor, the physical properties of one or several aqueous phases (*i.e.*, chemical makeup, ionic concentrations, and liquid water pH) will influence the interfacial reactions at the surface.[33, 34] For instance, increasing the ionic content of a brine can affect ionic or charge gradients near a surface, causing protonation or deprotonation of the surface. Increased cation concentrations can also neutralize more negatively charged regions of a surface, thereby affecting its physicochemical properties, energy, and affinity toward fouling.[12] In general, a surface with low energy is beneficial against fouling and has been established as a general requirement to combat all

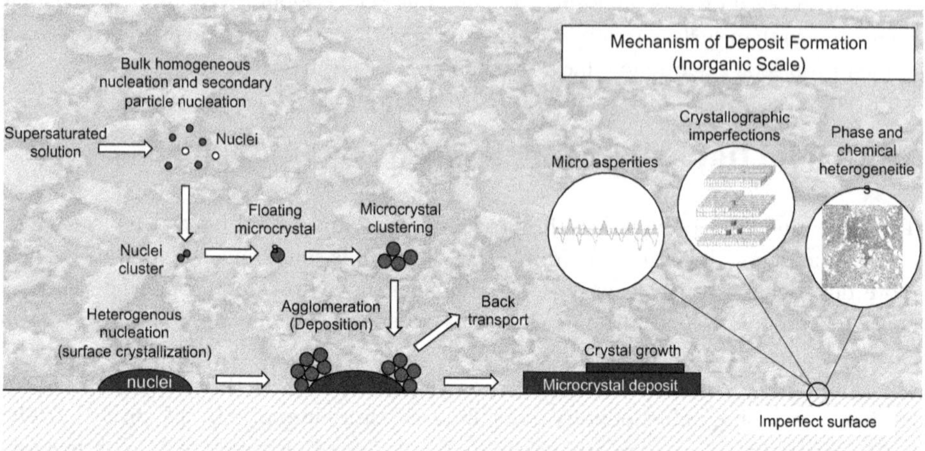

Figure 19.2. The deposition and adhesion of an inorganic scale broken down in homogeneous and heterogeneous nucleation, growth, and attachment in addition to prominent features of a surface that locally raise its energy as well as its tendency to fouling. A low nucleation rate in the bulk solution and/or the surface rate can be related to an extended induction time to scale deposition.

types of fouling.[3, 4, 6, 7, 12] Such a surface is conducive to acquiring less mass, and this mass is expected to be more removable because of weaker adhesion. For calcium sulfate, Table 19.1 illustrates that for metals such as copper, aluminum, iron (carbon steel), and brass, the surface energy directly influences the induction time of the sulfate scale, that is, the minimum time for any deposition to be observed on the designated materials.[35] For the metals in Table 19.1, a low surface energy simply postpones scaling. Interestingly, the listed DLC coating delays scale deposition the best, already suggesting the potential against fouling of this broad family of coatings.

Table 19.1: The influence of surface properties on calcium sulfate scaling kinetics.

Material	Surface Free Energy (nM/m)	Mean Roughness (mm)	Induction Time (h)
Copper	50	1	28.7
Aluminum	48	1	50.4
Steel	40	5	139
Brass	35	1	177
DLC coating	37	5	>475
PTFE	18	11	89

With fouling now explained, technologies for its control can be better understood. For instance, chemical inhibitors against inorganic scales are designed to kinetically prevent deposition by adsorbing on crystal nuclei and making these nuclei less appealing to the surface. Coatings and surface treatments function differently. They passively intervene by repelling nuclei or grown crystals. The literature suggests that heterogeneous nucleation over the surface is the principal mechanism influencing the extent of deposition.[13, 20] With heterogeneous nucleation in mind, as described in Figure 19.2, a coating or treated surface with both weak chemical affinities to saturated fluids and limited nucleation opportunities is expected to have strong antifouling characteristics. Because such coatings may not share a common behavior toward all deposits, a surface engineering solution developed or optimized against a specific deposit always remains at risk of not transferring its effectiveness toward a different deposit, as illustrated in this chapter.

Hydrodynamic factors such as flow velocity, flow direction, and flow regime (laminar vs. turbulent) influence fouling. In general, the precipitation and deposition of solids are more significant in turbulent areas, unless high-shear forces or momentum transfers develop along the surface.[13] Figure 19.3 mechanistically describes the effect of flow velocity on the shear force (stress) required to dislodge an adherent deposit over a hypothetical surface.[36] Removal of a deposit by tangential viscous flow conveying a surface shear force develops, when this force exceeds the adhesion force between the deposit and surface.[19] The minimum flow rate required to separate the deposit from the surface may be seen as a type of critical erosion flow rate, analogous to the critical erosion velocities of API RP14E.[37] Reducing the adhesion force between a deposit and a surface lowers the critical flow rate and expands the flow rate envelope in which the deposit will not attach to the surface. Figure 19.3 describes the positive effect of a coating on reducing the adhesion of a deposit. The more effective the anti-fouling from such coatings, the wider Region II becomes, as caused by a reduction in viscous shear. Without such a coating, deposition starts in Region I and continues throughout Region II. Similar to such valuable coatings, for inorganic scales, the presence of oil or gas has the beneficial effect on decreasing inorganic deposits by inserting a new medium between the water and the surface. With an oleophilic surface, which is keen on adsorbing nonpolar fluids such as oil, inorganic scales decrease.[15, 17, 36] An oleophilic surface such as an oil film is inclined to interfere with ions in water, weakening and making any deposition more removable by small shear forces, thus resulting in low critical velocities. In contrast, hydrocarbon products such as asphaltenes are driven to further segregate along this surface.[29–31] Thus, a universal strategy to broadly mitigate fouling requires accounting for different types of deposits, and hydrophobicity alone is lacking as a single coating and surface property.

Even though the exact set of properties defining a strong antifouling surface vs. another remain uncertain, surfaces with a low surface energy and a hydrophobic behavior tend to be less adherent to deposits. Developing such hydrophobicity can be achieved with low-surface-energy coatings and by controlling the surface topography, texture, or roughness at a small scale. Figure 19.4 shows the water contact angles of different coatings and surfaces, including polymers, fluoropolymers, ceramics, metals, and particularly thin DLC coatings.[19, 28, 38] Note that most coatings, specifically the fluorinated and DLC coatings, were applied on surfaces with the normal finish (roughness) of sealed bore components, implying surface roughness effects were secondary. In Figure 19.4, the contact angle values range between about 60° and 120° and identify 20 coatings with improved hydrophobicity over the two included metal surfaces. As expected, the fluoropolymers were the most hydrophobic, whereas the reported epoxies were the most hydrophilic. Among the listed surfaces, includ-

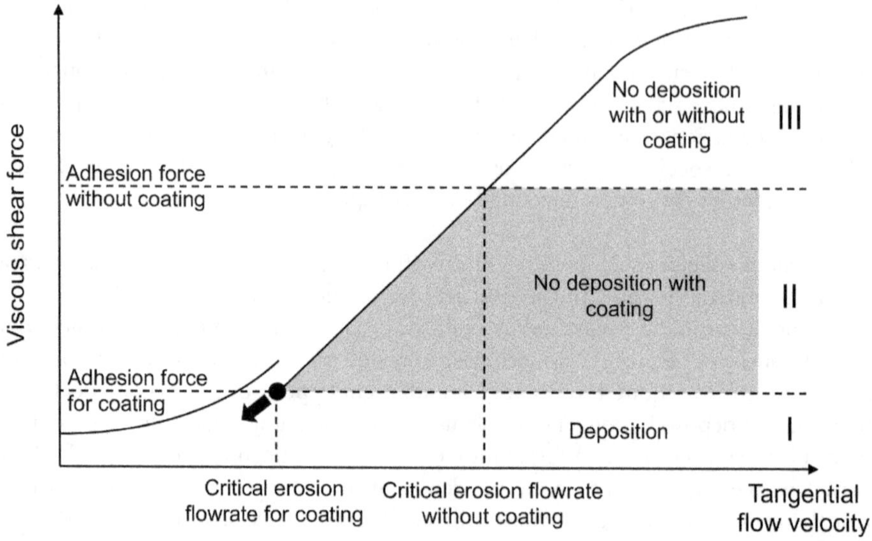

Figure 19.3. The effect of flow velocity on the viscous shear (adhesion) force to strip off a deposit as well as the beneficial effect of a properly engineered surface that promotes low adhesion and resists flow (abrasion and erosion mainly) to provide durable anti-fouling.

ing coatings from the same family, a considerable range of contact angles can exist, as well exemplified by the ceramic coatings, specifically the 10 DLC coatings referred to as DLC A-J, and the reported nickel and phosphorous (Ni-P) electroless platings. Thus, the intricate differences between surfaces, such as composition and structure, have a major effect on the water contact angle. If the water contact angle or surface energy were to simply control the deposition of inorganic or organic materials, the polymeric surfaces of Figure 19.4 would produce the least fouling; however, both experimental results and field history indicate that anti-fouling is more complicated. Fundamentally, the surface energy can be broken down into a polar component and a nonpolar (dispersive) component, with the smaller polar component usually known as the dominant factor influencing inorganic scaling.[6] In studying sulfate formation, Azimi *et al.* also established that a polar component less than 15% of the total surface energy greatly reduces scaling.[6] Surface energy is also partially associated with the electrical nature of a surface, and the right electrical attributes are essential to offer improved antifouling and corrosion advantages. As such, a noncharged surface may supply fewer heterogeneous nucleation sites and lower the probability of deposition. For polymers, Natsi *et al.* showed that the ionization of functional groups such as –OH, –COOH, $-SO_4H$ promotes the formation of inorganic scales, thereby highlighting the role of surface charges.[39] Lauer, in reporting the negative charging of asphaltenes, also suggested that polymers with negative charging attributes can noticeably mitigate asphaltene deposition.[24]

In nature, even a surface perceived as flat and smooth presents some degree of heterogeneous structure or texture. Texturing is also always associated with patterning down to the micro- and nanoscale to allow fluids such as water to drip without wetting. Examples of such surfaces include the lotus leaf, flower petals, fish scales, and insect wings.[40, 41] Mimicking nature by reproducing a suitable surface texture has been an elegant approach to the management of fouling, particularly for nonpressure applications such as heat exchangers. For calcium carbonate, various investigations have concluded that the adhesion of scales is influenced by surface topography, polar surface free energy (simplistically, contact angles), or both.[2, 20] Kumar *et al.* found that micro- and nanopatterning can decrease water wetting by providing contact angles as high as 150°; however, the authors did not observe higher anti-carbonate scaling due to texturing also acting as scale nucleation sites. In their studies, a fluoropolymer with 110° water contact angle outperformed all other surfaces. The effect of surface roughness has been convincingly established by Kesar *et al.* using steel samples exposed to carbonates.[42] In all practicalities, the authors identified that the stress required to pull the carbonate scale off its substrate is approximately proportional to the surface roughness.[42] This force is also about seven times greater for a machined surface than for a superfinished surface, clearly highlighting the benefit of a smooth surface despite frequent impracticalities. In line with other investigators, the authors also identified that carbonate deposits on smooth surfaces had lower bulk densities and different morphologies, suggesting that these deposits are also structurally weaker (likely more porous) and can be removed more readily.[42, 43] The increase in scale adhesion has been explained by the interfacial contact area, which naturally increases with rougher and more interlocking surfaces. Multiple investigations relying on statistical analyses have also highlighted interactions between surface energy and surface roughness, two main and common parameters of antifouling publications, as reexamined in subsequent sections.[10, 12, 13, 20]

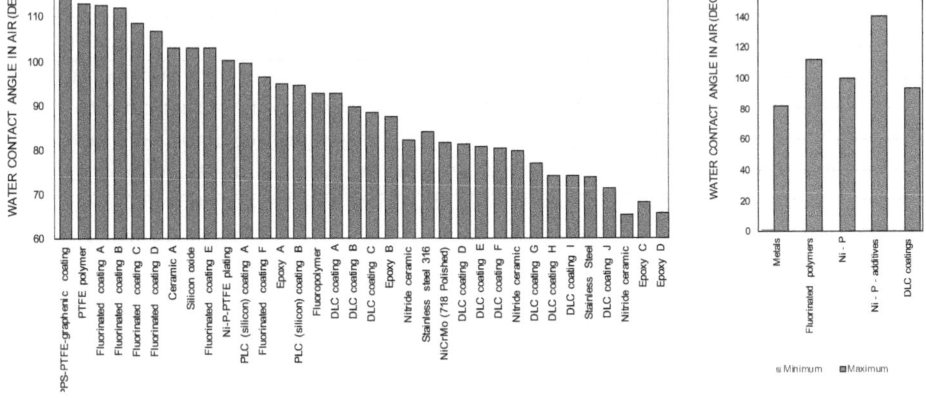

Figure 19.4. On the left, water contact angles in air ranked from the most to the least hydrophobic surface. On the right, the typical range of water contact angles is disclosed for distinct types of surfaces, among metals, fluoropolymers, platings, and carbon-rich films such as diamond-like carbon and polymer-like coatings.

19.3 Coatings and Surface Treatment Engineering

Oil- and gas-protective coatings can be divided into internal and external coatings. Since oil and gas fluids contain traveling solids at velocities up to 60 m/s (200 ft/s) for valve chokes, the impact of these solids induces wear (principally erosion), corrosion, or synergistically a combination of erosion and corrosion. Coatings and surface treatments designed for internal surfaces must be engineered to withstand these environments for decades at an attractive initial cost. Several case studies have proven that fouling can be successfully mitigated by coatings and less by surface treatments.[10, 20] The latter can be explained by what a surface treatment is and what it produces. It is both a process and a modified surface by which the outermost layer of a metal is metallurgically altered, as opposed to coatings that are applied as an added layer or topcoat. Common failure modes for coatings and surface treatments include blistering, cracking, spalling, and delamination.[23] Thicker, more defectuous, and more permeable surface treatments typically provide fewer durable solutions. The imperfections at the surface, if not enabling fluids to pass through the surface treatment, also tend to form nucleation sites for deposition.

19.3.1 Engineering Requirements

In contrast to well tubulars, major differences exist in the technical requirements of coatings and surface treatments for more advanced well completion equipment. The available construction materials also vary, with carbon and low-alloy steels in tubulars often accompanied by thick nonmetallic coatings, whereas the completion equipment more regularly includes corrosion-resistant alloys (CRAs) that allow only thin coatings.[7] With carbon and low-alloy steels, for which greater risks of weight loss corrosion prevail, coatings may be required to have a protective function. With CRAs, the thin coatings in use are primarily to provide lubrication and control friction, including prevent galling, and have been extended against fouling.[7, 28] For example, SSSVs may be requested with complementary anti-scaling characteristics to ensure safe activation in case of emergency, or worse, a blowout. For these valves, sticky deposits on internal wall surfaces can hinder actuation by substantially increasing the operating force or simply preventing actuation. The spring cavities, if filled with solids, can impede valve closure and thus pose catastrophic regulatory and safety risks. Unlike tubular products, coatings for SSSVs are thin, and yet they must endure time and be fully compatible with carbon and low-alloy steels, as well as a broad range of CRAs among stainless steels and nickel-based alloys. Given their thickness, when experiencing elevated liquid and gas pressures, these coatings are typically highly permeable, and thus may rapidly experience corrosion from underneath, which may lead to their failure. With proper alloy selection, particularly with CRAs selected to have a high pitting resistance equivalent number, corrosion under so-called insulation can be eliminated.[44] Completion equipment, such as sand screens, slotted liners, and ICDs, can also be subject to sticky deposits impeding the closure of small yet essential passageways. Such localized plugging may lead to collapses of base pipes, breakdown of the well equipment, and eventually major and costly damage to the production facilities. For screens, and in general filters, erosion is an added concern that demands surfaces much harder than polymers and therefore involves metallic and ceramic coatings and surface treatments. Unlike coatings, surface treatments chemically transform the metal surface but unnoticeably alter the dimensional tolerances. Nitrocarburizing (including by salt bath such as the Quench Polish Quench process, commonly known as QPQ), carbonitriding, nitriding, carburizing, and boronizing (or boriding) are just examples of diffusion treatments that are applied to well equipment, primarily to control friction and actuation, and reduce risk of galling. Thermal diffusion of zinc, aluminum, or chromium are also common surface treatments; however, since rarely encountered on oilfield equipment due to limited performance, these surface treatments are not addressed and solely mentioned. Their performance toward oilfield deposits may be expected to compare that of bare metals and thermal spray coatings, discussed in Section 3.3.

19.3.2 Polymer Coatings

Among the nonmetallics, the polymer coatings have served as thick liners for the rehabilitation of existing tubulars as well as for new products. Generally, they consist of mixtures of a binder with pigments, fillers, solvents, and additives with the binder representing the main component of the coating. For tubulars, epoxies, phenolics, urethanes, nylons, and fluoropolymers are the basis of many coatings, with the fluoropolymer-based thermoplastics favored above 550 bars (8,000 psi) and 175°C (350°F) and in the most severe environmental conditions (*e.g.*, 30% H_2S).[25] Due to their inertness and smooth finish, the fluoropolymer coatings have been widely used against organic and inorganic deposits in oil and gas wells.[23] For completion products, where the coating thickness is normally less than 0.076 mm (3 mil), fluoropolymers have been incorporated in chemically and thermally resistant resins such as PPS or PEEK. For lubrication, PTFE, PFA, and FEP along with solid lubricants have been available for decades in high-pressure applications demanding both low friction and scale mitigation.

In recent times, research has been conducted to include the so-called 2D carbon materials, among carbon nanotubes and more recently graphene particles. These material additives present advantages for wear resistance, but their anti-fouling resistance is not yet much documented.[41, 45] These 2D materials, among others, have been incorporated into polymer thin coatings applied by spray and bake processes, following a simple anchor patterning to improve substrate adhesion. These thin coatings are primarily composed of a single layer, whereas typical coatings for OCTGs tend to be multilayered to withstand fluid permeation and corrosion.[23] The latter may be visualized as layered systems consisting of interface bonding layers, chemical primer layers, and permeation barrier layers, with a topcoat. For a low frictional behavior and a smooth surface finish, the topcoat usually consists of a highly fluidic resin with at least a fluoropolymer component; for instance, PFA or PTFE that upon being remelted during a final manufacturing bake, leaves a lubricating topcoat over the surface.[23] Figure 19.5 shows a set of thin coatings, among DLC and polymer, after a carbonate scale test.[44] On each sample surface, the flaky white layer is calcium carbonate, specifically calcite.[44] Unlike the green and red polymer coatings, the black coatings blistered extensively despite a beneficial scale resistance. The DLC coating of the last image also resisted scale deposition, while remaining immune to structural damage. With thin polymer coatings or defective coatings with surface discontinuities (*e.g.*, pinholes, cracks), rapid permeation with gases such as carbon dioxide can induce debonding and coating failure.[44] Any gases, including light hydrocarbons, hydrogen, and hydrogen sulfide, remain a source of blistering under rapid pressure drops. This debonding between the metal substrate and the coating is promoted by impervious coatings and is manageable with engineered multilayer coatings, which are sparsely available to completion equipment.[46] Thin and lubricious polymer coatings are common for ESPs, although their long-term performance is controversial. Scale accumulation on the surfaces of ESP motors has been reported to impair motor cooling, causing pump failures due to the thermally insulative properties of scales.[47]

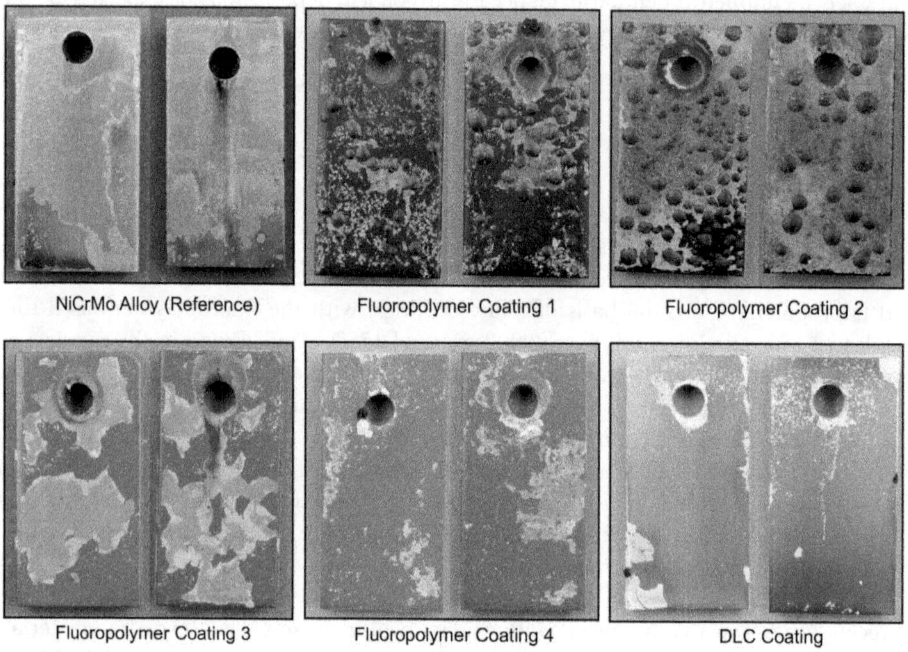

Figure 19.5. Laboratory reproduced carbonate scales on four thin polymer coatings, with blistering developing on the two black coatings. All coatings have one or more fluoropolymer components among polytetrafluoroethylene, and perfluoro alkoxy, in addition to carbon black, ceramic additives, and graphene.

19.3.3 Metallic Coatings

In the oilfields, only a few types of metallic coatings are commonly in use, as broadly categorized as follows: (1) electroless plating and electroplating; (2) thermal spray coatings (*e.g.*, detonation gun, arc spray, flame spray, high-velocity oxygen fuel [HVOF], and high-velocity air fuel); and (3) claddings and overlays (achieved by a welding process, commonly arc welding and laser beam welding). With rare exceptions, these coatings are not employed over the extended lengths of OCTGs, and thus are mainly unheard-of beyond a few feet in length. However, they are generally found in components of wellheads and completion equipment, where their primary functions remain lubrication and anti-galling, including for all critical sealing surfaces. For the latter, corrosion resistance is also important, and the claddings and overlays provide more industry accepted solutions. Except claddings and overlays, typically reproducing corrosion-resistant alloys at surfaces, several of these coatings have also been reported to minimize scale in fields, particularly in land applications such as SAGD, where they have seemingly been mostly evaluated.[20, 21]

Electroplating refers to the deposition of a metal on another metal upon the application of a direct electric current. The most common electroplated metals are nickel, copper, zinc, gold, silver, and chromium, with nickel, nickel-cobalt, and composites thereof also available and reportedly in use in oil and gas applications. Electroplatings can include wear-resistant phases and solid lubricants such as boron nitride, silicon carbide, diamond, PTFE, or even 2D carbon phases such as graphene oxide.[7, 41] Electroless plating, most commonly with nickel and phosphorus (Ni-P), involves an autocatalytic chemical reduction that leaves a uniformly thick layer of amorphous or semi-amorphous nickel-phosphorus alloy on a surface, metallic and conductive, or not.[48] The coating is produced with masking, where not desired, and by dipping into a solution of nickel salt with a reducing phosphorus agent (sodium hypophosphite) without any external current. The phosphorus content governs the plating microstructure, ductility, rate of plating, and level of additives, with content limitations for the more corrosion-resistant high-phosphorus nickel. In addition to steel tubulars, oilfield equipment, such as packers, ball valves, rod pumps, and couplings, represents a few examples, where electroless nickel has been used to take advantage of its chemical resistance. Variants of electroless nickel plating are also reported for ESP stages, where these coatings have been said to extend life expectancy in sandy conditions. Compared with low-alloy steels alone, laboratory and field tests reproducing SAGD conditions have shown that electroless nickel (Ni-P) provides significant antifouling properties, including those against asphaltene.[21, 22] For sand screens, the beneficial impact of electroless nickel over a low-alloy steel surface was also corroborated by Fattahpour *et al.*, who found that calcium carbonate and silica adhered less to the plating.[5, 22] Reduced scaling on electroless nickel was verified by Fermaniuk *et al.* with tubulars circulating for over 2 months in the fields.[11] Plated tubulars experienced limited scaling compared with uncoated tubulars. Zhu reported that the surface of electroless nickel is smooth despite spherical nodules, and certainly more so than a carbon steel surface experiencing corrosion.[22] Lawrence *et al.* found that electroless nickel enriched with boron nitride, diamond, and silicon carbide particles can provide suitable resistance to iron sulfide.[49] These authors also reported a proprietary dual-layer coating (EN-B) with an unusually elevated contact angle of 140°. Despite a lack of explanations, the near superhydrophobicity of this coating can be attributed to both its high-boron content and its texturized surface.[49] This coating, despite its nonwetting characteristics, performed marginally well by not offering any clear advantages. In a laboratory comparative evaluation of coatings against carbonate scales, Deepak *et al.* reported contact angles of 100° on a smooth electroless nickel-PFTE (Ni-P-PTFE) plating. The coating outperformed stainless steel, sprayed, and fluoropolymer coatings, and only an undisclosed doped DLC coating exhibited a superior antifouling performance.[3, 4] Considering low-alloy steel as the reference, Chen *et al.* showed through field tests in sour fields that a specially treated electroless nickel can reduce corrosion and sulfide deposition.[50] The authors also identified a tenacious microcrystalline Ni_3S_2 scale with low affinity for iron sulfide deposition, but expressed concerns over the long-term integrity of the coating due to blistering.

Thermal spraying encompasses various versatile deposition processes that involve the atomization of molten or semi-molten metal droplets by a supersonic atomization gas oriented to a prepared surface. Thermal spray processes, nowadays largely represented by HVOF, are line-of-sight processes well suited to outer surfaces and rarely employed for internal surfaces, particularly extended lengths. Several studies have evaluated such coatings against scales and have predominantly reported negative results.[51] As-deposited thermal spray coatings are characteristically rough and limited for anti-fouling applications. They are also subject to various forms of corrosion that lead to coating failures.[52] For geothermal applications, Sabard *et al.* evaluated Zn-Al thermal spray alloys and reported water contact angles between 110° and 135°. Despite this admirable hydrophobic behavior, which was partly attributed to submicron surface patterns, the authors concluded that carbonate deposition was far from avoided.[51, 53] Other investigations have also reported the potential benefits of oxidized metals produced by spray processes. For instance, sprayed copper oxide (CuO) exhibits high hydrophobicity.[53] When oxidized, such coatings may be characterized more as ceramics than metals. In field applications, they have not found widespread popular acceptance, because of questionable reliability in well equipment. With thermal spray coatings, the two main causes of failure are lack of bonding and porosity. Fluids, especially gases under pressure, can readily work their way through the porosity and other surface defects and cause the sprayed coating to react and eventually delaminate. The development of corrosion under coatings also causes failures in steel components.

19.3.4 Ceramic Coatings

Ceramic coatings are inorganic and nonmetallic, but they can still comprise metals as well as nonmetals in the form of inert compounds such as iron nitride or boron nitride. Ceramic coatings may be further divided according to their compositions, usually as oxide ceramics and nonoxide ceramics. These coatings are intrinsically hard and brittle; thus, they are abrasion-resistant and erosion-resistant under tangential flow, but are also usually restricted under high-angle impingement flows and for mechanical components experiencing localized yielding (plasticity). Ceramic coatings can be applied by various manufacturing processes. This section covers common oilfield ceramic coatings by surface treatments, specifically thermal diffusion, and then expands to more advanced and lesser known ceramic coatings. Diffusion coatings are true surface treatments in which a substrate material is alloyed, often with a light interstitial element such as nitrogen, carbon, or boron, and the result is a thin ceramic layer without noticeable variations in dimensions.

In oilfields, nitride ceramic coatings have found their place through processes referred to as nitrocarburizing, particularly the more scalable and economical salt-bath nitrocarburizing (QPQ process). In this process, an iron-containing alloy such as a carbon or low-alloy steel, a stainless steel, or a nickel-based alloy (with substantial iron and chromium) is converted into a nitride to reduce the risk of galling. For steel with little chromium, the process also provides some degree of atmospheric and seawater corrosion due to a surface iron nitride (white layer) and optionally an oxide layer.[54] Nitriding and nitrocarburizing, in their various forms among liquid, gas, and plasma or ion deposition routes, are primarily employed to counteract galling and control friction in oilfield equipment.[54] Baraka-Lokmane *et al.*, when comparing 18 coatings to a nitrided nickel alloy (by QPQ), found that the nitrided surface was subject to heavy carbonate and sulfate co-deposition, and this behavior of the nitrided surface was further explained by its high surface energy.[10] Unless superfinished after machining, thermal diffusion processes do not greatly modify the surface finish. Among nitriding processes, the QPQ process typically delivers dark and oxidized surfaces that can function as heterogeneous nucleation sites for subsequent scales. Furthermore, the valleys of the nitrided surfaces can shelter deposits from any flow-induced shear, thereby generally not offering the best options against solid fouling.

The performance of less common coatings, such as borides, transition metal nitrides, and/or other surface treatments such as micro-arc oxidation or plasma electrolytic oxidation, remains unreported. Perhaps ceramic compounds such as borides remain of interest, as evidenced by the reported hydrophobicity of the electroless nickel-boron (Ni-B) coating.[49] In general, ceramic thin coatings are applied by industrial processes such as sol-gel, physical vapor deposition (PVD), chemical vapor deposition (CVD), and related vacuum chamber processes. Sol-gel coatings feature a ceramic SiOx backbone, which, with and without organic and inorganic submicron additives, can be placed onto a component by conventional spray or dipping, followed by thermal treatment. Owing to their glassy nature, these coatings can also develop a fine texture with low surface energy for high repellency to fluids.[55] The antifouling resistance of sol-gel coatings with 8- to 10-µm thickness and about 100° water contact angles has been demonstrated for crude oils in exchangers, but it is largely unknown elsewhere.[55] In investigating sol-gel coatings applied to steel pipes for CO_2 service, Ramirez *et al.* found that blistering restricted the performance of these coatings, because of corrosive species diffusing through and reaching the base metals.[56] In geothermal brines, where sol-gel coatings appear to have caught more traction, Aristia *et al.* evaluated the addition of alumina particles (Al_2O_3) and identified protective benefits in both oxidizing and geothermal well environments.[57] For antifouling, alumina has been reported elsewhere as generally promoting deposition on the sol-gel coating.[57] Cheong *et al.* explored in laboratory conditions the carbonate scaling resistance of a hydrophobic chromium oxide (Cr_2O_3) ceramic coating produced by slurry and further compared it to PEEK and an undisclosed DLC coating.[40] The authors reported a low level of fouling, thereby confirming that some ceramics can help mitigate solid deposits. Unlike other coatings, sol-gel coatings are rarely encountered by third-party applicators of oil and gas equipment manufacturers. This alone remains a major limitation of coating technology. These thin coatings are also difficult to repair on-site and remain insufficiently proven to justify their use in oilfields. Even if further developed, the inherently thin sol-gel coatings would have to demonstrate advantages over other classes of coatings, including ceramic coatings such as the carbon-rich DLC and PLC coatings.

Among the physical, chemical, and related vapor deposition processes, diverse advanced engineered coatings may be found in completion equipment; however, few meet the full requirements of critical tubing-mounted parts. For instance, titanium and transition metal nitride coatings may be seen in oil and gas equipment but are scarce and unheard of on completion equipment surfaces. An exemption certainly exists for DLC coatings. The term DLC is generic and describes a range of thin amorphous films, including hydrogen-free DLC (a-C), hydrogenated DLC (a-C:H), tetrahedral amorphous carbon (ta-C), hydrogenated tetrahedral amorphous carbon (ta-C:H), and carbon-rich films with silicon and dopants.[58] DLC coatings exist with single- and multiple-layer configurations as well as with various percentages of sp^2 and sp^3 hybrid orbitals for graphite and diamond, respectively.[59] The results of incorporating a diamond-like phase among layered architectures and compositions enable several hardness, tribological properties, and environmental and antifouling characteristics.[59] The ratio of sp^3 to sp^2 in DLC coatings can vary greatly, specifically between about 10% and 90%, and tremendously affect the coating hardness.[58] This ratio is influenced by the deposition process, its parameters, the precursor gas, and the coating additives, including silicon (Si), fluorine (F), boron (B), and strong metal carbide formers such as tungsten (W), titanium (Ti), tantalum (Ta), or zirconium (Zr). In DLC coatings, the contact angle of water droplets changes with the presence of additives such as silicon and fluorine, which are well-known to favor a hydrophobic behavior.[60] DLC coatings can be produced on metallic components by PVD, CVD, and derived processes such as plasma-enhanced chemical vapor deposition, or hollow-cathode plasma deposition.[61] Few methods are metallurgically compatible and well suited to cylindrical oilfield parts and can be applied at low temperatures, as needed by long tubulars with mechanical and straightness requirements. Among these manufacturing processes, the hollow-cathode plasma process does not require a traditional vacuum chamber, because a hollow part or pipe precisely acts as the vacuum chamber.[61, 62] To maintain the part properties, straightness, and tolerances, the deposition occurs rapidly and at low temperatures, in contrast to typical CVD processes. A limitation of these low-temperature PVD processes may be the thickness and bonding of these coatings, which usually necessitate interlayers of controlled compositions and compositional gradients suitable for creating high coating interfacial strength and toughness.[59] Silicon doping improves coating adhesion, reduces internal residual stresses, and maintains hardness, and it has become quite common. Silicon, which is essential to the so-called PLC coatings, is important for a low-water-contact angle and has been explored among various antifouling coatings.[19, 28] At last, these thin carbon-rich film coatings often require improved protection to avoid in-shipping damage, particularly at component corners and edges, where the coatings are most vulnerable and prone to chipping. Like many nonmetallic coatings, the DLC coatings are not repairable in the fields. In fact, no ceramic coating is known to be serviceable on-site, a practical disadvantage that also restricts their use.

19.4 Coating Qualification Test and Qualification Procedures

The evaluation of antifouling benefits of coatings and surface treatments always underlies the complexity of deposition processes, and their numerous parameters are mostly absent in laboratory conditions. The laboratory tests are usually set to bridge as best as feasible and practical the gaps with field conditions. The hydrodynamics of the flow, liquid and gas compositions, oil and gas contents, water cut, formation water pH, fluid impurities, and presence of corrosion all affect fouling.[14, 17, 33, 34] In the literature, considerable testing has been reported in miscellaneous laboratory test conditions and field applications with new and traditional coatings and modified surfaces; however, the vast majority of these coatings lacks formulation details for proprietary reasons. While the general trends have already been explained, the typical procedures for qualifying coatings in laboratories are worth briefly examining.

First, the reader is invited to recognize that third-party contractors of oilfield manufacturers commonly produce coatings and surface treatments. Assuming that these coatings are applied according to stringent quality control processes, they are either "good-to-have" and therefore only optional add-ons, or critical to the normal operation of the equipment. When essential to maintain functionalities, these coatings and surface treatments must be properly evaluated for broad engineering and manufacturing requirements. These include (a) mechanical integrity requirements such as coating thickness, hardness, adhesion, friction, and wear resistance, including abrasion and erosion; (b) environmental compatibility requirements translating into an improved operational pressure-temperature envelope with liquids and gases; and (c) antifouling. Coatings and surface treatments may be qualified using standard test samples, which are typically small enough to accommodate test vessels, and/or at the first article level following larger scale tests, which are usually a step closer to reproducing field conditions. Standard organizations such as ASTM, NACE (now AMPP), and ISO propose standards and procedures that address different engineering requirements for specific coatings and coating test qualifications. Depending upon the coatings and testing of interest, the reader is invited to consult with these standards organizations, among others applicable in other regions.

To reproduce solid deposition on test sample surfaces, fluid or solution supersaturation must be established as the starting point, and the latter can be conducted under both static and hydrodynamic test laboratory conditions using a blend of two or more brines. Over time, these brines are set to precipitate solids with or without a change in pressure and/or temperature. For inorganic scales, these brines are normally selected as noncorrosive with relatively neutral hydronium potential (pH). At low-pH values below approximately 6, corrosion in carbon and low-alloy steels typically occurs, whereas at greater pH, scaling may dominate, for example, calcium sulfate ($CaSO_4$).[33] Testing, as well as qualification, is also always achieved with the inclusion of reference test materials, such as carbon and low-alloy steels and nickel-based alloys (*e.g.*, Alloy 718), or benchmarked coatings that are usually already in field use. These test samples were selected to have standardized dimensions and were kept small to fit test fixtures among glass cells and pressurized autoclaves. Even if only an approximation of field conditions is used, the testing and qualification of coatings and surface treatments should at least be comparative to allow a ranking of antifouling tendencies. As further described, this ranking is never universal, and a coating found effective for specific fouling conditions may not operate effectively in different conditions. A coating for use across reservoirs and regions should be qualified for various deposits, both organic and inorganic.

Before fouling tests, the test samples are characterized for the important parameters already discussed, starting with their surface roughness and contact angles, if not their surface energy. Surface roughness refers to irregularities in the surface texture produced by peaks and valleys. Specifically, the quantity Ra that is heavily referred to in the literature designates an arithmetic mean of the absolute departure over a mean line. The average roughness was calculated from measurements by contact profilometry, noncontact optical interferometry, or confocal microscopy.[63] Contact angle measurements of test samples are often performed using the sessile drop method in an open-air environment at ambient temperature. Polar liquids, such as pure or distilled water, and nonpolar fluids such as diiodomethane and specific oils, can be used. When using two complementary liquid media, the dispersive (nonpolar) and polar components of the surface energy can be determined from the measured contact angles. For the exact methods to determine the polar and nonpolar components of the surface energy, several references are available.[6, 13, 35]

The procedures for fouling tests vary depending on the types of deposits and are customizable according to the types of deposits expected in actual field environments. For carbonates and sulfates, NACE offers simple procedures at atmospheric pressure to induce precipitation over short periods.[64, 65] Typically, scale testing is accomplished in conditions that are more closely aligned to field conditions under high saturation ratios (SRs). The saturation of a brine can be calculated, for example, using Multiscale software, with an SR value exceeding unity and greater values for stronger scaling tendencies. For organic deposits, different procedures are used to best mimic the crude oils of production wells. Ideally, the pressure, temperature, and fluid flow can be adjusted and varied in the laboratory to explore the coating sensitivity to environmental changes. Flow loop testing, when accomplished on a larger scale with a custom-built apparatus, may be more suitable for measuring the performance of a coating in laminar and turbulent conditions in a variety of flow directions, with and without chokes along the fluid path. In such flow loops, several test samples may be mounted side by side for direct comparisons and in larger quantities than in either glass cells or autoclaves. Flow loop tests have been reported in several papers, particularly for complex scales and asphaltene deposits.[19, 30] Different procedures are available to ensure a good comparison of test samples and meaningful results relevant to field applications.

For test samples ready for quantitative and comparative analyses, various measurements are routinely performed and reported in the literature.[3–5, 10–13, 16, 19, 28, 38, 43, 49] These always include mass changes (mass or weight gains) drawn directly by weighting the samples on a precision scale before and after testing or alternatively thickness measurements.[19] To provide a visual ranking, some authors have conducted semiquantitative evaluation of images of surfaces with deposits. The percentage coverage and counts of crystal nuclei provide quantitative estimates that supplement the mass gain measurements.[43] For dense and sticky deposits, the adhesion stress or strength is also important to measure, because it provides an independent complementary measure that can downplay mass changes, if prone to major variations. It may be argued that scale adhesion measurement is more valuable than mass gains, even though it is less reported in the literature and not reported in NACE procedures.[64, 65] Adhesion can be measured using different methods, including mechanical and hydraulic methods that have been modified slightly to fit deposits as opposed to films, coatings, or adhesive joints.[7, 16, 28, 42] The simplest method for dense solid scales resembles ASTM D4541, ASTM C633, or ISO4624.[66–69] These tests use a suitable adhesive tape of known pull strength using a portable instrument or a tensile tester configured to record the maximum peel force required to remove the deposit.[66–69] An analogous equipment may be borrowed to measure the tensile shear force, as conducted in the qualification of adhesives and adhesive joints as per ASTM D1002 or ASTM F1044-05.[69, 70] These methods have been exploited successfully with carbonate as scales and wax deposits by several authors, although they are not easily transferable to semisolid materials such as asphaltene.[16] For solid deposits such as scales, adhesion tests can also be conducted using impingement jets, where a nozzle of suitable diameter is employed to deliver a saturated solution on the scaled surface.[19] When using this method, the impinging solution is saturated as such that it only causes the scale to detach and not to dissolve. Other investigators have employed finer tribological techniques typically in use for coatings and thin films, particularly using macro- and microscratch testers, and indentation techniques.[71–73] Microscratch testing uses a progressive load by a shaped indenter over a scaled surface in relative movement to define the critical load, stress, and work (energy) required to remove a deposit. This method for measuring scale adhesion also resembles another procedure for coatings, such as ASTM C1624-22.[71] Other common scratch tests are standardized by ASTM D7027-20 and ISO 19252:2008, and if used as per the standard, may lack sensitivity to highlight differences between scales and surfaces under measurements.[72, 73] All these tests rely on proper instrumentations and automatic scratch visualization to provide accurate measurements. Cheong *et al.* reported the use of a universal surface tester, with capabilities to measure surface interactions down to milli-Newtons.[38] For silica and asphaltene, several authors even extended the use of atomic force microscopy to measure adhesion, eventually concluding that coatings indeed successfully reduce adhesion.[5] For solid deposition accomplished in glass cells, the turbidity of the scaling solution is another independent characteristic that may be worth measuring, typically using a simple colorimeter. This supplementary measurement is useful for assessing the induction time to precipitation for inorganic scales.[32] Although not related to a coating or surface treatment, induction time provides valuable information on scaling kinetics and can provide additional insight, when faced with difficulties in comparing surfaces for fouling. Induction time has been reported in the literature based on indirect measurements of pH and electrolyte conductivity.[2]

Field tests are indispensable for validating the engineering selection of a coating or surface treatment for specific environmental conditions. A comparison of field tests with repeated laboratory tests usually provides a broader and more in-depth picture of the actual performance of a coating or surface treatment. Overall, several authors have reported laboratory and field test results in good agreement; however, in other cases, major differences have been observed and have often been explained by uncontrolled or unaccounted field parameters.[5, 13] The methodologies used to quantify deposits in the field are also subject to debate, and the lack of controlled conditions also usually initiates corrosion that may compete, or at least influence fouling. Measuring the weight of a pipe before and after testing is neither practical nor precise although this method has been used successfully.[74] Despite being already challenging, fouling that develops within surface facilities may be monitored remotely in real time by online nonintrusive sensors, among ultrasonics and radiation techniques. For well equipment, the challenge is considerably greater. Instead, field tests have been reported in test wells over extended periods, typically between a few months and a year.[5, 74] These tests consistently involve a series of connected short subs that are located within a zone of comparable pressure and temperature. By being next to each other along an identical flow, such tests are ideal for comparing antifouling surfaces in more realistic applications. The findings from such tests should outweigh any laboratory test results, if found conflicting, or at least call for additional in-depth analyses to bridge the gaps between different yet complementary tests. Other tests have been more intermediate between laboratory autoclave and field tests, for instance, involving mock-up valves testing under hydrodynamic scaling conditions.[8]

19.5 Coating Performances against Specific Solid Deposits

The literature is abundant on characterizing and ranking coatings according to their contact angles or surface energies, but few works have established the anti-scaling rankings for many coatings, surface treatments, and types of fouling. This section describes the performance against specific deposits of coatings and surface treatments, starting with inorganic scales and finishing with organic deposits. This section compares coatings for their performance against specific deposits. Note that none of the presented data were established using identical test procedures or under identical fouling conditions. Thus, it remains difficult to thoroughly compare all coatings and surface treatments. To overcome this challenge, this chapter incrementally discloses the literature to explore and analyze coatings or types of coatings that offer advantages against specific fouling.

Table 19.2 shows hardness, average friction coefficient, carbonate mass gains, and shear forces to dislodge a carbonate scale produced under hydrodynamic autoclave conditions (*i.e.*, with flow). Informative details on the coatings among DLC and polymers are also included. These include the inorganic fillers and their content, and the polymer present in the coatings. The polymers are either commercially available or have undergone modifications by various additives. These additives are among ceramics and contribute particularly to the mechanical properties (hardness) and color, as many also function as pigments. Stainless steel 410 (UNS S41000) was used as the benchmark. In Table 19.2, the R01, R02, and R03 coatings are commercial coatings that also provide references. In contrast, the C01 to C30 coatings are all exploratory and modified versions of commercial coatings incorporating various inorganic compounds, with a maximum content of 37 wt% for C11. For the polymer coatings, hardness varied considerably, with a minimal value of 5.4 for the R02 coating and a maximal value of 43.9 for the C11 FEP-PTFE-based coatings rich in inorganic fillers. The average dry friction coefficients of these coatings are shown to be between 0.06 and 0.11, with a slight increase in friction for the polymer coatings with inorganic additives. The inclusion of the friction coefficient, which is rarely discussed and is intuitively related to both surface energy and roughness, aims to explore whether the friction coefficient can be a meaningful and complementary engineering parameter for antifouling surfaces. The results in Table 19.2 already indicate that this is not the case. Furthermore, the carbonate deposition results in Table 19.2 may question the quality of the data, possibly the test repeatability since the data came from three identical autoclaves. The reported standard deviations represent a significant percentage of the average values and are occasionally as large as the reported averages (*e.g.*, R03). This scatter is common when qualifying surfaces against fouling. Notwithstanding the data quality, the results in Table 19.2 reveal a relative ranking among the groups of coatings. For ease of analysis, color coding was adopted for both the average mass gains and the shear forces, with the red color indicating unfavorable results with respect to antifouling and the green color indicating positive outcomes. Table 19.2 indicates that the two commercial coatings, R01 and R02, underperformed the newer polymer coatings with additives, whereas R03 showed very positive results. The DLC coatings also reveal uneven performance against carbonate scales, with the DLC-I coating performing remarkably well in stark contrast to the DLC-D coating. On the DLC-D coating, the mass gain is approximately three times smaller than that of stainless steel (UNS

SS41000), while it is substantially greater than that of the newer polymer coatings with inorganic additives. When considering only mass gain, the R03 coating may be seen as a clear winner in terms of scale deposition. However, when the shear force is also considered, other coatings take the spotlight. Practically, the ideal scale prevention coating would never allow the attachment of any scales, resulting in a zero-shear removal force. Among the disclosed coatings, several now outshine the R03 coating. From the top to bottom of Table 19.2, these coatings include DLC-I, C01, C02, and C11, among others. When hardness and dry friction are also considered, the identification of a single coating becomes difficult, because no coating simultaneously exhibits all desirable attributes.

Table 19.2: A comparison of polymer coatings, among traditional (R) and new (C), including two DLC coatings based on at least three repeats of each measurement.

Material	Surface free energy (nM/m)	Mean roughness (mm)	Induction time (h)
Copper	50	1	28.7
Aluminum	48	1	50.4
Steel	40	5	139
Brass	35	1	177
DLC coating	37	5	> 475
PTFE	18	11	89

While Table 19.2 introduces a meaningful subject, such as the antifouling test and data reproducibility, and describes the performance of many coatings, the information is only from a single source. This information is also restrictive by not including other relevant categories of materials, including additives. Figure 19.6 shows the relative performance of the additional coatings against an equivalent stainless steel. In agreement with Table 19.2, the data show both mass gains and adhesion stresses to provide a more comprehensive picture of anti-fouling. The results, despite different scaling conditions, prove that any of the selected coatings reduce deposition compared with bare stainless steel surfaces. As shown in Table 19.2, several coatings, including the DLC coating, microtextured polymers, and ultraviolet-curable silicone, show mass gain and adhesion stress values that complicate interpretation when viewed separately. Mass gains and adhesion values are simply not correlated, and truly provide independent pictures of anti-fouling. A new and important finding also relates to the chromium oxide (Cr_2O_3) ceramic coating, which was found to outperform all other coatings in both mass gain and adhesion values.

Complementary to Table 19.2 and Figure 19.6, Table 19.3 lists several paints, fluoropolymer coatings, and thin nitride coatings produced by the vapor deposition processes. Water contact angles, total surface energies (polar vs. nonpolar components), and mass gains for both carbonates and sulfates are included for laminar and turbulent flows.[74] In line with earlier results, the table color grouping shows that the more hydrophobic fluoropolymers and nitrides (closely behind based on contact angles) did not efficiently reduce either type of scale. Furthermore, the results for the carbonate and sulfate scales are not correlated,

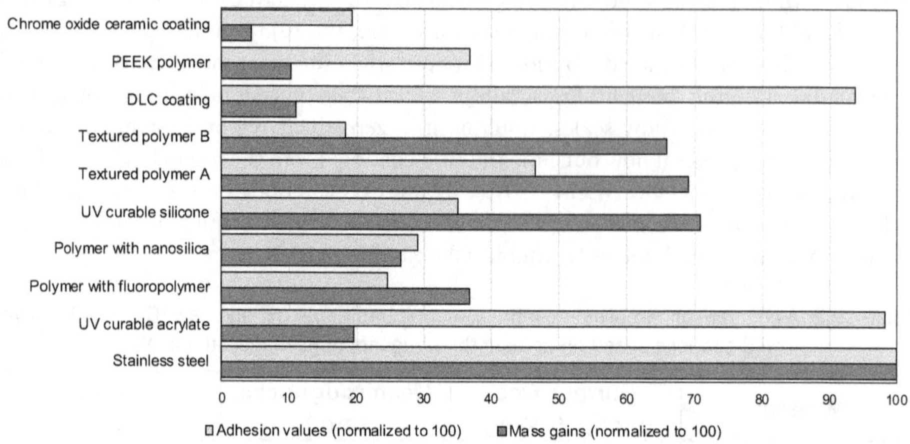

Figure 19.6. A comparison of coatings against carbonate scales, selected among various undisclosed polymers with additives, diamond-like carbon coatings, and a stainless steel reference. Produced with data extracted from Ref. [40].

illustrating that antifouling performance for one type of scale does not readily transfer to another type. As already discussed, turbulences can amplify mass deposition when the viscous shear stress values are low, as evidenced by the mass gains exceeding that of the laminar flow. The results in Table 19.1 and Figure 19.6 also highlight that mass gain alone is insufficient to unanimously converge onto a single preferred coating. In addition, neither the water contact angle nor the surface energy can be directly related to the mass gain values, as illustrated in Table 19.3. The data further exemplifies the complexity of scaling and reveals challenges in the testing and interpretation of laboratory test results to find an optimal antifouling coating. Azimi *et al.* reported that the polar component of the surface energy is a key parameter affecting the formation of scales, specifically calcium sulfate.[6] The data in Table 19.3 also reveal that surface energy, including its smaller polar component, is far from adequately explaining the results of Table 19.3. Deposition is a complex interplay of parameters among listed and unlisted in the table. Mass gain values constitute only surface or coating performance indicators and are thus insufficient to determine specific coating solutions for field applications.

Figure 19.7 shows the optical images of typical autoclave test samples before and after carbonate scale testing. All six test samples were concurrently tested and mounted in a rotating cage immersed in supersaturated scaling brine for 72 h. The untested samples, shown in the top row, are characterized by their initial black color for several DLC and carbon-filled polymer composite coatings, except for the disclosed PLC coating, which exhibits light diffraction fringes over a dark gray metallic surface. These coatings were selected, because they were reported in completion equipment and needed to be compared to recognize that they are not equally performing. The second row shows the same test samples after static scale testing (*i.e.*, with no flow).

Table 19.3: Average surface roughness, water contact angle, surface energy, and mass gains for two different scales deposited on a variety of surfaces with two flow conditions.

Name	Surface roughness, Ra (mm)	Water contact angle (deg)	Surface energy (mJ/m^2)			Mass gains (g)			
						Carbonate		Sulphate	
			Dispersive	Polar	Total	Turbulent	Laminar	Turbulent	Laminar
P-1 (Paint)	0.456 +/- 0.25	75.21	38.38	5.51	43.89	0.36	0.225	0.68	0.61
P-2 (Paint)	1.590 +/- 0.63	78.25	37.55	4.53	42.08	0.39	0.22	0.8	0.78
P-3 (Paint)	0.840 +/- 0.36	79.49	42.87	3.05	45.92	0.18	0.15	1.03	0.75
F-1 (Fluoropolymer)	0.420 +/- 0.17	109.95	18.29	0.31	18.6	0.52	0.27	1.615	0.63
F-2 (Fluoropolymer)	1.010 +/- 0.42	109.21	17.69	0.45	18.14	0.32	0.22	1.1	0.84
D-1 (DLC)	0.141 +/- 0.05	80.07	39.47	3.49	42.96	0.46	0.195	0.67	0.44
Stainless steel	0.080 +/- 0.03	92.28	38.23	0.73	38.96	0.56	0.345	0.48	0.53
G-1 (Ceramic)	0.063 +/- 0.02	102.39	19.39	1.27	20.66	0.525	0.355	1.22	0.72
C-1 (TiN and/or CrN)	0.119 +/- 0.05	96.33	27.63	1.21	28.84	0.78	0.25	0.63	0.32
C-2 (TiN and/or CrN)	0.148 +/- 0.05	88.21	33.13	2.22	35.35	0.48	0.2	0.54	0.25

Stainless steel (Reference)
Typically favorable and/or desirable
Typically unfavorable and/or undesirable

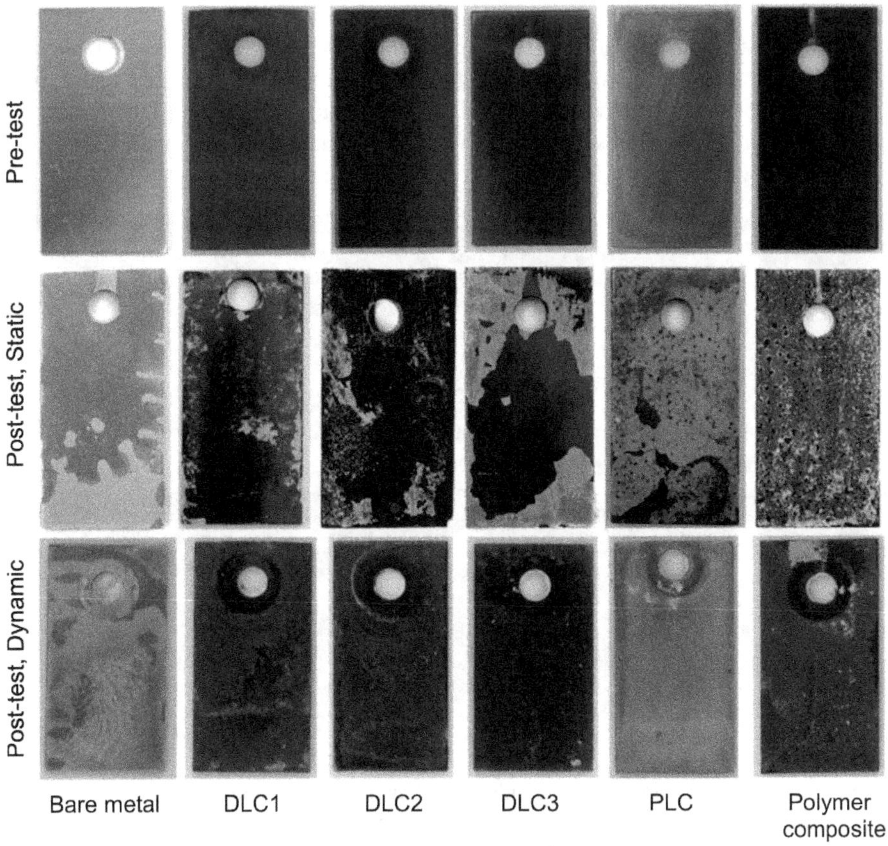

Figure 19.7. Set of optical images of several advanced coatings among diamond-like carbon, polymer-like carbon, and carbon-rich polymer composite showing different scaling behaviors in both static and hydrodynamic conditions with the same exact brines, pressure, and temperature cycles.

Most test samples are then characterized as having white, semicontinuous, flaky layers of scale over their surface. For the coatings specifically referred to as DLC1 and DLC2, the scale deposits are increasingly powdery and form more discontinuous patches. The PLC and polymer composite coatings are also covered by such scale over most of their surfaces. Scale deposits after hydrodynamic testing are depicted in the bottom row of Figure 19.7. Due to conditions altered by the flow, the uncoated substrate now exhibits a yellowish, rippled, and seemingly thick appearance. Compared with the static scale tests, in this test, the coatings were covered by less scale, as explained by the viscous shear removal stresses. With the hydrodynamic flow, both the PLC and the polymer-composite coating show a thin and continuous white layer on the scale. The images in Figure 19.7 consistently indicate that all coatings lowered the scaling tendency of all surfaces, as best evidenced by the tests under hydrodynamic conditions, where the bare metal surprisingly experienced heavy deposition. Figure 19.8 shows a set of SEM images comparing the carbonate deposits under identical magnification (500×). The scale produced under hydrodynamic conditions is only disclosed, with overall smaller crystals (*e.g.*, 2–3 μm) than in the static condition, particularly for the three disclosed DLC coatings. The scale deposit on the PLC coating consists of a combination of fine crystals and large crystal agglomerates, whereas the polymer-composite coating exhibits a high density of tiny crystals. Figure 19.8 demonstrates that scale deposition is partially controlled by preferred heterogeneous nucleation, and not all DLC coatings are equal. While the bare metal surface is fully covered by large carbonate crystals, for many coatings, scaling is still in the initial stages of development.

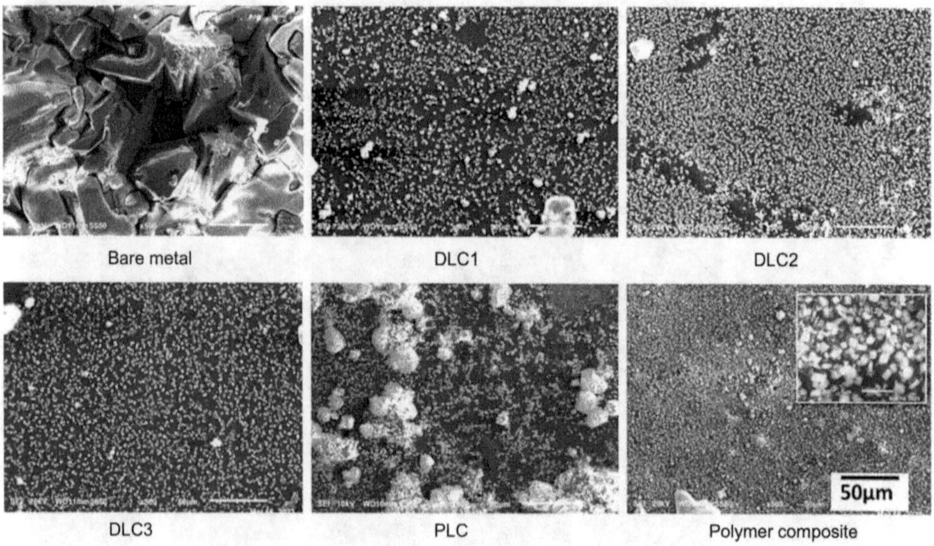

Figure 19.8. Scanning electron micrograph of calcium carbonate deposits on a bare metal (Alloy 718) and on different coatings, among three diamond-like carbon, a polymer-like carbon, and undisclosed polymer composite.

Continuing with the coatings shown in Figures 19.7 and 19.8 and adding repeated testing, Figure 19.9 discloses the average values of mass gain and adhesion strength for both static and hydrodynamic scaling conditions.[28] These results illustrate the variable effect of flow on scaling and, importantly, challenge the conclusions that would otherwise be reached with a single test and measurement method. In the exemplified hydrodynamic test, the coatings also reveal smaller mass gains than in the static tests, except for the bare metal, but greater minimum shear strengths for scale removal. The mass gains are attributed to the nucleation of calcium carbonate crystals and their growth on the surface, as well as the adhesion of scale particles from the bulk solution that adhere to previously nucleated particles on the surface. In this process of scale deposition, the initial layer is formed by the nucleation of calcium carbonate crystals, and subsequent deposition contributes to the growth of dense and more adherent scale. As seen previously, the bare metal with the largest mass gain and most adherent scale confirms that all disclosed coatings mitigate scaling. The selected polymer, a highly fluorinated composite, also performs admirably well with minimal values of both mass gain and adhesion stress. The DLC and PLC coatings, both thin films rich in carbon (with increased silicon content for the PLC coating), remain highly beneficial but are confirmed to be unequal based on their surface formulation (undisclosed).

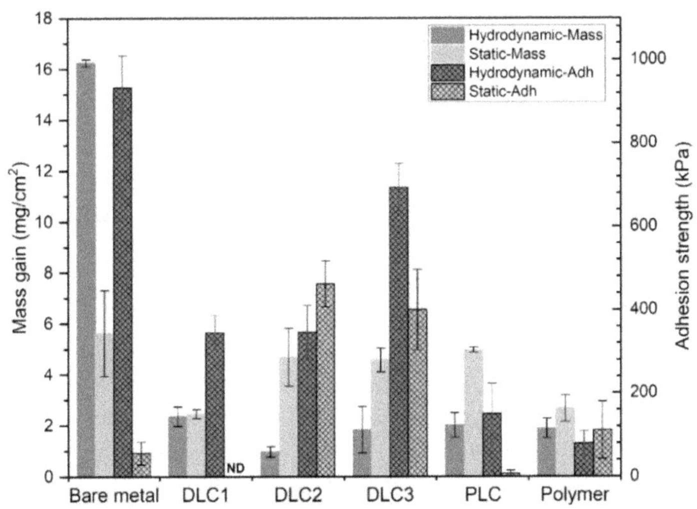

Figure 19.9. A comparison of calcium carbonate mass gain and adhesion strength values in static and hydrodynamic scaling conditions for three diamond-like carbon coatings, a polymer-like carbon coating, a polymer composite coating, and a bare metal (Alloy 718) surface (with similar average surface roughness, or Ra).

Table 19.4 presents a set of unpublished data specific to two undisclosed fluoropolymer coatings tested in various inorganic scaling conditions. These data, also chosen to be illustrative, further describe the effect of a proprietary chemical inhibitor and its interactions with various surfaces, including coatings. The standard coloring, green for generally beneficial, red for less desirable, and orange for all other in-between results, helps visualize the data. For all conducted scaling tests, both polymer coatings significantly outperformed the bare metal (Alloy 718) with all types of inorganic scale deposits. Considering testing uncertainties, the two fluoropolymers may be considered identical in performance compared with the bare metal surface. However, depending on the type of scale deposits, the mass gains are on average nine times smaller with the first polymer coating and 17 times smaller with the second coating, suggesting that the second coating reduces the various scales best. The data also demonstrate that the selected inhibitors are less effective than the coatings, with the latter providing by far the least mass gains. For the barium sulfate scale, the measured shear stresses also show a major reduction in adhesion stress because both coatings are almost indistinguishable.

Table 19.4: Summary table comparing the performances against four scaling conditions of two polymer coatings against a bare metal reference and highlighting the effect on a common scale inhibitor (Alloy 718).

			Alloy 718 (Reference)	Fluoropolymer composite coatings			
				#1	Ratio	#2	Ratio
Water contact angle (°)			NA	104		103	
Surface roughness (mm)			NA	0.88		0.93	
$BaSO_4$	Mass gain (g) with $BaSO_4$ precipitating brine	No inhibitor	6.20	1.00	6.2	1.40	4.4
	Critical shear stress (MPa) with $BaSO_4$ precipitating brine	No inhibitor	15.30	1.80	8.5	1.40	10.9
$CaCO_3$	Mass gain (g) with $CaCO_3$ precipitating brine	No inhibitor	5.55	0.52	10.7	0.45	12.3
		1mg/L inhibitor	5.08	1.20	4.2	0.10	50.8
		100 mg/L inhibitor	1.40	0.40	3.5	0.20	7.0
ZnS	Mass gain (g) with ZnS precipitating brine	No inhibitor	1.42	0.11	12.9	0.16	8.9
		1mg/L inhibitor	1.25	0.10	12.5	0.04	31.3
		100 mg/L inhibitor	1.25	0.04	31.3	0.05	25.0
$ZnS/PbS/CaCO_3$	Mass gain (g) with ZnS, PbS, $CaCO_3$ co-precipitating brine	No inhibitor	3.50	0.80	4.4	0.20	17.5
		1mg/L inhibitor	2.82	0.75	3.8	0.50	5.6
		100 mg/L inhibitor	5.10	0.64	8.0	0.35	14.6
Average (performance indicator for the coatings)			4.44	0.67	**9.62**	0.44	**17.12**

* All values are averages of at least 2 measurements

While Table 19.4 illustrates the advantages of engineered fluoropolymer composite coatings, a range of other coatings have been investigated specifically for sulfide scaling, particularly iron sulfide scaling. L. Kook *et al.* evaluated approximately 40 coatings, mostly selected among variants of electroless nickel plating with various additives, and compared their findings with those of an undisclosed DLC coating.[49] The authors first reported that the wetting of water on various electroless-nickel plating variants can vary between 80° and 145°, approximately correlating with the sulfide mass deposition. Specifically, the coatings with boron characterized by a nodular surface structure showed little scale accumulation, whereas no comment was made on the DLC coating and many other coatings with attractive results against iron sulfide scale.[49] In investigating the conventional mechanism of lead sulfide (PbS) fouling, W. Keogh *et al.* compared the performance of another undisclosed fluoropolymer coating with a SiO_x sol-gel coating, also against a common reference selected to be Alloy 718.[15] In line with previous separate investigations, the authors found that for all three tested and very different surfaces, mass gains and water contact angles were highly correlated, with the elevated water contact angle of the fluoropolymers minimizing lead sulfate deposition.[15]

For inorganic scales, internal coatings to mitigate both corrosion and asphaltene deposits have been successfully employed for tubular applications, particularly for oil and water wells.[27] These coatings have been proven through laboratory testing as well as field experience, where observed reductions in tubing cleanout frequency may be considered as the best evidence of their performance.[27] For asphaltene, paraffin, and scale deposition, internal plastic coatings are effective in diverse ways. By not experiencing corrosion due to proper coating characteristics (*i.e.*, formulation, layer construction, and total thickness), these thermoplastic coatings first prevent the formation of irregular corrosion products and barely offer any attachment sites for deposits. Their smooth nature, along with a low surface energy top layer, clearly helps mitigate deposits, both inorganic and organic. Asphaltenes and waxes have frequently been investigated as part of similar investigations.[29] The literature reports that smooth coatings possessing hydrophobic properties can also address asphaltene and wax flow assurance issues.[29, 36] For inorganic scales, coatings and surface treatments have thus proven their effects. Moradi *et al.* explored the correlation between asphaltene adhesion and wettability on industrial coatings.[75] The authors found that hydrophobic and epoxy resin coatings showed weak antifouling performance against asphaltene, yet reduced deposition. Hydrophilic surfaces, such as a normal steel surface or a phosphate surface, considerably reduce asphaltene deposition via the formation of a water film on the coated surface. Nonetheless, the fluorinated coatings, among PFA and PTFE, showed the best performance by reducing deposition by a staggering 99% under laboratory flow loop conditions. Saveri independently investigated the effects of the PTFE coating and compared it to nanosilica, which exhibits superhydrophobic behavior. The author concluded that PTFE is much more effective in reducing asphaltene deposition under both static and dynamic laboratory conditions.[76] The low surface energy of these coatings, along with the nature of the surface, are thus once again proven to be important.[24] Moradi *et al.* further concluded that the small polar component of the surface energy can be considered as the main reason behind the outstanding performance of fluoropolymers in general.[75] In an investigation of the adsorption of asphaltene on functional groups, Girard *et al.* confirmed that fluorinated surfaces promote low bonding to asphaltene, despite an initial fast monolayer interaction.[77] The findings well explained the effectiveness of fluorinated polymer coatings in OCTGs against asphaltene; though the latter deposits, it simply does not adhere and is removed by the shear-induced flow.

The performance of other coatings has been explored through additional literature, particularly field reports. Compared with low-alloy steels alone, laboratory and field tests for SAGD wells have shown that electroless nickel plating can significantly reduce asphaltene deposition.[21, 22] For both asphaltene and wax deposits, G.K. Bethke *et al.* proved the value of a novel hydrophobic thin carbon-rich film, either DLC or PLC, through field testing.[36] Specifically, the thin coating is reported to lower the scraping frequency by 75% compared with wells that simply have no coating. In a recent investigation by Zolfaghari *et al.*, the authors compared the same coatings as in Figures 19.7–19.9 with an asphaltene artificially deposited in laboratory autoclaves.[78] Even though more research would deepen the current understanding of the relationship between deposits and surface characteristics, the results captured in Figure 19.10 show that the DLC and PLC coatings that repel calcium carbonate still function against asphaltene.[78] Specifically, the PLC coating shows the least adhesion to both types of deposits, which is partly explained by its superhydrophobic behavior and inertness to water. In contrast, the DLC coatings that were not particularly hydrophobic, especially DLC2 with a contact angle near 70°, turned out to be as effective against the deposits as the superhydrophobic PLC coating. It follows that users of coatings should not overemphasize requirements on contact angles, surface finish, or other parameters difficult to specify and control in normal manufacturing and quality control. Instead, when specifying a coating for use in the field or a specific completion equipment, the emphasis should be on specifying a particular coating and applicator, if the coating is not internally manufactured.

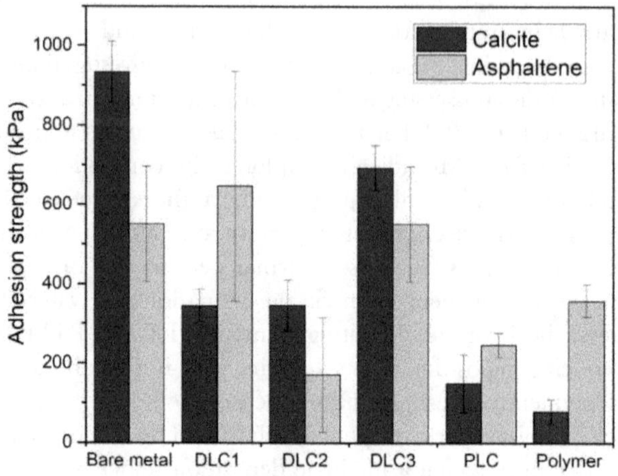

Figure 19.10. Adhesion strengths of bare metal, diamond-like carbon, a polymer-like carbon, and polymer composite coatings against carbonate and asphaltene deposition (all data in hydrodynamic conditions).

19.6 Conclusions

This chapter shows that among the coatings and surface treatments available for well tubulars and completion equipment, optimally managing fouling and balancing other service requirements is not straightforward. The main fouling-related challenge in considering coatings and surface treatments is exactly identifying the surface characteristics that prevent the deposition and accumulation of unwanted deposits under realistic conditions. The literature is abundant in test results applicable under static or quasi-static conditions. Laboratory tests show that hydrodynamic conditions have a profound effect on deposition. For instance, increases in the level of turbulence can amplify the amount of deposition on the surfaces or conversely reduce deposition in the presence of elevated surface shear stresses of erosive nature. As important as these parameters are, they are not the typical engineering requirements, but factors that influence deposition and affect the relative ranking of coatings and surface treatments. Varying these parameters is thus important to acquire a more complete picture of the performance of specific coatings and surface treatments. By default, the proper qualification of a coating or surface treatment should involve multiple and independent complementary tests in addition to multiple repeats to minimize any overlap between data. The data presented in this chapter were often based on multiple measurements, indicating that procedures to evaluate and qualify surfaces tend to produce non-negligible scatter, requiring proper design of experiments and statistical methods.

For tubular and completion equipment, engineering coatings of industrial relevance to combat fouling are attractive. Various field studies have reported the effectiveness of coatings in actual service conditions, particularly against simple carbon and low-alloy steels and thick coatings or liners. The data provided in this chapter indicate that coatings can outperform traditional inhibitors and are thus a crucial part of flow assurance strategies. Engineering coatings, depending on their end applications, either tubulars (OCTGs) or completion equipment, must fulfill engineering lifecycle requirements in addition to providing antifouling advantages. Upon analyzing the typical mechanisms of inorganic scale deposition, surface energy (basically represented by the water contact angle) and surface topography (simplistically summarized by the average roughness, Ra) are the two important parameters extensively covered in publications, but they are largely inexplicably linked. The selection of these two properties emphasizes deposits in which the surface energy, that is, hydrophobicity, is believed to be of fundamental importance. Through a comprehensive review of the literature, this chapter analyzes the relative performance of polymers, including fluoropolymers, and metals such as steels, stainless steels, and nickel alloys (*e.g.*, Alloy 718), plating, and ceramic coatings, including DLC and PLC coatings. To conduct quantitative comparisons of coatings and surface treatments, this chapter borrows test results from distinct and complementary laboratory tests. Given the complexity of fouling, this chapter also reviews the methods for qualifying coatings and surface treatments for anti-fouling. Most investigations have examined mass gains on standardized laboratory test samples and, more occasionally, on field components. This article illustrates that fouling must also be examined from the standpoint of adhesion, where a deposit attaches to a surface and requires a minimum force (work) for its removal. Unless determined by atomic force microscopy, this force is mainly produced and measured using mechanical or hydraulic means, such as tape pull tests, scratch tests, or impinging jets. Though only superficially cited, fouling should also be examined from the standpoint of induction time or the minimal time required for detecting deposition. Ideally, a surface that resists fouling should exhibit all three attributes under a variety of fouling conditions: (a) low mass gain, (b) low adhesion stress or strength, and (c) high induction time. The literature suggests that the results from laboratory tests can lead to different conclusions than those from larger field tests or field experience.

In general, surface treatments and thin ceramic coatings, such as nitrocarburizing or sol-gels, are not very reliable and effective in reducing scaling and are not reportedly used in the oilfields to prevent fouling. Among coatings, polymers that comprise a fluoropolymer component either as the main constituent or as an additive tend to produce coatings with the least surface energy and the least fouling. For inorganic scales and asphaltenes, surface energy remains a paramount parameter for reducing deposition. These coatings, which are inherently hydrophobic and occasionally super hydrophobic, perform well against deposits. However, super hydrophobicity should not be an engineering requirement for antifouling coatings for field use, because fouling is more than wettability and contact angles. Coatings such as DLC and PLC commonly offer water contact angles between only 60° and 100°, thus are on the low side of contact angles and similar to metals. Despite major variations in hydrophobicity, the DLC and PLC coatings can still reduce both inorganic and organic scales. With asphaltene, the mass gains may not consistently be among the lowest with these coatings, but they are lower than without, and above all, the adhesion strengths are lowered to levels approaching those of the fluoropolymers. With both inorganic and organic scales, particularly asphaltenes, fluoropolymer coatings (including nonsuperhydrophic) tend to outperform other surfaces in terms of mass deposition and adhesion.

Through its many examples, this chapter generally emphasizes the inverse relationship between the amount of deposition and surface energy, with the fluorinated surface outperforming all other surfaces. With a polymer, either fluorinated or having one or more fluorinated additives, the fact that these coatings flow over a surface by filling peaks and valleys smoothens the surface and overcomes its initial roughness, thus relegating it to an irrelevant parameter. In this entire chapter, smooth superfinished surfaces are rarely discussed, because they are deemed impractical and irrelevant to oilfield tubulars and completion equipment. Nonetheless, various coatings, in addition to polymers, DLC, and PLC coatings, have repeatedly proved beneficial and improved traditional field practices. Several studies have shown that electroless-nickel plating can be modified to produce many surface attributes. Electroless-nickel platings can be tailored for improved wetting, and similar to DLC and PLC coatings, these nickel platings can be designed to meet many engineering and field requirements. Unlike DLC and PLC coatings, their main applicability is to carbon and low-alloy steels, where if designed and processed correctly, these coatings may provide low-cost solutions to CRAs and help mitigate fouling. The reader should recognize that coatings are generally not designed to replace CRAs; coatings can be used to reduce corrosion of carbon and low-alloy steels, increase life expectancy, and reduce fouling. However, coatings should not be considered permanent solutions for the replacement of completion metallurgies, and their use as such remains restricted by standards such as NACE MR0175/ISO 15156.[79] In summary, given the multitude of requirements assigned to coatings for oilfield use, the right coating solution is based on the application and relative to an established performance. The most suitable coatings, either polymer, metallic, or ceramic, involve increasingly complex processes tailored to build up a coating architecture with more than a single layer of less complex formulation. For nonpolymeric coatings, the development of such coatings involves careful optimization of the mechanical requirements achievable by available manufacturing processes. With the increased use of coatings in downhole equipment and tubulars, their true impact on oilfields will eventually be fully established. With progress in the science and technology of coatings, challenges for the economical sourcing of CRAs, the need for low friction in electrically actuated completion equipment, and a push for greener chemicals, antifouling coatings presenting low frictional characteristics are certainly a part of the future of oilfield assets.

Nomenclature

AFM	Atomic force microscopy
AMPP	Association for Materials Protection and Performance
ASTM	American Society for Testing and Materials
BHA	Bottom hole assembly
CoF	Coefficient of friction
CRA	Corrosion-resistant alloys
CVD	Chemical vapor deposition
DLC	Diamond-like carbon
EOR	Enhanced oil recovery
ESPs	Electric submersible pumps

FCVs	Flow control valves
GLVs	Gas lift valves
HPHT	High pressure high temperature
HVAF	High-velocity air fuel
HVOF	High-velocity oxygen fuel
ICDs	Inflow control device and valves
ICVs	Inflow control valves
NACE	National Association of Corrosion Engineers
OCTG	Oil country tubular goods
PE-CVD	Plasma-enhanced chemical vapor deposition
PEEK	Polyether ether ketone
PFA	Perfluoro alkoxy
PLC	Polymer-like carbon
PPS	Polyphenylene sulfide
PREN	Pitting resistance equivalent number
PTFE	Polytetrafluoroethylene
PVD	Physical vapor deposition
QPQ	Quench polish quench
Ra	Average roughness
rms	Root mean square average for a surface roughness
SAGD	Steam-assisted gravity drainage
SEM	Scanning electron microscopy
SR	Saturation ratio
SSSV	Subsurface safety valves
UV	Ultraviolet rays
WAT	Wax appearance temperature

References

1. C.J. Hinrichsen, "Preventing Scale Deposition in Oil Production Facilities: An Industry Review," NACE-1998-61, Corrosion Annual Conference, held March 22–27, 1998 (San Diego, California, USA).
2. Y. Yan, T. Yu, H. Zhang, et al., "Co-Deposition Mechanisms of Calcium Sulfate and Calcium Carbonate Scale in Produced Water," *Crystals* 11 (2021): p. 1494.
3. D. Kumar, J. Welch, Z. Xu, "Reduction in Scale Build-Up From Sub-Surface Safety Valve Using Hydrophobic Material Coating," SPE-166218-MS, SPE Annual Technical Conference and Exhibition, held September 30–October 2, 2013 (New Orleans, Louisiana, USA).
4. D. Kumar, "Scale Resistant Production Tools," SPE-197994-MS, SPE Kingdom of Saudi Arabia Annual Technical Symposium and Exhibition, held April 24–27, 2017 (Dammam, Saudi Arabia).
5. V. Fattahpour, M. Mahmoudi, M. Roostaei, et al., "Evaluation of the Scaling Resistance of Different Coatings and Materials for Thermal Operations," SPE-193730-MS, SPE International Heavy Oil Conference and Exhibition, held December 10–12, 2018 (Kuwait).

6. G. Azimi, Y. Cui, A. Sabanska, et al., "Scale-Resistant Surfaces: Fundamental Studies of the Effect of Surface Energy on Reducing Scale Formation," *Applied Surface Science* 313 (2014): pp. 591–599.
7. M. Marya, V. Singh, A. Zolfaghari, et al., "Friction Reduction Coatings for Carbonate Scale Mitigation: A Comparison of Materials Technologies," NACE-2021-16533, Corrosion Annual Conference, held April 19–30, 2021 (Virtual).
8. G.M. Graham, N. Goodwin, E. Albino, et al., "Minimizing Scale Deposition Through Surface Enhancement in Downhole Tools," SPE-169781-MS, SPE International Oilfield Scale Conference and Exhibition, held May 14–15, 2014 (Aberdeen, UK).
9. S. Al-Ghamdi, A. Al Nazim, A. Bouyabes, et al., "A Novel Stimulation Approach for Scale Control in the Murrat Carbonate Reservoir—Case Studies From Joint Operations, PZ Kuwait," SPE-143649, SPE European Formation Damage Conference, held June 7–10, 2011 (Noordwijk, The Netherlands).
10. S. Baraka-Lokmane, J.V. Charpentier, A. Neville, et al., "Comparison of Characteristics of Anti-Scaling Coating for Subsurface Safety Valve for Use in Oil and Gas Industry," IPTC-17953-MS, International Petroleum Technology Conference, held December 10–12, 2014 (Kuala Lumpur, Malaysia).
11. B. Fermaniuk, M. Claerhout, M. Zhu, "In-Situ SAGD Thermal-Chemical Effects and Metal-Bond Coated Slotted Liner Design for Enhanced Sand Control," SPE-178465-MS, SPE Thermal Well Integrity and Design Symposium, held November 23–25, 2015 (Banff, Alberta, Canada).
12. Z. Wang, A. Neville, A.W. Meredith, "How and Why Does Scale Stick—Can the Surface be Engineered to Decrease Scale Formation and Adhesion," SPE-94993-MS, SPE International Symposium on Oilfield Scale, held May 11–12, 2005 (Aberdeen, UK).
13. M.M. Vazirian, J.V. Charpentier, M.O. Penna, et al., "Surface Inorganic Scale Formation in Oil and Gas Industry: As Adhesion and Deposition Processes," *Journal of Petroleum Science and Engineering* 137 (2016): pp. 22–32.
14. C. Crowe, S.B. McConnell, J.J. Hinkel, et al., "Scale Inhibition in Wellbores," SPE-27996-MS, University of Tulsa Centennial Petroleum Engineering Symposium, held August 29–31, 1994 (Tulsa, Oklahoma, USA).
15. W. Keogh, T. Charpentier, V. Eroini, et al., "Insights Into the Mechanism of Lead Sulfide PbS Fouling and the Influence of Light Distillate Fraction," SPE-190731-MS, SPE International Oilfield Scale Conference and Exhibition, held June 20–21, 2018 (Aberdeen, UK).
16. E. Brown, C. Koh, M. Pickerts, et al., "Deposition and Corrosion Mitigation Using Surface Treatments," NACE-2021-16729, Corrosion Annual Conference, held April 19–30, 2021 (Virtual).
17. J. McAdam, S.A. Parsons, "Calcium Carbonate Scale Formation and Control," *Reviews in Environmental Science and Biotechnology* 3 (2004):159–169.
18. M. Espinosa, J. Leal, T. Chen, et al., "Downhole Scale Removal in High Temperature Sour Gas Wells," AMPP-2024-20834, AMPP Annual Conference, held March 3–7, 2024 (New Orleans, Louisiana, USA).
19. M. Heydrich, A. Hammami, S. Choudhary, et al., "Impact of a Novel Coating on Inorganic Scale Deposit Growth and Adhesion," OTC-29218-MS, Offshore Technology Conference, held May 6–9, 2019 (Houston, Texas, USA).
20. T. Charpentier, A. Neville, S. Baraka-Lokmane, et al., "Evaluation of Anti-Fouling Surfaces for Prevention of Mineral Scaling in Sub-Surface Safety Valves," SPE-169750-MS, SPE International Oilfield Scale Conference and Exhibition, held May 14–15, 2014 (Aberdeen, UK).
21. D. Zhu, L. Gong, W. Hu, et al., "Application of Electroless Nickel Coating As a Scaling Resistant Alloy in Thermal Production," SPE-190749-MS, SPE International Oilfield Scale Conference and Exhibition, held June 20–21, 2018 (Aberdeen, UK).

22. D. Zhu, M. Leitch, J. Wang, et al., "Evaluation of an Advanced Metal Bonded Coating Technology for Improved SAGD Performance," SPE-181181-MS, SPE Latin America and Caribbean Heavy and Extra Heavy Oil Conference, held October 19–20, 2016 (Lima, Peru).
23. J.M. Brupbacher, J.E. Stiles, "Advanced Coating Lining Systems for Challenging Chemical Environments," NACE-97364, Corrosion Annual Conference, held March 9–14, 1997 (New Orleans, Louisiana, USA).
24. R.S. Lauer, "The Benefits of Using Internal Plastic Coatings on Chrome Tubulars," OTC-16026-MS, Offshore Technology Conference, held May 3–6, 2004 (Houston, Texas, USA).
25. L.W. McKeen, J. Hofmans, J. Nelissen, "Engineered Internal Downhole Coating Solutions for Corrosion and Deposition Control in Downhole Production Tubulars," SPE-1612-04, International Petroleum Conference and Exhibition, held November 11–14, 2012 (Abu Dhabi, UAE).
26. Z. Wang, X. Lin, L. Zhang, et al., "Opportunities for Fluorocarbons Interior Coating Technology in ASP Flooding EOR: Evaluation of Corrosion Protection and Drag Reduction," IPTC-18607-MS, International Petroleum Technology Conference, held November 14–16, 2016 (Bangkok, Thailand).
27. R. Edappillikulangara, M. Telang, R. Quttainah, et al., "First Time Worldwide Application of Glass Reinforced Epoxy Lined Tubing for Prevention of Asphaltene Deposition on Tubing in Oil Wells—A Case Study From Kuwait," IPTC-21441-MS, International Petroleum Technology Conference, held March 23–April 1, 2021 (Virtual).
28. V. Singh, A Zolfaghari, M. Marya, "A Comparative Parametric Investigation of Thin Diamond-Like and Polymer-Like Carbon Coatings to Explore Carbonate Anti-Scaling Performance," *Surface & Coatings Technology* 473 (2023): p. 130015.
29. M. Pickarts, E. Brown, J. Delgado-Linares, et al., "Deposition Mitigation in Flowing Systems Using Coatings," OTC-29380-MS, Offshore Technology Conference, held May 6–9, 2019 (Houston, Texas, USA).
30. M. Pickarts, J. Delgado-Linares, E. Brown, et al., "Evaluation of a Robust In-Situ Surface Treatment for Pipeline Solid Deposition Mitigation in Flowing Systems," OTC-30817-MS, Offshore Technology Conference, held May 4–7, 2020 (Houston, Texas, USA).
31. M. Pickarts, J. Delgado-Linares, E. Brown, et al., "Surface Treatment Strategies for Mitigating Gas Hydrate & Asphaltene Formation Growth, and Deposition in Flow Loops," OTC-31189-MS, Offshore Technology Conference, held August 16–19, 2021 (Houston, TX, USA).
32. E. Mavredaki, A. Neville, "Prediction and Evaluation of Calcium Carbonate Deposition at Surfaces," SPE-169796-MS, SPE International Oilfield Scale Conference and Exhibition, held May 14–15, 2014 (Aberdeen, UK).
33. M. Alahmad, "Factors Affecting Scale Formation in Sea Water Environments—An Experimental Approach," *Chemical Engineering & Technology* 31 (2008): pp. 149–156.
34. S.O. Adegoke, O.A. Falode, P.C. Nwankwo, "Understanding Oilfield Scale Deposition and Inhibition Mechanisms for Optimum Management: A Review," SPE-207133-MS, SPE Nigeria Annual International Conference and Exhibition, held August 2–4, 2021 (Lagos, Nigeria).
35. M. Forster, W. Augustin, M. Bonnet, "Influence of the Adhesion Force Between Crystal/Heat Exchanger Surface on Fouling Mitigation," *Chemical Engineering and Processing* 38 (1999): pp. 449–461.
36. G. Bethke, B. Snook, G. Herrera, et al., "A Novel Coating to Reduce Solids Deposition in Production Systems," OTC-28902-MS, Offshore Technology Conference, held April 30–May 3, 2018 (Houston, Texas, USA).
37. *API RP 14E, Recommended Practice for Design and Installation of Offshore Production Platform Piping Systems* (Washington D.C.: American Petroleum Institute, 1991).
38. W.C. Cheong, P.H. Gaskell, A. Neville, "Substrate Effect on Surface Adhesion/Crystallization of Calcium Carbonate," *Journal of Crystal Growth* 363 (2013): pp. 7–21.
39. P.D. Natsi, E. Zande, P.G. Koutsoukos, "Protective Coatings Against Formation of Scale Deposits," AMPP-2023-19528, AMPP Annual Conference, held March 19–23, 2023 (Denver, Colorado, USA).

40. W.C. Cheong, A. Neville, P. H. Gaskell, et al., "Using Nature to Provide Solutions to Calcareous Deposition," SPE-114082-MS, SPE International Oilfield Scale Conference, held May 28–29, 2008 (Aberdeen, UK).
41. R. Advincula, "Nanostructured and Superhydrophobic Coatings Against Corrosion," NACE-2019-13506, Corrosion Annual Conference, held March 24–28, 2019 (Nashville, Tennessee, USA).
42. S. Keysar, R. Semiat, D. Hasson, et al., "Effect of Surface Roughness on Morphology of Calcite Crystallization on Mild Steel," *Journal of Colloid, and Interface Science* 162 (1994): pp. 311–319.
43. V. Eroini, A. Neville, N. Kapur, et al., "Preventing Scale Formation Using Modified Surfaces," NACE-2011-11344, Corrosion Annual Conference, held March 13–17, 2011 (Houston, Texas, USA).
44. M. Marya, V. Singh, A. Zolfaghari, "Challenges and Solutions With Polymer, Polymer-Like Carbon, and Diamond-Like Carbon Coatings Against Geothermal Scaling and Corrosion," AMPP-2023-18808, AMPP Annual Conference, held March 19–23, 2023 (Denver, Colorado, USA).
45. P.D. Natsi, G. Ziomas, T. Kourassi, et al., "Is Surface Charge Important for Scaling on Polymeric Matrices," AMPP-2023-19524, AMPP Annual Conference, held March 19–23, 2023 (Denver, Colorado, USA).
46. M. Marya, "On the Corrosion of Commercial Coatings, Surface Treatments, and Carbides for Servicing Equipment Operating in Highly Sour Production Environments," NACE-2021-16532, Corrosion Annual Conference, held April 19–30, 2021 (Virtual).
47. A.I. Voloshin, V.V. Ragulin, N.E. Tyabayeva, et al., "Scaling Problems in Western Siberia," SPE-80407-MS, International Symposium on Oilfield Scale, held January 29–30, 2003 (Aberdeen, UK).
48. *ASTM B733-22, Standard Specification for Autocatalytic (Electroless) Nickel-Phosphorus Coatings on Metal* (West Conshohocken: ASTM International, 2022).
49. L. Kool, H. Zhu, Q. Wang, et al., "Development of Scaling-Resistant Coatings for High-Temperature Sour Gas Service," NACE-2018-11538, Corrosion Annual Conference, held April 15–19, 2018 (Phoenix, Arizona, USA).
50. T. Chen, F. Chang, F. Liang, et al., "Evaluation of a Hydrophobic Coating Material for Downhole Application, Lab vs. Field Evaluation," AMPP-2024-20756, AMPP Annual Conference, held March 3–7, 2024 (New Orleans, Louisiana, USA).
51. A. Sabard, S. Paul, B. Holmes, et al., "Coatings to Regulate Scaling in Geothermal Applications," AMPP-2022-17547, AMPP Annual Conference, held March 6–10, 2022 (San Antonio, Texas, USA).
52. H. Almasi, F. Namazi, A. Kamali, et al., "Comparison of Corrosion Resistance of HVOF Tungsten Carbide and Inconel 625 Coating on ENP Coating Carbon Steel in Inhibited HCl," NACE-2018-11549, Corrosion Annual Conference, held April 15–19, 2018 (Phoenix, Arizona, USA).
53. G. Mittal, I. Bhamji, F. Fanicchia, et al., "Development of Superhydrophobic Suspension Plasma Sprayed TiO_2, CuO, and Fe_3O_4 Coatings for Corrosion Mitigation in Geothermal Heat Exchangers," NACE-2021-16302, Corrosion Annual Conference, held April 19–30, 2021 (Virtual).
54. V. Singh, M. Marya, "Influence of Salt-Bath Nitrocarburizing on the Electrochemical Corrosion Properties of Stainless Steels and Nickel-Based Alloys in 3.5-wt% Solidum Chloride Solutions," *Journal of Materials Engineering and Performance* 31 (2022): pp. 6420–6434.
55. C. Bischoff, T. Poulsen, S. Sohoel, et al., "Development, Test, and Offshore Use of a Fouling Repellent Coating for Plate Heat Exchangers," SPE-169195-MS, SPE Seminar, held April 2, 2014 (Bergen, Norway).
56. J.J. Gonzales, M. Ramirez, M. Saenz, "Comparison of Different Techniques for the Improvement of the CO_2 Corrosion Behavior of Carbon Steels in Oil and Gas Industry," NACE-03348, Corrosion Annual Conference, held March 16–20, 2003 (San Diego, California, USA).

57. G. Aristia, R. Bäßler, L.Q. Hoa, et al., "Preliminary Study on Al_2O_3 Sol-Gel Coating for Corrosion Protection of Martensitic Stainless Steel in Artificial Geothermal Water," NACE-2021-16777, Corrosion Annual Conference, held April 19–30, 2021 (Virtual).
58. N. Ohtake, M. Hiratsuka, K. Kanda, et al., "Properties and Classification of Diamond-Like Carbon Films," *Materials* 14 (2021): p. 315.
59. L. Wang, J. Bailey, S. Rajagopalan, et al., "Development and Testing of Low Friction Coated Centralizers for Ultra-ERD Completions," SPE-193342-MS, International Petroleum Exhibition & Conference, held November 12–15, 2018 (Abu Dhabi, UAE).
60. J. Vetter, "60 Years of DLC Coatings: Historical Highlights and Technical Review of Cathodic Arc Processes to Synthetize Various DLC Types, and Their Evolution for Industrial Applications," *Surface and Coatings Technology* 257 (2014): pp. 213–240.
61. B. Boardman, K. Bonapolly, D. Upadhayaya, et al., "A Novel Corrosion Resistant Internal Method Using Hollow Cathode PECVD Technology With Improved Hardness," SPE-114020-MS, SPE Western Regional and Pacific Section AAPG Joint Meeting, held March 29–April 2, 2008 (Bakersfield, California, USA).
62. B. Boardman, K. Boinapally, T. Casserly, et al., "Corrosion and Mechanical Properties of Diamond-Like Carbon Films Deposited Inside Carbon," NACE-08032, held March 16–20, 2008 (New Orleans, Louisiana, USA).
63. ASME. *B46.1-2019, Surface Texture (Surface Roughness, Waviness, and Lay)* (New York: ASME, 2019).
64. *NACE Standard TM0179-2002, Standard Test Method, Laboratory Screening Test to Determine the Ability of Scale Inhibitors to Prevent the Precipitation of Barium Sulfate and/or Strontium Sulfate from Solution (For Oil and Gas Production Systems)* (Houston: NACE International, 2002).
65. *NACE Standard TM0374-2007, Standard Test Method, Laboratory Screening Test to Determine the Ability of Scale Inhibitors to Prevent The Precipitation of Calcium Sulfate and Calcium Carbonate From Solution (For Oil & Gas Production Systems)* (Houston: NACE International, 2007).
66. *ASTM D4541-22, Standard Test Method for Pull-Off Strength of Coatings Using Portable Adhesion Testers* (West Conshohocken: ASTM International, 2022).
67. *ASTM C633-13, Standard Test Method for Adhesion or Cohesion Strength of Thermal Spray Coatings* (West Conshohocken: ASTM International, 2021).
68. *ISO 4624:2016, Paint and Varnishes–Pulloff Test for Adhesion* (Geneva: International Organization for Standardization, 2016).
69. *ASTM D1002-10, Standard Test Method for Apparent Shear Strength of Single-Lap-Joint Adhesively Bonded Metal Specimens By Tension Loading (Metal-To-Metal)* (West Conshohocken: ASTM International, 2010).
70. *ASTM F1044-05, Standard Test Method for Shear Testing of Calcium Phosphate Coatings and Metallic-coatings* (West Conshohocken: ASTM International, 2018).
71. *ASTM C1624-22, Standard Test Method for Adhesion Strength, and Mechanical Failure Modes of Ceramic Coatings by Quantitative Single Point Scratch Testing* (West Conshohocken: ASTM International, 2022).
72. *ASTM D7027-20, Stndard Test Method for Evaluation of Scratch Resitance of Polymeric Coatings and Plastics Using an Intrumented Scratch Machine* (West Conshohocken: ASTM International, 2020).
73. *ISO19252:2008, Plastics–Determination of Scratch Properties* (Geneva: International Organization for Standardization, 2008).
74. M. Vazirian, M. Penna, A. Neville, "Assessing Surface Engineering Solutions for Oilfield Scale: Correlating Laboratory Tests to Field Trials," SPE-179862-MS, SPE International Oilfield Scale Conference and Exhibition, held May 11–12, 2016 (Aberdeen, UK).
75. S. Moradi, S. Amirjahadi, I. Danaee, et al., "Experimental Investigation on Application of Industrial Coatings for Prevention of Asphaltene Deposition in Well String," *Journal of Petroleum Science and Engineering* 181 (2019): p. 106095.

76. M.H. Savameri, S.N. Apourvani, A. Irannejad, et al., "Experimental Study and Modelling of Asphaltene Deposition on Metal Surfaces With Superhydrophobic and Low Sliding Angle Inner Coatings," *Nature Scientific Reports* 11 (2021): p. 16812.
77. H.L. Girard, P. Bourrianne, D. Chen, et al., "Asphaltene Adsorption on Functionalized Solids," *Langmuir* 36 (2020): pp. 3894–3902.
78. A. Zolfaghari, M. Marya, V. Singh, "On Defending Against Calcium Carbonate and Asphaltene Deposits Using Dual-Action Diamond-Like Carbon and Polymer Like Coatings," AMPP-2024-20896, AMPP Annual Conference, held March 3–7, 2024 (New Orleans, Louisiana, USA).
79. *NACE MR0175/ISO 15156, Selection and Qualification of Carbon and Low-Alloy Steels, Corrosion-Resistant Alloys, and Other Alloys for Service in Equipment in Oil and Natural Gas Production and NG Treatment Plants in H2S-Containing* (Houston: AMPP, 2015).

20

Hardware Technologies for Subsea Scale Control

Hua Guan[1] and Phillip A. Rice[1]
[1]SLB OneSubsea, Aberdeen, UK

20.1 Introduction

Over the past two decades, there has been a significant increase in the number of floating production, storage, and offloading (FPSO) vessels and subsea completions worldwide. Production from subsea fields is generally associated with unmanned platforms, wet trees, long tiebacks with large pressure and temperature variations, and limited chemical injection lines, while the completion strategies often involve long, horizontal, and complicated well shapes, multilaterals with inflow control devices and valves (ICDs and ICVs), and complicated sand control technologies. All of these have brought significant challenges to scale control and management, which is exacerbated by the uncertainties of scale risk evaluation because of the lack of reliable water compositions (*e.g.*, commingled water); in addition, inhibition chemicals are more vulnerable to degradation under harsh subsea environments, and well access and subsequent interventions are more complicated to successfully execute, making them more costly. These factors are the primary drivers in the development of laboratory technologies, analytical toolboxes, computer modeling tools, and subsea hardware systems that enable effective scale control and management in subsea fields.[1–8]

This chapter starts with an illustration of the journey that reservoir fluids take from downhole to the topside processing facility, a description of how inorganic scale can precipitate/accumulate on the production system network, and a discussion of how scale risk can change over the life of the field. This is followed by a review of hardware technologies used to tackle scale control and management challenges associated with subsea developments.

The focus of this chapter is the presentation of some key subsea hardware technologies that can be used for scale control and management in subsea fields. The functionalities and working mechanisms are discussed in detail to boost understanding of the hardware technologies, which are not familiar to most scale experts specializing in chemicals. The benefits of these innovative technologies vs. conventional techniques are illustrated by several successful field case histories. Moreover, for the subsea modular injection system, equipment flexibility and adaptability with other hardware is demonstrated by its integration with an open water downline (*e.g.*, coiled tubing or hose) to provide an optimal hydraulic intervention method that offers a cost-effective alternative to traditional riser-based techniques.

20.2 Challenges Associated with Subsea Fields

The study and characterization of produced fluids involves many disciplines, as shown by green boxes in Figure 20.1. The disciplines pertinent to this are pressure-volume-temperature (PVT) and fluid behavior, thermal hydraulics, scale, and production system architecture. In particular, scale must be controlled and managed to avoid production losses and improve the economic performance of subsea assets.

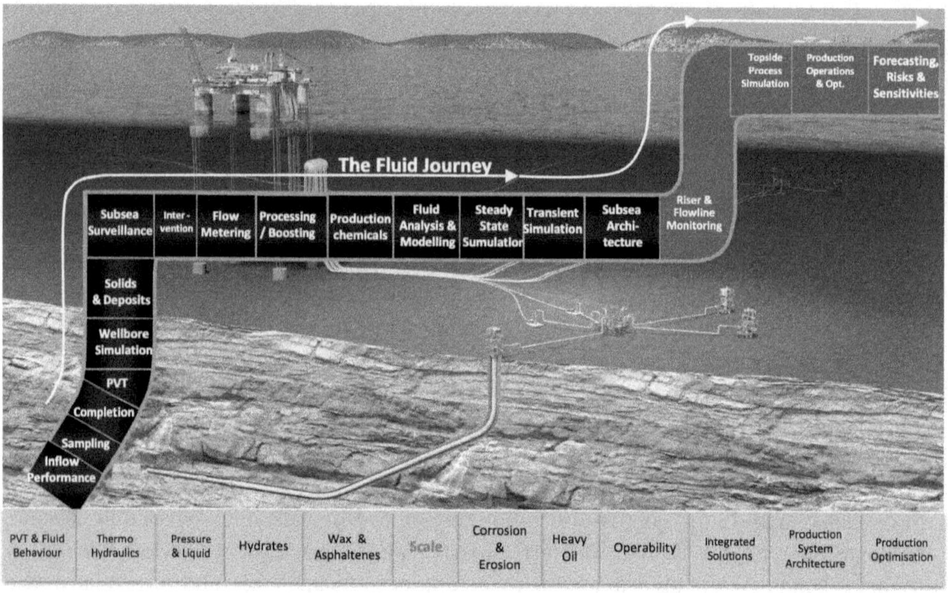

Figure 20.1. Produced fluid interacts with all disciplines across the fluid journey.

During the lengthy journey from the subsea reservoir to the surface processing facility, scale deposition of compounds such as barium sulfate ($BaSO_4$), calcium carbonate ($CaCO_3$), and barium carbonate ($BaCO_3$) occurs as a result of pressure/temperature and fluid composition changes over the life of the field.

Taking $CaCO_3$ scale as an example, it forms following the chemical reactions shown in Equations (1) and (2):

$$Ca^{2+} + CO_3^{2-} \leftrightarrow CaCO_{3(s)} \tag{1}$$

$$Ca^{2+} + 2HCO_3^- \leftrightarrow CaCO_{3(s)} + CO_{2(g)} + H_2O \tag{2}$$

Precipitation conditions are met downhole, where formation water is oversaturated with bicarbonate and calcium ions. During production, a pressure reduction along the production system causes CO_2 to leave the brine phase as gas, resulting in an increase in the solution pH and a reduction in solubility, which leads to $CaCO_3$ precipitation (Equation [2]).

Oversaturation is a factor influencing several variables, including water composition, temperature, pressure, and pH. In the case of carbonate scaling, hydrocarbon composition is also important, because CO_2 can partition in all three phases.[9]

As shown in Figure 20.2, during early field life (refer to black dotted line from reservoir to facility), the operational conditions are completely outside the scaling region, because pressures are high. However, in mid-to-late field life (refer to dark blue dotted line from reservoir to facility), for a given operating condition, when the pressure becomes depleted, it breaks below the equilibrium line and the operating condition enters the $CaCO_3$ scaling region. $SrCO_3$ (strontium carbonate) is another form of carbonate that can occur, when there is strontium present in the fluid.

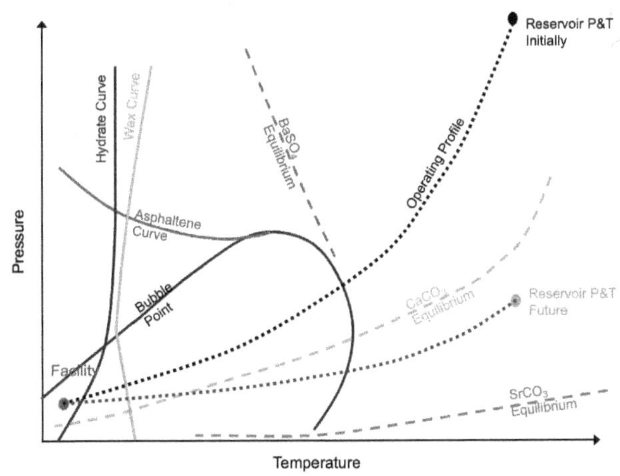

Figure 20.2. Fluid pressure (*P*) and temperature (*Y*), fluid behavior, and scale equilibrium along the fluid journey.

Thus, to monitor and understand a subsea asset's risk or better understand the causes of scale-induced production losses, it is beneficial to gather as much fluid data as practicable. One solution is subsea sampling. For these systems, it is preferable to capture multiphase fluids, maintain the collected fluids at a constant temperature and pressure during the sampling process, and use a sampling process that is simple to operate and efficient enough to perform at potentially frequent intervals.

In addition to changes in fluid compositions, temperatures, and pressures over the life of the field, fluids from different reservoir compartments are often commingled at the manifolds and mixed in the common flowlines. The increasing complexity of subsea infrastructure further complicates the correct understanding of produced fluids with adverse impacts on scale control and management:

- For unmanned floating production facilities, sampling, monitoring, and well access become more difficult and expensive and are sometimes impossible to execute.
- Chemical performance and monitoring are particularly challenged by the use of wet trees, long tiebacks, limited injection lines, and multiple manifolds.
- Different waters can be used in reservoir support, and the choice of injection water (aquifer water, seawater, desulfated water, produced water) has a direct effect on the severity of scaling.
- Following the development of advanced technologies, long horizontal and complicated wells such as snake and dragon wells, multilateral wells, and smart wells have been drilled, and different sand control techniques have been installed in subsea fields.

Figure 20.3 shows a typical scale management cycle which involves planning, execution, follow-up, and review and improvement. The associated subsea challenges bring many uncertainties during the scale management cycle:

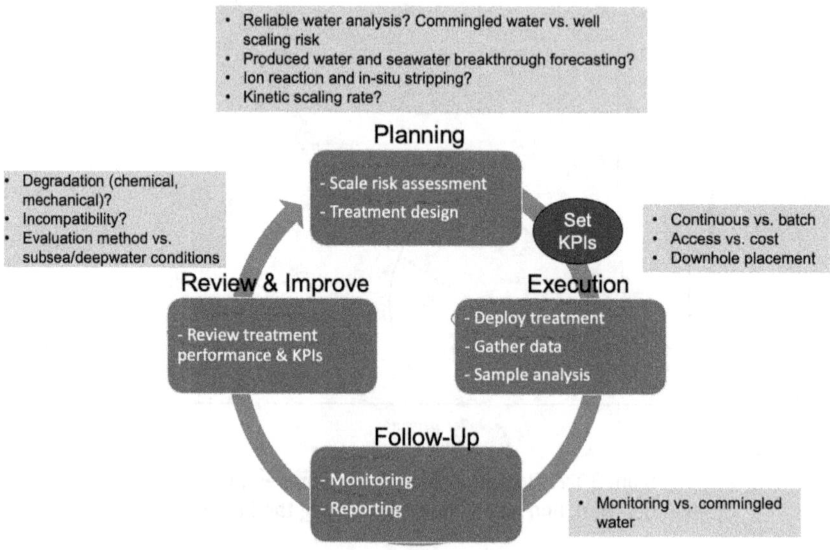

Figure 20.3. Uncertainties (oragen boxes) associated with the circled processes in a typical scale management cycle.

- During the planning stage, one key to successfully managing scale is the ability to identify the scaling threat over the life of the field. However, very often, scale problems can be overlooked or overtreated in subsea fields, because of the lack of reliable fluid sample analysis.
- During the execution stage, one of the primary concerns regarding squeeze deployment is well access and the associated costs. Placement can also be very difficult or problematic depending on the reservoir and completion.
- During the follow-up and review and improve stages, there are many uncertainties in monitoring from commingled fluids. Chemical performance in harsh subsea conditions is of significant concern.

All of the complexities in subsea production architecture affect capital expenditures (CapEx) vs. operational expenditures (OpEx) and total expenditures for a given field and its scale management strategy.

Inorganic scale risks vary with the fluid composition and temperature and pressure, which change with the field development strategy over the life of the field; the complexity of subsea architecture further affects the operational, contingency, and remediation scale management programs. If these changes can be monitored in real time, the scale risk can be updated and treated accordingly. However, it can be difficult to obtain a representative fluid sample and treat the scale efficiently in subsea fields, where well access can be problematic and expensive. Without representative fluid samples and easy access to each well, the scale control strategy is compromised, which can have a significant impact on production and the overall economics of the field.

To address these challenges, hardware and techniques have been developed, including solutions to obtain representative fluid samples from individual wells, which is essential to an effective scale control strategy. The ability to accurately and reliably control the scale inhibition chemical dosage to target an area of risk is crucial in managing operational cost, while the ability to easily access subsea systems to perform cost-effective scale squeeze interventions is necessary to realize the reservoir's potential, which can otherwise become compromised because of the typically significant cost of interventions in subsea environments.

20.3 Hardware Technologies for Scale Control

A variety of technologies have been developed and continue to evolve to address scale control for the subsea oil and gas industry. This section presents subsea hydraulic intervention systems, subsea representative sampling systems, subsea chemical injection metering valves (CIMVs), inflow control equipment, and how they seek to address scale control challenges.

As operators have developed deepwater prospects, surpassing 5,000 ft in 1997 and 9,000 ft in 2006, the need for cost-effective intervention methods has grown. Driven by deepwater rig costs and availability, riserless subsea well intervention is becoming a more widely used technique, with various wells around the world being worked on without a riser.[10]

The subsea well access system allows tools and/or fluids to be placed inside the well's tubulars, enabling various maintenance and/or repair operations. These systems can be either riser-based or riser-less. In many instances where the conditions of the reservoir and well are well known, perhaps through previous mechanical intervention (*i.e.*, when a conveyance method delivers tools inside the wellbore), an operator can proceed directly to a hydraulic intervention, which is one that only delivers fluids into the well or flow line without typically interfering with primary well barriers. Over the past 20 years, a number of operators have conducted a variety of hydraulic interventions in subsea wells, covering the following applications: acid stimulation, well kill, hydrate remediation, and scale squeeze.

Fluid properties play a crucial role throughout the life of the field, and many key economic and technical decisions (related to reservoirs) require reliable fluid property data. The acquisition of representative samples in situ (subsea) not only helps operators monitor their reservoirs and ongoing production but is also useful to assess fiscal allocation and evaluate reservoir communication. Representative fluid sampling and analysis that are updated over the life of the field also form the foundation of a reliable scale risk assessment and successful scale control strategy.

Chemical injection required in subsea wells for hydrates, scale, and corrosion calls for the continuous development of innovative technologies to control inhibitor dosages. Because of the high operator costs associated with downtime and unplanned interventions, subsea chemical injection methods must be highly reliable and available.

20.3.1 Subsea Access and Fluid Injection Systems

As noted previously, the well access required to perform subsea well scale squeezes can be problematic in terms of accessibility and cost. A multiple application reinjection system[11–13] can be used on existing or newly designed subsea trees to provide an injection point for scale inhibitor squeeze operations.

20.3.1.1 Subsea Well and Manifold Access Technology

The system gives access to the tree process flow via the choke body. By using a dual-bore multiaccess system insert in place of a production choke, process flow can be directed out with the tree and, when required, back through the dual-bore insert and onward to the pipeline. This enables tree-mounted process hardware to be retrofitted subsea and usually requires no tree modifications. Figure 20.4 demonstrates the working mechanism of the multiaccess technology used for connecting a process module (*e.g.*, CIMV, subsea sampling unit, or flowmeter).[14–16] By introducing a coaxial flow path insert into the tree, well fluids are directed externally to a flow loop that houses a processing technology. The flow is returned to the wellhead and the existing flowline infrastructure. Processing hardware is installed outboard of the existing isolation barriers, and it eliminates the need for expensive, high-risk well interventions, field shutdowns, or flowline decommissioning, while minimizing production losses.

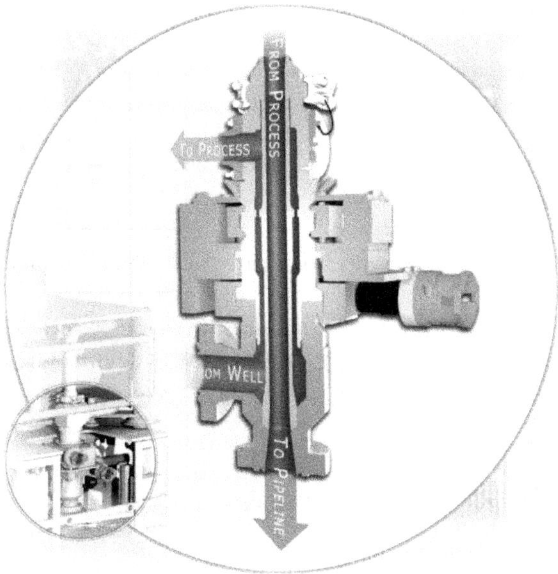

Figure 20.4. Multiaccess technology principle. Green arrows show production being diverted to an external process, before it rejoins the path toward the pipeline as before.

Alternatively, a multiaccess-system-enabled choke insert can be used to retain full choke functionality, while providing an external connection for scale squeeze or chemical injection operations. As shown in Figure 20.5, when the choke insert is upgraded to incorporate an injection flow path, the multiaccess technology can be used for scale squeezes, downhole chemical/fluid injection, pipeline flushing, hydrate management, and well kill.[12, 14, 15]

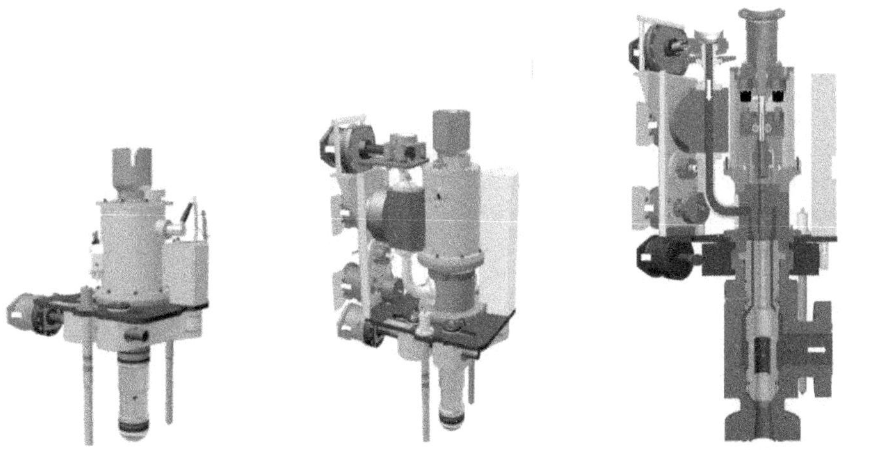

Figure 20.5. Multiaccess injection flow path to well for scale squeeze access.

By replacing the existing tree production choke with a multiaccess-system-enabled choke, a conduit is provided between an injection hub and the wellbore. The choke retains all of the standard production choke functionality and includes appropriate well barriers for the injection hub. Alternatively, a short-term multiaccess insert with no choke functionality can be fitted for the duration of the injection operations and replaced with a standard production choke after the work is complete.

The multiaccess system technology can also be fitted at a manifold, if required.

Alternatively, an operator may elect to have a directly mounted means of chemical injection (for interventions) on the tree itself. In this case, the additional subsea connection and isolation elements are installed on the tree and provide a ready-to-go access point, with a protection cap left in place during normal production operations.

20.3.1.2 Subsea Modular Injection System with Hydraulic Access

Other than having dedicated chemical injection lines from topside or a production facility running back to the subsea production system, which is typically associated with hydrate inhibition, the conventional solution to deliver a fluid treatment such as a scale squeeze to a reservoir is to use an intervention riser system deployed from a floating rig or a heavy well intervention vessel. By connecting it directly to a subsea tree, both mechanical and hydraulic access is achieved, enabling the conveyance of tools and fluids downhole. In scenarios that do not require the conveyance of tools downhole, a riserless light well intervention technique using a subsea intervention lubricator or similar provides a more cost-effective solution to riser-based intervention. For applications that require only the bullheading of fluids to a reservoir (*e.g.*, a scale squeeze treatment) without the need to convey tools downhole, the hydraulic intervention technique offers further optimization and cost reduction potential.

Hydraulic intervention and hydraulic access are the terms used to categorize the technique of delivering/receiving fluids to/from a subsea access point without the need to convey mechanical tooling of any kind. For bullheading treatments, hydraulic access offers a considerable OpEx advantage by negating the need for a rig or heavy intervention vessel, eliminating the need for an injection riser, and facilitating intervention operations on any tree, using either a multiaccess-system-enabled choke insert or an alternative access point.

As shown in Figure 20.6, a hydraulic intervention system typically comprises the following:

- A surface vessel (*e.g.*, a multipurpose service vessel or inspection, maintenance, and repair vessel) and topside equipment
- An open water conduit (*e.g.*, coiled tubing)
- A subsea safety module with integrated flexible jumper
- A subsea access point created by a multiaccess-system-enabled choke insert (or alternative access point)

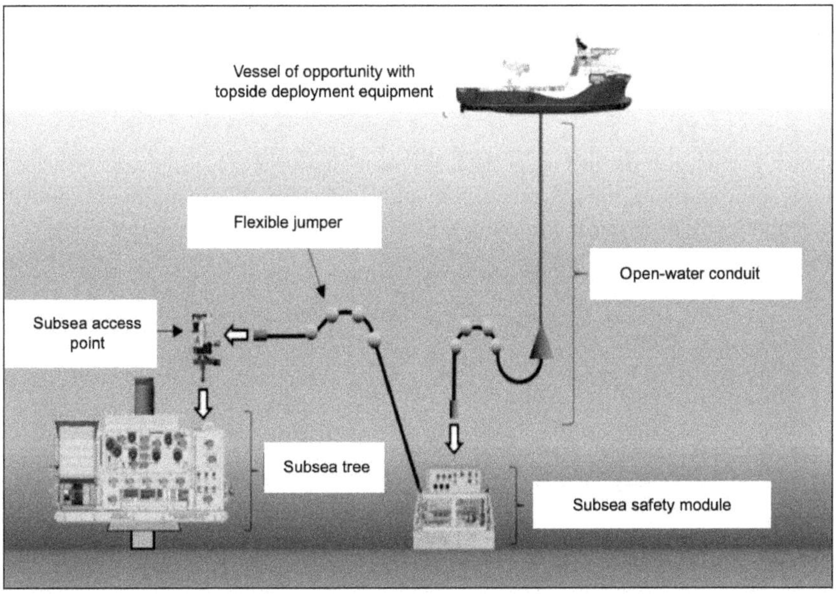

Figure 20.6. Typical hydraulic intervention system.

20.3.2 Subsea Sampling

Sample quality is critical to being able to accurately analyze samples to obtain reliable fluid properties. A representative sample in multiphase flow is an uncontaminated sample that has the same molecules in the same proportions as the flow at the sampling point, where all three phases are in equilibrium. Significant changes in pressure or temperature during sampling will result in mass transfer within phases, rendering the sample nonrepresentative of the production conditions. The subsea sampling system should therefore be as close as possible to isobaric and isothermal.

Additionally, because of the dynamic nature of flow, simply capturing a fluid sample mixture (oil, gas, and water) from a flowline can be ineffective, because the volume ratio of each phase in the mixture may not be sufficient to perform a detailed analysis of each phase (*e.g.*, gas phase in heavy oil or water phase in a low water cut). Multiphase sampling therefore becomes essential so that each phase is enriched and captured individually and sufficient sample volumes are collected from strategically chosen phase-rich sampling points, and no pressure or temperature drop occurs during sampling to maintain equilibrium between phases. Following sampling, each phase is analyzed and physically or mathematically recombined to obtain reservoir fluid properties.

The development of a subsea sampling system[17, 18] provides a new method of capturing a physically produced fluid sample from a subsea tree or manifold and recovering it for laboratory analysis. The sampling system uses the flow-through method of taking a sample, which involves diverting some of the production flow (refer to Figure 20.4) through a separator and returning the fluid downstream. After the separator has collected a sufficient volume of the target phase at equilibrium, the sample is transferred to a sample bottle, isolated, and recovered to the surface for analysis.

As shown in the example in Figure 20.7, the sampling system is physically attached to the host tree (a tie-in point) via a dedicated sampling interface. The sampling module includes an independently retrievable sampling skid and a saver sub and the necessary valving to provide sufficient isolation.

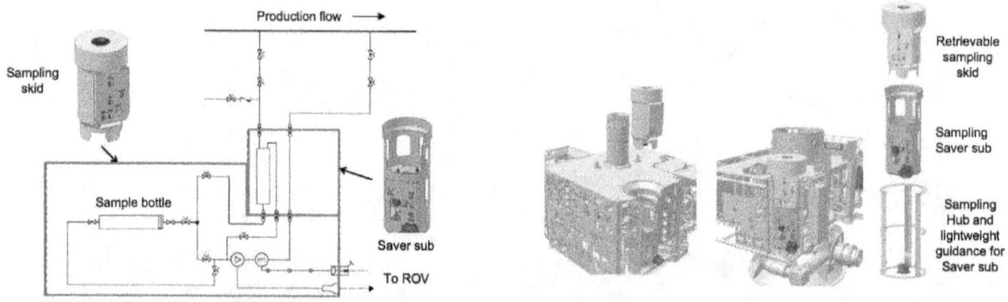

Figure 20.7. Subsea sampling system mounted on tree (schematic and key hardware).

Although various tie-in points (*e.g.*, tree or tree cap, flowline connection module, or manifold) can be used to attach the sampling interface, the principle remains the same: produced fluid is diverted into the sampling *module* and then returned via the dedicated connection. This enables sampling modules to be installed and recovered, providing additional functionality without the need to recover the subsea tree, completion, manifold, or flowline jumpers. Although a remotely operated vehicle (ROV) is required to perform specific tasks in operation and retrieval, such as the operation of manual valving and connection make-up/testing, the ROV is not tied to the subsea structure or sampling package nor does it take any hydrocarbons onboard. This reduces risks associated with ROV downtime and negates the need to collect produced fluids via other subsea equipment such as flexible hoses and hot stabs.

After the sampling module has been latched onto the sampling interface, the connection is pressure tested, and fluids are introduced into the sampling circuit. The circuit can be confined in a heated water bath, if required. The sampling circuit comprises:

- A multiphase displacement unit to maintain pressure
- A separator to segregate and enrich the phase
- Optical phase detectors to identify phases during enrichment and sampling
- Up to 12 transportable sample bottles

In a water sampling sequence, fluids are circulated from the top of the separator, enriching the water phase; gas and oil phases are simultaneously expelled, while the phase detectors are monitored through their electronic signals, which are sent back to the control system, until the chamber is full of water. Then, the contents of the separator are transferred into sample receptacles. For gas sampling, the flow is diverted from a gas-rich sampling point into the sampling circuit. As the gas phase accumulates at the top of the separator, the water and oil phases are expelled from the bottom of the separator, and the content is transferred into sample receptacles as illustrated in Figure 20.8. To capture an oil sample, after the gas phase is expelled from the top of the separator, the flow direction is switched to expel the water phase from the bottom, while the phase detectors are monitored continuously to isolate the oil phase. The system can also collect a multiphase fluid sample directly from the flow stream for liquid fraction measurements.

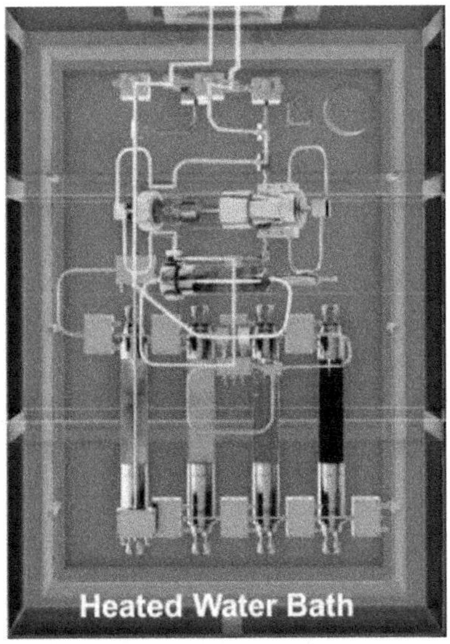

Figure 20.8. Subsea sampling system phase detection and collection assembly.

After the sampling process has been concluded, the complete sampling circuit is flushed with flushing fluids to eliminate any health, safety, and environmental concerns. The well is secured, and the sampling module is unlatched to return it to the surface for retrieval and to transport the samples to the laboratory for analysis.

Depending on the size of the sampling module, it can be either run on wire from a suitable vessel or flown in place by an ROV.[19–22] See Figure 20.9 for an example.

Figure 20.9. Remotely operated vehicle-deployed sampling system.

As illustrated in Figure 20.10, an integrated subsea sampling service enables the acquisition of representative samples under isothermal and isobaric conditions subsea using patented subsea hardware technology (sampling interface and sampling module) situated at the place of interest followed by transport and analysis of samples under representative conditions. This chain of sampling services supports reliable sample analysis, which then enables accurate scale prediction based on the data.[2]

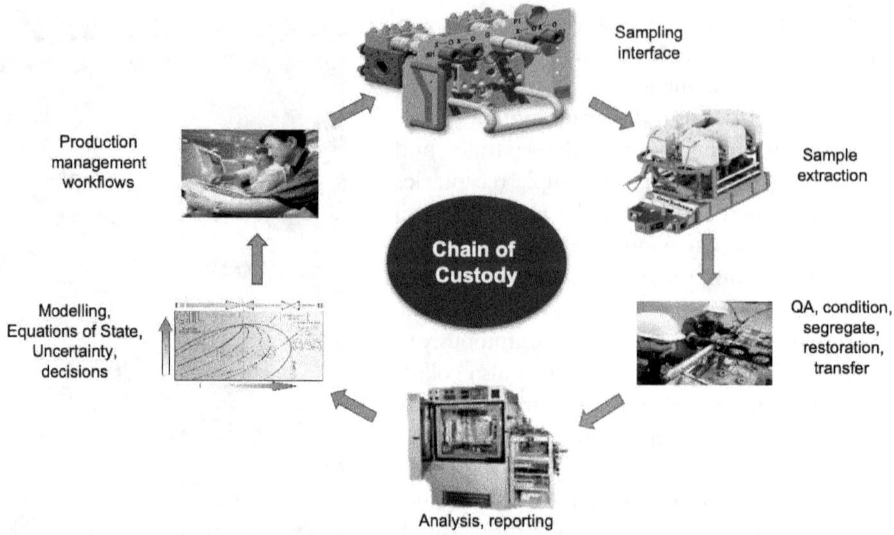

Figure 20.10. Integrated subsea sampling service from subsea access and transport to analysis.

The system has also been used to obtain produced water samples to evaluate the concentration of scaling ions, which is necessary to optimize scale squeeze well prioritization and fluid treatment volumes. In addition, the system has been used to detect the presence of tracers that have been injected into nearby wells, thus providing reservoir interconnectivity insights that can be used to help manage production or injection programs between multiple interrelated wells.

20.3.3 Control Subsea Chemical Injection Rates—Chemical Injection Metering Valves

For low-dose inhibitor (LDI) injection for long step-outs, deepwater tiebacks, or satellite wells, the subsea CIMV is a common solution (Figure 20.11). Pinpoint accuracy of injection is not only critical to achieving full inhibition, it can also save operators from pumping large volumes of excess chemical over the life of the field just to protect against chemical injection flowmeter inaccuracy. A typical field in which LDIs are injected at 2 L/h into 12 wells over a 20-year field life could become over injected by as much as 3–4 million liters of excess chemical based on the selection of CIMV technology.

In long-step-out deepwater projects with complex system architectures, LDI injection and control can be problematic. Subsea distributed chemical injection is often best suited to address the flow assurance challenges of these wells, but greater control and accuracy are required over the quantities of LDIs being added, at multiple points, into the flow.

Should the inhibitor dosage be over or under the optimal application rate, the repercussions can be problematic and costly. Reasons for under- or overdosing are often tied to chemical injection flowmeter accuracy, which can be heavily influenced by the properties of the injected chemicals. The under-injection of treatment chemicals can result in scale or paraffin buildup in well production strings and pipelines, lowering the production rate. If the scale or paraffin remains in the line for an extended period, the well may have to be shut in to undergo a batch treatment, incurring deferred production and intervention costs. Where corrosion inhibitors are being injected, subsea umbilicals, risers, and flowlines equipment can, in severe cases, be taken offline until failed components are replaced.

Figure 20.11. Chemical injection metering valve replacement on subsea tree by means of remotely operated vehicle intervention.

In the case of overdosing, excess chemical costs can be significant, and the additional chemical tanks take up valuable deck space on the platform. For instance, a company could spend more than U.S.$1 million to over inject just one well over the life of the field. Furthermore, excess levels of LDIs in the export crude could affect its value at the refinery. Over the life of a field, accurate flow measurement and control of LDI injection could reduce OpEx by tens of millions of dollars.

With LDI injection, particulate blockage is a recognized cause of CIMV failure. This type of blockage is partly caused by the extremely low required injection rates, which are sometimes less than 0.5 L/h, and the fact that particulate contamination can be introduced into the chemicals during transport, storage, and subsea distribution. These particulates can block moving parts in many flowmeter and CIMV designs.

Traditional flow measurement technologies used in subsea CIMVs typically use Venturi-type flow measurement. Inaccuracies in flow measurement can stem from particulate contamination and blockages in the CIMV and from the fact that CIMVs are engineered years before they are put into service, often with limited knowledge of the chemicals to be injected. As a result, CIMVs are not properly tailored for the chemicals being used, and ultimately, system underperformance can occur.

CIMV flowmeters can deliver an accuracy better than ±2% of reading for LDI injection compared with the industry-standard Venturi-type flowmeter, which may deliver a full-scale accuracy of only 5–10%. A generation of low-flow ultrasonic CIMVs that feature a microbore, nonintrusive line-of-sight ultrasonic flowmeter has been developed. Featuring no moving parts, the flowmeter delivers debris-tolerant flow measurement, is chemical-independent with a very low native pressure drop and does not require subsea filtration.

Developed to address the complete LDI chemical injection portfolio from 0.25 L/h to 600 L/h, this flowmeter is combined in closed-loop control with a throttling valve, providing a self-regulating device requiring only one user-defined input—low rate.

The CIMV's flowmeter addresses the key limitation of other LDI chemical injection technologies: sensitivity to blockages. The flowmeter is particulate-tolerant, meaning that contaminated fluid can easily pass through the unrestricted flowmeter tube. It also consistently provides high accuracy of reading independent of changes in chemical properties such as viscosity and reliably measures the chemical inhibitor flow rate. Real-time feedback from the flowmeter enables autonomous control of the throttling valve, maintaining a user-defined injection rate set point indefinitely regardless of up- or downstream system disturbances.

Packaged as an ROV-retrievable device with onboard diagnostics, the current-generation CIMVs enable full inhibition without the risk of under- or overdosing. Operators now have the option to reliably deliver LDIs via cost-efficient subsea distributed chemical injection systems with precision regardless of chemical properties or contamination, enabling them to make chemical decisions independent of the installed hardware.

As shown in Figure 20.12, a low-flow CIMV designed for the injection of LDIs (for issues such as wax, scale, and corrosion) uses a positive displacement flowmeter and a microneedle and seat throttling valve in closed-loop control. The example shown in Figure 20.11 has been engineered to handle flow rates from 0.25 L/h up to 600 L/h and packaged to be retrievable by ROV, whether mounted in either a horizontal or vertical orientation.[23]

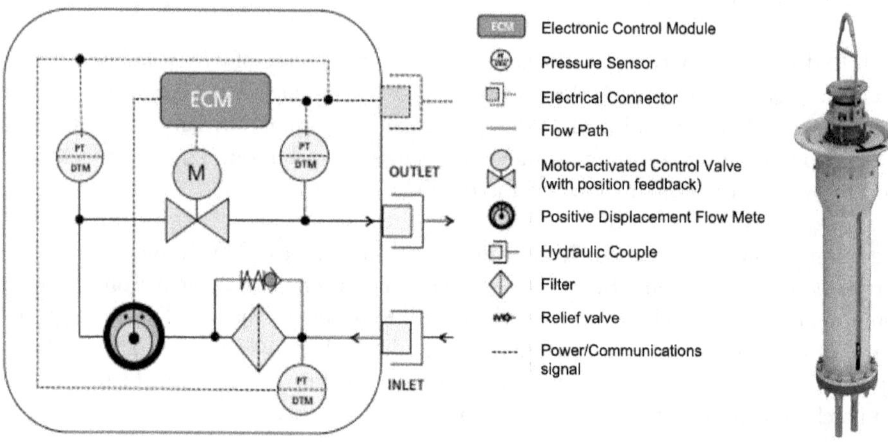

Figure 20.12. Chemical injection metering valve schematic and hardware.

A low-flow CIMV positive displacement flowmeter gives direct readings of flow. It operates using a rotary piston, of which each revolution transmits a discrete volume of fluid, so that each rotation can be directly translated into a flow rate based on the piston's frequency.

This improves the flow rate physical measurement accuracy. The low-flow CIMV uses a microneedle and seat throttling valve to regulate the flow through the system. The needle has an equal percentage flow curve, providing control at low flow rates. It is operated by an electric actuator, which provides very fine system adjustment, controlled by the feedback from the positive displacement flowmeter via the onboard closed-loop control algorithm.

20.4 Case Studies and Associated Benefits

To address the subsea scale management challenges in terms of subsea access, commingled fluids, and fluid property reliability, an innovative multiaccess-enabled insert is used to provide access to the production stream of any subsea tree using an inserted retrievable choke. This provides a subsea access point while maintaining proper environmental isolation when executing the following tasks:

- Scale removal, such as in an acid stimulation operation, where the treatment package is bullheaded from topside to subsea for an individual well
- Scale squeeze operations in which the treatment packages are bullheaded from topside to subsea for an individual well
- Diverting produced fluids into a subsea sampling module during a sampling operation
- Diverting chemicals into a CIMV for measurement during a chemical injection operation

For subsea scale treatments (either acid stimulation or scale squeezes) that will be carried out on an as-required basis, the treatments can be performed using either a rig or an intervention vessel to squeeze the well directly. Alternatively, a direct squeeze from the production facility can be achieved using an existing test or production flow line. Depending on the severity of the scaling issue, the system architecture, and the frequency of the operation, these options can have significant OpEx and production impacts.

The primary benefit of multiaccess-enabled hardware technology in these circumstances is that treatment fluids can be deployed from a multipurpose support vessel or an inspection, maintenance, and repair or other support vessel instead of from an expensive rig (*e.g.*, a mobile offshore drilling unit) or light well intervention vessel. When the multiaccess-enabled insert is installed, only the production from the well being treated is affected, thereby minimizing the production impact to a field.

The multiaccess-enabled hardware technology has been further evolved into an integrated subsea injection system, which includes both topside and subsea deployment elements that can be modularized and adjusted according to the specific requirement and field availability.

In 2008, an operator in the United Kingdom installed the first multiaccess-enabled subsea well intervention system to perform scale squeeze operations on a North Sea well in Bittern Field. Several additional units were later installed in other wells in this field and Pierce Field. In December 2009, an Angola-based operator successfully deployed multiaccess technology for scale treatment operations in the Lobito Field in 800-m water depth, and the treatment was deployed from the vessel available at the time[14]; in 2011, a multiaccess-enabled scale squeeze treatment was executed at 1,525-m water depth at Tahiti Field by a U.S.-based operator.

Detailed case studies describing two scale treatments are provided in the sections that follow and illustrate the benefits of using multiaccess-enabled hardware technology.

20.4.1 Case Study 1: Offshore Sabah, Malaysia

Asia's first multiaccess-enabled rigless subsea acid stimulation (to remove scale from production wells) was executed in 2018 by the operator, with intervention performed on three target wells offshore Sabah, Malaysia, at a water depth of approximately 1,400 m.[24]

20.4.1.1 Problem

The field came online in 2014 and, within a short period of time, experienced significant changes in productivity. After 8 months, the reservoir pressure had declined from 2.62×10^7 Pa to 1.45×10^7 Pa and experienced a significant change in fluid composition. Declining oil production and accelerating water cuts prompted an initial investigation, which confirmed calcium carbonate scale and fines migration within the reservoir. Physical sampling accompanied comprehensive reservoir simulations to diagnose the issues observed. The results highlighted a loss of horizontal permeability with increasing near-wellbore skin. It was suspected that this was a consequence of reservoir compaction resulting from the pressure decline, coupled with water breakthrough causing scale formation within the system.

20.4.1.2 Fluid Delivery Options

An intervention was proposed to treat the skin damage in the near-wellbore region and mitigate further scale deposition within the system. The potential modes of fluid delivery considered were hydraulic access, riserless light well intervention, and a full riser-based access solution using a rig.

The selected solution was a fully integrated acid stimulation and scale squeeze treatment delivered via a hydraulic access package using hybrid coiled tubing downline. This enabled bullheading of acid to the formation, with placement assisted by intermittent nitrified foam diversion stages, prior to a final scale inhibitor squeeze stage. Three target wells were identified for the proposed treatment.

The key benefit offered by this open water hydraulic access solution was the minimized financial exposure, because it had a lower cost compared with that of alternative intervention means. Because mechanical access to the wellbore was not required for the proposed treatment, an open water hydraulic access solution deployed from a multiservice vessel offered the functionality required to efficiently meet the objectives.

20.4.1.3 Subsea Modular Injection System with Hydraulic Access

As shown in Figure 20.13, the subsea modular injection system comprises standardized modules, including a topside deployment system, open water hybrid coiled tubing downline complete with circulation line, a subsea safety module complete with jumper, and a multi-access subsea tree choke access module.

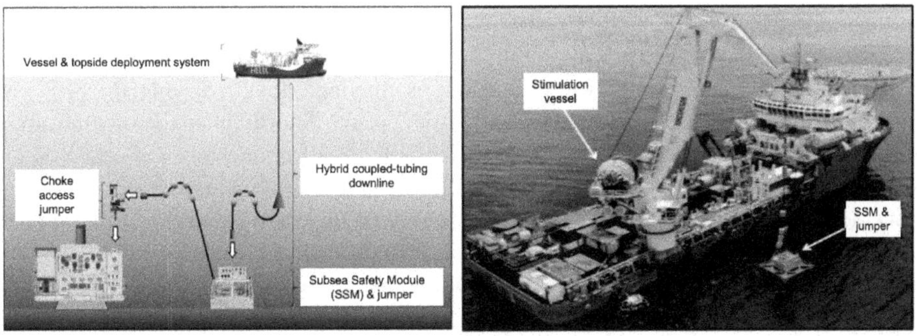

Figure 20.13. Sabah, Malaysia, subsea modular injection system (schematic and key hardware).

Treatment fluids were delivered subsea by an open water hydraulic access subsea modular injection system, using a hybrid coiled tubing downline. Access to the subsea trees was permitted via the multiaccess-enabled insert, enabling a flexible, OpEx-efficient, and low-risk intervention. The intervention system was installed on a multiservice vessel, with the downline deployed via the vessel moonpool. A second support vessel was used as required to provide additional fluid capacity without disturbing primary intervention operations.

20.4.1.4 Results

As shown in Table 20.1, poststimulation well testing confirmed that post-treatment productivity index has significantly increased for all of the three wells, with an average productivity index factor gain of 3.4. These results, combined with the efficient execution of the campaign, confirm the appropriateness of using open water hydraulic access with coiled tubing to perform cost-effective stimulations on complex subsea wells.

Table 20.1: Production performance results.

Well Name	Production Index		Productivity Index Factor	
	Prestimulation	Poststimulation	Actual	Target
Well A	0.3	1.8	6.0	5.0
Well B	2	6.5	3.0	3.0
Well C	3	3.8	1.3	1.5
Average gain	N/A		3.4	N/A

20.4.2 Case Study 2: Schiehallion Field

The well-published scale squeeze case for a U.K. operator's North Sea Schiehallion Field[25–28] has fully demonstrated the benefits of using multiaccess-enabled hardware technology in terms of flexibility, modularity, and providing significant cost savings.

20.4.2.1 Problem

Schiehallion Field (water depth of 500 m) first came online in 1998. With produced water injection as reservoir support, it was in constant production until 2013, when the FPSO was decommissioned and replaced in 2016 with a new vessel. Shortly after the commissioning of the new vessel, a significant reduction of injectivity on water injectors was observed on a subset of the wells such that insufficient rates were being achieved to maintain voidage replacement. Although the cost of obtaining in-well surveillance to validate the causal mechanism was prohibitive, compositional analysis of solids from normal operations showed a significant proportion of acid-soluble material. Hence, a decision was made to perform a low-cost chemical stimulation designed to prevent the most likely sources of damage in the injection wells as well as in affected producers.

20.4.2.2 Fluid Delivery Options

Acid stimulation was proposed to treat the skin damage in the near-wellbore region. The potential modes of fluid delivery considered were hydraulic access with a vessel of opportunity, riserless light well intervention, and a full riser-based access solution using a rig.

An open water hydraulic access system using an industrial hose downline from a vessel of opportunity was selected for the first campaign; this system was later moved onboard the stimulation vessel to provide the most cost-effective fluid delivery intervention to candidate wells. The key benefit was the minimized financial exposure, because the cost was significantly less than that of alternative intervention means.

Because the subsea wells were in water depths of approximately 450 m, the flexible downline chosen was an industrial polymer hose rather than the hybrid coiled tubing that is typically used in deeper water hydraulic access perforations. The use of industrial hose downline was suitable because of the tensile capacity of the hose for the given depth and operational requirements.

The vessel of opportunity provided additional cost savings; when the hose was later deployed directly from the stimulation vessel, the number of days the so-called intervention vessel was required on site was further reduced. The subsea construction vessel available at the time had full ROV and subsea capabilities, and it had been operating in Schiehallion Field for many years. The subsea vessel also had some basic pumping capability and fluid storage available. Therefore, there was an opportunity to perform a cost-effective bullheaded stimulation trial using committed contract days. Three target wells were identified for the proposed treatment.

20.4.2.3 Subsea Modular Injection System with Hydraulic Access

As shown in Figure 20.14, the subsea modular injection system comprises standardized modules, including a topside deployment system, an open water access industrial hose downline with an emergency quick disconnect connector, a subsea safety module complete with jumper, and a multiaccess subsea tree choke access module.

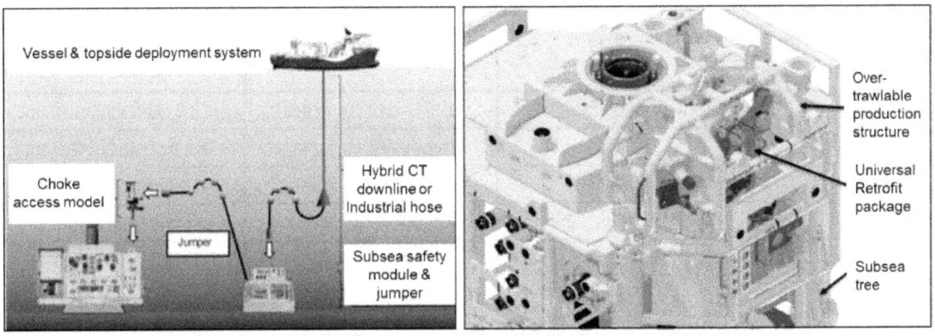

Figure 20.14. Schiehallion hydraulic access subsea modular injection system (schematic and tree with multiaccess insert).

Treatment fluids were delivered subsea by an open water hydraulic access system using an industrial hose downline. Access to the subsea trees was permitted via the multiaccess-enabled insert, enabling a flexible, OpEx-efficient, and low-risk intervention. The intervention system was installed on the available subsea construction vessel.

20.4.2.4 Results

The initial field trial took place on three wells—one producer and two injectors. As shown in Table 20.2, the producer had a 3-month increase in production, while the injectors had a 9- and 12-month injectivity increase following the treatment. In particular, a significant cleanup of near-well skin represented by the horizontal permeability K_h/lateral length improvement was observed for the two injectors, indicating additional lateral length had been accessed.

Table 20.2: Post-treatment performance results.

	Well A	Well B	Well C
Well type	Subsea oil producer	Subsea water injector	Subsea water injector
Lateral and screen lengths (m)	445 and 370	511 and 227	193 and 178
Completion type	Stand-alone screens	Stand-alone screens	Stand-alone screens
Pumping rate (bbl/min)	4.7	3.5	45
Operational duration (days)[(A)]	3.9	22.4	7.0
Acid type	Acetic/formic	15% Hydrochloric acid	15% Hydrochloric acid
Diverter type	N/A	Viscoelastic surfactant	Viscoelastic surfactant
Breaker type	N/A	U066 mutual solvent	U066 mutual solvent
Brine type	Potassium chloride	Ammonium chloride	Ammonium chloride
Productivity/matrix injectivity index increase (bbl/(D-psi)	From 3 to 12.3	From 9 to 18	From 7.5 to 22
Skin improvement	20 skin units	N/A	N/A
K_h/lateral length improvement (mD-ft)	N/A	>20,000	>2,500
Production increase duration (months)	3	N/A	N/A
Injection increase duration (months)	N/A	9	12

[(A)]Excluding mobilization and reloading.

20.4.3 Economical Considerations—Cost Analysis

Since the oil market downturn in 2014, the industry has experienced one of the most challenging times in its history; following the COVID-19 global pandemic outbreak in late 2019, oil prices were at historically low levels, with the price of West Texas Intermediate turning momentarily negative in April 2020. As a result, budget cuts, reduced outputs, and increased volatility became key trends in the energy industry. Technology development and optimization to drive down costs have become more important than ever.

The cost of hiring a vessel for an intervention operation (*e.g.*, a scale bullheading treatment) is one of the most significant expenditures. Using a multiaccess-enabled insert creates a subsea access point while maintaining environmental isolation. This eliminates the need for a rig or a light well intervention vessel. By choosing to deliver fluids to the well using a hydraulic access intervention system in place of a light well intervention package, the vessel specification is simplified and the resulting cost is further reduced.

Table 20.3 gives the estimated operational time for a typical scale squeeze treatment. The case without the use of multiaccess technology is estimated on the basis of published data,[29] while cases using the multiaccess system are estimated on the basis of the authors' experiences.

Table 20.3: Estimated operational time of a typical scale squeeze treatment.

Typical Vessel Required	Riser-Based Access Technology		Riserless Hydraulic Access
	Rig	Light Intervention Vessel	Support Vessel
Mobilize and position vessel (h)	72	24	24
Run intervention riser or hose (h)	48	24	24
Place inhibitor (h)	96	96	96
Pull riser or injection hose (h)	36	24	24
Demobilize vessel (h)	48	24	24
Total nonproductive time (NPT, h)	300	192	192
Total NPT (d)	12.5	8	8

Table 20.4 provides a cost comparison considering typical costs regarding CapEx, chemicals, vessel hire, and total deferred oil. For simplicity, only one estimate is given for the cost of chemicals and the CapEx required for the multiaccess system, although these costs may vary depending on the specific requirement. The oil production of the target well and the deferred well is assumed to be 1,000 bbl/d, while the oil price is assumed to be U.S.$50 per bbl. The vessel hire rate is based on a typical cost, although the actual rate may vary depending on the type of vessel used, its contract terms, and market conditions at the time.

Table 20.4: Estimated cost of a typical scale squeeze treatment.

Typical Vessel Required	Riser-Based Access Technology		Riserless Hydraulic Access
	Rig	Light Intervention Vessel	Support Vessel
CapEx for multiaccess system hardware (U.S.$)	0	50,000	50,000
Chemical cost (U.S.$)	50,000	50,000	50,000
Typical vessel cost (U.S.$ per day)	400,000	225,000	40,000
Total hire days	12.5	8	8
Total vessel hire cost (U.S.$)	5,000,000	1,800,000	320,000
Total NPT (d)	12.5	8	8
Total deferred oil (1,000 bbl/d)	12,500	8,000	8,000
Deferred oil cost at U.S.$50 per bbl (U.S.$)	625,000	400,000	400,000
Total cost per squeeze (U.S.$)	5,675,000	2,300,000	820,000

Although the estimates given in Tables 20.3 and 20.4 are provided only for the purposes of this analysis, they are intended to highlight the key differences between the access techniques and the relative cost savings that can be achieved. On the basis of Table 20.4, a traditional bullheading scale squeeze treatment without the use of the multiaccess system is two to three times more expensive than that using light intervention riser-based technology and seven to eight times more expensive than that using the hydraulic access technology. The multiaccess technology provides significant cost savings, with the hydraulic access technique being a compelling option in today's economic environment.

In addition to the cost savings, a greater pump rate (*e.g.*, 10 bbl/min) can be achieved using the multiaccess technology, while the maximum pump rate for the traditional bullheading squeeze without using the multiaccess insert is less than 5 bbl/min. It is well-known that a greater pump rate provides a better opportunity to effectively place scale squeeze treatments and conduct them within a shorter operational time.

20.4.4 Multiaccess-Enabled Subsea Sampling Case Study—Offshore Angola

A subsea sampling system was used in a challenging field in Angola for the first time, where low-water-cut wells commingled into a common flowline.

Lianzi Field (water depth ranging from 820 m to 1,070 m), which first came online in 2015, has a highly challenging fluid composition, and subsea sampling was identified as a solution to manage and mitigate various flow assurance challenges. Using water injection to maintain reservoir pressure increased the risk of scale buildup in the reservoir. Produced water analysis from individual wells was crucial to plan and optimize high-cost scale squeeze operations. Capturing water samples in the low water cut was a significant challenge that required enrichment of the water phase during sampling to ensure suitable quantities of the required water samples were collected for laboratory analysis.

Scale squeezes are very costly, and the ability to take samples subsea is crucial to determining the squeeze treatment frequency. Additionally, the water sampling information can be used to update the configuration files for the multiphase flowmeters and update the flow assurance strategy, specifically for the scale inhibitor dosage.

A subsea sampling system that included sampling interface quick-connect plates located on the trees, a subsea sampling module, and associated test and transportation equipment was deployed, as shown in Figures 20.15–20.17.

The first campaign took place in October 2016 (Figure 20.18). A total of 12 samples, with some shown in Figure 20.19, were successfully taken from three wells, with the first sample collected in approximately 6 h from deployment of the skid to retrieval to deck.

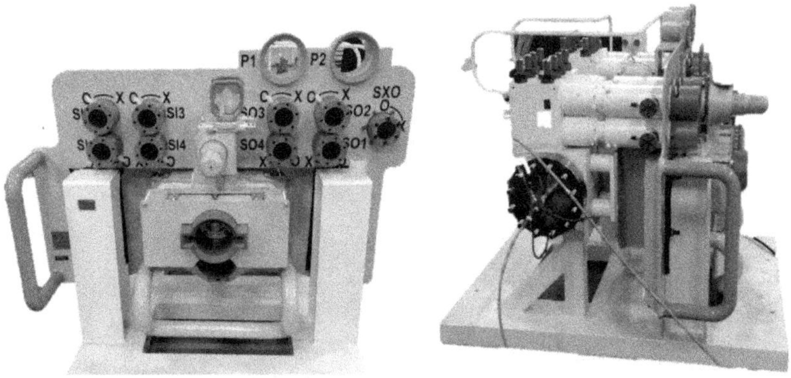

Figure 20.15. Subsea sampling interface and tree-mounted stab plate with couplers.

Figure 20.16. Subsea sampling interface installed on a tree.

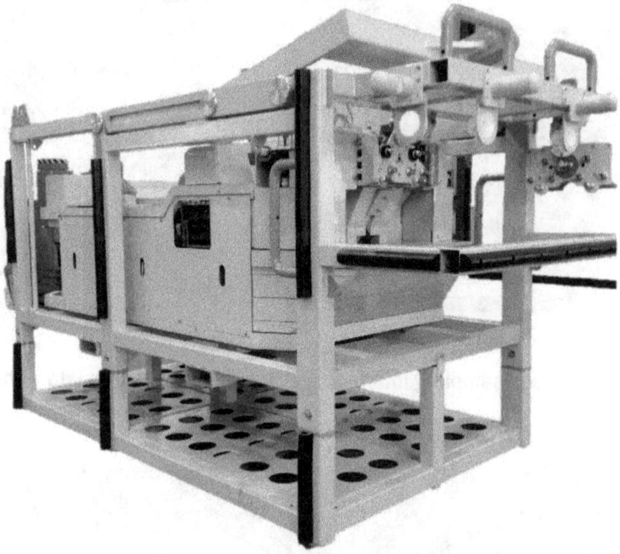

Figure 20.17. Subsea sampling module.

Figure 20.18. Subsea sampling system field and subsea operations.

Figure 20.19. Phase-enriched water samples captured during the operations.

With the information obtained from the samples, the squeeze treatment frequency was optimized, which resulted in cost savings estimated at U.S.$3 million per year.

To extend the applications for subsea sampling systems in existing fields, a concept was created for a multiaccess retrofit sampling choke assembly that could be used temporarily or permanently to act as a retrievable interface for sampling operations. Figure 20.20 (left) shows a sampling interface via a retrievable choke insert to be intervened on with a separate sampling skid and (right) a sampling choke system with fluid capture cylinders located directly on the choke insert.

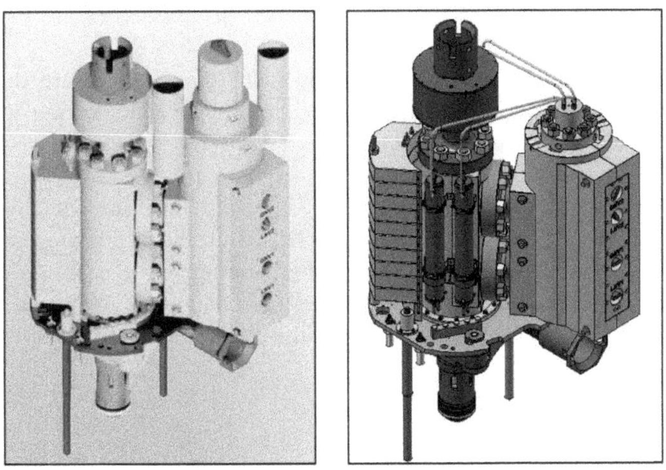

Figure 20.20. Concepts for sampling choke insert.

20.4.5 Inflow Control Equipment: Inflow Control Valve Scaling Case Study—Presalt Carbonate Reservoir

Inflow control is required because of the negative economic impact of excessive water production in subsea fields. For many, the optimal production control and surveillance method includes completions using sand control screens integrated with ICDs or ICVs. ICDs and ICVs are small nozzles installed along the length of the production tubing to control reservoir heterogeneity (different pressures and relative permeabilities) along horizontal wells; for example, ICDs and ICVs can choke back the flow from higher production compartments, balance out the water production, and increase overall recovery.[3, 29–31]

However, increased scale risks have been observed as a result of the increased shear stresses, turbulences, and pressure drops experienced across completions even under mild levels of oversaturation. Technical risks when installing ICDs and ICVs include sand production, equipment failure, system integrity failure, and scale formation. For example, medium $BaSO_4$ scaling upstream of ICVs can cause blockages in and malfunctioning of the ICVs themselves,[29] while $BaSO_4$ and $SrSO_4$ (strontium sulfate) scaling around ICD nozzles can occur, even when the scale inhibitor at the required minimum inhibitory concentration is applied.[30]

The ability to prevent and/or mitigate scale depends on the design and geometry of the completion. The failure of a functional ICD or ICV as a result of scale deposition can cause severe production impairment. As an example of recent measures, a coating material has been developed, which is applied to completion tools to prevent the scale formation and adhesion.[3, 31, 32]

A presalt carbonate reservoir indicated severe calcium carbonate scaling risk, which was expected to occur especially at the beginning of production. For optimal production control and surveillance in this field, deploying an ICV was considered to be the most favorable method at the time.

Using integrated multiphase flow simulation and scale chemistry software, the $CaCO_3$ scaling phase boundaries were calculated and plotted to provide an indication of the safe operating envelopes over the life of the field.

The early-life example is given in Figure 20.21. The solid line represents the scaling phase boundary, where no scaling is expected on the left-hand side of the line and scaling is expected on the right-hand side. The produced fluid journey proceeds from (1) the reservoir where it is at equilibrium to (2) the wellbore/perforation and then to the completion region where the ICV is located before traveling to (3) the tree, (4) the riser, and (5) topside.

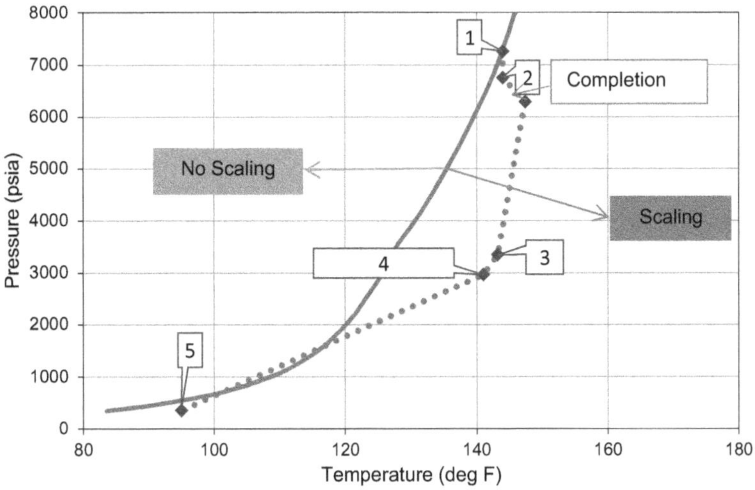

Figure 20.21. Safe operating envelope for early field production.

It is evident that the scaling tendency is greatest at the completion (*i.e.*, the ICV location). The scaling quantity is predicted in Table 20.5, where

- For the production lifetime (early, mid, late) evaluated, the calcium carbonate scaling tendency increases, as it goes from the reservoir/perforation (equilibrium), completion/ICV to the tree, and then decreases from the riser to the FPSO vessel. The greatest increases in scaling risks were found during early field life, and the risks varied from 1.2 (completion) to 1.5 (tree).
- Although the scaling tendency is mild, if scaling is permitted to occur, both the scaling mass and the daily accumulation rate are high. Specific to the completion/ICV condition, a maximum 120-mg/L scale precipitation in early life and a maximum accumulation rate of 32.7 kg/d in late life are expected to be experienced by the completion device.

Table 20.5: Summary of the scaling evaluation results for the ICV completion.

ICV Completion	Saturation Ratio	Maximum Precipitation (mg/L)	Maximum Accumulation Rate (kg/d)
Early life	1.2	120	1.8
Midlife	1.1	62	31.3
Late life	1.05	30	32.7

These prediction results necessitated the investigation of nonsticking surfaces for the ICV to prevent scale deposition from occurring in the field.

The selection of an appropriate surface finish for the ICDs/ICVs was crucial for this field. Several surface coatings were evaluated using a proper laboratory testing apparatus and testing conditions.[31] Intensive computational fluid dynamics (CFD) modeling was performed to assist the testing equipment design (including the autoclave reactor and mock valve) to replicate the field turbulence conditions and to validate suitable testing conditions for coupon testing in the laboratory.[3, 32]

A range of surface finishes—including 13% Cr steel, 25% Cr steel, tungsten carbide, and a patented coating—were tested in the laboratory. A review of the scanning electron microscope images indicated a few general trends, for example, calcium carbonate was present as rhombs on all coupons; chrome steels appeared to provide almost complete coverage, while the other alloys and tungsten carbide were patchy; and the patented coating had the lowest coverage, which was in line with the lowest mass of scale.[3, 31]

The noncoated surface (rough coupons) and coated surface (smooth coupons) were compared in CFD modeling (Figures 20.22 and 20.23). It was found that the velocity magnitude on the coupons' surface and the horizontal variation of the velocity were similar (with a maximum of approximately 4.8 m/s as shown in Figure 20.22) between the two surface configurations.

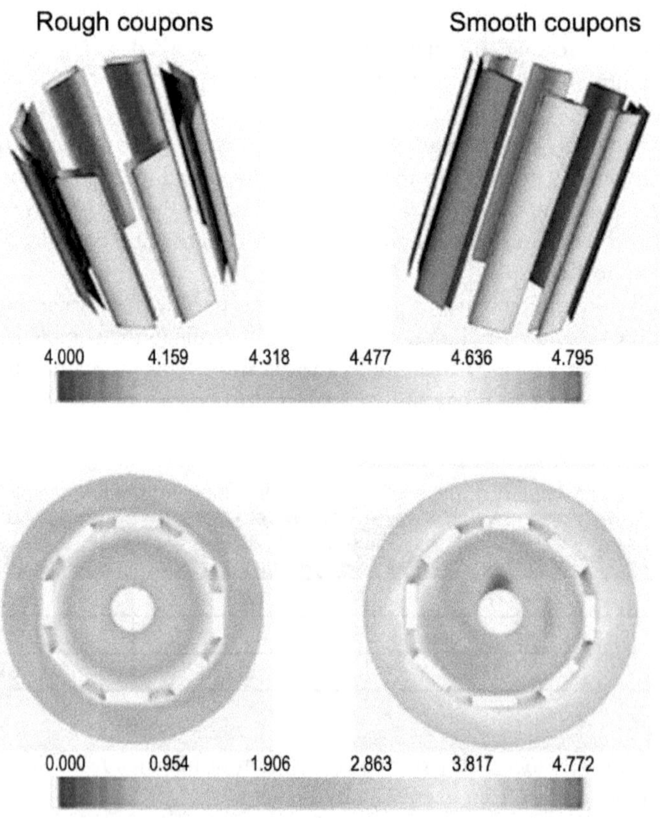

Figure 20.22. Computational fluid dynamics modeling results on velocity magnitude (m/s) for non-coated rough (left) vs. coated smooth (right) surfaces.

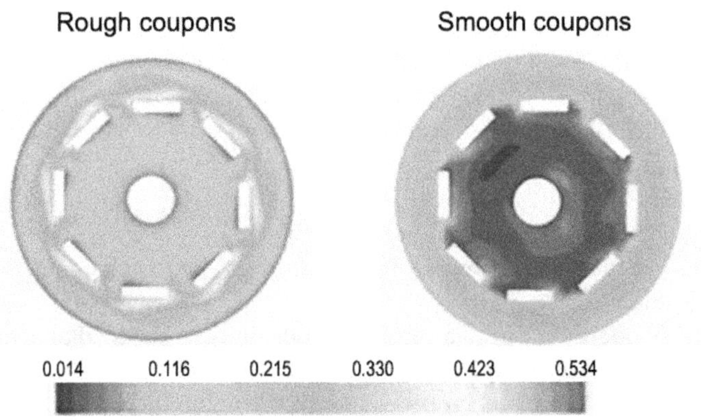

Figure 20.23. Computational fluid dynamics modeling results on turbulence kinetic energy (J/kg) for noncoated rough (left) vs. coated smooth (right) surfaces.

However, the noncoated coupons with a significant roughness showed much greater turbulence kinetic energy around their surfaces than the smooth coupons did (Figure 20.23). The tangential velocity plot showed a greater recirculation around the rough coupons, which suggested that the noncoated surfaces would be much more prone to scaling deposition and adhesion than the coated/smooth surfaces.[3, 32]

The CFD modeling results illustrate the mechanisms of scale formation and adhesion on different coupon surfaces; they indicate that the use of the patented coated surfaces in ICV is likely to reduce or eliminate the scale adhesion.

This case study shows the benefit of using a suitable coating on the ICV to prevent scale and highlights that the integration of scale prediction software with thermodynamic software can produce an overview of the scaling tendency along the produced fluid journey over the life of the field. Furthermore, the use of CFD technology can help provide understanding of the scaling mechanism in ICVs and guide laboratory qualification testing.

20.5 Conclusions

This chapter reviewed some specific scale management concerns in subsea fields, with a focus on subsea access challenges and the associated costs. A discussion of the fundamentals of inorganic scale formation resulting from changes in temperature/pressure and fluid properties along the fluid journey from the reservoir to topside revealed the importance of acquiring reliable fluid data through in situ subsea sampling, while dedicated CIMVs provide optimized dosing according to individual well risk.

In particular, the challenges of subsea access and the associated treatment costs have driven the innovation of access technology. To create easy subsea access without the need to use an expensive rig, multiaccess and related hardware technologies have been developed and applied for scale treatments.

Scale squeeze interventions have been further enhanced by the development of efficient subsea hydraulic access solutions with multiple connector options and selectable hydraulic access points to maximize system flexibility, reliability, and cost savings. By installing multiaccess-enabled choke inserts in existing or new trees, scale squeeze treatments have been optimized and provide the following benefits:

- Elimination of the requirement for a rig or divers
- Reduction of production shutdown time and, during installation, mobilization, and demobilization times
- Use of ROV operations from a vessel of opportunity as opposed to dedicated light well intervention assets, of which limited assets are readily available
- Reduction of overall vessel costs because of the ability to use a typical multipurpose service vessel equipped with an ROV and heave-compensated crane
- Ability to intervene in wells more frequently, because of the ability to conduct operations without needing to deploy a riser or remain directly over a target well (*i.e.*, the hydraulic access technique often enlarges the weather window).

In light of the recent oil industry downturn and market drivers, harnessing the power of hardware technology in scale control and management becomes more important. The recent evolution of open water hydraulic access technology provides further cost savings by using a riserless downline (either a hydraulic hose or coiled tubing), which enables the use of a suitable support vessel available at the time.

The number of field application experiences—including the two hydraulic access case studies summarized in this chapter—have demonstrated the benefits of multiaccess-enabled technologies, while the dedicated cost analysis example showed the magnitude of cost savings that can be achieved through the use of multiaccess technology. Traditional riser-based bullheading without using multiaccess technology can be as much as seven to eight times more expensive than a scale treatment using hydraulic access technology.

The benefit of integrating software technologies (*e.g.*, scale chemistry and multiphase flow simulation software and CFD modeling) with hardware technology (*e.g.*, surface coatings on the ICV) has been demonstrated by a case study in a presalt reservoir, where the scale tendency over the life of the field was quantified, and the scaling mechanism at the ICV was analyzed using several modeling techniques, which helped guide the laboratory validation of the anti-scale coating.

Nomenclature

CapEx	Capital expenditure
CFD	Computational fluid dynamics
CIMV	Chemical injection metering valve
CT	Coiled tubing
FPSO	Floating production, storage, and offloading

ICD	Inflow control device
ICV	Inflow control valve
LDI	Low-dose inhibitor
MARS	Multiple application reinjection system
NPT	Nonproductive time
OpEx	Operational expenditure
PVT	Pressure-volume-temperature
ROV	Remotely operated vehicle
SSM	Subsea safety module.
VES	Viscoelastic surfactant

References

1. H. Guan, P. Rice, M. Avery, "Harness the Power of Hardware Technologies for Scale Control and Management in Subsea Fields," SPE-209508-MS, SPE International Oilfield Scale Conference, held May 25–26, 2022 (Aberdeen, UK).
2. H. Guan, "Scale Deposition Control and Management in Subsea Fields," NACE-2015-5480, NACE Annual Conference, held March 15–19, 2015 (Dallas, Texas, USA).
3. H. Guan, R. Bouamra, T. Lindvig, et al., "Scale Risk Assessment and the Use of Novel Coating for Smart Completion Tools: Scale Simulation and CFD Modelling Assisted Approach," SPE-190730-MS, SPE International Oilfield Scale Conference, held June 20–21, 2018 (Aberdeen, UK).
4. H. Guan, A. Jenkins, "Testing Equipment and Procedures Suitable for Evaluation of Scale Inhibitors Applied in Deepwater Fields," NACE-11345, NACE-2015-5480, held March 13–17, 2011 (Houston, Texas, USA).
5. H. Guan, P. Farmer, "Evaluation of Scale Inhibitors Suitable for Deepwater Fields," SPE-131110-MS, SPE International Conference on Oilfield Scale, held May 26–27, 2010 (Aberdeen, UK).
6. H. Guan, J. Scheie, "The Effects of Chemical and Pipeline Insulation on Wax Deposition," 20th Oil Field Chemistry Symposium, held March 22–25, 2009 (Geilo, Norway).
7. G.M. Graham, I.R. Collins, "Assessing Scale Risks and Uncertainties for Subsea Marginal Field Developments," SPE-87460-MS, SPE International Symposium of Oilfield Scale, held May 26–27, 2004 (Aberdeen, UK).
8. M.M. Jordan, K. Sjuraether, I.R. Collins, et al., "Life Cycle Management of Scale Control Within Subsea Fields and its impact on Flow Assurance, Gulf of Mexico and the North Sea Basin," SPE-71557-MS, SPE Annual Technical Conference, held September 30–October 3, 2001 (New Orleans, Louisiana, USA).
9. H. Guan, "Carbonate Scaling Predictions: The Importance of Valid Data Input," *Journal of Materials Science and Engineering* 5 (2011): pp. 220–233.
10. N.C. Crawford, I.R. Still, "Riserless Subsea Well Intervention: Technology and Practice," OTC-20420-MS, Offshore Technology Conference, held May 3–6, 2010 (Houston, Texas, USA).
11. I. Donald, J. Reid, "Apparatus and Method for Processing Fluids From a Well," U.S. Patent 8297360B2, 2012.
12. G. Hall, R. Large, J.D. Trinder, "Well Intervention Tool and Method," U.S. Patent 9574420B2, 2017.
13. I. Donald, J. Reid, "Apparatus and Method for Recovering Fluids From a Well and/or Injecting Fluids Into a Well," U.S. Patent 8573306B2, 2013.

14. I. Donald, R. Sangster, "Using MARS (Multiple Application Re-Injection System) Technology for Improving Oil Recovery and Reducing Production Risk," SPE-132274-MS, CPS/SPE International Oil & Gas Conference, held June 8–10, 2010 (Beijing, China).
15. D. Petrone, "Subsea Well Fluid Intervention Using MARS (Multiple Application Reinjection System)," SPE-146267-MS, SPE Annual Technical Conference and Exhibition, held October 30–November 2, 2011 (Denver, Colorado, USA).
16. M. Avery, D. Morris, T. Morgan, et al., "Transitioning From Rig to Vessel Based Interventions to Maximize Economic Recovery," SPE-204402-MS, SPE/ICoTA Well Intervention Conference, held March 22–25, 2021 (Virtual).
17. G. Alexandre, "Apparatus and System for Passively Sampling Production Fluid From a Well," Patent GB2521294B, 2019.
18. G. Hall, A. Gordon, "Deepwater Multiphase Liquid Sampling Using Multiple Application Reinjection System (MARS)," SPE-143782-MS, Brasil Offshore Conference, held June 14–17, 2011 (Macaé, Brazil).
19. H. Guan, S. Rathod, "In-line Subsea Sampling: Non-Disruptive Subsea Intervention Technology for Production Assurance," Subsea Expo, held February 8, 2018 (Aberdeen, UK).
20. H. Guan, "Flow assurance risk evaluation and chemical management in a shale gas field," in *Industrial Water Treatment: New Trends, Challenges, and Solutions*, eds. Z. Amjad, T. Chen (Houston: NACE International, 2017), pp. 267–281.
21. H. Guan, S. Rathod, "Subsea Sampling on the Critical Path of Flow Assurance," DEVEX Conference, held May 19, 2016 (Aberdeen, UK).
22. A. Sbordone, G. Smith, M.T. Smith, et al, "Analysis of Subsea Sampling Applications and Drivers for Subsea Field Developments in Deep and Ultra-Deep Water," IBP-1836-12, Rio Oil & Gas Exp and Conference, held September 17–20, 2012.
23. D. Simpson, "Development & Qualification of CIMV for Gas Projects in the North Sea, South China Sea and Western Australia," Underwater Technology Conference, held June 19–20, 2013 (Bergen, Norway).
24. M. Avery, R. Large, H. Azhar, et al., "Asia's First Rigless Subsea Stimulation Using Patented Well Access Technology and a Fully Integrated Service Model," SPE-197073-MS, SPE Subsea Well Intervention Symposium, held August 13–15, 2019 (Galveston, Texas, USA).
25. A. Roy, A. Bird, S. Bremner, et al., "Pump Up the Volume—Massive Water Injection Increase Through Open Water Stimulations," SPE-205426-MS, SPE Offshore Europe Conference, held September 7–10, 2021 (Virtual).
26. B. Coupland, M. Avery, P. Rice, et al., "Subsea Asset Life Extension in Harsh Environments," OTC-30564-MS, Offshore Technology Conference, held May 4–7, 2020 (Houston, Texas, USA).
27. M. Avery, R. Greenaway, C. Bridge, et al., "Holistic Production Enhancement for Life of Field," OTC-30781-MS, Offshore Technology Conference, held May 4–7, 2020 (Houston, TX, USA).
28. G.M. Graham, I.R. Collins, T.J. Johnson, "Technical and Economic Analysis of the Downhole Scale Risks and Uncertainties for Subsea and Deepwater Field Developments," SPE-95049-MS, SPE International Symposium on Oilfield Scale, held May 11–12, 2005 (Aberdeen, UK).
29. N. Akram, S. Hicking, P. Blythe, et al., "Intelligent Well Technology in Mature Assets," SPE-71822-MS, held September 4–7, 2001 (Aberdeen, UK).
30. Ø. Gustavsen, O.M. Selle, F.H. Fadnes, et al., "Inflow-Control-Device Completion in a Scaling Environment: Findings and Experiences Obtained During Production Logging in the Heidrun Field," SPE-134994-MS, held September 19–22, 2010 (Florence, Italy).
31. G.M. Graham, M.C.M. Bezerra, N. Goodwin, et al., "Minimizing Scale Deposition Through Surface Enhancement in Downhole Tools," OTC-24530-MS, held October 29–31, 2013 (Rio de Janeiro, Brazil).
32. R. Bouamra, G. Carneiro, P. Machado, et al., "ScaleProTect—Scale Deposition Modeling in Pre-Salt Reservoir," OTC-29886-MS, held October 29–31, 2019 (Rio de Janeiro, Brazil).

www.ingramcontent.com/pod-product-compliance
Lightning Source LLC
LaVergne TN
LVHW080759011026
849160LV00015B/1491